19th International Conference on Strangeness in Quark Matter (SQM 2021)

EPJ Web of Conferences Volume 259 (2022)

New York, New York, USA
17 - 22 March 2021

Editors:

Gabor David
Prakhar Garg
Alexander Kalweit
Mukherjee Swagato
Thomas Ullrich
Zhangbu Xu
In-Kwon Yoo

ISBN: 978-1-7138-4387-0

Printed from e-media with permission by:

Curran Associates, Inc.
57 Morehouse Lane
Red Hook, NY 12571

Some format issues inherent in the e-media version may also appear in this print version.

For additional information, please contact EDP Sciences – Web of Conferences
at the address below.

EDP Sciences – Web of Conferences
17, Avenue du Hoggar
Parc d'Activité de Courtabœuf
BP 112
F-91944 Les Ulis Cedex A
France

Phone: +33 (0) 1 69 18 75 75
Fax: +33 (0) 1 69 28 84 91

contact-edps@webofconferences.org

Additional copies of this publication are available from:

Curran Associates, Inc.
57 Morehouse Lane
Red Hook, NY 12571 USA
Phone: 845-758-0400
Fax: 845-758-2633
Email: curran@proceedings.com
Web: www.proceedings.com

TABLE OF CONTENTS

Preface

The 19th International Conference on Strangeness in Quark Matter (SQM 2021) focuses on new experimental and theoretical developments on the role of strange and heavy-flavour quarks in high energy heavy-ion collisions and in astrophysical phenomena. Due to the ongoing pandemic, it was an online-only conference and serves as a bridge to the postponed SQM 2022 in South Korea taking place in June 13-18, 2022. The online conference is sponsored by BNL, SBU, CERN and PNU.

Overall, there were 41 plenary and 96 parallel talks at SQM 2021, including presentations in memory of Jean Cleymans, Jean Letessier, Dick Majka and Jack Sandweiss, who all made exceptional impacts on the field. Ninety-five articles were reviewed and included in the current proceedings.

SQM2021 Local Organizers:

Gabor David (SBU),
James Dunlop (co-chair, BNL),
Alexander Philipp Kalweit (CERN),
Berndt Mueller (BNL/Duke),
Swagato Mukherjee (BNL),
Rachid Nouicer (BNL),
Lijuan Ruan (BNL),
Thomas Ullrich (BNL),
Isaac Upsal (BNL),
Flemming Videbaek (BNL),
Zhangbu Xu (co-chair, BNL),
In-Kwon Yoo (PNU)

edp sciences

EPJ Web of Conferences

Statement of Peer review

In submitting conference proceedings to EPJ Web of Conferences, I certify to the Publisher that I adhere to the **Policy on Publishing Integrity** of the journal in order to safeguard good scientific practice in publishing.

1. All articles have been subjected to peer review administered by the proceedings editors.
2. Reviews have been conducted by expert referees, who have been requested to provide unbiased and constructive comments aimed, whenever possible, at improving the work.
3. Proceedings editors have taken all reasonable steps to ensure the quality of the materials they publish and their decision to accept or reject a paper for publication has been based only on the merits of the work and the relevance to the journal.

Title, date and place of the conference

The 19th International Conference of Strageness in Quark Matter,
May 17-22, 2021
(Online) New York, USA

Proceedings editor(s):

Gabor David
Prakhar Garg
Alexander Kalweit
Mukherjee Swagato
Thomas Ullrich
Zhangbu Xu
In-Kwon Yoo

Date and editor's signature

January 10, 2022

EPJ Web of Conferences **259**, 01001 (2022) https://doi.org/10.1051/epjconf/202225901001
SQM 2021

Highlights from the NA61/SHINE strong-interactions programme

Magdalena Kuich[1,*] for the NA61/SHINE Collaboration

[1]Faculty of Physics, University of Warsaw, Pasteura 5, 02-093 Warsaw, Poland

Abstract. NA61/SHINE is a multipurpose fixed-target facility at the CERN SPS. The main goals of the NA61/SHINE strong-interactions programme are to discover the critical point of strongly interacting matter as well as to study the properties of the onset of deconfinement. In order to reach these goals, a study of hadron production properties is performed in nucleus-nucleus, proton-proton and proton-nucleus interactions as a function of collision energy and size of the colliding nuclei. In this contribution, the NA61/SHINE results from a strong interaction measurement programme are presented. In particular, the latest results from different reactions $p+p$, Be+Be, Ar+Sc, and Pb+Pb on hadron spectra, as well as intermittency, higher-order moments of multiplicity fluctuations and spectator induced electromagnetic effects are discussed.

1 NA61/SHINE experiment

NA61/SHINE detection system is a large acceptance hadron spectrometer based on a set of eight Time Projection Chambers complemented by Time-of-Flight detectors. This setup allows for precise momentum reconstruction and identification of charged particles. The high-resolution forward calorimeter, the Projectile Spectator Detector (PSD), measures energy flow around the beam direction, so-called Forward Energy (E_F), which in nucleus-nucleus can be related to the collision centrality [1].

NA61/SHINE performed a two-dimensional scan in collision energy (13A-150A GeV/c and system size (p+p, p+Pb, Be+Be, Ar+Sc, Xe+La, Pb+Pb) to study the phase diagram of strongly interacting matter. Currently, NA61/SHINE pursues the physics programme focused on the open charm production measurements in order to determine the mechanism of open charm production, the impact of the onset of deconfinement on open charm production and investigate the impact of the QGP formation on J/Ψ production.

2 Study of the onset of deconfinement and onset of fireball

2.1 Particle production properties

The main motivation of the NA61/SHINE physics programme are the predictions of the Statistical Model of the Early Stage (SMES) [2]. The theory predicts a 1st order phase transition from the QGP to a hadron matter phase between top AGS and top SPS energies. In the transition region, constant temperature and pressure in the mixed-phase and an increase of

*e-mail: magdalena.kuich@cern.ch

the number of internal degrees of freedom is expected. SMES proposes several observables, which indicate the phase transition.

Fig. 1 presents the system-size dependence of K^+/π^+ ratio at mid-rapidity obtained by NA61/SHINE: *p+p* [5], Be+Be [6] and Ar+Sc (preliminary); and NA49: Pb+Pb [7]; at 150*A*/158*A* GeV/*c* beam momenta. The K^+/π^+ ratio was predicted within SMES as one

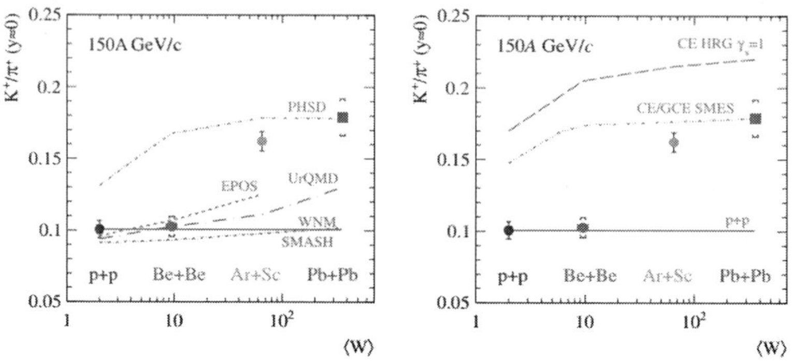

Figure 1. System size dependence of the K^+/π^+ ratio obtained at beam momenta of 150*A* GeV/*c* ($\sqrt{s_{NN}} \approx 17$ GeV) compared with dynamical (*left*) and statistical (*right*) models.

of the signatures of the onset of deconfinement. The results were compared with few dynamical models. Those without phase transition (EPOS [12], UrQMD [14] and SMASH [13]) agree with the results from small systems (*p+p* and Be+Be), while fail to describe the results from heavier systems (Ar+Sc and Pb+Pb). In contrast to the one with phase transition (PHSD [15, 16]), which follows the trend determined by the heaviest system (Pb+Pb) and overestimates the ratio for smaller systems. Similarly, testes statistical models: with (renormalised SMES [4]) and without (HRG [14]) the phase transition tends to overestimate the K^+/π^+ ratio especially in the small-mass region of colliding nuclei.

The left panel of fig. 2 presents the dependence of the ratio of the mean number of pions produced in a collision to the mean number of wounded nucleons $\langle\pi\rangle/\langle W\rangle$ versus the Fermi energy measure, $F = \left[\frac{(\sqrt{s_{NN}}-2m_N)^3}{\sqrt{s_{NN}}}\right]^{1/4} \approx \sqrt[4]{s_{NN}}$. NA61/SHINE results: *p+p* [8], Be+Be [9] and Ar+Sc [10] were compared with NA49 results from Pb+Pb [3] and others. $\langle\pi\rangle/\langle W\rangle$ for Ar+Sc reactions equals that for *N+N*[1] reactions at low SPS energies whereas it is consistent with that for central Pb+Pb reactions at high SPS energies. The behaviour of Ar+Sc creates a tension with to Be+Be measurements, which are close to the Pb+Pb results except for the top SPS beam energy. The steepening of the slope of the $\langle\pi\rangle/\langle W\rangle$, which is considered a signal of the onset of deconfinement within the SMES [2] is observed neither in *central* Ar+Sc nor in *central* Be+Be measurements.

The energy dependence of the sound velocities extracted from the data [8–10] are presented in fig. 2 (*right*). Results from central Pb+Pb and Au+Au collisions exhibit a clear minimum (usually called the softest point) at $\sqrt{s_{NN}} \approx 10$ GeV consistent with the reported onset of deconfinement [3, 7]. The energy range for results from Ar+Sc, Be+Be *central* collisions and inelastic $p+p$ interactions is too limited to allow a significant conclusion about a possible minimum.

[1]For *p+p* interactions the figure shows isospin symmetrized values [8] marked as *N+N*.

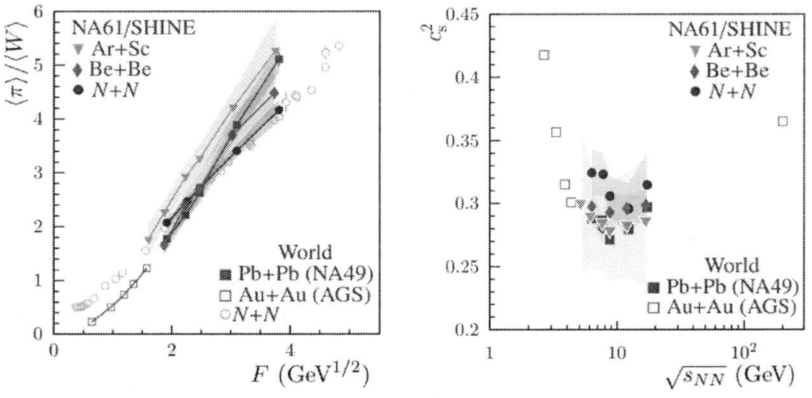

Figure 2. *Left*: Mean pion multiplicity to the mean number of wounded nucleons versus the Fermi energy measure. *Right*: The speed of sound c_s^2 as a function of center-of-mass collision energy as extracted from the data.

2.2 Flow

NA61/SHINE experiment reports results on anisotropic flow, measured in centrality selected Pb+Pb collisions at 13*A* and 30*A* GeV/*c* beam momentum. The experimental setup allows tracking and particle identification over a wide rapidity range. Flow coefficients were measured relative to the spectator plane estimated with PSD, which is unique for NA61/SHINE. Preliminary results on the directed flow of negatively charged pion are presented in the left panel of fig. 3. In mid-central collisions the flow of negatively charged pions has minimum at

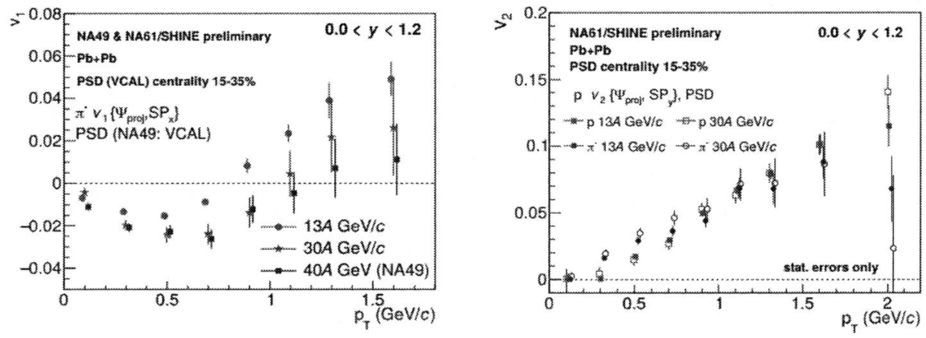

Figure 3. Preliminary results on negatively charged pions directed flow $v_1(p_T)$ and negatively charged pions and protons elliptic flow $v_2(p_T)$ for different collision energies.

$p_T \approx 0.2-0.6$ GeV/*c*. At the minimum the value v_1 is negative, while at $p_T \approx 0.7-1.3$ GeV/*c* it changes sign to positive. In the region on sign change v_1 values measured at 13*A* GeV/*c* differ from the ones measured at 30*A*/40*A* GeV/*c*. Dependence of $v_1(p_T)$ for protons produced in mid-central Pb+Pb collisions at 13*A* GeV/*c* beam momentum is also reported. The $v_1(p_T)$ values for protons are positive and increase with transverse momentum until $p_T \approx 1.5-2.0$ GeV/*c*. Preliminary results on the centrality dependence of dv_1/dy at mid-rapidity, measured in Pb+Pb collisions at 13*A* GeV/*c* were also reported. The slope of negatively charged pion

v_1 is always negative. In contrast, the slope of proton v_1 changes sign for centrality of about 20%. Preliminary results of elliptic flow, $v_2(p_T)$, for Pb+Pb collisions at 13*A* and 30*A* GeV/*c* are presented in the right panel of fig. 3. The p_T dependence of v_2 coefficient of negatively charged pions differs from the one measured for protons, but it does not show any energy dependence.

3 Search for critical point

One of the main goals of the NA61/SHINE experiment is to locate the critical point (CP) of strongly interacting matter. The expected signal of a CP is a non-monotonic dependence of various fluctuations (e.g. multiplicity, transverse momentum, baryon density fluctuations).

To study the fluctuations in systems of different sizes one should use quantities insensitive to system volume (so-called intensive quantities), for example constructed by division of cumulants κ_i of the measured distribution (up to fourth order), where i is the order of the cumulant. For second, third and fourth order cumulants intensive quantities are defined as: κ_2/κ_1, κ_3/κ_2 and κ_4/κ_2. Fig. 4 shows energy dependence of second (*left*), third (*middle*) and fourth (*right*) order cumulant ratio of negatively charged hadrons measured in *p*+*p* interactions as well as Be+Be and Ar+Sc *central* collisions. The second-order cumulant ratio

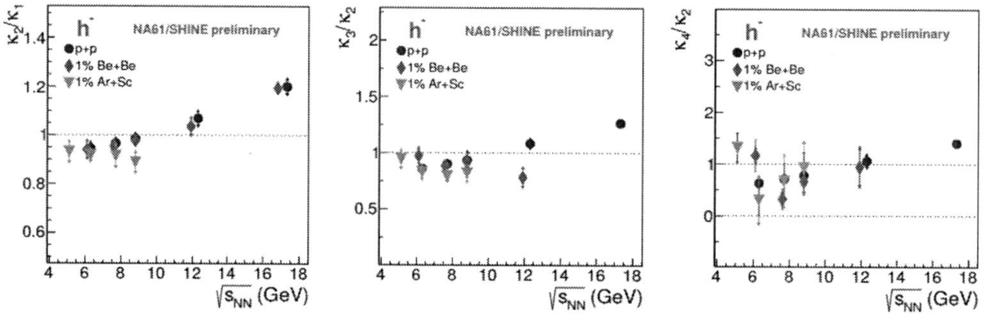

Figure 4. Preliminary results on the energy and system-size dependence of multiplicity fluctuations of negatively charged hadrons.

reveals an increasing difference between κ_2/κ_1 values from small systems (*p*+*p* and Be+Be) and a heavier system (Ar+Sc) with collision energy. The *p*+*p* results on third-order cumulant ratio tend to be higher than the ones from Be+Be at higher collision energies. Whereas, the fourth-order cumulant ratio indicates consistent values for all measured systems at given collision energy. More detailed studies are still needed, however, no structure indicating CP was observed so far.

Possible tool for search of CP is a proton intermittency. The proximity of CP is expected to manifest itself in local power-law fluctuations of the baryon density which can be searched for by studying the scaling behaviour of second factorial moments,

$$F_2(\delta) = \frac{\langle \frac{1}{M} \sum_{i=1}^{M} n_i(n_i - 1) \rangle}{\langle \frac{1}{M} \sum_{i=1}^{M} n_i \rangle^2}$$

, with the cell size or, equivalently, with the number of cells in $(p_x\ , p_y)$ space of protons at mid-rapidity [18–20]. NA61/SHINE measures $F_2(M)$ using statistically independent points and cumulative variables. Preliminary results on $F_2(M)$ of mid-rapidity protons measured in

0-20% most *central* Ar+Sc collisions at 150*A* GeV/*c* and 0-10% most central Pb+Pb collisions are presented in fig. 5 in left and right panels, respectively. The intermittency index ϕ_2

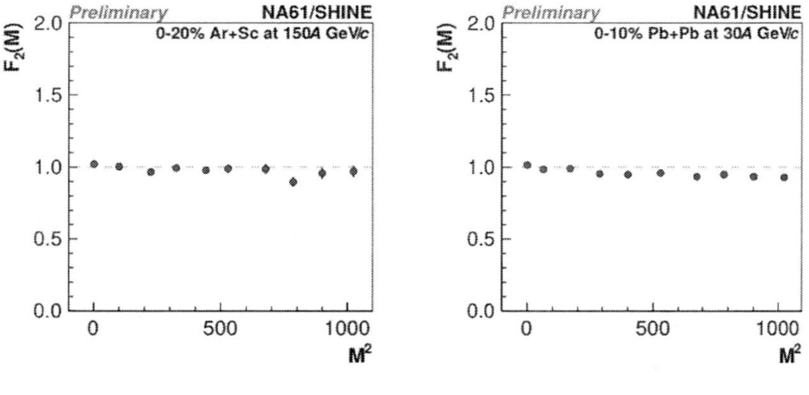

Figure 5. Preliminary results on $F_2(M)$ of mid-rapidity protons measured in 0-20% most central Ar+Sc collisions at 150*A* GeV/*c* (*left*) and 0-10% most central Pb+Pb collisions (*right*).

for a system freezing out at the QCD critical endpoint is expected to be $\phi_2 = 5/6$ assuming that the latter belongs to the 3-D Ising universality class. $F_2(M)$ of protons for Ar+Sc at 150*A* GeV/*c* and Pb+Pb at 30*A* GeV/*c* show no indication for power-law increase with a bin size.

4 Strangeness production in p+p interactions

NA61/SHINE provides unique high precision results on strange hyperon production in *p*+*p* interactions at 158 GeV/*c*: Ξ^-, $\overline{\Xi}^+$, $\Xi^0(1530)$ and $\overline{\Xi}^0(1530)$. The rapidity spectra were reported and compared to transport model predictions. Among tested models, EPOS describes the spectra of mentioned hyperons reasonably well. UrQMD agrees with Ξ^- spectra measured by NA61/SHINE. Additionally, the mean multiplicities of Ξ^- and $\overline{\Xi}^+$ produced in inelastic p+p interactions at 158 GeV/c was used to quantify the strangeness enhancement in A+A collisions at the same centre-of-mass energy per nucleon pair. More details can be found in [21, 22].

5 Spectator-induced electromagnetic effects

Charged spectators of the nuclear collision generate electromagnetic fields, which modify the trajectories of π^+ and π^-, which can result in charge splitting of directed flow. It is expected that the final state spectra π^+ and π^- are modified by those fields. For convenience the double differential π^+/π^- spectra ratios are studied [23, 24]. NA61/SHINE reports new data on spectator-induced electromagnetic effects in Ar+Sc collisions at 40*A* GeV/*c* beam momentum. The results are presented in fig. 6 for PSD selected event centrality classes: central (*left*), semi-central (*middle*) and semi-peripheral (*right*). The magnitude of the π^+/π^- spectra ratio modification increases with the peripherality of the collision. The effect is the strongest for pions moving at spectator velocity ($x_F \approx 0.15$), close to spectator trajectory ($p_T \approx 25$ MeV/*c*).

6 Summary

Presented experimental results on particle production properties reveal unexpected system-size dependence of K^+/π^+ ratio in mid-rapidity, which is not reproduced by any tested the-

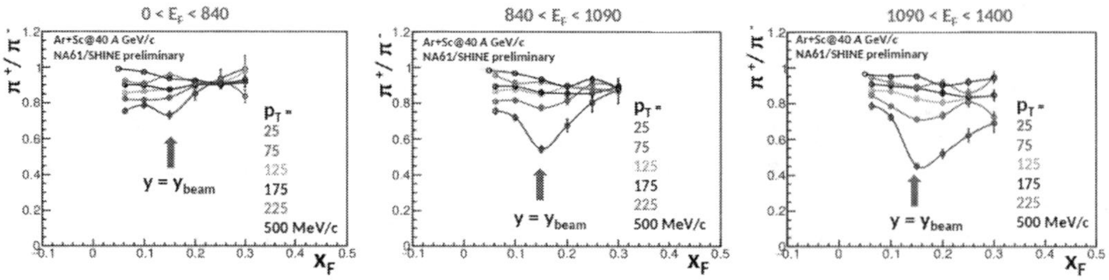

Figure 6. Dependence of π^+/π^- ratio in the final state on x_F measure and p_T measured in PDS selected central (*left*), semi-central (*middle*) and semi-peripheral (*right*) Ar+Sc at 40*A* GeV/*c* beam momentum.

oretical models. The velocity of sound extracted from the width of rapidity distribution of π mesons produced in *N*+*N* interactions and central Be+Be and Ar+Sc collisions is consistent with results for central Pb+Pb and Au+Au collisions. Measurements in a broader energy range are needed to conclude on a possible minimum of the sound velocity in small nuclei collisions. The results on the mean pion multiplicity to the number of wounded nucleons ratio and its collision energy dependence suggests an increase of the effective number of degrees of freedom already in *central* Ar+Sc collisions at the top SPS energies. Moreover, presented results show no indications of the CP in Ar+Sc collisions at 150*A* GeV/*c*. The measured directed flow of pions shows energy dependence, with the slope of negatively charged pions changing sign at different collision centralities in Pb+Pb collisions. Unique results on multi-strange baryon production in *p*+*p* interactions at 158 GeV/*c* beam momentum were discussed. Finally, the first observation of the electromagnetic effects in Ar+Sc at 40*A* GeV/*c* beam momentum was presented.

References

[1] N. Abgrall et al. [NA61/SHINE Collaboration], J. Inst. **9** (2014) P06005.
[2] M. Gazdzicki and M. Gorenstein Acta Phys.Polon. B**30** (1999) 2705.
[3] C. Alt et al. [NA49 Collaboration], Phys.Rev.C**77** (2008) 024903.
[4] R. Poberezhnyuk, M. Gaździcki, and M. Gorenstein, Acta Phys. Polon. B**46** (2015) 1991.
[5] A. Aduszkiewicz et al. [NA61/SHINE Collaboration], Eur. Phys. J. **C77** (2017) 10, 671.
[6] A. Acharya et al. [NA61/SHINE Collaboration], Eur. Phys. J. **C81** (2021) 1, 73.
[7] S. Afanasiev et al. [NA49 Collaboration], Phys. Rev. **C66** (2002) 054902.
[8] N. Abgrall et al. [NA61/SHINE Collaboration], Eur. Phys. J. **C74** (2014) 2794.
[9] A. Acharya et al. [NA61/SHINE Collaboration], Eur. Phys. J. **C80** (2020) 10, 961.
[10] A. Acharya et al. [NA61/SHINE Collaboration], Eur.Phys.J.C**81** (2021) 5, 397.
[11] M. Bleicher et al., EPJ Web of Conf. **52** (2013) 05001.
[12] K. Werner, Nucl. Phys. B**175** (2008) 81.
[13] J. Mohs et al., J. Phys. G**47** (2020) 6, 065101; and private communication;
[14] A. Motornenko et al., Phys. Rev. C**99** (2019) 3, 034909
[15] V. Kireyeu et al, Eur. Phys. J. A**56** (2020) 9, 223; and private communication.
[16] E. Bratkovskaya et al., arXiv:1908.00451.
[17] R. Stock, Phys. Rept. **135** (1986) 259–315.
[18] A. Bialas and R. B. Peschanski, Nucl. Phys. B**273** (1986) 703.
[19] N. Antoniou et al., PRL **97** (2006) 032002.

[20] L. Turko, Phys. Lett. B**227** (1989) 149.
[21] A. Aduszkiewicz et al. [NA61/SHINE Collaboration], Eur. Phys. J. C**80** (2020) 9, 833.
[22] A. Acharya et al. [NA61/SHINE Collaboration], arXiv:2105.09144.
[23] A.Rybicki., A. Szczurek, Phys. Rev. C**75** (2007) 054903.
[24] A. Rybicki, Acta Phys. Polon. B**42** (2011) 867.

Recent Highlights from the PHENIX Collaboration

Takahito Todoroki (for the PHENIX Collaboration)[1,*]

[1]Tomonaga Center for the History of the Universe, University of Tsukuba, Tsukuba, Ibaraki 305, Japan

Abstract. These proceedings present a set of PHENIX highlight results using recent high-luminosity data sets for p+p, p (d, ^{3}He)+A, and A+A(B) collisions at RHIC top energies with a focus on: (1) light-flavor particle production; (2) hard-probes such as jet-like correlations, direct photon, and heavy-quark productions; and (3) azimuthal anisotropy in small collision systems.

1 Introduction

During 16 years of operations, PHENIX collected a rich amount of data for 9 collision species at several collision energies. Now the focus of the PHENIX heavy-ion program has been complete analyses of recent high-luminosity data sets for p+p, p (d, ^{3}He)+Au(Al), as well as Au(Cu)+Au and U+U collisions at RHIC top energies. In these proceedings, we discuss the following recent highlight results from such an effort: (i) π^0, η, and ϕ meson productions; (ii) direct γ-hadron correlations and single direct photon production; (iii) inclusive J/ψ and $b\bar{b}$ pair productions; and (iv) azimuthal anisotropy in small collision systems.

2 Light Flavor Production

2.1 π^0 and η productions

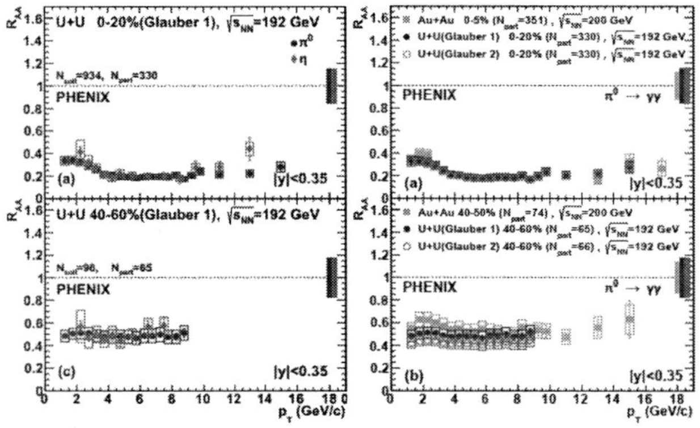

Figure 1. Nuclear modification factor R_{AA} as a function of p_T for π^0 and η in U+U collisions at $\sqrt{s_{NN}}$ = 192 GeV.

Figure 1 shows the first measurements of the nuclear modification factor R_{AA} for π^0 and η mesons in U+U collisions at $\sqrt{s_{NN}}$ = 192 GeV [1]. The R_{AA} of π^0 and η is measured to

*e-mail: todoroki@bnl.gov

 https://doi.org/10.1051/epjconf/202225901002

be consistent each other within uncertainties over the wide transverse momentum (p_T) range, indicating the similar suppressions of π^0 and η in QGP medium. A comparison of π^0 R_{AA} between U+U and Au+Au collisions made at similar numbers of participant nucleons shows the suppression depends only on the medium size regardless of collision systems.

2.2 ϕ production

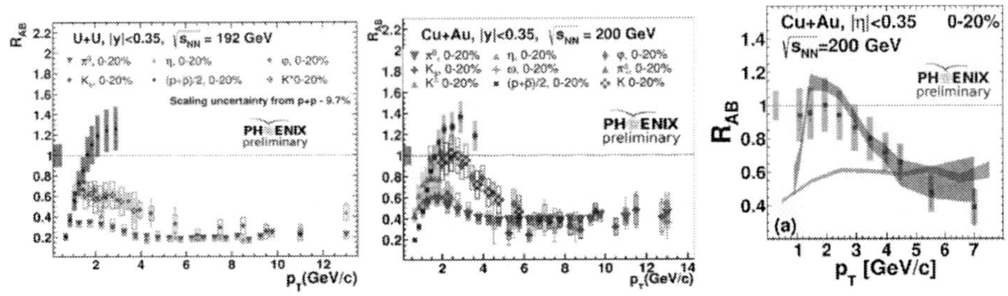

Figure 2. Nuclear modification factor R_{AA} for ϕ meson in (left) U+U collisions at $\sqrt{s_{NN}}$ = 192 GeV and (middle and right) Cu+Au collisions at $\sqrt{s_{NN}}$ = 200 GeV. The grey and purple bands in the right plot show AMPT with string melting and PYTHIA8 calculations, respectively.

Measurements of R_{AA} are now extended to ϕ meson for U+U and Cu+Au collisions as shown in Fig. 2. In both collisions, R_{AA} of various hadrons follows one universal curve at high-p_T probably due to flavor-independent energy loss in the medium. In contrast, at intermediate p_T, those hadrons show a hierarchy of $R_{AA}(p\bar{p}) > R_{AA}(\phi, K^*) > R_{AA}(\pi^0, \eta)$, which can be taken as an interplay of radial flow, strangeness enhancement, and quark coalescence effects. The importance of coalescence effect in ϕ meson production is supported by the better description of R_{AA} by AMPT with coalescence than PYTHIA without coalescence.

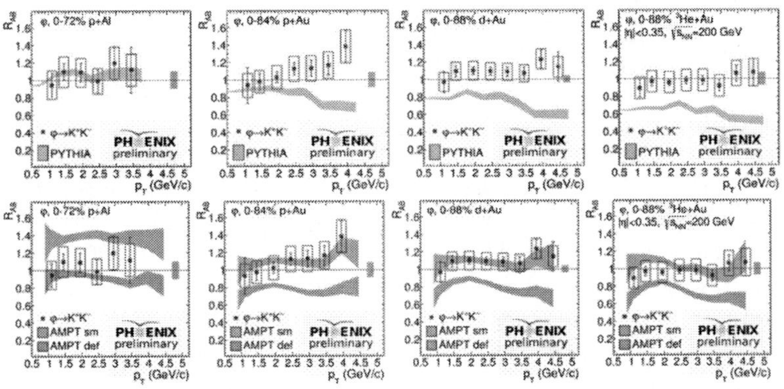

Figure 3. Nuclear modification factor R_{AB} for ϕ meson in p+Al, p+Au, d+Au, and ^{3}He+Au collisions compared with (top) PYTHIA and (bottom) AMPT calculations.

The ϕ meson production in small collision systems is investigated by measuring R_{AB} as presented in Fig. 3. From p+Au to ^{3}He+Au collisions, R_{AB} is well described by AMPT with coalescence. While, in p+Al collisions, PYTHIA and AMPT without coalescence effects better describe the data. This implies that the QGP medium does not form in p+Al collisions.

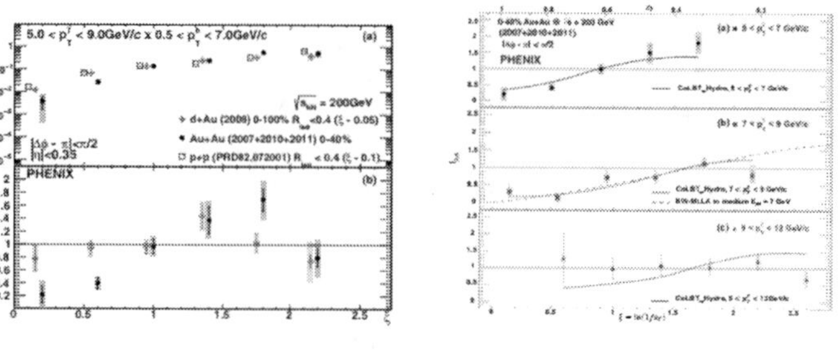

Figure 4. (Left-Top) Correlated yields and (Left-Bottom and Right) their ratios to the $p+p$ baseline I_{AA} as a function of $\xi = \ln(p^{\gamma}/p^{h\pm})$ in $p+p$, $d+$Au, and Au+Au collisions at $\sqrt{s_{NN}}$ = 200 GeV.

3 Hard Probes

3.1 γ-hadron correlations

In-medium parton energy-loss mechanism is studied via direct γ-hadron correlations in Au+Au collisions at $\sqrt{s_{NN}}$ = 200 GeV. Figure 4 shows the ratio of correlated yields I_{AA} between Au(d)+Au and $p+p$ collisions. In $d+$Au collisions, I_{AA} shows no modification from the $p+p$ baseline. However, in Au+Au collisions, I_{AA} shows a suppression at low $\xi = \ln(p^{\gamma}/p^{h\pm})$ (high p_T) and an enhancement at high ξ (low-p_T), which can be understood as parton energy loss at high p_T and its redistribution at low p_T. This observation can be quantitatively reproduced by the CoLBT_Hydro model calculation [3].

3.2 Direct photon production

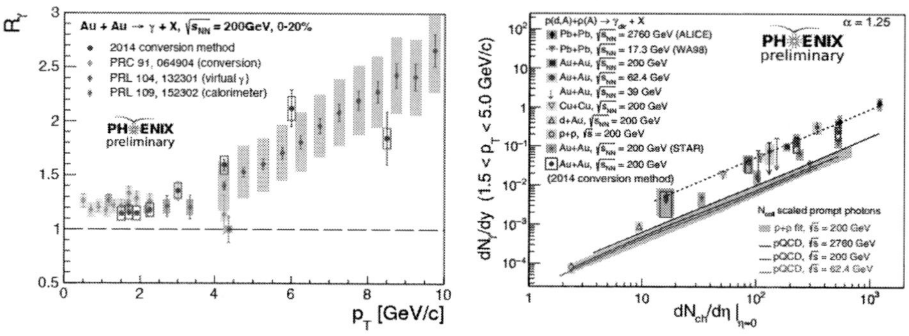

Figure 5. (Left) The direct photon fraction R_{γ} as a function of p_T and (Right) the integrated direct photon yield as a function of charged particle multiplicity in Au+Au collisions at $\sqrt{s_{NN}}$ = 200 GeV.

Single direct photon measurement has been updated using high luminosity Au+Au collision data set from the 2014 data taking period as shown in Fig. 5. The new measurement of direct photon fraction R_{γ} with the external conversion method reconfirms the previous measurements with different methods over the wide p_T range, significantly reducing the systematic uncertainties at high p_T. The integrated direct photon yield from this new measurement is also in excellent agreement with previous results.

3.3 Heavy quark production

Figure 6 shows the cross sections of inclusive J/ψ as a function of p_T in $p+p$ collisions at $\sqrt{s}$ = 200 and 510 GeV in mid- and forward-rapidities. In $p_T > 2$ GeV/c, the J/ψ production can be successfully described by the NLO+NRQCD calculations [6, 7] in both rapidi-

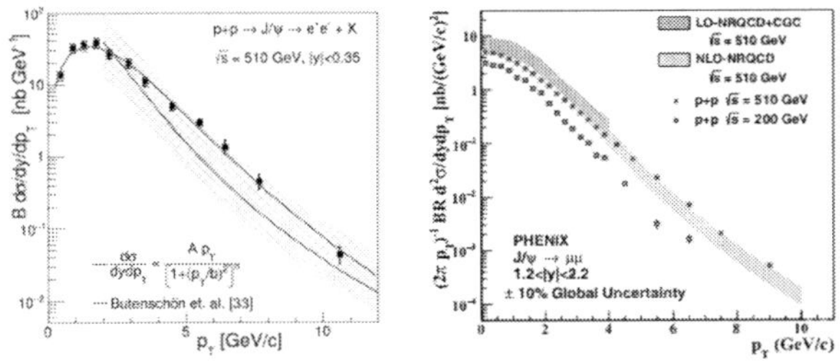

Figure 6. Cross section of J/ψ as a function of p_T and rapidity in p+p collisions at $\sqrt{s}$ = 200 and 510 GeV in (left) mid- and (right) forward-rapidities.

ties. Below p_T =2 GeV/c, the J/ψ production in forward-rapidity is overestimated by the LO+NRQCD+CGC calculations [7, 8].

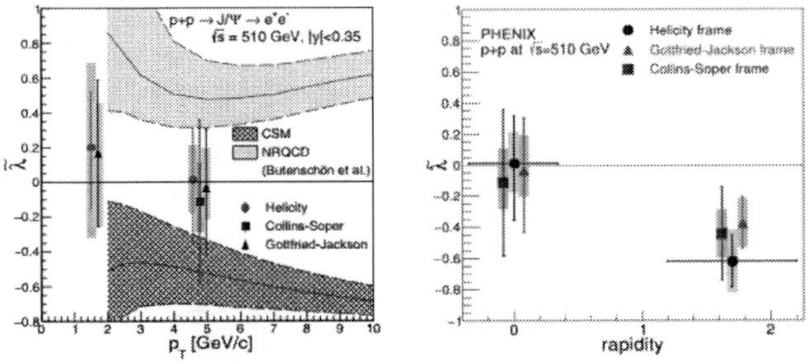

Figure 7. The frame invariant polarization $\tilde{\lambda}$ of J/ψ in the Helicity, Collins-Soper, and Gottfried-Jackson in p+p collisions at $\sqrt{s}$ = 510 GeV as a function of (left) p_T and (right) rapidity.

The J/ψ production mechanism has also been studied via measurements of the frame invariant polarization $\tilde{\lambda}$ in p+p collisions at $\sqrt{s}$ = 510 GeV as shown in Fig 7. At mid-rapidity, $\tilde{\lambda}$ is consistent with zero within large uncertainties over the entire p_T range. The calculations of NLO+NRQCD and color-singlet models are consistent with the data within large uncertainties. Unlike the mid-rapidity results, $\tilde{\lambda}$ shows a negative value in forward-rapidity, indicating longitudinal polarization.

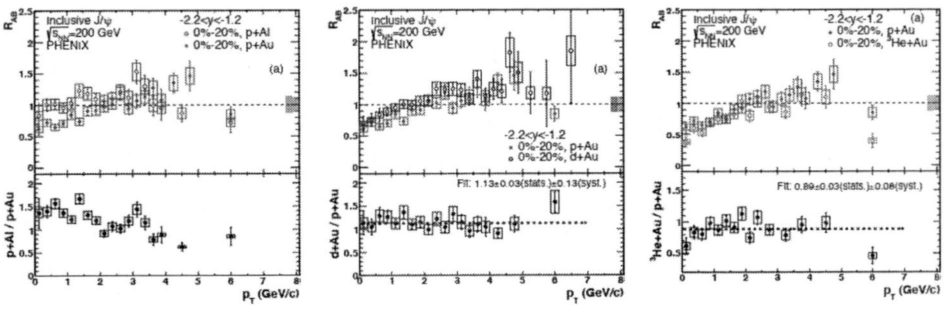

Figure 8. Nuclear modification factor R_{AB} of inclusive J/ψ in p+Al, p+Au, d+Au, and ^{3}He+Au collisions at $\sqrt{s_{NN}}$ = 200 GeV.

Cold nuclear matter effects in the inclusive J/ψ production are studied via the R_{AB} measurements in p+Al, p+Au, d+Au, and ^{3}He+Au collisions at $\sqrt{s_{NN}}$ = 200 GeV [9] as shown in Fig. 8. In p+Al collisions, J/ψ R_{AB} shows little suppression probably due to its small target size. From p+Au to ^{3}He+Au collisions, J/ψ R_{AB} is similar despite the increased energy density, indicating that cold-nuclear matter effects are dominant.

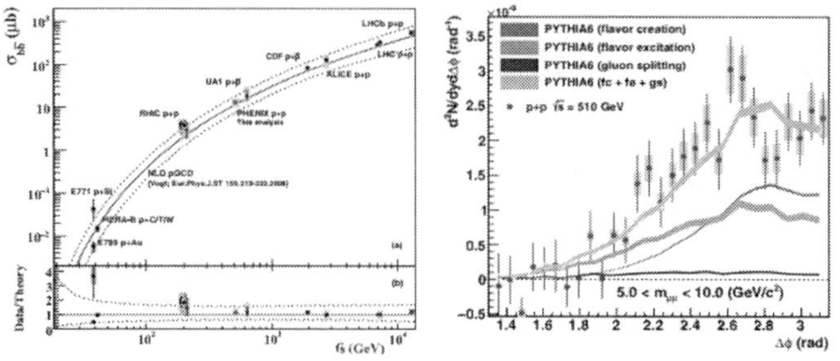

Figure 9. (Left) The production rate of $b\bar{b}$ pairs and (Right) relative angular distribution between $b\bar{b}$ pair quarks in p+p collisions at $\sqrt{s}$ = 510 GeV.

Shown in Fig. 9 is the production rate of $b\bar{b}$ pairs obtained via model dependent extrapolations from the yield of high-mass like-sign muons from neutral B meson oscillations in p+p collisions at $\sqrt{s}$ = 510 GeV [10]. The production rate was estimated to be $\sigma_{b\bar{b}}$=13.1±0.60(stat.)±1.5(sys.)±2.7(global). This new result follows the worldwide trend and is successfully described by pQCD calculations. A comparison of the relative angular distribution between $b\bar{b}$ pair quarks to PYTHIA calculations favors a dominant mixture of flavor creation and excitation over the gluon splitting.

4 Flow in small systems

As scientific due diligence, PHENIX revisited the measurements of second- and third-order azimuthal anisotropies v_2 and v_3 in high-multiplicity p+Au, d+Au, and ^{3}He+Au collisions at $\sqrt{s_{NN}}$ = 200 GeV using the 3×2PC method, taking the advantage of the event-mixing technique in canceling possible detector and beam optics effects [12]. In this 3×2PC method, we measure two-particle correlations of three different sets of pairs. The n-th order oscillation c_n extracted from those correlations is then combined to obtain single particle azimuthal anisotropy as $v_n^A = \sqrt{\frac{c_n^{AB} c_n^{AC}}{c_n^{BC}}}$, where A, B, and C stand for sub-events used to construct those correlations. As was done in the PHENIX results published in Nature Physics using the EP method [11], we used BBCS($-3.9 < \eta < -3.1$), FVTXS($-3 < \eta < -1$), and CNT($|\eta| <$ 0.35) detectors to measure mid-rapidity v_2 and v_3. Also as a systematic study, we employed FVTXS($-3 < \eta < -1$), CNT($|\eta| < 0.35$), and FVTXN($1 < \eta < 3$) detectors.

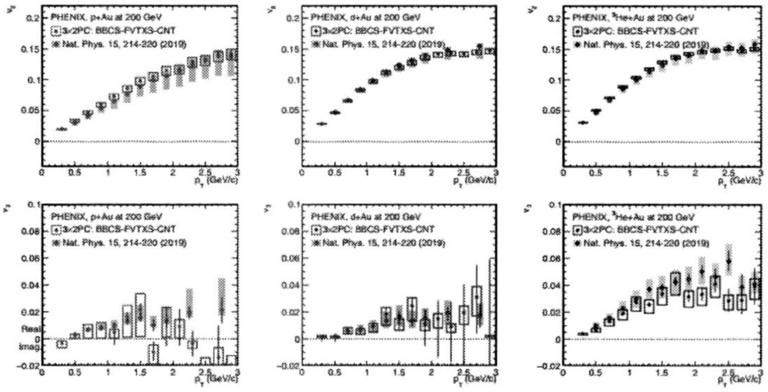

Figure 10. Measurements of (top panels) v_2 and (bottom panels) v_3 with the 3×2PC method by the BBCS-FVTXS-CNT detector combination in high-multiplicity p+Au, d+Au, and ^{3}He+Au collisions at $\sqrt{s_{NN}}$ = 200 GeV.

Figure 10 shows measurements of v_2 and v_3 with the 3×2PC method by the BBCS-FVTXS-CNT detector combination. Both v_2 and v_3 are in excellent agreement with PHENIX results published in Nature Physics, which reconfirms the robustness of the previous measurements.

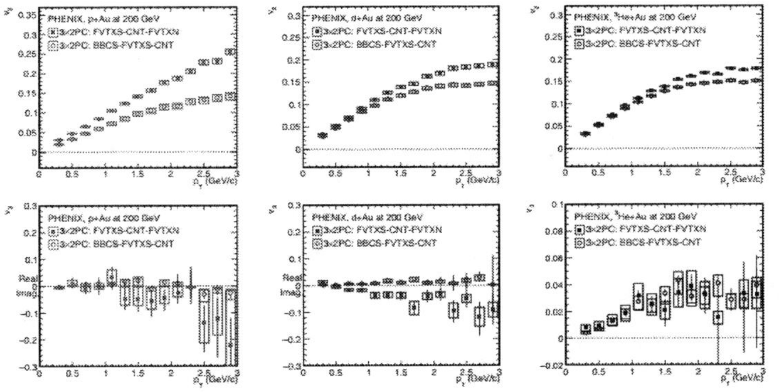

Figure 11. Measurements of (top panels) v_2 and (bottom panels) v_3 with the 3×2PC method by the FVTXS-CNT-FVTXN detector combination in high-multiplicity p+Au, d+Au, and ^{3}He+Au collisions at $\sqrt{s_{NN}}$ = 200 GeV.

In contrast, measurements with the 3×2PC method by the FVTXS-CNT-FVTXN detector combination show larger v_2 than that with the BBCS-FVTXS-CNT detector combination. This can be understood by stronger non-flow in the FVTN acceptance than BBCS acceptance due to smaller multiplicity in the FVTXN acceptance. Also, the wider rapidity gap between FVTXS and FVTXN than that between BBCS and FVTXS results in stronger event-plane decorrelations. Both of these changes by the kinematic selections push up v_2 values. This detector combination shows v_3 consistent with that with the BBCS-FVTXS-CNT detector combination in ^{3}He+Au collisions, while it shows mostly imaginary v_3 in p+Au and d+Au collisions. This is probably because the stronger non-flow and the event-plane decorrelation effects in these systems overwhelm the small v_3 signals.

These systematic measurements warrant more experimental and theoretical studies to understand the non-flow and event-plane decorrelation effects in small collision systems.

5 Summary

In summary, we have presented highlight results from recent ample data sets for p+p, small, and large system collisions such as: (i) π^0, η, and ϕ meson productions; (ii) direct γ-hadron correlations and single direct photon productions; (iii) inclusive J/ψ and $b\bar{b}$ pair productions; and (iv) azimuthal anisotropy in small collision systems. Many important analyses that use 36 billion events of Au+Au collisions at $\sqrt{s_{NN}}$ = 200 GeV from 2014 and 2016 operations are underway and many exciting physics results are yet to come from the PHENIX Collaboration.

References

[1] U. Acharya *et al.* (PHENIX Collaboration), Phys. Rev. C 102, 064905 (2020)
[2] U. Acharya *et al.* (PHENIX Collaboration), Phys. Rev. C 102, 054910 (2020)
[3] W. Chen *et al.*, Phys. Lett. B 777, 86-90 (2018)
[4] U. Acharya *et al.* (PHENIX Collaboration), Phys. Rev. D 101, 052006 (2020)
[5] U. Acharya *et al.* (PHENIX Collaboration), Phys. Rev. D 102, 072008 (2020)
[6] M. Butenschoen *et al.*, Phys. Lev. Lett. 106, 022003 (2011)
[7] Y. Ma *et al.*, Phys. Lev. Lett. 106, 042002 (2011)
[8] Y. Ma *et al.*, Phys. Lev. Lett. 113, 192301 (2014)
[9] U. Acharya *et al.* (PHENIX Collaboration), Phys. Rev. C 102, 014902 (2020)
[10] U. Acharya *et al.* (PHENIX Collaboration), Phys. Rev. D 102, 092002 (2020)
[11] C. Aidala *et al.* (PHENIX Collaboration), Nature Physics 15, 214 - 220 (2019)
[12] U. Acharya *et al.* (PHENIX Collaboration), arXiv:2107.06634 [hep-ex]

EPJ Web of Conferences **259**, 01003 (2022) https://doi.org/10.1051/epjconf/202225901003
SQM 2021

Overview of open heavy-flavour and quarkonia measurements with ALICE

Andrea Dubla[1,*] for the ALICE Collaboration

[1]GSI Helmholtzzentrum für Schwerionenforschung, Planckstraße 1, 64291 Darmstadt, Germany

Abstract.
Heavy-flavour hadrons, i.e. hadrons containing charm or beauty quarks, are effective probes to test perturbative-QCD (pQCD) calculations, to investigate the different hadronisation mechanisms, and to study the quark–gluon plasma (QGP) produced in relativistic heavy-ion collisions at the LHC. Measurements performed in pp and p–Pb collisions have recently revealed unexpected features not in line with the expectations based on previous measurements from e^+e^- and ep collisions, showing that charm fragmentation fractions are not universal. The investigation of initial-state effects such as shadowing in the collision of a proton with a heavy nucleus is also performed. Measurements of open heavy-flavour and quarkonia production in Pb–Pb collisions allow for testing the mechanisms of heavy-quark transport, energy loss, and coalescence effects during the hadronisation in the presence of a QCD medium. In this contribution, the most recent results on open heavy-flavour and quarkonia production in pp, p–Pb, and Pb–Pb collisions obtained by the ALICE Collaboration are discussed.

1 Introduction

The study of heavy-flavour hadron production in proton–proton (pp) collisions provides an important test for quantum chromodynamics (QCD) calculations. Heavy-flavour hadron production is usually computed with pQCD calculations as the convolution of the parton distribution functions (PDFs) of the incoming protons, the partonic cross section and the fragmentation functions that describe the transition from charm quarks into charm hadrons. The latter are typically parametrised from measurements performed in e^+e^- or ep collisions, under the assumption that the hadronisation of charm quarks into charm hadrons is a universal process independent of the colliding system. In the presence of a quark–gluon plasma (QGP), the strongly-interacting colour-deconfined state of matter created in ultra-relativistic heavy-ion collisions, an additional hadronisation mechanism, alternative to the in-vacuum string fragmentation, is considered. In this mechanism, known as coalescence, light quarks from the medium coalesce with the heavy quark to form a meson or baryon. By studying heavy-flavour production in nucleus–nucleus collisions, the relevance of coalescence in the medium for heavy quarks can be probed. Models that include hadronisation via coalescence for charm quarks qualitatively describe D-meson measurements in Pb–Pb collisions at the LHC. The possible occurrence of coalescence in pp collisions is one of the topics of investigation to provide a better description of heavy-baryon production.

*e-mail: andrea.dubla@cern.ch

2 Charm production and fragmentation in pp collisions

The large data samples collected during Run 2 of the LHC at $\sqrt{s}$ = 5.02 and 13 TeV allowed ALICE to measure the vast majority of charm quarks produced at midrapidity in the proton-proton (pp) collisions by reconstructing the decays of the ground-state charm hadrons down to very low transverse momenta (p_T), i.e. all charm-meson species (D^0, D^+, D_s^+ [1]) and charm-baryon species (Λ_c^+ [2], $\Xi_c^{0,+}$ [3, 4] and Ω_c^0). The ratios of different hadron species as function of the transverse momentum p_T help to study heavy flavour hadronisation. Both the prompt and non-prompt (coming from B-hadron decays) meson-to-meson yield ratios of D mesons are compatible with pQCD calculations [5–7] using fragmentation fractions extracted from e^+e^- collision data. Such calculations, however, significantly underpredict all the measured baryon-to-meson yield ratios, indicating that the fragmentation fractions are not universal. The Λ_c^+/D^0 ratio in pp collisions, reported in the left panel of Fig. 1, shows a clear decrease with increasing p_T and it is better described by models with an extension of colour reconnection beyond the leading colour approximation [8], models relying on hadronisation via coalescence [9, 10], or statistical hadronisation models with an augmented set of baryon states predicted by the relativistic quark model (RQM) [11]. Measurements of the heavier charm-strange baryons $\Xi_c^{0,+}$ and Ω_c^0 pose further important constraints to charm-quark hadronisation models. For the Ω_c^0 the absolute branching ratio (BR) of the decay channel $\Omega_c^0 \rightarrow \Omega^{\pm}\pi^{\mp}$ is not measured, hence only the BR multiplied with the cross section is reported. The yield ratios of these charm baryons to the D^0-meson one are presented in the middle and right panel of Fig. 1, and compared to predictions from the same models that describe the Λ_c^+/D^0 ratio. For the $\Xi_c^{0,+}/D^0$ and $BR\times\Omega_c^0/D^0$ all calculations, except the Catania model [10], which gets closer to the data and relies on hadronisation via coalescence, are underpredicting the measured ratios, hinting to an even stronger enhancement for charm-strange baryons. The predictions tuned on measurements in e^+e^- collisions underestimate all three baryon-to-meson yield ratios [12], providing further evidence that different processes are involved in charm hadronisation for elementary and hadronic collisions. These new charm-baryon measurements provide important constraints to models of charm-quark hadronisation in pp collisions, being, in particular, sensitive to the description of charm-strange baryon production in the colour reconnection approach, and to the possible contribution of coalescence to charm-quark hadronisation in pp collisions.

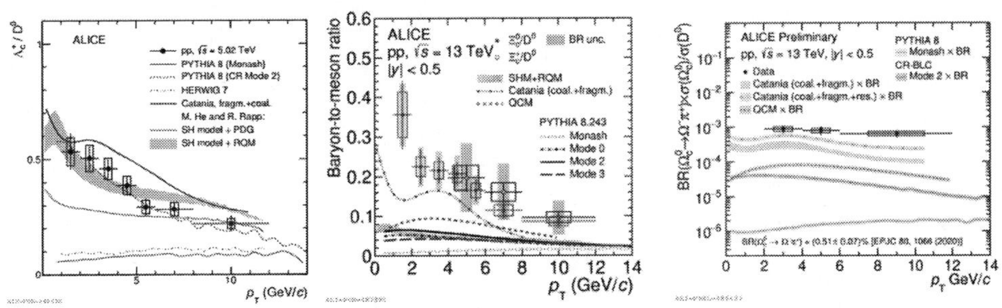

Figure 1. Left panel: The Λ_c^+/D^0 ratio measured in pp collisions at $\sqrt{s}$ = 5.02 TeV [2]. Middle panel: The Ξ_c^0/D^0 and Ξ_c^+/D^0 ratios as a function of p_T in pp collisions at $\sqrt{s}$ = 13 TeV [4]. Right panel: The $BR\times\Omega_c^0/D^0$ as a function of p_T in pp collisions at $\sqrt{s}$ = 13 TeV. All yield ratios are compared to theoretical model calculations.

The charm fragmentation fractions, $f(c \rightarrow H_c)$ (shown in the left panel of Fig. 2), which represent the probabilities of a charm quark to hadronise into a given charm hadron, were measured for the first time including baryon species at midrapidity in pp collisions at the LHC and are observed to be different from the ones measured in e^+e^- and ep collisions, showing evidence that the assumption of universality (collision-system independence) of charm-to-hadron fragmentation is not valid [13]. The charm quarks hadronize into baryons almost 40% of the time, which is four times more often than what was measured at colliders with electron beams. The fragmentation fraction for the Ξ_c^0 was measured for the first time in any collision system. The $c\bar{c}$ production cross section per unit of rapidity at midrapidity ($d\sigma^{c\bar{c}}/dy|_{|y|<0.5}$) was calculated by summing the cross sections of all measured ground-state charm hadrons (D^0, D^+, D_s^+, Λ_c^+, and Ξ_c^0 and their charge conjugates). The contribution of the Ξ_c^0 was multiplied by a factor of two, in order to account for the contribution of the Ξ_c^+. The resulting $c\bar{c}$ cross section per unit of rapidity at midrapidity is $d\sigma^{c\bar{c}}/dy|_{|y|<0.5} = 1165 \pm 44(\text{stat})^{+134}_{-101}(\text{syst})\ \mu\text{b}$. This measurement was done for the first time in hadronic collisions at the LHC including the charm baryon states [13]. The newly measured fragmentation fractions at midrapidity in pp collisions at $\sqrt{s} = 5.02$ TeV allowed the recomputation of the charm production cross sections per unit of rapidity at midrapidity in pp collisions at $\sqrt{s} = 2.76$ and 7 TeV. The charm cross sections were obtained by scaling the p_T-integrated D^0-meson cross section for the relative fragmentation fraction of a charm quark into a D^0 meson. The updated $c\bar{c}$ cross sections at $\sqrt{s} = 2.76$ and 7 TeV are about 40% higher than the previously published results, reflecting the differences in the fragmentation into charm baryons with respect to those measured in e^+e^- and pp collisions. In the right panel of Fig. 2, the measured $c\bar{c}$ production cross sections per unit of rapidity at midrapidity are compared with FONLL [5] and NNLO [14, 15] predictions as a function of the collision energy. The $c\bar{c}$ cross sections measured at midrapidity at the LHC lies at the upper edge of the theoretical pQCD calculations. The measurements described above not only provide constraints to pQCD calculations but are also important as references for the investigation of the charm-quark interaction with the medium created in heavy-ion collisions.

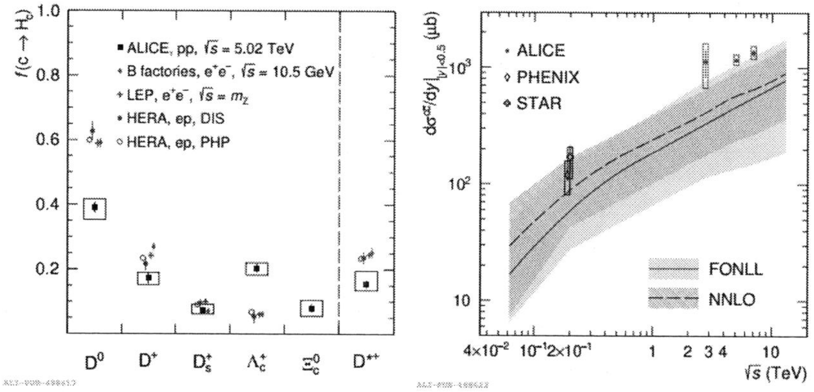

Figure 2. Left panel: Charm-quark fragmentation fractions into charm hadrons measured in pp collisions at $\sqrt{s} = 5.02$ TeV in comparison with experimental measurements performed in e^+e^- and in ep collisions [13]. Right panel: Charm production cross section at midrapidity per unit of rapidity as a function of the collision energy. The FONLL and NNLO calculations are also shown [5, 14, 15].

3 Radial flow, hadronisation, and shadowing in p–Pb collisions

Additional investigations on the charm hadronisation measuring charm baryon production are carried out in p–Pb collisions at $\sqrt{s_{NN}} = 5.02$ TeV [2]. The Λ_c^+/D^0 yield ratio as a function of

p_T in pp and p–Pb collisions is shown in the left panel of Fig. 3. For the first time at the LHC the Λ_c^+ has been measured down to $p_T = 0$. A clear decreasing trend with increasing p_T is seen in p–Pb collisions for $p_T > 2$ GeV/c. The ratios measured in pp and p–Pb collisions are qualitatively consistent with each other, although a larger Λ_c^+/D^0 ratio in $3 < p_T < 8$ GeV/c and a lower ratio in $1 < p_T < 2$ GeV/c are measured in p–Pb collisions with respect to pp collisions suggesting the presence of possible radial flow effect or of a further modification of the hadronisation mechanism in p–Pb collisions. In order to further investigate these effects and to study also cold nuclear matter effects (CNM), the nuclear modification factor (R_{pPb}) was measured (right panel of Fig. 3). For $p_T < 2$ GeV/c an R_{pPb} lower than unity is measured with a significance of 2.6σ, which provides a hint for a suppression of the yield in p–Pb with respect to pp collisions [2]. The suppression is described by model calculations that include modification of the parton distribution function in Pb nuclei (shadowing). For $p_T >$ 2 GeV/c the R_{pPb} is systematically above unity, with a maximum deviation from $R_{pPb} = 1$ reaching 2.2σ for $5 < p_T < 6$ GeV/c. The indication that Λ_c^+ production is suppressed at low p_T and enhanced at mid p_T in p–Pb collisions with respect to pp collisions could also be seen as possible radial flow effects present in p–Pb collisions. The current precision of the measurement is not sufficient to draw conclusions on the role of different CNM effects and the possible presence of final-state effects, like the radial flow.

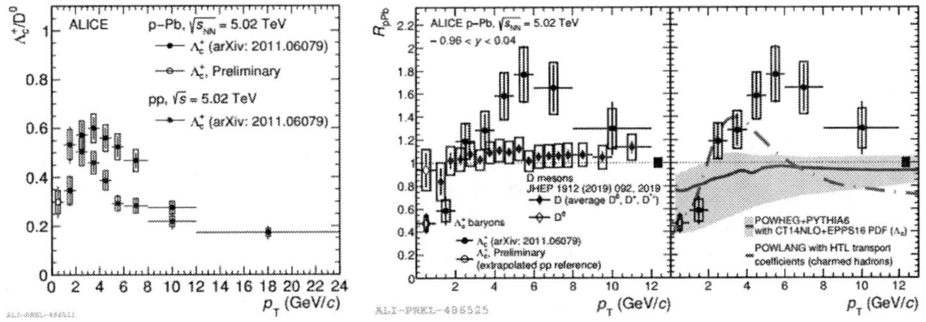

Figure 3. Left panel: The Λ_c^+/D^0 yield ratio as a function of p_T measured in pp and in p–Pb collisions $\sqrt{s_{NN}} = 5.02$ TeV. Right panel: R_{pPb} of prompt Λ_c^+ baryons in p–Pb collisions at $\sqrt{s_{NN}} = 5.02$ TeV as a function of p_T compared to D-meson R_{pPb} (left) and to model expectations (right) [2].

Recent measurements of inclusive, prompt, and non-prompt J/ψ production in p–Pb collisions at a nucleon–nucleon centre-of-mass energy of $\sqrt{s_{NN}} = 5.02$ TeV provide additional information on the role of CNM effects [16]. The J/ψ mesons are reconstructed in the di-electron decay channel at midrapidity down to a transverse momentum $p_T = 0$. Non-prompt J/ψ mesons, which originate from the decay of beauty hadrons, are separated from promptly produced J/ψ on a statistical basis for $p_T > 1$ GeV/c. In the left panel of Fig. 4, the R_{pPb} of prompt J/ψ is reported as a function of p_T in comparison with ATLAS results [17]. Given also the relatively small fraction of J/ψ from b-hadron decays for $p_T < 14$ GeV/c, the R_{pPb} of prompt J/ψ is comparable with that of the inclusive J/ψ. As shown in the left panel of Fig. 4, both trends indicate that the suppression observed at midrapidity is a low-p_T effect, concentrated at $p_T < 3$ GeV/c. The measurements are compared with results from model predictions which embed different CNM effects into prompt J/ψ production [18]. A calculation including the effects of coherent energy loss [19], with or without the introduction of nuclear shadowing effects according to EPS09 nPDF, provides a fairly good description of the measurements. The measured value of the p_T-integrated R_{pPb} of non-prompt J/ψ (shown in the right panel of Fig. 4) is 0.79 ± 0.11 (stat.)±0.13 (syst.)$^{+0.01}_{-0.02}$(extr.), suggesting the presence of nuclear effects also for the non-prompt component. Within uncertainties, the measurements are compatible with those of the LHCb collaboration [20] at both forward and backward rapidity. All charmonium measurements reported here are in agreement with the mild degree of suppression predicted by models employing nPDFs.

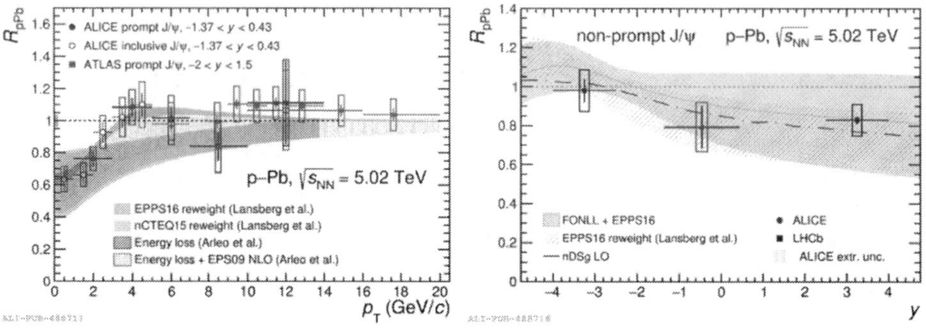

Figure 4. Left panel: R_{pPb} of prompt J/ψ as a function of p_T along with that of inclusive J/ψ at midrapidity. Right panel: Nuclear modification factor R_{pPb} of non-prompt J/ψ as a function of rapidity [16].

4 Charm and beauty production in heavy-ion collisions

The Λ_c^+/D^0 yield ratio in central Pb–Pb collisions at $\sqrt{s_{NN}}$ = 5.02 TeV extends the study of the hadronisation properties of charm baryons to the extreme conditions realized in central Pb–Pb collisions. This ratio is larger than the one measured in pp collisions in the interval 4 < p_T < 8 GeV/c. Figure 5 shows the p_T-differential Λ_c^+/D^0 yield ratio measured in central collisions compared to three different theoretical predictions: Catania [21], TAMU [22], and the GSI/Hd+BW statistical hadronisation model [23]. Theoretical calculations that model the charm-quark transport in the QGP and include hadronisation via both coalescence and fragmentation processes describe the data.

The production of non-prompt D^0 and D_s^+ mesons from b-hadron decays was measured at midrapidity ($|y|$ < 0.5) in Pb–Pb collisions at $\sqrt{s_{NN}}$ = 5.02 TeV. The R_{AA} of non-prompt D_s^+ is measured in the range 2 < p_T < 36 GeV/c. A significant suppression up to a factor of about three is observed for p_T > 5 GeV/c in the 10% most central collisions. The data are described by models that include both collisional and radiative processes in the calculation of beauty quark in-medium energy loss and quark coalescence as a hadronisation mechanism. A higher R_{AA} for non-prompt D_s^+ mesons than for prompt D_s^+ mesons is measured at p_T > 3 GeV/c. This is expected by models in which beauty quarks lose less energy than charm quarks in the QGP medium because of their larger mass. In the right panel of Fig. 5, the R_{AA} of non-prompt D_s^+ mesons was divided by that of non-prompt D^0 mesons and compared with the same ratio from the TAMU model [22]. The R_{AA} for non-prompt D_s^+ mesons is expected to be higher than the one of non-prompt D^0 mesons due to the enhanced production of B_s^0 mesons in heavy-ion collisions in case of beauty-quark hadronisation via coalescence. The data are in agreement with the TAMU predictions within the uncertainties.

Further studies on the evolution of heavy-flavour hadronisation across collision systems, including studies as a function of the event multiplicity for rare baryons, will be done in future LHC runs.

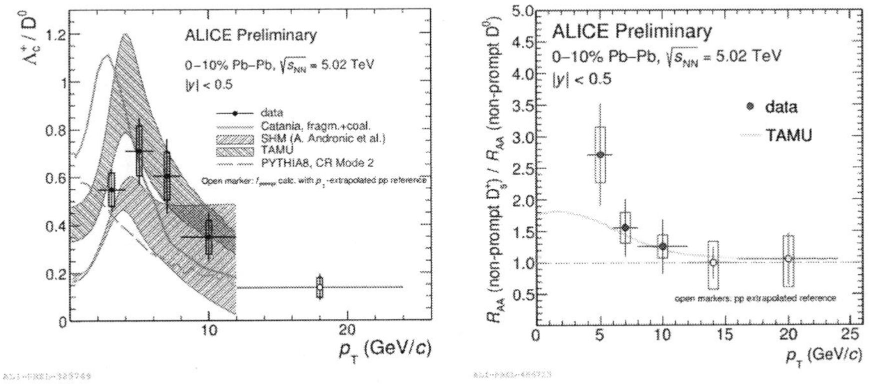

Figure 5. Left panel: The $p_{\rm T}$-differential Λ_c^+/D^0 yield ratio in pp collisions and in the10% most central Pb–Pb collisions at $\sqrt{s_{\rm NN}}$ = 5.02 TeV compared to the predictions of different theoretical calculations. Right panel: Ratio between the $R_{\rm AA}$ of non-prompt D_s^+ mesons and non-prompt D^0 mesons in Pb–Pb collisions at $\sqrt{s_{\rm NN}}$= 5.02 TeV for the 0–10% centrality class compared to the TAMU model [22].

References

[1] S. Acharya et al. (ALICE), JHEP **05**, 220 (2021), 2102.13601
[2] S. Acharya et al. (ALICE) (2020), 2011.06078
[3] S. Acharya et al. (ALICE) (2021), 2105.05616
[4] S. Acharya et al. (ALICE) (2021), 2105.05187
[5] M. Cacciari, S. Frixione, N. Houdeau, M.L. Mangano, P. Nason, G. Ridolfi, JHEP **10**, 137 (2012), 1205.6344
[6] G. Kramer, H. Spiesberger, Nucl. Phys. B **925**, 415 (2017), 1703.04754
[7] I. Helenius, H. Paukkunen, JHEP **05**, 196 (2018), 1804.03557
[8] J.R. Christiansen, P.Z. Skands, JHEP **08**, 003 (2015), 1505.01681
[9] J. Song, H.h. Li, F.l. Shao, Eur. Phys. J. C **78**, 344 (2018), 1801.09402
[10] V. Minissale, S. Plumari, V. Greco (2020), 2012.12001
[11] M. He, R. Rapp, Phys. Lett. B **795**, 117 (2019), 1902.08889
[12] P. Skands, S. Carrazza, J. Rojo, Eur. Phys. J. C **74**, 3024 (2014), 1404.5630
[13] S. Acharya et al. (ALICE) (2021), 2105.06335
[14] D. d'Enterria, A.M. Snigirev, Eur. Phys. J. C **78**, 359 (2018), 1612.08112
[15] M. Czakon, P. Fiedler, A. Mitov, Phys. Rev. Lett. **110**, 252004 (2013), 1303.6254
[16] S. Acharya et al. (ALICE) (2021), 2105.04957
[17] M. Aaboud et al. (ATLAS), Eur. Phys. J. C **78**, 171 (2018), 1709.03089
[18] A. Kusina, J.P. Lansberg, I. Schienbein, H.S. Shao, Phys. Rev. Lett. **121**, 052004 (2018), 1712.07024
[19] F. Arleo, R. Kolevatov, S. Peigné, M. Rustamova, JHEP **05**, 155 (2013), 1304.0901
[20] R. Aaij et al. (LHCb), JHEP **02**, 072 (2014), 1308.6729
[21] S. Plumari, V. Minissale, S.K. Das, G. Coci, V. Greco, Eur. Phys. J. C **78**, 348 (2018), 1712.00730
[22] M. He, R. Rapp, Phys. Rev. Lett. **124**, 042301 (2020), 1905.09216
[23] A. Andronic, P. Braun-Munzinger, M.K. Köhler, A. Mazeliauskas, K. Redlich, J. Stachel, V. Vislavicius, JHEP **07**, 035 (2021), 2104.12754

Heavy ion results from LHCb

Jiayin Sun (for the LHCb collaboration)[1,2,*]

[1]Università degli Studi di Cagliari, Cagliari, Italy
[2]INFN Sezione di Cagliari, Cagliari, Italy

Abstract. The LHCb detector is a single-arm spectrometer fully instrumented in the forward rapidity at the LHC. In this contribution we introduce some of the most recent results in heavy ion collisions, the upgrades for LHC Run3 and the future perspectives of the heavy ion program from the LHCb experiment.

1 Introduction

The LHCb detector [1] is a single-arm forward spectrometer fully instrumented in the pseudorapidity range $2 < \eta < 5$. The detector includes a high-precision tracking system providing excellent vertex and momentum resolution, two ring-imaging Cherenkov detectors for charged particle identification, a calorimeter system to identify photons, electrons and hadrons, and a muon system. The coverage in the forward rapidity region is complementary to other major experiments at the LHC, and allows unique access to Bjorken-x region $\sim 10^{-6} - 1$[2–4] inside the nucleus. With the introduction of the SMOG (System for Measuring overlap with gas) in 2015, the LHCb detector is equipped with the unique capability to operate in fixed-target configuration and study beam-gas collisions at lower energies using noble gas targets injected inside the primary LHC vacuum.

LHCb has collected a variety of heavy-ion data in recent years. In the standard beam-beam collision mode, pPb and Pbp collisions are recorded at $\sqrt{s_{\mathrm{NN}}} = 5.02$ TeV with a total luminosity of $\sim 1.6\ \mathrm{nb}^{-1}$ in year 2013, and at $\sqrt{s_{\mathrm{NN}}} = 8.16$ TeV with a total luminosity around $35\ \mathrm{nb}^{-1}$ in 2016. PbPb collisions at $\sqrt{s_{\mathrm{NN}}} = 5.02$ TeV are also collected in 2015 and 2018 with a luminosity of $\sim 10\ \mu\mathrm{b}^{-1}$ and $210\ \mu\mathrm{b}^{-1}$, respectively.

2 Recent heavy-ion results

2.1 Prompt charged hadron production in pp and pPb collisions at 5.02 TeV

The production of prompt charged hadrons is measured with precision in pp and pPb collisions at 5.02 TeV [5]. Since the LHCb detector covers only one direction in rapidity, two beam configures of pPb collisions are used. In the pPb or forward configuration, the proton beam travels from the vertex detector VELO to the muon chambers. In the Pbp or backward configuration, the directions of the proton and lead beams are switched. The double-differential cross-section of prompt charged hadrons as a function of transverse momentum, p_{T}, in the range of $0.2 < p_{\mathrm{T}} < 8\ \mathrm{GeV}/c$, is shown in Fig. 1. The pseudorapiditiy (η) acceptance

*e-mail: jiayin.sun@cern.ch

covers $2.0 < \eta < 4.8$ for pp, $1.5 < \eta < 4.3$ for pPb and $-5.2 < \eta < -2.5$ for Pbp, and is divided into 6 η bins in this measurement. The prompt charged hadrons are measured with the LHCb tracking system. Their yields are corrected for the reconstruction and selection efficiency, as well as contamination from ghost tracks and non-prompt particles. The total uncertainty of the production cross-section is about 3% for most (p_{T}, η) bins.

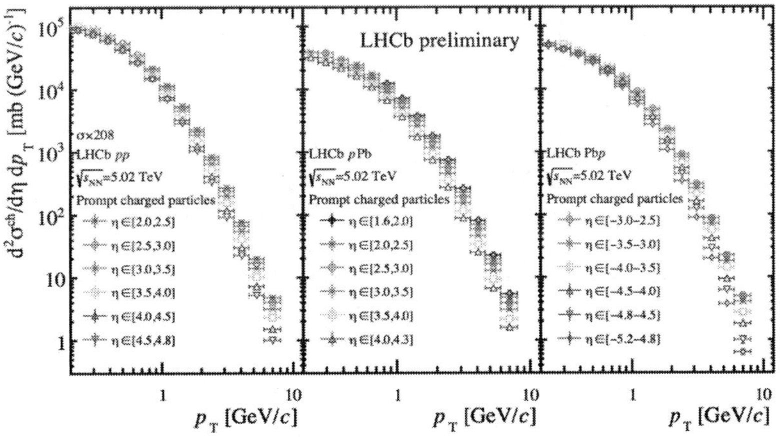

Figure 1. Double differential cross-section of prompt charged hadron production in pp (left), pPb (middle) and Pbp (right) collisions at $\sqrt{s_{\mathrm{NN}}} = 5.02$ TeV.

The nuclear modification factor $R_{p\mathrm{Pb}}$ is presented in Fig. 2. The five upper (lower) panels show $R_{p\mathrm{Pb}}$ as a function of p_{T} in the pPb (Pbp) configuration. A significant suppression is observed in the forward region. The $R_{p\mathrm{Pb}}$ value reaches ~ 0.3 at lowest p_{T} in the most forward pseudorapidity bin. In the backward region, an enhancement is found for $p_{\mathrm{T}} > 1.5\,\mathrm{GeV}/c$. The measurement is compared to calculations from three phenomenological models in the region $p_{\mathrm{T}} \geq 1.5\,\mathrm{GeV}/c$. The calculation using nuclear PDFs [6] agrees with the forward data within large nPDF uncertainties, but does not reproduce the enhancement in the backward region. The gluon saturation model [7] qualitatively describes the increasing trend in the forward rapidity. In the backward rapidity, the data are not well described by the pQCD calculation considering parton multiple scattering [8].

2.2 Prompt production ratio of χ_{c2}/χ_{c1} in pPb collisions at 8.16 TeV

The χ_{c2} and χ_{c1} charmonium production in nuclear collisions at LHC energies is first measured by the LHCb with pPb collisions at 8.16 TeV [9]. The χ_{cJ} states are reconstructed via the decay $\chi_c \to J/\psi(\to \mu^+\mu^-)\gamma$ The J/ψ candidates are reconstructed from a pair of oppositely charged muons. They are combined with a photon candidate to form a χ_c candidate. Two types of photons are used in this analysis: photons converted into electron positron pairs in the detector material and reconstructed by the tracking system (*converted photons*), and photons from their energy deposits in the electromagnetic calorimeter (*calorimetric photons*). The χ_{c1} and χ_{c2} signals reconstructed via converted photons are shown in the left panel of Fig. 3, while those reconstructed via calorimetric photons are shown in the right. The converted photon technique provides better mass resolution with well separated χ_{c1} and χ_{c2} peaks, whereas the calorimetric photon technique provides ~ 20 times more statistics. An upper limit of the pseudo-decay time, $t_z < 0.1$ ps, is imposed to remove non-prompt χ_c production from b-hadron decays.

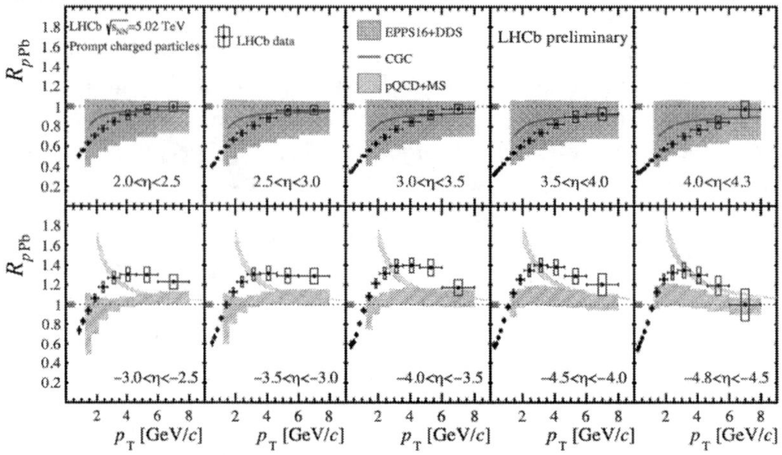

Figure 2. The nuclear modification factor R_{pPb} of prompt charged hadrons as a function of p_T in the forward (upper) and backward (lower) rapidity regions.

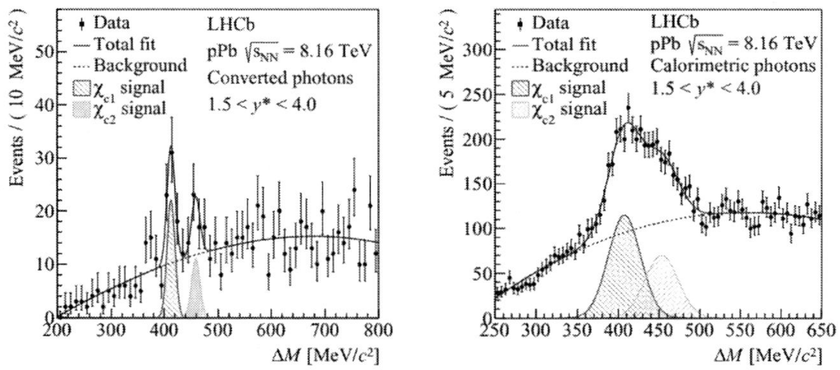

Figure 3. Invariant mass distribution of χ_{c1} and χ_{c2} candidates in pPb collisions reconstructed via the converted photon method (left) and via the calorimetric photon method (right).

As the χ_{c1} and χ_{c2} decays share nearly identical kinematics, various detector effects in the ratio of $\sigma\chi_{c2}/\sigma\chi_{c1}$, such as tracking and particle identification efficiencies, are cancelled, reducing the systematic uncertainty of the measurement. The measured ratio is presented in Fig. 4. The $\sigma\chi_{c2}/\sigma\chi_{c1}$ ratios measured with the two photon types are consistent with each other, and with unity in both the forward and backward rapidity regions. The significantly larger calorimetric photon sample allows for a more precise measurement. The cross-section ratio in pPb data is also compared with a similar measurement from pp collisions at $\sqrt{s} = 7$ TeV [10]. The results are consistent within two standard deviations. This implies that nuclear effects, if any, have the similar impacts on both states.

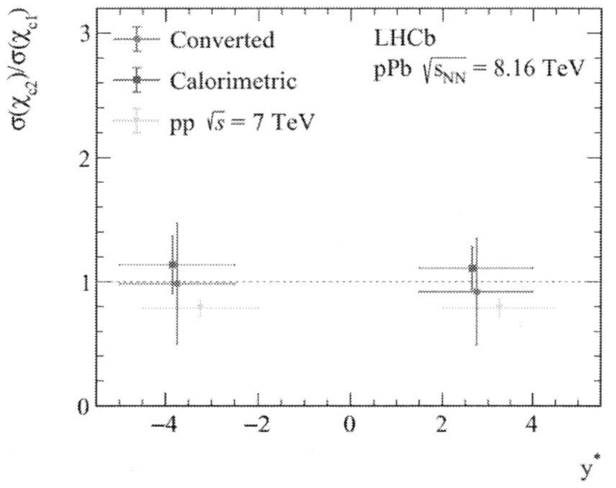

Figure 4. Production cross-section ratio $\sigma\chi_{c2}/\sigma\chi_{c1}$ measured in the forward and backward rapidity regions. The error bars reflect the total uncertainties.

2.3 Coherent J/ψ production in peripheral PbPb collisions at $\sqrt{s_{NN}}$ = 5.02 TeV

One of the first results of AA collisions from LHCb, the photo-production of J/ψ at low transverse momentum is studied in peripheral lead-lead collisions at $\sqrt{s_{NN}} = 5.02$ TeV, using a data sample collected by LHCb in 2018 with an integrated luminosity of 210 μb^{-1} [11]. The J/ψ meson candidates are reconstructed through the $J/\psi \rightarrow \mu^+\mu^-$ decay channel. The dimuon invariant mass distribution of the selected candidates is shown in the left of Fig. 5 for J/ψ candidates with $p_T < 15.0$ GeV/c and $2.0 < y < 4.5$ in a peripheral centrality interval corresponding to $\langle N_{part} \rangle = 10.6 \pm 2.9$. The prompt J/ψ candidates are selected using the requirement on the pseudo decay-time $t_z < 0.3$ ps. The photo-produced J/ψ mesons and the hadronically produced J/ψ mesons are separated with an unbinned maximum likelihood fit to the dimuon $ln(p_T^2)$ spectrum, as shown in the right of Fig. 5. The photo-produced J/ψ candidates (red dotted line) are visible in the range $0 < p_T < 250$ MeV/c.

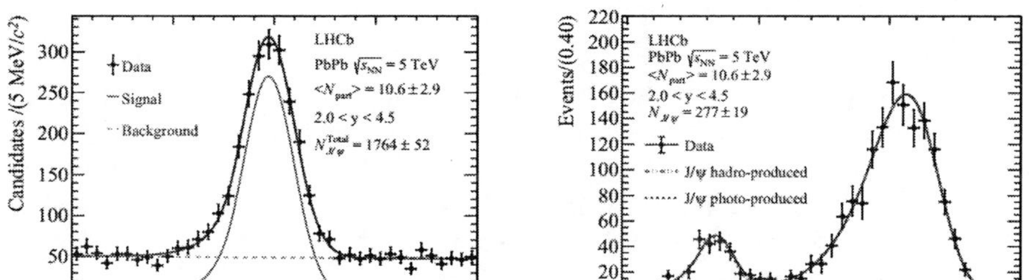

Figure 5. Left: invariant mass distribution of J/ψ candidates in $p_T < 15.0$ GeV/c and $2.0 < y < 4.5$, with $\langle N_{part} \rangle = 10.6 \pm 2.9$. Right: the lnp_T^2 distribution of J/ψ candidates after background subtraction for the same kinematic interval.

Figure 6 presents the measured production yield of photo-produced J/ψ mesons as a function of p_T (left), and rapidity, y, (right), determined for the first time at the LHC. The mean

p_T of the coherent J/ψ is estimated to be $\langle p_T \rangle = 64.9 \pm 2.4$ MeV/c. Theoretical calculations [12, 13] are drawn in open circles, and are qualitatively in agreement with the data in the p_T and y shape.

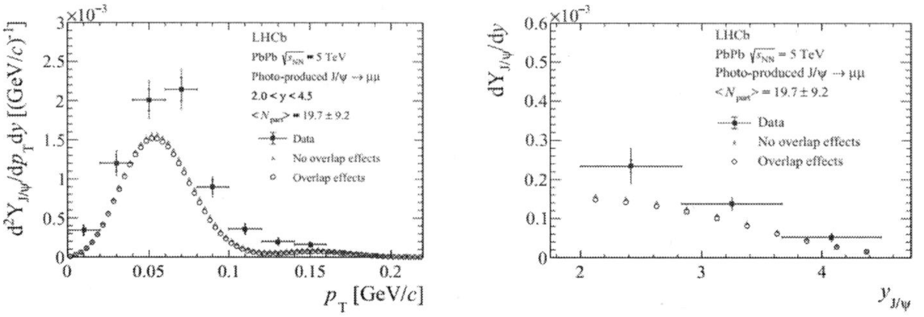

Figure 6. Invariant yields of photo-produced J/ψ mesons as a function of p_T (left) and rapidity (right), for the centrality interval $\langle N_{part} \rangle = 19.7 \pm 9.2$. The blue error bars represent the statistical uncertainty and the red error bars the total uncertainty. Theoretical model predictions [12, 13] are shown in open circles in black and green.

3 Upgrade and future plans

During the LHC Long Shutdown 2 the LHCb is undergoing a major upgrade [14] to meet the challenge of the increased luminosity in the LHC Run3. A new software-only trigger system with 40 MHz data acquisition and real time data reconstruction will replace the current trigger system using a hardware-based trigger step. The detector readout scheme is being upgraded to allow event processing at 40 MHz rate. Upgrades of the sub-detectors include a new Vertex Locator (VELO) with pixel segmentation, new tracking stations and new RICH optics and PMTs. The upgrade will reduce the occupancy limitation in PbPb collisions, and allow access to mid-central PbPb collisions up to 30% in centrality.

The upgrade of the LHCb's fixed target program, SMOG2, is the most important upgrade for heavy-ion physics [15]. Depending on the beam energy, the centre-of-mass energies of the fixed-target collisions range from 40 to 115 GeV, between the SPS and RHIC energies. The forward acceptance of the LHCb geometry covers the middle and backward rapidity regions, providing access to high Bjorken-x region in the target nuclei. The installation of a gas storage cell upstream of the interaction point together with a new Gas Feed System for precise luminosity determination opens the door to many future measurements of great interest. The design of the SMOG2 storage cell is shown in the left panel of Fig. 7, the right panel shows a photo of the installed storage cell inside the LHCb detector together with the vertex detector VELO. With the new storage cell, more gas species such as H_2, D_2, N_2, O_2. Kr and Xe can be injected in addition to He, Ne and Ar. A well-defined interaction region of beam-gas collisions allows the possibility to record beam-gas and beam-beam collisions simultaneously without affecting the performance in the collider mode. The gas target areal density can be increased by up to 100 times, leading to higher luminosity for fixed-target collisions.

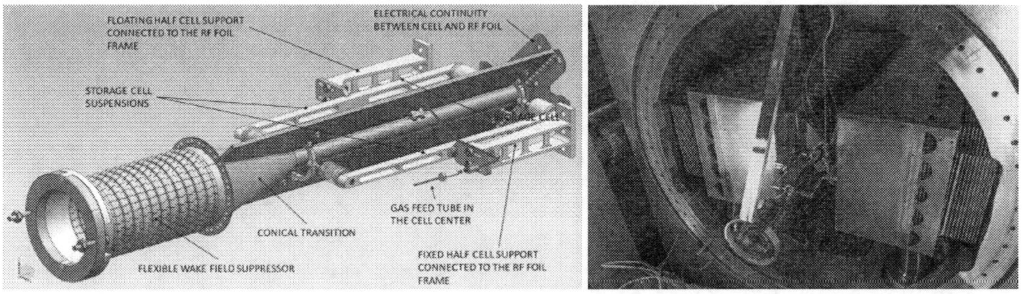

Figure 7. Left: design of the SMOG2 storage cell. Right: installed SMOG2 storage cell together with the VELO detector.

4 Conclusion and outlook

LHCb has a rich heavy ion program with excellent detector performance and unique kinematic coverage. The high statistical *p*Pb datasets offer valuable opportunities to perform precision measurements, study rare probes and investigate small Bjorken-*x* physics. First results from the PbPb datasets also demonstrate the capabilities of LHCb in studying nuclear effects in different systems. With the current upgrade, significantly increased statistics of *p*Pb and PbPb data samples in Run3 will allow for more precision measurements and open the possibilities for many new analyses. The upgrade will also unlock PbPb collisions up to mid-central events, enabling QGP studies with the LHCb detector. SMOG2 will bring the current fixed-target system to a new level with high statistics without detector saturation in unexplored energy and kinematic regions.

References

[1] A.A. Alves Jr. et al. (LHCb collaboration), JINST **3**, S08005 (2008)
[2] A. Kusina, J.P. Lansberg, I. Schienbein, H.S. Shao, Phys. Rev. Lett. **121**, 052004 (2018)
[3] A. Kusina, F. Lyonnet, D. Clark, E. Godat, T. Jezo, K. Kovarik, F. Olness, I. Schienbein, J. Yu, Eur. Phys. J. C **77**, 488 (2017), 1610.02925
[4] K.J. Eskola, P. Paakkinen, H. Paukkunen, C.A. Salgado, Eur. Phys. J. C **77**, 163 (2017), 1612.05741
[5] R. Aaij et al. (LHCb collaboration) (2021), in preparation
[6] I. Helenius, K.J. Eskola, H. Paukkunen, Journal of High Energy Physics **2014** (2014)
[7] T. Lappi, H. Mäntysaari, Phys. Rev. D **88**, 114020 (2013)
[8] Z.B. Kang, I. Vitev, E. Wang, H. Xing, C. Zhang, Physics Letters B **740**, 23 (2015)
[9] R. Aaij et al. (LHCb collaboration), Phys. Rev. **C103**, 064905 (2021), 2103.07349
[10] R. Aaij et al. (LHCb collaboration), JHEP **10**, 115 (2013), 1307.4285
[11] R. Aaij et al. (LHCb collaboration) (2020), in preparation
[12] W. Zha, S.R. Klein, R. Ma, L. Ruan, T. Todoroki, Z. Tang, Z. Xu, C. Yang, Q. Yang, S. Yang, Phys. Rev. C **97**, 044910 (2018)
[13] W. Zha, L. Ruan, Z. Tang, Z. Xu, S. Yang, Phys. Rev. C **99**, 061901 (2019)
[14] *Framework TDR for the LHCb Upgrade: Technical Design Report* (2012)
[15] *LHCb SMOG Upgrade* (2019)

Recent milestones from STAR: new developments and open questions

Rongrong Ma[1,*] (For the STAR Collaboration)

[1]Brookhaven National Laboratory, Upton, NY 11973, USA

Abstract. In these proceedings, an overview of recent STAR results on selected topics is presented. These results utilize Au+Au collisions at various energies, and are aimed at understanding the properties of QED and QCD, characterizing the quark-gluon plasma, as well as searching for the possible critical point in the QCD phase diagram. Specifically, following measurements are discussed: global hyperon polarization, net-proton fluctuations, low transverse momentum dimuon pair production, hyperon-baryon correlations, and $f_0(980)$ elliptic flow.

1 Introduction

The Quark-Gluon Plasma (QGP), consisting of deconfined quarks and gluons, can be created through a phase transition from ordinary nuclear matter in which quarks and gluons are confined within hadrons. This transition is realized at the Relativistic Heavy Ion Collider (RHIC) in head-on collisions of high energy Au beams. At top center-of-mass energy per nucleon-nucleon pair, i.e. $\sqrt{s_{\rm NN}} = 200$ GeV, the main physics goal of the RHIC program is to confirm the QGP formation and study its properties. Additionally, the Beam Energy Scan (BES) program, consisting of two phases, is designed to explore the QCD phase diagram and search for a possible critical point, using Au+Au collisions down to 7.7 GeV (3.0 GeV) in collider (fixed-target) mode. In recent years, heavy-ion collisions and the QGP are also used to carry out research beyond traditional heavy-ion physics, such as coherent photon-photon processes, hyperon-nucleon interactions, etc.

In these proceedings, recent developments from the Solenoidal Tracker at RHIC (STAR) experiment [1] are presented. Physics implications of these results and remaining open questions are discussed along the way.

2 QGP and critical point search

2.1 Global hyperon polarization

In 2017, a new phenomenon of global polarization for both Λ and anti-Λ hyperons, shown as open and filled circles in the left panel of Fig. 1, was discovered in mid-central Au+Au collisions [2]. Such a global polarization is believed to be induced by the vorticity of the medium, which aligns the spins of Λ and anti-Λ along the system's intital orbital momentum direction. A hint of a rising trend of the global polarization with decreasing collision energy is seen, whose confirmation requires measurements of better precision at low energies.

*e-mail: marr@bnl.gov

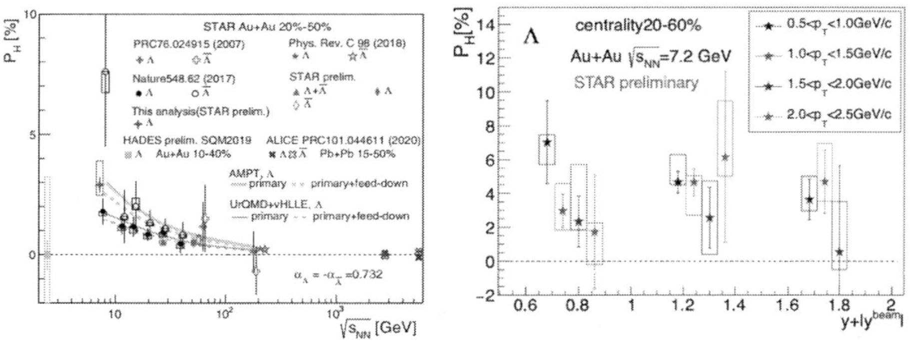

Figure 1. Left: collision energy dependence of global polarization for Λ and anti-Λ in mid-central Au+Au collisions. Right: rapidity dependence of the Λ global polarization in 20-60% central Au+Au collisions at $\sqrt{s_{NN}} = 7.2$ GeV.

The latest measurement of the Λ global polarization in 7.2 GeV Au+Au collisions is shown as the red star in Fig. 1, left panel. Its value is consistent with previous measurement at 7.7 GeV but with better statistical precision, and also consistent with increasing polarization at lower energies. To better understand how the system's angular momentum couples with the hyperon spin, studying the rapidity dependence of the global polarization is of great interest. Dominance of the initial local orbital angular momentum driven by the collision geometry would lead to an increase of the global polarization with rapidity [3], while local thermal vorticity and hydrodynamic evolution predict a decreasing trend or weak dependence with rapidity [4]. The right panel of Fig. 1 shows the Λ global polarization as a function of rapidity in 20-60% central Au+Au collisions at $\sqrt{s_{NN}} = 7.2$ GeV. Within the large experimental uncertainties, no obvious rapidity dependence is seen. The data set taken during BES-II with more statistics and a larger rapidity coverage will significantly improve the precision of rapidity dependence measurement.

Since the Λ global polarization arises from the coupling between the system's angular momentum and particle spin, other hyperons should show similar behavior. Figure 2 shows the measurements of global polarizations for Ξ and Ω in 20-80% central Au+Au collisions at $\sqrt{s_{NN}} = 200$ GeV [5]. Results for particles and anti-particles are combined. The Ξ polarization

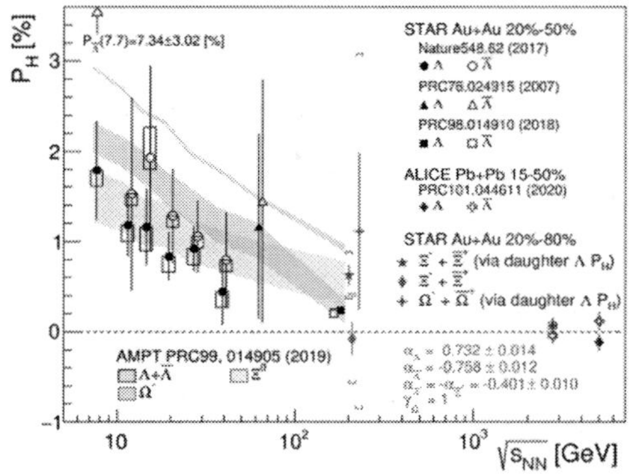

Figure 2. Measurements of global polarizations for Ξ and Ω in 20-80% central Au+Au collisions at $\sqrt{s_{NN}} = 200$ GeV [5]. They are compared with similar measurements for Λ at both RHIC and the LHC.

is measured in two ways, via analysis of the angular distribution of the daughter particles (magenta diamond) and via measuring the polarization of the daughter Λ-hyperon (red star). The average of the two gives Ξ global polarization of $\langle P_\Xi \rangle = 0.47 \pm 0.10(\text{stat.}) \pm 0.23(\text{syst.})\%$.

The Ω global polarization is measured to be $\langle P_\Omega \rangle = 1.11 \pm 0.87(\text{stat.}) \pm 1.97(\text{syst.})\%$. Despite the large uncertainties, the positive values are consistent with the underlying mechanism that the system's vorticity induces the observed global polarization.

2.2 Search for the critical point

Searching for the possible critical endpoint in the QCD phase diagram is one of the main goals of the BES program at RHIC. One promising tool is to study fluctuations of conserved quantities, such as the baryon number, whose correlation lengths diverge at the critical point. Experimentally, ratios of high-order moments of net-proton multiplicity distribution, i.e. number of protons minus that of anti-protons, are used, and results are shown in Fig. 3 [6]. The left and right panels show the skewness times variance ($S\sigma = C_3/C_2$) and kurtosis times variance square ($\kappa\sigma^2 = C_4/C_2$) as a function of collision energy, respectively using BES-I data, where C_n is the nth moment of the net-proton distribution. Results from 0-5% (filled circles) and

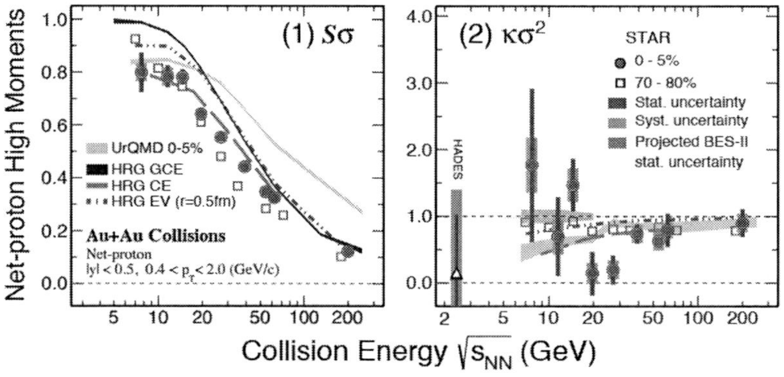

Figure 3. Collision energy dependence of net-proton $S\sigma$ (left) and $\kappa\sigma^2$ (right) in 0-5% (filled circles) and 70-80% (open squares) Au+Au collisions [6].

70-80% (open squares) Au+Au collisions are shown, where the latter is used as a reference for the case of no critical point. A non-monotonic behavior is seen for $\kappa\sigma^2$ in 0-5% central collisions with a significance of 3.1σ, consistent with the expectation of the presence of the critical point, while the 70-80% result shows little dependence on collision energy.

Collision energy dependence of even higher-order moment ratio (C_6/C_2), presumably more sensitive to the correlation length, is shown in Fig. 4. Theoretical calculations predict negative C_6/C_2 values when a phase transition between hadronic and partonic matter occurs [7]. The measured values are positive for 70-80% peripheral collisions, while they are negative for 0-40% central collisions at most energies. Enhanced statistics from BES-II is of great importance to these measurements.

3 Explore new physics with heavy-ion collisions

3.1 Photon-photon process

In ultra-relativistic Au+Au collisions, large fluxes of linearly polarized photons are emitted by incoming highly-charged nuclei. Consequently, coherent photon-photon interactions would occur, producing excess dilepton pairs above known hadronic sources at very low transverse momentum (p_T). Such a process has been observed in the dielectron channel

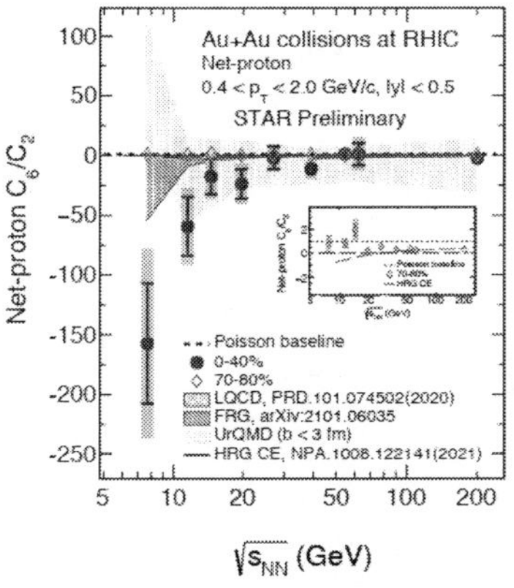

Figure 4. Collision energy dependence of net-proton C_6/C_2 in 0-40% (filled circles) and 70-80% (open diamonds) Au+Au collisions.

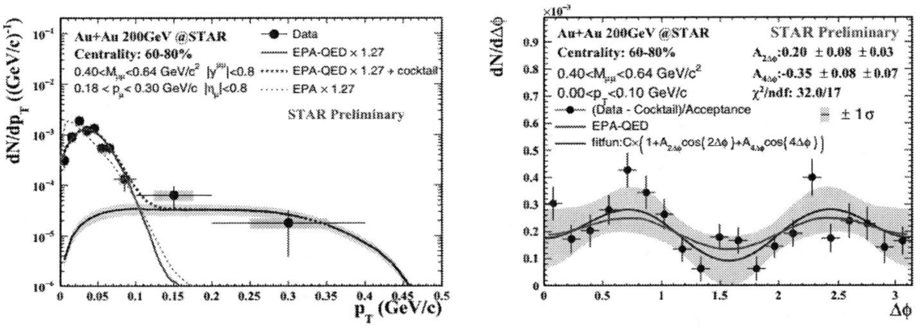

Figure 5. p_T (left) and azimuthal (right) distributions of dimuon pairs in 60-80% peripheral Au+Au collisions at $\sqrt{s_{NN}}$ = 200 GeV. See text for the definition of the azimuthal angle.

[8, 9], and new complimentary measurements via the dimuon channel are shown in Fig. 5. The left panel shows the p_T distribution of dimuon pairs with the pair mass between 0.40 and 0.64 MeV/c^2 in 60-80% peripheral Au+Au collisions at $\sqrt{s_{NN}}$ = 200 GeV. Compared to the known hadronic sources (black solid curve), a clear excess at $p_T < 0.1$ GeV/c is seen, which is a signature of coherent photon-photon interactions. Theoretical calculations based on QED are consistent with the data within uncertainties [10].

Since colliding photons are linearly polarized, $\cos(2\Delta\varphi)$ and $\cos(4\Delta\varphi)$ modulations are expected for the produced dilepton pairs [11], where $\Delta\varphi$ is the angle between the momenta of the lepton pair and the positively charged lepton. The $\cos(4\Delta\varphi)$ modulation has been observed in the dielectron channel [9], while the $\cos(2\Delta\varphi)$ modulation is easier to be measured in the dimuon channel since it is suppressed by the power of m^2/p_T^2 for the electron channel. The $\Delta\varphi$ distribution for low-p_T dimuon pairs is shown in the right panel of Fig. 5. A fit to the distribution yields a compatible $\cos(4\Delta\varphi)$ modulation to the dielectron result, and a first observation of $\cos(2\Delta\varphi)$ modulation with a significance of 2.3σ.

3.2 Hyperon-baryon interactions

Understanding interactions between hyperons and baryons will provide important inputs to resolving the "hyperon puzzle" in neutron star research. Such measurements can be carried out in heavy-ion collisions as hyperons and baryons are copiously produced in these collisions. The left panel of Fig. 6 shows the ratio of the p-Ξ correlations in peripheral and

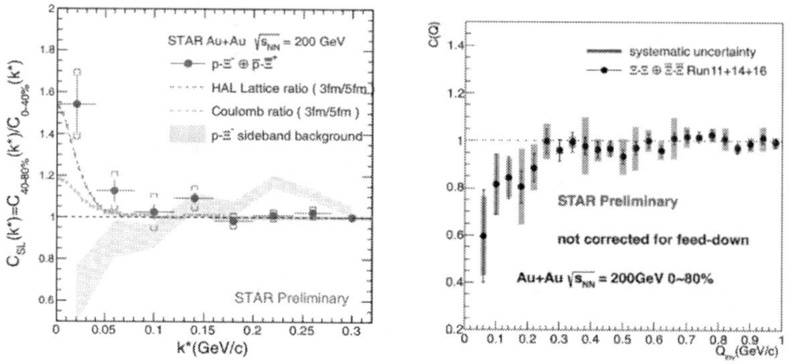

Figure 6. Left: ratio of p-Ξ correlation between peripheral and central Au+Au collisions at $\sqrt{s_{\text{NN}}}$ = 200 GeV as a function of half of relative momentum between p-Ξ pair in the pair rest frame. Right: correlation of Ξ-Ξ as a function of relative momentum in 0-80% Au+Au collisions at $\sqrt{s_{\text{NN}}}$ = 200 GeV.

central Au+Au collisions at $\sqrt{s_{\text{NN}}}$ = 200 GeV as a function of k^*, which is half of the relative momentum between the p-Ξ pair in the pair rest frame. Compared to the baseline calculation taking into account the Coulomb interaction (orange dashed line) [12], the data indicate additional attractive strong interaction between p and Ξ for $k^* < 0.05$ GeV/c, which is consistent with lattice QCD calculation (green dashed line) [13]. Furthermore, the correlation between $\Xi - \Xi$ pairs is studied in 0-80% 200 GeV Au+Au collisions, and the result is shown in the right panel of Fig. 6. A hint of negative correlation at small relative momentum is seen. Additional studies on feed-down contribution and Coulomb effect are needed to better interpret the data.

3.3 f_0(980) quark content

Despite being observed in 1970's, the exact quark content of f_0(980) meson remains an open question. Different scenarios include standard $q\bar{q}$, tetraquark, meson-meson bound state, or gluon ball, or a superposition of these states. In Au+Au collisions at RHIC, particle's elliptic flow (v_2) is seen to empirically follow the number of constituent quark (NCQ, n_q) scaling, i.e. v_2/n_q collapses to a common curve for different particles when plotted against $(m_T - m_0)/n_q$, where $m_T = \sqrt{m_0^2 + p_T^2}$ and m_0 is the particle's rest mass. This is because v_2 is mainly developed during the partonic phase before hadronization. In turn, one can use this scaling property to infer the quark content of f_0(980) by examing its v_2. Results are shown in Fig. 7, in which both two- and four-quark assumptions are used. Using the common trend by fitting the results for other particles, the inferred number of constituent quarks is $3.0 \pm 0.7(\text{stat.}) \pm 0.5(\text{syst.})$, which is inconclusive due to large uncertainties. Additional data taking of 200 GeV Au+Au collisions planned for 2023 and 2025 will be beneficial in reducing uncertainties of this measurement.

4 Summary

These proceedings present recent developments from the STAR experiment at RHIC. Significant global polarization for Λ is measured in 7.2 GeV Au+Au collisions without any strong rapidity dependence. For Ξ and Ω, positive global polarizations are also seen in 200 GeV

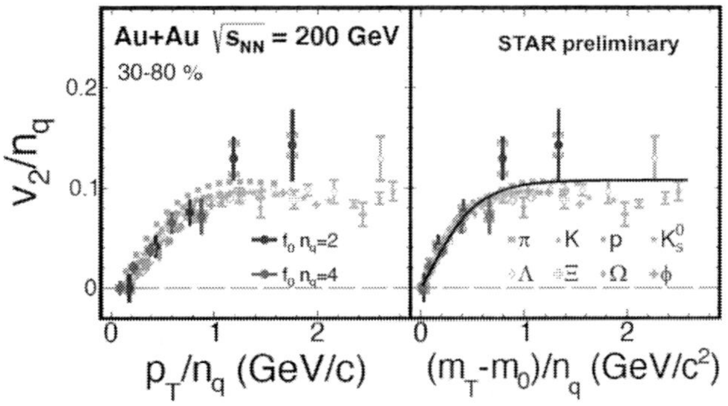

Figure 7. $f_0(980)$ v_2/n_q as a function of quark number scaled p_T (left) and transverse kinetic energy (right). Measurements for other hadrons are shown for comparison.

Au+Au collisions. From the BES program, net-proton kurtosis exhibits a non-monotonic behavior against collision energy, consistent with the expectation of a QCD critical point. Furthermore, low-p_T dimuon pair production, arising from coherent photon-photon interactions, is observed, and reveals a $\cos(2\Delta\varphi)$ modulation due to the linear polarization of the incoming photons. An indication of attractive strong interaction between p and Ξ at low relative momenta is seen, which provides important inputs to neutron star studies. Last but not least, the $f_0(980)$ v_2 is checked for NCQ scaling, but a firm conclusion requires better precision. Datasets taken during BES-II and future data-taking at top energy will greatly improve the precision of these measurements.

References

[1] K.H. Ackermann et al. (STAR), Nucl. Instrum. Meth. A **499**, 624 (2003)
[2] L. Adamczyk et al. (STAR), Nature **548**, 62 (2017), 1701.06657
[3] Z.T. Liang, J. Song, I. Upsal, Q. Wang, Z.B. Xu, Chin. Phys. C **45**, 014102 (2021), 1912.10223
[4] W.T. Deng, X.G. Huang, Phys. Rev. C **93**, 064907 (2016), 1603.06117
[5] J. Adam et al. (STAR), Phys. Rev. Lett. **126**, 162301 (2021), 2012.13601
[6] J. Adam et al. (STAR), Phys. Rev. Lett. **126**, 092301 (2021), 2001.02852
[7] W.j. Fu, X. Luo, J.M. Pawlowski, F. Rennecke, R. Wen, S. Yin (2021), 2101.06035
[8] J. Adam et al. (STAR), Phys. Rev. Lett. **121**, 132301 (2018), 1806.02295
[9] J. Adam et al. (STAR) (2019), 1910.12400
[10] W. Zha, J.D. Brandenburg, Z. Tang, Z. Xu, Phys. Lett. B **800**, 135089 (2020), 1812.02820
[11] C. Li, J. Zhou, Y.J. Zhou, Phys. Lett. B **795**, 576 (2019), 1903.10084
[12] K. Morita, A. Ohnishi, F. Etminan, T. Hatsuda, Phys. Rev. C **94**, 031901 (2016), [Erratum: Phys.Rev.C 100, 069902 (2019)], 1605.06765
[13] T. Hatsuda, K. Morita, A. Ohnishi, K. Sasaki, Nucl. Phys. A **967**, 856 (2017), 1704.05225

 https://doi.org/10.1051/epjconf/202225901006

Updates on Flavor Production from STAR

Sooraj Radhakrishnan[1,*]

[1]Kent State University, Kent, OH, 44240

Abstract. In these proceedings, some of the latest results from the STAR experiment on heavy, strange and light flavor hadron production and their flow are summarized. Over the years STAR has collected heavy ion collision data over a wide range of collision energies from $\sqrt{s_{\rm NN}}$ = 200 GeV down to $\sqrt{s_{\rm NN}}$ = 3 GeV. In high energy heavy ion collisions, heavy flavor hadrons are a valuable probe to study properties of the Quark Gluon Plasma (QGP) created in these collisions. Measurements of strange and light flavor hadrons at lower collision energies help understand the properties of the QCD matter at finite baryon densities and the phase structure of the QCD matter. Furthermore, the high statistics data collected at $\sqrt{s_{\rm NN}}$ = 3 GeV allows measurements of hypernuclei production and flow, which can provide insights into the hyperon - nucleon interactions and hyperon contribution to the nuclear equation of state.

1 Introduction

Measurements of elliptic flow (v_2) of D mesons and electrons from heavy flavor hadron decays (HFE) in heavy-ion collisions at top RHIC energy and LHC energies show comparable values to those of light flavor hadrons, indicating that charm quarks interact strongly with the Quark Gluon Plasma (QGP) at these energies [1, 2]. These measurements have helped constrain the charm quark diffusion coefficient in the QGP [3]. Measurements of HFE v_2 at lower collision energies can help constrain the temperature dependence of charm quark transport in the QGP. The observed J/ψ nuclear modification factor (R_{AA}) in heavy-ion collisions at RHIC and LHC has contributions from a regeneration component, in addition to that from in-medium dissociation [4]. Better precision measurements at lower collision energies can also help constrain these different contributions.

The Beam Energy Scan Phase II (BES-II) program at STAR, following the BES-I, aims to conduct high statistics measurements of various observables in the high baryon density region (200 < μ_B < 720 MeV) to understand the phase structure of QCD matter. As part of BES-II, STAR has collected a high statistics dataset of Au+Au collisions at $\sqrt{s_{\rm NN}}$ = 3 GeV (μ_B = 720 MeV), using a fixed-target setup. The detailed measurements of the collective flow of various identified light and strange flavor hadrons with this dataset can improve our understanding of the nature and Equation of State (EoS) of the QCD matter in the high μ_B region. In addition, the high statistics 3 GeV dataset allows the measurement of strange and multi-strange hadron production, providing insights into the canonical suppression of strangeness production at low collision energies [6]. The same dataset also enables measurements of the yields, lifetime and flow of hypernuclei, owing to their enhanced cross-sections at these low energies, which can

*e-mail: skradhakrishnan@lbl.gov

help understand the hyperon - nucleon interactions and hyperon contribution to the nuclear EoS.

In these proceedings, measurements of HFE v_2 in Au+Au collisions at $\sqrt{s_{NN}}$ = 54.4 and 27 GeV and R_{AA} of J/ψ in Au+Au collisions at $\sqrt{s_{NN}}$ = 54.4 GeV are presented. Measurements of v_2 of identified hadrons, yield ratios of strange and multi-strange hadrons, and yields, lifetime and directed flow (v_1) of hypernuclei ${}^{\Lambda}_{3}H$ and ${}^{\Lambda}_{4}H$ in Au+Au collisions at $\sqrt{s_{NN}}$ = 3 GeV are also presented. The results are compared to model calculations and the physics implications are discussed.

2 Heavy flavor measurements at $\sqrt{s_{NN}}$ = 54.4 and 27 GeV

STAR had previously carried out measurements of v_2 of HFE at $\sqrt{s_{NN}}$ = 62.4 and 39 GeV [2]. The measured HFE v_2 values were found to be smaller, although consistent within the large uncertainties, than the values measured at 200 GeV. In 2017 and 2018 STAR collected 10 times more statistics at $\sqrt{s_{NN}}$ = 54.4 GeV and 27 GeV compared to those used for previous measurements at 62.4 and 39 GeV. The left panel of Fig. 1 shows the measured v_2 of HFE at 54.4 and 27 GeV as a function of transverse momentum (p_T), compared to the measurements at 200 GeV. The measured v_2 values at 54.4 GeV are non-zero and comparable to those at 200 GeV, indicating strong interactions of the charm quarks with the QGP medium. The measurements at 27 GeV show a hint of lower values, albeit with large uncertainties. The right panel of Fig. 1 shows the comparison of the measured HFE v_2 values at 54.4 GeV to different theoretical model calculations. The TAMU [7] and PHSD [8] model calculations, which incorporate charm quark diffusion in the QGP, describe the data at high p_T, but are lower than the measured values at low p_T.

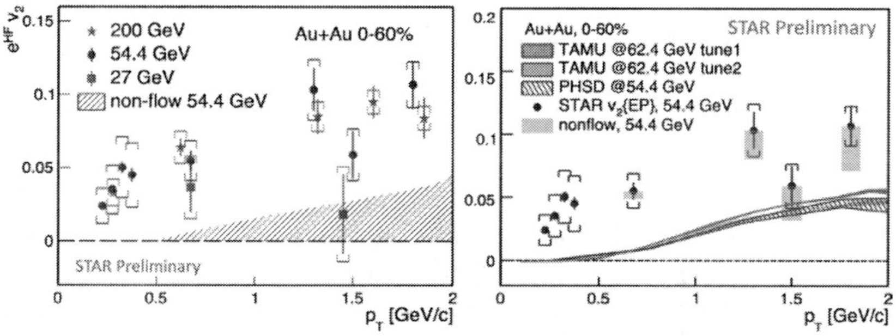

Figure 1. (Left) The v_2 of HFE as a function of p_T in 0-60% Au+Au collisions compared between $\sqrt{s_{NN}}$ = 54.4, 27 and 200 GeV. The error bars are statistical uncertainties and brackets indicate systematic uncertainties. The blue hatched area indicates the estimated non-flow contribution to the measured v_2 values. (Right) The v_2 of HFE as a function of p_T in 0-60% Au+Au collisions at $\sqrt{s_{NN}}$ = 54.4 GeV compared to different model calculations [7, 8].

The in-medium dissociation of J/ψ, due to Debye like color screening in the QGP medium, has been proposed as an evidence of the formation of the QGP and a valuable tool to study its properties, particularly the medium temperature [9]. Suppression of J/ψ production has been observed at both RHIC and the LHC. However, the observed suppression also gets contribution from regeneration of J/ψ from deconfined charm quarks in the medium [4, 10]. Measurements of the J/ψ suppression at a lower collision energy of $\sqrt{s_{NN}}$ = 54.4 GeV, where the charm quark production cross-section and thus the regeneration contribution are lower, can help better understand the in-medium J/ψ dissociation. The left panel of Fig. 2 shows the nuclear modification factor R_{AA}, the ratio of yields in A+A collisions to that in p+p collisions scaled by the number of binary nucleon-nucleon collisions N_{coll}, as a function of number of participating nucleons in the collision (N_{part}) for different collision energies. The right panel shows the R_{AA} in central collisions as a function of collision energy. The new measurements at 54.4 GeV are shown in red markers, and have significantly improved precision compared to previous low energy STAR measurements at 62.4 and 39 GeV [11]. The measured values show no significant collision energy dependence from 17.2 to 200 GeV within uncertainties. A model calculation [10] that takes into account both in-medium dissociation and regeneration, along with cold nuclear matter effects, can reproduce the measured values.

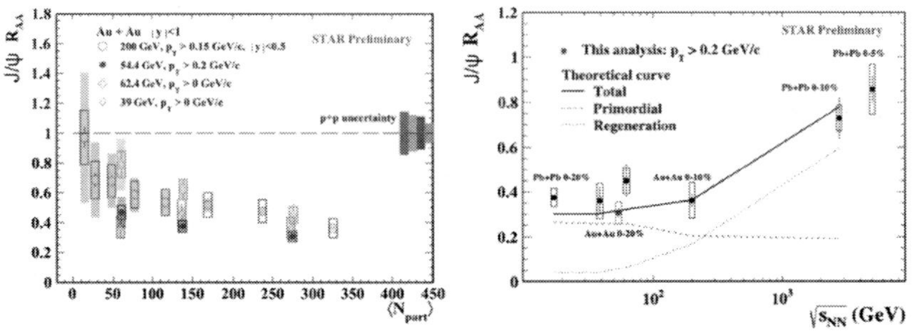

Figure 2. (Left) The R_{AA} of J/ψ as a function of N_{part} for different collision energies. The error bars, boxes and shaded bands on the data points denote statistical, systematic and N_{coll} uncertainties respectively. The shaded bands at 1 indicate the uncertainties from the p+p baseline. (Right) The R_{AA} of J/ψ in central collisions as a function of collision energy, including measurements at SPS and LHC [12]. The error bands indicate statistical uncertainties and boxes indicate the combined systematic, N_{coll} and p+p baseline uncertainties.

3 Strange and light flavor production

3.1 Disappearance of partonic collectivity at $\sqrt{s_{NN}}$ = 3 GeV

The large positive values of v_2 and its number of constituent quark (NCQ) scaling observed for identified hadrons in high energy heavy-ion collisions have been argued as evidence for the creation of a strongly interacting QGP [13]. At lower collision energies, below $\sqrt{s_{NN}}$ = 27 GeV, where the passage time of the colliding nuclei becomes comparable to or larger than the time scale of parton production, a negative 'squeeze out' contribution from the shadowing by spectator nucleons is expected for the v_2, in addition to the contribution from partonic collectivity [14]. At even lower collision energies, $\sqrt{s_{NN}} \sim 5$ GeV, hadronic transport and baryonic mean field interactions among the colliding nuclei also contribute to the observed flow harmonics [15]. In 2018 STAR collected about 2.6×10^8 minimum bias Au+Au collisions at $\sqrt{s_{NN}}$ = 3 GeV. The high statistics dataset allows to carry out detailed measurements of the v_2 of various identified hadrons at mid-rapidity in this interesting region of QCD phase space.

Figure 3 shows the NCQ scaled v_2 of identified hadrons as a function of the scaled transverse kinetic energy in Au+Au collisions at $\sqrt{s_{NN}}$ = 3 GeV. The figure also shows measurements at $\sqrt{s_{NN}}$ = 4.5 GeV [5] and from the new datasets at $\sqrt{s_{NN}}$ = 54.4 and 27 GeV. The v_2 values at 54.4 and 27 show NCQ scaling. Positive values and scaling within uncertainties are also observed at $\sqrt{s_{NN}}$ = 4.5 GeV. At $\sqrt{s_{NN}}$ = 3 GeV, the measured v_2 values turn negative and the NCQ scaling breaks down. Figure 4 shows the centrality dependence of the v_2 values of different identified hadrons at $\sqrt{s_{NN}}$ = 3 GeV. Calculations from the UrQMD model [16] at $\sqrt{s_{NN}} = 3$ GeV in the cascade mode, that incorporates hadronic transport and spectator shadowing, and in the meanfield mode, which includes additional baryonic meanfield interactions, are also shown. The UrQMD cascade mode cannot describe the measured v_2 values and are of opposite sign. The UrQMD meanfield mode calculations, with a value of the incompressibility $\kappa = 380$ MeV, can reproduce the trends in the data, except for K^+. These measurements and comparisons show that a medium dominated by hadronic and baryonic interactions, rather than partonic interactions, is produced in Au+Au collisions at $\sqrt{s_{NN}} = 3$ GeV.

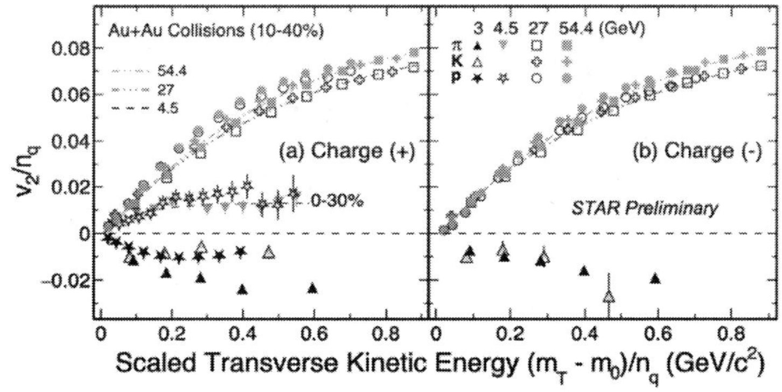

Figure 3. The v_2 scaled by the number of constituent quarks n_q as a function of the scaled transverse kinetic energy at different collision energies for negative (left) and positive particles (right). Statistical and systematic uncertainties are indicated by error bars and shaded bands on data points, respectively. The dashed lines denote the NCQ scaling curves at different energies.

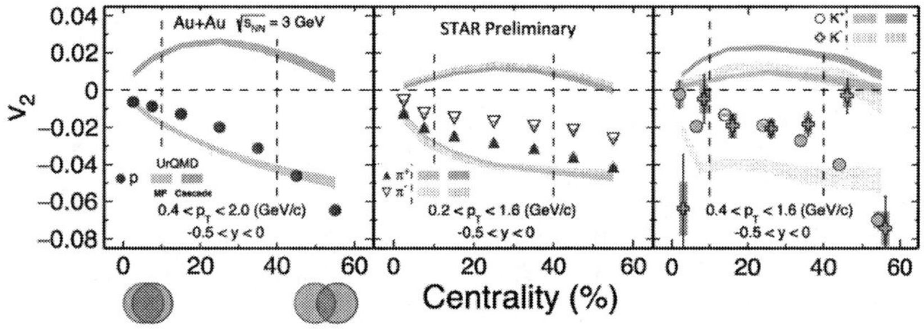

Figure 4. The v_2 of protons (left), pions (middle) and kaons (right) as a function of centrality for Au+Au collisions at $\sqrt{s_{NN}} = 3$ GeV. Statistical and systematic uncertainties are indicated by error bars and shaded bands on data points, respectively. UrQMD calculations in cascade and meanfiled modes are shown as gray and golden bands respectively.

3.2 Canonical suppression of strangeness production

Statistical thermal models have been successful in describing the hadron yields in heavy-ion collisions [17]. At low collision energies and in small collision systems, as the strangeness production is rare, it has been argued that the strangeness number needs to conserved locally and on an event by event basis. Therefore a Canonical Ensemble (CE) is more suitable to describe the statistical production of strange hadrons as opposed to a Grand Canonical Ensemble (GCE). This would cause progressive suppression of yields of strange hadrons with non-zero strangeness numbers [6, 18]. Figure 5 shows the yield ratios of ϕ mesons to that of K^- on the left and yield ratios of ϕ relative to Ξ^- on the right as a function of collision energy. The new measurements at 3 GeV are shown as colored markers. The GCE calculations are strongly disfavored, being lower than the measured ratios at 5σ and 4σ significances respectively, in the 0-10% most central collisions. The CE calculation with a strangeness correlation length $(r_c) \sim 3.2$ fm can describe the ϕ/K^- yield ratio. There is a hint of a centrality dependence of the ratios at $\sqrt{s_{\mathrm{NN}}} = 3$ GeV. Future higher precision measurements can provide better insights into the centrality dependence. Transport model calculations from a modified UrQMD [19] and SMASH [20] that take into account high mass resonances that decay into ϕ and Ξ^- can also describe the ratios at 3 GeV.

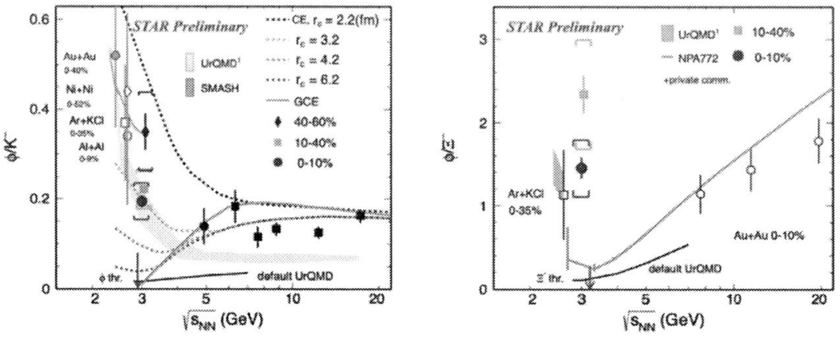

Figure 5. The yield ratios of ϕ to that of K^- (left) and of ϕ to that of Ξ^- (right) as a function of collision energy. The new measurements at 3 GeV are shown in colored markers and at other energies [21] in solid black and open gray markers. The error bars and brackets on the 3 GeV data points denote the statistical and systematic uncertainties respectively. The gray solid line indicates the GCE calculation while the dashed lines indicate CE calculations with different r_c values [6, 18, 22]. The default UrQMD [16] calculations are shown in solid black line, while the modified UrQMD [19] and SMASH [20] calculations are shown in the green and red bands respectively.

4 Hypernuclei production at $\sqrt{s_{\mathrm{NN}}}$ = 3 GeV

Hypernuclei production cross-sections in heavy-ion collisions are large at colliding energies ~ 3 GeV [23]. This provides an opportunity to measure the lifetime, yield and collective flow of hypernuclei with the $\sqrt{s_{\mathrm{NN}}} = 3$ GeV Au+Au collision dataset. The lifetimes of ${}^{\Lambda}_{3}H$ and ${}^{\Lambda}_{4}H$ measured using the 3 GeV dataset are 232 ± 29 (stat) $\pm$ 37 (sys) and 218 ± 8 (stat) $\pm$ 12 (sys), respectively. The ${}^{\Lambda}_{4}H$ lifetime measurement obtained is the most precise to date. The two left panels of Fig. 6 show the yields within rapidity $|y| < 0.5$ of the hypernuclei ${}^{\Lambda}_{3}H$ and ${}^{\Lambda}_{4}H$, along with model calculations, as a function of collision energy. Thermal model calculations [23] can describe the measured ${}^{\Lambda}_{3}H$ yield at 3 GeV and also at the LHC energy [24]. The thermal model calculations are done with a CE for strangeness production. The model however under-predicts the yield of ${}^{\Lambda}_{4}H$ at 3 GeV. Coalescence model calculations [25] overestimate the ${}^{\Lambda}_{3}H$ yield while under-predict the ${}^{\Lambda}_{4}H$ yield at 3 GeV. A hybrid UrQMD calculation over-predicts yields of both ${}^{\Lambda}_{3}H$ and ${}^{\Lambda}_{4}H$. The right panel of Fig. 6 shows the slope of directed flow at midrapidty, $dv_1/dy|_{y=0}$, of various light nuclei and hypernuclei as a function of their mass for Au+Au collisions at $\sqrt{s_{\mathrm{NN}}} = 3$ GeV. A mass scaling of directed flow slope has been observed for light nuclei which is attributed to the coalescence production of light nuclei [26]. The hypernuclei also show a mass scaling of v_1 slope, albeit with large uncertainties, suggesting coalescence production also for hypernuclei. These measurements will help better constrain the hypernucleus production models and also the hyperon contribution to nuclear EoS.

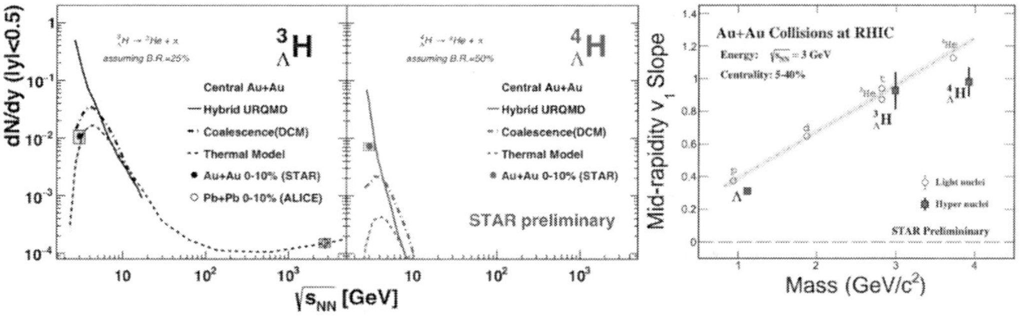

Figure 6. The yield in $|y| < 0.5$ of $^{\Lambda}_{3}H$ (left) and $^{\Lambda}_{4}H$ (middle), along with model comparisons, as a function of collision energy. The boxes indicate the total uncertainties. (Right) The measured slope, $dv_1/dy|_{y=0}$, in Au+Au collisions at $\sqrt{s_{NN}} = 3$ of different light nuclei and hypernuclei as a function of mass. Error bars indicate total uncertainties. The yellow line shows a linear fit to the light nuclei measurements. Combined results from the 2-body and 3-body decay channels are used for the $^{\Lambda}_{3}H$ v_1 measurements, while for $^{\Lambda}_{4}H$ v_1 measurements and for yield measurements of both the hypernuclei results from only 2-body decay channel are used.

5 Summary

We have presented some of the newest measurements from STAR on heavy, strange and light flavor hadron production and collective flow. Measurement of HFE v_2 with a much improved precision in Au+Au collisions at $\sqrt{s_{NN}} = 54.4$ GeV, compared to previous measurements at 62.4 and 39 GeV, shows non-zero values comparable to the v_2 of HFE measured at 200 GeV. This indicates that charm quarks interact strongly with the QGP medium produced in 54.4 GeV collisions. The R_{AA} of J/ψ measured in Au+Au collisions at $\sqrt{s_{NN}} = 54.4$ GeV has improved precision compared to previous measurements at 62.4 and 39 GeV. The measurements show no significant collision energy dependence of R_{AA} within uncertainties compared to that at 200 GeV and other lower energy measurements. Elliptic flows of identified hadrons in Au+Au collisions at $\sqrt{s_{NN}} = 3$ GeV show negative values and a breakdown of the NCQ scaling. Hadronic transport model calculations from UrQMD with baryonic meanfield interactions can describe the trends seen in data. These observations indicate the formation of a medium dominated by hadronic interactions and a nuclear EoS in Au+Au collisions at 3 GeV. Strange and multi-strange hadron yield measurements in Au+Au collisions at $\sqrt{s_{NN}}$ = 3 GeV strongly disfavor GCE calculations and show evidence of canonical suppression in strangeness production. Thermal model calculations are consistent with the measured $^{\Lambda}_{3}H$ yield, while under-predct the $^{\Lambda}_{4}H$ yield. Coalescence model calculations fail to describe yields of both the hypernuclei. Measurements of directed flow slope, $dv_1/dy|_{y=0}$, of light nuclei and hypernuclei show mass scaling within uncertainties. The mass scaling is consistent with a picture of coalescence production for the light nuclei and hypernuclei.

References

[1] L. Adamczyk et al. (STAR), Phys. Rev. Lett. **118**, 212301 (2017); S. Acharya et al. (ALICE), Phys. Rev. Lett. **120**, 102301 (2018)
[2] L. Adamczyk et al. (STAR), Phys. Rev. C **95**, 034907 (2017)
[3] X. Dong, Y.-J. Lee, R. Rapp, Ann. Rev. Nucl. Part. Sci. **69** 417–445 (2019)

[4] B. Abelev et. al (ALICE), Phys. Lett. B **734**, 314-327 (2014); J. Adam et al. (STAR), Phys. Lett. B **797**, 134917 (2019)
[5] J. Adam et al. (STAR), Phys. Rev. C **103** 34908 (2021)
[6] K. Redlich and A. Tounsi, Eur. Phys. J. C **24**, 589 (2002)
[7] M. He et al., Phys. Rev. C **91**, 024904 (2015)
[8] T. Song et al., Phys. Rev. C **92**, 014910 (2015); T. Song et al., Phys. Rev. C. **96**, 014905 (2017)
[9] T. Matsui, H. Satz, Phys. Lett. B **178** 416–422 (1986).
[10] X. Zhao, R. Rapp, Phys. Rev. C **82**, 064905 (2010)
[11] L. Adamczyk et al. (STAR), Phys. Lett. B **771**, 13 (2017)
[12] L. Kluberg, Eur. Phys. J. C **43**, 145 (2005); B. Abelev et. al (ALICE), Phys. Lett. B **734**, 314 (2014); X. Bai (for ALICE Collaboration), Nucl. Phys. A **1005**, 121769 (2021)
[13] U. Heinz and R. Snellings, Annu. Rev. Nucl. Part. Sci. **63** 123 (2013); B. I. Abelev et al. (STAR), Phys. Rev. Lett. **99**, 112301 (2007)
[14] C. Pinkenburget et al.(E895), Phys. Rev. Lett. **83**, 1295 (1999)
[15] P. Danielewicz et al, Science **298**, 1592-1596, (2002)
[16] M. Bleicheret et al., J. Phys. G **25**, 1859 (1999)
[17] A Andronic et al. Nature **561**, 321 (2018)
[18] A. Andronic et al. Nucl. Phys. A **772** 167-199 (2006)
[19] J. Steinheimer and M. Bleicher, J. Phys. G: Nucl. Part. Phys. **43**, 015104 (2015).
[20] V. Steinberg et al., Phys. Rev. C **99**, 064908 (2019)
[21] M. S. Abdallah et al. (STAR), arXiv:2108.00924 (2021)
[22] S. Wheaton et al., Comput. Phys. Commun. **180**, 84 (2009)
[23] A. Andronic et al., Phys. Lett. B **697**, 203 (2011)
[24] J. Adam et al. (ALICE), Phys. Lett. B **754**, 360 (2016)
[25] J. Steinheimer et al. Phys. Lett. B **714**, 47985 (2012)
[26] L. Adamczyk et al. (STAR), Phys. Rev. C **94**, 034908 (2016)

EPJ Web of Conferences **259**, 01007 (2022) https://doi.org/10.1051/epjconf/202225901007
SQM 2021

HADES Overview

Recent results from Ag+Ag collisions at $\sqrt{s_{NN}}$ = 2.55 GeV measured by HADES

Simon Spies[1,*] for the HADES collaboration

[1]Institut für Kernphysik, Goethe-Universität, 60438 Frankfurt, Germany

Abstract. In March 2019 the HADES experiment recorded 14 billion Ag+Ag collisions at $\sqrt{s_{NN}}$ = 2.55 GeV as a part of the FAIR phase-0 physics program. With the capabilities to measure and analyze particles forming the bulk matter, namely pions, protons and light nuclei, as well as rare probes like dilepton decays of vectormesons and strange hadrons, the HADES experiment allows to study the properties of matter at high densities in great detail. In this contribution a special focus is put on the reconstruction of weakly decaying strange hadrons.

1 Introduction

In heavy-ion collisions at bombarding energies of a few GeV as measured by HADES, matter is compressed to densities above its groundstate density. At the same time, it is heated to temperatures of 70 MeV extracted using dilepton radiation [1]. Similar densities and temperatures are also expected in merging neutron stars [2, 3]. Therefore, the study of heavy-ion collisions in the few GeV regime allows to deduce information on the microscopic composition and ultimately also macroscopic properties like for example the equation of state of astronomic objects like merging neutron stars.

1.1 Strangeness Production

In this energy regime, strange hadrons are produced close to or even below their free nucleon nucleon threshold energy which results in a steep rise of their excitation function with increasing $\sqrt{s_{NN}}$. Thus, even slight effects connected to the surrounding medium significantly influence their production [4].

In previous measurements of HADES a stronger than linear rise of the production rates of strange hadrons with the average amount of nucleons participating in the collision $\langle A_{Part} \rangle$ [5] was observed and might be connected to these medium effects. Furthermore, the strong excess of baryons in the fireball leads to different potentials for strange and anti-strange mesons [6–8], which was studied by multiple experiments [9–14].

*e-mail: s.spies@gsi.de

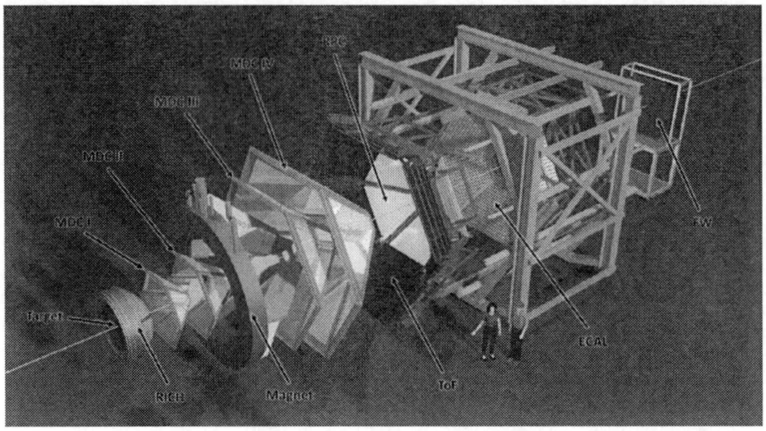

Figure 1. Schematic expanded view of the HADES detector setup.

2 HADES Experiment

The HADES experiment, which is schematically depicted in an expanded view in figure 1, is a multi-purpose fixed-target magnet spectrometer operated with the SIS18 accelerator at GSI, Darmstadt, Germany. Besides heavy-ion A+A collisions, also elementary p+A and p+p as well as pion induced π+A collisions are measured, e.g. [15–17]. Starting from the segmented 15-fold silver target, the particles traverse a Ring-Imaging CHerenkov (RICH) detector for electron/positron identification, two low-mass Mini Drift Chamber (MDC) tracking stations, a toroidal magnetic field, two further MDC tracing stations and depending on the polar angle either a Resistive Plate Chamber (RPC) or a Time of Flight (ToF) scintillation detector for time of flight measurement. The setup is completed by an Electromagnetic CALorimeter (ECAL) for photon detection, a Forward Wall (FW) hodoscope for measurement of the event plane and a diamond reaction time and trigger detector (START) located in front of the target. Besides the RICH, the FW and the START detector, all detectors are splitted into six independent sectors covering almost the entire azimuthal angle. More details on the HADES experiment are given in [18].

3 Weak Decay Recognition

In contrast to the charged kaons which have a mean flight length $c\tau$ of few meters and are thereby measured directly in the detectors, the Λ hyperon and K^0_S meson have a $c\tau$ of only few centimeters and thereby need to be reconstructed from their decay daughters. Due to the large amounts of protons and pions not associated with weak decays, a method to strongly separate the signals from the background is required. As a result of the rather high $c\tau$ values of weakly decaying particles, a spacial displacement between the primary event vertex and the secondary decay vertex can be observed. This displacement results in a so-called *Off-Vertex-Decay-Topology* which is quantified using the following parameters:

- Distance between the primary and secondary vertex (VDX).
- Distance of closest approach (DCA) between the daughter tracks and the primary vertex. (Dau1VD and Dau2VD)
- DCA between reconstructed mother track and primary vertex. (MotVD)
- DCA between the two daughter tracks. (MTD)
- Opening angle between the two daughter tracks. (A)

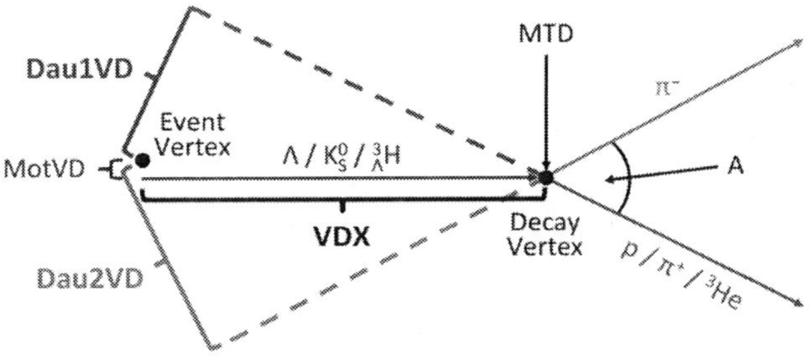

Figure 2. Schematic depiction of the *Off-Vertex-Decay-Topology* of Λ, K_S^0 and ${}^3_\Lambda H$ decays.

In addition, the method proposed by R. Armenteros and J. Podolanski [19] is used to take the kinematic constraints of decays into account. In the ultrarelativistic limit, these constraints reduce to the ellipse equation 1 with the parameters defined by equation 2.

$$\left(\frac{\alpha - \alpha_0}{r_\alpha}\right)^2 + \frac{p_\perp^2}{p_{cms}^2} = 1 \tag{1}$$

$$p_{cms} = p_{cms}^{\pm} = \sqrt{\left(\frac{M^2 + m_\pm^2 - m_\mp^2}{2M}\right)^2 - m_\pm^2} \quad \text{and} \quad \alpha_0 = \frac{m_+^2 - m_-^2}{M^2} \quad \text{and} \quad r_\alpha = \frac{2p_{cms}}{M} \tag{2}$$

Since the ultrarelativistic limit is not reached at typical HADES collision energies, we adapt the original formalism by introducing an artificial boost of $\beta = 0.99$ in longitudinal direction on the measured Lorentz vectors before calculating the Armenteros-Podolanski variables. Next, the characteristic Armenteros-Podolanski ellipses are interpreted in polar coordinates by assigning them a radial and an azimuthal component in the α-$p_\perp$-plane according to equation 3.

$$r_{\alpha p_\perp} = \sqrt{\left(\frac{\alpha - \alpha_0}{r_\alpha}\right)^2 + \frac{p_\perp^2}{p_{cms}^2}} \quad \text{and} \quad \phi_{\alpha p_\perp} = \arctan\left(\frac{p_\perp}{p_{cms}} \cdot \frac{r_\alpha}{\alpha_0 - \alpha}\right) \tag{3}$$

The radial component $r_{\alpha p_\perp}$ is strongly correlated to the invariant mass of the daughter particles and thereby inappropriate to separate signal and background. However, the azimuthal component $\phi_{\alpha p_\perp}$, in the following *APAngle*, is at most weakly correlated to the invariant mass and thereby well suited as a further parameter to reject background. To our knowledge, such a modification has not been discussed in literature yet.

Finally, the rejection of background is enhanced further by using a Multi-Layer-Perceptron (MLP) artificial Neural Network (aNN) from the Toolkit for Multivariate Data Analysis (TMVA) [20]. Therefore, the Off-Vertex-Decay-Topology parameters Dau1VD, Dau2VD, MotVD, VDX and MTD as well as the azimuthal angle of the Armenteros-Podolanski ellipse APAngle are used as input parameters and the aNN is trained using simulated signal samples and mixed-event background samples.

In previous analyses [5, 21, 22] it was shown that the use of an aNN to reject background in the reconstruction of weak decays significantly improves the reconstructed signals compared to pure hard-cut analyses.

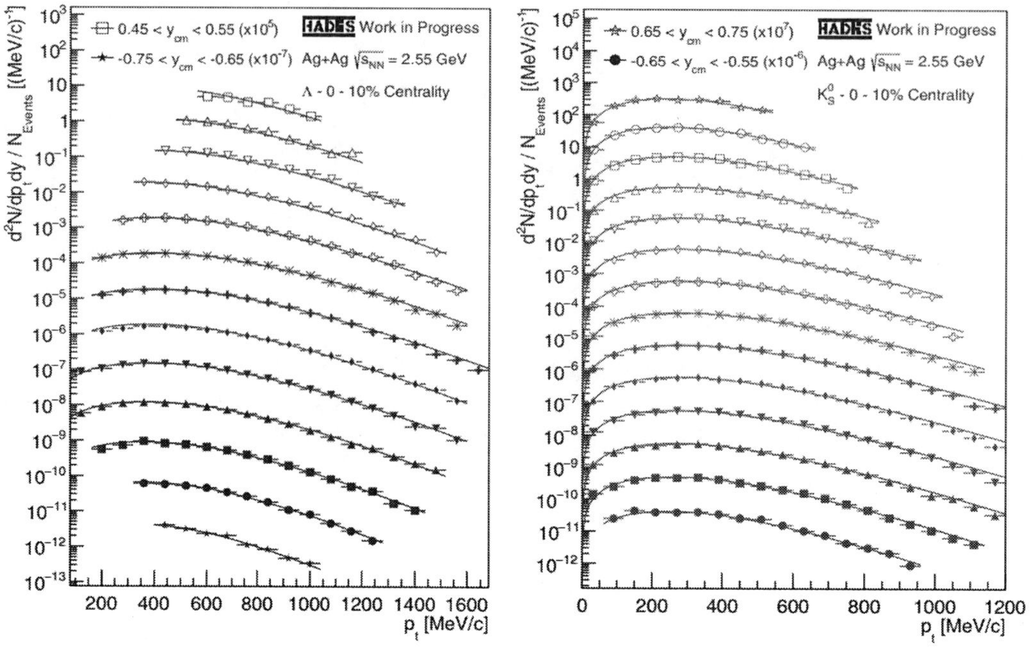

Figure 3. Transverse momentum spectra of Λ hyperons (left) and K^0_S mesons (right) from the 0-10% most central Ag(1.58A GeV)+Ag events measured by HADES.

4 Multi-Differential Analysis

The production of strange hadrons is analyzed multi-differentially in transverse momentum, rapidity and centrality. Figure 3 shows on the left side the transverse momentum spectra of Λ hyperons in 13 rapidity intervals and on the right side the transverse momentum spectra of K^0_S mesons in 14 rapidity intervals both for the 0-10% most central Ag(1.58A GeV)+Ag collisions. Thanks to the excellent acceptance of the HADES detector, both analyses cover a large proportion of the forward and backward hemisphere of the total phase-space which allows for in-depth studies of the production mechanisms of strangeness at SIS18 energies.

The measured transverse spectra are extrapolated to the regions of the total phase-space not covered by the analysis assuming the particles to be produced with Boltzmann distributed momenta. The corresponding fit functions, that are also depicted in figure 3, describe the measured data over the entire range despite the simplicity of the underlying assumption.

Furthermore, the global polarization is measured using Λ hyperons as probes. Likewise, this measurement is performed multi-differentially in transverse momentum, rapidity and centrality. In both heavy-ion collision systems measured by HADES, Au(1.23A GeV)+Au and Ag(1.58A GeV)+Ag, a significant global polarization was measured unlike at high collision energies where the global polarization vanishes. The HADES global polarization measurement is described in more detail in the contribution [23] of this conference.

In addition to the Λ hyperons and K^0_S mesons, also the ${}^3_\Lambda H$ was successfully reconstructed by HADES via its weak decay into a ${}^3He$ and a π^- in Ag(1.58A GeV)+Ag collisions. This is the first time the ${}^3_\Lambda H$ is observed at mid-rapidity at such low collision energies. As a hypernucleus, the ${}^3_\Lambda H$ is sensitive to the medium due to both production of strangeness in collective processes and the formation of nuclei from the baryon dominated matter. Furthermore, the

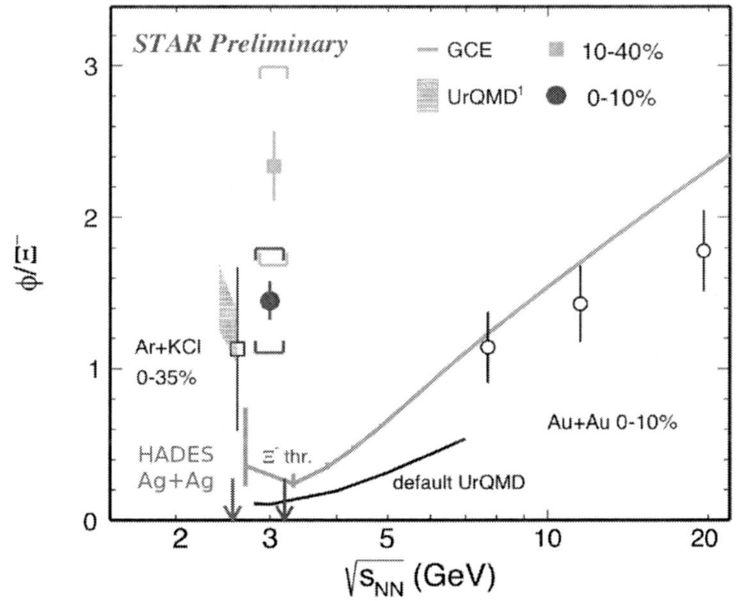

Figure 4. ϕ/Ξ^- ratio in dependence of $\sqrt{s_{NN}}$ as presented by the STAR collaboration on this conference [14]. The red arrow on the x-axis depicts the free nucleon nucleon threshold energy of Ξ^- production and the blue arrow depicts the $\sqrt{s_{NN}}$ of the Ag(1.58A GeV)+Ag collisions measured by HADES. This emphasizes the need for more data in this energy regime.

study of hypernuclei allows to gain more information on hyperon nucleon interaction which are crucial to understand the influence of hyperons on the nuclear equation of state.

5 Summary and Outlook

In this contribution we presented first data from the HADES Ag(1.58A GeV)+Ag measurement campaign conducted as a part of the FAIR phase-0 physics program. The combination of the high amount of recorded events and the sophisticated analysis methods allows for in-depth studies of the production of (rare) strange hadrons. Furthermore, the ${}^{3}_{\Lambda}H$ was measured for the first time at midrapidity at such low collision energies. In combination with the STAR fixed-target program in which Au+Au collisions down to $\sqrt{s_{NN}} = 3$ GeV were measured, the high baryon density region of the phase-diagram of QCD matter is explored in more detail than ever before.

Figure 4, which was presented by the STAR collaboration on this conference [14], shows the ϕ/Ξ^- ratio in dependence of $\sqrt{s_{NN}}$. The plot shows results from the HADES Ar(1.76A GeV)+KCl measurement campaign and the STAR Au+Au fixed target program as well as theoretical predictions. In the HADES Au(1.23A GeV)+Au measurement campaign the ϕ was successfully measured [13], however, the Ξ^- could not be reconstructed successfully. In the HADES Ag(1.58A GeV)+Ag measurement campaign, thanks to the increased amount of recorded events, the increased energy and the decreased amount of background nucleons, we expect to reconstruct and analyze the ϕ mesons and the Ξ^- hyperons at lower energies than ever before.

In parallel to the ongoing analysis of the Ag(1.58A GeV)+Ag data, the HADES collaboration is implementing the Kalman-Filter based KF-Particle-Finder [24] package to reconstruct decaying particles. This approach is already used by STAR, ALICE and the future CBM experiment to increase the reconstruction efficiencies of decaying particles. Thereby, it will strongly enhance the reconstruction and analysis of strange hadrons, in particular the double strange Ξ hyperons.

SIP JUC Cracow, Cracow (Poland), 2017/26/M/ST2/00600; TU Darmstadt, Darmstadt (Germany), VH-NG-823, DFG GRK 2128, DFG CRC-TR 211, BMBF:05P18RDFC1; Goethe-University, Frankfurt (Germany) and TU Darmstadt, Darmstadt (Germany), ExtreMe Matter Institute EMMI at GSI Darmstadt; NRNU MEPhI Moscow, Moscow (Russia), RFBR funding within the research project no. 18-02-00086, Fundamental properties of elementary particles and cosmology No 0723-2020-0041; JLU Giessen, Giessen (Germany), BMBF:05P12RGGHM; IPN Orsay, Orsay Cedex (France), CNRS/IN2P3; NPI CAS, Rez, Rez (Czech Republic), MSMT LTT17003, MSMT LM2018112, MSMT OP VVV CZ.02.1.01/0.0/0.0/18_046/0016066;

References

[1] J. Adamczewski-Musch *et al.* HADES Collaboration, Nature Phys. **15** 1040-1045 (2019).
[2] M. Hanauske, J. Steinheimer, L. Bovard, A. Mukherjee, S. Schramm *et al.*, J. Phys. Conf. Ser. **878** 012031 (2017).
[3] C. Ecker, M. Järvinen, G. Nijs and W. van der Schee, Phys. Rev. D **101** 103006 (2020).
[4] V. Steinberg, J. Staudenmaier, D. Oliinychenko, F. Li, Ö. Erkiner and H. Elfner, Phys. Rev. C **99**, 064908 (2019).
[5] J. Adamczewski-Musch *et al.* HADES Collaboration, Phys. Lett. B **793** 457-463 (2019).
[6] D.B. Kaplan and A.E. Nelson, Phys. Lett. B **175** 57-63 (1986).
[7] C.B. Dover and G.E. Walker, Phys. Rept. **89** 1-177 (1982).
[8] C.H. Lee, G.E. Brown, D.P. Min and M. Rho, Nucl. Phys. A **585** 401-449 (1995).
[9] F. Laue *et al.* KaoS Collaboration, Phys. Rev. Lett. **82** 1640 (1999).
[10] M. Menzel *et al.* KaoS Collaboration, Phys. Lett. B **495** 26 (2000).
[11] A. Förster *et al.* KaoS Collaboration, Phys. Rev. Lett. **91** 152301 (2003).
[12] A. Forster, F. Uhlig, I. Bottcher, D. Brill *et al.*, Phys. Rev. C **75** 024906 (2007).
[13] J. Adamczewski-Musch *et al.* HADES Collaboration, Phys. Lett. B **778** 403-407 (2018).
[14] G. Xie *et al.* STAR Collaboration, SQM2021 conference talk (2021).
[15] G. Agakishiev *et al.* HADES Collaboration, Eur. Phys. J. A **50** 81 (2014).
[16] J. Adamczewski-Musch *et al.* HADES Collaboration, Phys.Rev.C **95** 015207 (2017).
[17] J. Adamczewski-Musch *et al.* HADES Collaboration, Phys. Rev. C **102** 024001 (2020).
[18] G. Agakishiev *et al.* HADES Collaboration, Eur. Phys. J. A **41** 243-277 (2009).
[19] J. Podolanski and R. Armenteros, Phil. Mag. **45** 13-30 (1954).
[20] A. Hocker, P. Speckmayer, J. Stelzer, J. Therhaag, E. von Toerne *et al.*, TMVA - Toolkit for Multivariate Data Analysis, CERN-OPEN-2007-007 (2007).
[21] S. Spies, Master Thesis Goethe Universität Frankfurt (2018).
[22] S. Spies, J. Phys. Conf. Ser. **1667** 012041 (2020).
[23] F. Kornas *et al.* HADES Collaboration, these proceedings.
[24] S. Gorbunov and I. Kisel, CBM-SOFT-note-2007-003 (2007).

Dynamic modeling for heavy-ion collisions

Chun Shen[1,2,*]

[1]Department of Physics and Astronomy, Wayne State University, Detroit, Michigan, 48201, USA
[2]RIKEN BNL Research Center, Brookhaven National Laboratory, Upton, NY 11973, USA

Abstract. Recent theory progress in (3+1)D dynamical descriptions of relativistic nuclear collisions at finite baryon density are reviewed. Heavy-ion collisions at different collision energies produce strongly coupled nuclear matter to probe the phase structure of Quantum Chromodynamics (QCD). Dynamical frameworks serve as a quantitative tool to study properties of hot QCD matter and map collisions to the QCD phase diagram. Outstanding challenges are highlighted when confronting theoretical models with the current and forthcoming experimental measurements from the RHIC beam energy scan program.

1 Introduction

Quantifying the phase structure of hot and dense QCD matter is one of the primary goals in relativistic nuclear physics. Lattice QCD calculations [1, 2] provide conclusive information that hot nuclear matter at zero net baryon density transits from the hadron gas to the Quark-Gluon Plasma (QGP) phase via a smooth crossover at the pseudo-critical temperature $T_{pc} = 156.5 \pm 1.5$ MeV [3]. Moreover, the same type of phase structure remains for small-to-moderate values of the net baryon chemical potential μ_B [3]. It was conjectured that the quark-hadron transition turns from a crossover to a first-order phase transition at some finite baryon chemical potential, suggesting the existence of a critical point [4, 5]. However, the first principles lattice QCD calculations can only provide limited guidance on the existence and location of the QCD critical point in the $T - \mu_B$ phase diagram because of the fermion sign problem at finite net baryon density [6]. Meanwhile, exploration of the dense quark matter properties at large net baryon density is of particular importance since the gravitational waves detection, emerging from black hole/neutron star mergers, now give stringent constraints on the properties of the compact stars, including the nuclear matter equation of state (EoS) [7–10].

To establish quantitative connections between the QCD phase structure and measurements from relativistic heavy-ion collisions over an extensive collision energy range, we need to model the entire dynamical evolution of the heavy-ion collisions. Relativistic hydrodynamics, incorporated with nuclear matter EoS, viscosity, and initial state fluctuations, has been a precision tool to understand the macroscopic dynamics of the strongly coupled QGP. The fluid dynamic description is connected with hadronic transport approaches to microscopically describe the system's evolution in the dilute phase. Such a hydrodynamics + hadronic transport hybrid theoretical framework has successfully described and even predicted various

*e-mail: chunshen@wayne.edu

flow correlation measurements with remarkable precision [11–14]. Phenomenological studies with precise flow measurements of the hadronic final state from the Beam Energy Scan (BES) program can elucidate the collective aspects of the baryon-rich QGP and constrain the QGP EoS and transport properties, such as its viscosity and charge diffusion coefficients [15, 16]. Strange quarks produced in relativistic nuclear collisions are an interesting probe for studying the evolution of the collisions. Since the collision system is strangeness neutral, the $s\bar{s}$ pair production mechanism is sensitive to the properties of strongly-interacting matter, especially at the early time [17]. The partonic and hadronic production channels are very different and may signal the onset of deconfinement and the quark-gluon plasma [18]. Furthermore, unstable resonance states of strange baryons can provide detailed information about hadronic reactions in the late-stage of heavy-ion collisions [19]. This proceeding will highlight the recent developments of the hybrid dynamical frameworks, emphasizing the strangeness-related observables.

2 Flowing through the QCD crossover region

Relativistic heavy-ion collisions with collision energies $\sqrt{s_{\rm NN}} \gtrsim 20$ GeV can probe the nuclear matter properties with a net baryon chemical potential up to ~ 300 MeV, where the QGP and hadronic phases are connected with a smooth crossover [20, 21]. At $\sqrt{s_{\rm NN}} \sim O(10)$ GeV, heavy-ion collisions violate longitudinal boost invariance, and the overlapping time for the two nuclei to pass through each other is significant compared to the total collision lifetime [22–24]. Therefore, it is crucial to develop initial conditions with non-trivial longitudinal dynamics. The complex 3D collision dynamics can be approximated by parametric energy depositions based on the collision geometry [24–27]. Non-trivial dynamics could be included by modeling the energy loss in individual nucleon-nucleon collisions based on either string deceleration [23, 28] or full transport simulations [22, 29, 30]. Further theory developments exist to understand the early-stage baryon stopping from the Color Glass Condensate-based approaches in the fragmentation region [31, 32] and a holographic approach at intermediate couplings [33]. In the meantime, dynamical initialization schemes have been developed to interweave initial-state and hydrodynamics on a local basis to model the extended interaction region between the two colliding nuclei at low collision energies [23, 30, 34, 35]. These schemes have also been applied to study small systems and jet-medium interactions [36, 37].

Figure 1 shows a series of snapshots of one central Au+Au collision at $\sqrt{s_{\rm NN}} = 39$ GeV flowing through the QCD phase diagram. The numerical simulation is based on 3D MC-Glauber with string deceleration model [23]. The first row presents the early dynamical initialization stage $\tau \leq 1.5$ fm/c, during which the two colliding nuclei pass through each other. The collisions among nucleons continuously deposit energy-momentum currents to form the strongly coupled QGP. Because most of the baryon charge sources are at the end of strings, their deposition times are later than those of the energy-momentum currents. Hence, the hydrodynamic medium first heats up at $\mu_B \sim 0$. Around $\tau \sim 1.2$ fm/c, the baryon charges start to be doped into the system and drive the system to flow to a large μ_B region in the phase diagram. For central collisions at $\sqrt{s_{\rm NN}} = 39$ GeV, the duration of energy-momentum and net baryon charge depositions last for about 1.5-2 fm/c. During this phase, the fireball reaches $T_{\rm max} \sim 250$ MeV and $\mu_{B,{\rm max}} \sim 200$ MeV at the mid-rapidity. For fireball in the forward space-time rapidity region $1 < \eta_s < 1.8$, the peak temperature of the medium drops to about 200 MeV while the medium reaches out to a larger μ_B region, $\mu_B \sim 300$ MeV. For $\tau \geq 2$ fm/c, the fireball expands hydrodynamically in 3D and lives up to 10.5 fm/c before all the fluid cells convert to hadrons in the dilute hadronic phase.

Because most hybrid dynamical frameworks assume grand-canonical ensemble (GCE) at the particlization, the hadron chemistry is mainly controlled by the distributions of tem-

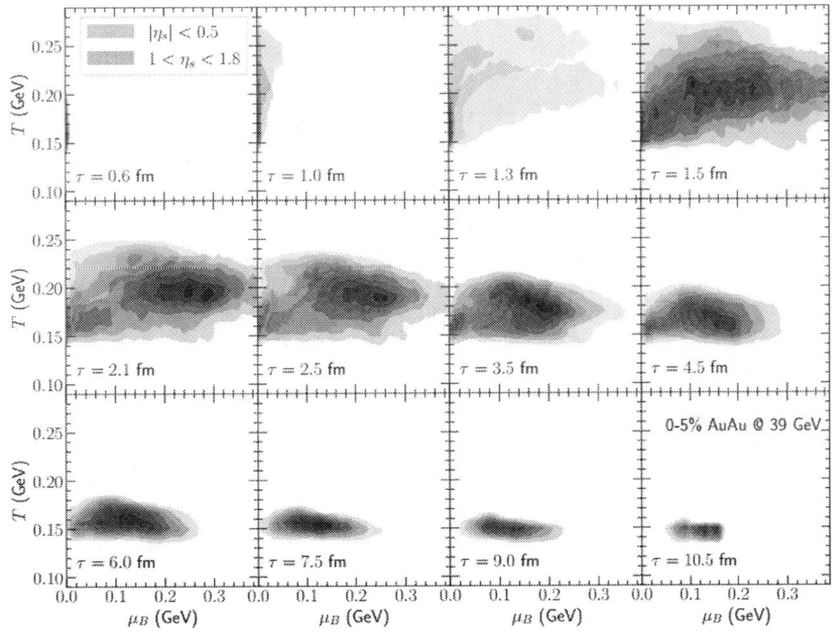

Figure 1. The QCD phase diagram is dynamically probed by a central Au+Au collision at $\sqrt{s_{\rm NN}} =$ 39 GeV. The red and blue contours indicate the time snapshots of fluid cell $T - \mu_B$ distributions in the mid-rapidity $|\eta_s| < 0.5$ and a forward space-time rapidity region $1 < \eta_s < 1.8$, respectively.

perature and chemical potentials on the constant energy density particlization hyper-surface. Furthermore, the late-stage hadronic transport model can further modify the relative particle abundances via inelastic scatterings and baryon-anti-baryon annihilation. Therefore, the yield ratios of identified particles provide strong constraints on the particlization condition in hybrid simulations [38, 39]. Assuming GCE, one can map the net electric charge (Q), strangeness (S), and baryon (B) chemical potentials to identified particle ratios as follows, $\pi^-/\pi^+ \propto \exp(-2\mu_Q)$, $K^-/K^+ \propto \exp[-2(\mu_Q + \mu_S)]$, and $\bar{p}/p \propto \exp[-2(\mu_Q + \mu_B)]$. The ratios $K^-/\pi^- \propto \exp(-\mu_S)$ and $K^+/\pi^+ \propto \exp(+\mu_S)$ provide direct information about the strangeness chemical potential. The yield ratios of multi-strange baryons to their antiparticles $\bar{\Lambda}/\Lambda \propto \exp[-2(\mu_B - \mu_S)]$, $\bar{\Xi}^+/\Xi^- \propto \exp[-2(\mu_B - 2\mu_S - \mu_Q)]$, and $\bar{\Omega}/\Omega \propto \exp[-2(\mu_B - 3\mu_S)]$ contain different weights on μ_S. Figure 2 shows the comparison between the preliminary calculations from the 3D string deceleration model [23] and the identified particle yield ratios in central Au+Au collisions at the RHIC BES energies. The model calculations with a constant switching energy density $e_{\rm sw} = 0.26$ GeV/fm^3 reproduce the collision energy dependence of the data from 7.7 GeV to 200 GeV. The key theoretical ingredients that lead to this good description are initial-state baryon stopping plus the constraints $n_S = 0$ and $n_Q = 0.4 n_B$ for Au+Au collision [38, 46]. The strangeness neutrality condition $n_S = 0$ results in a linear correlation between the strangeness chemical potential and μ_B, $\mu_S \sim \mu_B/3$ [46]. The collision energy dependence of μ_S can be constrained with the K^+/π^+ and K^-/π^- ratios. Below $\sqrt{s_{\rm NN}} = 5$ GeV, the K^+/π^+ ratio becomes smaller because of the canonical suppression of the strange quark production [20, 47]. The small μ_Q from the constraint $n_Q = 0.4 n_B$ quantitatively reproduces the π^-/π^+ ratios. The model calculation underestimates the ratios among multi-strangeness baryons for $\sqrt{s_{\rm NN}} < 40$ GeV, which hints that these strangeness baryons

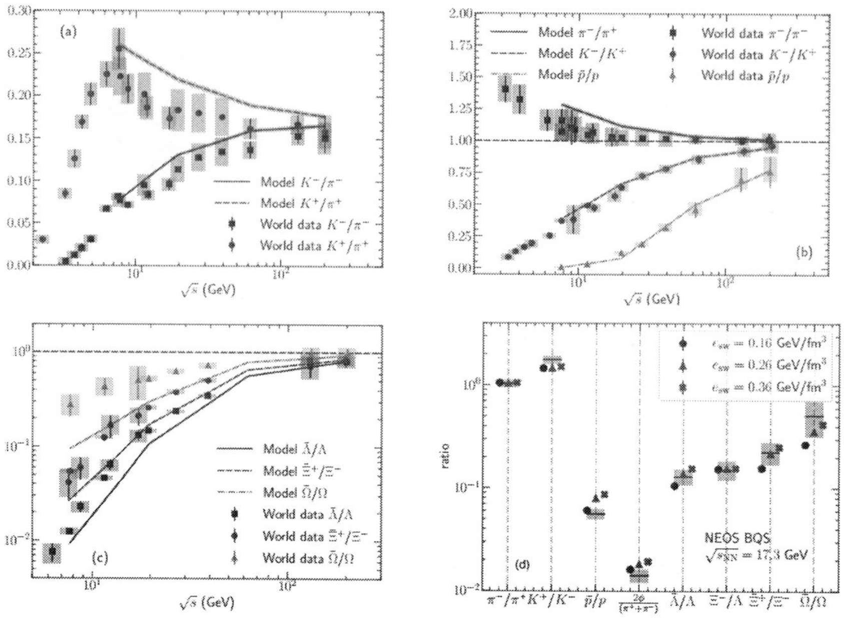

Figure 2. Panels (a, b, c): Preliminary calculations of identified particle ratios as functions of the collision energy in central Au+Au collisions compared with the world data [40–45]. Panel (d): Identified particle ratios at three switching energy densities for central Pb+Pb collisions at the top SPS energy [38].

could have a higher chemical freeze-out energy density than the non-strange hadrons [48] illustrated in Fig. 2d.

3 Outlook and challenges

Over the past decade, the RHIC Beam Energy Scan program has excited a wave of theory developments towards a (3+1)D paradigm of dynamical modeling of relativistic nuclear collisions. The advancements in 3D initial-state models, equation of state at finite net baryon density, dynamical initialization schemes enable us to quantify initial baryon stopping and study the collectivity of the QGP in a baryon-rich environment. To confront precision flow measurements from the upcoming RHIC BES program phase II, our community has adopted the Bayesian Inference method as a standard approach to systematically constrain the QGP's thermodynamic and transport properties and initial-state fluctuation spectrum [14, 49, 50]. Meanwhile, the increasing theoretical and numerical complexities in 3D hybrid models could pose significant challenges in future high-dimensional Bayesian analysis.

The current RHIC fixed-target experiment and the future FAIR/NICA experimental programs push the exploration of the QCD phase diagram beyond μ_B = 400 MeV, where guidance from lattice QCD on EoS is very limited. Searching for experimental signals of critical point and first-order phase transition requires the theoretical frameworks to be calibrated by particle production and flow measurements to provide reliable baseline expectations without critical phenomena [51–54]. Further studying the effects of out-of-equilibrium evolution of critical fluctuations [55–59] on net proton high-order cumulants and light nuclei productions [60] as functions of collision energy and rapidity intervals would shed light on the QCD phase structure at large baryon density.

Acknowledgments

This work is support in part by the National Science Foundation (NSF) under grant number PHY-2012922 and by the U.S. Department of Energy, Office of Science, Office of Nuclear Physics, under contract number DE-SC001346 and within the framework of the Beam Energy Scan Theory (BEST) Topical Collaboration. The calculations in this work used resources provided by the Open Science Grid [61, 62], which is supported by the NSF award 2030508.

References

[1] T. Bhattacharya et al., Phys. Rev. Lett. **113**, 082001 (2014), 1402.5175
[2] A. Bazavov et al., Phys. Rev. D **85**, 054503 (2012), 1111.1710
[3] A. Bazavov et al. (HotQCD), Phys. Lett. B **795**, 15 (2019), 1812.08235
[4] M. Asakawa, K. Yazaki, Nucl. Phys. A **504**, 668 (1989)
[5] A. Bzdak, S. Esumi, V. Koch, J. Liao, M. Stephanov, N. Xu, Phys. Rept. **853**, 1 (2020), 1906.00936
[6] P. de Forcrand, PoS **LAT2009**, 010 (2009), 1005.0539
[7] B.P. Abbott et al. (LIGO Scientific, Virgo), Phys. Rev. Lett. **119**, 161101 (2017), 1710.05832
[8] B.P. Abbott et al. (LIGO Scientific, Virgo), Phys. Rev. Lett. **121**, 161101 (2018), 1805.11581
[9] G. Baym, T. Hatsuda, T. Kojo, P.D. Powell, Y. Song, T. Takatsuka, Rept. Prog. Phys. **81**, 056902 (2018), 1707.04966
[10] V. Dexheimer, J. Noronha, J. Noronha-Hostler, C. Ratti, N. Yunes, J. Phys. G **48**, 073001 (2021), 2010.08834
[11] C. Shen, Z. Qiu, H. Song, J. Bernhard, S. Bass, U. Heinz, Comput. Phys. Commun. **199**, 61 (2016), 1409.8164
[12] J.E. Bernhard, J.S. Moreland, S.A. Bass, J. Liu, U. Heinz, Phys. Rev. C **94**, 024907 (2016), 1605.03954
[13] B. Schenke, C. Shen, P. Tribedy, Phys. Rev. C **102**, 044905 (2020), 2005.14682
[14] D. Everett et al. (JETSCAPE), Phys. Rev. C **103**, 054904 (2021), 2011.01430
[15] G.S. Denicol, C. Gale, S. Jeon, A. Monnai, B. Schenke, C. Shen, Phys. Rev. C **98**, 034916 (2018), 1804.10557
[16] E. McLaughlin, J. Rose, T. Dore, P. Parotto, C. Ratti, J. Noronha-Hostler (2021), 2103.02090
[17] S. Pratt, C. Plumberg, Phys. Rev. C **99**, 044916 (2019), 1812.05649
[18] C. Blume, EPJ Web Conf. **171**, 03001 (2018), 1710.07508
[19] D. Oliinychenko, C. Shen (2021), 2105.07539
[20] A. Andronic, P. Braun-Munzinger, K. Redlich, J. Stachel, Nature **561**, 321 (2018), 1710.09425
[21] L. Adamczyk et al. (STAR), Phys. Rev. C **96**, 044904 (2017), 1701.07065
[22] I.A. Karpenko, P. Huovinen, H. Petersen, M. Bleicher, Phys. Rev. C **91**, 064901 (2015), 1502.01978
[23] C. Shen, B. Schenke, Phys. Rev. C **97**, 024907 (2018), 1710.00881
[24] C. Shen, S. Alzhrani, Phys. Rev. C **102**, 014909 (2020), 2003.05852
[25] T. Hirano, U.W. Heinz, D. Kharzeev, R. Lacey, Y. Nara, Phys. Lett. B **636**, 299 (2006), nucl-th/0511046
[26] P. Bozek, W. Broniowski, Phys. Rev. C **97**, 034913 (2018), 1711.03325

[27] A. Sakai, K. Murase, T. Hirano, Phys. Rev. C **102**, 064903 (2020), 2003.13496
[28] A. Bialas, A. Bzdak, V. Koch, Acta Phys. Polon. B **49**, 103 (2018), 1608.07041
[29] L. Pang, Q. Wang, X.N. Wang, Phys. Rev. C **86**, 024911 (2012), 1205.5019
[30] L. Du, U. Heinz, G. Vujanovic, Nucl. Phys. A **982**, 407 (2019), 1807.04721
[31] M. Li, J.I. Kapusta, Phys. Rev. C **99**, 014906 (2019), 1808.05751
[32] L.D. McLerran, S. Schlichting, S. Sen, Phys. Rev. D **99**, 074009 (2019), 1811.04089
[33] M. Attems, Y. Bea, J. Casalderrey-Solana, D. Mateos, M. Triana, M. Zilhão, Phys. Rev. Lett. **121**, 261601 (2018), 1807.05175
[34] C. Shen, G. Denicol, C. Gale, S. Jeon, A. Monnai, B. Schenke, Nucl. Phys. A **967**, 796 (2017), 1704.04109
[35] Y. Akamatsu, M. Asakawa, T. Hirano, M. Kitazawa, K. Morita, K. Murase, Y. Nara, C. Nonaka, A. Ohnishi, Phys. Rev. C **98**, 024909 (2018), 1805.09024
[36] M. Okai, K. Kawaguchi, Y. Tachibana, T. Hirano, Phys. Rev. C **95**, 054914 (2017), 1702.07541
[37] Y. Kanakubo, Y. Tachibana, T. Hirano, Phys. Rev. C **101**, 024912 (2020), 1910.10556
[38] A. Monnai, B. Schenke, C. Shen, Phys. Rev. C **100**, 024907 (2019), 1902.05095
[39] P. Alba, V.M. Sarti, J. Noronha-Hostler, P. Parotto, I. Portillo-Vazquez, C. Ratti, J.M. Stafford, Phys. Rev. C **101**, 054905 (2020), 2002.12395
[40] J.L. Klay et al. (E895), Phys. Rev. Lett. **88**, 102301 (2002), nucl-ex/0111006
[41] S.V. Afanasiev et al. (NA49), Phys. Rev. C **66**, 054902 (2002), nucl-ex/0205002
[42] C. Alt et al. (NA49), Phys. Rev. Lett. **94**, 192301 (2005), nucl-ex/0409004
[43] C. Alt et al. (NA49), Phys. Rev. C **78**, 034918 (2008), 0804.3770
[44] L. Adamczyk et al. (STAR), Phys. Rev. C **96**, 044904 (2017), 1701.07065
[45] J. Adam et al. (STAR), Phys. Rev. C **102**, 034909 (2020), 1906.03732
[46] A. Monnai, B. Schenke, C. Shen, Int. J. Mod. Phys. A **36**, 2130007 (2021), 2101.11591
[47] M. Abdallah et al. (STAR) (2021), 2108.00924
[48] R. Bellwied, J. Noronha-Hostler, P. Parotto, I. Portillo Vazquez, C. Ratti, J.M. Stafford, Phys. Rev. C **99**, 034912 (2019), 1805.00088
[49] J.E. Bernhard, J.S. Moreland, S.A. Bass, Nature Phys. **15**, 1113 (2019)
[50] G. Nijs, W. van der Schee, U. Gürsoy, R. Snellings, Phys. Rev. C **103**, 054909 (2021), 2010.15134
[51] W. Zhao, C. Shen, C.M. Ko, Q. Liu, H. Song, Phys. Rev. C **102**, 044912 (2020), 2009.06959
[52] D. Oliinychenko, C. Shen, V. Koch, Phys. Rev. C **103**, 034913 (2021), 2009.01915
[53] S. Pratt, R. Steinhorst, Phys. Rev. C **102**, 064906 (2020), 2008.08623
[54] V. Vovchenko, V. Koch, C. Shen (2021), 2107.00163
[55] K. Rajagopal, G. Ridgway, R. Weller, Y. Yin, Phys. Rev. D **102**, 094025 (2020), 1908.08539
[56] L. Du, U. Heinz, K. Rajagopal, Y. Yin, Phys. Rev. C **102**, 054911 (2020), 2004.02719
[57] T. Dore, J. Noronha-Hostler, E. McLaughlin, Phys. Rev. D **102**, 074017 (2020), 2007.15083
[58] M. Nahrgang, M. Bluhm, Phys. Rev. D **102**, 094017 (2020), 2007.10371
[59] L. Du, X. An, U. Heinz (2021), 2107.02302
[60] K.J. Sun, L.W. Chen, C.M. Ko, Z. Xu, Phys. Lett. B **774**, 103 (2017), 1702.07620
[61] R. Pordes et al., J. Phys. Conf. Ser. **78**, 012057 (2007)
[62] I. Sfiligoi, D.C. Bradley, B. Holzman, P. Mhashilkar, S. Padhi, F. Wurthwrin, WRI World Congress **2**, 428 (2009)

Overview of recent lattice QCD results: phase diagram, fluctuations and strangeness

Jana N. Guenther[1,*]

[1]Aix-Marseille Université

Abstract. In recent years there has been much progress on the investigation of the QCD phase diagram with lattice QCD. This work will focus on the developments in the last few years with physical parameters and low chemical potential which are obtained by extrapolations from zero or imaginary chemical potential.

1 Introduction

The investigation of the QCD phase diagram has been a very active topic for many years. Lattice QCD is a very effective, non perturbative tool to solve QCD. Therefore, it is an obvious choice to turn to when instigating the QCD phase diagram. Simulations at finite baryon chemical potential are, however, hindered by the sign problem. Therefore, other methods are necessary to gain insight at finite μ_B physics from lattice QCD. This work will review recent results obtained from lattice QCD simulations with physical parameters at low finite μ_B by extrapolations from zero or imaginary μ_B. A more detailed review, including more parameter spaces, can be found in Ref. [1].

Since at $\mu_B = 0$ the transition is a crossover (Ref. [2–7]), some observables can be discribed by an analytic function in the vicinity of zero. This fact can be exploited, by using results at imaginary or zero chemical potential. A very common technique for extrapolation is the so called Taylor method. The pressure is prametrized as

$$\frac{p}{T^4} = \sum_{j=0}^{\infty} \sum_{k=0}^{\infty} \frac{1}{j!k!} \chi_{jk}^{BS} \hat{\mu}_B^j \hat{\mu}_S^k \tag{1}$$

with $\hat{\mu} = \frac{\mu}{T}$. The χ_{jk}^{BS} can be measured on zero μ lattices. Another expansion from which results will be discussed in this work, is the so called fugacity expansion or sector method. Here the parametrization

$$\frac{p}{T^4} = \sum_{j=0}^{\infty} \sum_{k=0}^{\infty} P_{jk}^{BS} \cosh(j\hat{\mu}_B - k\hat{\mu}_S) \tag{2}$$

is used. Here the expansion coefficients P_{jk}^{BS} have to be determined either from the χ_{jk}^{BS} or from imaginary chemical potential. In general for simulations with imaginary chemical potential arbitrary fit functions are possible to describe the data at $\mu^2 < 0$. The choice of

*e-mail: jana.gunther@univ-amu.fr

 https://doi.org/10.1051/epjconf/202225902002

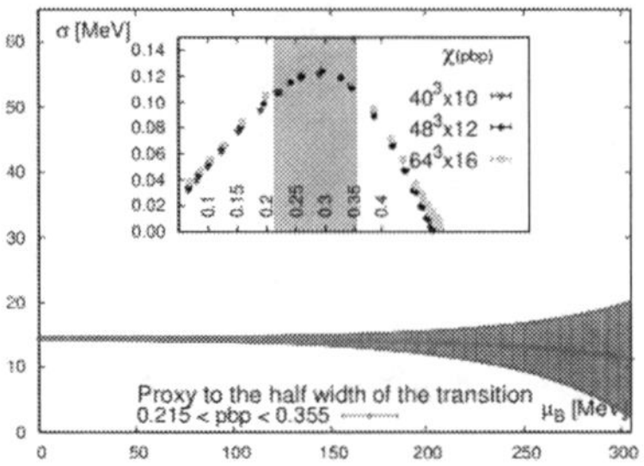

Figure 1. (Ref. [11]) Extrapolation of the half width of the transition to finite baryon chemical potential as defined in equation (3). The insert shows the chiral suceptibility χ as a function of the chiral condensate $\langle\bar{\psi}\psi\rangle$. The shaded region illustrates $\langle\bar{\psi}\psi\rangle_c \pm \Delta\langle\bar{\psi}\psi\rangle/2$.

fit function is usually guided by the fit quality, sometimes complemented by known physical features. An advantage of the fugacity expansion is its rapid convergence in the hadronic phase and the added information on the particle contend from different sectors. On the other hand the Taylor expansion convergences rapidly in the high temperature limit.

2 The phase diagram

The crossover nature of the transition and it temperature has been determined since 2006 with a variety of observables (Ref. [2–9]). More recently the transition temperature has been determined with increased precision in Ref. [10] and Ref. [11]. By now the error on the transition temperature at $\mu_B = 0$ is much smaller than the width σ of the crossover. Therefore it can be determined seperatly. A possible definition, that has been used in Ref. [11] is

$$\langle\bar{\psi}\psi\rangle(T_c \pm \sigma/2) = \langle\bar{\psi}\psi\rangle_c \pm \Delta\langle\bar{\psi}\psi\rangle/2. \tag{3}$$

The result, extrapolated to the continuum from three lattice spacings, can be seen in figure 1 up to $\mu_B = 300$ MeV.

The behaviour of the transition temperature in the μ_B-T-plane can be parameterized by the Taylor expansion as

$$\frac{T_c(\mu_B)}{T_c(0)} = 1 - \kappa_2 \left(\frac{\mu_B}{T_c}\right)^2 - \kappa_4 \left(\frac{\mu_B}{T_c}\right)^4 + O(\mu_B^6). \tag{4}$$

The odd powers of $\frac{\mu_B}{T}$ vanish due to the isospin symmetry. The coefficients κ_2 has been determined in several computations. The more recent ones from Refs. [10–14] are compared in the right side of figure 2. The green points were obtained from extrapolations with imaginary chemical potential while the blue points were obtained by the Taylor method. Both method yield agreeing results. In addition to κ_2 new results for κ_4 are available from Ref. [10, 11]. While it is clear that $\kappa_4 \ll \kappa_2$ the relative error of κ_4 is more than 100% and the sign therefore still undetermined.

The extrapolation of the transition temperature to finite μ_B for the choice of $n_S = 0$ and $n_Q = 0.4n_B$, to match the condition in heavy ion collisions from Refs. [10, 11] is shown in figure 3. As a comparison several results for the freeze-out temperature (Refs. [15–21] are shown. In a addition, on the right side, a result from Dyson-Schwinger-Equation calculations (Ref. [22]) is shown. The curvature agrees well with the lattice result, while the absolute value is set to previous lattice value, which was determined by a different observable and with a larger error. Therefore the difference does not imply a contradiction between the two calculations.

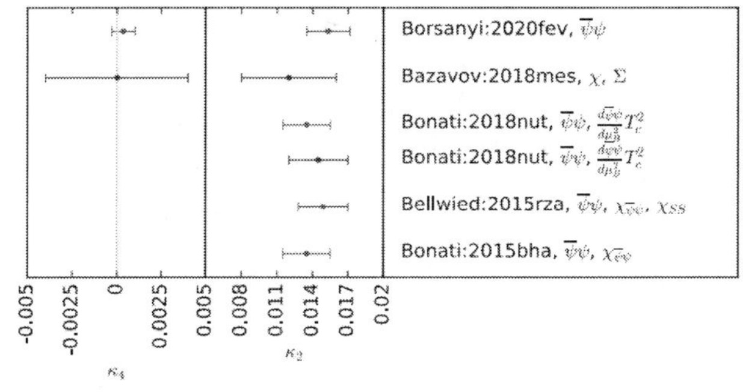

Figure 2. (Ref. [11]) Overview of the different determinations of the κ_2 and κ_4 coefficients as defined in equation (4). The values are taken from Ref. [10–14]. Green points correspond to determinations from imaginary chemical potential, while resuls shown in blue were obtained by the Taylor method.

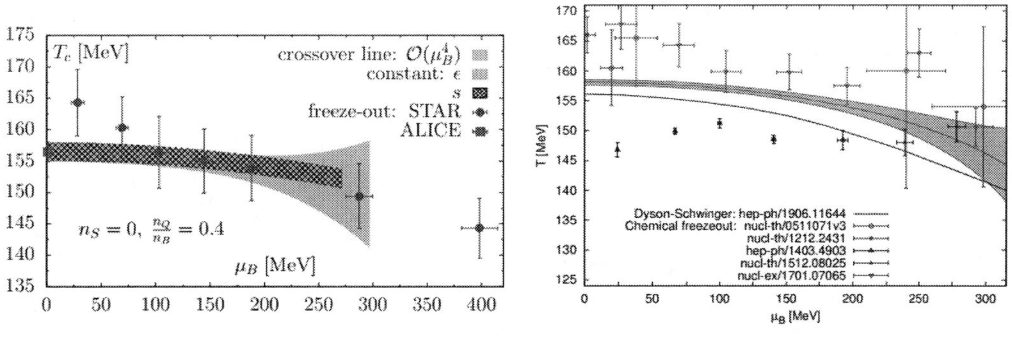

Figure 3. Left: (Ref. [10]) The extrapolation of the transition from the Taylor method compared with the freeze out temperatures from Ref. [20, 21], Right: (Ref. [11]) A comparison of the extrapolated transition temperature from simulations at imaginary chemical potential (green band) and Dyson-Schwinger-Equations (Ref. [22]). In addition the freeze-out temperatures obtained from heavy ion collisions in Ref. [15–19] are shown.

3 Fluctuations

Fluctuations are computed as the derivatives of the pressure with respect to various chemical potentials:

$$\chi_{i,j,k}^{B,Q,S} = \frac{\partial^{i+j+k}(p/T^4)}{(\partial\hat{\mu}_B)^i(\partial\hat{\mu}_Q)^j(\partial\hat{\mu}_S)^k}\,,\ \hat{\mu}_i = \frac{\mu}{T} \tag{5}$$

While fluctuations to various order have previously published for example in Ref. [23–27], now new continuum extrapolated results are available in Ref. [28]. The precision of these results is high enough is allow a comparison to different models with detailed studies on inclusion or exclusion of various states for example in a Hadron Resonance Gas model.

The ratios of various fluctuations can be used to express the cumulants of the Baryon number distribution. This offers an observable for comparisons with heavy ion collision measurements of the proton number distribution. At the current precision level this can only be a rough comparison. If the precision is increased in the future, other effects should be taken into account, like the continuum limit on the lattice side, or volume fluctuations and on equilibrium effects on the experimental side.

Figure 4 shows the ratios $R_{31}^B(T,\mu_B) = \chi_3^B/\chi_1^B = S_B\sigma_B^3/M_B$, $R_{42}^B(T,\mu_B) = \chi_4^B/\chi_2^B = \kappa_B\sigma_B^2$, $R_{51}^B(T,\mu_B) = \chi_5^B/\chi_1^B$ and $R_{62}^B(T,\mu_B) = \chi_6^B/\chi_2^B$ as a function of $R_{12}^B(T,\mu_B) = M_B/\sigma_B^2$ evaluated along the transition line. While the two lower order ratios are a continuum estimate from $N_t = 8$ and $N_t = 12$ lattices $R_{51}^B(T,\mu_B)$ and $R_{62}^B(T,\mu_B)$ are computed only on an $N_t = 8$ lattice.

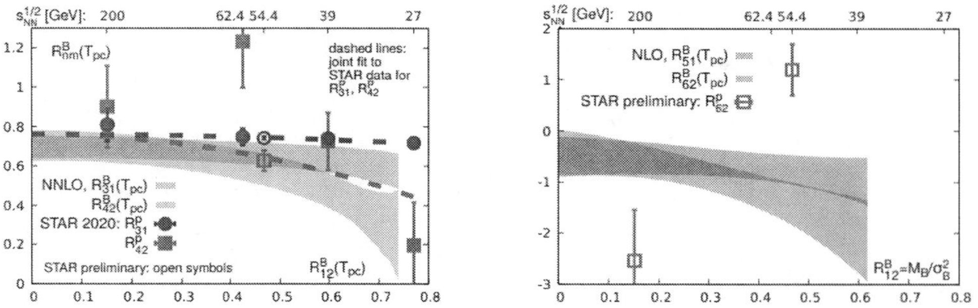

Figure 4. (Ref. [27]) Left: The ratios $R^B_{31}(T,\mu_B) = S_B\sigma^3_B/M_B$ and $R^B_{42}(T,\mu_B) = \kappa_B\sigma^2_B$ as a function of $R^B_{12}(T,\mu_B) = M_B/\sigma^2_B$ evaluated along the transition line in comparison to the data from the STAR collaboration (Ref. [29, 30]). Right: The ratios $R^B_{51}(T,\mu_B)$ and $R^B_{62}(T,\mu_B)$ as a function of $R^B_{12}(T,\mu_B)$ in comparison to the data from the STAR collaboration (Ref. [30]).

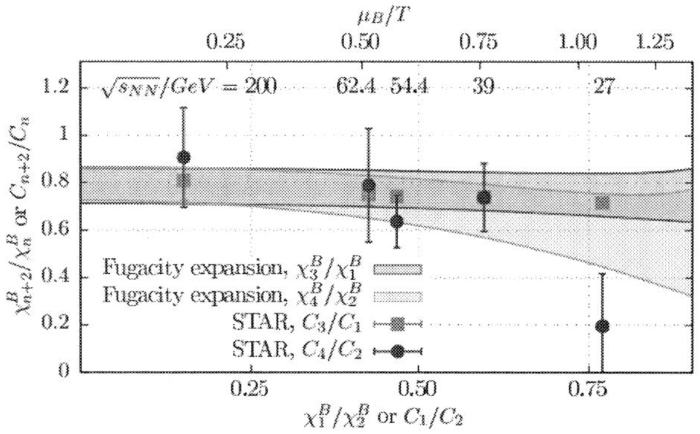

Figure 5. (Ref. [31]) The ratios $R^B_{31}(T,\mu_B) = S_B\sigma^3_B/M_B$ and $R^B_{42}(T,\mu_B) \equiv \kappa_B\sigma^2_B$ as a function of $R^B_{12}(T,\mu_B) = M_B/\sigma^2_B$ evaluated along the transition line in comparison to the data from the STAR collaboration (Ref. [29]).

A corresponding result from Ref. [31] is shown in figure 5. Here the results are a continuum estimate obtained with a different action on $N_t = 8, 10$ and 12 lattices. The analysis is based on a two dimensional fugacity expansion with imaginary μ_B and μ_S (see equation (2)).

To include strange particles like K or Λ in the comparison with lattice calculations, suitable observables have to be constructed. Ref. [32] used the Hadron Resonance Gas (HRG) model to compare different proxies (see left of figure 6). The proxy $\sigma^2_\Lambda/(\sigma^2_\Lambda + \sigma^2_K)$ for the fluctuation ratio $-\frac{\chi^{BS}_{11}}{\chi^S_2}$ was further investigated both by comparing the HRG and the experimental (see middle of figure 6) results, as well as lattice calculations from both the Taylor and sector method (see right of figure 6).

4 The equation of state

Another quantiy that has been investigated for long time is the equation of state. In the following some progress on the baryon number n_B will be discussed. When extrapolated to finite μ_B with a Taylor expansion up to μ_B^6 is shows a surprising behaviour around the transition temperature (see left and middle of figure 7). This behaviour has been observed by different groups and on different data sets (Refs. [25**?**]). Ref. [34] could link it to the general convergence of the Taylor series in this region. A possible expansion for the difficulties of the convergence might be that the extrapolation has to cover both the hadronic and the quark gluon plasma phase. Therefore a new extrapolation scheme from imaginary μ_B was proposed that shows a smooth behaviour as can be seen on the right side of figure 7.

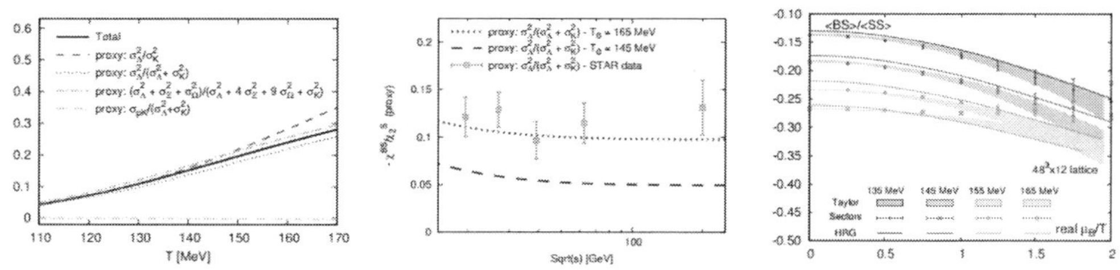

Figure 6. (Ref. [32]) Left: Calculations for different combinations of praticle number cumulants (proxis) that could be measurable in heavy ion collision expermients and the total of $-\frac{\chi_{11}^{BS}}{\chi_2^S}$ that is accessible in lattice QCD calculations with the Hadron Resonance Gas (HRG) model. Middle: Comparison of the proxy $\sigma_\Lambda^2/(\sigma_\Lambda^2 + \sigma_K^2)$ for two temperatures in the HRG model and experimental results from Ref. [33] Right: Comparison between the Taylor (see equation (1)) and the sector (see equation (2)) method for $-\frac{\chi_{11}^{BS}}{\chi_2^S}$ on an $48^3 \times 12$ lattice, as well as results from the HRG model.

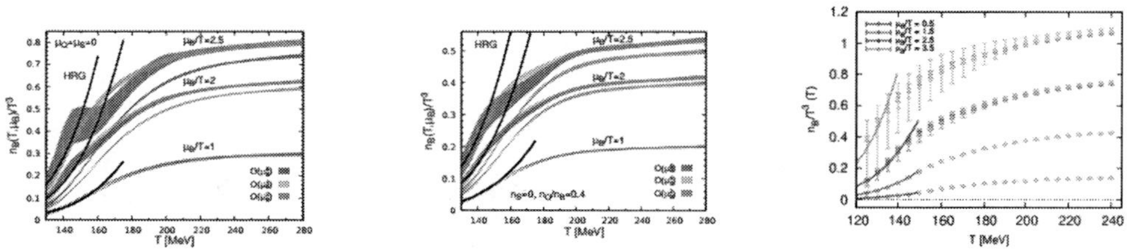

Figure 7. (Ref. [25]) The extrapolation to finite baryon number n_B done by the Taylor method (see equation (1)): Left: $\mu_Q = \mu_S = 0$, Middle: $n_S = 0$ and $n_Q/n_B = 0.4$, Right: (Ref. [34]) The pressure and the baryon number from the resummed extrapolation. The lighter bars show the increased error by the inclusion of a higher order term.

Acknowledgments

The project leading to this publication has received funding from Excellence Initiative of Aix-Marseille University - A*MIDEX, a French "Investissements d'Avenir" programme, AMX-18-ACE-005.

References

[1] J.N. Guenther, Eur. Phys. J. A **57**, 136 (2021), 2010.15503

[2] Y. Aoki, G. Endrodi, Z. Fodor, S.D. Katz, K.K. Szabo, Nature **443**, 675 (2006), hep-lat/0611014

[3] Y. Aoki, Z. Fodor, S. Katz, K. Szabo, Phys. Lett. B **643**, 46 (2006), hep-lat/0609068

[4] Y. Aoki, S. Borsanyi, S. Durr, Z. Fodor, S.D. Katz, S. Krieg, K.K. Szabo, JHEP **06**, 088 (2009), 0903.4155

[5] S. Borsanyi, Z. Fodor, C. Hoelbling, S.D. Katz, S. Krieg, C. Ratti, K.K. Szabo (Wuppertal-Budapest), JHEP **09**, 073 (2010), 1005.3508

[6] T. Bhattacharya et al., Phys. Rev. Lett. **113**, 082001 (2014), 1402.5175

[7] A. Bazavov et al., Phys. Rev. **D85**, 054503 (2012), 1111.1710

[8] M. Cheng et al., Phys. Rev. D **74**, 054507 (2006), hep-lat/0608013

[9] A. Bazavov et al., Phys. Rev. D **80**, 014504 (2009), 0903.4379
[10] A. Bazavov et al. (HotQCD), Phys. Lett. **B795**, 15 (2019), 1812.08235
[11] S. Borsanyi, Z. Fodor, J.N. Guenther, R. Kara, S.D. Katz, P. Parotto, A. Pasztor, C. Ratti, K.K. Szabo (2020), 2002.02821
[12] C. Bonati, M. D'Elia, F. Negro, F. Sanfilippo, K. Zambello, Phys. Rev. D **98**, 054510 (2018), 1805.02960
[13] R. Bellwied, S. Borsanyi, Z. Fodor, J. Guenther, S.D. Katz, C. Ratti, K.K. Szabo, Phys. Lett. **B751**, 559 (2015), 1507.07510
[14] C. Bonati, M. D'Elia, M. Mariti, M. Mesiti, F. Negro, F. Sanfilippo, Phys. Rev. **D92**, 054503 (2015), 1507.03571
[15] A. Andronic, P. Braun-Munzinger, J. Stachel, Nucl. Phys. **A772**, 167 (2006), nucl-th/0511071
[16] F. Becattini, M. Bleicher, T. Kollegger, T. Schuster, J. Steinheimer, R. Stock, Phys. Rev. Lett. **111**, 082302 (2013), 1212.2431
[17] P. Alba, W. Alberico, R. Bellwied, M. Bluhm, V. Mantovani Sarti, M. Nahrgang, C. Ratti, Phys. Lett. **B738**, 305 (2014), 1403.4903
[18] V. Vovchenko, V.V. Begun, M.I. Gorenstein, Phys. Rev. **C93**, 064906 (2016), 1512.08025
[19] L. Adamczyk et al. (STAR), Phys. Rev. **C96**, 044904 (2017), 1701.07065
[20] A. Andronic, P. Braun-Munzinger, K. Redlich, J. Stachel, Nature **561**, 321 (2018), 1710.09425
[21] L. Adamczyk et al. (STAR), Phys. Rev. C **96**, 044904 (2017), 1701.07065
[22] P. Isserstedt, M. Buballa, C.S. Fischer, P.J. Gunkel, Phys. Rev. **D100**, 074011 (2019), 1906.11644
[23] C. Schmidt (BNL-Bielefeld), Nucl. Phys. A **904-905**, 865c (2013), 1212.4278
[24] M. D'Elia, G. Gagliardi, F. Sanfilippo, Phys. Rev. **D95**, 094503 (2017), 1611.08285
[25] A. Bazavov et al., Phys. Rev. **D95**, 054504 (2017), 1701.04325
[26] S. Borsanyi, Z. Fodor, J.N. Guenther, S.K. Katz, K.K. Szabo, A. Pasztor, I. Portillo, C. Ratti (2018), 1805.04445
[27] A. Bazavov et al., Phys. Rev. D **101**, 074502 (2020), 2001.08530
[28] D. Bollweg, J. Goswami, O. Kaczmarek, F. Karsch, S. Mukherjee, P. Petreczky, C. Schmidt, P. Scior (2021), 2107.10011
[29] J. Adam et al. (STAR) (2020), 2001.02852
[30] T. Nonaka (STAR), *Measurement of the Sixth-Order Cumulant of Net-Proton Distributions in Au+Au Collisions from the STAR Experiment*, in *28th International Conference on Ultrarelativistic Nucleus-Nucleus Collisions* (2020), 2002.12505
[31] R. Bellwied, S. Borsanyi, Z. Fodor, J.N. Guenther, S.D. Katz, P. Parotto, A. Pasztor, D. Pesznyak, C. Ratti, K.K. Szabo (2021), 2102.06625
[32] R. Bellwied, S. Borsanyi, Z. Fodor, J.N. Guenther, J. Noronha-Hostler, P. Parotto, A. Pasztor, C. Ratti, J.M. Stafford, Phys. Rev. D **101**, 034506 (2020), 1910.14592
[33] L. Adamczyk et al. (STAR), Phys. Lett. B **785**, 551 (2018), 1709.00773
[34] S. Borsányi, Z. Fodor, J.N. Guenther, R. Kara, S.D. Katz, P. Parotto, A. Pásztor, C. Ratti, K.K. Szabó, Phys. Rev. Lett. **126**, 232001 (2021), 2102.06660

https://doi.org/10.1051/epjconf/202225902003

Global Polarization Theory Overview

Jian-Hua Gao[1,*]

[1]Shandong Provincial Key Laboratory of Optical Astronomy and Solar-Terrestrial Environment, Institute of Space Sciences, Shandong University, Weihai, Shandong 264209, China

Abstract. We give a brief overview about theory development of spin polarization in relativistic heavy ion collisions, which includes how the polarization could be generated by single scattering, what the polarization could be in equilibrium, how to address some recent puzzles in spin polarization in heavy ion collisions and how much progress we have made in spin hydrodynamics and spin kinetic theory. We will also discuss the possible helicity polarization in relativistic heavy ion collisions.

1 Introduction

Spin freedom plays an important role to understand the nuclei's structure in low energy nuclear physics. For example, nuclei exhibit extra stability when either the number of proton or neutron is equal to the magic number 2, 8, 20, 28, 50, 82 or 126. The magic numbers can only be explained by adding a spin-orbital coupling term in nuclear shell model. Another example is high spin states in heavy ion fusion evaporation reaction. When two nuclei collide at low energy, they can combine together, exhibit a collective rotation and becomes a high spin state. Such high spin state can be regarded as a rigid rotating body and collective rotation can align the nucleon's angular momentum along the direction of the angular velocity due to Coriolis force, termed as angular momentum alignment. At high-energy heavy ion collisions, the colliding nuclei can not fusion into high spin states like a rigid rotating body any more. Instead, Liang and Wang in [1] first pointed out that the initial orbital angular momentum in non-central heavy ion collisions could be transferred into the momentum or rapidity gradient along the transverse direction. In local comoving frame of the fluid, the colliding particles will carry a local orbital angular momentum along the initial global orbital angular momentum. A natural question is: could this local orbital angular momentum be transformed into the spin angular momentum — spin polarization? As such polarization is always along the initial global orbital angular momentum, it has been given a specific name as global polarization.

2 Global spin polarization

Global polarization was first estimated by single scattering in static potential model [1] and then was generalized to realistic two particle scattering with hard thermal loop propagator [2]. The scattering with initial local angular momentum of two particles can be described

*e-mail: gaojh@sdu.edu.cn

by the cross section in impact parameter space. In general,the cross section for two particles with impact parameter $\mathbf{x}_T$ can be decomposed into two parts

$$\frac{d\sigma_s}{d^2\mathbf{x}_T} = \frac{d\sigma}{d^2\mathbf{x}_T} + s\frac{d\Delta\sigma}{d^2\mathbf{x}_T}, \qquad \frac{d\Delta\sigma}{d^2\mathbf{x}_T} \propto \mathbf{n}\cdot(\mathbf{x}_T \times \mathbf{p}) \tag{1}$$

where $s = \pm 1$ denotes the particle's spin index. It is obvious that when the polarization direction $\mathbf{n}$ is along the orbital angular momentum, the polarization reaches to maximal value:

The estimation for global polarization at RHIC energy 200 GeV from static potential model is about -30% while the observed global polarization from STAR at this energy is consistent with zero within the precision of the measurements [3]. When we generalize the static potential model to hard thermal loop propagator and go beyond the small angle approximation, the estimation can be consistent with the data in the Bjorken scenario at least qualitatively [2]. The most important thing about single scattering model is that it provided us a very intuitive way to understand how the orbital angular momentum can be transformed into spin angular momentum.

Single scattering is only a naive limit, we need consider the multiple scattering process for the realistic heavy ion collisions at high energy. For simplicity, we can consider another naive limit — what the polarization would be if the particles experience multiple scattering so that the system reaches global equilibrium with angular momentum. The spin polarization vector $S^\mu(x, p)$ in phase space for spin $1/2$ particles in such case can be directly calculated from the density operator at global equilibrium with given angular momentum and the result is given by [4]

$$S^\mu(x,p) = -\frac{1}{8m}(1-n_F)\epsilon^{\mu\nu\rho\sigma}p_\nu\varpi_{\rho\sigma}, \qquad \varpi_{\rho\sigma} = -\frac{1}{2}\left[\partial_\rho(\beta u_\sigma) - \partial_\sigma(\beta u_\rho)\right], \quad \beta = \frac{1}{T} \tag{2}$$

where n_F is Fermi-Dirac distribution, m is the particle's mass and $\varpi^{\rho\sigma}$ is thermal vorticity tensor. We note that the polarization is determined by the thermal vorticity

Although this result is derived from global equilibrium, we expect it should give dominant contribution at local equilibrium. With such assumption, the spin polarization vector at freeze-out hyper-surface in heavy ion collisions can be calculated from

$$S^\mu(p) = -\frac{1}{8m}\epsilon^{\mu\nu\rho\sigma}p_\nu\frac{\int d\Sigma_\lambda p^\lambda \varpi_{\rho\sigma} n_F(1-n_F)}{\int d\Sigma_\lambda p^\lambda n_F} \tag{3}$$

In non-central heavy ion collisions, the initial orbital angular momentum could be transferred into global vorticity along this direction which further induce the global polarization. This conclusion is consistent with the result from single scattering model. The calculations for the global polarization based on this expression within different models are all quantitatively consistent with the recent experimental data. The averaged polarization indicate a vorticity of $\omega \approx (9+1)\times 10^{21}$ [5], which is much greater than the vorticity of all other known fluids.

3 Local spin polarization

In relativistic heavy ion collisions, the hot system produced after collisions display more complicated vorticity structure than the global vorticity along the direction of initial orbital angular momentum. The simulation indicates that some very interesting local vorticity structure could exist in heavy ion collisions, for example the transverse vorticity exhibits a smoke-ring distribution along the bean momentum axis and the longitudinal vorticity exhibits a quadruple structure in the transverse plane [6, 7]. These local vorticity structure will lead to the local spin polarizaiton, which can be detected by the azimuthal angle dependence of the polarization. However the experimental data shows that both transverse and longitudinal polarization are opposite to the hydrodynamic prediction based on the expression (4). This is so-called local spin puzzle in relativistic heavy ion collisions. This inconsistency can not be solved by the feed-down effect from more heavy hyperons [8, 9]. Recently, some collaborations has proposed that the shear tensor could possibly contribute to the spin polarization. In Refs. [10, 11], the authors derive the shear-induced spin polarization (SIP) from chiral kinetic theory or linear response theory

$$A^{\mu}_{\rm SIP} = -\beta n_F(1-n_F)\frac{p_\perp^2}{u\cdot p}\epsilon^{\mu\nu\alpha\rho}u_\nu Q_\alpha{}^\lambda\sigma_{\rho\lambda}, \qquad \sigma_{\rho\lambda} = \frac{1}{2}\left(\partial^\perp_\rho u_\lambda + \partial^\perp_\lambda u_\rho + \frac{2}{3}\partial\cdot u\Delta_{\rho\lambda}\right) \tag{4}$$

where $p^\mu_\perp = \Delta^{\mu\nu}p_\nu$, $\partial^\mu_\perp = \Delta^{\mu\nu}\partial_\nu$, and $Q^{\mu\nu} = -p^\mu_\perp p^\nu_\perp/p^2_\perp + \Delta^{\mu\nu}/3$. Similarly, the authors in Refs [12, 13]. also derive the spin polarization from spin-thermal shear coupling by stationary non-equilibrium density operator. With the assumption of isothermal local equilibrium (ILE), they obtain

$$S^\mu_\xi = -\frac{1}{4m}\epsilon^{\mu\nu\sigma\tau}\frac{p_\tau p^\rho}{p\cdot\hat{t}}\frac{\int_\Sigma d\Sigma\cdot p\, n_F(1-n_F)\hat{t}_\nu\xi_{\sigma\rho}}{\int_\Sigma d\Sigma\cdot p\, n_F}, \qquad \xi_{\sigma\rho} = \frac{1}{2}\left[\partial_\sigma(\beta u_\rho) + \partial_\rho(\beta u_\sigma)\right] \tag{5}$$

where $\hat{t}$ is the unit vector normal to the hypersurface Σ_B which corresponds to the time direction in the QGP frame. Although these two methods are slightly different but both shear tensor contribution always shows the same azimuthal angle dependence as experimental result. In the "strange memory" scenario of the first method and with the assumption of isothermal local equilibrium of the second method, the total results from shear tensor and vorticity tensor give the right trend with the experimental tensor.

Another spin puzzle in heavy ion collisions is relevant to spin alignment of vector meson. In the recombination scenario, naive estimation gives the spin alignment should be squared order of hyperon's global polarization $P_H^2 \sim 10^{-4}$ while the experiment result is the order 10^{-1}. Besides, the ϕ meson can take a significant positive deviation from 1/3 which is very different from K^{*0} meson. It has been proposed that the existence of a mean field of the ϕ meson generated in heavy-ion collisions can give a possible mechanism to result in positive deviation from 1/3 [14]. It has been also found that the spin alignment does not necessary signal the global polarization but may also originate from local spin polarization [15]. The strong alignment measured by STAR and ALICE which is not consistent with hyperon polarization might be due to the fact that the global polarization is very small while local polarization fluctuation is very large.

4 Spin hydrodynamics and spin kinetic theory

Single scattering and global equilibrium are two opposite limits, spin hydrodynamics or the spin kinetic theory can connect these two limits and gives more realistic description for the spin polarization in heavy ion collisions. In recent years, a lot of important progresses have been made in this direction.

In the chiral system, the particle's spin always coincides with the momentum and the spin is not a dynamical variable, spin hydrodynamics can be reduced into anomalous hydrodynamics [16, 17],which had been established very well,

$$\partial_\mu T^{\mu\nu} = F^{\nu\lambda}j_\lambda, \quad \partial_\mu j^\mu = 0, \quad \partial_\mu j^\mu_5 = CE\cdot B, \tag{6}$$

where the information for spin has been encoded in the axial currrent j_5^μ. Similarly, the spin kinetic theory for chiral particles can be reduced into chiral kinetic theory in which only right-handed or left-handed particle distribution function f is independent [18, 19],

$$p \cdot \partial f = \int_{BCD} C_{ABCD}[f]\left(1 - \int_{B'C'D'} \frac{\partial}{\partial f} C_{AB'C'D'} \frac{\bar{\Delta} \cdot \bar{n}'}{p \cdot \bar{n}'}\right) \tag{7}$$

where the collision kernel C_{ABCD} is defined by the difference between the rate of collisions $C_{ABCD} \equiv W_{CD \to AB} - W_{AB \to CD}$, $\bar{n}'$ is the no-jump frame in the collison $AB' \leftrightarrow C'D'$ and $\bar{\Delta}$ denotes tunnelinglike motion in nonlocal collisions. The recent simulation [20] from chiral kinetic equation has shown that the non-local collision from angular momentum conservation plays a very important role to reproduce the consistent result of local polarization.

When the particle has a finite mass, the spin will become a dynamical variable and we need extra variables describe the spin evolution. In spin hydrodynamics, spin chemical potential $\omega^{\mu\nu}$ has been introduced into relativistic hydrodynamics to describe the spin evolution. In addition to the conservations of energy-momentum and charge, spin dynamics is determined by the conservation of angular momentum [21],

$$\partial_\mu T^{\mu\nu} = 0, \quad \partial_\mu (n u^\mu) = 0, \quad \partial_\lambda S^{\lambda,\mu\nu} = 0. \tag{8}$$

In the ideal spin hydrodynamics, the spin tensor $S^{\lambda,\mu\nu}$ is proportional to spin chemical potential $S^{\lambda,\mu\nu} \propto u^\lambda \omega^{\mu\nu}$. However, from effective field theory, even ideal spin hydrodynamics requires non-dissipative dynamics at higher order than second order, which indicates very non-triviality in spin hydrodynamics. In the regime of ideal spin hydrodynamics, we cannot deal with the spin generation and dissipation. In order to address this problem, we must go beyond ideal spin hydrodynamics and use viscous spin hydrodynamics. Viscous spin hydrodynamics can be formulated from entropy-current analysis and constrain possible constitutive equations of a spin current and the antisymmetric part of canonical energy-momentum tensor [22].

$$\partial_\lambda S^{\lambda,\alpha\beta} = T^{\beta\alpha} - T^{\alpha\beta} \equiv -2T_A^{\alpha\beta} \tag{9}$$

where antisymmetric part of energy momentum tensor $T_A^{\alpha\beta}$ is given by

$$T_A^{\alpha\beta} = -2\lambda\left(-u \cdot \partial u^{[\mu} + \beta \partial_\perp^{[\mu} \beta^{-1} + 4 u_\rho \omega^{\rho[\mu}\right) u^{\nu]} - 2\gamma\left(\partial_\perp^{[\mu} u^{\nu]} - 2\Delta_\rho^\mu \Delta_\sigma^\nu \omega^{\rho\sigma}\right) \tag{10}$$

where λ and γ are new transport coefficients associated with spin transport. The viscous spin hydrodynamics can be also derived from classical kinetic equation at relaxation time approximation and the spin chemical potential satisfies

$$\dot{\omega}^{\mu\nu} = D_\Pi^{\mu\nu}(\partial \cdot u) + \left(\nabla^\alpha \frac{\mu}{T}\right) D_n^{[\mu\nu]}{}_\alpha + D_\pi^{[\nu}{}_\lambda \sigma^{\lambda\mu]} + D_{\Sigma 1}^\alpha \nabla^{[\mu} \omega^{\nu]}{}_\alpha + D_{\Sigma 2}^{[\mu\nu]\alpha} \nabla^\lambda \omega_{\alpha\lambda} \tag{11}$$

where the explicit expressions for D coefficient are given by [23]. The spin hydrodynamics has also been derived from chiral kinetic theory with realistic collision term [24] and the results is more complicated than these simplified version.

The spin kinetic equation for massive fermions in background field approximation without collision term can be directly derived from Wigner function approach. However the kinetic equation can take different forms which depends on the different independent distribution function. Here we will only present one form given in [25] in which scalar distribution $\mathcal{F}$ and spin vector distribution $\boldsymbol{\mathcal{A}}$ have been chosen as independent variables.

$$(\nabla_t + \mathbf{v} \cdot \boldsymbol{\nabla})\mathcal{F} = -\frac{\hbar}{2mE_p}[(\mathbf{B} + \mathbf{E} \times \mathbf{v})\nabla_\mathbf{v} - (\mathbf{B} \cdot \mathbf{v})\nabla_\mathbf{v}\mathbf{v}] \cdot \boldsymbol{\mathcal{A}}, \tag{12}$$

$$(\nabla_t + \mathbf{v} \cdot \boldsymbol{\nabla})\boldsymbol{\mathcal{A}} = \mathbf{B} \times \boldsymbol{\mathcal{A}} - \mathbf{E}(\mathbf{v} \cdot \boldsymbol{\mathcal{A}}) - \frac{\hbar}{2mE_p}(\mathbf{B} + \mathbf{E} \times \mathbf{v})\nabla_\mathbf{v}\mathcal{F}, \tag{13}$$

where $\mathbf{v} = \mathbf{p}/E_p$, $\nabla_t = \partial_t + \mathbf{E}\cdot\boldsymbol{\nabla}_p$, $\boldsymbol{\nabla} = \boldsymbol{\nabla}_x + \mathbf{B}\times\boldsymbol{\nabla}_p$ and $\nabla_{\mathbf{v}} = \mathbf{v}\cdot\boldsymbol{\nabla} + E_p\overleftarrow{\boldsymbol{\nabla}}_x\cdot\boldsymbol{\nabla}_p$. For spin kinetic theory with collision term derived from quantum field theory, some attempt has been made and the final result is much more complicated than the spin kinetic equation without collision term. Further work is needed for numerical simulation. However the non-local collision term can be derived explicitly [26]

$$p\cdot\partial\tilde{f}(x,p,s) = \int d\Gamma_1 d\Gamma_2 d\Gamma'\tilde{W}\Big[\tilde{f}(x+\Delta_1,p_1,s_1)\tilde{f}(x+\Delta_2,p_2,s_2)$$
$$-\tilde{f}(x+\Delta,p,s)\tilde{f}(x+\Delta',p',s')\Big] + \int d\Gamma_2 dS_1(p)M\tilde{f}(x+\Delta_1,p_1,s_1)\tilde{f}(x+\Delta_2,p_2,s_2) \quad (14)$$

where Δ, Δ', Δ_1 and Δ_2 are position shifts which exhibit non-local collision term explicitly,

5 Helicity polarization in relativistic heavy ion collisions

In addition to the global or local spin polarization along the global orbital angular momentum or beam momentum, another possible spin polarization along the momentum of final hadrons — helicity polarization is also discussed recently [27, 28]. The helicity polarization can be induced by both helicity charge and vorticity tensor,

$$\mathcal{A}^\mu(x,p) = \frac{p^\mu}{E_p}\sum_\lambda \lambda f_\lambda - \frac{\epsilon^{\mu\nu\rho\sigma}p_\nu\varpi_{\rho\sigma}}{4E_p}\sum_\lambda f_\lambda(1-f_\lambda), \quad (15)$$

where the first term is pure helicity charge contribution and the second term is usual vorticity contribution. It is easy to show that only the spatial components of vorticity tensor are involved for the helicity polarization. Even with the initial zero helicity charge, the helicity charge can be redistributed due to the chiral separate effect from vorticity for the viscous fluid

$$\partial_\mu(n_5 u^\mu) = -\frac{T}{6s}\epsilon_{\mu\lambda\rho\sigma}u^\lambda\partial^\rho u^\sigma\partial_\nu\pi^{\mu\nu}. \quad (16)$$

where n_5 is helicity charge density, $\pi^{\mu\nu}$ is dissipative tensor and s is entropy density.

In relativistic heavy ion collisions, the initial helicity charge could exist due to classical color longitudinal fields or QCD sphaleron transitions in the quark gluon plasma. Such initial helicity charge indicates the local parity violation in QCD. Recently it has been proposed that this initial helicity charge can be measured by the correlation of the hyperons' helicity polarization [27, 28]. The helicity alignment for vector meson from helicity charge can contribute to the spin alignment along x, y and z axis,

$$\bar{\rho}^x_{00} = \frac{1}{3} + \frac{1}{4}\Delta + \frac{3}{4}v_2\Delta,\ \ \bar{\rho}^y_{00} = \frac{1}{3} + \frac{1}{4}\Delta - \frac{3}{4}v_2\Delta,\ \ \bar{\rho}^z_{00} = \frac{1}{3} - \frac{1}{2}\Delta \quad (17)$$

where $\Delta = \rho_{00} - 1/3$ is deviation from $1/3$ for the helicity density matrix element ρ_{00} and v_2 is elliptic flow of the momentum distribution of vector meson.

6 Summary and outlook

Spin plays an important role in nuclear structure at low energy. The observation of global polarization opens a new avenue to understand the quark gluon plasma at high energy. With the upcoming running of NICA, FAIR, J-PARK and HIAF, we can expect that a big picture of spin in nuclear physics would be coming. Local spin polarization for hyperons and spin alighment for vector mesons are both still in debate and quantitative prediction by spin hydrodynamics or spin kinetic theory is needed. Helicity polarization would give us another new probe to detect the spin polarization in heavy ion collisions.

7 Acknowledgments

This work was supported in part by the National Natural Science Foundation of China under Grant Nos. 11890710, 11890713 and 11475104.

References

[1] Z. T. Liang and X. N. Wang, Phys. Rev. Lett. **94**, 102301 (2005) [erratum: Phys. Rev. Lett. **96**, 039901 (2006)]

[2] J. H. Gao, S. W. Chen, W. t. Deng, Z. T. Liang, Q. Wang and X. N. Wang, Phys. Rev. C **77**, 044902 (2008)

[3] B. I. Abelev *et al.* [STAR], Phys. Rev. C **76**, 024915 (2007) [erratum: Phys. Rev. C **95**, no.3, 039906 (2017)]

[4] F. Becattini, V. Chandra, L. Del Zanna and E. Grossi, Annals Phys. **338**, 32-49 (2013)

[5] L. Adamczyk *et al.* [STAR Collaboration], *Nature* **548**, 62 (2017).

[6] F. Becattini and I. Karpenko, Phys. Rev. Lett. **120**, no.1, 012302 (2018)

[7] X. L. Xia, H. Li, Z. B. Tang and Q. Wang, Phys. Rev. C **98**, 024905 (2018)

[8] X. L. Xia, H. Li, X. G. Huang and H. Z. Huang, Phys. Rev. C **100**, no.1, 014913 (2019)

[9] F. Becattini, G. Cao and E. Speranza, Eur. Phys. J. C **79**, no.9, 741 (2019)

[10] S. Y. F. Liu and Y. Yin, JHEP **07**, 188 (2021)

[11] B. Fu, S. Y. F. Liu, L. Pang, H. Song and Y. Yin, [arXiv:2103.10403 [hep-ph]].

[12] F. Becattini, M. Buzzegoli and A. Palermo, Phys. Lett. B **820**, 136519 (2021)

[13] F. Becattini, M. Buzzegoli, A. Palermo, G. Inghirami and I. Karpenko, [arXiv:2103.14621 [nucl-th]].

[14] X. L. Sheng, L. Oliva and Q. Wang, Phys. Rev. D **101**, no.9, 096005 (2020)

[15] X. L. Xia, H. Li, X. G. Huang and H. Zhong Huang, Phys. Lett. B **817**, 136325 (2021)

[16] D. T. Son and P. Surowka, Phys. Rev. Lett. **103**, 191601 (2009)

[17] S. Pu, J. h. Gao and Q. Wang, Phys. Rev. D **83**, 094017 (2011)

[18] M. A. Stephanov and Y. Yin, Phys. Rev. Lett. **109**, 162001 (2012)

[19] J. Y. Chen, D. T. Son and M. A. Stephanov, Phys. Rev. Lett. **115**, no.2, 021601 (2015)

[20] S. Y. F. Liu, Y. Sun and C. M. Ko, Phys. Rev. Lett. **125**, no.6, 062301 (2020)

[21] W. Florkowski, B. Friman, A. Jaiswal and E. Speranza, Phys. Rev. C **97**, no.4, 041901 (2018)

[22] K. Hattori, M. Hongo, X. G. Huang, M. Matsuo and H. Taya, Phys. Lett. B **795**, 100-106 (2019)

[23] S. Bhadury, W. Florkowski, A. Jaiswal, A. Kumar and R. Ryblewski, Phys. Rev. D **103**, no.1, 014030 (2021)

[24] S. Shi, C. Gale and S. Jeon, Phys. Rev. C **103**, no.4, 044906 (2021)

[25] J. H. Gao and Z. T. Liang, Phys. Rev. D **100**, no.5, 056021 (2019)

[26] N. Weickgenannt, E. Speranza, X. l. Sheng, Q. Wang and D. H. Rischke, Phys. Rev. Lett. **127**, no.5, 052301 (2021)

[27] F. Becattini, M. Buzzegoli, A. Palermo and G. Prokhorov, [arXiv:2009.13449 [hep-ph]].

[28] J. H. Gao, [arXiv:2105.08293 [hep-ph]].

Energy Dependence of Strangeness Production in Heavy-ion Collision

Md Nasim[1,*]

[1]Department of Physics,
Indian Institute of Science Education and Research (IISER), Berhampur,
Postal Code-760010, Odisha, India.

Abstract. An experimental overview of the energy dependence of strangeness production is presented. The strange hadrons are considered a good probe to study the QCD matter created in relativistic nucleus-nucleus collisions. The heavy-ion experiments at SPS, RHIC, and LHC have recorded a wealth of data in proton-proton, proton-nucleus and nucleus-nucleus collisions at several beam energies. In this proceeding, I discuss the invariant yield and azimuthal anisotropy measurement of strange hadrons in nucleus-nucleus collisions at SPS, RHIC, and LHC.

1 Introduction

The relativistic heavy-ion collision provides a unique opportunity to study the properties of QCD matter at various temperatures and densities. In order to map out the phase diagram of the QCD matter, experimental programs started in the early 1990s at the Brookhaven Alternating Gradient Synchrotron (AGS) and the CERN Super Proton Synchrotron (SPS) followed by Relativistic Heavy Ion Collider (RHIC) at Brookhaven and recently at Large Hadron Collider (LHC) at CERN. These experiments allowed us to study the QCD matter over a large range of baryonic chemical potential (or net-baryon density) by varying center-of-mass energy($\sqrt{s_{NN}}$) of the two colliding nucleus.
Strange hadron production in nucleus-nucleus collisions is believed to be an important tool to study the dynamics of the produced QCD matter. Due to the large production cross-section of strange quarks in the QGP, an enhancement in the production of strange hadrons in QGP compared to non-QGP medium is expected [1]. Due to low energy density, the matter produced in the nucleon-nucleon (p+p) collisions is considered as a baseline for the non-QGP medium. Production of strange hadrons is being measured in nucleus-nucleus (A+A) collisions and compared with that in p+p collisions to look for any deviation from the non-QGP baseline. Any deviation from the non-QGP baseline would mean the formation of deconfined matter of quarks and gluons in A+A collisions.
Among all strange hadrons, multi-strange hadrons (e.g. ϕ, Ω) are expected to have relatively small hadronic interaction cross-sections and therefore they can carry the information directly from the chemical freeze-out stage with almost no distortion due to late-stage hadronic interaction [2, 3]. The measurement of production yield and azimuthal anisotropy

*e-mail: nasim@iiserbpr.ac.in

of multi-strange hadrons are considered as a useful probe to study the hadronization and collective properties of the QGP medium.

2 Strange Hadron Production

2.1 Transverse momentum (p_T) integrated particle ratio

Test of thermal model:
Figure 1 (the left panel) shows K/π ratios as a function of $\sqrt{s_{NN}}$ in central A+A collision measured at AGS, SPS, and RHIC [4–9]. The K/π ratio is of interest, as it reflects the strangeness content relative to entropy in heavy-ion collisions. The K^+/π^+ and K^-/π^- ratios are shown by red and black symbols, respectively. The observed peak position in the energy dependence of K^+/π^+ has been suggested to be a signature of a change in degrees of freedom [10, 11]. The thermal model calculations are shown as yellow band for K^+/π^+ and green band for K^-/π^- [12–14]. It is shown that the thermal model calculation explains the measured K/π ratios fairly well.
Cannonical Ensemble at high-baryon density:
The ϕ/K^- ratios as a function of $\sqrt{s_{NN}}$ is presented in the right panel of Fig. 1 [15, 16]. The thermal model calculations, based on the Grand Canonical Ensemble (GCE) and Canonical Ensemble (CE) for strangeness with several different choices of strangeness correlation length are shown by different line colors [13, 14]. We can see for $\sqrt{s_{NN}} > 5$ GeV, both GCE and CE calculations explain the data, however for $\sqrt{s_{NN}} \leq 3.0$ GeV GCE calculation under-predicts the measured ϕ/K^- ratios. Calculation using Canonical Ensemble (CE) with strangeness correlation length $\sim$ 3 fm can explain the data measured by STAR and HADES experiment.
Evidence of hadronic rescattering in central A+A collision:
Figure 2 shows K^{*0}/K and ϕ/K ratios as a function of multiplicity in p+Pb and Pb+Pb collisions at 2.76 and 5.02 TeV measured by the ALICE experiment [17]. We see that K^{*0}/K ratios decrease with increasing multiplicity. Whereas the ϕ/K ratio is almost independent of multiplicity. This observation can be understood considering the re-scattering of daughter particles in the hadronic phase. The lifetime of K^{*0} resonance is short ($\sim$ 4 fm/c) and therefore they decay inside the fireball. The daughter particles of K^{*0} undergo re-scattering in the hadronic medium and their momentum may get changed. This would lead to a loss in the reconstruction of the parent resonance. The lifetime of the ϕ meson is $\sim$ 42 fm/c. Because of the longer lifetime, the ϕ meson will mostly decay outside the fireball and therefore its daughters will not have much time to rescatter in the hadronic phase. We can also see from Fig. 2 that thermal model calculations, which do not include hadronic rescattering, explain ϕ/K ratios and over-predict K^{*0}/K at the central Pb+Pb collisions. This is a clear piece of evidence of the hadronic rescattering effect on short-lived resonance in central Pb+Pb collisions.

2.2 Particle ratio as a function of p_T

Probing hadronization through baryon-to-meson ratio:
The baryon-to-meson ratios have been predicted to be sensitive to the parton dynamics of the collision system. Figure 3 shows the $\bar{\Lambda}/K^0_S$ ratio as a function of p_T for different centralities in Au+Au collisions at $\sqrt{s_{NN}}$ = 7.7–54.4GeV [18]. The enhancement of $\bar{\Lambda}/K^0_S$ ratios at intermediate p_T in central Au+Au collisions compared to peripheral Au+Au at $\sqrt{s_{NN}} \geq$19.6 GeV is interpreted as a consequence of hadron formation through parton recombination.We do not see any significant difference between central and peripheral collisions for $\sqrt{s_{NN}} \leq$

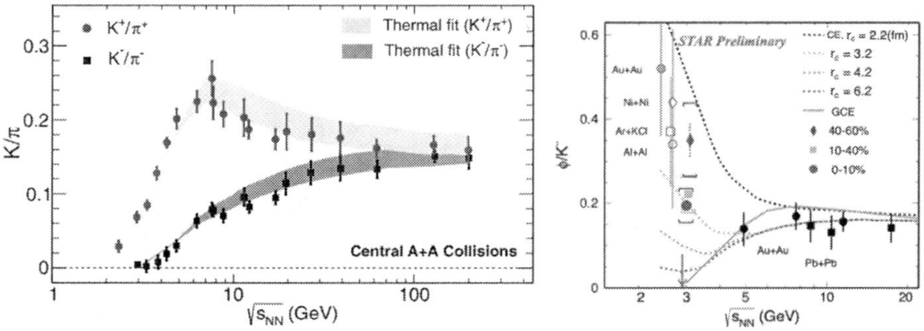

Figure 1. Left panel: The K^+/π^+ and K^-/π^- ratios as a function of $\sqrt{s_{NN}}$ in central A+A collision [4–9]. Right panel: The ϕ/K^- ratios as a function of $\sqrt{s_{NN}}$ [15, 16].

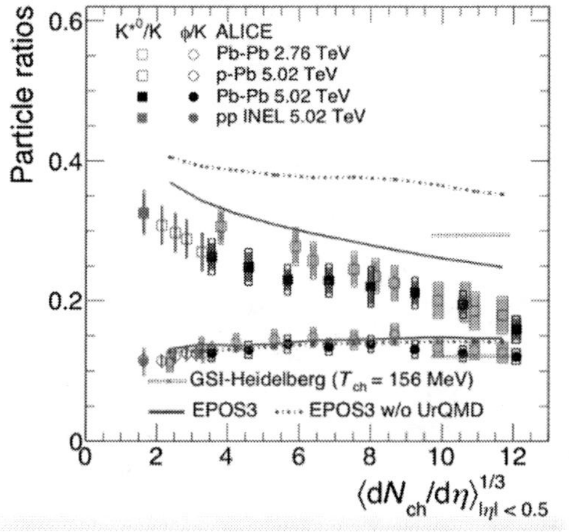

Figure 2. The K^{*0}/K and ϕ/K ratios as a function of multiplicity in p+Pb and Pb+Pb collisions at 2.76 and 5.02 TeV measured by the ALICE experiment [17].

11.5 GeV. This may indicates formation of hadronic interaction dominated matter at $\sqrt{s_{NN}} \leq$ 11.5 GeV.

2.3 Nuclear modification factor

Probing QGP through parton energy loss:
The nuclear modification factor (R_{CP}) is defined as the ratio of the particles yield in the central to peripheral collisions normalized by the number of binary collisions (Nbin). The value of Nbin is calculated from the Monte Carlo Glauber simulation. If R_{CP} is equal to one, then the nucleus-nucleus collision is simply the superposition of nucleon-nucleon collisions. Deviation of R_{CP} from the unity would imply contribution from the nuclear medium effects. The measured R_{CP}, which is less than one at high p_T, at top RHIC and LHC energies indicates the energy loss of partons traversing through QGP medium. Figure 4 show R_{CP} of strange hadrons measured in Au+Au collisions at $\sqrt{s_{NN}}$ = 7.7–54.4GeV [18]. From the Fig. 4 one can see, at the intermediate p_T , R_{CP} goes above unity with the decrease in beam energy. This indicates that at lower beam energy the parton energy loss effect could be less important.

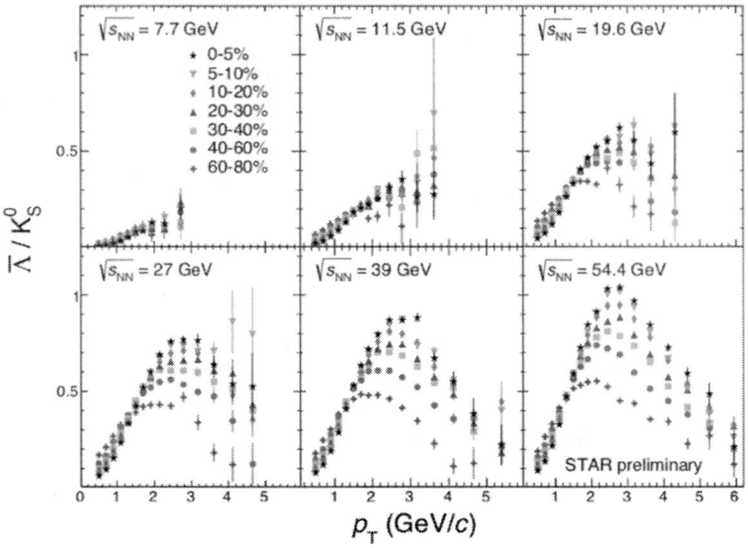

Figure 3. The $\bar{\Lambda}/K^0_S$ ratio as a function of p_T for different centralities in Au+Au collisions at $\sqrt{s_{NN}}$ = 7.7–54.4GeV [18].

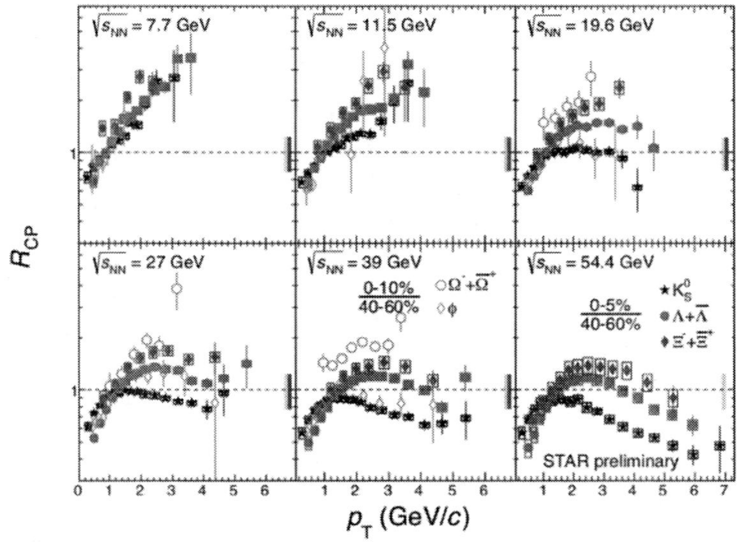

Figure 4. The nuclear modification factor of strange hadrons measured in Au+Au collisions at $\sqrt{s_{NN}}$ = 7.7–54.4GeV [18].

3 Collective Flow of Strange Hadrons

Partonic collectivity at top RHIC and LHC energies:
Figure 5 shows the v_2 as a function of p_T for π, p, ϕ, and Ω for 0–80% centrality in Au+Au collisions at $\sqrt{s_{NN}}$ = 200 GeV and for ϕ, and Λ in 30–40% central Pb+Pb collisions at $\sqrt{s_{NN}}$ = 5.02 TeV [19, 20]. We see mass ordering in v_2 at low p_T and particle type dependence at the intermediate p_T. The observed mass ordering can be explained by the hydrodynamics

model whereas particle type dependence at the intermediate p_T can be explained through the parton recombination model. It is clear from Fig. 5 that the v_2 of hadrons consisting of only strange quarks (ϕ and Ω) is similar to that of non-strange hadrons π and p. As we know that the ϕ and Ω do not participate strongly in the hadronic interactions unlike π and p, therefore Fig. 5 suggests that the major part of collectivity is developed during the partonic phase at top RHIC energy. This is also true for LHC energies as shown by the ALICE experiment [20].

Turn-off of QGP and change of equation of states ?

The energy dependence of ϕ-meson v_2 measured at intermediate p_T is shown in Fig. 6 [21]. We can see almost constant magnitude (~0.1) of ϕ v_2 for $\sqrt{s_{NN}} \geq 19.6$ GeV. However, for $\sqrt{s_{NN}}$ =11.5 and 7.7 GeV, magnitude of ϕ v_2 is small or close zero within large uncertainties. Small ϕ v_2 is considered as a signature of formation of hadronic interaction dominated matter as ϕ mesons are expected to have small hadronic interaction cross sections. The energy dependence of the slope of ϕ-meson directed flow (v_1) is shown in the right panel of Fig. 6 [22].We see there is indication of ϕ-meson v_1 slope changes sign below $\sqrt{s_{NN}}$ = 10 GeV. This could be related to the change of equation of states at the low beam energy. The uncertainties in the current measurement are large, upcoming BES-II program at RHIC is expected to provide precision measurement of ϕ-meson v_1 and v_2.

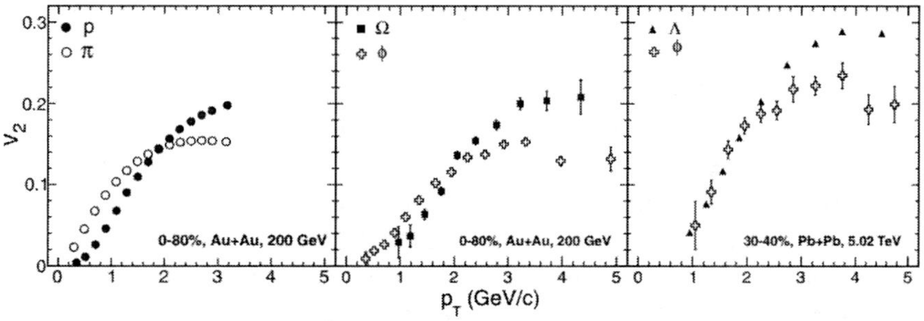

Figure 5. The elliptic flow as a function of p_T for π, p, ϕ, and Ω for 0%–80% centrality in Au+Au collisions at $\sqrt{s_{NN}}$ = 200 GeV and for ϕ, and Λ in 30%–40% central Pb+Pb collisions $\sqrt{s_{NN}}$ = 5.02 TeV [19, 20].

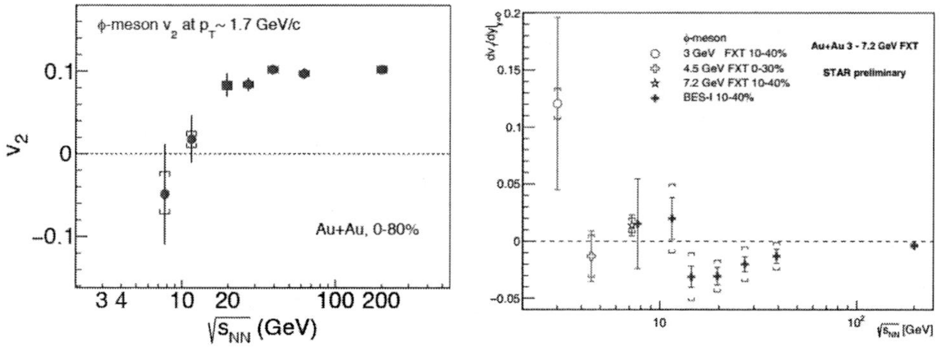

Figure 6. The energy dependence of ϕ-meson v_2 at intermediate p_T and slope of directed flow (dv_1/dy) measured by the STAR experiment [21, 22].

4 Summary

The energy dependence of strange hadron production in heavy-ion collisions is presented. It is found that the thermal model explains the measured particle ratios (e.g. K/π and ϕ/K) and its energy dependence. It has been found that the ϕ/K ratio is a sensitive probe to study the strangeness canonical suppression. We observed thermal model calculations over-predicts K^{*0}/K at the central Pb+Pb collisions. Also, the K^{*0}/K ratio decreases with increasing event multiplicity. This clearly indicates the effect of hadronic re-scattering on short-lived resonance in central A+A collisions. The measured elliptic flow of strange hadrons found to be similar to that of non-strange hadrons at top RHIC and LHC energies, indicating formation of partonic collectivity.
The measure ϕ v_2 at intermediate p_T is found to be close zero within large uncertainties at $\sqrt{s_{NN}}$ =11.5 and 7.7 GeV. The ϕ-meson v_1 slope changes sign below $\sqrt{s_{NN}}$ = 10 GeV. The nuclear modification factor is greater that unity at the intermediate p_T for $\sqrt{s_{NN}} \leq 19.6$ GeV. The baryon-to-meson ($\bar{\Lambda}/K^0_S$) ratios as a function of p_T is found to centrality independent for $\sqrt{s_{NN}}$ =11.5 and 7.7 GeV. These results indicates formation of hadronic interaction dominated matter below $\sqrt{s_{NN}}$ = 20 GeV.

References

[1] J. Rafelski and B. Muller, Phys. Rev. Lett. 48, 1066 (1982).
[2] J. Adam et al. (STAR), Nucl. Phys. A 757, 102 (2005).
[3] A. Shor, Phys. Rev. Lett. 54, 1122 (1985).
[4] L. Ahle et al. (E866, E917), Phys. Lett. B 476, 1 (2000).
[5] L. Ahle et al. (E-802, E-866), Phys. Rev. C 60, 044904 (1999).
[6] L. Ahle et al. (E866, E917), Phys. Lett. B 490, 53 (2000).
[7] S. Afanasiev et al. (NA49), Phys. Rev. C 66, 054902 (2002).
[8] C. Alt et al. (NA49), Phys. Rev. C 77, 024903 (2008).
[9] L. Adamczyk et al. (STAR), Phys. Rev. C 96, 044904 (2017) and sQM2021 proceedings.
[10] M. Gazdzicki and M. I. Gorenstein, Acta Phys. Polon. B 30, 2705 (1999).
[11] J. Cleymans, et al. Phys. Lett. B 615, 50 (2005).
[12] J. Randrup and J. Cleymans, Phys. Rev. C 74, 047901 (2006).
[13] A. Andronic, et al. Nucl. Phys. A 772, 167 (2006)
[14] J. Cleymans et al. Phys.Lett. B 603, 146 (2004).
[15] G. Agakishiev et al. (HADES) Phys. Rev. C 80, 025209 (2009)
[16] M. S. Abdallah et al. (STAR), arXiv:2108.00924 and sQM2021 proceedings.
[17] S. Acharya, et al. (ALICE) PLB 802, 135225 (2020).
[18] J. Adam et al. (STAR), Phys. Rev. C 102, 34909 (2020) and sQM2021 proceedings.
[19] L. Adamczyk et al. (STAR), Phys. Rev. Lett. 116, 062301 (2016).
[20] S. Acharya, et al. (ALICE) JHEP 09, 006 (2018).
[21] L. Adamczyk et al. (STAR), Phys. Rev. Lett. 110, 142301 (2013)
[22] L. Adamczyk et al. (STAR), Phys. Rev. Lett. 120, 62301 (2018) and sQM2021 proceedings.

Strange Hadron Production in Au+Au Collisions at $\sqrt{s_{NN}}$ = 54.4 GeV

Yan Huang[1,*] for the STAR collaboration

[1]Engineer Physics Department, Tsinghua University

Abstract. The RHIC Beam Energy Scan (BES) program was proposed to study the QCD phase transition. As a sensitive probe to the transition from the hadron gas to the Quark-Gluon Plasma, the particle ratios of strange hadrons are measured to study the properties of nuclear matter. Recently, the production of strange hadrons ($K^0_s, \Lambda, \Xi, \Omega$, and ϕ) at mid-rapidity (|y|<0.5) in Au+Au collisions at $\sqrt{s_{NN}}$ = 54.4 GeV were measured by the STAR experiment at RHIC. In this paper nuclear modification factor, baryon-to-meson ratio, as well as the overall integrated yields of these strange hadrons at $\sqrt{s_{NN}}$ = 54.4 GeV are presented. In particular, the multi-strange baryon-to-meson ratio $N(\Omega^- + \bar{\Omega}^+)/[2N(\phi)]$ is studied and compared to previous measurements from other BES energies. The strange-hadron-to-pion ratios versus charged hadron multiplicity are also studied and compared to the measurements in heavy-ion collisions at higher energies.

1 Introduction

The main goal of the RHIC BES Program is to map the quantum chromodynamics (QCD) phase diagram, including locating the onset of deconfinement, searching the critical point and the phase boundary of the first order phase transition [1]. The multi-strange hadrons are sensitive probes to study the initial stage of the collision as they freeze out early and are expected to have a small hadronic interaction cross section [2, 3]. The central-to-peripheral nuclear modification factor R_{cp} is less than unity at high p_T for various particles at top RHIC energy [4, 5], indicating that there is a significant energy loss of the scattered partons in QGP. Different trends in baryon/meson ratio at intermediate $p_T \approx 2.5$ GeV/c can give information about the parton recombination [6, 7]. The enhancement of Ω/ϕ ratio at intermediate p_T indicates hadron formation through parton recombination [8], therefore the study of Ω and ϕ particles offers the knowledge of the transition from partonic to hadronic matter.

The Solenoidal Tracker At RHIC (STAR) is a multi-purpose detector at the RHIC collider at Brookhaven National Laboratory. It provides measurements with full azimuthal coverage, uniform acceptance and excellent particle identification capability [9, 10]. This work presents strangeness production in Au+Au collisions at $\sqrt{s_{NN}}$ = 54.4 GeV, collected by the STAR experiment during the first phase of RHIC BES program in 2017.

*e-mail: h-y12@tsinghua.org.cn

2 Results and discussion

The strange hadrons $K_s^0, \Lambda(\bar{\Lambda}), \Xi^-(\bar{\Xi}^+), \Omega^-(\bar{\Omega}^+)$, and ϕ are reconstructed using the invariant mass technique with the corresponding decay channels:

$K_s^0 \rightarrow \pi^+ + \pi^-$ $(B.R.\ 69.20\%)$

$\Lambda(\bar{\Lambda}) \rightarrow p(\bar{p}) + \pi^-(\pi^+)$ $(B.R.\ 63.90\%)$

$\Xi^-(\bar{\Xi}^+) \rightarrow \Lambda(\bar{\Lambda}) + \pi^-(\pi^+)$ $(B.R.\ 99.887\%)$

$\Omega^-(\bar{\Omega}^+) \rightarrow \Lambda(\bar{\Lambda}) + K^-(K^+)$ $(B.R.\ 67.8\%)$

$\phi \rightarrow K^+ + K^-$ $(B.R.\ 49.1\%)$

Weak decay feed-down correction was carried on for Λ hyperons. The daughter particles of these strange hadrons, π, K, and p are identified by dE/dx in TPC to reconstruct the secondary vertex. More details of the strange particle reconstruction can be found in Ref. [7]. In the following, the particle yields, R_{cp}, Ω/ϕ ratio, and particles/pions ratio of the strange hadrons are presented and discussed.

2.1 Particle yields

The integrated yield dN/dy per half of average number of participating nucleon ($<N_{part}>/2$) as a function of $<N_{part}>$ at $\sqrt{s_{NN}} = 54.4$ GeV are plotted in Fig. 1. We can see the normalized yields of K_s^0, Ξ, Ω, and ϕ increase with $<N_{part}>$ and energy. The feed down corrected $\bar{\Lambda}$ normalized yield has weak centrality dependence similar to other BES energies, which can be understood by the possible annihilation processes on anti-baryon production in more central collisions.

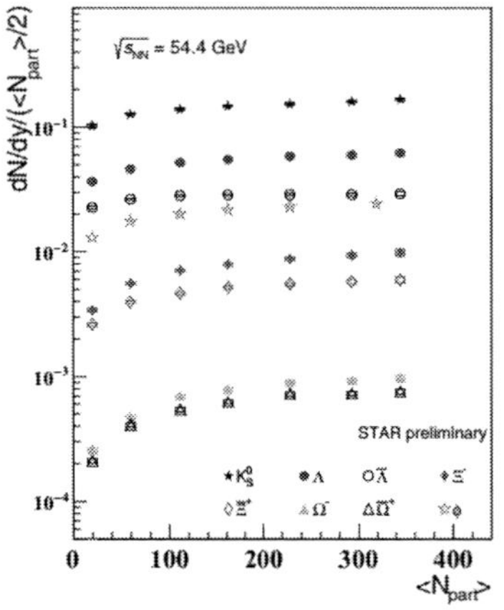

Figure 1: The integrated yield dN/dy per half of average number of participating nucleon ($<N_{part}>/2$) as a function of $<N_{part}>$. The box on each data point denotes the systematic error.

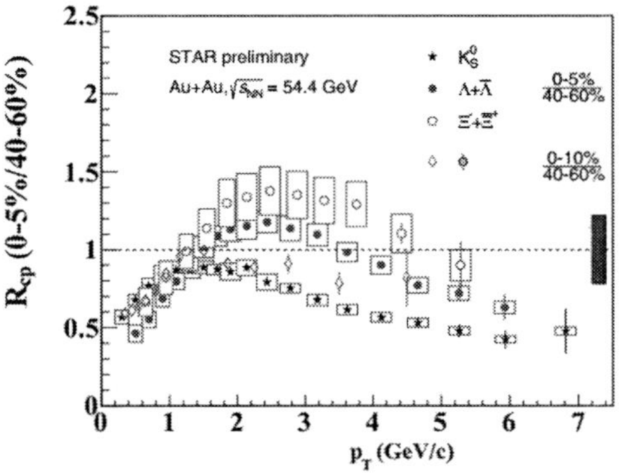

Figure 2: R_{cp}(0-5%)/(40-60%) of K_s^0, Λ, Ξ, and R_{cp}(0-10%)/(40-60%) of ϕ at mid-rapidity $|y|<0.5$ at $\sqrt{s_{NN}} = 54.4$ GeV. The box on each data point of K_s^0, Λ, and Ξ denotes the systematic error. There are only statistical error for ϕ.

2.2 Nuclear modification factor R_{cp}

The nuclear modification factor R_{cp} is defined as the ratio of $(dN/dp_T)/<N_{coll}>$ in central collisions to that in peripheral ones. Here N_{coll} is determined from Glauber Monte Carlo simulations [7]. R_{cp} of strange hadrons at $\sqrt{s_{NN}}$ = 54.4 GeV are presented in Fig .2. There is a strong suppression at high p_T for all particles indicating energy loss effect of scattered partons in QGP. A separation of baryon/meson at intermediate p_T suggests the hadron formation through parton recombination. Considering the R_{cp} results at other BES energies shown in Fig. 28 from Ref. [7], one can see that there is no suppression at highest measured p_T at $\sqrt{s_{NN}} \leq$ 11.5 GeV, more statistics are needed for high p_T and results from BES phase -II program are expected.

2.3 Ω/ϕ ratio

Figure 3 (a) shows the centrality dependence of $N(\Omega^- + \bar{\Omega}^+)/[2N(\phi)]$ ratio as a function of transverse momentum at mid-rapidity (|y|<0.5) at $\sqrt{s_{NN}}$ = 54.4 GeV, the ratios keep increasing from peripheral to central collisions at intermediate p_T, this can be interpreted as a consequence of hadron formation through parton recombination and parton collectivity in central collisions. Figure 3 (b) presents the Ω/ϕ ratio for different energies in central collision (0-10%) [8]. An enhancement at intermediate p_T at $\sqrt{s_{NN}}$ = 54.4 GeV is apparently observed, indicating hadron formation through parton recombination. More statistics are needed for $\sqrt{s_{NN}}$ = 11.5 GeV to make a conclusion and this is one of the aims of BES-II program.

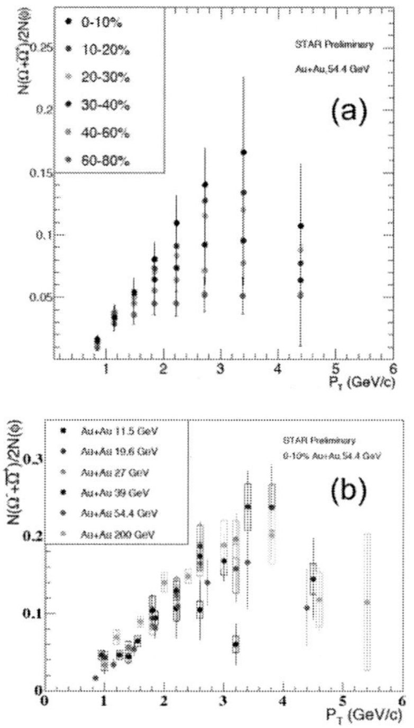

Figure 3: Centrality dependence of Ω/ϕ ratio at $\sqrt{s_{NN}}$ = 54.4 GeV (a), and energy dependence of that for central collisions 0-10% (b). Statistical and systematic errors are shown.

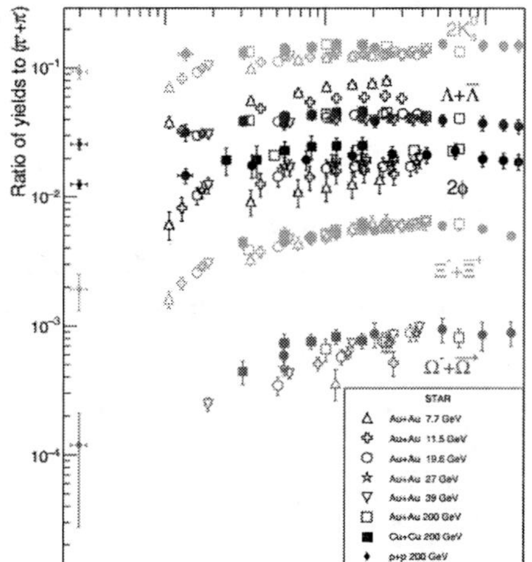

Figure 4: The ratio of yields of strange particles to that of pions [12, 13]. The errors are statistical and systematic.

2.4 Strange particles/pions ratio

The measurements by the ALICE experiment show particle ratios in $p+p$ and $p+Pb$ are identical at the same $dN_{ch}/d\eta$ [11], indicating that the final-state particle density might indeed be a good scaling variable between different systems. Figure 4 presents the ratio of yields of strange particles to that of pions [12, 13]. Here the charged-particle multiplicity density $dN_{ch}/d\eta$ is the sum of $k^{\pm}, \pi^{\pm}, p$, and $\bar{p}$ $dN/d\eta$. The ratios with $dN_{ch}/d\eta$ are system and energy independent, except for Λ and ϕ at lower energies. More statistics are needed for a firm conclusion for Ω.

3 Conclusion

The yield measurement of strange hadrons ($K_s^0, \Lambda, \Xi, \Omega$, and ϕ) at mid-rapidity (|y|<0.5) in Au+Au collisions at $\sqrt{s_{NN}}$ = 54.4 GeV has been presented. The normalized yields of K_s^0, Ξ, Ω, and ϕ increase with average number of participating nucleons and energy. The corrected $\bar{\Lambda}$ normalized yield has weak centrality dependence, indicating the possible annihilation processes on anti-baryon production. The strong suppression at high p_T of R_{cp} can be interpreted by the energy loss of parton in QGP. The separation of baryon and meson R_{cp} and the enhancement in Ω/ϕ ratio at intermediate p_T give information about parton recombination. This study on strange particles/pions ratio shows that $dN_{ch}/d\eta$ is a good scaling independent of system and energy for K_s^0 and Ξ.

4 Acknowledgement

This work is supported by the Ministry of Science and Technology under GrantNo. 2020YFE0202001, 2018YFE0205200, 2016YFA0400100. The work is also supported by National Natural Science Foundation of China under Grant No. 11927901, 11420101004, 11735009, 11890710, 11890712, U1832118.

References

[1] STAR white paper, https://drupal.star.bnl.gov/STAR/starnotes/public/sn0598
[2] J. Rafelski and B. Muller, Phys.Rev.Lett. **56**, 2334 (1986).
[3] STAR, NPA **757**, 102(2005).
[4] STAR, Phys. Rev. Lett. **98**, 062301 (2004).
[5] STAR, Phys. Rev. Lett. **92**, 052302 (2004).
[6] STAR, Phys. Rev. Lett. **108**, 072301 (2012).
[7] STAR, Phys. Rev. C **102**, 034909 (2020).
[8] STAR, Phys. Rev. C **93**, 021903(R) (2016).
[9] STAR, Nucl. Instrum.Meth. A **499**, 624 (2003).
[10] M. Anderson et al., Nucl. Instrum. Meth. A **499**, 659 (2003).
[11] ALICE, Nature Physics volume **13**, 535-539 (2017).
[12] STAR, Phys. Rev. C **96**, 044904 (2017).
[13] ALICE, Phys. Rev. C **88**, 044910 (2013).

Light and strange hadron production and anisotropic flow measurement in Au+Au collisions at $\sqrt{s_{NN}} = 3\,\mathrm{GeV}$ from STAR

Guannan Xie (for the STAR Collaboration)[1,*]

[1]Lawrence Berkeley National Laboratory

Abstract. In this proceeding, we present on our first measurements of identified particle (π, K, p, K_s^0, Λ, ϕ, Ξ^-) production and anisotropic flow (v_1, v_2) in Au+Au collisions at $\sqrt{s_{NN}} = 3\,\mathrm{GeV}$. Various models including thermal and transport model calculations are compared to data, these results imply that the matter produced in the 3 GeV Au+Au collisions is considerably different from that at higher energies.

1 Introduction

Relativistic heavy ion physics is aiming at the detailed investigation of phase structures of strongly interacting matter under extreme conditions. Searching for the onset of Quark-Gluon Plasma (QGP), studying the properties of the produced QCD matter, and locating the possible QCD phase boundary are the focus of the RHIC Beam Energy Scan (BES) Program [1].

Particle production and collective flow have been used to investigate the properties of the QCD matter produced in heavy-ion collisions. The RHIC BES program covers a wide range of energies, including the transition from a partonic dominated area to hadronic dominated area. Of particular interest is the high baryon density region which is accessible through production and flow measurements of particles including light and strange hadrons in the STAR fixed-target program. In these proceedings, invariant yields and rapidity density distributions of π, K, ϕ mesons and Ξ^- hyperons as well as the directed/elliptic flow of π, K, p, K_s^0, Λ, ϕ are presented.

2 Experiment

The dataset used in this analysis consists of Au+Au collisions at $\sqrt{s_{NN}} = 3\,\mathrm{GeV}$ collected by the STAR experiment under the fixed target (FXT) configuration in the year of 2018. The single beam was provided by RHIC with total energy equal to 3.85 GeV/nucleon. The thickness of the gold target is about 0.25 mm, corresponding to a 1% interaction probability to minimize the pileup and energy loss effect in target. The target is located at 200 cm to the west of the center of the STAR detector, it is installed inside the vacuum pipe, 2 cm below the center of the beam axis. The main detectors used for this analysis are the Time Projection Chamber (TPC), the Time of Flight (TOF) detector, and the Beam-Beam Counter (BBC) [2]. The trigger is provided by the signal in the east BBC detector and at least five hits in the TOF

*e-mail: GuannanXie@lbl.gov

detector. Tracking and particle identification (PID) are done using the energy loss (dE/dx) information from TPC and time of flight ($1/\beta$) information from TOF. In total, approximately 2.6×10^8 minimum bias triggered events are used in this analysis. Reconstruction of short lived particles, $K_S^0 \rightarrow \pi^+ + \pi^-$, $\Lambda \rightarrow p + \pi^-$ and $\Xi^- \rightarrow \Lambda + \pi^-$ is performed using the KF Particle Finder package based on the Kalman Filter method [2]. Those combinatorial backgrounds are obtained by rotating daughter tracks. The ϕ mesons are reconstructed through the hadronic decay channel, $\phi \rightarrow K^+ + K^-$, where combinatorial background is estimated using the mixed event technique. After corrected for the acceptance and efficiency, the 4π acceptance yield (dN) and anisotropic flow (v_1, v_2) are obtained.

3 Light and Strange Hadron Yields

Figure 1 left shows the K^-/K^+ ratio along with collision energy $\sqrt{s_{\mathrm{NN}}}$. The relative yield of K^- is much smaller compared to K^+ at 3 GeV which demonstrates the importance of K^+ production in association with the Λ ($N+N \rightarrow N+\Lambda+K$). Right plot shows the K^+/π^+ ratio along with $\sqrt{s_{\mathrm{NN}}}$. The horn structure around 7 GeV was proposed as a possible signal of onset of deconfinement in early days, but later on the data are described well by the statistics models. The measured K^-/K^+ and K^+/π^+ ratios at 3 GeV follow with the world trend [3].

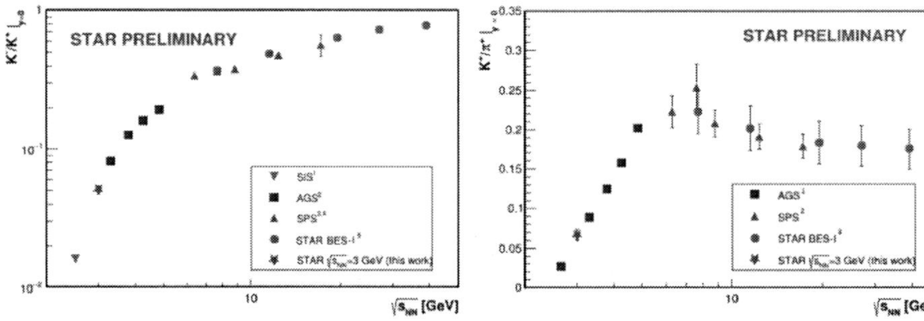

Figure 1. K^-/K^+ (left) and K^+/π^+ (right) ratio as a function of $\sqrt{s_{\mathrm{NN}}}$. The red star shows the measurements at 3 GeV with statistical and systematic uncertainties added together, while other markers are used for data from other energies [3].

Statistical thermal models have often been used to characterize the thermal properties of the produced media. In low energies, the strangeness production is rare, therefore it has been argued that strangeness number needs to be conserved locally on an event-by-event basis described by the Canonical Ensemble (CE), which leads to a reduction in the yields of hadrons with non-zero strangeness number. Figure 2 shows the ϕ/K^- (left) and ϕ/Ξ^- (right) ratios from our measurements in different centralities as a function of $\sqrt{s_{\mathrm{NN}}}$. The measured ϕ/K^- and ϕ/Ξ^- ratios at 3 GeV are slightly higher than the values at high energies for $\sqrt{s_{\mathrm{NN}}} \geqslant 7$ GeV. There is a hint for both the measured ϕ/K^- and ϕ/Ξ^- ratios in mid-central collisions are larger than that in central collisions, and this need further statistics to systematically study the centrality dependence in detail. The Grand Canonical Ensemble (GCE) underestimates our data with ~5σ effect for ϕ/K^- and ~4σ effect for ϕ/Ξ^-, while the CE calculation with strangeness correlation length (r_c) ~3.2 fm can reasonably describe our measurements. The precise determination of the thermal parameters (including T_{ch}, μ_B and r_c) needs a global thermal model fit with all the particle yields at 3 GeV. The modified transport models (UrQMD and SMASH) calculations by including high mass resonances decay to ϕ and Ξ^- can also reasonably describe the data at low energies [4]. In the Au+Au

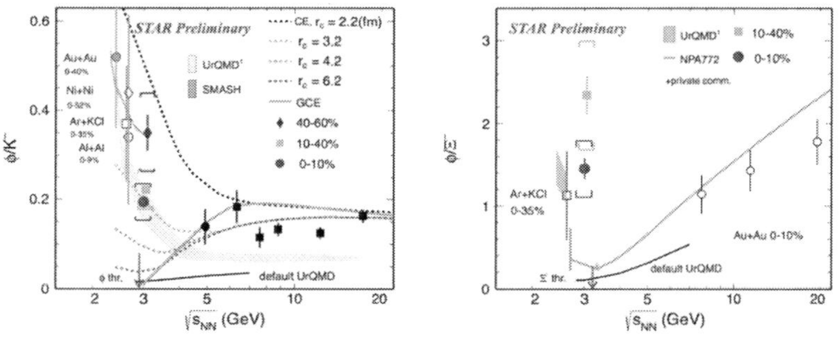

Figure 2. ϕ/K^- (left) and ϕ/Ξ^- (right) ratio as a function of $\sqrt{s_{NN}}$. The solid color markers show the measurements at 3 GeV from different centrality bins with vertical lines for statistical and bracket symbols for systematic errors, while other markers are used for data from various other energies and/or collision systems [4].

collisions at 3 GeV, the observed strangeness production mechanism may be different from that at high energy, and this may indicate a change of EoS at this low energy.

4 Light and Strange Hadron Anisotropic Flow

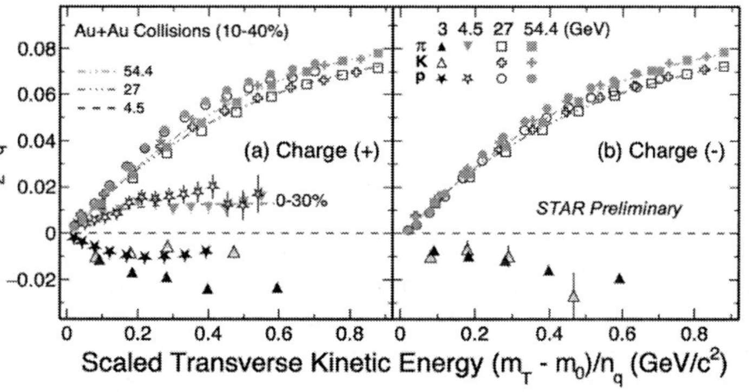

Figure 3. NCQ scaled v_2 as a function of $(m_T - m_0)/n_q$ for pions, kaons and protons from Au+Au collisions in 10-40% centrality for positive (left) and negative (right) charged particles.

Directed flow v_1 and elliptic flow v_2 are sensitive to the stiffness of the nuclear Equation of State (EoS) at high baryon density region [5]. The number of constituent quark (NCQ) scaling of v_2 is observed for particle or anti-particle for collision energy $\geq$ 7.7 GeV and it is often argued as an evidence for the formation of a QGP phase with partonic degrees of freedom. For 3 GeV collisions, the elliptic flow of all charged hadrons measured at midrapidity are negative and the NCQ scaling is absent as shown in Fig. 3, especially for positive charged particles. The slope of the midrapidity directed flow of all measured hadrons are found to be positive as shown in Fig. 4 [5]. This is the first time observed of positive v_1 slop for kaons and ϕ mesons. In addition, the UrQMD model calculations with the baryonic mean-field potential qualitatively reproduce our observations. Due to the formation of the QGP, at $\sqrt{s_{\mathrm{NN}}}$ larger than 10 GeV, we observed the opposite behaviors, namely all hadrons' v_2 are positive while all slopes of v_1 are negative. In the Au+Au collisions at 3 GeV, the observed opposite behavior implies the vanishing of partonic collectivity and a new EoS dominated by the baryonic interactions in these region.

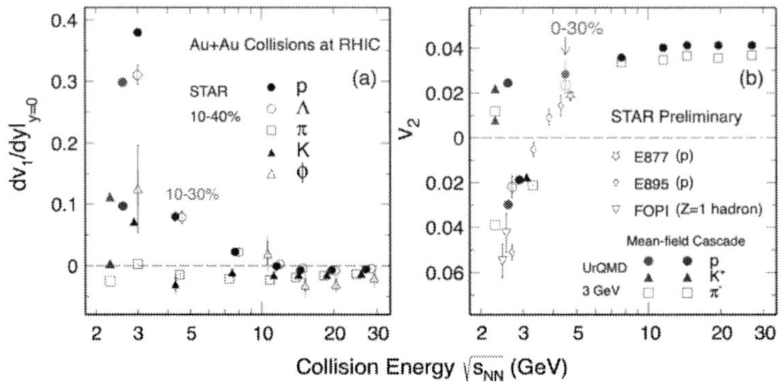

Figure 4. v_1 slope $dv_1/dy|_{y=0}$ (left) and v_2 (right) along with $\sqrt{s_{\mathrm{NN}}}$ in heavy-ion collisions [5].

5 Summary

In Au+Au collisions at 3 GeV, our result suggests that the strangeness production mechanism may be different compared to that in high energies. The collective flow behaviors show an opposite trend to that in high energies. These results imply that the matter produced in the 3 GeV Au+Au collisions is considerably different from those at higher energies.

References

[1] Y. Akiba *et al.*, (2005) arXiv:1502.02730 [nucl-ex]; W. Busza, *et al.*, Ann. Rev. Nucl. Part. Sci. **68**, 339 (2018).

[2] K. H. Ackermann *et al.*, Nucl. Instrum. Meth. **A 499**, 624 (2003); M. Anderson, Nucl. Instrum. Meth. **A 499**, 659 (2003); W. J. Llope, Nucl. Instrum. Meth. **A 661**, S110 (2012); I. Kisel (CBM), J. Phys. Conf. Ser.**1070**, 012015 (2018).534

[3] A. Foerster *et al.*, (KaoS), J. Phys. **G 28**, 2011 (2002); L. Ahle *et al.*, (E866/E917), Phys. Lett. **B 490**, 53 (2000); Phys. Lett. **B 476**, 1 (2000); C. Alt *et al.* (NA49), Phys. Rev. **C 77**, 024903 (2008); Phys. Rev. **C 66**, 054902 (2002); L. Adamczyk *et al.*, (STAR), Phys. Rev. **C 96**, 044904 (2017); (Theory) Acta Phys. Pol. **B 30**, 2705 (1999); Phys. Lett. **B, 673**, 142 (2009).

[4] B. B. Back *et al.*, (E917), Phys. Rev. **C 69**, 054901 (2004); C. Alt *et al.*, (NA49), Phys. Rev. **C 78**, 044907 (2008); J. Adam *et al.*, (STAR), Phys. Rev. **C 102**, 034909 (2020); P. Gasik *et al.*, Eur. Phys. J. **A 52**, 177 (2016); K. Piasecki *et al.*, (FOPI), Phys. Rev. **C 91**, 054904 (2015); G. Agakishiev *et al.*, (HADES), Phys. Rev. **C 80**, 025209 (2009); Phys. Lett. **B 778**, 403 (2018); Phys. Rev. Lett. **103**, 132301 (2009).

[5] J. Brachmann *et al.*, Phys. Rev. **C 61**, 024909 (2000); L. Adamczyk *et al.*, (STAR) Phys. Rev. **C 88** 14902 (2013); Phys. Rev. **C 103**, 034908 (2021); J. Barrette *et al.*, (E877) Phys. Rev. **C 56**, 3254-3264 (1997); P. Chung *et al.*, (E895), Phys. Rev. Lett. **85**, 940 (2000); A. Andronic *et al.*, (FOPI) Phys. Lett. **B 612**, 173 (2005).

Momentum and multiplicity dependence of strangeness and nuclei production

Alberto Calivà[1,2,*]

[1]GSI Helmholtzzentrum für Schwerionenforschung GmbH, Planckstraße 1, 64291 Darmstadt, Germany
[2]Physikalisches Institut, University of Heidelberg, 69120 Heidelberg, Germany

Abstract. Prominent phenomena observed in high-energy hadronic collisions, such as the strangeness enhancement in small collision systems with increasing multiplicity and the production of loosely bound states in collisions where extreme temperatures are reached, are still mysterious and at the center of the experimental programs of several scientific Collaborations. In this contribution, recent experimental results from the ALICE and STAR Collaborations on these two fronts are presented and discussed.

1 Strangeness enhancement

One of the most intriguing observations in the research field dedicated to the study of strangeness production in high-energy hadronic collisions is the continuous evolution of the ratio between (multi)strange hadron yields and pion yields with increasing charged-particle multiplicity at mid-rapidity $\langle dN_{ch}/d\eta \rangle_{|\eta|<0.5}$ [1, 2]. The measured ratios show an increase for $\langle dN_{ch}/d\eta \rangle_{|\eta|<0.5} \lesssim 50$ which is followed by a saturation for higher multiplicities, up to the most central nucleus–nucleus collisions at the LHC. This behavior seems to be independent of the collision system and the center-of-mass energy of the collision. A larger increase is observed for hadrons with larger strangeness content. The origin of strangeness enhancement is still not fully understood from the theoretical point of view and different approaches are used to describe the experimental observations. Three of the most used models are presented in the following.

In the Canonical Statistical Model (CSM) [3], the multiplicity dependence of strange hadron to pion ratios emerges from the exact conservation of electric charge, baryon number, and strangeness in the so-called correlation volume V_c. In the perspective of this model, the strangeness enhancement with increasing multiplicity is rather seen as a strangeness suppression with respect to the grand-canonical limit, reached in central nucleus–nucleus collisions, going to lower multiplicities. Recent developments of this model include a multiplicity-dependent chemical freeze-out temperature T_{chem} and the assumption of an incomplete chemical equilibration described by a multiplicity-dependent strangeness saturation parameter γ_S [3]. This model provides a good description of the available data except for the p/π ratio, which is overestimated by approximately 2σ at all multiplicities [3]. In the two-component "core-corona" model [4], the smooth evolution from the string fragmentation regime in pp collisions to statistical hadronization in nucleus–nucleus collisions is interpreted as an increase of the relative contribution from the high-density "core" with increasing multiplicity.

*e-mail: alberto.caliva@cern.ch

Strangeness enhancement can also be described in the context of the Lund string fragmentation model using high-density color strings (ropes) and the color reconnection mechanism [5].

1.1 Canonical suppression

The role of canonical suppression becomes relevant also at low center-of-mass energy collisions. This is nicely highlighted by the measurements of ϕ/K^- and ϕ/Ξ^- as a function of centrality and center-of-mass energy by the STAR Collaboration (Fig. 1). The data indicate an increasing suppression of open strange hadron yields, relative to the hidden-strangeness meson ϕ yield, for a fixed $\sqrt{s_{\mathrm{NN}}}$ going from central to peripheral collisions. The suppression is stronger for hadrons with larger strangeness content ($\phi/\Xi^- > \phi/K^-$). The $\sqrt{s_{\mathrm{NN}}}$ dependence of these ratios originates from an interplay between the $\sqrt{s_{\mathrm{NN}}}$ dependence of the chemical freeze-out temperature and baryochemical potential μ_{B}, and the effect of canonical suppression, as described by statistical hadronization models [6, 7]. Canonical suppression becomes dominant for $\sqrt{s_{\mathrm{NN}}} < 4$, as indicated by the data.

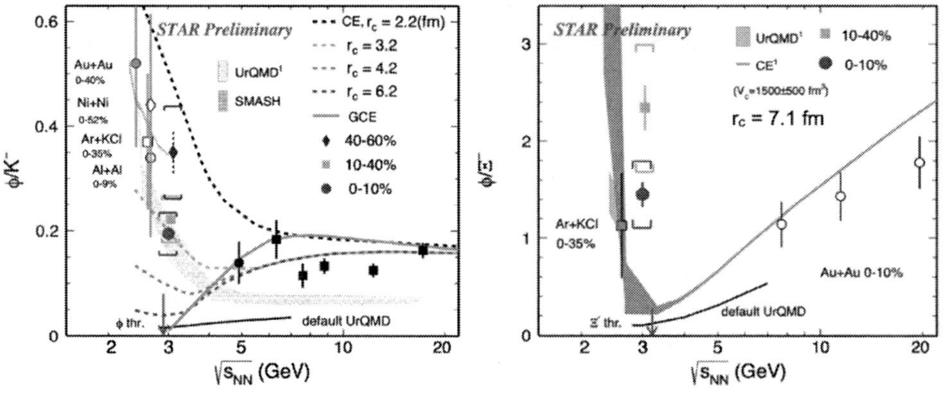

Figure 1. ϕ/K^- and ϕ/Ξ^- measured as a function of the center-of-mass energy in different collision systems by the STAR Collaboration.

1.2 Strangeness production and effective energy

Strangeness enhancement is further studied by the ALICE Collaboration as a function of effective energy. This is the energy effectively available for particle production at mid-rapidity considering that a fraction of the initial center-of-mass energy is carried away by baryons emitted at forward rapidity. Their energy is measured using the Zero-Degree Calorimeters (ZDC) and the effective energy of the collision is obtained as $E_{\mathrm{eff}} = \sqrt{s_{\mathrm{NN}}} - E_{\mathrm{ZDC}}$. Preliminary results from the ALICE Collaboration indicate that the strange hadron yields measured as a function of multiplicity in events with low and high effective energy are consistent within the uncertainties (Fig. 1). Strangeness production is therefore independent of the effective energy and is mainly driven by the final-state multiplicity.

1.3 Strangeness production in and out of jets

Strangeness production is further characterized by the ALICE Collaboration by measuring the yields of strange hadrons in jets and in the underlying event (UE) in pp collisions at $\sqrt{s}$

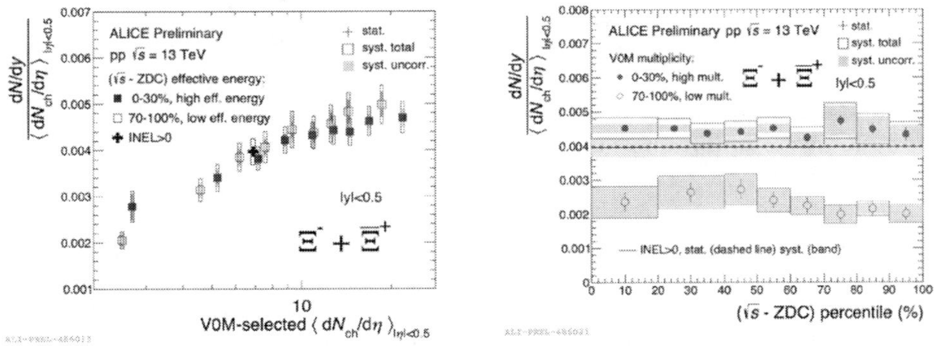

Figure 2. Ξ yields, normalized by its average, shown as a function of multiplicity for low and high effective energy (left) and as a function of the E_{eff} percentile for different multiplicities (right).

= 13 TeV. The two-particle correlation method is used in this measurement to distinguish the different contributions: a charged particle with the highest p_{T} and $p_{\text{T}} > 3$ GeV/c, called "trigger particle", is used as a proxy for the jet. The "in-jet" and "out-of-jet" regions are defined based on $\Delta\eta$ and $\Delta\phi$ of strange hadrons with respect to the trigger particle. The measured yield of strange hadrons in the UE rises faster with multiplicity than in the jet region, as shown in Fig. 3 left. The inclusive Ξ/K^0_{S} measured as a function of the event multiplicity is consistent with that measured in the UE, which implies a smaller relative contribution of strangeness production in jets. The measured ratio increases as a function of multiplicity also for the jet region with a slope similar to that in the UE. These results suggest that strangeness enhancement originates both from jets and the UE.

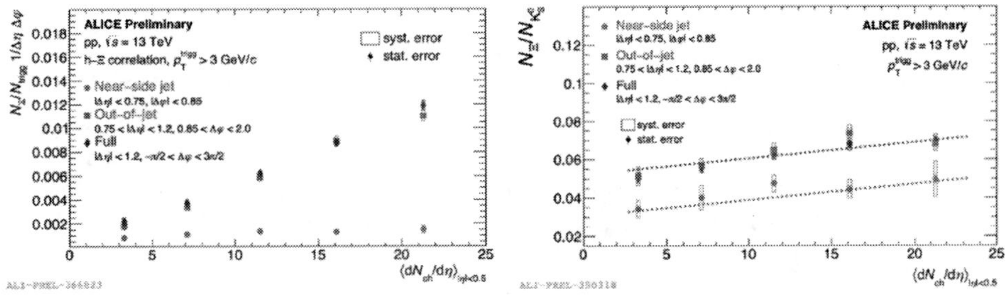

Figure 3. Ξ yield density (left) and Ξ/K^0_{S} as a function of multiplicity (right) for different regions.

2 (Hyper)nuclei producton

The production mechanism of light (anti)(hyper)nuclei in high-energy hadronic collisions is still not understood and currently under intense debate in the scientific community. The experimental data are typically described using two different phenomenological models: the statistical hadronization model (SHM) and the baryon coalescence approach. In the statistical hadronization model [8–13], light (anti-)(hyper-)nuclei, as well as other hadron species, are assumed to be emitted by a source in local thermal and hadrochemical equilibrium with their abundances being fixed at chemical freeze-out, at a temperature of $T_{\text{chem}} = 156 \pm 4$ MeV

[12]. Multi-baryon states are assumed to be produced as compact multi-quark systems, which evolve towards the final configuration with typical timescales of 5 fm/c or longer [12]. This model provides an excellent description of the measured hadron yields in central nucleus–nucleus collisions [12]. In the coalescence model [14–19], multi-baryon states are assumed to be formed by coalescence of baryons that are close in phase space at kinetic freeze-out. In the state-of-the-art implementations of the coalescence approach, the quantum-mechanical properties of baryons and bound states are taken into account and the coalescence probability is calculated from the overlap between the wave functions of individual baryons and the Wigner density of the final-state cluster.

From systematic studies of light (anti-)nuclei production in hadronic collisions at different multiplicities, it seems that the dominant production mechanism evolves smoothly from small to large collision systems with the charged-particle multiplicity. A smooth transition from the low-multiplicity regime towards the grand-canonical limit is observed in the yield ratios d/p and ^{3}He/p. While the multiplicity evolution of the d/p ratio is fairly well reproduced by the coalescence model [19], some tensions are observed in the intermediate multiplicity region for the ^{3}He/p ratio. The canonical statistical model, with fixed $T_{\mathrm{chem}} = 155$ MeV for all multiplicities [11], provides only a qualitative description of these data. A continuous evolution with multiplicity is also observed for the coalescence parameter, defined as the ratio between the invariant yield of a nucleus with a mass number A and that of protons to the power of A. In particular, the coalescence parameters for a fixed p_{T}/A continuously decrease going from pp to central nucleus–nucleus collisions. In the coalescence approach, this behavior is connected to the multiplicity evolution of the particle-emitting source size [16]: nucleons with fixed p_{T}/A are, on average, closer in space in small systems as compared to large systems, hence the coalescence probability in small systems is larger.

2.1 Deuteron production in and out of jets

The deuteron production is studied by the ALICE Collaboration in jets and in the UE in pp collisions at $\sqrt{s} = 13$ TeV [20]. The p_{T}-differential jet-associated deuteron production is measured using the two-particle correlation method as described in Sect. 1.3. The p_{T} spectrum of deuterons inside the jets is found to be consistent with predictions based on PYTHIA coupled to a simple coalescence afterburner. In the latter, deuteron formation is assumed to happen if a proton and neutron have a momentum difference $\Delta p < p_0$, with $p_0 = 110$ MeV, thus ignoring the space coordinates in the calculation of the coalescence probability.

The deuteron p_{T} spectra are also studied in the jet region and in the UE as a function of the EU activity. The latter is quantified by the ratio $R_{\mathrm{T}} = N^{\mathrm{T}}_{\mathrm{ch}}/\langle N^{\mathrm{T}}_{\mathrm{ch}}\rangle$, where $N^{\mathrm{T}}_{\mathrm{ch}}$ is the charged-particle multiplicity measured in the transverse region with respect to the trigger particle. Preliminary results from the ALICE Collaboration show similar coalescence parameters both in the toward and transverse regions for fixed R_{T}. The toward region contains contributions from jets and UE, while the transverse region is dominated by the UE. The data suggest a larger B_2 in the toward region at low R_{T} (low UE activity), as shown in Fig. 4. PYTHIA 8.3 describes the ordering and p_{T} dependence of B_2 but not the magnitude.

2.2 Hypernuclei production and their structure

The hypertriton is a weakly bound state composed of a proton, a neutron, and a Λ. This system has a halo structure with a "deuteron" core surrounded by an extended Λ probability cloud. The average distance between them is connected to the Λ separation energy: the smaller B_{Λ} the larger their distance. The widely-accepted measurement is $B_{\Lambda} = 130 \pm 50$ KeV [21]. The hypertriton lifetime is at the center of a long-standing puzzle which consists

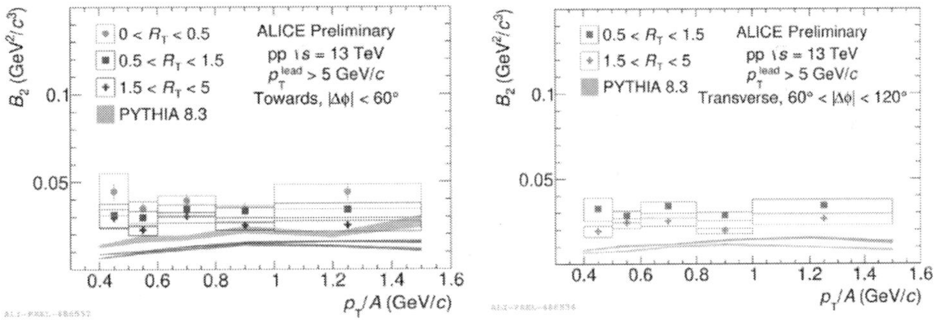

Figure 4. Coalescence parameter B_2 measured in the toward (left) and transverse (right) region for different R_T intervals.

in the difficulty to reconcile the small value of the Λ separation energy with the relatively short lifetime measured by the experiments. In fact, most of the theory calculations, which use the Λ separation energy as input, predict a hypertriton lifetime very close to that of the vacuum Λ, with a maximum deviation of a few %. The current world average, instead, is about 30% shorter with significant tension between the published ALICE and STAR results. The most recent preliminary measurements of the hypertriton lifetime from the two Collaborations give now consistent results, which are closer to the Λ lifetime, as shown in Fig. 5. In addition, the ALICE Collaboration has measured a Λ separation energy consistent with zero. These new results depict a consistent picture of the hypertriton structure and the solution to the lifetime puzzle seems to be imminent.

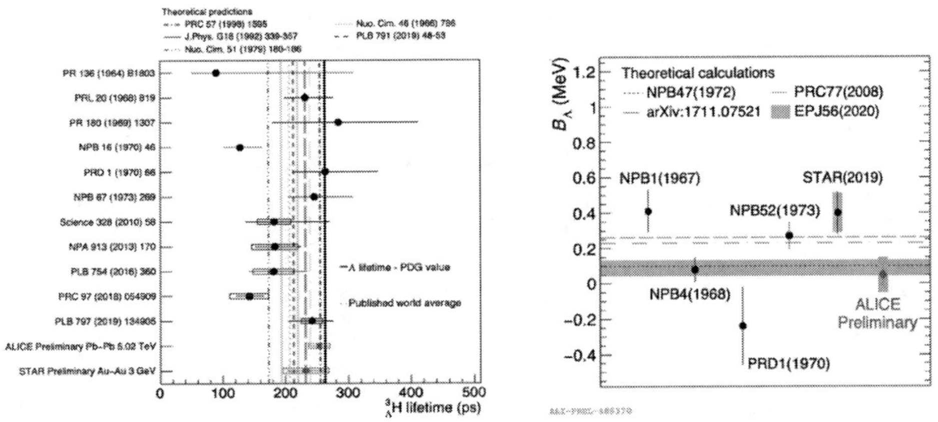

Figure 5. Hypertriton lifetime (left) and Λ separation energy (right).

The radial extension of the hypertriton wave function, constrained by the lifetime and Λ separation energy measurements, is a relevant parameter in the coalescence model which enters explicitly into the coalescence probability [19]. On the contrary, the predictions from the statistical hadronization model are independent of the system size and depend only on the mass and spin of the bound state [8–13]. The hypertriton yield measured in central nucleus–nucleus collisions, both at low and high center-of-mass energy, are consistent with the thermal model (Fig. 6 left). On the other hand, recent measurements of the hypertriton production in small collision systems by the ALICE Collaboration favor the two-body coalescence model. The preliminary measurement of the ${}^{4}_{\Lambda}$H – a bound state with two protons, a neutron and a Λ – in central Au–Au collisions at 3 GeV challenges both models (Fig. 6 right).

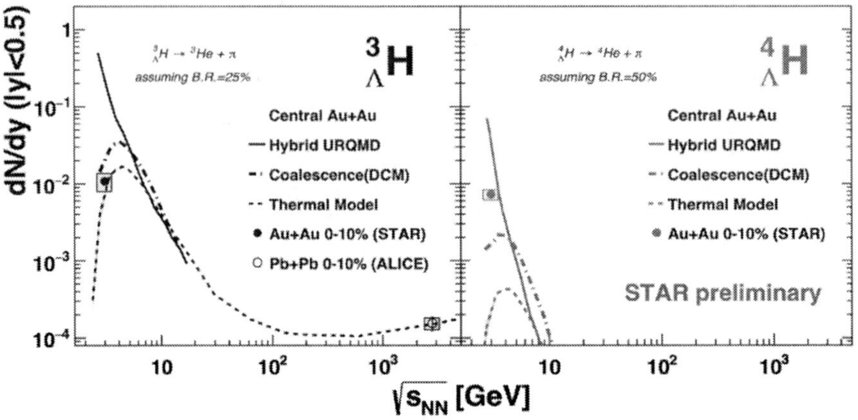

Figure 6. Hypertriton (left) and ${}^{4}_{\Lambda}$H yield (right) as a function of the center-of-mass energy.

References

[1] J. Adam *et al.* (The ALICE Collaboration), Nature Phys. **13**, 535-539 (2017)
[2] S. Acharya *et al.* (The ALICE Collaboration), Eur. Phys. J. C **80**, 167 (2020)
[3] V. Vovchenko, B. Dönigus, H. Stöcker, Phys. Rev. C **100**, 054906 (2019)
[4] Y. Kanakubo, Y. Tachibana, T. Hirano, Phys. Rev. C **101**, 024912 (2020)
[5] R. Nayak, S. Pal, S. Dash, Phys. Rev. D **100**, 074023 (2019)
[6] J. Cleymans *et al.*, Phys. Lett. B **603**, 146-151 (2004)
[7] A. Andronic, P. Braun-Munzinger, J. Stachel, Nucl. Phys. A **772**, 167-199 (2006)
[8] J. Cleymans *et al.*, Phys. Rev. C **84**, 054916 (2011)
[9] A. Andronic, P. Braun-Munzinger, J. Stachel, H. Stöcker, Phys. Lett. B **697**, 203-207 (2011)
[10] F. Becattini, M. Bleicher, E. Grossi, J. Steinheimer, R. Stock, Phys. Rev. C **90**, 054907 (2014)
[11] V. Vovchenko and H. Stöcker, Phys. Rev. C **95**, 044904 (2017)
[12] A. Andronic, P. Braun-Munzinger, K. Redlich, J. Stachel, Nature **561**, 321-330 (2018)
[13] N. Sharma, J. Cleymans, B. Hippolyte, M. Paradza, Phys. Rev. C **99**, 044914 (2019)
[14] S. T. Butler and C. A. Pearson, Phys. Rev. **129**, 836 (1963)
[15] J. I. Kapusta, Phys. Rev. C **21**, 1301-1310 (1980)
[16] R. Scheibl and U. W. Heinz, Phys. Rev. C **59**, 1585-1602 (1999)
[17] W. Zhao, L. Zhu, H. Zheng, C. M. Ko, H. Song , Phys. Rev. C **98**, 054905 (2018)
[18] K. Blum and M. Takimoto, Phys. Rev. C **99**, 044913 (2019)
[19] K. J. Sun, C. M. Ko, B. Dönigus, Phys. Lett. B **792**, 132-137 (2019)
[20] S. Acharya *et al.* (The ALICE Collaboration), Phys. Lett. B **819**, 136440 (2021)
[21] D.H. Davis, Nucl. Phys. A **754** (2005)

Bottomonium suppression and flow in heavy-ion collisions

Michael Strickland[1,*]

[1]Department of Physics, Kent State University, Kent, OH 44242

Abstract. The strong suppression of bottomonia production in ultra-relativistic heavy-ion collisions is a smoking gun for the creation of a deconfined quark-gluon plasma (QGP). In this proceedings contribution, I review recent work that aims to provide a more comprehensive and systematic understanding of bottomonium dynamics in the QGP through the use of pNRQCD and an open quantum systems approach. This approach allows one to evolve the heavy-quarkonium reduced density matrix, taking into account non-unitary effective Hamiltonian evolution of the wave-function and quantum jumps between different angular momentum and color states. In the case of a strong coupled QGP in which $E_{\rm bind} \ll T, m_D \ll 1/a_0$, the corresponding evolution equation is Markovian and can therefore be mapped to a Lindblad evolution equation. To solve the resulting Lindblad equation, we make use of a stochastic unraveling called the quantum trajectories algorithm and couple the non-abelian quantum evolution to a realistic 3+1D viscous hydrodynamical background. Using a large number of Monte-Carlo sampled bottomonium trajectories, we make predictions for bottomonium R_{AA} and elliptic flow as a function of centrality and transverse momentum and compare to data collected by the ALICE, ATLAS, and CMS collaborations.

1 Introduction

Bottomonium states that propagate through a deconfined quark-gluon plasma (QGP) are suppressed due to both Debye-screening, which modifies the real part of the heavy-quark potential, and in-medium transitions/breakup, which are encoded in the imaginary part of the heavy-quark potential. In recent years there have significant advances in our understanding of heavy-quark bound state dynamics in the QGP stemming from a combination of potential non-relativistic QCD (pNRQCD) and open quantum systems (OQS) [1–4]. In this effective field theory treatment one relies on a separation scales between the inverse size of the bound states $1/\langle r \rangle$, the effective temperature of the system T, and the binding energy of the states E. Through the inclusion of hard-thermal-loop effects one can also include the effect of the induced Debye mass, m_D, of the system on the heavy-quark potential. In a strongly coupled QGP, due to the large coupling, there is not a strict ordering of the temperature and Debye mass scales. Instead one typically has $\pi T \sim m_D$ and the relevant hierarchy of scales is $1/a_0 \gg \pi T, m_D \gg E$, where a_0 is the Bohr radius. In this case it can be shown that the quantum evolution is Markovian, with the local medium relaxation time being shorter than

*e-mail: mstrick6@kent.edu

both the time scale associated with internal transitions and the probe (heavy quarkonium) relaxation time [5–7]. As a consequence, the dynamical equation which governs the evolution of the heavy quarkonium reduced density matrix is of Lindblad form [1, 2].

The Lindblad equation for the reduced density is challenging to solve numerically. This stems from the fact that one must decompose states in angular momentum and color quantum numbers followed by discretization of the underlying wave-functions on a lattice of size N, resulting in a reduced density matrix with the number of elements proportional to $(l_{\max}+1)^2N^2$, where $l_{\max}$ is the largest angular momentum quantum number considered. For reliable computation one must take both a large number of lattice points and a large angular momentum cutoff, which makes directly solving the Lindblad equation numerically prohibitive, with memory size scaling like $(l_{\max}+1)^2N^2$ and per step evolution times scaling like $(l_{\max}+1)^4N^4$ [8].[1] When faced with a problem with such high dimensionality it is frequently beneficial to make use of Monte-Carlo methods. A particularly well-suited algorithm is provided by the quantum trajectories algorithm [9–14]. In this approach, one maps the solution of the Lindblad equation to solution of an ensemble of a independent one-dimensional Schrödinger equation evolutions that are subject to stochastic color and angular momentum transitions (quantum jumps) [14]. When averaged over the ensemble, one obtains a solution to the three-dimensional Lindblad equation. Due to the fact that each quantum trajectory is independent, the quantum trajectories algorithm lends itself to massive parallel computation. Additionally, since between quantum jumps the evolution proceeds with fixed color and angular momentum quantum numbers, one does not have to place a cutoff on the magnitude of the angular momentum. Herein, we present results obtained using an open-source code called QTraj, which implements the quantum trajectories algorithm for bottomonium states [8].

In this proceedings contribution, we briefly review the theoretical underpinnings of the method and present comparisons between the phenomenological results obtained using a realistic hydrodynamical background and experimental data for bottomonium observables collected by the ALICE, ATLAS, and CMS collaborations in 5 TeV Pb-Pb collisions.

2 Methodology

Using pNRQCD and OQS, in Refs. [1, 2] the authors obtained a set of master equations for heavy quarkonium in a strongly coupled QGP. When the temperature is much larger than the binding energy, $T \gg E$, one can make an expansion in E/T, finding at leading order an evolution equation of Lindblad form [15, 16]

$$\frac{d\rho(t)}{dt} = -i[H,\rho(t)] + \sum_n \left(C_n\rho(t)C_n^\dagger - \frac{1}{2}\left\{C_n^\dagger C_n, \rho(t)\right\}\right), \tag{1}$$

where, for N_c colors, the reduced density matrix and Hamiltonian are given by

$$\rho(t) = \begin{pmatrix} \rho_s(t) & 0 \\ 0 & \rho_o(t) \end{pmatrix}, \tag{2}$$

$$H = \begin{pmatrix} h_s & 0 \\ 0 & h_o \end{pmatrix} + \frac{r^2}{2}\gamma \begin{pmatrix} 1 & 0 \\ 0 & \frac{N_c^2-2}{2(N_c^2-1)} \end{pmatrix}, \tag{3}$$

with $\rho_s(t)$ and $\rho_o(t)$ being the singlet and octet reduced density matrices, respectively, and $h_{s,o} = \mathbf{p}^2/M + V_{s,o}$ being the singlet or octet Hamiltonians. The jump (collapse) operators C

[1]For application to heavy-quarkonium dynamics in the QGP one needs $N \sim 4000$ and $l_{\max} \sim 5$.

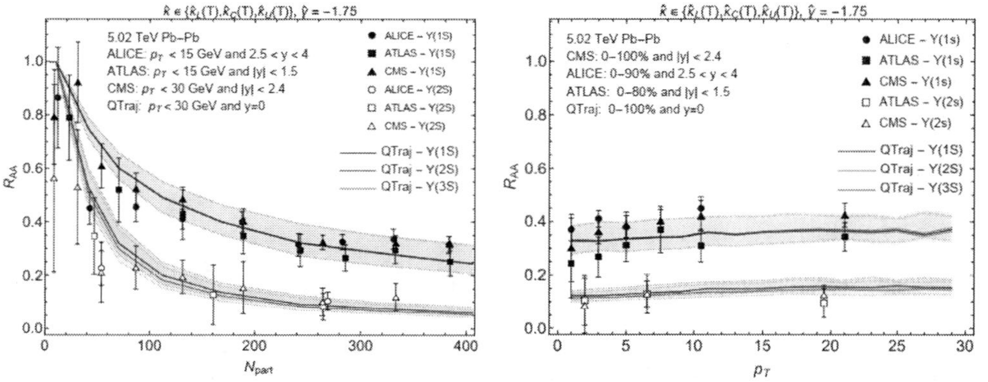

Figure 1. The nuclear modification factor R_{AA} of the $\Upsilon(1S)$, $\Upsilon(2S)$, and $\Upsilon(3S)$ as a function of N_{part} compared to experimental measurements from the ALICE [23], ATLAS [24], and CMS [25] collaborations. The bands indicate variation with respect to $\hat{\kappa}(T)$ (left) and $\hat{\gamma}$ (right). The central curves represent the central values of $\hat{\kappa}(T)$ and $\hat{\gamma}$, and the dashed and dot-dashed lines represent the lower and upper values, respectively, of $\hat{\kappa}(T)$ and $\hat{\gamma}$.

appearing in Eq. (1) are

$$C_i^0 = \sqrt{\frac{\kappa}{N_c^2 - 1}} r^i \begin{pmatrix} 0 & 1 \\ \sqrt{N_c^2 - 1} & 0 \end{pmatrix}, \quad C_i^1 = \sqrt{\frac{(N_c^2 - 4)\kappa}{2(N_c^2 - 1)}} r^i \begin{pmatrix} 0 & 0 \\ 0 & 1 \end{pmatrix}, \tag{4}$$

with the transport coefficients κ and γ given by the real and imaginary parts of the chromoelectric correlator

$$\kappa = \frac{g^2}{18} \int_0^\infty dt \left\langle \left\{ \tilde{E}^{a,i}(t, \mathbf{0}), \tilde{E}^{a,i}(0, \mathbf{0}) \right\} \right\rangle, \quad \gamma = -i \frac{g^2}{18} \int_0^\infty dt \left\langle \left[\tilde{E}^{a,i}(t, \mathbf{0}), \tilde{E}^{a,i}(0, \mathbf{0}) \right] \right\rangle. \tag{5}$$

3 Results

We solve (1) using the quantum trajectories algorithm. For the transport coefficients κ and γ, we use lattice quantum chromodynamics measurements to constrain them. For the results presented herein we will vary κ over three temperature-dependent parameterizations $\hat{\kappa}(T) = \kappa(T)/T^3 \in \{\hat{\kappa}_L(T), \hat{\kappa}_C(T), \hat{\kappa}_U(T)\}$ with the lower, central, and upper limits determined using the results of Ref. [17]. For γ, less is known. To allow for all current indirect lattice extractions, we take $\hat{\gamma} = \gamma/T^3 = \{-3.5, -1.75, 0\}$ [18–22]. We note that the coefficient κ sets the magnitude of the imaginary part of the potential, while γ sets the magnitude of the modification of the real part of the potential.

To compute the nuclear suppression in AA collisions we compute the survival probability for a large number of physical trajectories with Monte-Carlo-sampled initial production points and transverse momenta. Along each physical trajectory we additionally average over a set of quantum trajectories in which different quantum evolutions are sampled. Once the survival probability for each state under consideration is computed, we then perform late-time excited state feed down using a feed down matrix F constructed from the measured branching ratios and cross sections for bottomonium states [3].

With this one can compute the nuclear suppression of state i from

$$R_{AA}^i(c, p_T, \phi) = \frac{\left(F \cdot S(c, p_T, \phi) \cdot \vec{\sigma}_{\text{direct}}\right)^i}{\vec{\sigma}_{\text{exp}}^i}, \tag{6}$$

https://doi.org/10.1051/epjconf/202225904001

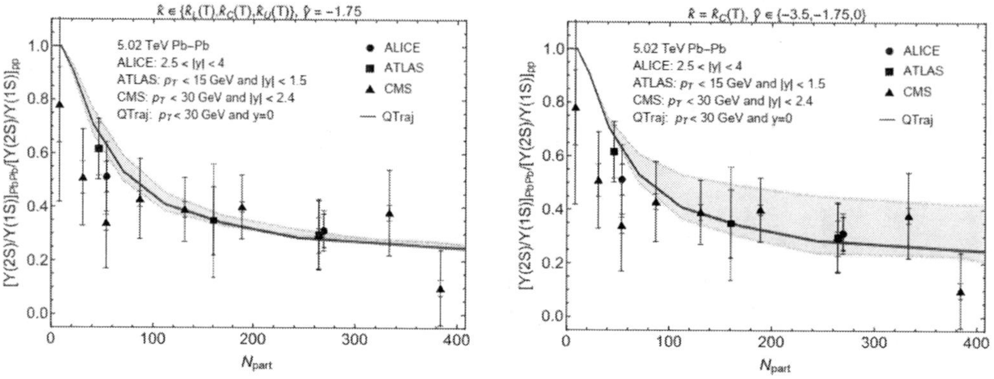

Figure 2. (Color online) The double ratio of the nuclear modification factor $R_{AA}[\Upsilon(2S)]$ to $R_{AA}[\Upsilon(1S)]$ as a function of $N_{\rm part}$ compared to experimental measurements of the ALICE [23], ATLAS [24], and CMS [31] collaborations. The bands indicate variation of $\hat{\kappa}(T)$ and $\hat{\gamma}$ as in Fig. 1. The black and red experimental error bars represent statistical and systematic uncertainties, respectively.

where $\vec{\sigma}_{\rm direct}$ and $\vec{\sigma}_{\rm exp}$ are the pre-feeddown and experimentally observed pp cross sections for bottomonium production, respectively. In this expression c, p_T, and ϕ correspond to the centrality, transverse momentum, and azimuthal angle bins being considered. The survival probability $S(c, p_T, \phi)$ is a diagonal matrix containing each state's survival probablity along the diagonal. For the background evolution, we use a three-dimensional anisotropic hydrodynamics code (aHydro3p1) that has been tuned to soft-observables such as the pion, kaon, and proton spectra and elliptic flow [26]. The aHydro3p1 code implements quasiparticle anisotropic hydrodynamics, including a lattice-based equation of state [27–30].

In practice we averaged over millions of sampled physical trajectories using a large one-dimensional lattice to solve for the stochastic evolution. For the precise lattice spacing, time steps, initial conditions used, etc., we refer the reader to Ref. [4]. In the left panel of Fig. 1 we plot the nuclear suppression factor as a function of the number or participants. The band in the figure indicates our theoretical uncertainty in R_{AA} stemming from the uncertainty in the transport coefficient κ. Similar results are obtained when varying γ, however, the band is somewhat larger (see Ref. [4]). As can be seen from Fig. 1, our predictions for the dependence of R_{AA} on $N_{\rm part}$ agree quite well with the reported experimental data. In the right panel of Fig. 1 we present our predictions for the p_T-dependence of R_{AA} and compare to experimental data. Once again we see good agreement given current theoretical and experimental uncertainties.

In Fig. 2 we present our predictions for the double ratio $[\Upsilon(2S)/\Upsilon(1S)]_{\rm PbPb}/[\Upsilon(2S)/\Upsilon(1S)]_{\rm pp}$. In the left and right panels, the bands result from variation of κ and γ, respectively. As the left panel demonstrates, the dependence on κ largely cancels when considering the double ratio; however, in the right panel we see that this ratio has a stronger dependence on γ. Similar behavior was seen in the 3S to 1S double ratio [4]. This offers some hope that, with increased experimental statistics, one can use 2S/1S and 3S/1S double ratio data to constrain the transport coefficient γ.

Finally, in Fig. 3 we present our predictions for the elliptic flow of the $\Upsilon(1S)$ in three different centrality classes along with a centrality-integrated measurement in the right subpanel of both figures. The left and right figures in Fig. 3 correspond to varying κ and γ, respectively. In the case of $v_2[\Upsilon(1S)]$, the variation with κ and γ are similar and our predictions are, again, in reasonable agreement with available data. We note that if one focuses on the integrated

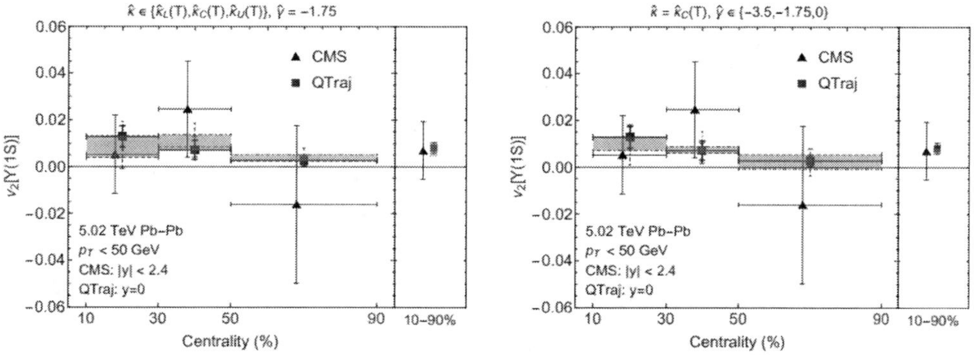

Figure 3. (Color online) The elliptic flow v_2 of the $\Upsilon(1S)$ as a function of centrality compared to experimental measurements of the CMS [32] collaboration. The QTraj bands represent uncertainties as in Fig. 1. QTraj error bars indicate the statistical uncertainty of our extraction.

results shown in the right sub-panel of each figure, our predictions have small systematic and statistical uncertainties and are consistent with CMS measurements within uncertainties.

4 Conclusions and outlook

In this proceedings contribution, I highlighted recent work that uses the framework of OQS together with pNRQCD to make phenomenological predictions for bottomonium suppression in the quark-gluon plasma. The code used to generate the results presented herein has been released as an open-source package along with detailed documentation in Ref. [8]. Additional theory/data comparisons and discussions can be found in Ref. [4]. Our results demonstrate that it is possible to understand experimental measurements of bottomonium suppression based on a first-principles approach that is fully quantum and incorporates features that are unique to non-abelian QCD, such as singlet-octet transitions.

Using pNRQCD and OQS methods, at leading-order one must solve an evolution equation of Lindblad form for the reduced density matrix. Due to the difficulty of solving such a large matrix evolution equation, we made use of an algorithm originally developed for quantum optics applications, called the quantum trajectories algorithm. With this method, solution of the 3D non-abelian Lindblad equation could be reduced to the solution of a 1D Schrödinger equation with a non-Hermitian Hamiltonian subject to stochastic internal transitions (angular momentum and color) dubbed quantum jumps. This algorithm allows one to solve the 3D non-abelian Lindblad equation in a massively parallel manner due to the fact that each quantum trajectory is independent.

Looking to the future, it is necessary to include next-to-leading order corrections in the binding energy over the temperature, E/T, in order to assess the convergence of the high-temperature expansion used to obtain the leading-order Lindblad equation. This will help to reduce the theoretical uncertainty in the low-temperature limit associated with the leading-order truncation and will be allow us to include drag/recoil effects.

Acknowledgments

M.S. has been supported by the U.S. Department of Energy, Office of Science, Office of Nuclear Physics Award No. DE-SC0013470. M.S. also thanks the Ohio Supercomputer Center for support under the auspices of Project No. PGS0253.

References

[1] N. Brambilla, M.A. Escobedo, J. Soto, A. Vairo, Phys. Rev. **D96**, 034021 (2017), 1612.07248

[2] N. Brambilla, M.A. Escobedo, J. Soto, A. Vairo, Phys. Rev. **D97**, 074009 (2018), 1711.04515

[3] N. Brambilla, M.A. Escobedo, M. Strickland, A. Vairo, P. Vander Griend, J.H. Weber, JHEP **05**, 136 (2021), 2012.01240

[4] N. Brambilla, M.A. Escobedo, M. Strickland, A. Vairo, P.V. Griend, J.H. Weber (2021), 2107.06222

[5] Y. Akamatsu (2020), 2009.10559

[6] A. Rothkopf, Nucl. Phys. A **1005**, 121922 (2021), 2002.04938

[7] X. Yao (2021), 2102.01736

[8] H.B. Omar, M.A. Escobedo, A. Islam, M. Strickland, S. Thapa, P.V. Griend, J.H. Weber (2021), 2107.06147

[9] J. Dalibard, Y. Castin, K. Molmer, Phys. Rev. Lett. **68**, 580 (1992)

[10] K. Mølmer, Y. Castin, J. Dalibard, J. Opt. Soc. Am. B **10**, 524 (1993)

[11] M.B. Plenio, P.L. Knight, Rev. Mod. Phys. **70**, 101 (1998), quant-ph/9702007

[12] H. Carmichael, *Statistical methods in quantum optics : master equations and fokker-planck equations* (Springer, New York, 1999), ISBN 978-3-540-54882-9

[13] U. Weiss, *Quantum Dissipative Systems* (WORLD SCIENTIFIC, 1993)

[14] A.J. Daley, Adv. Phys. **63**, 77 (2014), 1405.6694

[15] G. Lindblad, Commun. Math. Phys. **48**, 119 (1976)

[16] V. Gorini, A. Kossakowski, E.C.G. Sudarshan, J. Math. Phys. **17**, 821 (1976)

[17] N. Brambilla, V. Leino, P. Petreczky, A. Vairo, Phys. Rev. D **102**, 074503 (2020), 2007.10078

[18] N. Brambilla, M.A. Escobedo, A. Vairo, P. Vander Griend, Phys. Rev. D **100**, 054025 (2019), 1903.08063

[19] S. Kim, P. Petreczky, A. Rothkopf, JHEP **11**, 088 (2018), 1808.08781

[20] G. Aarts, C. Allton, S. Kim, M.P. Lombardo, M.B. Oktay, S.M. Ryan, D.K. Sinclair, J.I. Skullerud, JHEP **11**, 103 (2011), 1109.4496

[21] R. Larsen, S. Meinel, S. Mukherjee, P. Petreczky, Phys. Rev. D **100**, 074506 (2019), 1908.08437

[22] S. Shi, K. Zhou, J. Zhao, S. Mukherjee, P. Zhuang (2021), 2105.07862

[23] S. Acharya et al. (ALICE) (2020), 2011.05758

[24] Songkyo Lee (ATLAS Collaboration), *Quarkonium production in Pb+Pb collisions with ATLAS*, Quark Matter 2020 https://indico.cern.ch/event/792436/contributions/3535775/ (2017)

[25] A.M. Sirunyan et al. (CMS), Phys. Lett. B **790**, 270 (2019), 1805.09215

[26] M. Alqahtani, M. Strickland (2020), 2008.07657

[27] M. Martinez, M. Strickland, Nucl. Phys. **A848**, 183 (2010), 1007.0889

[28] W. Florkowski, R. Ryblewski, Phys.Rev. **C83**, 034907 (2011), 1007.0130

[29] M. Alqahtani, M. Nopoush, M. Strickland, Phys. Rev. **C92**, 054910 (2015), 1509.02913

[30] M. Alqahtani, M. Nopoush, M. Strickland, Prog. Part. Nucl. Phys. **101**, 204 (2018), 1712.03282

[31] A.M. Sirunyan et al. (CMS), Phys. Rev. Lett. **120**, 142301 (2018), 1706.05984

[32] A.M. Sirunyan et al. (CMS), Phys. Lett. B **819**, 136385 (2021), 2006.07707

Heavy-quark effects on cold quark matter and self-bound stars

José C. Jiménez[1,*] and *Eduardo* S. Fraga[2,**]

[1]Instituto de Física, Universidade de São Paulo,
Rua do Matão, 1371, Butantã, 05508-090, São Paulo, SP, Brazil
[2]Instituto de Física, Universidade Federal do Rio de Janeiro,
Caixa Postal 68528, 21941-972, Rio de Janeiro, RJ, Brazil

Abstract. The heavy-quark effects on the equation of state for cold and dense quark matter are obtained from perturbative QCD, yielding observables parametrized only by the renormalization scale. In particular, we investigate the thermodynamics of charm quark matter under the constraints of β equilibrium and electric charge neutrality in a region of densities where perturbative QCD is, in principle, much more reliable. Finally, we analyze the stability of charm stars, which might be realized as a new branch of ultradense hybrid compact stars, and find that such self-bound stars are unstable under radial oscillations.

1 Introduction

The presence of heavy flavors in cold quark matter might have non-trivial effects in matter subject to extreme conditions of density. For instance, the particular case of charm quarks immeaditely lead us to consider the possibility of charm stars, i.e. self-bound strange stars with some amount of charm quarks at their cores. However, although usual constraints forbid their existence [1, 2], they might still exist in ultradense hybrid neutron stars. Besides, perturbative quantum chromodynamics (pQCD) can be used much more confidently at these very high densities [3] when building the equation of state (EoS) for charm matter[1] in electroweak equilibrium. In this proceeding, we investigate cold quark matter with heavy quarks using in-medium pQCD. Specifically, we build our framework using the formalism developed in Ref. [6] for $N_f = N_l + 1$ flavors, i.e. N_l is the number of massless quarks and "1" the massive flavor, to include any number of massive flavors and determine the associated EoS. As an application of our framework, we discuss charm matter in β equilibrium and also being electrically neutral. Interestingly, this allow us to discuss the existence charm quark stars from the pQCD viewpoint. In the past, these exotic stars were investigated only within the bag model being unstable against radial oscillations [7]. By applying the Gondek's method [8] for a couple of first-order differential equations for the Lagrangian radial displacements, our calculations indicate that such self-bound stars are unstable.

*e-mail: jimenez@if.usp.br
**e-mail: fraga@if.ufrj.br

[1]This complements advancements in pQCD [4] and lattice QCD at high temperatures [5].

2 Cold and dense pQCD with heavy quarks

The calculation of the EoS for cold quark matter within pQCD has a long history [3]. The state-of-the-art three loop result of Ref. [6] includes renormalization-group effects of the mass and strong coupling parameters. More especifically, this framework deals with $N_f = N_l + 1$ quark flavors, i.e., N_l massless plus 1 massive, e.g. massless up and down and massive strange. The corresponding pQCD thermodynamic potential is composed by a massless contribution plus a massive term, together with the mixed vacuum-matter (VM) diagrams and the corresponding ring terms, i.e. $\Omega = \Omega^{m=0}(\vec{\mu}) + \Omega^{m}(\tilde{\mu}, m) + \Omega^{x}_{\rm VM}(\vec{\mu}, m) + \Omega_{\rm ring}(\vec{\mu}, \tilde{\mu}, m)$ with mass m and $\tilde{\mu}$ corresponds to the massive flavor $\vec{\mu} \equiv (\mu_1, ..., \mu_{N_l})$ is the vector chemical potential for the massless quarks. Besides, this Ω depends on a renormalization scale parameter $\bar{\Lambda}$, which is a pQCD scale and α_s is $\alpha_s(\bar{\Lambda}) = (4\pi/\beta_0 L)\left(1 - (2\beta_1/\beta_0^2)\ln L/L\right)$ where $\beta_0 = 11 - 2N_f/3, \beta_1 = 51 - 19N_f/3$, and $L = 2\ln\left(\bar{\Lambda}/\Lambda_{\overline{\rm MS}}\right)$ with $\Lambda_{\overline{\rm MS}}$ as the $\overline{\rm MS}$ point scale. Usually the renormalization scale is defined as $\bar{\Lambda} = X\sum_i \mu_i/N_f$, where the sum runs over the active quark flavors, and the dimensionless parameter X sits between 1 and 4 [6].

On the other hand, at high densities one expects that light quarks could be accompanied by heavy flavors. In this situation, the total number of flavors are $N_f = \sum_{i=1}^{N_m}(N_l + 1)^{(i)}$, where N_m is the number of massive quarks. This restrict us to add at least one massless quark for each massive flavor. Notice that further physical conditions are required in order to control when a heavy partner appears actively. In fact, such conditions affect the renormalization scale $\bar{\Lambda}$, depending on the chosen heavy quark to be added. Thus, the Ω at $O(\alpha_s^2)$ for N_l massless and N_m massive quarks is $\bar{\Omega}[N_f] = \sum_{i=1}^{N_m}\left\{\Omega[N_l^{(i)}] + \Omega[1^{(i)}]\right\}$, choosing first the massless flavors when adding a massive one, so that $\Omega[N_l^{(i)}] \equiv (\Omega^{m=0}(\vec{\mu}))^{(i)}$ is the massless contribution and $\Omega[1^{(i)}] \equiv (\Omega^m + \Omega^x_{\rm VM} + \Omega_{\rm ring})^{(i)}$ is the mixed massive contribution, where $\vec{\mu}^{(i)} = (\mu_1, ..., \mu_i)$ is the massless vector chemical potential, $\tilde{\mu}^{(i)}$ the massive (heavy) quark chemical potentials, and $m^{(i)}$ their corresponding masses. Here, $\Omega[...]$ indicates just the implicit parameter dependence (e.g. on N_f), whereas $\Omega(...)$ represents an explicit function dependence.

3 Charm matter thermodynamics

We now apply the aforementioned formalism to charm quark matter for which pQCD is more reliable. The EoS for charm matter is built as follows. Below the charm threshold, the usual strange matter conditions are the same [9]. Above the charm threshold, the reactions $u + d \leftrightarrow c + d$ yield $\mu_c = \mu_u$. Furthermore, electric neutrality is $(2/3)n_u + (2/3)n_c - (1/3)n_d - (1/3)n_s - n_e - n_\mu = 0$, with muons having being degenerate satisfying $\mu_\mu = \mu_e$. Then, all chemical potentials are functions of μ_s. The strange and charm quark mass parameters (up to $O(\alpha_s^2)$) are [10] $m_s(\bar{\Lambda}) = \hat{m}_s\,(\alpha_s/\pi)^{4/9}\left(1 + 0.895062\,(\alpha_s/\pi) + 1.37143\,(\alpha_s/\pi)^2\right)$ and $m_c(\bar{\Lambda}) = \hat{m}_c\,(\alpha_s/\pi)^{12/25}\left(1 + 1.01413\,(\alpha_s/\pi) + 1.38921\,(\alpha_s/\pi)^2\right)$, respectively. The $\{\hat{m}_q\}$ are the invariant quark masses. By noting that α_s depends on N_f, this gives different values of $\Lambda_{\overline{\rm MS}}$. Choosing $\alpha_s(\bar{\Lambda} = 1.5\ {\rm GeV},\ N_f = 3, 4) = 0.336^{+0.012}_{-0.008}$ [11] gives $\Lambda^{2+1}_{\overline{\rm MS}} = 343^{+18}_{-12}$ MeV and $\Lambda^{2+1+1}_{\overline{\rm MS}} = 290^{+18}_{-12}$ MeV, thus defining $\alpha_s^{2+1}(\bar{\Lambda})$ and $\alpha_s^{2+1+1}(\bar{\Lambda})$, respectively. Fixing the strange quark mass at $m_s(2\ {\rm GeV}, N_f = 3, 4) = 92.4(1.5)$ MeV [12] give $\hat{m}_s^{2+1} \approx 246.2$ MeV when using α_s^{2+1}, and $\hat{m}_s^{2+1+1} \approx 243.7$ MeV with α_s^{2+1+1}. Additionally, fixing the charm quark mass at $m_c(3{\rm GeV}, N_f = 4) = 0.9851(63)\ {\rm GeV} \equiv m_c^0$ [12], gives $\hat{m}_c^{2+1+1} \approx 3.0895$ GeV when using α_s^{2+1+1}, being m_c^0 the vacuum charm mass. Furthermore, charm quarks must satisfy $\mu_c = \mu_s - \mu_e > m_c^{\rm medium} > m_c^0$ being $m_c^{\rm medium}$ the unknown in-medium charm mass and the renormalization scale parameter below and above the charm threshold is $\bar{\Lambda} = X(\mu_u + \mu_d + \mu_s + 0)/3$ for $\mu_s \lesssim m_c^0$ and $X^*(\mu_u + \mu_d + \mu_s + \mu_c)/3$ for $\mu_s \gtrsim m_c^0$, where the

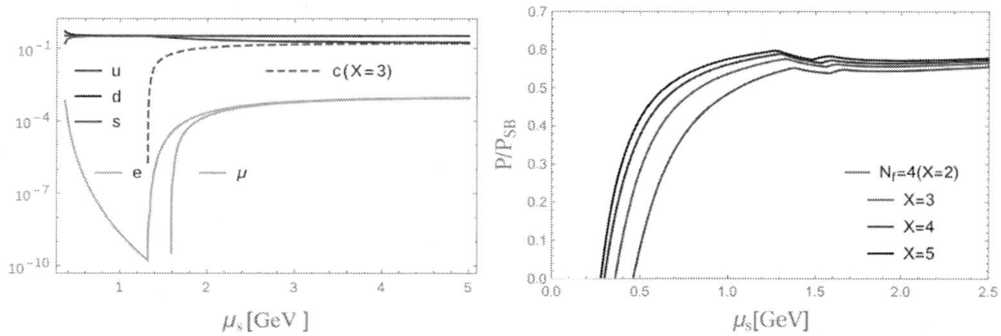

Figure 1. Left panel: Quark and lepton populations, n_i/n, present in charm matter for $X = 3$. Above the charm threshold, the lepton fractions increase to ensure electric charge neutrality at high densities. Right panel: Total pressure of $N_f = 2+1+1$ quarks plus leptons in β equilibrium and electrically charge neutral normalized by the Stefan-Boltzmann gas with $N_f = 4$. The case $X = 5$ is shown only to verify how the EoS depends on X when including one additional heavy flavor. Taken from Ref. [14].

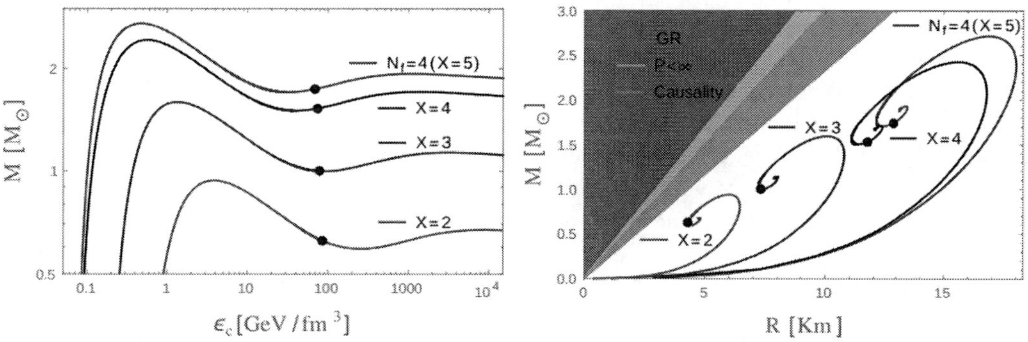

Figure 2. Left panel: Gravitational masses vs central energy densities for a system with $N_f = 2 + 1 + 1$ quarks plus leptons. Right panel: Mass-radius diagram for the associated quark stars. The black dots indicate the appearance of the charm quarks. Stars not satisfying the general-relativistic (GR, magenta), causality (green), and finite pressure ($P < \infty$, light blue) limits are excluded [13]. Taken from Ref. [14].

approximations in the inequalities come from noticing that before the threshold point the electron chemical potential takes its lowest value compared to the strange one, thus allowing us to make the approximation $\mu_c \approx \mu_s$. To build the EoS, we choose $N_l^{(1)} = 1$ for the up, $N_l^{(2)} = 1$ for the down, and $N_m = 2$ for the strange and charm quarks, in the thermodynamic potential $\bar{\Omega}[N_f = 2+1+1] = \left\{\Omega[N_l^{(1)} = 1] + \Omega[1^{(1)}]\right\} + \left\{\Omega[N_l^{(2)} = 1] + \Omega[1^{(2)}]\right\}$ so that the flavors are counted as $N_f = (1+1)^{(1)} + (1+1)^{(2)} = (u+c)^{(1)} + (d+s)^{(2)}$. We define the total quark number density for charm matter, for a given X, as $n_q \equiv n_u + n_d + n_s + n_c$ and the total particle density as $n = n_q + n_L$, where $n_L = n_e + n_\mu$. In Fig. 1, we show the behavior of the relative particle populations in β equilibrium and electrically neutral charm quark matter in the case of $X = 3$. The total pressure and energy density are obtained by keeping thermodynamic consistency systematically. This allows us to build numerically the EoS, $P = P(\epsilon, X)$. In Fig. 1, the total pressure is displayed for charm matter normalized by a Stefan-Boltzmann gas of quarks with $N_f = 4$ as a function of the strange quark chemical potential. Besides, the charm EoS is largely softened, generating an apparent instability which could have astrophysical effects, e.g. the possibility of charm stars.

4 Results and Conclusion

The possibility of bare charm quark stars was studied in Ref. [7] using the simplest version of the bag model. Their conclusion was that charm stars would be unstable against radial perturbations. We revisit this question using our first-principle pQCD results for the EoS for charm quark matter. We also choose the range $X \geq 3$, which satisfies the Bodmer-Witten hypothesis of strange quark matter. First of all, although the structure equations ensure their relativistic hydrostatic equilibrium, they must also satisfy the condition $\partial M/\partial \epsilon_c \geq 0$ [9]. The maximal masses are identified where $\partial M/\partial \epsilon_c = 0$ (see Fig. 2). However, we note that for the case $X = 2$, this condition is not satisfied when charm quarks appear. On the other hand, for $X > 3$, this condition is satisfied when charm quarks are present, which would correspond to stable charm stars. In Fig. 2, we show the mass-radius diagram for quark stars made of $N_f = 2 + 1 + 1$ quarks plus electrons and muons with varying X. Performing the dynamical stability analysis against radial pulsations using the method of Gondek *et al.* [8], where amplitudes oscillate harmonically when the frequencies are such that $\mathrm{Re}(\omega_n) > 0$ and $\mathrm{Im}(\omega_n) = 0$, or increase exponentially if $\mathrm{Re}(\omega_n) \geq 0$ and $\mathrm{Im}(\omega_n) > 0$. Besides, if $\mathrm{Im}(\omega_0) > 0$ [and $\mathrm{Re}(\omega_0) = 0$] from some value of central energy density ϵ_c, then all the higher modes will become complex too, representing the onset of the instability. Our calculations show that for $X = 3$ the amplitudes of the stellar configurations increase even in the region where charm stars are expected, thus making them dynamically unstable. Since the same behavior was obtained for larger values of X, one can conclude from a perturbative QCD analysis that charm stars are unstable. In this work, we have extended the pQCD $N_f = N_l + 1$ formalism to include heavy quarks in the EoS for cold and dense quark matter. Their effects at low densities are non trivial through the renormalization scale parameter $\bar{\Lambda}$. After performing a dynamical analysis, it was found that charm stars are unstable. In the future, it would be interesting to study the presence of charm quarks in neutron-star mergers.

Acknowledgement

This work was partially supported by INCT-FNA (Process No. 464898/2014-5). J. C. J. acknowledges support from FAPESP (Processes No. 2020/07791-7 and No. 2018/24720-6). E. S. F. is partially supported by CAPES (Finance Code 001), CNPq, and FAPERJ.

References

[1] J. M. Lattimer and M. Prakash, arXiv:1012.3208.
[2] J. M. Lattimer, Universe **5**, 159 (2019).
[3] J. I. Kapusta and C. Gale, *Finite-Temperature Field Theory: Principles and Applications* (Cambridge University Press, Cambridge, England, 2006).
[4] M. Laine and Y. Schroder, Phys. Rev. D **73**, 085009 (2006).
[5] S. Borsanyi *et al.*, Nature (London) **539**, 69 (2016).
[6] A. Kurkela, P. Romatschke, and A. Vuorinen, Phys. Rev. D **81**, 105021 (2010).
[7] C. Kettner *et al.*, Phys. Rev. D **51**, 1440 (1995).
[8] D. Gondek, P. Haensel and J. L. Zdunik, Astron. Astrophys. **325**, 217 (1997)
[9] N. K. Glendenning, *Compact Stars* (Springer, New York, 2000).
[10] J. A. M. Vermaseren, S. A. Larin, and T. van Ritbergen, Phys. Lett. B **405**, 327 (1997).
[11] A. Bazavov *et al.*, Phys. Rev. D **90**, 074038 (2014).
[12] B. Chakraborty *et al.*, Phys. Rev. D **91**, 054508 (2015).
[13] J. M. Lattimer and M. Prakash, Phys. Rep. **442**, 109 (2007).
[14] J. C. Jiménez and E. S. Fraga, Phys. Rev. D **102**, 034015 (2020)

EPJ Web of Conferences **259**, 04003 (2022) https://doi.org/10.1051/epjconf/202225904003
SQM 2021

From lattice QCD to in-medium heavy-quark interactions via deep learning

Shuzhe Shi[1,*], *Kai* Zhou[2], *Jiaxing* Zhao[3], *Swagato* Mukherjee[4], and *Pengfei* Zhuang[3]

[1]Department of Physics, McGill University, Montreal, Quebec H3A 2T8, Canada.
[2]Frankfurt Institute for Advanced Studies, Ruth Moufang Strasse 1, D-60438, Frankfurt am Main, Germany.
[3]Department of Physics, Tsinghua University, Beijing 100084, China.
[4]Physics Department, Brookhaven National Laboratory, Upton, New York 11973, USA.

Abstract. Bottomonium states are key probes for experimental studies of the quark-gluon plasma (QGP) created in high-energy nuclear collisions. Theoretical models of bottomonium productions in high-energy nuclear collisions rely on the in-medium interactions between the bottom and antibottom quarks, which can be characterized by real ($V_R(T,r)$) and imaginary ($V_I(T,r)$) potentials, as functions of temperature and spatial separation. Recently, the masses and thermal widths of up to $3S$ and $2P$ bottomonium states in QGP were calculated using lattice quantum chromodynamics (LQCD). Starting from these LQCD results and through a novel application of deep neural network (DNN), here, we obtain model-independent results for $V_R(T,r)$ and $V_I(T,r)$. The temperature dependence of $V_R(T,r)$ was found to be very mild between $T \approx 0-330$ MeV. Meanwhile, $V_I(T,r)$ shows rapid increase with T and r, which is much larger than the perturbation theory based expectations.

In-medium modifications of quarkonium states are sensitive probes of the QGP produced in high energy nuclear collisions [1–13]. Sequential suppression patterns among the $\Upsilon(1S)$, $\Upsilon(2S)$ and $\Upsilon(3S)$ states have been observed in heavy-ion collision experiments [14–17]. Theoretical understanding of these experimental observations rely on effective field theories (EFT), which naturally lead to an open quantum system based treatment of both open and hidden bottom states in QGP (for a recent review, see [12]). If interactions between the color-singlet and color-octet states are neglected then the pNRQCD reduces to a theoretical description of quarkonia solely based on a potential between the heavy quark and antiquark. A potential based description allows studies of quarkonia by employing Schrödinger-type equations [18–21]. One-loop hard thermal loop (HTL) perturbative QCD calculations [22, 23], and later on pNRQCD calculations [24, 25], show that at high temperatures heavy quark potential becomes complex with a nonvanishing imaginary part. However, it is difficult to provide satisfactory descriptions of bound states arising out of strong interactions solely using perturbative expansions and a nonperturbative treatment, such as the LQCD, is called for. In the static limit, the heavy quark potential can be extracted from the spectral functions of the thermal Wilson loop using nonperturbative LQCD calculations [26–29]. To extract the heavy quark potential from the Euclidean-time Wilson loop spectral functions, these LQCD calculations rely on weak-coupling motivated functional forms of the spectral functions. On the

*e-mail: shuzhe.shi@mcgill.ca

other hand, recent LQCD studies have led to quantification of the masses, thermal widths, and Bethe–Salpeter amplitudes (BSA) of up to $3S$ and $2P$ bottomonium states in QGP [30–32]. As we shall see later, one-loop HTL motivated functional forms of $V_R(T, r)$ and $V_I(T, r)$ are not compatible with the recent LQCD results. This observation calls for a model-independent nonperturbative extraction of the in-medium heavy quark potential.

In our recent work [33], we have developed a model-independent DNN-based method and determine the r and T-dependence of the in-medium heavy quark potential starting from the LQCD results [31] for the masses and thermal widths of up to $3S$ and $2P$ bottomonium states at various temperatures. The underlying idea is as follows: At a fixed T, various bottomonium states differ in sizes and their wavefunctions probe different distances. Knowledge of the masses and thermal widths of multiple bottomonium states, thereby, provide constraints on not only the strength of the real and imaginary parts of the bottom-antibottom interactions in QGP but also its r-dependence. Thus, LQCD results for the masses and thermal widths of multiple bottomonium states at different temperature can be used to extract $V_R(T, r)$ and $V_I(T, r)$ and, presently, DNN is probably the best tool achieve this in an unbiased fashion. According to the universal approximation theorem [34, 35], DNN can generally provide an unbiased, yet flexible enough, parameterization to approximate arbitrary functional relations. We exploited the DNNs to represent the real and imaginary potentials,

$$V_R(T, r) = V_{R,\mathrm{DNN}}(\boldsymbol{b}_R, \boldsymbol{W}_R; T, r), \qquad V_I(T, r) = V_{I,\mathrm{DNN}}(\boldsymbol{b}_I, \boldsymbol{W}_I; T, r), \tag{1}$$

where $\boldsymbol{b}$ and $\boldsymbol{W}$ — called *bias* and *weight*, respectively — are the DNN parameters to be determined by fitting the LQCD masses and thermal widths [31]. In this Proceeding, we will briefly summarize our method and results. We will further discuss the possible interpretation and phenomenological consequence, as the complement of [33].

In typical machine-learning problems, one usually knows some direct informations of the target function — *e.g.*, its value at discrete points — and is able to tune the DNN parameters by directly compare its output to the known knowledge. In our network, however, we cannot compare Eq. (1) to the LQCD masses and thermal widths and fix the parameters. There is an extra step to map the DNN output (potential) to observables (masses and thermal widths) — the reduced complex-valued two-body time-independent Schrödinger equation,[1]

$$-\frac{\nabla^2}{m_b}\psi_n + \left[V_R(T, r) + i \cdot V_I(T, r)\right]\psi_n = E_n\psi_n\,, \tag{2}$$

where the wavefunction ψ_n and the energy eigenvalues E_n for in-medium bottomonia are complex-valued. Here, $V_I(T = 0, r) = 0$, $\mathrm{Re}[E_n] = m_n - 2m_b$ and $\mathrm{Im}[E_n] = -\Gamma_n$, where m_n and Γ_n are the mass and thermal width of the nth bottomonium state, respectively. We solved Eq. (2) with potential provided by DNN (1) and obtain the masses and thermal widths. The DNN parameters ($b_i^{(l)}$ and $W_{ij}^{(l)}$) are tuned by minimizing the cost function

$$J = \frac{\lambda}{2}\sum_{l,i}\left(b_i^{(l)}\right)^2 + \frac{\lambda}{2}\sum_{l,i,j}\left(W_{ij}^{(l)}\right)^2 + \frac{1}{2}\sum_{T,n}\left(\frac{m_{T,n} - m_{T,n}^{\mathrm{LQCD}}}{\delta m_{T,n}^{\mathrm{LQCD}}}\right)^2 + \left(\frac{\Gamma_{T,n} - \Gamma_{T,n}^{\mathrm{LQCD}}}{\delta\Gamma_{T,n}^{\mathrm{LQCD}}}\right)^2, \tag{3}$$

Here, the $\propto \lambda$ terms are regularizers in DNN to avoid over-fitting. The summation runs over six temperature points, $T \in \{0, 151, 173, 199, 251, 334\}$ MeV, and five bottomonium states, $n \in \{1S, 2S, 3S, 1P, 2P\}$ and the LQCD values were taken from Ref. [31]. We used gradient descent with Back-Propagation optimization technique, which is based on the derivatives of

[1]By taking the Cornell potential, we can well reproduce the vacuum masses of up to $3S$ and $2P$ bottomonium states [36] and the corresponding vacuum BSA obtained from LQCD calculation [32], (see [33] for more details).

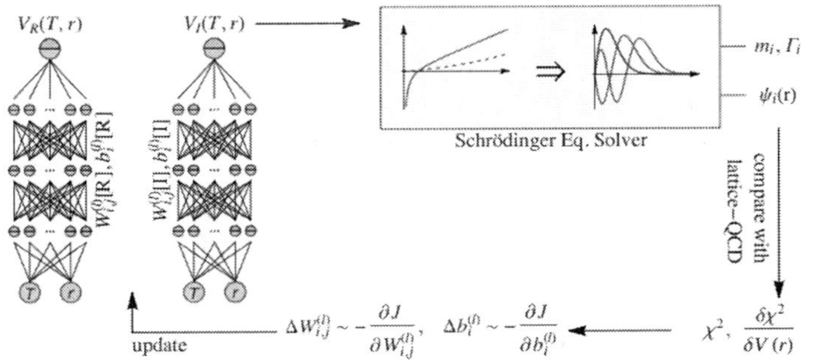

Figure 1. Flow chart of the potential reconstruction scheme — using generalized back-propagation to optimize parameters in the deep neural networks coupled with a Schrödinger equation.

the cost function with respect to the network parameters. We overcame the challenge of gradients evaluation of such implicit functions through perturbative solution of the Schrödinger equation with respect to small change of $V(T, r)$. Moreover, we invoked Bayesian inference for uncertainty quantification, whereby the posterior distribution of the network parameters was evaluated. In Fig. 1 we show the flow chart of our methodology of the potential reconstruction with DNNs coupled to a Schrödinger equation. To the best of our knowledge, the current method is developed for the first time here. More details on the method is provided in [33], along with a closure test to justify our methodology and assess its reliability.

Results and Discussions — We begin with pointing out the inadequacy of weak-coupling motivated functional form of the potential to consistently describe the LQCD masses and thermal widths. We chose the functional form proposed in Ref. [37], which incorporates one-loop HTL based functional forms of V_I and of color-electric screening, in addition to a vacuum potential satisfying Gauss's law. Taking this functional form for the potential, we fix α, σ, and B by their vacuum values, and tune μ_D at different temperatures to fit the finite-temperature bottomonia masses and widths. As shown by the open squares in Fig. 2, one-loop HTL motivated functional form of V_I and color-electric screening in V_R fail to simultaneously reproduce the LQCD results for the mass shifts and the thermal widths of bottomonium. Even if allowing an extra magnification factor for V_I, one would still miss the state-dependence of the thermal width (see open diamonds which take magnification factor to be four). The failure of the only known analytic form to describe the LQCD results necessitates a model-independent extraction of $V(T, r)$ using an adequate unbiased parameterization. We devised

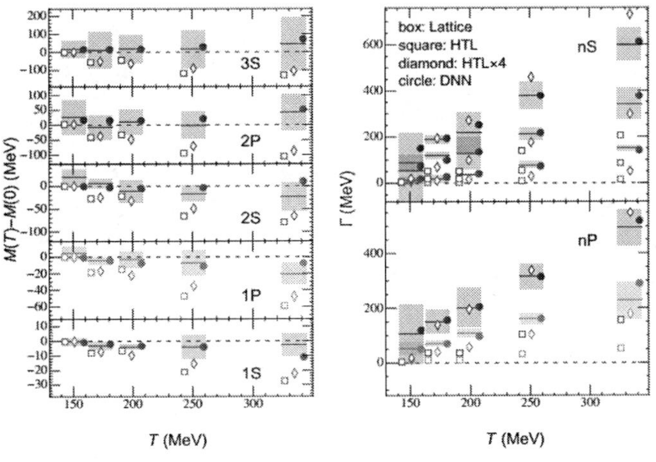

Figure 2. In-medium mass shifts (left) and the thermal widths (right) of different bottomonium states obtained from fits to LQCD results [31] (lines and shaded bands) using HTL functional forms [37] (open symbols) and DNN based optimization (filled circle). The points are shifted horizontally for better visualization. $\Upsilon(1S)$, $\chi_{b_0}(1P)$, $\Upsilon(2S)$, $\chi_{b_0}(2P)$ and $\Upsilon(3S)$ states are represented by red, orange, green, blue, and purple symbols, respectively.

the above outlined method by coupling Schrödinger equation with DNNs and achieved good agreement with the LQCD results [31] (see solid symbols in Fig. 2).

The T- and r-dependence of the extracted real and imaginary potentials are shown in Fig. 3. We see signs that with increasing temperature $V_R(T,r)$ becomes flatter at large r, as expected from color screening effect. However, the temperature dependence of $V_R(T,r)$ is very mild between $T \approx 150-330$ MeV, and closely approximating its vacuum counterpart. In the same temperature range, we observed that $V_I(T,r)$ monotonically increases both with temperature and distance. Whereas V_I varies a lot as temperature changes, the scaled imaginary potential, V_I/T as function of $r \times T$, is insensitive to the change of temperature, (see Right panel of Fig. 3).

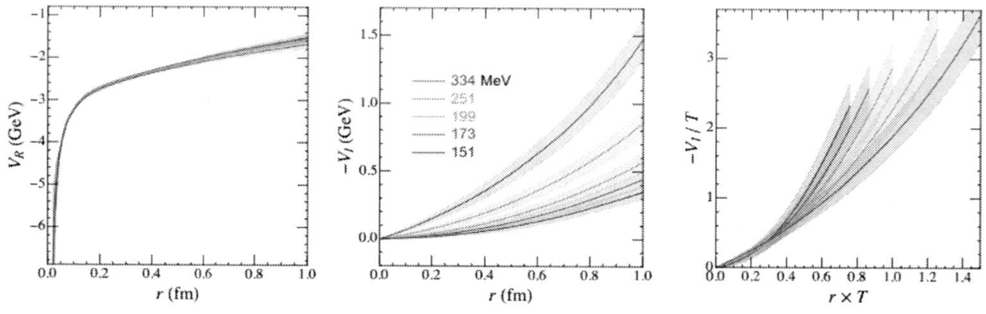

Figure 3. The DNN reconstructed real (Left) and imaginary (Middle) parts of the heavy quark potential at temperatures T = 151(purple), 173(blue), 199(green), 251(orange), and 334 MeV(red). The uncertainty bands represent the 68%(1σ) confident region. Right panel is the same as Middle, but for x- and y-axis scaled by temperature T.

The heavy quark potential obtained here is based on LQCD calculations of bottomonium state using 2+1 flavor dynamical gauge field background with nearly physical values of up, down, and strange quark masses. Whereas the extant LQCD calculations of V_R and V_I are for infinitely heavy static quarks and mostly in pure SU(3) gauge theory without dynamical quarks [26–29]. Direct quantitative comparison, thus, is difficult. However, our results for the heavy quark potential is qualitatively different from the weak-coupling motivated extractions of the static quark potentials [26–29]. Unlike the previous studies, the V_R obtained in this work show very little signs of color-electric Debye screening for $r \lesssim 1$ fm for the entire temperature range $T \in [0, 334]$ MeV. The V_I here is much larger in magnitude and increases more rapidly both with T and r than the one-loop HTL motivated extractions. On the other hand, it is reassuring that the preliminary results on the static quark potential from very recent LQCD calculations and without using the one-loop HTL motivated forms are quite similar to the potential obtained here [38, 39].

Based on our results, one might speculate that, for phenomenologically relevant temperatures $T \lesssim 330$ MeV, bottomonia are approximately vacuum-like excitation but very short lifetime that are inversely proportional to their large thermal widths. This is different from the conventional picture of the bottomonium in-medium behavior. In the traditional picture, both the finite temperature free-energy and interaction potential becomes flat at large distance. As the temperature increases, the platform decreases in height, thus the binding energy of the boundstate decreases, while its average size increase. At a certain temperature, T_m, the binding energy approaches to zero, and the average size diverges [18, 19, 40]. T_m is referred to as the melting temperature, above which a boundstate no longer exists. Similar behaviors are observed for potentials with non-vanishing imaginary part [22, 23, 37, 41, 42]. According to the recent lattice QCD calculation for static quark potential [37], the $\Upsilon(3S)$ state melts at $T \sim 0.2$ GeV, while $\Upsilon(2S)$ disappears at $T \gtrsim 0.35$ GeV. In our new picture, on the contrary, all the five bottomonium states can survive up to $T = 334$ MeV, and their masses are insensitive to the temperature. On the other hand, the imaginary potential induced thermal widths, which characterize the dynamical dissociation and increase with temperature, are significantly larger than previous results [37]. The inverse of thermal width can be interpreted as lifetime of state at finite temperature. For instance, at $T = 334$ MeV, we find $\Gamma(1S) \sim 0.15$ GeV, $\Gamma(2S) \sim 0.3$ GeV, and $\Gamma(3S) \sim 0.6$ GeV, correspond to lifetime 1.3 fm, 0.7 fm, and 0.3 fm, respectively. In such a new continuous dissociation picture, the enhancement of the thermal width compensates the vanishing of the melting effect in the thermal dissociation.

This is important for studying the Quarkonium dynamics in the rapidly expanding QGP in heavy ion collisions. The suppression of the Quarkonia production rate in relativistic heavy-ion collisions is caused by both the color-screening melting and the dynamical dissociation, the new picture suggests a reduction of the former and an enlargement of the latter. Noting that such two effects exhibit different temperature dependencies, phenomenological consequence might be expected in the experimental measurements of azimuthal angle distribution of the production rate, especially for the excitations.

Conclusion — In this work, a bias-free in-medium heavy quark potential with DNN representation is determined from the recently obtained LQCD results [31] for the masses and thermal widths of up to $3S$ and $2P$ bottomonium states in QGP. By coupling Schrödinger equation to DNN, we introduced a novel method for unbiased extractions of the real and imaginary parts of the heavy quark potential, and invoked Bayesian inference to quantify the potential uncertainties in a non-local fashion. We obtained model-independent results for $V_R(T,r)$ and $V_I(T,r)$ for $r \lesssim 1$ fm and $T \lesssim 330$ MeV. The $V_R(T,r)$ has very mild T dependence and closely resembles the vacuum potential. On the other hand, $V_I(T,r)$ is large and rises rapidly with T and r. These results are qualitatively different from the static quark potential obtained using one-loop HTL perturbative calculations. It would be very interesting to see the phenomenological consequences [13] of this heavy quark potential, model-independently extracted from the non-perturbative LQCD calculations.

The authors thank Min He, Matthew Heffernan, Rasmus Larsen, Simon Mak, Peter Petreczky, Ralf Rapp, Alexander Rothkopf, and Nan Su for many insightful discussions. This material is based upon work supported by: (i) The NSFC under grant Nos. 11890712 and 12075129 and Guangdong Major Project of Basic and Applied Basic Research No. 2020B0301030008 (J.Z. and P.Z.); (ii) The Natural Sciences and Engineering Research Council of Canada (S.S.); (iii) The Fonds de recherche du Québec - Nature et technologies (FRQNT) through the Programmede Bourses d'Excellencepour Étudiants Étrangers (PBEEE) scholarship (S.S.); (iv) The BMBF funding under the ErUM-Data project and the AI grant at FIAS of SAMSON AG, Frankfurt (K.Z.); (v) The GPU Grant of the NVIDIA Corporation (K.Z.); (vi) The U.S. Department of Energy, Office of Science, Office of Nuclear Physics through the Contract No. DE-SC0012704 (S.M.); (vii) The U.S. Department of Energy, Office of Science, Office of Nuclear Physics and Office of Advanced Scientific Computing Research, within the framework of Scientific Discovery through Advance Computing (SciDAC) award Computing the Properties of Matter with Leadership Computing Resources (S.M.).

References

[1] T. Matsui, H. Satz, Phys. Lett. B **178**, 416 (1986)
[2] F. Karsch, M.T. Mehr, H. Satz, Z. Phys. C **37**, 617 (1988)
[3] J.P. Blaizot, J.Y. Ollitrault, Phys. Rev. Lett. **77**, 1703 (1996)
[4] P. Braun-Munzinger, J. Stachel, Phys. Lett. B **490**, 196 (2000)
[5] S. Digal, P. Petreczky, H. Satz, Phys. Rev. D **64**, 094015 (2001)
[6] L. Grandchamp, R. Rapp, G.E. Brown, Phys. Rev. Lett. **92**, 212301 (2004)
[7] T. Song, K.C. Han, C.M. Ko, Phys. Rev. C **84**, 034907 (2011)
[8] X. Du, R. Rapp, Nucl. Phys. A **943**, 147 (2015)
[9] Y. Liu, B. Chen, N. Xu, P. Zhuang, Phys. Lett. B **697**, 32 (2011)
[10] K. Zhou, N. Xu, Z. Xu, P. Zhuang, Phys. Rev. C **89**, 054911 (2014)
[11] R. Katz, P.B. Gossiaux, Annals Phys. **368**, 267 (2016)
[12] X. Yao, W. Ke, Y. Xu, S.A. Bass, B. Müller, JHEP **21**, 046 (2020)
[13] A. Islam, M. Strickland, Phys. Lett. B **811**, 135949 (2020)
[14] S. Chatrchyan et al. (CMS), Phys. Rev. Lett. **107**, 052302 (2011)
[15] S. Chatrchyan et al. (CMS), Phys. Rev. Lett. **109**, 222301 (2012), [Erratum: Phys.Rev.Lett. 120, 199903 (2018)]
[16] V. Khachatryan et al. (CMS), Phys. Lett. B **770**, 357 (2017)
[17] A.M. Sirunyan et al. (CMS), Phys. Rev. Lett. **120**, 142301 (2018)
[18] H. Satz, J. Phys. G **32**, R25 (2006)
[19] J. Zhao, K. Zhou, S. Chen, P. Zhuang, Prog. Part. Nucl. Phys. **114**, 103801 (2020)
[20] H.W. Crater, J.H. Yoon, C.Y. Wong, Phys. Rev. D **79**, 034011 (2009)
[21] X. Guo, S. Shi, P. Zhuang, Phys. Lett. B **718**, 143 (2012)
[22] M. Laine, O. Philipsen, P. Romatschke, M. Tassler, JHEP **03**, 054 (2007)
[23] A. Beraudo, J.P. Blaizot, C. Ratti, Nucl. Phys. A **806**, 312 (2008)
[24] N. Brambilla, J. Ghiglieri, A. Vairo, P. Petreczky, Phys. Rev. D **78**, 014017 (2008)
[25] N. Brambilla, M.A. Escobedo, J. Ghiglieri, J. Soto, A. Vairo, JHEP **09**, 038 (2010)
[26] A. Rothkopf, T. Hatsuda, S. Sasaki, Phys. Rev. Lett. **108**, 162001 (2012)
[27] Y. Burnier, O. Kaczmarek, A. Rothkopf, Phys. Rev. Lett. **114**, 082001 (2015)
[28] Y. Burnier, O. Kaczmarek, A. Rothkopf, JHEP **12**, 101 (2015)
[29] D. Bala, S. Datta, Phys. Rev. D **101**, 034507 (2020)
[30] R. Larsen, S. Meinel, S. Mukherjee, P. Petreczky, Phys. Rev. D **100**, 074506 (2019)
[31] R. Larsen, S. Meinel, S. Mukherjee, P. Petreczky, Phys. Lett. B **800**, 135119 (2020)
[32] R. Larsen, S. Meinel, S. Mukherjee, P. Petreczky, Phys. Rev. D **102**, 114508 (2020)
[33] S. Shi, K. Zhou, J. Zhao, S. Mukherjee, P. Zhuang (2021), 2105.07862
[34] M. Leshno, S. Schocken, Neural Networks **6**, 861 (1993)
[35] A. Kratsios, Annals of Mathematics and Artificial Intelligence **89**, 435 (2021)
[36] M. Tanabashi et al. (Particle Data Group), Phys. Rev. D **98**, 030001 (2018)
[37] D. Lafferty, A. Rothkopf, Phys. Rev. D **101**, 056010 (2020)
[38] J.H. Weber, *Complex static potential in lattice qcd* (2021), https://indico.cern.ch/event/838970/contributions/4258563/
[39] R. Larsen, *Excited states of bottomonia in qgp from lattice qcd* (2021), https://indico.cern.ch/event/985652/contributions/4302187/
[40] A. Mocsy, P. Petreczky, Phys. Rev. Lett. **99**, 211602 (2007)
[41] M. Ali-Akbari, D. Giataganas, Z. Rezaei, Phys. Rev. D **90**, 086001 (2014)
[42] L. Thakur, U. Kakade, B.K. Patra, Phys. Rev. D **89**, 094020 (2014)

Strangeness in astrophysics: Theoretical developments

Veronica Dexheimer[1,*] and *Krishna* Aryal[1,**]

[1]Department of Physics, Kent State University,Kent, OH 44242

Abstract. In this conference proceeding, we review important theoretical developments related to the production of strangeness in astrophysics. This includes its effects in supernova explosions, neutron stars, and compact-star mergers. We also discuss in detail how the presence of net strangeness affects the deconfinement to quark matter, expected to take place at large densities and/or temperatures. We conclude that a complete description of dense matter containing hyperons and strange quarks is fundamental for the understanding of modern high-energy astrophysics.

1 Historical Background

The history of strangeness in dense matter astrophysics started in the early 80's with a paper from Glendenning describing the role played by hyperons in neutron stars [1]. In this seminal work, it is discussed how as "the density of matter rises, the Fermi energy of the neutrons will eventually exceed the threshold for decay, first to protons and then to the hyperons" and that these decays are not forbidden, since astrophysical timescales are orders of magnitude larger than the one characteristic of the weak interaction. The baryon octet is described by a modified Wallecka model and strange particle populations are shown to occupy most of the neutron star radii for massive stars, which could in this approach reach 1.8 $M_\odot$. A few years latter, a very famous paper was written by Witten describing the possibility of stable strange quark stars [2]. In this work, the quarks are described as a free gas of massless quarks stabilized by a vacuum pressure, which is introduced to take into account the non-perturbative effects of QCD. Under an adequate choice of choice bag constant, 2 solar mass stars could easily be reproduced. In the 90's, the possibility of having kaons (and their further condensation) in neutron stars were also studied [3]. It was found that, it is hard to support anything larger than 1.44 $M_\odot$ neutron stars in the presence of a large strangeness content.

The numbers stated above immediately point out to what later became know as the hyperon puzzle [4]. This relates to the fact that the softening caused by the new strange degrees of freedom turns the equation of state softer and reproduces lower neutron-star masses. Initial attempts to parametrize very stiff equations of state in order to compensate for the softening due to the strange degrees of freedom reproduced larger neutron-stars, in disagreement with laboratory and observational data. In reality, equations of state do not have to be overall soft or stiff, but can be a mixture of both by being fitted to different data at different densities. This can be constructed by tuning the strength of the attractive and repulsive interactions, and also exploring the nature of different couplings [5, 6]. In addition, it was shown that including not

*e-mail: vdexheim@kent.edu
**e-mail: karyal@kent.edu

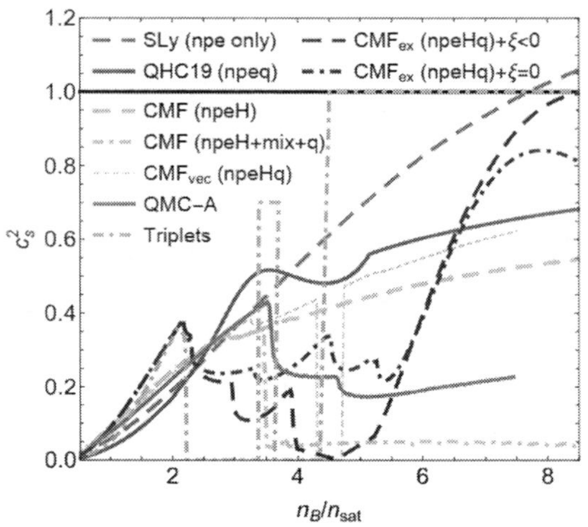

Figure 1: Speed of sound squared for the following models as a function of baryon number density in nuclear saturation units: SLy [20], QHC19 [10], CMF [6, 14–16], QMC-A EoS [11, 12], and Triplets [13]. Figure modified from Ref. [9].

only hyperons, but also their respective negative parity states could still be consistent 2 $M_\odot$ neutron stars (and other laboratory and observational data), if a crossover transition to quark matter is allowed [7, 8].

2 Different Scenarios

The aforementioned softening of the equation of state caused by the appearance of strange hadronic and quark degrees of freedom can be seen for a few different models in Fig. 1, which was modified from Ref. [9]. Note that equations of state constructed from state of-the-art models for neutron stars generically lead to characteristic non-trivial, non-monotonic structure in the speed of sound [6, 10–16]. In the particular case of Ref. [16], a phase transition associated with the amount of strangeness was found to be associated with the appearance of twin stars. But is is very important to keep in mind that observational constraints for neutron stars can still be fulfilled by models that present "bumps" in this speed of sound. This is illustrated in Fig. 2, where parametric equations of state [17] produce mass-radius diagrams and tidal deformability-mass diagrams showing curves that are in agreement with NICER and LIGO/VIRGO data. Furthermore, bumps that present an early raise in the speed of sound (before 2 times saturation density) can generate massive stars beyond 2.5 $M_\odot$, which points out to the possibility of the secondary object observed gravitationally in the merger GW190814 [18] being a neutron star (and not a black hole). Other approaches have also been able to describe the secondary object in GW190814, in spite of its strange content (see for example Refs. [6, 19]).

In supernova explosions, the presence of s-quarks were shown to turn non-explosive massive-star 3-dimensional simulations into successful ones, as strangeness affects neutrino interactions, reducing neutrino opacity, and generating higher luminosities and mean energies of neutrino emission [21]. Hyperons were also shown to affect both the density and temperature profiles inside massive exploding stars [22]. In neutron-star mergers, it was shown that similar changes in neutrino luminosity are not expected to be distinguishable before high temperature regions are swallowed into the black hole, but the changes in gravitational wave

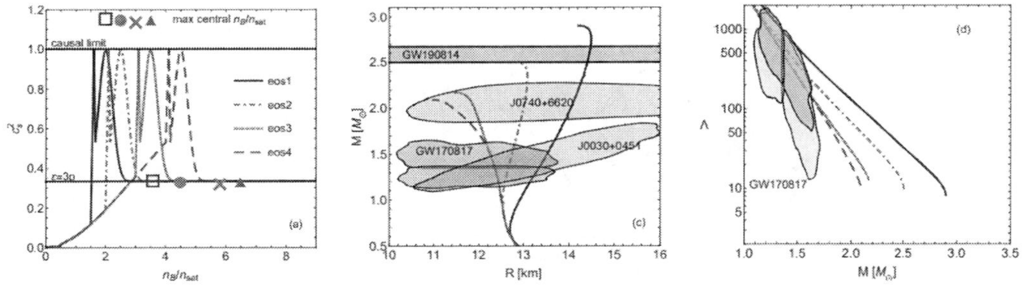

Figure 2: Speed of sound, mass-radius diagram, and tidal deformability for a family of equations of state with double peaks in the speed of sound at different locations. Symbols show the central density for the most massive star of the sequence. Figure modified from Ref. [9]

frequency and amplitude before and after merger could signal hyperons [23], that could lead to a faster collapse of hypermassive stars into black holes [24]. The densities and temperatures found in mergers when allowing for exotic (hadronic and quark) degrees of freedom [25] are actually ideal conditions to generate hadronic and quark strangeness, as shown in Fig. 6 of Ref. [26].

The final scenario we describe is the matter generated in particle collisions. Although net strangeness (difference between content of strange particles and strange anti-particles) is not produced in the time frame involved in these collisions, strange and anti strange particles are produced and measured. For example, recent results from ALICE experiment in CERN give very good estimates for the Λ, Σ, and Ξ hyperon potentials in medium [27]. Using these results, it was shown that that neutron stars with masses larger than 2.1 $M_\odot$ can still be reproduced when the isovector coupling is rescaled [28, 29].

Concerning deconfinement to (strange) quark matter, the presence of net strangeness can affect the position of the phase transition, as shown in the left panel of Fig. 3. A calculation using the Chiral Mean Field (CMF) [15, 30–32] model showed that the effect of net strangeness can affect the position of deconfinement (with respect of free energy per baryon $\widetilde{\mu} = \mu_B + Y_Q\mu_Q + Y_S\mu_S$ or baryon chemical potential μ_B) by more than 40 MeV, depending on the temperature T and charge fraction Y_Q (or isospin fraction Y_I), being larger for larger temperatures. Note that the free energy per baryon is only equal to the baryon chemical potential when either the charge fraction or charge chemical potential μ_Q and either the strangeness fraction Y_S or chemical potential μ_S are zero. For zero and low temperature, the effect of strangeness is larger for lower charge (or more negative isospin fraction), corresponding to conditions achieved in chemically equilibrated neutrons stars or their mergers. Larger charge or zero/slightly positive isospin fraction are achieved in situations where beta-equilibrium hasn't taken the system away from a more energy favorable (under normal conditions of density and temperature) isospin symmetric case, corresponding to the case of particle collisions or supernova explosions. Note that the relation between charge and isospin chemical potential is not trivial (meaning always differing by exactly 0.5) in the presence of strangeness, as seen in the deconfinement curves shown in the right panel of Fig. 3 for the zero temperature case. The relation between isospin and charge fractions depends on the strangeness fraction, which again is larger for lower charge/isospin fractions $Y_I = Y_Q - \frac{1}{2} + \frac{1}{2}Y_S$.

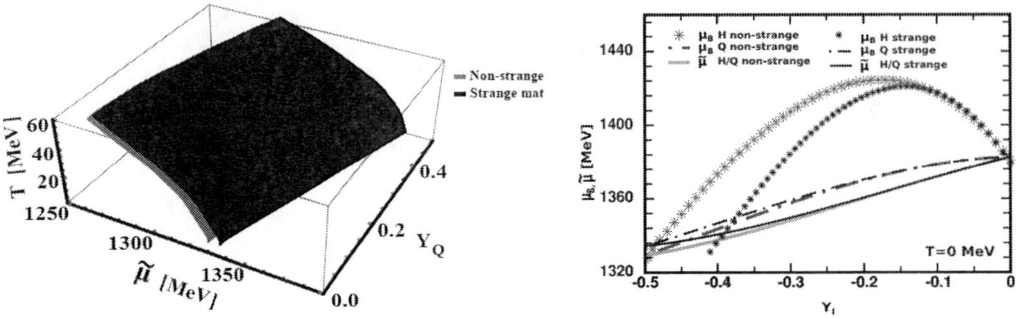

Figure 3: Phase diagrams for deconfinement to quark matter calculated from the CMF model for non-strange (color), and strange matter (black) calculated for charge fraction varying from zero to $Y_Q = 0.5$. Left panel shows temperature (only until 60 MeV), free energy per baryon $\tilde{\mu}$, and charge fraction Y_Q. Right panel shows free energy per baryon $\tilde{\mu}$, baryon chemical potential μ_B (in either side of the phase transition), and isospin fraction Y_I. Figure modified from Ref. [30].

2.1 Conclusions

Give our current understanding of astrophysical environments, strangeness is not ruled out in dense matter and it is expected to appear in significant quantities already in supernova explosions, remaining present in cold catalyzed neutron stars, and generated in large amounts in hot neutron-star mergers. In this way, strangeness is expected to be important for several dynamical processes and could generate neutrinos, electromagnetic and gravitational signals expected to be measured in the near future.

Although comparisons with particle collisions could give us insight into high-energy matter, the conditions produced in this case are quite different from astrophysical scenarios, even considering that low-energy collisions overlap in temperature with neutron star mergers. These differences include the amount of charge/isospin and strangeness, both of which affect the deconfinement to quark matter. In this sense, deconfinement signals from neutron-star mergers could give us additional information about the amount of strangeness inside compact stars.

Acknowledgements

Support for this research comes from the National Science Foundation under grant PHY-1748621 and PHAROS (COST Action CA16214).

References

[1] N.K. Glendenning, Physics Letters B **114**, 392 (1982)
[2] E. Witten, Physical Review D **30**, 272 (1984)
[3] V. Thorsson, M. Prakash, J.M. Lattimer, Nuclear Physics A **572**, 693 (1994)
[4] N. Chamel, P. Haensel, J.L. Zdunik, A.F. Fantina, Int. J. Mod. Phys. E **22**, 1330018 (2013)

[5] V. Dexheimer, R. de Oliveira Gomes, S. Schramm, H. Pais, J. Phys. G **46**, 034002 (2019)
[6] V. Dexheimer, R.O. Gomes, T. Klähn, S. Han, M. Salinas, Phys. Rev. C **103**, 025808 (2021)
[7] A. Mukherjee, S. Schramm, J. Steinheimer, V. Dexheimer, Astron. Astrophys. **608**, A110 (2017)
[8] A. Motornenko, J. Steinheimer, V. Vovchenko, S. Schramm, H. Stoecker, Phys. Rev. C **101**, 034904 (2020)
[9] H. Tan, T. Dore, V. Dexheimer, J. Noronha-Hostler, N. Yunes (2021), `arXiv:2106.03890 [astro-ph.HE]`
[10] G. Baym, S. Furusawa, T. Hatsuda, T. Kojo, H. Togashi, shys. J. **885**, 42 (2019)
[11] P.A.M. Guichon, K. Saito, E.N. Rodionov, A.W. Thomas, Nucl. Phys. A **601**, 349 (1996)
[12] J.R. Stone, V. Dexheimer, P.A.M. Guichon, A.W. Thomas, S. Typel, Mon. Not. Roy. Astron. Soc. **502**, 3476 (2021)
[13] M.G. Alford, A. Sedrakian, Phys. Rev. Lett. **119**, 161104 (2017)
[14] V. Dexheimer, S. Schramm, Astrophys. J. **683**, 943 (2008)
[15] V.A. Dexheimer, S. Schramm, Phys. Rev. C **81**, 045201 (2010)
[16] V. Dexheimer, R. Negreiros, S. Schramm, Phys. Rev. C **91**, 055808 (2015)
[17] H. Tan, J. Noronha-Hostler, N. Yunes, Phys. Rev. Lett. **125**, 261104 (2020)
[18] R. Abbott et al. (LIGO Scientific, Virgo), Astrophys. J. Lett. **896**, L44 (2020)
[19] I.A. Rather, U. Rahaman, V. Dexheimer, A.A. Usmani, S.K. Patra, Astrophys. J. **917**, 46 (2021)
[20] E. Chabanat, P. Bonche, P. Haensel, J. Meyer, R. Schaeffer, Nucl. Phys. A **635**, 231 (1998), [Erratum: Nucl.Phys.A 643, 441–441 (1998)]
[21] T. Melson, H.T. Janka, R. Bollig, F. Hanke, A. Marek, B. Müller, The Astrophysical Journal **808**, L42 (2015)
[22] P. Char, S. Banik, D. Bandyopadhyay, The Astrophysical Journal **809**, 116 (2015)
[23] Y. Sekiguchi, K. Kiuchi, K. Kyutoku, M. Shibata, Phys. Rev. Lett. **107**, 211101 (2011)
[24] A. Perego, S. Bernuzzi, D. Radice, Eur. Phys. J. A **55**, 124 (2019), `1903.07898`
[25] E.R. Most, L.J. Papenfort, V. Dexheimer, M. Hanauske, S. Schramm, H. Stöcker, L. Rezzolla, Phys. Rev. Lett. **122**, 061101 (2019)
[26] E.R. Most, L. Jens Papenfort, V. Dexheimer, M. Hanauske, H. Stoecker, L. Rezzolla, Eur. Phys. J. A **56**, 59 (2020)
[27] L. Fabbietti, V.M. Sarti, O.V. Doce (2020), `arXiv:2012.09806 [nucl-ex]`
[28] S. Weissenborn, D. Chatterjee, J. Schaffner-Bielich, Phys. Rev. C **85**, 065802 (2012)
[29] N. Hornick, L. Tolos, A. Zacchi, J.E. Christian, J. Schaffner-Bielich, Phys. Rev. C **98**, 065804 (2018)
[30] K. Aryal, C. Constantinou, R.L.S. Farias, V. Dexheimer, Phys. Rev. D **102**, 076016 (2020)
[31] V. Dexheimer, K. Aryal, C. Constantinou, J. Peterson, R.L.S. Farias, J. Phys. Conf. Ser. **1602**, 012013 (2020)
[32] V. Dexheimer, K. Aryal, .M. Wolf, .C. Constantinou, R.L.S. Farias, Astron. Nachr. **342**, 347 (2021)

 https://doi.org/10.1051/epjconf/202225905002

Strange Baryons in Nuclei and Neutron Stars

Laura Tolos[1,2,3,4,*]

[1]Institute of Space Sciences, Campus UAB, Carrer de Can Magrans, 08193 Barcelona, Spain
[2]Institut d'Estudis Espacials de Catalunya, 08034 Barcelona, Spain
[3]Faculty of Science and Technology, University of Stavanger, 4036 Stavanger, Norway
[4]Frankfurt Institute for Advanced Studies, Ruth-Moufang-Str. 1, 60438 Frankfurt am Main, Germany

Abstract. In this contribution we review the dynamics of hyperons with nucleons and nuclear matter, paying a special attention to hypernuclei. We also discuss the presence of hyperons in the inner core of neutron stars and the consequences for the structure of these compact stars.

1 Introduction

The dynamics of hyperons with nucleons and nuclear matter has been object of high interest in connection with the formation of hypernuclei as well as the probable existence of hyperons in neutron stars [1]. In order to understand this dynamics, first we need to address the features of the hyperon-nucleon (YN) and hyperon-hyperon (YY) interactions, so as to comprehend how few-body systems such as hypernuclei are formed and how hyperons interact with other nucleons and hyperons, in particular in the core of neutron stars.

In this contribution we briefly review the status of YN and YY interactions, paying a special attention to hypernuclei. We then address hyperons in dense nuclear matter within the chiral effective theory (χEFT) framework. The final goal is to discuss the probable presence of hyperons in the core of neutron stars and the consequences for the structure of these compact objects. The present contribution represents a brief summary of the review of Ref. [1].

2 YN and YY interactions from a theoretical perspective

Over the years a lot of effort has been made from the theoretical point of view to describe the YN and YY interactions. The theoretical approaches can be grouped in meson-exchange models, χEFT approaches, computations on the lattice (LQCD), low-momentum schemes and quark-model potentials.

The interaction between two baryons in *meson-exchange models* is mediated by the exchange of mesons. Starting from the nucleon-nucleon (NN) meson-exchange model, $SU(3)_{\rm flavor}$ symmetry is assumed to determine the YN and YY interactions. Those schemes include the Jülich [2] and Nijmegen potentials [3].

Within chiral effective field theory, the Jülich-Bonn-Munich group has built the YN and YY interactions starting from their previous χEFT approach for the NN interaction [4–6].

*e-mail: tolos@ice.csic.es

As for LQCD, Monte Carlo techniques are used to solve the QCD path integral over the quark and gluon fields at each point of a four-dimensional space-time grid. This effort has been lead by the HALQCD [7] and the NPLQCD [8] collaborations.

And, finally, other schemes for the description of the YN and YY interactions include *low-momentum interactions* and *quark-model potentials*. Whereas the former aims at determining a universal effective low-momentum potential for YN and YY by means of renormalization-group methods [9], the latter constructs the YN and YY interactions within constituent quark models [10].

3 Hypernuclei

Hypernuclei are bound systems composed of neutron, protons, and one or more hyperons. Over the years their properties have been addressed experimentally and theoretically, so as to relate them with the underlying YN and YY interactions [11].

In particular, in the last years the determination of the hypertriton ($^3_\Lambda$H) lifetime has received a lot of attention. The separation energy of the Λ in this hypernucleus is about 130 keV, and this results in an RMS radius of 10.6 fm for the hypertriton. A very low binding energy suggests a small change of the Λ wave function in a nucleus and, hence, the lifetime of the hypertriton should be very close to that of the free Λ hyperon. However, several measurements reported a shorter lifetime for $^3_\Lambda$H and this gave rise to the so called *hypertriton lifetime puzzle*.

The recent measurement of the STAR collaboration [12] suggests a value of the binding energy about three times larger than the original benchmark [13]. However, measurements of the hypertriton lifetime by ALICE [14] give a small value of the hypertriton binding energy, indicating consistency with the free Λ lifetime.

4 Hyperons in dense matter within χEFT framework

During the past decade the YN interaction has been extensively studied within the χEFT. In this framework the YN interaction has been obtained within the chiral expansion, improving calculations systematically by moving to higher orders in the Weinberg power counting [4–6].

Using, as kernel, the LO and NLO contributions for YN, the properties of Λ and Σ in dense matter have been obtained within the Brueckner-Hartree-Fock framework [15, 16]. Whereas the Σ-nuclear potential is found to be repulsive, the Λ single-particle potential is in good qualitative agreement with the empirical values extracted from hypernuclear data, becoming repulsive about two to three-times saturation density. The inclusion of the three-body forces for the Λ-nuclear interaction in dense matter [17] gives an extra repulsion at large densities. As we will discuss in the next section, this might help to solve the so-called *hyperon puzzle*.

5 Hyperons inside Neutron Stars

Neutron stars are one of the most compact known stellar objects and, therefore, turn out to be a unique laboratory for testing dense matter [18, 19]. The mass and radius of these stellar objects depend on the properties of matter in their interior and, thus, are strongly correlated to the equation of state (EoS) of the core. Very accurate observations of $2M_\odot$ neutron stars have been determined [20–22], whereas both mass and radius for certain neutron stars have been obtained very recently by NICER (Neutron star Interior Composition ExploreR) [23–26].

The composition of the core of neutron stars is determined imposing equilibrium against weak interaction processes, the so-called β-stability. Whereas the core of neutron stars has been traditionally modelled by a uniform fluid of neutron rich matter in β-equilibrium, other degrees of freedom could be expected, such as hyperons. This is due to the fact that, on the one hand, the value of the nucleon chemical potential increases rapidly with density and, on the other hand, the density in the core of neutron stars is very high.

The probable appearance of hyperons in the core of neutron stars would affect the structure of these compact objects. Indeed, the EoS and, hence, the pressure becomes softer with respect to the case when only nucleons are present. The reason behind is that the addition of one specie opens a set of new available low-energy states that can be filled, hence lowering the total energy (and pressure) of the system. As the total pressure of the star is lowered, the gravitational pull is reduced in order to maintain the hydrostatic equilibrium. As a consequence, the less pressure, the less mass a neutron star needs to sustain, leading to masses for neutron stars below the $2M_\odot$ observations. This is usually referred as *the hyperon puzzle*.

Several solutions have been put forward so as to obtain $2M_\odot$ neutron stars with hyperons in the core. Those solutions move from stiffer YN and YY interactions, stiffening induced by hyperonic three-body forces, the appearance of Δ baryons, meson condensates or a phase transition to quark matter below the hyperon onset to the existence of dark matter in the core of neutron stars or the need of modified gravity theories for the description of neutron stars (see [1, 27] and references therein).

6 Summary

We have briefly reviewed the dynamics of hyperons with nucleons and nuclear matter, paying a special attention to hypernuclei and, in particular, to the hypertriton lifetime puzzle. We have also discussed the hyperon puzzle in the context of neutron stars, as hyperons would appear in the core of neutron stars, softening the equation of state and leading to neutron stars with masses below the $2M_\odot$ observations.

Acknowledgements

The author wishes to acknowledge support from the Spanish Ministerio de Ciencia e Innovación and the European Regional Development Fund (ERDF) under contract PID2019-110165GB-I00 (Ref.10.13039/501100011033); the Consejo Superior de Investigaciones Científicas under Project Ref. 202050I008; the Deutsche Forschungsgemeinschaft (DFG, German research Foundation) under Project Nr. 411563442 and Project Nr. 315477589 - TRR 211 (Strong-interaction matter under extreme conditions); the EU STRONG-2020 project under the program H2020-INFRAIA- 2018-1, grant agreement no. 824093; and the Generalitat Valenciana under contract PROMETEO/2020/023.

References

[1] L. Tolos, L. Fabbietti, Prog. Part. Nucl. Phys. **112**, 103770 (2020), 2002.09223

[2] J. Haidenbauer, U.G. Meissner, Phys. Rev. C **72**, 044005 (2005), nucl-th/0506019

[3] T.A. Rijken, M.M. Nagels, Y. Yamamoto, Prog. Theor. Phys. Suppl. **185**, 14 (2010)

[4] H. Polinder, J. Haidenbauer, U.G. Meissner, Nucl. Phys. A **779**, 244 (2006), nucl-th/0605050

[5] J. Haidenbauer, S. Petschauer, N. Kaiser, U.G. Meissner, A. Nogga, W. Weise, Nucl. Phys. A **915**, 24 (2013), 1304.5339

[6] J. Haidenbauer, U.G. Meißner, A. Nogga, Eur. Phys. J. A **56**, 91 (2020), 1906.11681
[7] https://inspirehep.net/experiments/1405371?ui-citation-summary=true (2021)
[8] https://www.ub.edu/nplqcd/index.html (2021)
[9] B.J. Schaefer, M. Wagner, J. Wambach, T.T.S. Kuo, G.E. Brown, Phys. Rev. C **73**, 011001 (2006), nucl-th/0506065
[10] Y. Fujiwara, Y. Suzuki, C. Nakamoto, Prog. Part. Nucl. Phys. **58**, 439 (2007), nucl-th/0607013
[11] A. Gal, E.V. Hungerford, D.J. Millener, Rev. Mod. Phys. **88**, 035004 (2016), 1605.00557
[12] J. Adam et al. (STAR), Nature Phys. **16**, 409 (2020), 1904.10520
[13] M. Juric et al., Nucl. Phys. B **52**, 1 (1973)
[14] S. Acharya et al. (ALICE), Phys. Lett. B **797**, 134905 (2019), 1907.06906
[15] J. Haidenbauer, U.G. Meißner, Nucl. Phys. A **936**, 29 (2015), 1411.3114
[16] S. Petschauer, J. Haidenbauer, N. Kaiser, U.G. Meißner, W. Weise, Eur. Phys. J. A **52**, 15 (2016), 1507.08808
[17] D. Gerstung, N. Kaiser, W. Weise, Eur. Phys. J. A **56**, 175 (2020), 2001.10563
[18] A.L. Watts et al., Rev. Mod. Phys. **88**, 021001 (2016), 1602.01081
[19] A.L. Watts et al., Sci. China Phys. Mech. Astron. **62**, 29503 (2019)
[20] P. Demorest, T. Pennucci, S. Ransom, M. Roberts, J. Hessels, Nature **467**, 1081 (2010), 1010.5788
[21] J. Antoniadis et al., Science **340**, 6131 (2013), 1304.6875
[22] H.T. Cromartie, E. Fonseca, S.M. Ransom, P.B. Demorest, Z. Arzoumanian, H. Blumer, P.R. Brook, M.E. DeCesar, T. Dolch, J.A. Ellis et al., Nature Astronomy **4**, 72 (2020), 1904.06759
[23] M.C. Miller et al., Astrophys. J. Lett. **887**, L24 (2019), 1912.05705
[24] T.E. Riley et al., Astrophys. J. Lett. **887**, L21 (2019), 1912.05702
[25] T.E. Riley et al. (2021), 2105.06980
[26] M.C. Miller et al. (2021), 2105.06979
[27] I. Vidaña, Proc. Roy. Soc. Lond. A **474**, 0145 (2018), 1803.00504

Hyperon-nucleon femtoscopy, nuclear production and bearing on astrophysics

Valentina Mantovani Sarti[1,*]

[1]Physik Department E62, Technische Universität München, James-Franck-Str. 1, 85748, Garching, Germany

Abstract. The presence of hyperons in the inner core of neutron stars is still an open issue in astrophysics and the so-called hyperon puzzle is far away from being solved. A complete understanding of how the interaction between hyperons and nucleons, at the level of two- and three-body, behaves in vacuum and at finite baryon density is a key requirement in order to build a realistic Equation of State for matter inside the core of neutron stars. Recently, measurements of hadron-hadron correlation functions with the femtoscopy technique in small colliding systems delivered a large amount of precise data on several hyperon-nucleon pairs. Several interactions relevant for the physics of neutron stars, such as pΛ and pΞ, have been compared to the femtoscopic data and the effects on the corresponding Equation of State have been investigated. The extension to the three-body forces with the measurement of correlations involving triplets and proton-deuteron pairs will provide for the first time access to the genuine three-body interaction and shed light on the hyperon-puzzle.

1 Introduction

The interaction of hyperons, hadrons with one of more strange quarks, with nucleons is one of the key ingredients needed to understand the composition of the most dense objects in our universe: neutron stars (NS) [1]. These kinds of stars are characterized by large masses ($M \approx 1.5-2.2 M_{\odot}$) and small radii ($R \approx 10-15$ km) [2–4], leading to the possibility to achieve densities in the core up to several times the nuclear saturation density $\rho_0 = 0.17\ fm^{-3}$. The measurements of masses and radii have been so far fundamental in the investigation and modeling of the Equation of State (EoS), namely on how nuclear matter behaves in the inner core of NS. Recently, the experimental database has been increased by the addition of precise radii measurements from the NICER Collaboration [1, 5] and by the determination of the tidal deformability extracted from gravitational waves events involving NS systems [6, 7]. The recent increase of experimental data on NS, in particular the improved precision in the measurements of the radius, opens a new chapter towards the understanding of the main constituents in the EoS and how these constituents interact in a dense environment as the core of NS.

This high density environment ($\rho \approx 4\rho_0$) supposed to occur in the interior of NS leads to an increase in the Fermi energy of the nucleons, translating into the appearance of new degrees of freedom such as hyperons [8]. The energetically favored production of strange hadrons

*e-mail: valentina.mantovani-sarti@tum.de

induces a softening of the EoS which cannot be reconciled with the current observations of heavy NS, beyond $2M_\odot$. The solution of this so-called hyperon puzzle is a precise understanding of the interaction between hyperons and nucleons, at the two-body (YN) and three-body (YNN) level. At the moment, however, experimental data on the genuine three-body interactions with hyperons are not yet present and the precision on the knowledge of YN interactions from scattering data and hypernuclei is mostly limited to the strangeness $|S| = 1$ sector [8, 9]. In order to build a realistic EoS, precise data on two- and three-body interactions with hyperons are needed. The measurements of correlations in the momentum space between hyperons and nucleons, performed in ultrarelativistic proton-proton (pp) collisions at the Large Hadron Collider (LHC) by the ALICE Collaboration, provided the most precise data for several YN interactions. In this contribution, the main experimental results on YN and YNN interactions, obtained by means of the femtoscopy technique, useful for the physics of NS and its EoS, will be presented.

2 Femtoscopy at the Large Hadron Collider

The main observable in femtoscopy is the correlation function $C(k^*)$. In Fig. 1, the theoretical definition given by the Koonin-Pratt formula [10] is shown in (a). One of the two main ingredients is the particle emitting source $S(r^*)$ (in orange in (c)) which depends on the relative distance r^* among particles. The pair of interest is composed of particles emitted from the source which are close in the relative momentum k^*, evaluated in the pair rest frame (see [11] for details). As the particles fly towards the detector they will likely undergo final-state interaction. This is embedded in the second ingredient needed to evaluate the theoretical $C(k^*)$: the modulus squared of the relative wavefunction $\psi(\mathbf{k}^*, \mathbf{r}^*)$ obtained from the solution of the Schrödinger equation [12] (in blue in (c)). Depending on the nature of the interaction, the correlation function has a specific shape, as seen in (b) and (d). In particular, in the absence of interaction the correlation function equals unity, a repulsive interaction leads to a depletion at low k^*, an attractive interaction leads to an enhancement and a bound state is presented differently based on the size of the emitting source.
From an experimental point of view, the $C(k^*)$ can be obtained by measuring the k^* distribution of pairs within the same event ($N_{same}(k^*)$) divided by pairs from different (mixed) collisions ($N_{mixed}(k^*)$). The factor $\xi(k^*)$ includes the corrections due to experimental effects (e.g momentum resolution) [13]. In order to study the strong hadron-hadron interaction, typically acting at distances of 1 fm, the particles composing the pair must be emitted at similar relative distances.

3 Accessing the strong interaction with femtoscopy in small colliding systems

At LHC, the size of the emitting source can be varied experimentally by changing the colliding system. In the ALICE experiment, source sizes ranging from $\approx 1-1.8$ fm up to $\approx 1.8-2.5$ fm (ultra-peripheral and peripheral) for pp and p–Pb collisions and $\approx 4-10$ fm (semi-central and central) for Pb–Pb collisions can be probed. The study of the short-range interaction between hadrons can be directly accessed by performing femtoscopic measurements in small colliding systems as pp and p–Pb, as will be shown in Sec. 4.
In the left panel of Fig. 2 a typical Gaussian source distribution of similar size to the one obtained experimentally is shown and a typical nuclear potential is presented for comparison. As can be seen, the range of the strong interaction is relevant up to ≈ 2 fm. In this range a

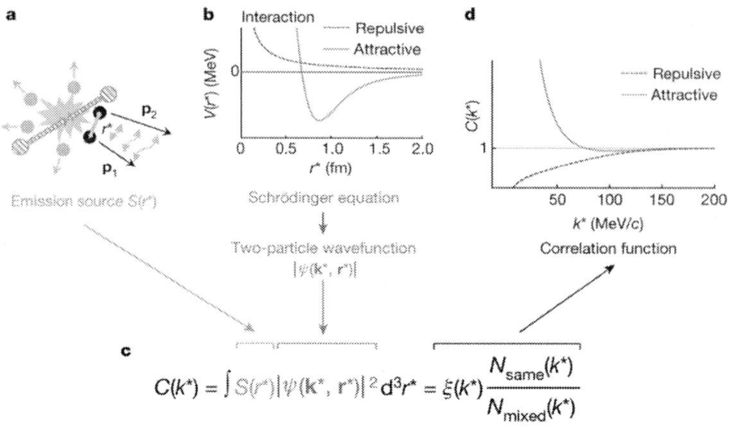

Figure 1. Figure taken from [13]. Schematic representation of the femtoscopy technique. In (a), a sketch of the source from which the pair of interest is emitted is shown, while in (b) two examples of interacting potential are presented. In (c) both the theoretical and the experimental definition of $C(k^*)$ are given, and finally in (d) the corresponding correlations for the two interactions in (b) are shown.

large amount of particle pairs with small k^* can be produced, as shown by the blue shaded area.
The measurement of $C(k^*)$ in small colliding systems provides not only a quantitative insight into the strong interaction but it also acts as a doorway to access baryon density above ρ_0.

In order to investigate the strong interaction, a solid knowledge on the emitting source

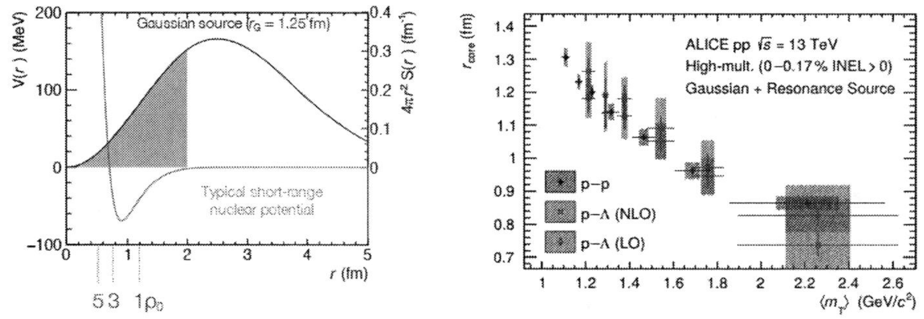

Figure 2. Left: the Gaussian source distribution (blue) for a source size of $r_G = 1.25$ fm as a function of the relative distance, compatible with measurements in pp and p–Pb collisions at LHC. In green, a typical short-range nuclear potential as a function of the relative distance. The blue shaded area represents the amount of particle pairs sensitive to the short-range interaction that can be emitted for the given source. Right: Figure taken from [14]. Source radius r_{core}, for p–p (blue) and p–Λ pairs (red [15], green [16]) as a function of average m_T using a core Gaussian source with strong resonances contributions.

in small colliding systems is required. The ALICE Collaboration delivered a data-driven analysis on the modeling of the source distribution in high-multiplicity pp collisions at $\sqrt{s} = 13$ TeV which quantitatively showed the existence of a common core Gaussian source for baryon-baryon pairs, once contributions from strongly decaying resonances are properly taken into account [14]. In the right panel of Fig. 2, the extracted core r_{core} is shown as a function of the average pair´s transverse mass. The analysis is performed for p–p and p–Λ pairs since the interaction in these two systems is already well known from scattering experiments. A common scaling of the core radius is observed indicating the presence of a universal emitting source for baryons in pp collisions. These results provide a data-driven method which allows to investigate the strong interaction of any hadron-hadron pair by determining the r_{core} at the corresponding m_{T} from the p–p results and by including the strong resonance contributions. The total source can be approximated by an effective Gaussian [14], which for the presented measurements in Sec. 4 has an effective size in the range $\mathrm{r_{eff}} \approx 1 - 1.3$ fm.

4 Results

In this section we present the results obtained from the femtoscopic measurements of the correlation functions of YN pairs with different strangeness content.

|S| = 1: pΛ interaction and the role of the NΣ coupling

- The dynamics in the coupled system NΛ–NΣ originates from the same quantum numbers and the similar mass that these two system share. The NΛ–NΣ coupling deeply affects the behaviour of the Λ hyperons at finite density and it has implications for the ΛNN interaction. Recent Chiral Effective field theory (χEFT) calculations with a weaker NΛ–NΣ coupling strength (NLO19) [9], still reproducing scattering and hypernuclei data, predict a more attractive single-particle potential U_Λ at large densities in pure neutron matter. The current experimental constraints, mostly from scattering experiments, cannot discriminate between these different chiral potentials.
The measurement of p–Λ pairs in high-multiplicity pp collisions at $\sqrt{s}$ = 13 TeV by the ALICE Collaboration [17] is shown in Fig. 3. These data were able to extend the kinematic range in which the interaction is tested and to achieve the highest precision currently available on this interaction ($\approx$ 1%). This measurement also provided the first direct observation of the opening of the NΣ channel in a two-body reaction, seen in the middle panel of Fig. 3 as a cusp-structure occurring at $k^* \approx 290$ MeV/c. The data favours a weaker NΛ–NΣ coupling as predicted by NLO19 calculations. Additionally, a better agreement with the measured correlation is achieved if the residual p–Σ^0 contribution ($\approx$ 16% of the total signal) is assumed flat (right panel in Fig. 3), in agreement with the results obtained from the direct measurements of p–Σ^0 pairs [18].

|S| = 2: pΞ^- interaction and test of Lattice QCD potentials

- In the left panel of Fig. 4, the p–Ξ^- correlation function measured in high-multiplicity pp collisions at $\sqrt{s}$ = 13 TeV by the ALICE Collaboration is shown [13]. The data are above the Coulomb-only prediction (green line), clearly indicating the first observation of a strong attractive interaction in this system. The assumption, along with Coulomb, of a lattice potential [19], calculated at the physical quark masses, for the strong part of the interaction shows an nice agreement with the data (pink band). This scenario has also been confirmed by the identical measurement performed in p–Pb at $\sqrt{s}$ = 5.02 TeV [20]. Calculations at finite density based on the same lattice potentials, validated in vacuum by the above-mentioned femtoscopic measurements, provide a slightly repulsive $U_\Xi \approx 6$ MeV in pure neutron matter [21], which can push the production of Ξ hyperons in the core of NS to higher densities and consequently stiffen the EoS. In the left panel of Fig. 4, the obtained mass-radius relation from an EoS based on mean-field calculations [22, 23] tuned

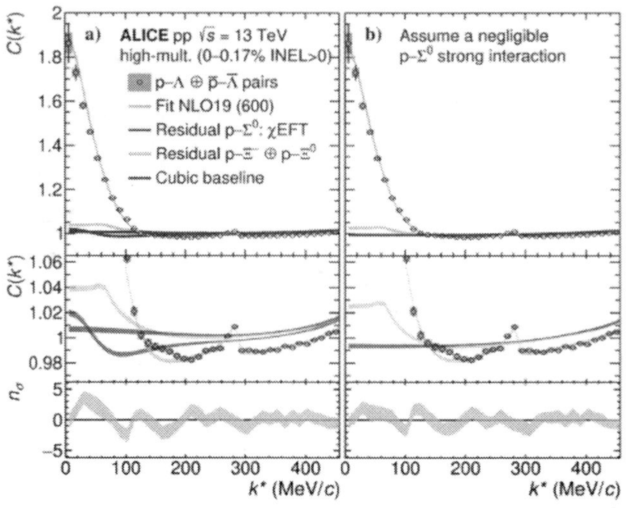

Figure 3. Figure taken from [17]. Measured p–Λ correlation and zoom in the k^* region with the cusp-structure due to the NΛ–NΣ coupling at $k^* = 289$ MeV/c. Lower panel: deviation between data and model in terms of n_σ. The fit using NLO19 [9] chiral potential is shown. Left: fit assuming a χEFT potential [15] for the p-Σ^0 residual. Right: fit assuming a flat p-Σ^0 residual.

to reproduce the lattice constraint on U_Ξ in medium is shown [11]. A maximum mass of $M = 2.13 M_\odot$, compatible with the recent measurements [2–4] is obtained.

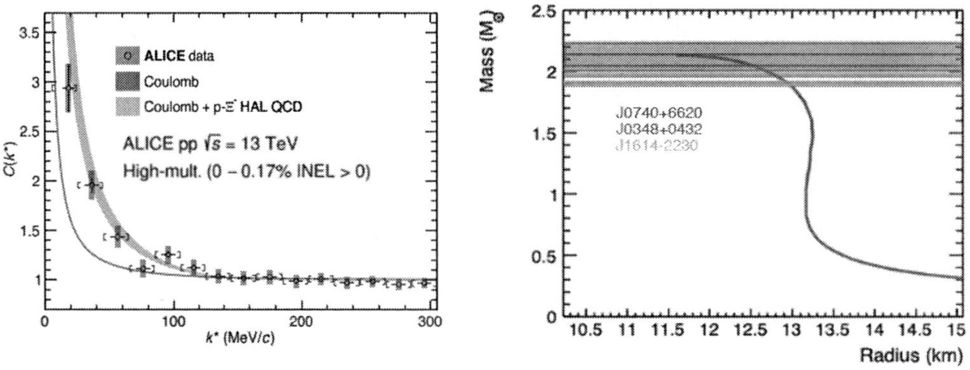

Figure 4. Left: Figure taken from [13]. Measured p–Ξ^- correlation function in high multiplicity pp collisions at $\sqrt{s}$ = 13 TeV by ALICE Collaboration. The fits assuming Coulomb-only (green) and Coulomb + strong HAL QCD potentials (pink) are shown [19]. Right: mass-radius relation obtained by assuming the predicted HAL QCD Ξ interaction in medium [21]. Recent astrophysical measurements of massive NS are shown in blue, green and orange [2–4].

The increased statistics expected for the upcoming Runs 3 and 4 of the LHC [24] will provide additional insights on the YN interaction and it will extend the studies to three-particle correlations in order to investigate the role played by three-body interactions, of great relevance in NS.
The access to the three-body interaction in femtoscopy will occur through a direct three-

particle correlation measurements, by extending the two-body formalism to triplets, and through an indirect measurement, by looking at correlations between protons and composite objects such as deuterons. Recently, a projector method has been developed [25], which will allow to remove the two-body contributions from the three-body correlations. This technique will grant direct access, for the first time, to the genuine three-body interaction in p–p–p and p–p–Λ systems.

References

[1] T.E. Riley et al., Astrophys. J. Lett. **887**, L21 (2019), 1912.05702

[2] P. Demorest, T. Pennucci, S. Ransom, M. Roberts, J. Hessels, Nature **467**, 1081 (2010), 1010.5788

[3] J. Antoniadis et al., Science **340**, 6131 (2013), 1304.6875

[4] H.T. Cromartie et al., Nature Astron. **4**, 72 (2019), 1904.06759

[5] M.C. Miller, F.K. Lamb, A.J. Dittmann, S. Bogdanov, Z. Arzoumanian, K.C. Gendreau, S. Guillot, A.K. Harding, W.C.G. Ho, J.M. Lattimer et al., The Astrophysical Journal **887**, L24 (2019)

[6] B.P. Abbott et al. (LIGO Scientific, Virgo), Phys. Rev. Lett. **119**, 161101 (2017), 1710.05832

[7] B. Abbott, R. Abbott, T. Abbott, F. Acernese, K. Ackley, C. Adams, T. Adams, P. Addesso, R. Adhikari, V. Adya et al., Physical Review Letters **121** (2018)

[8] L. Tolos, L. Fabbietti, Prog. Part. Nucl. Phys. **112**, 103770 (2020), 2002.09223

[9] J. Haidenbauer, U.G. Meißner, A. Nogga, Eur. Phys. J. A **56**, 91 (2020), 1906.11681

[10] M.A. Lisa, S. Pratt, R. Soltz, U. Wiedemann, Ann. Rev. Nucl. Part. Sci. **55**, 357 (2005)

[11] L. Fabbietti, V.M. Sarti, O.V. Doce (2020), 2012.09806

[12] D. Mihaylov, V. Mantovani Sarti, O. Arnold, L. Fabbietti, B. Hohlweger, A. Mathis, Eur. Phys. J. C **78**, 394 (2018), 1802.08481

[13] S. Acharya et al. (ALICE), Nature **588**, 232 (2020)

[14] S. Acharya et al. (ALICE), Phys. Lett. B **811**, 135849 (2020), 2004.08018

[15] J. Haidenbauer, S. Petschauer, N. Kaiser, U.G. Meißner, A. Nogga, W. Weise, Nucl. Phys. A **915**, 24 (2013), 1304.5339

[16] H. Polinder, J. Haidenbauer, U.G. Meißner, Nucl. Phys. A **779**, 244 (2006), nucl-th/0605050

[17] S. Acharya et al. (ALICE) (2021), 2104.04427

[18] S. Acharya et al. (ALICE), Phys. Lett. B **805**, 135419 (2020), 1910.14407

[19] K. Sasaki et al. (HAL QCD), Nucl. Phys. A **998**, 121737 (2020), 1912.08630

[20] S. Acharya et al. (ALICE), Phys. Rev. Lett. **123**, 112002 (2019), 1904.12198

[21] T. Inoue (LATTICE-HALQCD), PoS **INPC2016**, 277 (2016), 1612.08399

[22] S. Weissenborn, D. Chatterjee, J. Schaffner-Bielich, Phys. Rev. C **85**, 065802 (2012), [Erratum: Phys.Rev.C 90, 019904 (2014)], 1112.0234

[23] N. Hornick, L. Tolos, A. Zacchi, J.E. Christian, J. Schaffner-Bielich, Phys. Rev. C **98**, 065804 (2018), 1808.06808

[24] S. Acharya et al. (ALICE Collaboration), **ALICE-PUBLIC-2020-005, CERN-LHCC-2020-018** (2020)

[25] R. Del Grande, L. Šerkšnytė, L. Fabbietti, V.M. Sarti, D. Mihaylov (2021), 2107.10227

Global and local polarization of Λ ($\overline{\Lambda}$) hyperons in Pb–Pb collisions in ALICE at the LHC

Debojit Sarkar[1,*] (for the ALICE Collaboration)

[1]Wayne State University, 666 W Hancock St, Detroit, MI-48201, USA

Abstract. The global polarization of the Λ and $\overline{\Lambda}$ hyperons ($P_{\rm H}$) has been measured in Pb–Pb collisions at $\sqrt{s_{\rm NN}}$ = 2.76 TeV and 5.02 TeV in ALICE at the Large Hadron Collider (LHC). The hyperon global polarization is found to be consistent with zero at both collision energies. The local polarization of the Λ and $\overline{\Lambda}$ hyperons along the beam (z) direction, P_z, has also been measured in Pb–Pb collisions at $\sqrt{s_{\rm NN}}$ = 5.02 TeV. The P_z, measured as a function of the hyperon emission angle relative to the second harmonic symmetry plane, exhibits a second harmonic sine modulation, as expected due to elliptic flow. The measurements of global and local hyperon polarization are reported for different collision centralities and as a function of transverse momentum in semicentral collisions. These results show the first experimental evidence of a non-zero hyperon P_z in Pb–Pb collisions at the LHC.

The initial large orbital angular momentum of the colliding nuclei in a relativistic non-central nucleus–nucleus collision might be partially transferred to the system created in such collision [1]. This system, known to behave almost like an ideal fluid [2], would exhibit a vortical behavior with vorticity on average directed perpendicular to the reaction plane defined by the impact parameter vector and the beam direction. Due to spin-orbit interactions, all emitted final state particles may become polarized along the system's orbital angular momentum, the phenomenon termed as global polarization [1, 3, 4]. As the spin of a particle cannot be measured directly, the parity violating weak decays of $\Lambda \rightarrow {\rm p} + \pi^-$ and $\overline{\Lambda} \rightarrow \overline{\rm p} + \pi^+$ in which the momentum of the daughter (anti)proton is correlated with the spin of the hyperon, are used to measure the polarization in the system. The polarization vector $P_{\rm H}$ generally depends on the hyperon kinematics, namely the transverse momentum $p_{\rm T}$, rapidity $y_{\rm H}$, and its azimuthal angle with respect to the reaction plane, $\varphi - \Psi_{\rm RP}$, as well as the collision centrality [5]

$$P_{\rm H}(p_{\rm T}, y_{\rm H}, \varphi) = -\frac{8}{\pi\alpha_{\rm H}}\langle\sin(\varphi_{\rm p}^* - \Psi_{\rm RP})\rangle, \quad (1)$$

where $\varphi_{\rm p}^*$ is the azimuthal angle of the (anti)proton direction in the hyperon rest frame and α_H is the hyperon decay parameter [6]. The reaction plane angle $\Psi_{\rm RP}$ is estimated using the spectator plane angle $\Psi_{\rm SP}$, characterizing the deflection direction of the spectator neutrons [5]. In addition to the vorticity due to the orbital angular momentum of the entire system, other physics processes, such as anisotropic flow, jet energy deposition, deviation from longitudinal boost invariance of the transverse velocity fields, generate vorticity along different directions depending on the location of the fluid elements in the created system [7–10]. In particular, in non-central nucleus-nucleus collisions, the strong elliptic flow would generate a non-zero

*e-mail: debojit.sarkar@cern.ch

vorticity component along the beam axis (z) [8, 9]. The vorticity and the corresponding polarization exhibits a quadrupole structure in the transverse plane. This polarization can be characterized by the second harmonic sine component in the Fourier decomposition of the polarization along the beam axis (P_z) as a function of the particle azimuthal angle relative to the second harmonic symmetry plane [11]

$$P_{z,s2} = \langle P_z \sin(2\varphi - 2\Psi_2)\rangle, \quad (2)$$

$$P_z(p_T, y_H, \varphi) = \frac{\langle \cos\theta_p^* \rangle}{\alpha_H \langle (\cos\theta_p^*)^2 \rangle}, \quad (3)$$

where θ_p^* is the polar angle of the (anti)proton direction in the hyperon rest frame and Ψ_2 is the second harmonic symmetry plane angle. The factor $\langle (\cos\theta_p^*)^2 \rangle$ serves as a correction for finite acceptance along the longitudinal direction. The sign of $P_{z,s2}$ determines the phase of the P_z modulation relative to the second harmonic symmetry plane.

The analyzed data samples were recorded by ALICE in 2010 and 2011 for Pb–Pb collisions at $\sqrt{s_{NN}}$ = 2.76 TeV, and in 2015 and 2018 at $\sqrt{s_{NN}}$ = 5.02 TeV [5, 11]. The centrality is determined using the sum of the charge deposited in the V0A ($2.8 < \eta < 5.1$) and the V0C ($-3.7 < \eta < -1.7$) scintillators [5, 11]. After removing the pile up contributions, events that pass central, semi-central, or minimum-bias trigger criteria with a z-component of the reconstructed event vertex (V_z) within ±10 cm are selected. The Λ and $\overline{\Lambda}$ hyperons are reconstructed inside the Time Projection Chamber (TPC) using the decay topology of $\Lambda \rightarrow p + \pi^-$ and $\overline{\Lambda} \rightarrow \overline{p} + \pi^+$ (64% branching ratio) [12] as described in Ref. [13]. The daughter tracks are assigned the identity of a pion or a (anti)proton based on the charge and particle identification using the specific energy loss (dE/dx) measurement in the TPC. The tracks of the daughter pions and (anti)protons are selected within the pseudorapidity range of $|\eta| < 0.8$ inside the TPC. The Λ and $\overline{\Lambda}$ candidates having invariant mass M_{inv} within the interval 1.103 –1.129 GeV/c^2 with $p_T > 0.5$ GeV/c and $|y_H| < 0.5$ are considered in this measurement [5, 11].

The event-plane method is used for the polarization measurement [14]. The spectator plane angle (Ψ_{SP}) is reconstructed with a pair of neutron Zero Degree Calorimeter (ZDC) detectors [5]. The second harmonic event plane angle (Ψ_2) is reconstructed using the TPC tracks and signals in the V0A and V0C scintillators [11]. A recentering procedure [14] is applied run-by-run as a function of the event centrality, event vertex position (V_x, V_y, V_z), and duration of the run in order to compensate for various effects (e.g., imperfect detector acceptance, varying beam conditions) [5, 11].

The P_H and $P_{z,s2}$ are measured using the invariant mass method [5, 11] by calculating $Q = \langle \sin(\varphi_p^* - \Psi_{SP}) \rangle$ for global and $Q = \langle \cos\theta_p^* \sin(2\varphi - 2\Psi_2)\rangle$ for local polarization for all hyperon candidates as a function of the invariant mass and fitting it with the expression

$$Q(M_{inv}) = f^S(M_{inv})Q^S + f^{BG}Q^{BG}(M_{inv}), \quad (4)$$

where f^S and $f^{BG} = 1 - f^S$ are the signal and background fraction of the Λ ($\overline{\Lambda}$) candidates estimated from the invariant mass yields, respectively. The parameter Q^S estimates the signal and $Q^{BG}(M_{inv})$ estimates the possible contribution from the combinatorial background of Λ ($\overline{\Lambda}$) hyperons in the measured polarization. Figure 1 shows the measured hyperon global polarization (P_H) as a function of centrality in Pb–Pb collisions for two collision energies [5]. The P_H at the LHC is found to be consistent with zero within the experimental uncertainties for all studied centrality classes. No signal was observed as a function of rapidity or p_T either [5]. Recent measurements at RHIC show a significant global polarization of Λ and $\overline{\Lambda}$ hyperons in Au–Au collisions at $\sqrt{s_{NN}}$ = 7.7 – 200 GeV with the polarization magnitude monotonically decreasing with increasing $\sqrt{s_{NN}}$ [15, 16]. The decrease in the P_H at

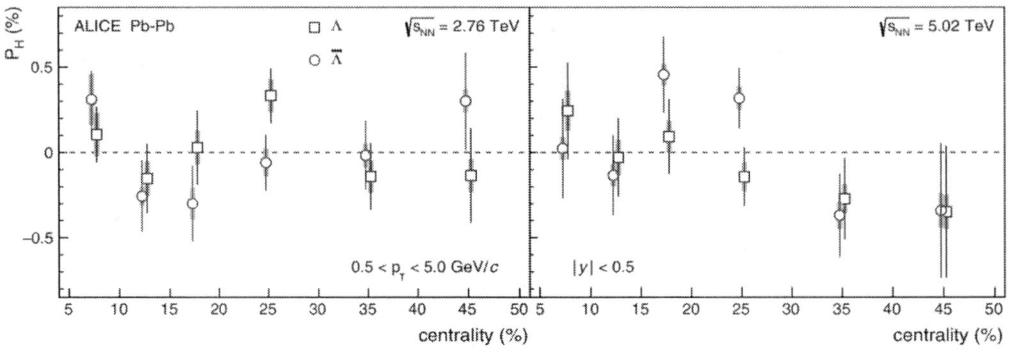

Figure 1. (color online) The global hyperon polarization (P_H) as function of centrality for Pb–Pb collisions at $\sqrt{s_{NN}}$ = 2.76 TeV (left) and 5.02 TeV (right) [5].

midrapidity with collision energy is usually attributed to a decreasing role of the baryon stopping [17] in the initial velocity distributions. The ALICE measurements are consistent with hydrodynamical model calculations for the LHC energies and empirical estimates based on the collision energy dependence of the directed flow due to the tilted source [1, 8, 18].

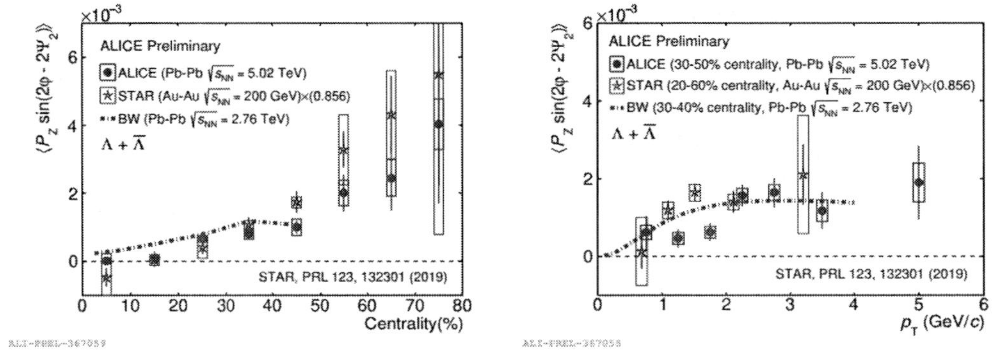

Figure 2. (color online) Centrality dependence of $\langle P_z \sin(2\varphi - 2\Psi_2)\rangle$ averaged for Λ and $\overline{\Lambda}$ in Pb–Pb collisions at $\sqrt{s_{NN}}$ = 5.02 TeV and it's comparison with the RHIC results in Au–Au collisions at $\sqrt{s_{NN}}$ = 200 GeV. The Blast-Wave model calculations [19] for Pb–Pb collisions at $\sqrt{s_{NN}}$ = 2.76 TeV are shown by dashed-dotted line.

The centrality and p_T dependences of $P_{z,s2}$ in Pb–Pb collisions at $\sqrt{s_{NN}}$ = 5.02 TeV and its comparison with the STAR measurements in Au–Au collisions at $\sqrt{s_{NN}}$ = 200 GeV [20] are shown in Fig. 2. The $P_{z,s2}$ decreases towards more central collisions, similar to the elliptic flow. For centralities larger than 60%, the large uncertainties prevent a firm conclusion on the exact mature of its centrality dependence [11]. The $P_{z,s2}$ also shows an increase with p_T up to $p_T \approx 2.0$ GeV/c. For $p_T > 2.0$ GeV/c, the data is consistent with being constant but the uncertainty in the measurement does not allow for a strong conclusion. As the STAR results were obtained with $\alpha_H = 0.642$ whereas the ALICE measurement uses updated values $\alpha_H = 0.750$ (Λ) and -0.758 ($\overline{\Lambda}$), the STAR results are rescaled with a factor 0.856 for a proper comparison [11]. The $P_{z,s2}$ in Pb–Pb collisions at $\sqrt{s_{NN}}$ = 5.02 TeV is similar in magnitude for the central collisions with somewhat smaller value in the semi-central collisions and has same sign compared to the top RHIC energy. At lower transverse momenta (p_T < 2.0 GeV/c), $P_{z,s2}$ at the LHC is smaller than that at the top RHIC energy as shown in the right panel of Fig. 2. The $P_{z,s2}$ does not exhibit a significant dependence on rapidity as shown in Ref. [11]. The sign of the $P_{z,s2}$ is positive at both RHIC and the LHC and in disagreement with hydrodynamic and AMPT models estimations accounting only for the thermal vorticity [21, 22]. Surprisingly, a simple Blast-Wave model, which accounts only for the kinematic vorticity, tuned to reproduce transverse momentum spectra, elliptic flow, and femtoscopic radii, describes the ALICE and STAR results reasonably well [16]. However, the recent hydrodynamical calculations of hyperon polarization based on fluid shear and thermal vorticity coupled with additional assumptions on the hadronization temperature or mass of the spin carrier reproduce the experimentally observed positive $P_{z,s2}$ at RHIC and the LHC energies [11, 21, 22]. These studies indicate that longitudinal polarization is sensitive to the hydrodynamic gradients as well as the dynamics of the spin degrees of freedom through the different stages of the evolution of the system created in heavy-ion collisions. The upcoming Run 3 at the LHC will provide much larger data samples for more differential and precision measurements of local and global hyperon polarization and will further constrain the models aiming to explain the vorticity and the particle polarization in heavy-ion collisions.

References

[1] F. Becattini, G. Inghirami, V. Rolando, A. Beraudo, L. Del Zanna, A. De Pace, M. Nardi, G. Pagliara, V. Chandra, Eur. Phys. J. C **75**, 406 (2015), [Erratum: Eur.Phys.J.C 78, 354 (2018)], 1501.04468
[2] P. Braun-Munzinger, V. Koch, T. Schäfer, J. Stachel, Phys. Rept. **621**, 76 (2016), 1510.00442
[3] Z.T. Liang, X.N. Wang, Phys. Rev. Lett. **94**, 102301 (2005), [Erratum: Phys.Rev.Lett. 96, 039901 (2006)], nucl-th/0410079
[4] S.A. Voloshin (2004), nucl-th/0410089
[5] S. Acharya et al. (ALICE), Phys. Rev. C **101**, 044611 (2020), 1909.01281
[6] M. Ablikim et al. (BESIII), Nature Phys. **15**, 631 (2019), 1808.08917
[7] B. Betz, M. Gyulassy, G. Torrieri, Phys. Rev. C **76**, 044901 (2007), 0708.0035
[8] S.A. Voloshin, EPJ Web Conf. **171**, 07002 (2018), 1710.08934
[9] F. Becattini, I. Karpenko, Phys. Rev. Lett. **120**, 012302 (2018), 1707.07984
[10] X.L. Xia, H. Li, Z.B. Tang, Q. Wang, Phys. Rev. C **98**, 024905 (2018), 1803.00867
[11] S. Acharya et al. (ALICE) (2021), 2107.11183
[12] P.A. Zyla et al. (Particle Data Group), PTEP **2020**, 083C01 (2020)
[13] S. Acharya et al. (ALICE), JHEP **09**, 006 (2018), 1805.04390
[14] A.M. Poskanzer, S.A. Voloshin, Phys. Rev. C **58**, 1671 (1998), nucl-ex/9805001
[15] L. Adamczyk et al. (STAR), Nature **548**, 62 (2017), 1701.06657
[16] J. Adam et al. (STAR), Phys. Rev. C **98**, 014910 (2018), 1805.04400
[17] H. Sorge, A. von Keitz, R. Mattiello, H. Stoecker, W. Greiner, Nucl. Phys. A **525**, 95C (1991)
[18] B. Abelev et al. (ALICE), Phys. Rev. Lett. **111**, 232302 (2013), 1306.4145
[19] T. Niida, private communication, 2020. For the description of the Blast-Wave model see [20]
[20] J. Adam et al. (STAR), Phys. Rev. Lett. **123**, 132301 (2019), 1905.11917
[21] F. Becattini, M. Buzzegoli, A. Palermo, G. Inghirami, I. Karpenko (2021), 2103.14621
[22] B. Fu, S.Y.F. Liu, L. Pang, H. Song, Y. Yin (2021), 2103.10403

Global polarization and spin alignment measurements

Takafumi Niida[1,3,*]

[1]University of Tsukuba, 1-1-1 Tennoudai, Tsukuba, Ibaraki 305-8571, JAPAN

Abstract. Recent experimental results on global polarization of hyperons and spin alignment of vector mesons in heavy-ion collisions are reviewed, especially in context of the energy dependence and particle species dependence. New results on local polarization along the beam direction at the LHC are discussed with previous measurements at RHIC. Future outlook of these measurements and possible future directions for better understanding of local vorticity field are briefly discussed.

1 Introduction

In non-central heavy-ion collisions, an initial orbital angular momentum carried by two colliding nuclei is partially converted to the spin angular momentum of particles produced in the collisions [1–3]. This phenomenon is called global polarization, i.e. produced particles are globally polarized along the direction of the initial orbital angular momentum which is perpendicular to reaction plane. The observation of Λ global polarization by the STAR Collaboration [4], indicating the creation of vortical fluid, opens new direction to study a nuclear matter not only under high temperature and/or baryon density but also under the fastest vorticity field [5, 6] as well as possibly to study the initial strong magnetic field ($B \sim 10^{13}$ T) and its lifetime which has large uncertainty [7].

While results on the average global polarization are well reproduced by theoretical models [8–12] in a wide range of collision energies, differential measurements and local polarization measurements [13–15] found discrepancies with models which still remain open questions. In this proceedings, recent experimental updates on global and local polarization measurements of hyperons as well as spin alignment measurements of vector mesons are reviewed and their physics implications are discussed.

2 Hyperon global polarization

2.1 Λ global polarization

Polarization can be studied using hyperons' weak decay, by analyzing angular distribution of decay product: $dN/d\cos\theta^* \propto 1 + \alpha_H P_H \cos\theta^*$ where θ denotes a polar angle of daughter baryon relative to the direction of hyperon polarization P_H, α is decay parameter ($\alpha_\Lambda \approx 0.732$ [16]), and the asterisk indicates analysis in the hyperon rest frame. Global polarization of Λ hyperons was first observed by STAR Collaboration in the beam energy scan of Au+Au collisions at $\sqrt{s_{NN}}$ = 7.7–39 GeV [4], and later confirmed at 200 GeV with better

*e-mail: niida@bnl.gov

precision [17]. As shown in Fig. 1, the results show an energy dependence, increasing toward lower energies, which could be understood by rapidity dependence of the initial shear flow field and baryon stopping at lower energies with finite detector acceptance. Longer lifetime of the system at higher energy may also dilute the polarization, leading to the similar energy dependence. Various theoretical models [8–12] such as hydrodynamic and transport models can explain well the measured polarization averaged over phase-space as shown in Fig. 1, while there exists open questions in differential measurements as mentioned, e.g. discrepancy in azimuthal angle dependence between the data and models and unknown rapidity dependence. Based on phenomenological and theoretical expectations [18, 19], the polarization at the LHC energy would be of the order of 0.1%. Results in Pb+Pb collisions at $\sqrt{s_{NN}}$ = 2.76 and 5.02 TeV from ALICE [20] are consistent with zero and the current statistical uncertainty is still in the level of the expected signal, which can be explored with large statistics in LHC-Run3.

The polarization at lower energy is also of particular interest since the system would turn to hadronic matter from partonic matter when decreasing the energy. How does the initial orbital angular momentum couple with spins of partons or hadrons and any difference there? What is the relaxation time for spin-orbit coupling compared to the system lifetime? New preliminary result at 7.2 GeV from STAR fixed-target program [21] is found to follow the global trend as shown in Fig. 1. Furthermore, new preliminary results in Ag+Ag 2.55 GeV and Au+Au 2.4 GeV collisions from HADES experiment indicate comparable or even larger polarization compared to that at 7.2 GeV. A transport model (UrQMD) [22] predicts that the kinetic vorticity becomes maximum around 3 GeV and then decreases when approaching $\sqrt{s_{NN}} = 2m_N$ where m_N is the mass of nucleon. It is worth to mention that after SQM2021

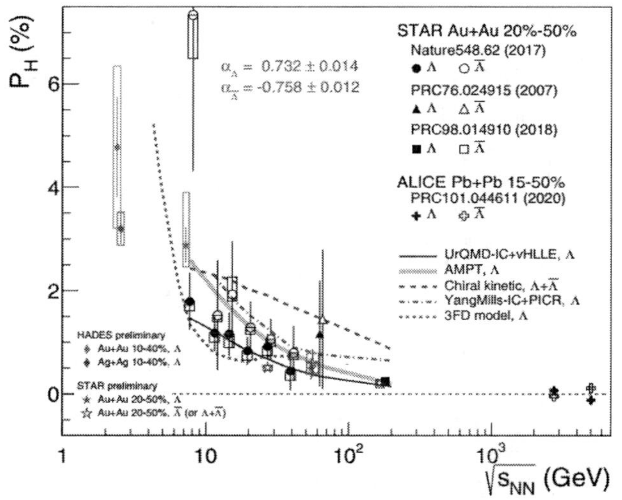

Figure 1. Energy dependence of global polarization of Λ hyperons in heavy-ion collisions, comparing with various theoretical calculations [8–12]. The experimental data are taken from Refs. [4, 17, 20, 21, 23] and rescaled with new decay parameter [16] shown in the figure.

conference the STAR Collaboration has reported new result on Λ global polarization at 3 GeV [24], which is comparable to or possibly larger than the HADES preliminary results. Note that different detector acceptance lead to difference in the measured polarization if there is rapidity and/or p_T dependence which seems to be not significant with current uncertainties.

2.2 Ξ and Ω global polarization

As mentioned in previous section, there still exists open questions and discrepancy between the data and models. Further experimental inputs, especially polarization measurements with other particles (e.g. different spin), will help to better understand the nature of the polarization in heavy-ion collisions. With the assumption of local thermal equilibrium of the system, the polarization of these hyperons P is determined by the local thermal vorticity ω as $P = (s+1)\omega/T$ where s is the spin of the particle and T is the temperature.

The STAR Collaboration has recently extended the measurement to Ξ (spin-1/2) and Ω (spin-3/2) hyperons [25]. One should note that the decay parameter is different for each hyperons: α_{Ξ}=-0.401 and $\alpha_{\Omega} = 0.0157$ [16]. Smaller magnitude of the decay parameter makes the measurement difficult, especially for case of Ω it is practically impossible to measure Ω polarization in this way. Instead one can utilize the following relation to measure the polarization of Ξ and Ω hyperons: $\mathbf{P}_P^* = C_{PD}\mathbf{P}_D^*$ where $\mathbf{P}_{P(D)}$ denotes the polarization of parent (P) and its daughter (D) particles in the parent rest frame and C_{PD} is the polarization transfer factor in the decay from the parent to the daughter. Results on Ξ and Ω global polarization (particles and antiparticles combined) at 200 GeV from STAR [25] are shown in Fig. 2, together with previous results for Λ and $\bar{\Lambda}$. The Ξ polarization was measured with the two independent methods: by analysis of daughter Λ distribution in Ξ rest frame and by measuring polarization of daughter Λ taking into account the polarization transfer of $C_{\Xi\Lambda} = +0.944$. Both results are consistent within uncertainties and the combined result show a slightly larger polarization than inclusive Λ, although the difference is not significant with the current uncertainties. Similarly, Ω polarization was measured via daughter polarization but there is

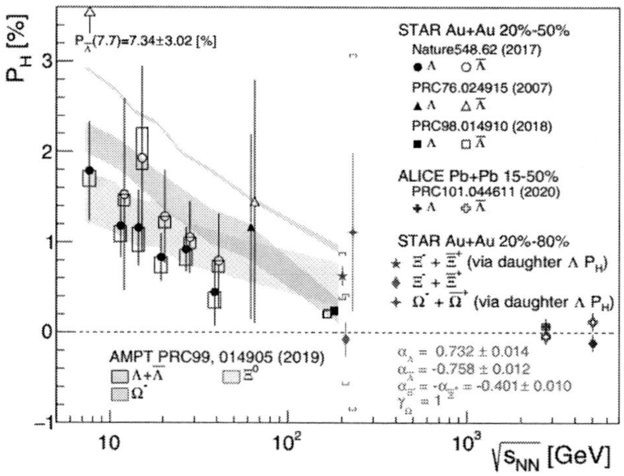

Figure 2. Ξ and Ω global polarization at 200 GeV with previous results for Λ ($\bar{\Lambda}$) as a function of collision energy. Calculations from AMPT model [26] are also shown with shaded bands. Previous results for Λ ($\bar{\Lambda}$) polarization are rescaled with new decay parameter [16] shown in the figure.

an uncertainty in $C_{\Omega\Lambda} = (1 + 4\gamma_{\Omega})/5$ due to the unmeasured decay parameter γ_{Ω} which is likely either +1 or -1. The result based on $\gamma_{\Omega} = +1$ is shown in Fig. 2. The new results on Ξ and Ω polarization, consistent with theoretical expectations as shown in the figure, reinforce the fluid vorticity and global polarization picture in heavy-ion collisions. More precise measurements are anticipated at RHIC and the LHC, especially the uncertainty of the γ_{Ω} could be clarified assuming the vorticity picture and furthermore the measurement of Ω^- and $\bar{\Omega}^+$ separately would be interesting to look for the possible difference due to the initial magnetic field since Ω has a factor of three larger magnetic moment compared to Λ and Ξ.

3 Spin alignment of vector mesons

Vector mesons (spin-1) can be also used to study the particle polarization. Unlike hyperons, they decay via strong interaction where the parity is conserved, therefore one cannot know the direction of polarization. The spin state of a vector meson can be described by 3×3 spin density matrix where there exists only one independent diagonal element ρ_{00}. The diagonal element ρ_{00} represents the probability to have spin projection onto a quantization axis to be zero. It can be studied by measuring angular distributions of decay products of vector mesons: $dN/d\cos\theta^* \propto [1 - \rho_{00} + (3\rho_{00} - 1)\cos^2\theta^*]$ where θ^* is angle of decayed daughter relative to the polarization direction in the decay rest frame. If the particles are unpolarized, $\rho_{00} = 1/3$ on average. The deviation from $1/3$ indicates spin alignment of vector mesons.

Figure 3 left panels show ρ_{00} of K^{*0} and ϕ mesons from ALICE experiment [27]. The results relative to the event plane show a negative deviation from $1/3$ at $p_T < 2$ GeVc, while the results of K_s^0 mesons (spin-0) are consistent with $1/3$ as expected. Also, the results relative to random event plane serving as a baseline confirm the significance of the deviation. Preliminary results from STAR experiment show similar behavior for K^{*0} ρ_{00} at $\sqrt{s_{NN}} = 54.4$ and 200 GeV [28] as shown in the top-right panel of Fig. 3, while ρ_{00} of ϕ mesons [29] show positive values at low p_T as opposed to the negative deviation at the LHC. Note that the contribution to the spin alignment from the vorticity is a second-order effect, i.e. $\rho_{00} \sim 1/[1 + (\omega/T)^2]$. Therefore the observed deviation of ρ_{00} cannot be explained by the vorticity picture based on the results on the Λ global polarization. Also there seems no significant energy dependence in K^{*0} ρ_{00} unlike the Λ global polarization. Such a large deviation should also contribute to the elliptic flow [2]. Mean field of ϕ mesons may play a role but likely not for K^{*0} ρ_{00} [30]. Further investigation is needed for better understanding the spin alignment measurements.

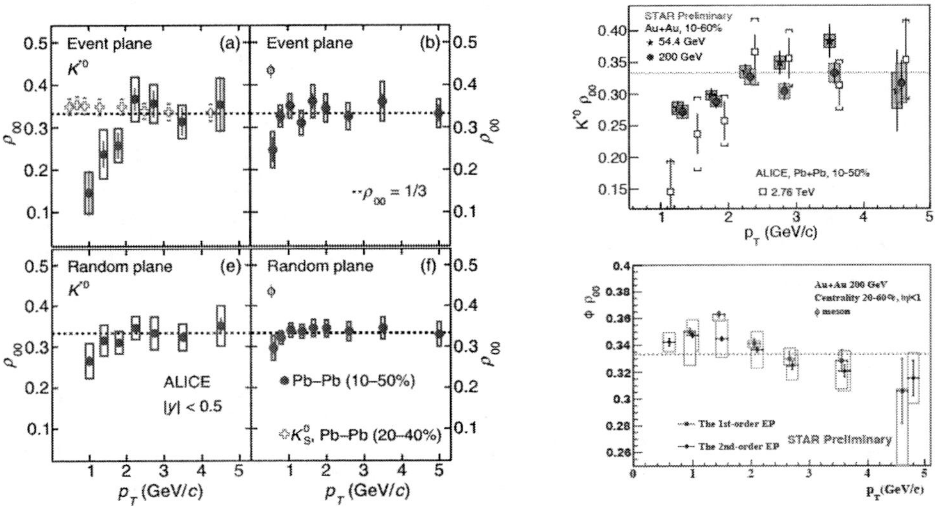

Figure 3. Average spin density matrix element ρ_{00} of K^{*0} and ϕ mesons in $\sqrt{s_{NN}} = 2.76$ TeV Pb+Pb collisions at the LHC (left) [27] and in $\sqrt{s_{NN}} = 54.4$ and 200 GeV Au+Au collisions at RHIC (right) [28, 29].

4 Local polarization

The vorticity field may be more complicated varying with phase-space due to density fluctuations, energy deposit from jets, and collective flow. Theoretical calculations predict that the vorticity along the beam direction might be created due to the azimuthal anisotropic flow [18, 31]. The STAR experiment observed the polarization of Λ ($\bar{\Lambda}$) along the beam direction (P_z) in Au+Au collisions at $\sqrt{s_{NN}}$ = 200 GeV, where the sign of P_z changes with azimuthal angle as expected from the elliptic flow in origin. However there is discrepancy in the sign between the data and models, even among various model calculations [15].

In this conference, new results in Pb+Pb collisions at $\sqrt{s_{NN}} = 5.02$ TeV [32] were reported from the ALICE Collaboration as shown in Fig. 4. The results show similar behaviors and are

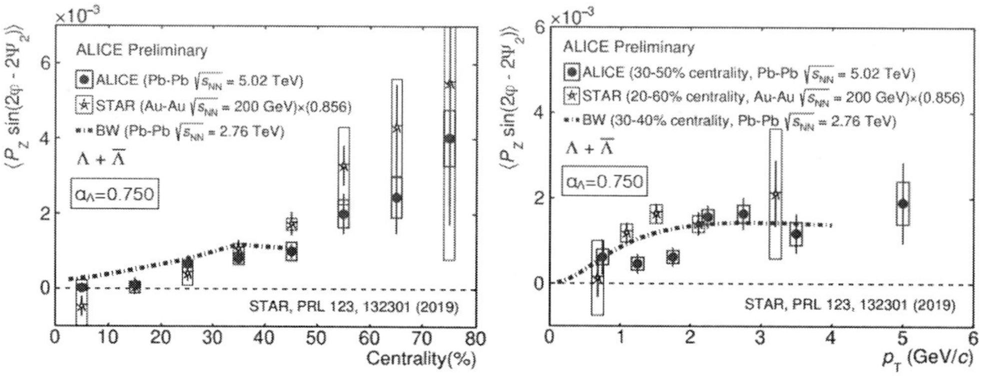

Figure 4. Fourier sine coefficient of $\Lambda + \bar{\Lambda}$ polarization along the beam direction as a function of centrality (left) and transverse momentum p_T (right) in Pb+Pb collisions at $\sqrt{s_{NN}}$ = 5.02 TeV [32] compared to those in Au+Au collisions at $\sqrt{s_{NN}}$ = 200 GeV [13].

comparable to the results from STAR; the sine modulation of P_z increases towards peripheral collisions (Fig. 4 left) and shows a hint of p_T dependence at $p_T < 2$ GeV/c (Fig. 4 right). The blast-wave model calculations reproduce the data well in their sign and magnitude unlike more realistic hydrodynamic models. It should be noted that recent theoretical developments that consider contribution from shear tensor to the polarization in addition to thermal vorticity may explain the experimental data [33, 34].

Similar to the elliptic flow, one can naively expect that the longitudinal polarization may arise from higher-order flow, which can be explored with high statistics data in the future data taking. Furthermore troidal structure of vorticity field is expected in central asymmetric collisions such as Cu+Au, d+Au, and p+Au collisions [18, 35]. In this case, the direction of the polarization is along the azimuthal angle $\vec{\phi}$ of a hyperon. Future data with high statistics will allow us to study this phenomena.

5 Summary

This paper has presented recent experimental results on hyperon polarization and spin alignment of vector mesons. New results on global polarization of Λ hyperons indicate that the polarization still increases towards $\sqrt{s_{NN}}$ = 2.4-7 GeV. With given large uncertainties, it is still not clear whether the polarization starts to decrease as expected from theoretical calculations. The first measurements on Ξ and Ω global polarization reinforce the vorticity picture in heavy-ion collisions. To confirm the possible mass and/or spin dependence of the polarization, more precise measurements are needed. Polarization of Λ hyperons along the beam direction as well as significant deviation of ρ_{00} from 1/3 were observed for K^{*0} and ϕ mesons both at RHIC and the LHC but there still exists open questions. Future runs at RHIC and the LHC as well as the data of the beam energy scan program phase-II from STAR will hopefully clarify those questions together with theoretical developments.

The author thanks the organizers of Strangeness in Quark Matter 2021 for the invitation to give this plenary talk. The author thanks S. Voloshin, S. Esumi, M. Lisa, F. Becattini, Y. Karpenko, and X.-G. Huang for fruitful discussions.

References

[1] Z.T. Liang, X.N. Wang, Phys. Rev. Lett. **94**, 102301 (2005), [Erratum: Phys.Rev.Lett. 96, 039901 (2006)]
[2] S.A. Voloshin (2004), `nucl-th/0410089`
[3] F. Becattini, F. Piccinini, J. Rizzo, Phys. Rev. C **77**, 024906 (2008)
[4] L. Adamczyk et al. (STAR), Nature **548**, 62 (2017)
[5] Y. Jiang, J. Liao, Phys. Rev. Lett. **117**, 192302 (2016)
[6] Y. Fujimoto, K. Fukushima, Y. Hidaka, Phys. Lett. B **816**, 136184 (2021)
[7] L. McLerran, V. Skokov, Nucl. Phys. A **929**, 184 (2014)
[8] I. Karpenko, F. Becattini, Eur. Phys. J. C **77**, 213 (2017)
[9] H. Li, L.G. Pang, Q. Wang, X.L. Xia, Phys. Rev. C **96**, 054908 (2017)
[10] Y. Sun, C.M. Ko, Phys. Rev. C **96**, 024906 (2017)
[11] Y. Xie, D. Wang, L.P. Csernai, Phys. Rev. C **95**, 031901 (2017)
[12] Y.B. Ivanov, V.D. Toneev, A.A. Soldatov, Phys. Rev. C **100**, 014908 (2019)
[13] J. Adam et al. (STAR), Phys. Rev. Lett. **123**, 132301 (2019)
[14] T. Niida (STAR), Nucl. Phys. A **982**, 511 (2019)
[15] F. Becattini, M.A. Lisa, Ann. Rev. Nucl. Part. Sci. **70**, 395 (2020)
[16] P.A. Zyla et al. (Particle Data Group), PTEP **2020**, 083C01 (2020)
[17] J. Adam et al. (STAR), Phys. Rev. C **98**, 014910 (2018)
[18] S.A. Voloshin, EPJ Web Conf. **171**, 07002 (2018)
[19] I. Karpenko, F. Becattini, Nucl. Phys. A **982**, 519 (2019)
[20] S. Acharya et al. (ALICE), Phys. Rev. C **101**, 044611 (2020)
[21] K. Okubo (STAR), these proceedings
[22] X.G. Deng, X.G. Huang, Y.G. Ma, S. Zhang, Phys. Rev. C **101**, 064908 (2020)
[23] F. Kornas (HADES), these proceedings
[24] M.S. Abdallah et al. (STAR) (2021), `arXiv:2108.00044`
[25] J. Adam et al. (STAR), Phys. Rev. Lett. **126**, 162301 (2021)
[26] D.X. Wei, W.T. Deng, X.G. Huang, Phys. Rev. C **99**, 014905 (2019)
[27] S. Acharya et al. (ALICE), Phys. Rev. Lett. **125**, 012301 (2020)
[28] S. Singha (STAR), Nucl. Phys. A **1005**, 121733 (2021)
[29] C. Zhou, Nucl. Phys. A **982**, 559 (2019)
[30] X.L. Sheng, L. Oliva, Q. Wang, Phys. Rev. D **101**, 096005 (2020)
[31] F. Becattini, I. Karpenko, Phys. Rev. Lett. **120**, 012302 (2018)
[32] D. Sarker (ALICE), these proceedings
[33] F. Becattini, M. Buzzegoli, A. Palermo, G. Inghirami, I. Karpenko (????), `arXiv:2103.14621`
[34] B. Fu, S.Y.F. Liu, L. Pang, H. Song, Y. Yin (2021), `arXiv:2103.10403`
[35] M.A. Lisa, J.G.P. Barbon, D.D. Chinellato, W.M. Serenone, C. Shen, J. Takahashi, G. Torrieri, Phys. Rev. C **104**, 011901 (2021)

Measurement of global polarization of Λ hyperons in Au+Au $\sqrt{s_{NN}}$ = 7.2 GeV fixed target collisions at RHIC-STAR experiment

Kosuke Okubo (for the STAR Collaboration)*

[1]University of Tsukuba, 1-1-1 Tenno-dai, Tsukuba, Ibaraki 305-8571, Japan

Abstract. In non-central nuclear collisions, the created matter should exhibit strong vorticity, which induces spin polarization. Global polarization is a good tool to investigate the dynamics of the system. Global polarization of Λ and $\bar{\Lambda}$ hyperons has been measured at $\sqrt{s_{NN}}$ = 2.4 GeV - 5.02 TeV in various experiments. We have measured the global polarization of Λ hyperons in Au+Au collisions with fixed-target configuration at $\sqrt{s_{NN}}$ = 7.2 GeV at RHIC-STAR experiment. The non-zero signal is observed and it follows the global trend of Λ global polarization that increases towards lower collision energies. The global polarization is larger in more peripheral collisions but it does not show any significant rapidity dependence as well as transverse momentum dependence within uncertainties.

1 Introduction

In non-central heavy-ion collisions, the created matter should exhibit a rotational motion because of the conservation of the initial orbital angular momentum carried by the two nuclei, which is perpendicular to the reaction plane. Particle's and anti-particle's spins are aligned due to the spin-orbit coupling [1]. The strong magnetic field would appear in the initial state and its direction is almost the same as the initial angular orbital momentum. Particle's and anti-particle's spins also could get aligned in the opposite direction by the magnetic field due to the opposite signs of their magnetic moments. Therefore, global polarization is a good tool to investigate the dynamics of the system. Global polarization can be measured with hyperons through their parity-violating weak decays. The STAR Collaboration reported the global polarization of Λ and $\bar{\Lambda}$ hyperons at $\sqrt{s_{NN}}$ = 7.7 - 200 GeV and observed positive signals [2–4]. Additionally, the ALICE Collaboration measured global polarization of Λ and $\bar{\Lambda}$ hyperons at $\sqrt{s_{NN}}$ = 2.76, 5.02 TeV [5]. These measurements show that global polarization becomes larger at lower collision energies. The STAR fixed-target program provides an opportunity to extend such measurements at even lower energies. We here present the measurement of global polarization of Λ hyperons in Au+Au collisions at $\sqrt{s_{NN}}$ = 7.2 GeV with fixed-target configuration.

2 Analysis details

The good minimum bias data of Au+Au collisions have been collected by the STAR detector with fixed-target configuration in 2018 and 209 million events were analyzed in this study.

*e-mail: kokubo@rcf.rhic.bnl.gov

The gold target has been installed inside the vacuum pipe at 2.0 meters away from the center of the STAR detector. The decay channel of $\Lambda \rightarrow p + \pi$ (BR: 63.9% [6]) was utilized for Λ identification. The daughter particles were identified using TPC dE/dx and TOF and then Λ hyperons were reconstructed based on the invariant mass, applying cuts on decay topology. In this parity-violating weak decay, the daughter proton preferentially decays along the Λ's spin direction. Furthermore, since the system angular momentum is perpendicular to the reaction plane, the global polarization can be measured via the distribution of the azimuthal angle of the daughter proton in the Λ's rest frame with respect to the reaction plane as follows [2]:

$$P_H = \frac{8}{\pi \alpha_H} \frac{\langle \sin(\Psi_1 - \phi_p^*) \rangle}{\mathrm{Res}(\Psi_1)} \tag{1}$$

where α_H is Λ's decay parameter ($\alpha_\Lambda = 0.732 \pm 0.014$ [6]), Ψ_1 is the first order event plane, and ϕ_p^* is the azimuthal angle of the daughter proton in the Λ's rest frame. The event plane reconstructed by Event Plane Detector (EPD) and the resolution estimated by three subevent method [7].

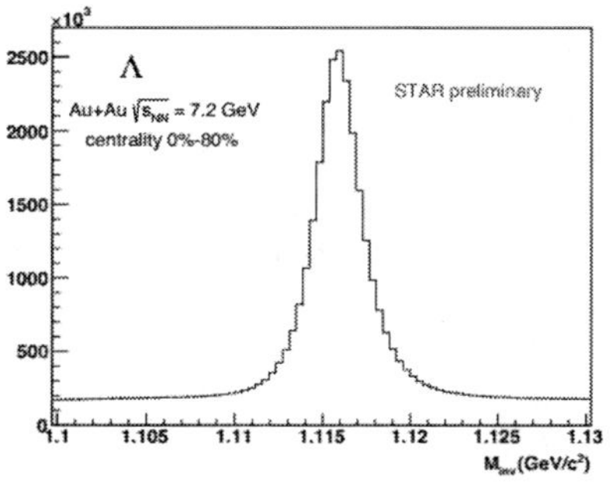

Figure 1: Invariant mass distribution of Λ hyperons in 0%-80% centrality for Au+Au collisions at $\sqrt{s_{\mathrm{NN}}} = 7.2$ GeV.

3 Result

Figure 2 shows global polarization of Λ and $\bar{\Lambda}$ as a function of collision energy. Solid markers indicate Λ and open markers indicate $\bar{\Lambda}$ global polarization. Because the Λ and $\bar{\Lambda}$'s decay parameter were updated recently, the previous results were rescaled by using the new values ($\alpha_{\mathrm{old}}/\alpha_{\mathrm{new}}$). The result at 7.2 GeV shows a non-zero positive signal. It follows the global trend of Λ global polarization that increases towards lower collision energies.

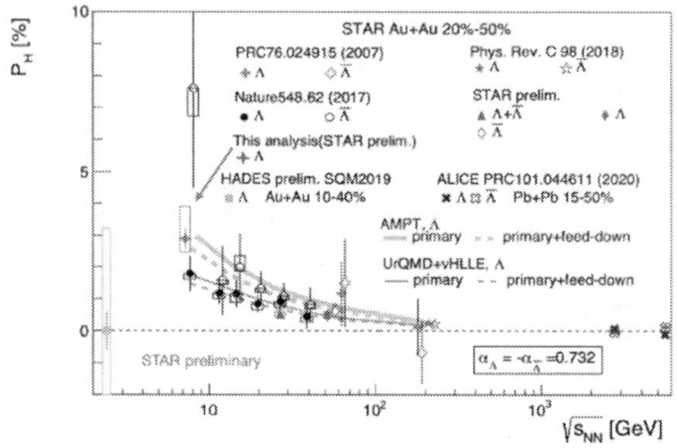

Figure 2: Global polarization of Λ (solid markers) and $\bar{\Lambda}$ (open markers) as a function of collision energy. Bold lines show the AMPT model calculations [8] and thin lines show calculations by a 3+1D cascade + viscous hydrodynamic model (UrQMD + vHELL) [9]. Primary Λ with and without the feed-down effect are indicated by dashed and solid lines, respectively.

We have also performed differential measurements such as centrality, rapidity, and transverse momentum dependence of global polarization. Figure 3 shows centrality dependence of Λ global polarization. Λ global polarization increases in more peripheral collisions as expected from the initial collision geometry.

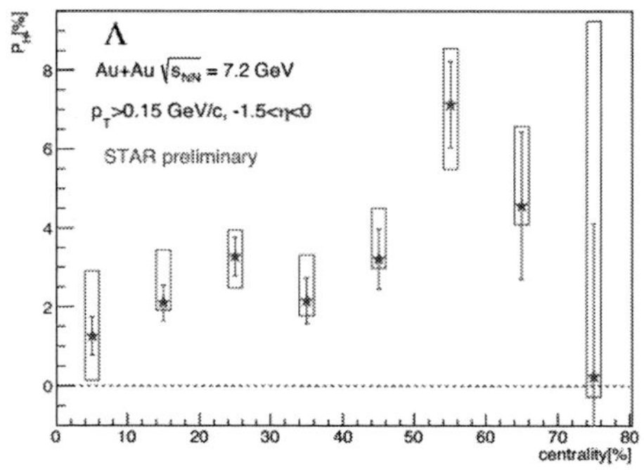

Figure 3: Global polarization of Λ hyperons as a function of centrality for $p_T > 0.15$ GeV/c, -1.5 $< \eta <$ 0.

Theoretical models predict that the polarization depends on rapidity, although the prediction is different among the models [10–13]. Figure 4 shows rapidity dependence of Λ global polarization for 20%-60% centrality, shown for 4 different p_T regions. Λ global polarization is consistent with being constant within uncertainties within the acceptance of the STAR detector. STAR experiment has recently installed a new detector inner Time Projection Chamber (iTPC) and will upgrade the forward detectors (2023+2025). With these upgrades, global polarization in large rapidity regions can be explored in the future. The Λ global polarization is expected to decrease at low p_T due to the smearing effect caused by scattering at the later stage of the collision. On the other hand, it may also decrease at high p_T due to the jet fragmentation. Figure 5 shows Λ global polarization as a function of transverse momentum for 20%-60% centrality. The Λ global polarization does not show significant p_T dependence within uncertainties.

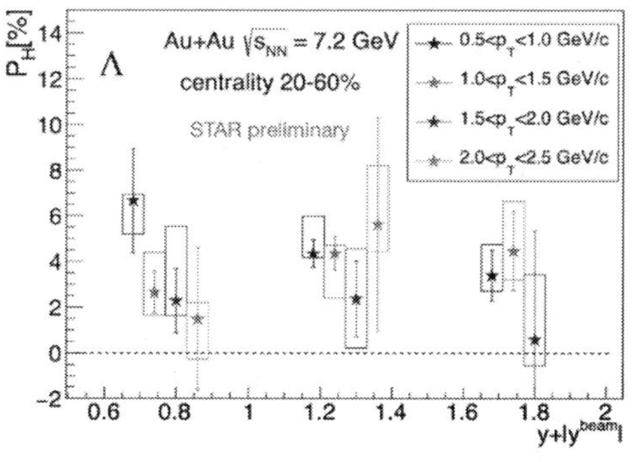

Figure 4: Global polarization of Λ hyperons as a function of rapidity for 20%-60% centrality each p_T region.

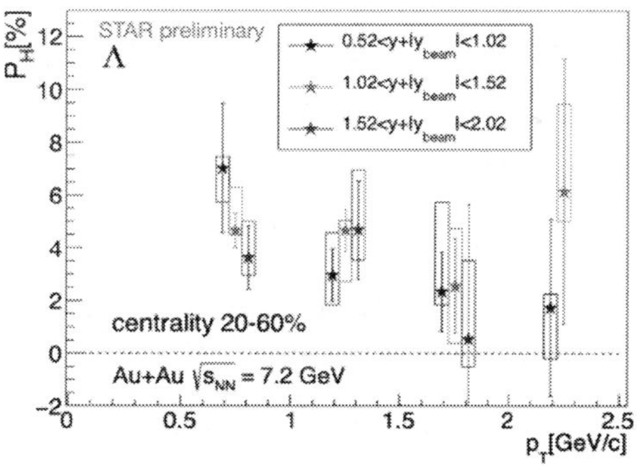

Figure 5: Global polarization of Λ hyperons as a function of transverse momentum for 20%-60% centrality each rapidity region.

4 Summary

We present the first measurement of Λ global polarization in Au+Au collisions at $\sqrt{s_{NN}} = 7.2$ GeV with STAR fixed-target configuration. The positive non-zero signal has been observed and it follows the global trend of energy dependence. We have also performed differential measurements of global polarization. The increasing trend of polarization towards peripheral collisions is observed. We have observed no significant rapidity and p_T dependence within uncertainties. With newly upgraded iTPC and future upgrade of forward detectors in 2023-2025, the global polarization will be explored in more forward regions.

References

[1] Z. T. Liang and X. N. Wang, Phys. Rev. Lett. 94, 102301 (2005).
[2] B. I. Abelev *et al.* (STAR), Phys, Rev. C 76 024915 (2007).
[3] L. Adamczyk *et al.* (STAR), Nature 548 62 (2017).
[4] J. Adam *et al.* (STAR), Phys. Rev. C 98 14910 (2018).
[5] S. Acharya *et al.* (ALICE), Phys. Rev. C 101 4, 044611 (2020).
[6] P. A. Zyla *et al.* (Particle Data Group), Prog. Theor. Exp. Phys. 2020, 083C01 (2020).
[7] A. M. Poskanzer and S. A. Voloshin, Phys. Rev. C 58, 1671 (1998).
[8] H. Li, L. Pang, Q. Wang, and X. Xia, Phys. Rev. C 96, 054908 (2017).
[9] I. Karpenko and F. Becattini, Eur. Phys. J. C 77, 213 (2017).
[10] W. T. Deng and X. -G Huang, Phys. Rev. C 93, 064907 (2016).
[11] Z. T. Liang *et al.*, Chin. Phys. C 45 1, 014102 (2021).
[12] D. X. Wei, W. T. Deng and X. G. Huang, Phys. Rev. C 99.014905 (2019).
[13] H. Z. Wu *et al.*, Phys. Rev. Research 1, 033058 (2019).

From Nuggets of SQM to Domains of Broken Symmetry (In memoriam of Jack Sandweiss)

Evan Finch[1,*]

[1]Southern Connecticut State University

Abstract. For the last 20 years of his remarkable career, Jack Sandweiss was an enthusiastic leader in heavy-ion physics, especially excited about two main topics: the search for stable or metastable strange quark matter and the possibility of local parity violation in heavy-ion collisions. I take a look back at some of Jack's contributions in the context of the current state of both of these very active areas.

1 Jack's early career

It was an honor for me to be able to give this talk remembering Jack Sandweiss at the SQM '21 conference. I had the great pleasure of working in Jack's group at Yale for about twenty years and have extremely fond memories of working with him, Dick Majka and Alexei Chikanian among the many great personalities of the 5th floor of Gibbs Lab. Anyone who worked with Jack remembers his truly extraordinary broad and deep knowledge of physics, but what stands out to many of us even more was his relentless enthusiasm, not just for his field or for physics in general, but for just about everything he was involved with, and just how much he enjoyed talking to other people about what they were excited about to broaden his own interests. John Harris, who was chairing this session at SQM, shared how Jack's presence was one of the major draws for John coming to Yale, and also remembered Jack's as a "true academic", a term that really fit Jack with his vast interests and the involvements he shared with his wife Letha as it does few others.

Jack's graduate studies were done at the rad lab in Berkeley, and he started out right in the thick of particle physics, where he worked in one of the groups searching for the antiproton, and though their technique didn't win the race due to the unexpectedly large annihilation cross-section, they were able to make one of the first measurements of that cross section [1]. He did a bit of work [2] that contributed to the understanding of the "$\tau - \theta$" kaon puzzle that helped lead to the discovery of weak parity violation. He did it all - he soon moved to Yale where he worked with one of his first students Charlie Baltay as they became experts on building RF-field separated beams at Brookhaven to isolate beams of kaons and hyperons. And one of the accomplishments he often spoke of most fondly was the design and construction of the high-resolution streamer chamber for charm quark measurements at Fermilab [3]. He became an eminently respected physicist; helping to steer the field by chairing the HEPAP advisory panel in the 80s, and succeeding his friend Bob Adair as chief editor of Physical Review Letters through the 90s and 2000s.

*e-mail: FinchL3@southernCT.edu

Figure 1. Jack at CERN. AIP Emilio Segre Visual Archives, Segre Collection

2 Searches for SQM

Jack's interest in searches for stable Strange Quark Matter (SQM) began in the mid-80s. SQM, made up of roughly 1/3 up, down, and strange quarks in a single hadronic state with baryon number $A \geq 2$, was suggested almost 50 years ago as the possible true ground state of nuclear matter [4, 5]. That quark matter with only up and down quarks could be the true ground state of hadronic matter at zero temperature and pressure is easily dismissed by the fact that normal matter doesn't decay into it, but this argument doesn't rule out the stability of SQM because a similar transition would require a prohibitive number of simultaneous weak decays. Theoretical calculations, chiefly with the MIT bag model, give some support to the idea of SQM stability for a respectable chunk of parameter space, but the question of stability is only answerable experimentally (or observationally), with the experimental signatures being hadronic objects with anomalously low charge to mass ratios or that are considerably more massive than stable nuclei.

Strangelets (small chunks of SQM) may also be metastable and decay with weak lifetimes, but Jack's chief interest was generally in the idea of stable cold SQM. Always part of his talks on the subject was a slide about the exciting implications of stability: the extensions of chemistry and materials with stable nuclei up to charge of $Z \approx 1000$ (the surface electric field strength would prevent higher charge) and superdense materials, the potential as a convenient clean energy source [6]. His interest led him to join Yale colleague Shiva Kumar as part of Experiment 814 at the Brookhaven AGS, where his group performed an initial strangelet search and started their work in heavy-ion physics. He (with Dick Majka as co-spokesperson) then led the much more specialized E864, a large-acceptance open-geometry spectrometer with a particularly notable hadronic calorimeter and "Late Energy Trigger" system that facilitated a highly sensitive search [7] that reached a sensitivity of around 10^{-8} per central AGS heavy-ion collision, for positive, negative and neutral(!) strangelets with $A > 5$ and lifetime $\tau > 50ns$.

No strangelets were found by E864, but like many AGS experiments it was great place to be a student-John Lajoie, Zhangbu Xu, Ken Barish and Jamie Nagle (from Shiva's group) are among the Yale PhD students from E864 who are still very active in the field. In addition, measurements were made of production yields (and limits) for stable and unstable light nuclei, antinuclei and hypernuclei, as summarized in the left panel of Figure 2: and detailed in Refs [8, 9]. Among other things, these measurements strongly suggested that cold SQM wasn't going to be formed in collisions as the "penalties" for coalescing additional nucleons (a factor of ≈ 48, as shown by the fit on the positive A side of the Figure) and strangeness (another factor of a couple, though with a large uncertainty) into clusters was too steep. The search for cold SQM would need other avenues...

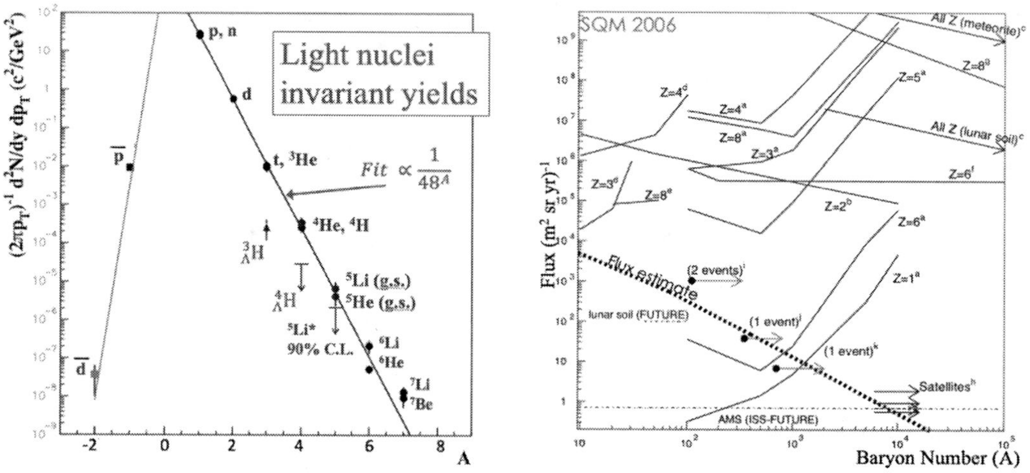

Figure 2. Left panel shows a compilation of some of the banner results from E864: yields and limits for light nuclei, antinuclei, hypernuclei at the AGS, with implications for coalescing strangelets described in text. Right panel gives an overview circa 2006 of searches relevant to strangelets in cosmic rays as strange star relics along with the implied flux. The lunar soil search is long completed [12] and the first published AMS results are recently in [13], to supercede those "future" lines in the figure with sensitivities noted in the text.

After getting reeled in by various interests in heavy-ions at RHIC (see section 3), Jack returned to his interest in strangelets following the 1998 hypothesis of the existence of a "color-flavor-locked" phase of SQM that pointed Jack down the road of astrophysical searches to investigate if "neutron stars" could in fact all be strange stars. (The possibility additional stable SQM phases removed the objection that observed pulsar glitches would not be consistent with fully SQM stars). Under that assumption, theorist Jes Madsen computed a flux estimate [11] for a strangelet component of cosmic rays incident on Earth. Jack then spent some time researching whether that estimated flux could be ruled out by various balloon or satellite-based studies, or relic searches with terrestrial or lunar material (the landscape hasn't changed much since our group looked into it - a still-almost-current plot is in Figure 2 from Ref [10]. As Jack emphasized in his contribution proceedings to the 2003 SQM conference: "...the question as to (stable SQM's) existence remains open. A new experiment, the Alpha Magnetic Spectrometer (AMS), to be mounted on the International Space Station offers the possibility of a definitive answer to the question." The promise of AMS is a tremendously high-class detector for cosmic ray measurements sitting largely above the influence of the geomagnetic field. For a strangelet search, it operates in principle much as E864 did, with redundant time-of-flight measurements (to measure speed β) and precise tracking (for a rigidity R) that could be combined to measure the charge to mass ratio of each cosmic ray. Because the launch of AMS was delayed and for a time uncertain in the aftermath of the Columbia accident, another clever idea of Jack's was to use lunar soil (applied for and granted from among NASA's Apollo 11 samples) as a beam source in Yale's tandem Van-de-graff, which operated as a mass spectrometer for a strangelet search. In particular, this was a follow-up on some interesting events recorded by the AMS-01 prototype flight. The sensitivities achieved by the analysis of Ke Han [12] from Jack's group translate to a flux of around 200 per (m^2 sr yr) over a somewhat narrow bite in mass of $40 < A < 70$, and so were sufficient to begin address the flux predictions and the AMS-01 events, but not at the level that AMS would promise. And at last in 2021, the long-awaited first results from AMS on strangelets are published [13]- no observation of strangelets, with a flux sensitivity of around 40 per (m^2 sr yr) over a large range of charge and mass. The limit sits below the expected flux in the right panel of Figure 2 for a very significant mass range. With the large uncertainties attached to the expected flux, one shouldn't yet argue that this has succeeded at Jack's expectation being able to rule out the idea of stable SQM, but improvements in the AMS search sensitivity still hold promise to do that (or, indeed, for discovery).

3 Local Parity Violation

Leading E864 heightened Jack's interest in the larger heavy-ion physics world, and the arrival at Yale of John Harris drew him the rest of the way in as he joined the STAR experiment. As the RHIC collider was nearing its first run, Jack also had a very lovely idea (as he would point out with a broad smile, that became tenable only after he found a minor error in a paper by some excellent theorists) to measure the charged particle stopping (and through this estimate the energy density achieved) at RHIC via a measurement of Bremstrahlung photons. I really enjoyed this description of the design that I ran across on Ben Monreal's twitter feed soon after Jack's passing: "Start with a cute, elegant bit of classical E&M phenomenology everyone else overlooked, realize it produces a tiny photon signal that's swamped by pion backgrounds, and concoct a specialized bit of apparatus that enables a signal on/signal off background subtraction. A tiny work of art ". The bremstrahlung experiment didn't go forward at RHIC, bit it has recently seen renewed interest at LHC energy [14]

Though he worked on other topics in heavy-ions, his main interest started with the 1998 paper [15] suggesting that parity and CP symmetries might be broken "locally" in regions generated in heavy-ion collisions. He led the charge to push on what could be done with this idea experimentally, in the early days helping to organize many joint meetings with theorists and experimentalists to try to understand the types of observables that could be formed from combinations of particles' final-state momenta to be sensitive [16], and the methods to extract a reliable measurement given the nature of this "local" violation to reverse handedness from region to region, and later working on complementary methods such as measurements of correlations of Λ-helicities [17].

While Jack's attention was largely back on strangelets in the mid 2000s, the leap was made forward to understand that this Local Parity Violation should combine with the large magnetic field in heavy-ion events to cause charge separation across the event reaction plane (known as the "Chiral Magnetic Effect") [18] and the methods for observing this were developed [19], and Jack rejoined the effort full-bore as STAR made initial measurements that found signals qualitatively consistent with many aspects of the CME expectations [20]. This was a big step, but as the New York Times noted: "Dr. Sandweiss cautioned that it was still possible that some other effect could be mimicking the parity violation". Backgrounds, chiefly from flowing two-particle clusters or similar sources, were known early-on to be a potential issue and have proven very difficult to convincingly separate from CME-related signal despite a great deal of work on the subject.

Perhaps the most promising method to separate signal from background is through the use of isobar collisions that were run by RHIC in 2018. $^{96}_{44}$Ru+$^{96}_{44}$Ru and $^{96}_{40}$Zr+$^{96}_{40}$Zr collision systems are expected to have differences in CME-related signal by around 15% due to the larger magnetic field in Ru, but to have similar backgrounds (with background differences that to first order can be estimated by measuring flow in the two systems). The results from a large-scale effort of several groups within STAR, all performing blind analyses, are due to be released to the public by September 2021, and the field is eagerly awaiting these results. While this work has been going on, there have been some very nice developments in other CME-analysis techniques. One particularly promising method involves utilizing the difference of measured CME signal and flow with respect to the reaction plane in the central region (which will maximize backgrounds) and the event plane measured in forward detectors (which will maximize CME-related signal). Depending on assumptions about how the CME signal should scale between Au+Au to the isobar systems, it can be argued that this method has similar or even higher sensitivity than the isobar method. STAR has recently submitted results from this method for publication [21]. These results are reproduced in Figure **??** and they imply that while the fraction of measured charge separation signal originating from CME is small ($< 10\%$), it is 1-3 sigma away from zero, though some concerns about non-flow backgrounds remain. It will be very interesting to see what the isobar results bring to light! Looking to the future, another Jack-inspired idea is to correlate the CME-related charge separation with other parity-odd signals such as net lambda helicity[22]. Such a study holds the promise of eliminating flow-related background but how much the signal will be present in the strange quarks is a significant question.

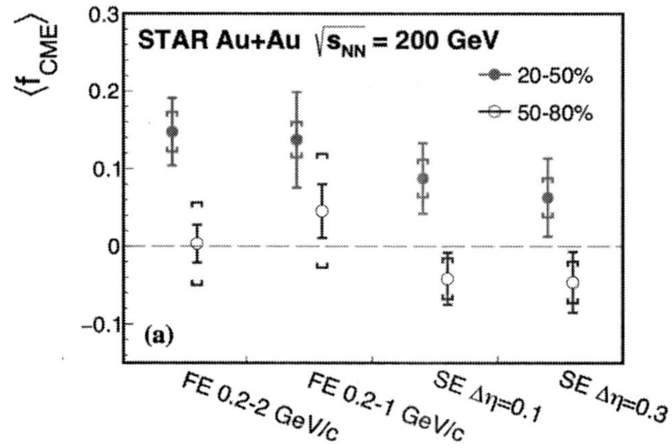

Figure 3. STAR results for the fraction of charge separation due to CME (f_{CME} utilizing the difference in various signals with respect to forward and central reaction planes. Different columns represents variations in the methods, using full (FE) or sub-events (SE) for the analysis. The resulting signal is consistent with zero in peripheral collisions but 1-3 sigma away from zero in mid-central collisions. There are many important nuances here-see [21] for details.

4 Summary

The intention of the organizers was that we remember Jack by discussing some of the physics that particularly delighted him, and it is appropriate that we're talking about him as the STAR isobar results near completion and the AMS strangelet limits are released. We keenly miss having him with us to discuss these new results that he would have been eagerly awaiting, but that's truly only a small piece of what we miss about Jack!

References

[1] D H Stork, R W Birge, R P Haddock, L T Kerth, J Sandweiss and M N Whitehead, Phys Rev **105**, 729 (1957)
[2] R W Birge, R P Haddock, L T Kerth, J Sandweiss, D H Stork and M N Whitehead, Il Nuovo Cimento **4**, 351 (1956)
[3] J Sandweiss, Physics Today **31**, 10 (1978)
[4] A R Bodmer, Phys Rev D **4**, 1601 (1971)
[5] E Witten, Phys Rev D **30**, 272 (1984)
[6] G L Shaw, M Shin, R Dalitz and M Desai, Nature **337**, 436 (1989)
[7] T A Armstrong et. al. (E864 Collaboration), Phys Rev C **63**, 054903 (2001)
[8] T A Armstrong et. al. (E864 Collaboration), Phys Rev Lett **83**, 5431 (1999)
[9] T A Armstrong et. al. (E864 Collaboration), Phys Rev C **67**, 014902 (2004)
[10] J Sandweiss, J. Phys G, **30**, S51 (2004), E Finch, J. Phys G, **32**, S251 (2006)
[11] J Madsen, Phys Rev D **71**, 014026 (2005)
[12] K Han et. al. (LSSS Collaboration), Phys Rev Lett **103**, 092302 (2009)
[13] M Aguilar et. al. (AMS Collaboration), Physics Reports **894**, 1 (2021)
[14] S Park and U Wiedemann, arXiv:2107.05129-hep-ph (2021)
[15] D Kharzeev, R D Pisarski, M H G Tytgat, Phys Rev Lett **81**, 512 (1998)
[16] L E Finch, A Chikanian R S Longacre, J Sandweiss and J H Thomas, Phys Rev C **65**, 014908 (2002)
[17] F Du, L E Finch and J Sandweiss, Phys Rev C **78**, 044908 (2008)
[18] K Fukushima, D E Kharzeev and H J Warringa, Phys Rev D **78**, 074033 (2008)
[19] S Voloshin, Phys Rev C **70**, 057901 (2004)
[20] B I Abelev et al (STAR Collaboration), Phys Rev Lett **103**, 251601 (2009)
[21] M Abdallah et al (STAR Collaboration), arXiv:2106.09243-nucl-ex (2021)
[22] L E Finch and S J Murray, Phys Rev C **96**, 044911 (2017)

Exact strangeness conservation in heavy ion collisions*

Krzysztof Redlich[1,**] and *Natasha* Sharma[2,***]

[1]Institute of Theoretical Physics, University of Wrocław, 50-204 Wrocław, Poland
[2]Indian Institute of Science Education and Research (IISER) Berhampur, 760010 Odisha, India

Abstract. We investigate the increase in strangeness production with charged particle multiplicity (dN_{ch}/dy) seen by the ALICE collaboration at CERN in p-p, p-Pb and Pb-Pb collisions using the hadron resonance gas model. The strangeness canonical ensemble is used taking into account the interactions among hadrons using S-matrix corrections based on known phase shift analyses. We show the essential role of constraints due to the exact conservation of strangeness which is instrumental to describing observed features of strange particle yields and their scaling with dN_{ch}/dy. Furthermore, the results on comparing the hadron resonance gas model with and without S-matrix corrections, are presented. We observe that the interactions introduced by the phase shift analysis via the S-matrix formalism are essential for a better description of the yields data independent of collision system. This work is based on our long-lasting collaboration and most recent publications with Jean Cleymans****.

1 Introduction

The experimental particle yeild data from heavy-ion collisions across different experiments and broad range of energies have been shown to originate from a thermal production [1–7]. These data have been successfully explained by the hadron gas model (HRG) with a common freezeout temperature T_f and chemical potentials $\vec{\mu}_f$ associated with the conserved charges. It has been shown that the freezeout temperature obtained from the HRG conincides with the chiral crossover temperature from lattice QCD (LQCD) at vanishingly small finite μ_B [8].

It is argued that the S-matrix scheme can improve the HRG model in approximating the QCD partition function in the hadronic phase and hence can provide a more accurate description of the measured particle yields [9–14]. In this work, we apply the S-matrix extended HRG model to analyze data obtained by the ALICE collaboration on charged-particle multiplicity in p-p collisions at 7 TeV [15] as well as 13 TeV [16], in p-Pb collisions at 5.02 TeV [7, 17] and in Pb-Pb collisions at 2.76 TeV [18–20], all in the central region of rapidity.

We estimate the yields of (multi-)strange hadrons and discuss strangeness suppression as a function of $dN_{ch}/d\eta$ and strangeness content of hadrons. We formulate the HRG model in the canonical ensemble of strangeness conservation and account for differences between the fireball volume at midrapidity V_A and the correlation volume V_C required for exact global strangeness conservation [21–24]. We include the S-matrix corrections to proton production

*In memory of Jean Cleymans
**e-mail: krzysztof.redlich@uwr.edu.pl
***e-mail: natasha.sharma@cern.ch
****https://home.cern/news/obituary/cern/jean-willy-andre-cleymans-1944-2021

by using the empirical phase shifts of πN scattering and the contribution of hyperons to proton yield and corrections to the hyperon yields, employing an existing coupled-channel study involving $\pi\Lambda$, and $\pi\Sigma$ interactions in the $S = -1$ sector [25].

2 Strangeness production and HRG model

In the framework of HRG model with the grand canonical ensemble, the quantum numbers are conserved on average and are implemented using the corresponding chemical potentials $\vec{\mu}$ linked to conserved charges $\vec{Q}$. At the LHC energies, all chemical potentials $\vec{\mu}$ are set to zero, i.e. the system is charge neutral. Such a model has been applied to quantify thermalization and particle production in most central heavy-ion collisions at the LHC. At low multiplicity events or low collisions energies, a thermal description requires exact implementation of charge conservation which is usually described in the C-ensemble [21, 22, 26–30].

We focus on the canonical ensemble with exact strangeness conservation (total strangeness $S = 0$) and assume all other quantum numbers are conserved on average in the GC ensemble with $\vec{\mu} = 0$. The strangeness canonical partition function can be expressed as a series of Bessel functions [21, 27, 30, 31],

$$Z^C_{S=0} = \sum_{n,p=-\infty}^{\infty} I_n(S_2) I_p(S_3) I_{-2n-3p}(S_1). \tag{1}$$

The mean multiplicity in the canonical ensemble of particle k carrying strangeness s in a given experimental acceptance A, are obtained as

$$\langle N^s_k \rangle_A = V_A\, n^s_k(T)\, \frac{1}{Z^C_{S=0}} \times \sum_{n,p=-\infty}^{\infty} I_n(S_2) I_p(S_3) I_{-2n-3p-s}(S_1). \tag{2}$$

The particle yields are obtained using the temperature (T) and two volume parameters: the volume of the system in the acceptance V_A and the correlation volume V_C of global strangeness conservation which appear in the arguments of the Bessel functions. To get the total multiplicity of a given strange particle in the C ensemble resonance decays have to be added. To identify the contribution and dependence of various terms in Eq. (2) on strange particle quantum number s, only leading terms can be considered ($p = n = 0$), so

$$\langle N^s_k \rangle_A \simeq V_A\, n^s_k(T)\, \frac{I_s(S_1)}{I_0(S_1)}. \tag{3}$$

The ratio of $I_s(S_1)/I_0(S_1)$ is the suppression factor which decreases with increasing s and with decreasing thermal phase-space occupied by strange particles. These correspond to the strangeness canonical suppression that have been introduced [21] to describe thermal production of multi(strange) hadrons in heavy-ion collisions. In this work, we apply the HRG model with the C-ensemble to quantify the production of (multi-)strange hadrons and their behavior with charged particle multiplicity. We modify the HRG model with a more complete implementation of interactions within the S-matrix formalism.

3 S-matrix formalism and HRG

The various interactions among hadrons modify the density of states (DOS) of a thermal system and hence the thermal abundances of hadron states [9, 10, 32, 33]. In the scattering matrix (S-matrix) formalism, an effective spectral function $B(M)$ (M represents center-of-mass energy of the system) describing the DOS can be computed from the S-matrix [9]. It summarizes the various interactions among the scattering channels. As the energy increases, new interaction channels open and scattering becomes inelastic. The channel-specific spectral function $B_a(M)$ describes the energy dependent component of the full $B(M)$ for a given channel a. The channel yield, e.g., from a resonance decaying into multiple final states, can be obtained as

$$n_a(T) = \int_{m_{th}}^{\infty} \frac{dM}{2\pi} B_a(M)\, n^{(0)}(T, M), \tag{4}$$

where $n^{(0)}$ is the ideal gas formula for the particle density.

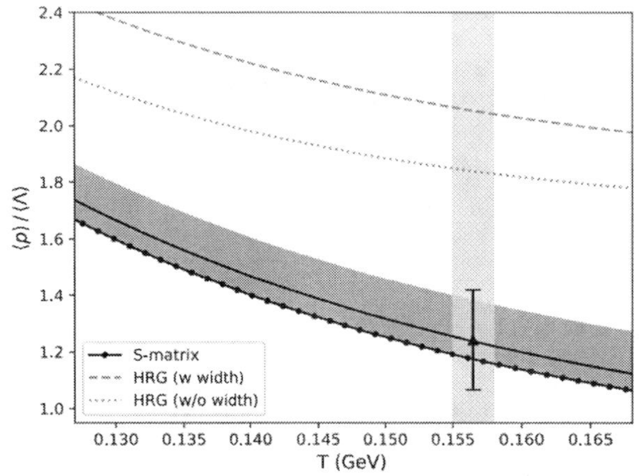

Figure 1. Proton to $\Lambda + \Sigma^0$-baryons yield ratio is shown as a function of temperature. The predictions from the S-matrix scheme and from the HRG scheme (with and without width) are presented as line, dashed line and dotted line respectively. The blue solid triangle represents the experimental value measured by the ALICE collaboration for the most central Pb-Pb collisions (at 2.76 TeV).

We focus on computing the hadron yields of protons and $\Lambda + \Sigma^0$-baryons. Following previous studies for protons, we employ the empirical phase shifts from the GWU/SAID PWA on the πN scattering [34]. In addition to this, we implement a $\pi\pi N$ background contribution based on the LQCD computation of the baryon charge-susceptibility χ_{BQ}. For the strange ($|S| = 1$) baryon system, we employ an existing coupled-channel model involving $\bar{K}N$, $\pi\Lambda$, and $\pi\Sigma$ interactions. The S-matrix scheme for $|S| = 1$ hyperons is based on Ref. [35] without any extra tuning.

It is observed that the HRG model overestimates the yields of protons and underestimates the yields of $\Lambda + \Sigma^0$-baryons. As a result, the ratio of the yields of protons to $\Lambda + \Sigma^0$-baryons is larger than the measured experimental ratio. Figure 1 shows the comparison of ratio of protons to $\Lambda + \Sigma^0$ yields measured by the ALICE collaboration for the most central Pb-Pb collisions (at 2.76 TeV), with the corresponding predictions from the HRG model and from the S-matrix scheme. It is observed that the result obtained from the measured hadron yields favor the S-matrix prediction and hence illustrates the robustness of this method.

4 Comparison of results with ALICE data

We have used the latest version of THERMUS [36] and extended it to include the S-matrix approach. As a first step, we made a fit for each multiplicity bin of pp collisions at $\sqrt{s} = 7$ TeV [15] by keeping the number of parameters to a minimum. Two temperature values were chosen: $T_f = 156.5$ MeV and $T_f = 160$ MeV based on the results from LQCD [37] and the recent HRG model analysis of ALICE data for central Pb-Pb collisions [4]. The strangeness suppression factor γ_s is fixed to 1 ($\gamma_s = 1$) as motivated by fits in central Pb-Pb collisions,. The chemical potentials due to conservation of baryon number and electric charge are set to zero for LHC energy. Thus, in the SCE only two parameters remained, the volume of the system in the experimental acceptance V_A and the canonical volume V_C which quantifies the range of exact strangeness conservation.

Figure 2 (left) shows the hadrons yields calculated using the SCE for different charged particle multiplicities $dN_{ch}/d\eta$ for a single volume $V = V_A \simeq V_C$. For comparison, the experimental results from the ALICE collaboration are also shown as symbols. The SCE model qualitatively explains the hadron yields data even with a single volume parameter. For large $dN_{ch}/d\eta > 100$ all hadron yields, as well as extracted V, depend linearly on charged particle rapidity density. However, for lower $dN_{ch}/d\eta$ this dependence is clearly non-linear for strange particles due to strangeness canonical suppression which increases with the strangeness content of particles. At the quantitative level, however, as seen in Fig. 2 (left), using a single volume leads to too much suppression at small charged particle multiplicities, particularly for $S = -2$ and $S = -3$ baryons. This result is consistent with the previous observation, that a single volume canonical model implies too strong strangeness suppression in low multiplicity events [17].

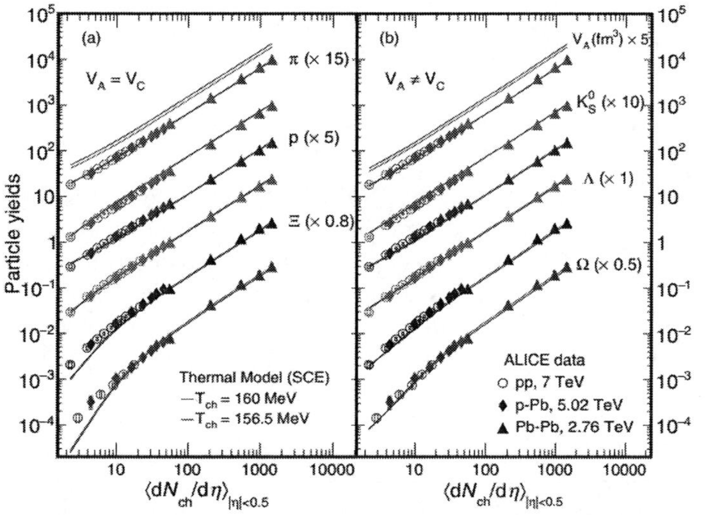

Figure 2. Left: Yields for $V_A = V_C$. Right: Yields for $V_A \neq V_C$, The two top lines are the fitted volumes V_A (x5) in fm^3. The particle yields are indicated together with the multiplicative factor used to separate the yields. The solid blue lines have been calculated for $T = 156.5$ MeV while the solid red lines have been calculated for $T = 160$ MeV. The values of the volumes used have been parametrized empirically (see text for more details).

We have also performed the SCE model fit to data with two independent volume parameters as shown in Fig. 2 (right). In general, strangeness conservation relates to the full phase-space whereas particle yields are measured in some acceptance window. Thus, the strangeness canonical volume parameter V_C can be larger than the fireball volume V_A, restricted to a given acceptance. The resulting yields exhibit much better agreement with data by decreasing strangeness suppression at lower multiplicities due to larger value of V_C than V_A.

The pion yields data are slightly underestimated by the model points while the kaons are overestimated. This has implications for the kaon to pion ratio discussed further below. The

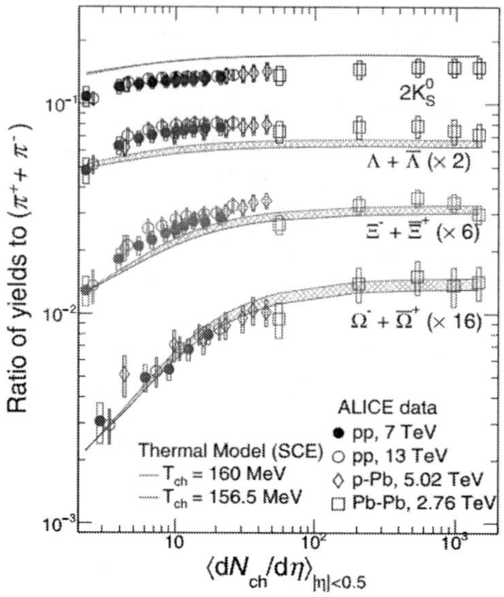

Figure 3. Strange and multi-strange particles to pions yield ratios versus charged particle multiplicity. The symbols show experimental data and the lines represent the SCE model results obtained using $V_A \neq V_C$.

strangeness suppression effect and its CSE model description are important to describe the dependence of particle yields on the fireball volume V_A. The ratio of strange particle and pion yields, is shown in Fig. 3. This ratio has been discussed prominently by the ALICE collaboration [15] where a comparison with other model calculations was presented. The SCE model introduced here compares very favorably to the ones discussed in [15]. The underestimation of the pion yield is responsible for the larger discrepancy in the kaon to pion ratio.

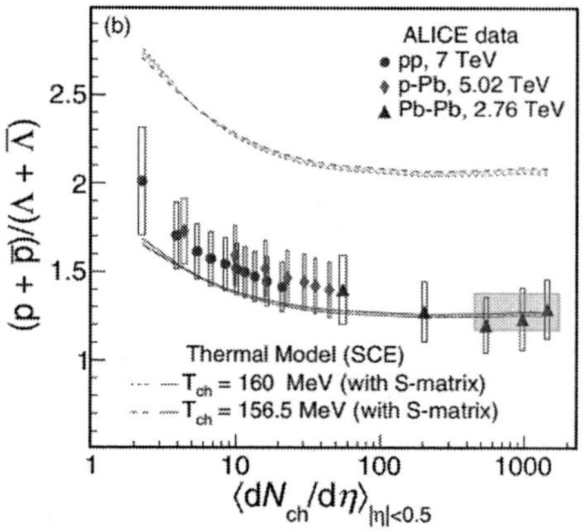

Figure 4. Proton to Lambda ratio as a function of charged particle multiplicity is shown. The lines represent predictions from the HRG model including the S-matrix corrections while the dotted lines represent predictions from the HRG model without S-matrix corrections at temperature T = 160 MeV and 156.5 MeV. The solid symbols represent the experimental data measured by the ALICE collaboration[7, 15–20].

The importance of the S-matrix description of proton and hyperon yields can be seen with data from the central Pb-Pb collisions by comparing measured p/Λ ratio with the S-matrix results as shown in Fig. 4. The lines represent predictions from the HRG model including the S-matrix corrections while the dotted lines represent predictions from the HRG model without S-matrix corrections at temperature T = 160 MeV and 156.5 MeV. The shaded region (in gray) shows the theoretical predictions spanned by different stages of the S-matrix improvement: from including only elastic scatterings of ground state hadrons (upper limit) to the full list of interactions (lower limit). The solid symbols represent the experimental data measured by the ALICE collaboration[7, 15–20]. The ratio is very sensitive to the S-matrix corrections which amount to a reduction of proton yield by a factor of 0.75 and a 1.24 enhancement of the Λ yield. The thermal model without S-matrix corrections exhibits large deviations from the p/Λ ratio data. We also found that the ratios of strange and multi-strange particles from SCE model describe the ALICE data very accurately.

5 Summary

We have studied the effect of global strangeness quantum number conservation on strangeness production in heavy-ion and elementary collisions in a given acceptance region using the hadron resonance gas (HRG) model in the canonical ensemble. The HRG model is extended with the S-matrix corrections to the yields of protons and hyperons. The S-matrix calculation is based on the empirical phase shifts of πN scattering, an estimate of the $\pi\pi N$ background constrained by Lattice QCD results of baryon-charge susceptibility, and an existing coupled-channel model describing the $|S| = 1$ strange baryons.

It is demonstrated that an accurate description of the widths of resonances and the non-resonant interactions in a thermal model leads to a reduction of the proton yield relative to the HRG baseline (by $\approx 25\%$). Including the protons from strong decays of $|S| = 1$ hyperons, which constitute $\approx 6\%$ of the total yields, does not alter this conclusion. Such a reduction is also crucial for resolving the proton anomaly in the LHC data. For the hyperons, the S-matrix scheme predicts an increase in the $\Lambda + \Sigma^0$ yields relative to the HRG baseline by $\approx 23\%$. This is consistent with the data from the ALICE collaboration. Furthermore, the S-matrix prediction on the ratio of yields of proton to $\Lambda + \Sigma^0$ is in good agreement with the measured values by the ALICE collaboration in Pb-Pb collisions in the events with the largest multiplicities ($dN_{ch}/d\eta$). The evolution of (multi-)strange baryons to Λ yields with $dN_{ch}/d\eta$ calculated in the present thermal model in the C-ensemble follows the measured values within two standard deviations.

A good description was obtained for the variation of the strangeness content in the final state as a function of the number of charged hadrons at mid-rapidity with the same freezeout temperature $T_f \sim 156.5$ MeV. This further supports the idea that at LHC energies (independent of colliding system), the freezeout temperature coincides with the chiral crossover as calculated in LQCD. An exact conservation of strangeness is to be imposed in the full phase-space rather than in the experimental acceptance at mid-rapidity. Consequently, the correlation volume parameter where strangeness is exactly conserved was found to be larger than the fireball volume at mid-rapidity.

The S-matrix formulation discussed here accurately describes the measured hadron yields and supports the interpretation of the LQCD results. It will be useful to employ this in analyzing the multi-strange baryon sectors and the light mesons.

6 Acknowledgments

K.R. acknowledges partial support by the Polish National Science Center (NCN) under OPUS Grant No. 2018/31/B/ST2/01663, and by the Polish Ministry of Science. N.S. acknowledges the support of SERB Research Scientist research grant (D.O. No. SB/SRS/2020-21/48/PS) of the Department of Science and Technology, Government of India. We also acknowledge fruitful collaboration with Pok Man Lo.

References

[1] A. Andronic, P. Braun-Munzinger, K. Redlich, J. Stachel, Nature **561**, 321 (2018), 1710.09425
[2] A. Andronic, Int. J. Mod. Phys. A **29**, 1430047 (2014), 1407.5003
[3] F. Becattini, P. Castorina, A. Milov, H. Satz, Eur. Phys. J. **C66**, 377 (2010), 0911.3026
[4] A. Andronic, P. Braun-Munzinger, B. Friman, P.M. Lo, K. Redlich, J. Stachel, Phys. Lett. B **792**, 304 (2019), 1808.03102
[5] S. Chatterjee, S. Das, L. Kumar, D. Mishra, B. Mohanty, R. Sahoo, N. Sharma, Adv. High Energy Phys. **2015**, 349013 (2015)
[6] S. Das, D. Mishra, S. Chatterjee, B. Mohanty, Phys. Rev. C **95**, 014912 (2017), 1605.07748
[7] B.B. Abelev et al. (ALICE), Phys. Lett. B **728**, 25 (2014), 1307.6796
[8] A. Bazavov et al., Phys. Rev. D **95**, 054504 (2017), 1701.04325
[9] R. Dashen, S.K. Ma, H.J. Bernstein, Phys. Rev. **187**, 345 (1969)
[10] R. Venugopalan, M. Prakash, Nucl. Phys. A **546**, 718 (1992)
[11] W. Weinhold, B. Friman, W. Norenberg, Phys. Lett. B **433**, 236 (1998), nucl-th/9710014
[12] P.M. Lo, B. Friman, K. Redlich, C. Sasaki, Phys. Lett. B **778**, 454 (2018), 1710.02711
[13] A. Dash, S. Samanta, B. Mohanty, Phys. Rev. C **99**, 044919 (2019), 1806.02117
[14] A. Dash, S. Samanta, B. Mohanty, Phys. Rev. **C97**, 055208 (2018), 1802.04998
[15] J. Adam et al. (ALICE), Nature Phys. **13**, 535 (2017), 1606.07424
[16] S. Acharya et al. (ALICE), Eur. Phys. J. C **80**, 167 (2020), 1908.01861
[17] J. Adam et al. (ALICE), Phys. Lett. B **758**, 389 (2016), 1512.07227
[18] B. Abelev et al. (ALICE), Phys. Rev. C **88**, 044910 (2013), 1303.0737
[19] B.B. Abelev et al. (ALICE), Phys. Rev. Lett. **111**, 222301 (2013), 1307.5530
[20] B.B. Abelev et al. (ALICE), Phys. Lett. B **728**, 216 (2014), [Erratum: Phys.Lett.B 734, 409–410 (2014)], 1307.5543
[21] S. Hamieh, K. Redlich, A. Tounsi, Phys. Lett. B **486**, 61 (2000), hep-ph/0006024
[22] R. Hagedorn, K. Redlich, Z. Phys. C **27**, 541 (1985)
[23] H. Satz, Acta Phys. Polon. Supp. **10**, 711 (2017)
[24] P. Castorina, H. Satz, Int. J. Mod. Phys. E **23**, 1450019 (2014), 1310.6932
[25] J. Cleymans, P.M. Lo, K. Redlich, N. Sharma, Phys. Rev. C **103**, 014904 (2021), 2009.04844
[26] R. Hagedorn (1971)
[27] J. Cleymans, K. Redlich, E. Suhonen, Z. Phys. **C51**, 137 (1991)
[28] J. Cleymans, H. Oeschler, K. Redlich, Phys. Rev. C **59**, 1663 (1999), nucl-th/9809027
[29] C. Ko, V. Koch, Z.W. Lin, K. Redlich, M.A. Stephanov, X.N. Wang, Phys. Rev. Lett. **86**, 5438 (2001), nucl-th/0010004
[30] P. Braun-Munzinger, K. Redlich, J. Stachel, *Particle production in heavy ion collisions*, Quark–Gluon Plasma (World Scientific, 2003), nucl-th/0304013
[31] P. Braun-Munzinger, J. Cleymans, H. Oeschler, K. Redlich, Nucl. Phys. A **697**, 902 (2002), hep-ph/0106066

[32] P.M. Lo, Eur. Phys. J. C **77**, 533 (2017), 1707.04490
[33] P.M. Lo, Phys. Rev. D **102**, 034038 (2020)
[34] R. Workman, R. Arndt, W. Briscoe, M. Paris, I. Strakovsky, Phys. Rev. C **86**, 035202 (2012), 1204.2277
[35] C. Fernandez-Ramirez, I. Danilkin, D. Manley, V. Mathieu, A. Szczepaniak, Phys. Rev. D **93**, 034029 (2016), 1510.07065
[36] S. Wheaton, J. Cleymans, M. Hauer, Comput. Phys. Commun. **180**, 84 (2009), hep-ph/0407174
[37] A. Bazavov et al. (HotQCD), Phys. Lett. B **795**, 15 (2019), 1812.08235

Charming, equilibrated, and not at all strange: My memories of Jean Letessier and his work

Giorgio Torrieri[1,*]

[1]IFGW,Unicamp

Abstract. This is a tribute to Jean Letessier, some memories of my PhD research in collaboration with him, a summary of the results we have achieved, and some personal thoughts as to the significance of these results to the field of heavy ion physics as well as the future of our field.

Theoretical physics has developed a cult of "assertiveness". To be noticed, you must be loud, confident, and willing to tell the whole seminar room or conference hall that you are correct and the rest of us do not have a clue. We all know, deep down, that the correlation of competence and arrogance is spurious, but somehow we always end up believing it is not.

Jean Letessier is a primary counter-example going against this prejudice. One would generally not notice him engaging in this behavior. And yet brilliant and competent he certainly was, without any doubt. His untimely departure means the loss of one of the pioneers of our field.

Figure 1. Jean Letessier, 1938-2020

I[1] met Jean soon after I started my PhD, under the supervision of Johann Rafelski [1]. I got to extensively appreciate Jean's brilliance, competency and helpfulness as I spent years studying how to make a sold link between the relativistic statistical mechanics of quark-gluon plasma and oodles of experimental data haphazardly organized in literature and electronic data-bases, via the subtlety of fitting algorithms.

*e-mail: torrieri@ifi.unicamp.br

[1]This, and the references in this work, are meant to be a personal memory, not a comprehensive overview of the field. The reader is advised to go to look through the talks and proceedings of [3] for such a thorough list of perspectives

The topic, strangeness enhancement in heavy ion collisions, is one of the pillars of the phenomenology of the quark-gluon plasma phase transition, mainly because of it's spectacular experimental confirmation [2, 3].

Yet, the physical interpretation of this phenomenon was and to a point still is up for grabs. Jean and Johann Rafelski made the important realization that, because of the proximity of the mass of the strange quark to Λ_{QCD} strangeness equilibration could function as a "clock", measuring the degree of chemical equilibration and the global properties of the quark-gluon plasma. This is both because in a Quark-Gluon plasma strangeness would equilibrate *faster* and because strangeness content would be *greater* than in the corresponding hadronic phase. A hadronization via quark coalescence would enhance both effects on the yield of multi-strange particles.

While fast equilibration is generally accepted within the field, most of the heavy ion community believed, and still does, that strangeness content in the quark-gluon plasma relative to a hadron gas is largely irrelevant to freeze-out dynamics because the system remains in chemical equilibrium, $\mu_{q,s} = -\mu_{q,s}$, throughout the hadronization phase transition. Considering that the chemical equilibrium is maintained if and only if

$$\frac{\chi_{q,s}}{T\rho_{q,s}} \times \frac{d\rho_{q,s}}{d\tau} \times \tau_{q,s} \ll 1 \tag{1}$$

Here, the first factor is the susceptibility χ relative to density ρ, the second the hydrodynamic expansion speed and the third is the chemical equilibration time.

considering $\chi_{q,s}$ and bulk viscosity [4] jump over the phase transition [2] and we do not know the behavior of the equilibration time $\tau_{q,s}$, it could just be that the system at freeze-out is in thermal equilibrium but maintains the chemical content of the quark gluon plasma phase. In [2], a particular scenario was proposed where this was realized via an explosive hadronization from a super-cooled plasma phase. While super-cooling necessitates a first order phase transition, other mechanisms, such as jumps in bulk viscosity [4], could produce the same result within a cross-over.

These non-equilibrium effects could be parametrized by phase-space factors γ_i ($i = q$ for light quarks, s for strange ones), with the effective chemical potential μ

$$\mu_{i,\bar{i}} \to \pm\mu_i \ln(\gamma_i) \tag{2}$$

and, if freeze-out is fast, we would expect $\gamma_q \simeq 1.5$ and $\gamma_s \simeq 1.5 - 2$.

This was the theory. One needed considerable computational efforts to, on one side, see what the properties of the quark gluon plasma would say about $\gamma_{q,s}$, on te other try to infer $\gamma_{q,s}$ from experimental data. Jean Letessier was the undisputed leader in the first effort, and one of the primary participants of the second one, which also formed the bulk of my thesis work.

The bulk of our effort was to write a now well-known computer code [5–7] capable of inferring chemical parameters from experimental data. The results were intriguing but inconclusive: The fits [8, 9] pointed to *exactly* the parameter space required by the arguments in [2]: Above a thrershold in energy and system size, associated with the onset of nearly equilibrated deconfinement, freeze-out temperature drops to $\sim 140MeV$ and $\gamma_{q,s} \geq 1.5$. Below this thershold, $T \sim 170$ MeV and $\gamma_{q,s} \leq 1$, reflecting a "hot" but under-equilibrated system that just "hadronizes" without significant partonic dynamics. Note that the enhancement of the ϕ meson from small to large systems [8] can be interpreted as to mean that strangeness enhancement is driven by chemical phase space γ_s^2, rather than kinematic conservation laws ("canonical suppression" [2]).

However, the statistical significance of $\gamma_{q,s} > 1$ was not enough to make this into a fully experimentally-supported statement. Since the difference between equilibrium and non-equilibrium goes with quark number, the penta-quark was the ideal candidate to look for phase space enhancement [11]. Unfortunately, it turned out to be a spurious state.

Of course we did not give up. An interesting opportunity was provided by the experimental study, in heavy ion collisions, of short-lived hadronic resonances, such as

$$Y^* = \Lambda(1520), K^*(892), \Sigma^*(1385), \Xi^*(1530), \Delta(1232), \rho, ...$$

and others. These have the same quark composition as the corresponding stable hadrons

$$Y = \Lambda, K, \Xi, p, \pi, ...$$

, but a larger mass. Hence, if freeze-out temperature is lower than the one expected for chemical freeze-out, as expected if the plasma is over-equilibrated [2], one would expect that their abundance, measurable via ratios such as Y^*/Y , is depleted. Of course, a long hadronic lifetime out of chemical equilibrium can also deplete such resonances, so more than one particle is necessary [12]. It is fair to say that the effect of rescattering, regeneration and lack of chemical equilibrium still has to be conclusively disentangled.

The next step was to examine particle fluctuations. Bose-Einstein statistics, which enhance mesonic and baryonic yields, enhance even more fluctuations of π mesons, whose mass is comparable to freeze-out temperature. The lower freeze-out temperature associated with over-equilibrated freeze-out provides a further boost to this enhancement by eliminating correlations due to resonances. Centrality fluctuations can be eliminated by considering fluctuations of particle ratios, preferably ratios where numerator and denominator are correlated by resonance decays, such as K/π (which probes the presence, at freeze-out,of K^*),p/π (which probes the Δ) and Λ/π (which probles Σ^*). Efforts in such a comprehensive analysis are on-going [6], but, to our knowledge, the model in [6], in the over-equilibrated regime, is the only one that can describe both yields and event-by-event fluctuations with a single set of parameters.

This physical picture, where chemical composition, analyzed via the non-equilibrium ansatz, provides clues to the dynamics and global properties of the hot quark-gluon plasma stage, also gives a special role to the charm quark. This quark is certainly and unquestionably out of equilibrium. It is dominantly produced within the initial hard scatterings, and then it's current is well-described by a conservation law equation driven by collective flow u_μ

$$\partial_\mu (\rho_c u^\mu) = 0 \quad , \quad \rho_c \equiv \gamma_c F\left[g, m, T\right] \tag{3}$$

where F is the relativistic Boltzmann factor, in terms of the spin degeneracty g and mass of the hadron m and the modified Bessel function $K_2(x)$

$$F\left[g, m, T\right] = \frac{4\pi g}{(2\pi)^2} m^2 T K_2\left(\frac{m}{T}\right) \tag{4}$$

In an equilibrated quark-gluon plasma, even at LHC energies, ρ_c would be thoroughly negligible w.r.t. $\rho_{q,s}$. Unlike strangeness and light-quark abundance, that depend respectively on the strangeness susceptibility and entropy content of the quark-gluon plasma phase, γ_c is sensitive to $N_{coll}\sigma_{pp\to cX}$, due to the fact that after the initial production charm quarks are effectively conserved.

Charmed hadrons from a quark-gluon plasma will however be created via a coalescence type process, and hence will, just as light and strange quarks, directly be sensitive to the

Table 1. The regimes of interest of heavy ion physics. See [2] for details

$(R\sqrt{s})$ / $\sqrt{s}/\Lambda_{QCD}$	$\ll 1$	~ 1	$\gg 1$
$\ll 1$	"Hadrons"	transport	"Gas"
~ 1	Regge?	String gas?	Hagedorn? Criticality?
$\gg 1$	"Partons"	"Glasma"?	QGP

charm composition.

$$n_i = \gamma_q^{q_i}\gamma_s^{s_i}\gamma_c^{c_i}\exp\left[\frac{\sum_{k=q,s,c}(k-\overline{k})\mu_k}{T}\right]F(g_i,m_i,T) \tag{5}$$

At LHC energies this will lead to a dramatic enhancement of charm and multi-charm baryons and mesons, by the same coalescence mechanism of strangeness enhancement described in [2], with the effect being quantifiable according to the publically available computer code in [7].

As a scientific conclusion, we are at the dawn of not one but two new eras: On the one hand, the increased luminosity of the LHC will allow experimentalists to look for multiply heavy baryon and meson states: $\Lambda_c, \Xi_c, B_s, csq$ states etc. etc. etc. as well as possible tetra and pentaquark states. This will be done both for "large" AA systems and for "small" pA and pp systems. On the other, low energy accellerators such as FAIR and NICA will explore the onset of deconfinement by looking for rare probes (ϕ,multi-strange states, D-mesons, J/Ψ) at low energy A-A collisions.

The variety of experimental data we will obtain should allow us to perform conclusively a thorough scan,with light, strange and charmed hadrons, of the relevant regimes (in system size R and energy $\sqrt{s}$ and intrinsic QCD scale $\Lambda_{QCD} \sim T_c \sim m_s$) to probe the onset of deconfinement (table 1) The systems shown in the table are all qualitatively different, requiring their own effective theory. "Hadron" refers to a small hadronic system, best studied by effective theory in the scattering regime. "Gas" refers to an out-of-equilibrium hadron gas, amenable to either hadronic transport simulations or thermal effective field theory. Partons would be an out of equilibrium, hadronizing burst of partons while QGP is the "holy grail", a deconfined system where statistical mechanics applies. The boundaries of these regimes are, respectively, equilibration and deconfinement. In the middle, non-trivial and fundamentally not understood critical phenomena, such as a Hagedorn/stringy regime and critical dynamics, could lurk [2].

Despite the enormous amount of results obtained, papers published, workshops and so on these boundaries have as yet to be firmly established. The tools of analysis of particle abundances outside of chemical equilibrium, pioneered by Jean and Jan, could be instrumental in achieving this mapping. Conclusively established jumps in $T, \gamma_{q,s,c}$, might be instrumental in identifying and characterizing transition regimes, as well as pointing out where, in energy and system size, could these transitions occur.

As Jean left this world, our field entered into a phase where the questions he spent a good part of his professional life investigating could be experimentally answered, by the methods he helped develop.

Collaborating with Jean Letessier was an absolute joy. His quiet demeanor provided a counter-balance to J.Rafelski's exuberance. One would however be gravely mistaken to confuse his quiet demanour with lack of involvement. Every result mentioned in this review would not exist without his contribution. Our field will miss him, and his friends and collegues will sorely miss his company. On a personal level, I cherish his helpfulness while I struggled with understanding the computational and statistical aspects of performing the analyses that went into the core of my thesis work. PhD work is often professionally lonely and daunting, and finding someone who will readily and meaningfully ask questions can make all the difference on a doctoral career. R.I.P. Jean and thanks for your collaboration, help and friendship.

G.T. acknowledges support from Bolsa de produtividade CNPQ 306152/2020-7, Bolsa de pesquisa FAPESP 2021/01700-2 and Partecipation to Tematico FAPESP, 2017/05685-2.

References

[1] G. Torrieri, [arXiv:nucl-th/0405026 [nucl-th]].

[2] J. Letessier and J. Rafelski,

[3] SQM2021, Online and organied by BNL
https://indico.cern.ch/event/985652/
Proceedings in press, EPJ Web of Conference

[4] G. Torrieri, B. Tomasik and I. Mishustin, Phys. Rev. C **77**, 034903 (2008) doi:10.1103/PhysRevC.77.034903 [arXiv:0707.4405 [nucl-th]].

[5] G. Torrieri, S. Steinke, W. Broniowski, W. Florkowski, J. Letessier and J. Rafelski, Comput. Phys. Commun. **167**, 229-251 (2005) doi:10.1016/j.cpc.2005.01.004 [arXiv:nucl-th/0404083 [nucl-th]].

[6] G. Torrieri, S. Jeon, J. Letessier and J. Rafelski, Comput. Phys. Commun. **175**, 635-649 (2006) doi:10.1016/j.cpc.2006.07.010 [arXiv:nucl-th/0603026 [nucl-th]].

[7] M. Petran, J. Letessier, J. Rafelski and G. Torrieri, Comput. Phys. Commun. **185**, 2056-2079 (2014) doi:10.1016/j.cpc.2014.02.026 [arXiv:1310.5108 [hep-ph]].

[8] J. Rafelski, J. Letessier and G. Torrieri, Phys. Rev. C **72**, 024905 (2005) doi:10.1103/PhysRevC.72.024905 [arXiv:nucl-th/0412072 [nucl-th]].

[9] J. Letessier, J. Rafelski and G. Torrieri, [arXiv:nucl-th/0411047 [nucl-th]].

[10] J. Letessier, G. Torrieri, S. Hamieh and J. Rafelski, J. Phys. G **27**, 427-438 (2001) doi:10.1088/0954-3899/27/3/322 [arXiv:nucl-th/0011048 [nucl-th]].

[11] J. Letessier, G. Torrieri, S. Steinke and J. Rafelski, Phys. Rev. C **68**, 061901 (2003) doi:10.1103/PhysRevC.68.061901 [arXiv:hep-ph/0310188 [hep-ph]].

[12] J. Rafelski, J. Letessier and G. Torrieri, Phys. Rev. C **64**, 054907 (2001) [erratum: Phys. Rev. C **65**, 069902 (2002)] doi:10.1103/PhysRevC.64.054907 [arXiv:nucl-th/0104042 [nucl-th]].

Hypernuclei and Antihypernuclei Production in Heavy-Ion Collisions

Yue-Hang Leung[1,*]

[1]Nuclear Science Division, Lawrence Berkeley National Laboratory

Abstract. In these proceedings, an overview of recent hypernuclei measurements in heavy ion collisions is presented. These results include the lifetime, Λ binding energy, production yield, and directed flow in heavy ion collisions. These results provide constraints on the hyperon-nucleon interaction, enhance our understanding in hypernuclear structure, and provide insight on the production mechansims of hypernuclei in heavy ion collisions.

1 Introduction

Hypernuclei are nuclei containing at least one hyperon. Hypernuclei provide access to the hyperon-nucleon (Y-N) interaction, which is an important ingredient in the equation of state of high density nuclear matter, such as the core of neutron stars or the hadronic phase of a heavy ion collision. More than forty Λ-hypernuclei have been discovered, and through the systematic study of their Λ separation energies (B_Λ), it can be inferred that the $\Lambda - N$ interaction is attractive, with a depth of approximately 30 MeV. More precise measurements on the lifetime, B_Λ, etc. can help us further constrain the $\Lambda - N$ interaction. On the other hand, $\Xi - N$, $\Sigma - N$ or $\Lambda - \Lambda$ interactions are not well understood due to the few double s hypernuclei being discovered and studied.

Traditionally, hypernuclei are studied through techniques such as emulsion, γ-ray or strangeness exchange reactions. The first observations of hypernuclei and antihypernuclei from heavy ion collisions are from E864 [1] in 2004 and STAR [2] in 2009 respectively. There are few advantages of studying hypernuclei in heavy ion collisions. Firstly, light nuclei are generally produced in copious amounts, thus enabling precise measurements on the hypernuclei structure, such as binding energy and lifetime. Also, the production mechanisms of light nuclei may be studied, providing input to the Y-N interaction as well as properties of the matter formed. Finally, it has been proposed that exotic hypernuclei may be produced in heavy ion collisions [3]. Recent measurements of hypernuclei in heavy ion collisions provide new input to the Y-N interaction.

2 The Internal Structure of Hypernuclei

2.1 Hypertriton Lifetime Puzzle

The hypertriton is a weakly bound object, leading to small overlap between the Λ and the deuteron core. This leads to the theoretical expectation that the ${}^3_\Lambda$H lifetime is close to the

*e-mail: yhleung@lbl.gov

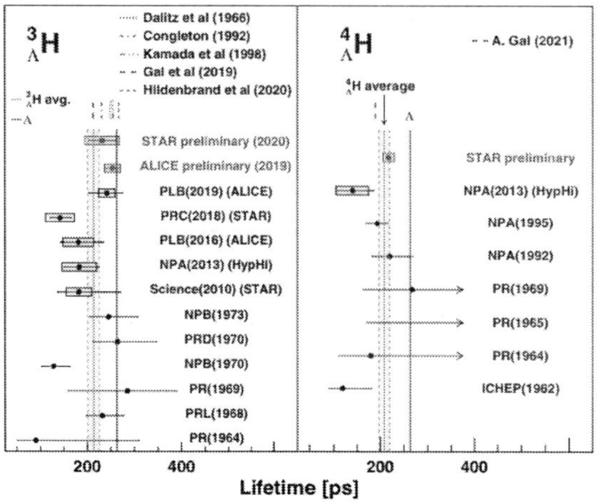

Figure 1. $^{3}_{\Lambda}$H (a) and $^{4}_{\Lambda}$H (b) measured lifetimes, compared to previous measurements, theoretical calculations and the free Λ lifetime. The experimental average lifetimes and the corresponding uncertainty are also shown as orange bands.

free Λ lifetime τ_Λ. However, STAR and HypHI has reported lifetimes 30 – 40% shorter than τ_Λ. This is commonly known as the hypertriton lifetime puzzle.

Recently, STAR [4] and ALICE [5] have presented preliminary results of the $^{3}_{\Lambda}$H lifetime. The results are shown in Fig. 1a. Taking into account of these results, the current experimental averaged $^{3}_{\Lambda}$H lifetime is 81 ± 5%τ_Λ. Future measurements using high statistics data sets from STAR, ALICE and HADES [6] are expected to further reduce the experimental uncertainty. On the theoretical side, recent studies from Gal et al. [7] indicate that pion final state interactions are attractive and may decrease the $^{3}_{\Lambda}$H lifetime by ~ 10%. Predictions incorporating these considerations are consistent with the world average. In addition, the possible relationship between B_Λ and lifetime has been studied in [8, 9]. There is active progress both experimentally and theoretically towards a very precise understanding of the $^{3}_{\Lambda}$H lifetime.

2.2 Hypertriton B_Λ

The Λ separation energy of the $^{3}_{\Lambda}$H is defined as $B_\Lambda(^{3}_{\Lambda}\mathrm{H}) = (\mathrm{m_d} + \mathrm{m_\Lambda} - \mathrm{m}_{^{3}_{\Lambda}\mathrm{H}})\mathrm{c}^2$, where m_d, m_Λ and $m_{^{3}_{\Lambda}\mathrm{H}}$ are the deuteron mass, the Λ mass and the $^{3}_{\Lambda}$H mass. In 2019, STAR measured the B_Λ of $^{3}_{\Lambda}$H and $^{3}_{\bar{\Lambda}}\bar{\mathrm{H}}$ using data taken by the Heavy Flavor Tracker. The B_Λs measured are consistent with each other, and the average is 0.41 ± 0.12(stat.) ± 0.11(syst.)[MeV], seemingly in tension with the result from emulsion experiments, B_Λ = 0.13 ± 0.05 [MeV]. On the other hand, studies from [10] showed that recalibrating the results from emulsion measurements gives an updated value of B_Λ = 0.27 ± 0.08 [MeV], consistent with the STAR measurement. ALICE reported a precise new preliminary result of B_Λ = 0.05 ± 0.06(stat.) ± 0.10(syst.)[MeV] [13], further reducing the experimental uncertainty in $B_\Lambda(^{3}_{\Lambda}\mathrm{H})$. Precise measurements is important since it has a strong influence in the coalescence calculation of production yield in heavy ion collisions, as well as potential impact on lifetime calculations as well.

2.3 Measurements of A=4 Hypernuclei

Besides $^{3}_{\Lambda}$H , measurements on the lifetime and B_Λ for $A = 4$ hypernuclei have been reported by the STAR collaboration in this conference, using data from Au+Au collisions at $s_{\mathrm{NN}} = 3$ GeV taken in 2018. The lifetime of $^{4}_{\Lambda}$H is measured to be 217 ± 8(stat.) ± 12(syst.)[ps], consistent with theoretical calculations from [7]. The B_Λ for $^{4}_{\Lambda}$H and $^{4}_{\Lambda}$He have been measured. A non-zero $\Delta B_\Lambda = B_\Lambda(^{4}_{\Lambda}\mathrm{He}) - \mathrm{B}_\Lambda(^{4}_{\Lambda}\mathrm{H})$ is commonly known as charge symmetry breaking. The B_Λ difference in the ground state is measured to be 130 ± 130(stat.) ± 70(syst.) keV, and

 https://doi.org/10.1051/epjconf/202225908001

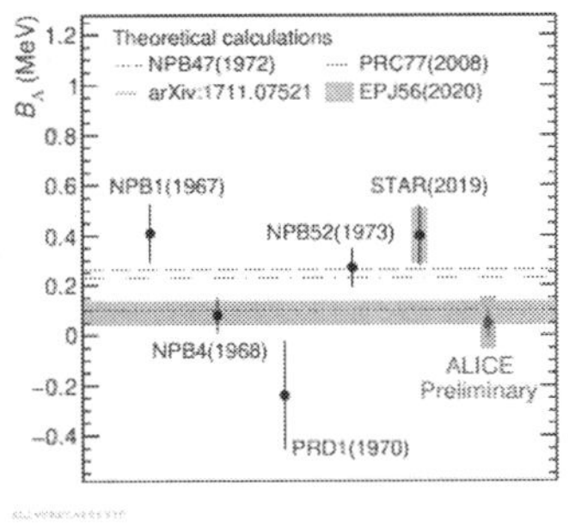

Figure 2. Preliminary $B_\Lambda(^3_\Lambda\mathrm{H})$ measurement from ALICE (red), compared with theoretical predictions and previous measurements. The values reported for previous measurements are recalibrated values reported in [10].

is compared to theoretical calculations in Fig. 3. Combining the ground state result with transition energies measurements [11, 12], the B_Λ difference in the excited state is also reported, and is consistent with model calculations. STAR detector upgrades and future data sets will reduce the statistical uncertainty, providing stronger constraints to theoretical calculations.

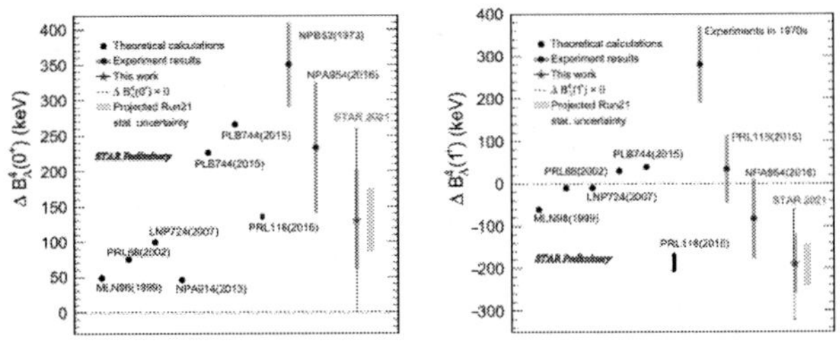

Figure 3. The B_Λ difference between $^4_\Lambda\mathrm{H}$ and $^4_\Lambda\mathrm{He}$ in ground states (left) and in excited states (right) compared with theoretical model calculations (black dots) and previous measurements (blue dots). The green bands are projected statistical uncertainties from STAR run 2021 3 GeV data.

3 Production of Hypernuclei in Heavy-Ion Collisions

In heavy ion collisions, light hypernuclei are expected to be abundantly produced at low energies due to the high baryon density. However, the production mechanisms of hypernuclei in heavy ion collisions are not well unuderstood, due to the scarcity of data. Coalescence has been proposed as a possible production mechanism, particularly in central heavy ion collisions, while in peripheral collisions or small systems, production via the absorption of hyperons in the spectator fragments has been suggested.

Hypernuclei yields have been measured by ALICE [16], STAR [17], and HypHI [18] in the past. The $^3_\Lambda\mathrm{H}$ yield at mid-rapidity from Pb+Pb collisions at 2.76 TeV measured by ALICE are consistent with thermal model predictions [19] and UrQMD [20]. Multi-differential measurements in the production yield of hypernuclei are in different systems are essential in advancing our understanding in the field.

3.1 $^3_\Lambda\mathrm{H}$ and $^4_\Lambda\mathrm{H}$ Production from 3 GeV Au+Au Collisions at STAR

STAR has recently measured the $^3_\Lambda\mathrm{H}$ and $^4_\Lambda\mathrm{H}$ production yield at mid-rapidity in Au+Au collisions with $\sqrt{s_{\mathrm{NN}}} = 3$ GeV, using data taken in 2018. The p_T spectra are obtained, and the p_T integrated yields are estimated by extrapolating to 0 p_T. Systematic uncertainties are estimated by using different functional forms for extrapolation. The rapidity distributions of $^3_\Lambda\mathrm{H}$ and $^4_\Lambda\mathrm{H}$ are shown in Fig. 4. For $^4_\Lambda\mathrm{H}$, different trends in the rapidity distribution in central $(0-10\%)$ and mid-central $(10-50\%)$ collisions are observed. This is likely due to differences in collision geometry, possibly due to spectator reactions playing a larger role in non-central collisions.

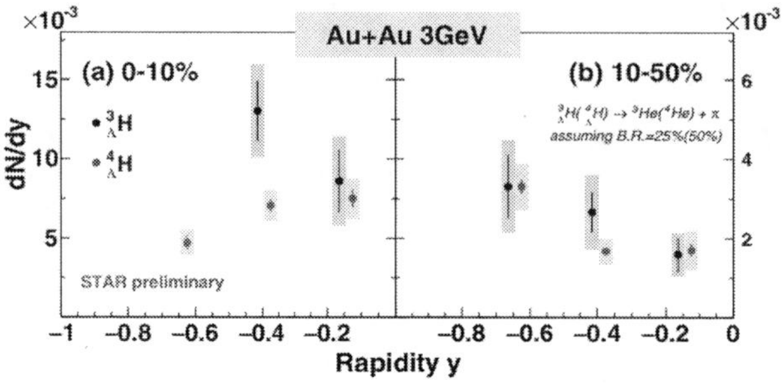

Figure 4. dN/dy as a function of rapidity y for ${}^{3}_{\Lambda}$H (black) and ${}^{4}_{\Lambda}$H (red) for (a) $0-10\%$ centrality and (b) $10-50\%$ centrality Au+Au collisions at 3 GeV.

The mid-rapidity $|y| < 0.5$ yield in central heavy ion collisions is plotted as a function of beam energy in Fig. 5, and are compared to ALICE measurements at 2.76 [16] and 5.02 TeV [5] Pb+Pb collisions, as well as theoretical calculations. The statistical thermal model [19] which incorporates canonical suppression through the introduction of the canonical volume paramter V_C, can successfully describe the ${}^{3}_{\Lambda}$H yield over few orders of magnitude of $\sqrt{s_{\mathrm{NN}}}$, while underestimating the ${}^{4}_{\Lambda}$H yield at 3 GeV. Similarly, the coalescence (DCM) [20] also describes the ${}^{3}_{\Lambda}$H yield at 3 GeV while underestimating ${}^{4}_{\Lambda}$H . This discrepancy may possibly due to the coalescence parameters adopted in [20]; the parameters for hyperons are assumed to be identical for nucleons and not constrained by data. The Hybrid URQMD model [20], although can describe the ${}^{3}_{\Lambda}$H yield at LHC energies, overestimate the ${}^{3}_{\Lambda}$H and ${}^{4}_{\Lambda}$H yield at 3 GeV by an order of magnitude.

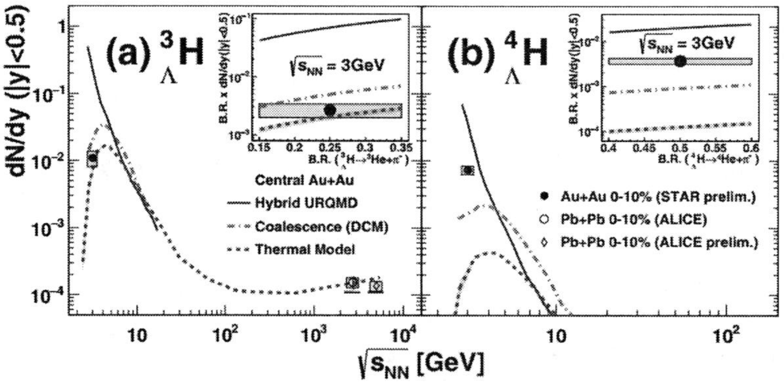

Figure 5. (a)${}^{3}_{\Lambda}$H and (b)${}^{4}_{\Lambda}$H yields at $|y| < 0.5$ as a function of beam energy in central heavy ion collisions. The symbols represent measurements while the lines represent different theoretical calculations. The data points assume a branching ratio of 25(50)% for ${}^{3}_{\Lambda}$H(${}^{4}_{\Lambda}$H) $\rightarrow^{3}$ He(^{4}He) + π^{-}. The insets show the (a)${}^{3}_{\Lambda}$H and (b)${}^{4}_{\Lambda}$H yields at $|y| < 0.5$ times the branching ratio as a function of the branching ratio.

Besides the production yield, the directed flow v_1 as a function of rapidity, for ${}^{3}_{\Lambda}$H and ${}^{4}_{\Lambda}$H in $5-40\%$ Au+Au collisions have been measured. The slope at mid-rapidity is obtained via a linear fit through the origin, and are shown in Fig. 6. The hypernuclei dv_1/dy are compared to that of light hypernuclei. For light nuclei, a scaling behavior as a function of mass is observed, i.e. baryon number scaling. Since the v_1 slope of hypernuclei is similar to that of light nuclei with similar mass, the results are qualitatively consistent with hypernuclei production from coalescence of hyperons and nucleons at mid-rapidity.

In summary, these results provide new quantitative input on the production mechanisms of loosely bound objects in high baryon density region.

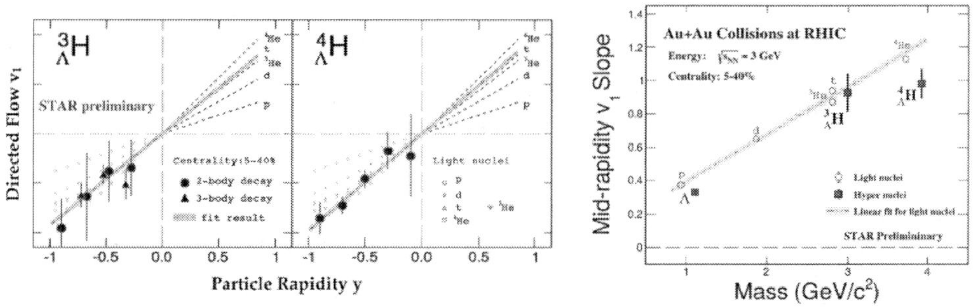

Figure 6. (left)$^3_\Lambda$H and $^4_\Lambda$H v_1 as a function of rapidity. The red line represents a linear fit through the origin. (right) $^3_\Lambda$H and $^4_\Lambda$H mid-rapidity v_1 slope as a function of mass. The dv_1/dy for light nuclei is shown for comparison.

3.2 $^3_\Lambda$H and $^3_{\bar{\Lambda}}\bar{\text{H}}$ Production in Small Systems at ALICE

$^3_\Lambda$H and $^3_{\bar{\lambda}}\bar{\text{H}}$ signals have been observed by ALICE in $0-40\%$ p+Pb collisions at $\sqrt{s_{\text{NN}}} = 5.02$ TeV and high multiplicity p+p collisions at $\sqrt{s} = 2.76$ TeV, both at $\sim 5\sigma$ significance [13]. The ratio $(^3_\Lambda\text{H} + ^3_{\bar{\Lambda}}\bar{\text{H}})/2\Lambda$ and the strangeness population factor $S_3 = (^3_\Lambda\text{H}/^3\text{He}) \times (\text{p}/\Lambda)$ are extracted and shown in Fig. 7. The p+p data is consistent with 2-body coalescence [14] and in tension with the statistical hadronization model (SHM) [15]. For the p+Pb data, the measurement excludes certain configurations of the statistical model, while being consistent with 2-body coalescence. These new measurements give insight on the production mechanisms of hypernuclei in small systems, possibly with enough precision to distinguish between statistical hadronization and coalescence.

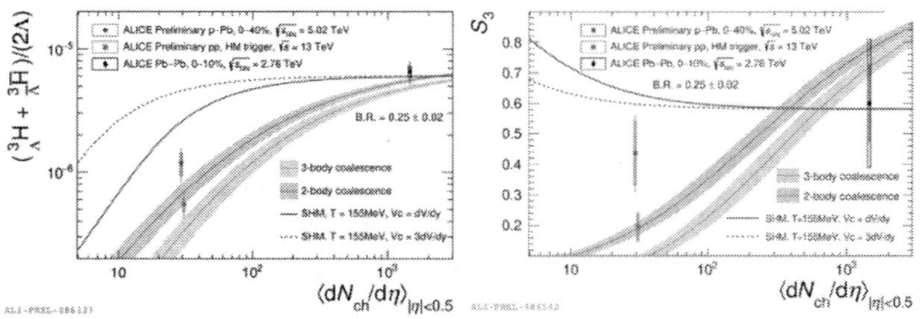

Figure 7. $^3_\Lambda$H (left) and S_3 (right) measurements in p+p (orange), p+Pb (green) and Pb+Pb collisions(indigo) as a function of mean charged-particle multiplicity. The measured ratios are compared with expectations for the canonical statistical hadronization [15] and coalescence models [14].

4 Summary and Experimental Outlook

These proceedings present recent hypernuclei measurements from the STAR experiment and the ALICE experiment. The most precise $^3_\Lambda$H and $^4_\Lambda$H lifetime measurements to date, as well as measurements on B_Λ of $^3_\Lambda$H , $^4_\Lambda$H and $^4_\Lambda$He are presented, providing strong constraints to theoretical models. The $^3_\Lambda$H and $^4_\Lambda$H rapidity distributions at $\sqrt{s_{NN}}$ = 3 GeV Au+Au collisions are presented. Thermal model and coalescence calcaulations are consistent with the

${}^{3}_{\Lambda}$H yield, while no model can describe the ${}^{4}_{\Lambda}$H yield. These new results provide insight to hypernuclei in production in the high baryon density region. Directed flow measurements are also presented, and the results are qualitatively consistent with expectations from production via coalescence. Finally, the ${}^{4}_{\Lambda}$H yield is measured in p+p collisions at 13 TeV and p+Pb collisions at 5.02 TeV. The results exclude certain configurations of the statistical hadronization model, and are consistent with coalescence models.

Future data from LHC can also shed light on the production mechanisms in smaller systems, while future data from STAR BES-II can help us understand the properties of high baryon density matter. Besides these current experiments, in the near future, data from heavy ion collisions at FAIR (CBM, SIS), JPARC, NICA, HIAF, etc. are expected to contribute to our understanding of the Y-N and Y-Y interactions through the study of hypernuclei. These experiments cover a huge range in collision energy, allows the systematic study of hypernuclei production as a function of energy. Lifetime and binding energy of various hypernuclei can be measured to great precision, further enhancing our understanding of hypernuclear structure and the Y-N interaction.

Similar to Λ hypernuclei providing constraints to the Y-N interaction, the Y-Y interaction can be accessed through double-Λ hypernuclei. Currently the $\Lambda - \Lambda$ potential is not well understood due to the lack of data. Only few double-Λ hypernuclei have been discovered. According to thermal model predictions, the modest production rate of light double-Λ hypernuclei, such as ${}^{5}_{\Lambda\Lambda}$H, ${}^{5}_{\Lambda\Lambda}$He etc., gives discovery potential to heavy ion experiments. A discovery of these bound states, along with a measurement of the binding energy, are expected have a high impact on our understanding of the Y-Y interaction, which has important implications on nuclear and astrophysics.

References

[1] B. I. Abelev et al. (E864 collaboration), Phys. Rev. C **70**, 024902(2004)
[2] T. A. Armstrong et al. (STAR collaboration), Science **328**, 58-62(2010)
[3] D.E. Lanskoy et al., EPJ Web of Conferences **222**, 03012(2019)
[4] C. Hu (STAR collaboration), these proceedings
[5] E. Bartsch (ALICE collaboration) Nuclear Phys. A, **1005**, 121791(2021)
[6] S. Spies (HADES collaboration), these proceedings
[7] A. Gal et al, Phys. Lett. B **791**, 48(2019), A. Gal, these proceedings
[8] A. Pérez-Obiol et al, Phys. Lett. B **811**, 135916(2020)
[9] F. Hildenbrand et al, Phys. Rev. C **102**, 064002(2020)
[10] P. Liu et al, Chin. Phys. C **43**, 124001(2019)
[11] T. O. Yamamoto et al. (J-PARC E13 collaboration), Phys. Rev. Lett. **115**, 222501(2015)
[12] M. Bedjidian et al, Phys. Lett. B **62**, 467(1976)
[13] P. Fecchio (ALICE collaboration), these proceedings
[14] K.-J. Sun et al., Phys. Lett. B **792**, 132(2019)
[15] V. Vovchenko et al., Phys. Lett. B **785**, 171(2018), Phys. Rev. C **100**, 054906(2019)
[16] A. Jaroslav et al. (ALICE collaboration), Phys. Lett. B **754** 360(2016)
[17] B. I. Abelev et al (STAR collaboration), Science **328** 58(2010)
[18] C. Rappold et al, (HypHI collaboration), Phys. Lett. B **747** 129(2015)
[19] A. Andronic et al, Phys. Lett. B **697** 203(2011), A. Andronic (updated, preliminary)
[20] J. Steinheimer et al, Phys. Lett. B **714** 85(2012)

Recent progress on few-body hypernuclei

Avraham Gal[1,⋆]

[1] *Racah Institute of Physics, the Hebrew University, Jerusalem 91904, Israel*

Abstract. Few-body Λ hypernuclei provide valuable information towards understanding strange matter. Recent experimental progress by the STAR Collaboration at the RHIC facility and by the ALICE Collaboration at the LHC has been matched by theoretical progress reviewed here: (i) lifetimes of the hypertriton ${}^3_\Lambda$H, ${}^3_\Lambda$n if particle-stable, ${}^4_\Lambda$H and ${}^4_\Lambda$He and their charge symmetry breaking, and (ii) the onset of $\Lambda\Lambda$ hypernuclear binding.

1 Introduction

Single- and double-Λ hypernuclei provide unique extension of nuclear physics into strange hadronic matter [1]. Experimental data on Λ and $\Lambda\Lambda$ hypernuclei are unfortunately poorer both in quantity and quality than the data available on normal nuclei. Nevertheless, the few dozen Λ separation energies B_Λ of single-Λ hypernuclei (${}^A_\Lambda$Z) determined across the periodic table from A=3 to 208, and the three $\Lambda\Lambda$ hypernuclei (${}^A_{\Lambda\Lambda}$Z) firmly established so far [2], provide useful testground for the role of strangeness in dense hadronic matter, say in neutron star matter. Particularly meaningful tests of hyperon-nucleon and hyperon-hyperon strong-interaction models are possible in light Λ and $\Lambda\Lambda$ hypernuclei, $A \leq 6$, which the topics reviewed below are concerned with.

2 ${}^3_\Lambda$H lifetime

Measurements of the ${}^3_\Lambda$H lifetime in emulsion or bubble-chamber experiments during the 1960s and early 1970s gave conflicting and puzzling results. Particularly troubling appeared a conference report by Block *et al.* claiming a lifetime of $\tau({}^3_\Lambda\mathrm{H})=(95^{+19}_{-15})$ ps [3], to be compared with a free Λ lifetime $\tau_\Lambda=(236\pm6)$ ps [4] measured in the same He bubble chamber (BC). However, the He BC experiment [5] concluding that era, coauthored by the same Block, reported a value of $\tau({}^3_\Lambda\mathrm{H})=(246^{+62}_{-41})$ ps, in agreement with a Λ lifetime of (263±2) ps. Given a weakly bound Λ, $B_\Lambda({}^3_\Lambda\mathrm{H})=0.13\pm0.05$ MeV from emulsion studies [6], it was anticipated that $\tau({}^3_\Lambda\mathrm{H})\approx\tau_\Lambda$. Recent measurements of $\tau({}^3_\Lambda\mathrm{H})$ in relativistic heavy ion collision experiments renewed interest in the ${}^3_\Lambda$H lifetime problem. The first round of results from STAR [7] and ALICE [8] suggested that $\tau({}^3_\Lambda\mathrm{H})$ is shorter than τ_Λ by as much as (30±17)%. Whereas STAR's most recent published lifetime, from Au-Au collisions at $\sqrt{s_{\mathrm{NN}}}$=7.7 to 200 GeV, is even shorter than that: $\tau({}^3_\Lambda\mathrm{H})=142^{+24}_{-21}\pm29$ ps [9], ALICE latest published lifetime, from Pb-Pb collisions at $\sqrt{s_{\mathrm{NN}}}$=5.02 TeV, is close to τ_Λ: $\tau({}^3_\Lambda\mathrm{H})=242^{+34}_{-38}\pm17$ ps [10], or 254±15±17 ps as reported in ICHEP 2020 [11]. The latest news is that this STAR-ALICE apparent divergence has been

⋆avragal@savion.huji.ac.il, invited talk at the 19th International Conference on Strangeness in Quark Matter (SQM2021), Brookhaven National Laboratory, Upton, New York, USA (online), May 2021.

resolved by new lifetime measurements in Au-Au collisions at $\sqrt{s_{NN}}$=3 GeV reported by STAR at SQM2021: $\tau({}^{3}_{\Lambda}\mathrm{H})$=231±29±37 ps [12].

Table 1. Calculated hypertriton weak decay rates $\Gamma({}^{3}_{\Lambda}\mathrm{H}_{g.s.})$ in units of the free-Λ decay rate Γ_{Λ}, using ${}^{3}_{\Lambda}\mathrm{H}(\frac{1}{2}^{+})$ wavefunctions that satisfy $B_{\Lambda}({}^{3}_{\Lambda}\mathrm{H})$=0.13±0.05 MeV from emulsion studies [6], and two-body branching ratios $R_3=\Gamma({}^{3}_{\Lambda}\mathrm{H}\to\pi^-+{}^3\mathrm{He})/\Gamma({}^{3}_{\Lambda}\mathrm{H}\to\pi^-+\mathrm{all})$. The underlined R_3 value is the BC world-average experimental value [5].

Source	Method	πFSI	R_3	$\Gamma({}^{3}_{\Lambda}\mathrm{H})/\Gamma_{\Lambda}$
RD (1966) [13]	closure-Λpn	no	–	1.14
Congleton (1992) [14]	closure-Λd	no	0.33±0.02	1.15
Kamada (1998) [15]	Faddeev-Λpn	no	0.379	1.06
GG (2019) [16]	closure-Faddeev-Λpn	no	0.362	1.11±0.01
HH (2020) [17]	$\not\pi$EFT(LO)-Λd	no	0.37±0.05	0.98±0.15
GG (2019) [16]	closure-Faddeev-Λpn	yes	0.357	1.23±0.02
POGFG (2020) [18]	πEFT(LO) Λpn+ΣNN	yes	<u>0.35±0.04</u>	$1.38^{+0.18}_{-0.14}$

A taste of what Theory has to say about the hypertriton lifetime is demonstrated by a representative selection of ${}^{3}_{\Lambda}\mathrm{H}_{g.s.}(\frac{1}{2}^{+})$ lifetime calculations assembled in Table 1. Its upper part lists works that disregard pion final-state interaction (πFSI). The closure approximation was used in Refs. [13, 14, 16], whereas the other two works [15, 17] accounted microscopically for the outgoing nucleon phase space and FSI. The lifetimes $\tau({}^{3}_{\Lambda}\mathrm{H})=1/\Gamma({}^{3}_{\Lambda}\mathrm{H})$ derived from this upper part are shorter than τ_{Λ} by less than 13%. In contrast, the two works listed in the lower part of the table suggest ${}^{3}_{\Lambda}\mathrm{H}$ lifetimes shorter than τ_{Λ} by more than 20%, owing primarily to πFSI where outgoing-pion plane waves are superseded by realistic pion distorted waves. This enhances $\Gamma({}^{3}_{\Lambda}\mathrm{H})$ by about 10% [16], or more realistically by 15% [18]. Interestingly, the recent πEFT work [18] is the only one that bothered to address the small ΣNN components, of order ≲1% probability, in the ΛNN dominated ${}^{3}_{\Lambda}\mathrm{H}$. A relatively large reduction of $\Gamma({}^{3}_{\Lambda}\mathrm{H})$ by ≈10% was found, owing to interference between the $\Sigma\to N\pi$ and $\Lambda\to N\pi$ weak-decay amplitudes. This effect was disregarded by Kamada *et al.* [15], thereby rendering their widely cited lifetime questionable.

The ${}^{3}_{\Lambda}\mathrm{H}$ lifetime results shown in Table 1 hold generally for ${}^{3}_{\Lambda}\mathrm{H}$ wavefunctions corresponding to given B_{Λ} values within the emulsion $B_{\Lambda}({}^{3}_{\Lambda}\mathrm{H})$=0.13±0.05 MeV interval. The B_{Λ} dependence of $\tau({}^{3}_{\Lambda}\mathrm{H})$ was studied in two of these works, with almost B_{Λ}-independent lifetime calculated in LO $\not\pi$EFT [17] and a rather strong B_{Λ} dependence LO πEFT [18], as shown in Fig. 1. The three red points in the figure correspond to three distinct values of $B_{\Lambda}({}^{3}_{\Lambda}\mathrm{H})$ reached by considering ultraviolet (UV) cutoff Λ_{UV} values smaller than Λ_{UV}=1200 MeV, beginning at which convergence is assured. This amounts to slightly varying the πEFT short-range input. The resulting B_{Λ} dependence of the computed two-body decay rate $\Gamma({}^{3}_{\Lambda}\mathrm{H}\to{}^3\mathrm{He}+\pi^-)$ is satisfied by the blue points, all for Λ_{UV}=1200 MeV, obtained by varying systematically within allowed uncertainties some of the NN and YN chiral fit data [19]. Of the three red points, the middle one is that highlighted in Table 1 while the left one corresponds to $B_{\Lambda}({}^{3}_{\Lambda}\mathrm{H})$=0.069 MeV and $\tau({}^{3}_{\Lambda}\mathrm{H})$=234±27 ps, a lifetime compatible with both ALICE and STAR new preliminary lifetime values [11, 12].

Given the strong correlation found in the πEFT calculation [18] between $B_{\Lambda}({}^{3}_{\Lambda}\mathrm{H})$ and $\tau({}^{3}_{\Lambda}\mathrm{H})$, a ${}^{3}_{\Lambda}\mathrm{H}$ lifetime as close to τ_{Λ} supports ALICE preliminary value $B_{\Lambda}({}^{3}_{\Lambda}\mathrm{H})$=0.05±0.06±0.10 MeV [20] over STAR's published value 0.41±0.12±0.11 MeV [21]. As demonstrated in Ref. [18], going to as high values of $B_{\Lambda}({}^{3}_{\Lambda}\mathrm{H})$ would lead to considerably shorter values of $\tau({}^{3}_{\Lambda}\mathrm{H})$, which are compatible perhaps with STAR's published value [9], but are incompatible with STAR's new preliminary value [12].

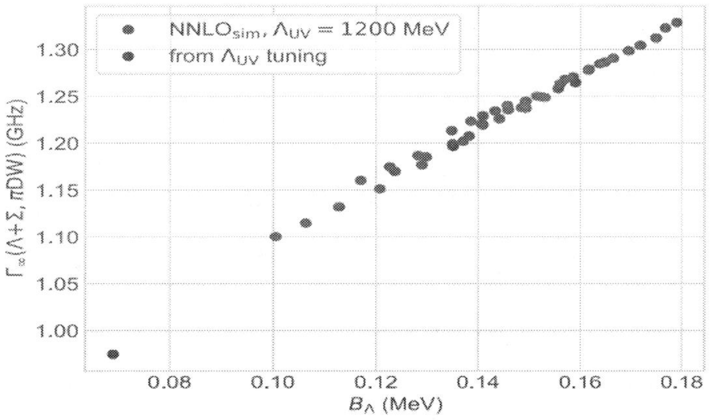

Figure 1. $\Gamma({}^3_\Lambda\mathrm{H}\to\pi^-+{}^3\mathrm{He})$ vs. $B_\Lambda({}^3_\Lambda\mathrm{H})$ in πEFT [19]. The three red points are from Ref. [18], see text.

3 ${}^3_\Lambda$n lifetime

${}^3_\Lambda$n was conjectured by the HypHI GSI Collaboration [22] to be particle stable, while unstable unanimously in recent theoretical calculations [23–25]. In ${}^3_\Lambda$n decays induced by $\Lambda \to p + \pi^-$, where ${}^3_\Lambda$n neutrons are spectators, the ${}^3_\Lambda\mathrm{n} \to (pnn) + \pi^-$ weak decay rate is given in the closure approximation essentially by the $\Lambda \to p + \pi^-$ free-space weak-decay rate, whereas in $\Lambda \to n + \pi^0$ induced decays the production of a third low-momentum neutron is suppressed by the Pauli principle, so this ${}^3_\Lambda$n weak decay branch may be disregarded up to perhaps a few percents. Hence $\Gamma({}^3_\Lambda\mathrm{n})/\Gamma_\Lambda \approx 1.114 \times 0.641 = 0.714$ [16], where the factor 1.114 follows from a difference between recoil energies in the ${}^3_\Lambda$n and Λ phase space factors, and the factor 0.641 is the free-space $\Lambda \to p+\pi^-$ fraction of the total $\Lambda \to N + \pi$ weak decay rate. This corresponds to a ${}^3_\Lambda$n lifetime estimate of $\tau({}^3_\Lambda\mathrm{n})\approx 368$ ps, considerably longer than $181^{+30}_{-24}\pm 25$ ps or $190^{+47}_{-35}\pm 36$ ps deduced from the $nd\pi^-$ and $t\pi^-$ alleged ${}^3_\Lambda$n decay modes [22, 26], and thereby questioning the conjectured stability of ${}^3_\Lambda$n.

4 ${}^4_\Lambda$H and ${}^4_\Lambda$He lifetimes

A preliminary STAR lifetime value for ${}^4_\Lambda$H was reported in this meeting: $\tau({}^4_\Lambda\mathrm{H})=218\pm8\pm12$ ps [12]. Do we understand this value in simple terms? A back-of-the-envelope estimate gives the following approximate expressions for ${}^4_\Lambda$H (Z=1) and ${}^4_\Lambda$He (Z=2) decay rates relative to the free Λ decay:

$$\Gamma({}^4_\Lambda\mathrm{Z})/\Gamma_\Lambda \approx (1 + \eta(\bar{q})) \times (\alpha_Z \times 0.7 + 1 \times 0.3) + 0.20, \quad (1)$$

where $\eta(\bar{q}) \approx 0.50 \pm 0.05$ is an exchange contribution to the dominant s-wave pionic weak decay rate, $\alpha_Z = \frac{2}{3}$ $(\frac{1}{3})$ for Z=1 (Z=2) by applying the $\Delta I{=}\frac{1}{2}$ rule to the π^4He two-body decay modes, R_4=0.7 is the pionic two-body decay branching ratio [27], with $1-R_4$=0.3 standing for the pionic multi-body part. Additional factors arising from recoil kinematics enhancement and p-wave decay suppression largely cancel out. Finally, the factor 0.20 stands for the observed $\Lambda N \to NN$ non-mesonic (n.m.) decay fraction $\Gamma_{\mathrm{n.m.}}/\Gamma_\Lambda$ [28, 29]. The resulting $\tau({}^4_\Lambda\mathrm{Z})=1/\Gamma({}^4_\Lambda\mathrm{Z})$ lifetime estimates

$$\tau_{\mathrm{th}}({}^4_\Lambda\mathrm{H}) = 195 \pm 10 \text{ ps}, \quad \tau_{\mathrm{th}}({}^4_\Lambda\mathrm{He}) = 263 \pm 13 \text{ ps}, \quad (2)$$

with a 5% assigned theoretical uncertainty, are in good agreement with $\tau_{\rm exp}(^4_\Lambda{\rm H})$=193±25 ps and with $\tau_{\rm exp}(^4_\Lambda{\rm He})$=250±19 ps as measured at KEK [28] and confirmed, for $^4_\Lambda$He, at BNL [29]. A preliminary value $\tau_{\rm exp}(^4_\Lambda{\rm H})$=180±7 ps (stat. only) was derived in a test run of E73 at J-PARC [30].

5 Charge symmetry breaking

A special feature of the $^4_\Lambda$H–$^4_\Lambda$He mirror hypernuclei is the particularly strong charge symmetry breaking (CSB) reflected in their spectra, as shown in Fig. 2. Although one-pion exchange (OPE) does not contribute directly to the ΛN strong interaction owing to isospin invariance, it does contribute as pointed out by Dalitz and von Hippel (DvH) through a CSB potential $V^{\rm OPE}_{\rm CSB}$ generated by admixing the SU(3) octet $\Lambda_{I=0}$ and $\Sigma^0_{I=1}$ hyperons in the physical Λ hyperon [34]. For the $^4_\Lambda$H–$^4_\Lambda$He g.s. levels built on the mirror ^{3}H–^{3}He cores, and using A=4 wavefunctions generated within a LO πEFT no-core shell-model calculation [35], the regularized Yukawa tail of $V^{\rm OPE}_{\rm CSB}$ gives rise to

$$\text{OPE (DvH)}: \quad \Delta B^{J=0}_\Lambda \approx 175 \pm 40 \text{ keV}, \quad \Delta B^{J=1}_\Lambda \approx -50 \pm 10 \text{ keV}, \tag{3}$$

where $\Delta B^J_\Lambda \equiv B^J_\Lambda(^4_\Lambda{\rm He}) - B^J_\Lambda(^4_\Lambda{\rm H})$. Remarkably, the large $\Delta B^{J=0}_\Lambda$ OPE CSB (central plus tensor) contribution to the splitting of the $^4_\Lambda$H–$^4_\Lambda$He mirror g.s. levels roughly agrees with the observed value $\Delta B^{J=0}_\Lambda$=233±92 keV shown in Fig. 2 which is considerably larger than the ≈70 keV CSB part of the Coulomb-dominated $\Delta B(^3{\rm H}-^3{\rm He})$=764 keV in the mirror core nuclei, driven apparently by short-range ρ^0-ω mixing. Considerably smaller hypernuclear CSB contributions were found in ΛNNN calculations by Coon *et al.* [36]: $\Delta B^{J=0}_\Lambda(\pi^0\eta + \rho^0\omega) \approx -20$ keV, $\Delta B^{J=0}_\Lambda(\pi^0\eta + \rho^0\omega) \approx -10$ keV.

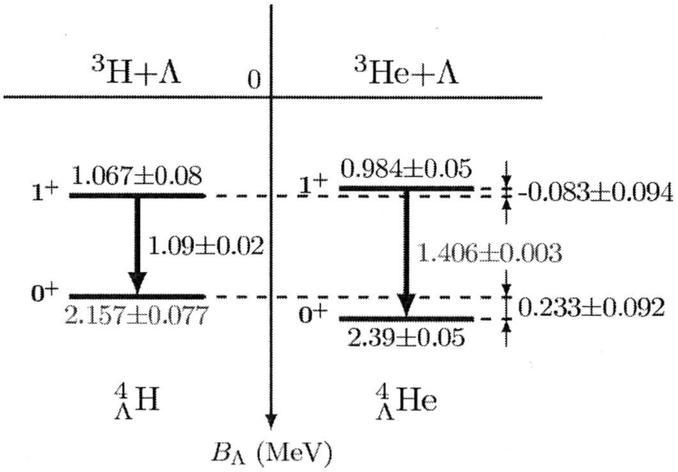

Figure 2. A = 4 hypernuclear level scheme. Recent measurements of $^4_\Lambda$He(1$^+$) excitation energy [31] and of $^4_\Lambda$H($0^+_{\rm g.s.}$) binding energy [32] are marked in red. CSB splittings are shown to the right of the $^4_\Lambda$He levels. Preliminary values from STAR are $\Delta B^{J=0}_\Lambda$=0.13±0.13±0.07 MeV, $\Delta B^{J=1}_\Lambda$=−0.19±0.13±0.07 MeV [33].

6 Onset of ΛΛ hypernuclear binding

Reliable data on ΛΛ hypernuclei are scarce. the Nagara emulsion event [37, 38] identified unambiguously as $^{\ 6}_{\Lambda\Lambda}$He, with $\Delta B_{\Lambda\Lambda}(^{\ 6}_{\Lambda\Lambda}{\rm He}) = B_{\Lambda\Lambda}(^{\ 6}_{\Lambda\Lambda}{\rm He}) - 2B_\Lambda(^5_\Lambda{\rm He})$=0.67±0.17 MeV [38], is the lightest

https://doi.org/10.1051/epjconf/202225908002

particle-stable $\Lambda\Lambda$ hypernucleus found so far. Are there lighter particle-stable $\Lambda\Lambda$ species? This question has been addressed recently in a LO $\not\pi$EFT calculation [39], sketched below, and in a full coupled-channel πEFT calculation at NLO [40], both concluding that the onset of $\Lambda\Lambda$ hypernuclear binding is given by the isodoublet ${}^{5}_{\Lambda\Lambda}$H–${}^{5}_{\Lambda\Lambda}$He.

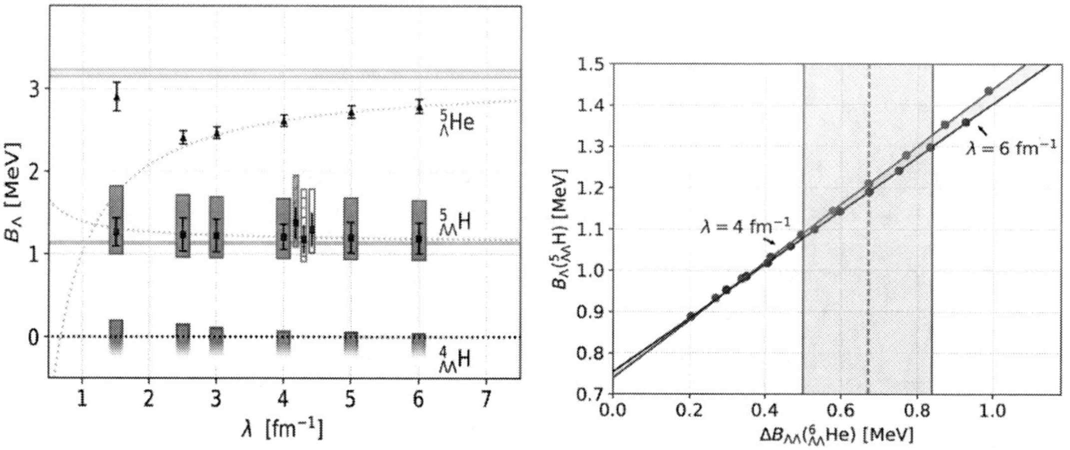

Figure 3. Left: B_Λ(${}^{5}_{\Lambda}$He, ${}^{4}_{\Lambda\Lambda}$H, ${}^{5}_{\Lambda\Lambda}$H) from $\not\pi$EFT calculations [39]. Black error bars reflect given uncertainties in ${}^{3}_{\Lambda}$H, ${}^{4}_{\Lambda}$H, ${}^{4}_{\Lambda}$H*, ${}^{6}_{\Lambda\Lambda}$He B_Λ input data, red rectangles arise by varying $a_{\Lambda\Lambda}$ between −0.5 to −1.9 fm. Thin dotted lines show extrapolations to $\lambda \to \infty$ limits marked by gray horizontal bands. Right: Tjon lines calculations of $B_\Lambda({}^{5}_{\Lambda\Lambda}$H) vs. $\Delta B_{\Lambda\Lambda}({}^{6}_{\Lambda\Lambda}$He), with vertical straight lines marking the experimental uncertainty of $\Delta B_{\Lambda\Lambda}({}^{6}_{\Lambda\Lambda}$He).

Λ separation energy values $B_\Lambda({}^{5}_{\Lambda\Lambda}$H) from $\not\pi$EFT calculations [39] are shown in the left panel of Fig. 3. Several representative values of the $\Lambda\Lambda$ scattering length were used from a broad range of values suggested by analyses of $\Lambda\Lambda$ correlations derived recently in relativistic heavy-ion collisions and by analyzing the KEK-PS E522 [41] invariant mass spectrum in the reaction ^{12}C$(K^-, K^+)\Lambda\Lambda X$ near the $\Lambda\Lambda$ threshold; see Ref. [39] for references. Here the choice of $a_{\Lambda\Lambda}$ determines the one $\Lambda\Lambda$ low-energy constant (LEC) required at LO, while the $\Lambda\Lambda N$ LEC was fitted to the $\Delta B_{\Lambda\Lambda}({}^{6}_{\Lambda\Lambda}$He)=0.67±0.17 MeV datum [38]. The reported calculations used several ΛN scattering-length combinations, demonstrating that the ΛN model dependence is weak when it comes to double-Λ hypernuclei, provided B_Λ values of single-Λ hypernuclei for $A < 5$ are fitted to generate the necessary ΛNN LECs. With values of $B_\Lambda({}^{5}_{\Lambda\Lambda}$H)~1 MeV, calculated over a broad range of cutoff values λ, it is clear that the particle stability of ${}^{5}_{\Lambda\Lambda}$H is robust, in contrast to ${}^{4}_{\Lambda\Lambda}$H for which $B_\Lambda({}^{4}_{\Lambda\Lambda}$H) comes out negative over most of the permissible parameter space in these calculations. Finally, Tjon-line correlations [42] between $B_\Lambda({}^{5}_{\Lambda\Lambda}$H) and $B_\Lambda({}^{6}_{\Lambda\Lambda}$He) are demonstrated in the right panel of Fig. 3. We conclude that the isodoublet ${}^{5}_{\Lambda\Lambda}$H–${}^{5}_{\Lambda\Lambda}$He marks the onset of $\Lambda\Lambda$ hypernuclear binding. J-PARC Experiment E75 will search for ${}^{5}_{\Lambda\Lambda}$H [43].

Acknowledgments

Discussions with my good colleagues coauthoring Refs. [16, 18, 35, 39] are gratefully acknowledged. This work is part of a project funded by the EU Horizon 2020 Research & Innovation Programme under grant agreement 824093.

References

[1] A. Gal, E.V. Hungerford, D.J. Millener, Rev. Mod. Phys. **88**, 035004 (2016)
[2] A. Gal, D.J. Millener, Phys. Lett. B **701**, 342 (2011)
[3] M.M. Block *et al.*, Proc. Sienna Int'l. Conf. on Elementary Particles, Vol. **1**, 63 (Bologna 1963)
[4] M.M. Block *et al.*, Phys. Rev. **130**, 766 (1963)
[5] G. Keyes, J. Sacton, J.H. Wickens, M.M. Block, Nucl. Phys. B **67**, 269 (1973)
[6] D.H. Davis, Nucl. Phys. A **754**, 3c (2005)
[7] B.I. Abelev *et al.* (STAR Collaboration), Science **328**, 58 (2010)
[8] J. Adam *et al.* (ALICE Collaboration), Phys. Lett. B **754**, 360 (2016)
[9] L. Adamczyk *et al.* (STAR Collaboration), Phys. Rev. C **97**, 054909 (2018)
[10] S. Acharya *et al.* (ALICE Collaboration), Phys. Lett. B **797**, 134905 (2019)
[11] F. Mazzaschi on behalf of the ALICE Collaboration, PoS **390** (ICHEP2020), 486 (2021)
[12] C. Hu on behalf of the STAR Collaboration, presented at SQM2021
[13] M. Rayet, R.H. Dalitz, Nuovo Cimento **46A**, 786 (1966)
[14] G. Congleton, J. Phys. G **18**, 339 (1992)
[15] H. Kamada, J. Golak, K. Miyagawa, H. Witala, W. Glöckle, Phys. Rev. C **57**, 1595 (1998)
[16] A. Gal, H. Garcilazo, Phys. Lett. B **791**, 48 (2019)
[17] F. Hildenbrand, H.-W. Hammer, Phys. Rev. C **102**, 064002 (2020)
[18] A. Pérez-Obiol, D. Gazda, E. Friedman, A. Gal, Phys. Lett. B **811**, 135916 (2020)
[19] A. Pérez-Obiol, D. Gazda, E. Friedman, A. Gal, in preparation (2021)
[20] P. Fecchio on behalf of the ALICE Collaboration, presented at SQM2021
[21] J. Adam *et al.* (STAR Collaboration), Nat. Phys. **16**, 409 (2020)
[22] C. Rappold *et al.* (HypHI Collaboration), Phys. Rev. C **88**, 041001(R) (2013)
[23] H. Garcilazo, A. Valcarce, Phys. Rev. C **89**, 057001 (2014)
[24] E. Hiyama, S. Ohnishi, B.F. Gibson, Th.A. Rijken, Phys. Rev. C **89**, 061302(R) (2014)
[25] A. Gal, H. Garcilazo, Phys. Lett. B **736**, 93 (2014)
[26] T. Saito *et al.* (HypHI Collaboration), Nucl. Phys. A **954**, 199 (2016)
[27] D. Bertrand *et al.*, Nucl. Phys. B **16**, 77 (1970)
[28] H. Outa *et al.*, Nucl. Phys. A **639**, 251c (1998)
[29] J.D. Parker *et al.*, Phys. Rev. C **76**, 035501 (2007)
[30] Y. Ma (J-PARC E73 status report), in https://kds.kek.jp/event/38719/
[31] T.O. Yamamoto *et al.* (J-PARC E13 Collaboration), Phys. Rev. Lett. **115**, 222501 (2015)
[32] F. Schulz *et al.* (MAMI A1 Collaboration), Nucl. Phys. A **954**, 149 (2016)
[33] T. Shao on behalf of the STAR Collaboration, presented at SQM2021
[34] R.H. Dalitz, F. von Hippel, Phys. Lett. **10**, 153 (1964)
[35] D. Gazda, A. Gal, Phys. Rev. Lett. **116**, 122501 (2016), Nucl. Phys. A **954**, 161 (2016)
[36] S.A. Coon, H.K. Han, J. Carlson, B.F. Gibson, arXiv:nucl-th/9903034
[37] H. Takahashi *et al.*, Phys. Rev. Lett. **87**, 212502 (2001)
[38] J.K. Ahn *et al.* (KEK-PS E373 Collaboration), Phys. Rev. C **88**, 014003 (2013)
[39] L. Contessi, M. Schäfer, N. Barnea, A. Gal, J. Mareš, Phys. Lett. B **797**, 134893 (2019)
[40] H. Le, J. Haidenbauer, U.-G. Meißner, A. Nogga, Eur. Phys. J. A **57**, 217 (2021)
[41] C.J. Yoon *et al.* (KEK-PS E522 Collaboration), Phys. Rev. C **75**, 022201(R) (2007)
[42] J.A. Tjon, Phys. Lett. B **56**, 217 (1975)
[43] H. Fujioka *et al.*, Few-Body Syst. **62**, 47 (2021)

Status and Perspectives of the CBM experiment at FAIR

Norbert Herrmann[1,2,*] for the CBM collaboration

[1]Physics Institute, Heidelberg University, Germany
[2]GSI Helmholtzzentrum für Schwerionenforschung GmbH, Darmstadt, Germany

Abstract. The CBM experiment, a fixed target experiment, designed to explore the QCD matter phase diagram in the baryon rich domain, is currently being constructed at the FAIR facility at GSI. The status of the construction towards the startup configuration of CBM and the initial physics goals are described. An outlook to the future physics perspectives is given.

1 Overview

The Compressed Baryonic Matter (CBM) experiment aims to study the QCD phase diagram in the baryon rich domain accessible by the SIS100 synchrotron of FAIR. The SIS100 machine with its rigidity of 100 Tm will provide beams of protons and heavy nuclei up to U. This enables measurements of heavy reaction systems like Au+Au up to a center-of-mass energy of $\sqrt{s_{NN}} = 4.9\,\text{GeV}$. The energy range accessible by CBM will allow the detailed study of a possible first order phase transition as well as the location of the critical end point (CEP) according to recently published predictions [2, 3]. Of particular interest to CBM is the high beam current and superb beam quality of SIS100 allowing the design of an apparatus that is capable of measuring rare probes with high statistics.

*e-mail: N.Herrmann@gsi.de

Figure 1. Setup of the CBM and HADES experiments sharing a common SIS100 beamline with different interaction regions.

 https://doi.org/10.1051/epjconf/202225909001

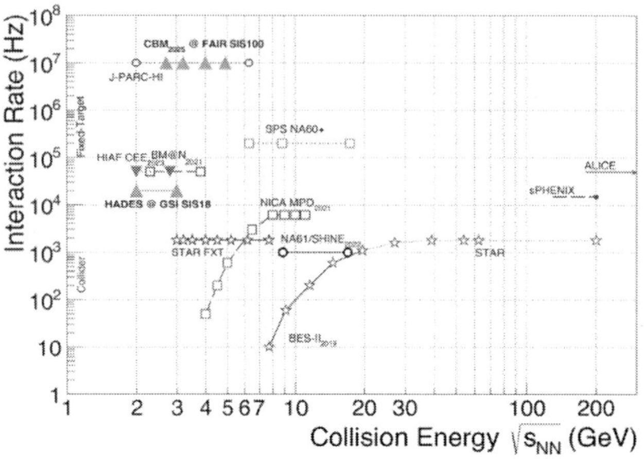

Figure 2. Comparison of the rate capability of experiments, Fig. updated from [4, 5]

The layout of CBM with its different subsystems is shown in Fig.1 together with the HADES experiment [1] which will be installed along the same beamline but with a different target location. CBM is a general purpose forward spectrometer with a polar angle acceptance of $2.5^o < \theta < 25^o$ and full azimuthal coverage. CBM can be operated in different configurations that are referred to as Hadron-Electron (HE) -, Muon (M) -, and Hadron-only (H) - setup. Common to all setups is the Silicon-Tracking-System (STS) that comprises 8 layers of double-sided micro strip Si-counters with a strip pitch of 58 μm operated in a dipole magnetic field with a bending power of 1 Tm. This common STS tracker can be complemented by different subsystems enabling particle identification or optimizing topological signatures. These subsystems are shown in Fig. 1: Micro Vertex Detector (MVD), Ring Imaging Cerenkov Counter (RICH), MUon CHambers (MUCH), Transistion Radiation Detector (TRD) , Time-Of-Flight system (TOF) and Projectile Spectator Detector (PSD). For combined electron and hadron identification the HE - setup makes use of MVD, STS, RICH, TRD, TOF and PSD, while the measurement of di-muons is done with the M - setup that implements an instrumented absorber and uses the subsystems STS, MUCH, TRD, TOF and PSD. Highest luminosity for identified hadron measurements is enabled by the H - setup employing STS, RICH, TRD and TOF. An important component not included in the figure is the data acquisition system of CBM. Since no hardware trigger is capable of reducing the event rate to a manageable level the full system is employing a trigger-less readout scheme. All detector signals are time stamped and forwarded by a high speed network to a high performance computing center where the event reconstruction and selection is done in software.

CBM and its subsystems are being designed to cope with interaction rates of heavy systems like Au + Au up to a peak rate of 10 MHz with full event reconstruction. This feature is so far not reached by any other experiment, a comparison to other facilities is shown in Fig. 2.

2 Site preparation and experiment status

With the restart of FAIR in 2017, the progress in construction and civil engineering is impressive as can be seen from Fig.3 [6]. After a rebaselining process in 2021 first beams from SIS100 are now expected for 2026. The timeline for CBM has been adjusted accordingly. The CBM building will be finished in 2022 and equipped with heavy infrastructure like the magnet foundation and the rail system. In 2023 the magnet will be delivered and all the services

Figure 3. Status of civil construction of the CBM cave as of April 2021. The walls and the beam dump are finished, the roof is planned to be done in Q3, 2021. Aerial photography was extracted from a video of a drone fly over the FAIR construction site in April 2021 [6].

will be installed. Detector installation is planned for 2024 ending with the commissioning phase that is anticipated for 2025.

Almost all the CBM subsystems have concluded their R&D phases and either have already built the entire detector (PSD) or are preparing for series production (STS, TRD, RICH, MUCH and TOF). The MVD comprising 288 sensors with 148 M pixels in MAPS technology based on the Tower Jazz 180 nm process is also at its final evaluation steps and recently submitted its Technical Design Report (TDR). Many of the subsystems have had major milestones successfully completed recently. For the magnet, the acceptance of Production Readiness Review (PRR) for the yoke and the power supply marks the start of production. The STS has already all its sensors produced and tested for acceptance. The current focus is on integration of prototype modules and ladders with the final readout ASIC (STS-XYter 2.2). In addition thermal stability issues are now addressed since the sensors will be operated below a temperature of -5°C. MUCH has procured all the necessary GEM foils and is on the track for the Engineering Design Review (EDR) of its mechanics. With a successful operation of their MAPMT sensors in the HADES experiment, RICH is now focussing on the camera design and its mechanics. The TRD subsystem passed its PRR for the outer chamber production and at the same time submitted a TDR for a modified design of the inner readout chambers which will enhance the tracking capability by a 2-dimensional position measurement and will increase the CBM acceptance for charged particles to vanishing transverse momenta at mid-rapidity. The TOF system employing MRPC detectors is pushing for the determination of the rate limitations of their counters that use different technologies for the floating electrodes (see section 5). In addition aging test have been performed at the ISRAM facility in Bucharest. The PSD subsystem is most advanced along its timeline as demonstrated by the delivery of their support structure to FAIR. All calorimeter modules are produced and used in the framework of FAIR phase-0 in the BM@N and NA61 experiments.

Further details about the detector technologies of the subsystems can be found in the TDRs that are publicly available [7]. In summary, prototype or pre-series modules are available for all subsystems. Some CBM detectors are already used in running physics experiments, PSD modules are operated in NA61 (CERN) and BM@N (JINR) and TOF modules are contributing to BES-II data taking in STAR (BNL). All of the subsystems are present in CBM's own demonstrator, the mini-CBM (mCBM) experiment (see section 5).

3 Physics program

The overall physics program of CBM is outlined in [4] and addresses all topics relevant for the characterization of dense baryonic matter. As mentioned above a highlight of CBM is the possibility to correlate bulk observables of the reactions with the abundance of rare probes. According to current full system simulations, bulk properties like centrality and

 https://doi.org/10.1051/epjconf/202225909001

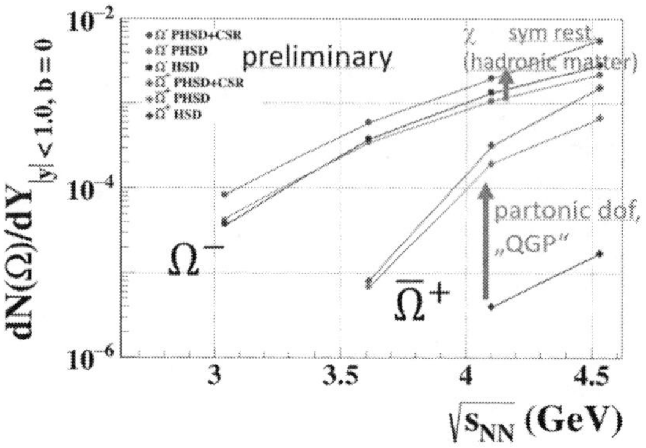

Figure 4. Prediction of Ω and $\bar{\Omega}$ production in the transport models HSD and PHSD [8, 9], for details see text.

collective flow can be well reconstructed in the forward hemisphere of the reaction. The rare probes whose phase space distribution appear to be accessible to CBM at SIS100 are multi-strange baryons, di-electrons, di-muons, J/Ψ and single and double Λ - hypernuclei. With this menu of observables, new insights are expected for the Equation-Of-State (EOS) of dense baryonic matter with its relevance for the stability of neutron stars, the phase structure of the QCD phase diagram, the existence of a CEP, chiral symmetry restoration and in-medium modifications of hadrons, strange matter and the interaction of heavy quarks with cold dense matter. In addition, the experiment is well suited to search for exotic QCD states like multi-strange di-baryons, strange penta-quarks, multi-Λ* systems, double Λ hypernuclei and the like. To which extent the production of these objects is enabled or favored by the dense baryonic environment is an interesting research topic that will be pursued in the future. All of these programs will need systematic measurements in terms of excitation functions and characterization of centrality and reaction plane orientation. To obtain significant results, a high rate experiment like CBM is indispensable.

4 Day-1 measurements

Since the ultimate rates, which the experiment is designed for, will not be available at the beginning of SIS100 operation the experiment will start in 2025 with its so-called day-1 configuration, the HE - and the M - setup. Despite an initial rate limitation of 10^5 interactions per second due to the readout speed of the MVD, important measurements may be obtained from these setups. For example, the excitation function of the production probability of multiple-strange baryons and the one excitation of di-lepton production may be measured. Note that the CBM day-1 rate exceeds the ones of all currently running experiments (c.f. Fig.2).

As an example, Fig. 4 depicts the expectations of Ω and $\bar{\Omega}$ production yields for a possible outcome of the measurement in the framework of a transport model [8] where different assumptions on the underlying physics were implemented. It can be seen that a strong increase of the $\bar{\Omega}$ is obtained when partonic degrees of freedom (string contributions) are allowed. While the effect of partonic degrees of freedom on the Ω production is rather small, both species are affected by the hadronic phase that is modeled to include chiral symmetry restoration (CSR). The measurement of the strange baryon rates with sufficient statistics will allow the discrimination of the scenarios. Full system simulations were performed and a highly efficient particle reconstruction package ("KFParticle") was developed and applied. Reconstruction efficiencies for Ω ($\bar{\Omega}$) in the Au + Au reaction at an incident energy of $\sqrt{s_{NN}} = 4.5$ GeV were determined to 5% (3%). Within a week of data taking about 10^6 (10^5) Ω ($\bar{\Omega}$) parti-

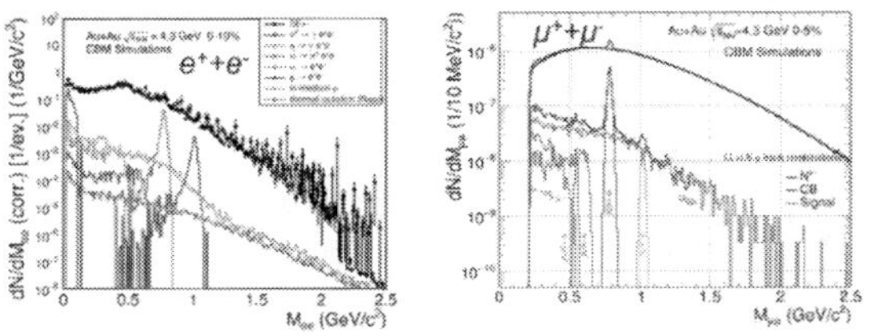

Figure 5. Simulated invariant mass spectra of e^+e^- (left) and $\mu^+\mu^-$ (right) at an incident energy of $\sqrt{s_{NN}} = 4.3\,\mathrm{GeV}$ obtained with the corresponding day-1 setups, HE and M, respectively

cles will be reconstructed at this energy if the scenario "PHSD" is assumed. This allows a significant distinction from the "HSD" scenario which predicts only $2 \cdot 10^3$ reconstructed $\bar{\Omega}$ - baryons. A high statistics measurement will be able to differentiate among the various assumptions that are admittedly different in different models.

Figure 5 presents the current status of simulations for the invariant mass of lepton pairs at an incident energy of $\sqrt{s_{NN}} = 4.3\,\mathrm{GeV}$. Both setups show clear peaks of the ω - meson and a similar signal-to-background ratio. An advantage of the SIS100 energy range for the study of di-leptons with respect to higher beam energies is the negligible contribution from Drell-Yan process and from open charm decays. With the current material budget and detector performance, the most demanding goal to determine the slope of the invariant mass spectrum in the intermediate mass region (IMR), i.e. invariant mass range from 1.1 GeV to 3 GeV, seems to be possible. This region is specifically interesting since the slope of the spectra provides access to the initial fireball temperature in the dense phase of the reaction and the yield is carrying information about chiral symmetry [10]. However, such a measurement needs large statistics to cope with the small signal-over-background ratio. To reach a statistical accuracy of 10% for the slope parameter with the HE - setup about 10^{11} events have to be recorded for which a total time of about 20 days will be needed. The comparison of the spectra from di-electron and di-muons will allow the systematic uncertainty to be assessed. CBM is therefore well prepared to do systematic measurements with these penetrating probes to diagnose the in-medium properties of hadrons and eventually quantify chiral symmetry restoration or even find a signal for a phase transition [4].

5 mCBM experiment

In order to prepare for its challenging program, CBM has constructed and installed a setup called mini CBM (mCBM) that allows tests of pre-series and prototype detectors under realistic conditions using high intensity SIS18 beams. In addition, mCBM also serves a demonstrator and development site of the free streaming data acquisition and online data analysis chain including necessary tools for controls and monitoring. A picture of the setup is shown in Fig. 6. For a trigger-less data acquisition system as under development for CBM, it is mandatory that all front-end devices that provide time stamps to the data messages are and stay synchronized even in the presence of high load from misbehaving detector components throughout the data taking. This important step could be verified during test beam campaigns in 2020 and 2021, the successful time synchronization is visible in Fig. 6 (right, bottom) where the time difference to a common reference is shown. Building on the synchronized data stream spatial correlations of the detector subsystems MUCH, RICH and TRD with reference tracks provided by the TOF subsystems (as depicted in the upper right panel of Fig. 6) have been observed and are currently being analyzed to extract detection efficiencies. Interaction rate of up to 1 MHz and charged particle fluxes of up to $20\,\mathrm{kHz/cm^2}$ at a distance of 3 m from the target have been reached and could be handled by the detectors. As a next step

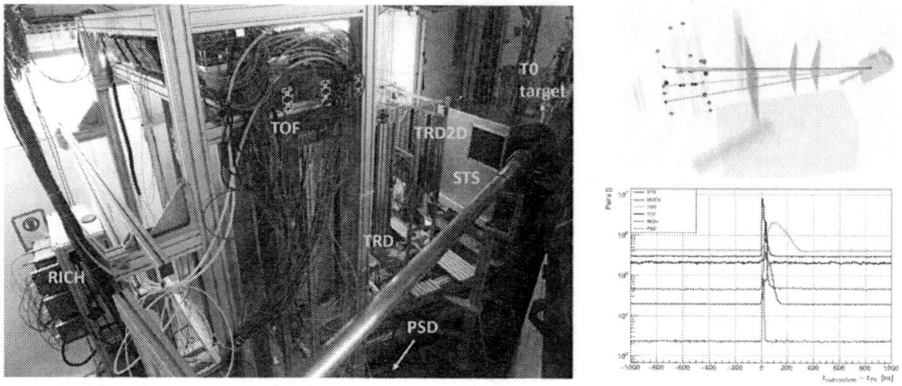

Figure 6. Photo of the mCBM setup with subsystem labels (left), event display with reconstructed TOF tracks (right, top) and time correlation of all subsystems with respect to the beam counter (right, bottom)

for full mCBM the reconstruction of Λ - baryons is planned with the perspective to measure the excitation function in the sub-threshold region.

6 Conclusion and Outlook

CBM as a high rate experiment has a unique physics potential in mapping out the properties of rare probes originating from the compressed phase of baryonic matter produced in heavy-ion reactions. The preparation of the experiment is advancing well towards the day-1 setups that will be ready to take first collision data in 2025.

Acknowledgments

This work is supported by funding agencies of Germany (BMBF 05P15VHFC1, 05P19PMFC1, 05P19PXFCA, 05P19RFFC1, 05P19RGFCA, DFG 405580046, GSI, HIC for FAIR, HGS-HIRe, HFH FAIR, FIAS), European Union (Horizon 2020 no.871072 CREMLINplus), Russia (MSHE of RF 0723-2020-0041, RFBR 18-02-40086, RSF 17-72-20234), India (VECC EHEP and A-Group, University of North Bengal), Poland (WUT IDUB-POB-FWEiTE-1), China (NNSF 11927901, 11420101004, 11735009, U1832118, MST 2018YFE0205200, 2016YFA0400100), Czech Republic (CZ.02.1.01/0.0/0.0/16 013/0001677, LM2018112), and Romania (MRID NUCLEU PN19060103, FAIR-03/16.11.2020). The mCBM experiment in carried out within the framework of the FAIR phase-0 program.

References

[1] G. Agakichiev et al., Eur. Phys. J. A 41 (2009) 243
[2] W. Fu, J. Pawlowski, F. Rennecke, Phys.Rev. D 101 (2020) 5
[3] F. Karsch, arXiv:1905.03936
[4] T. Ablyazimov et al. (CBM Collaboration), Eur. Phys. J. A 53 (2017) 60
[5] T. Galatyuk, Nucl.Phys. A 982 (2019) 163
[6] https://www.youtube.com/watch?v=btwRHqcVW4A

[7] https://www.cbm.gsi.de/documents
[8] W. Cassing et al., Phys.Rev. C 93 (2016) 014902
[9] I. Vassiliev, priv. comm.
[10] R. Rapp, H. v.Hees, Phys.Lett. B 753 (2016) 586

The NICA Complex and the MPD experiment at JINR: status and physics potential

Adam Kisiel[1,2,*]

[1]Joint Institute for Nuclear Research, Joliot-Curie 6, 141980 Dubna. Moscow Region, Russia
[2]Faculty of Physics, Warsaw University of Technology, Koszykowa 75, 00-662 Warsaw, Poland

Abstract. The Multi-Purpose Detector is under construction at the Joint Institute for Nuclear Research in Dubna, as part of the NICA Accelerator Complex. The MPD Collaboration aims to study the phase diagram of QCD matter at maximum baryonic density, determine the nature of the phase transition between the deconfined and hadronic matter and search for the critical point. The combination of significant luminosity at NICA, collider geometry of the experiment and collision energies $\sqrt{s_{NN}}$ spanning the range of 4 to 11 GeV, where the transition from baryon to meson dominated matter is expected will allow MPD to provide unique insight into these questions. The status of the assembly of the detector in MPD Hall, as well as the progress in the production of MPD subdetectors will be reported on.
The design physics performance of the detector will be discussed with the emphasis on the measurements in the strangeness sector. Tracking and particle identification procedures will enable precise determination of spectra of identified hadrons, including hyperons and hypernuclei, with the possibility of differential measurements and total yield extraction. The quality of directed and elliptic flow determination will be discussed, with comparison to model expectations. The sensitivity of event-by-event fluctuations to the nature of the phase transition and the presence of a critical point will be given. The femtoscopic analysis for charged pions and kaons has shown sensitivity to first order and cross-over phase transitions in EoS seen for the vHLLE model. All those studies will be put in context of addressing outstanding questions on the properties of the region of the QCD phase diagram probed by heavy-ion collisions in the NICA energy range.

1 Introduction

The Nuclotron-base Ion Collider fAcility (NICA) is a major science project realized by the Joint Institute for Nuclear Research (JINR). Its aims to investigate phase diagram of QCD matter in the region of maximum baryonic density by studying (heavy-)ion collisions in the energy range from 2.5 to 11 GeV/u. The Baryonic Matter at Nuclotron (BM@N) and Multi-Purpose Detector (MPD) experiments will collect data at the facility. This work describes the status of construction of NICA and the MPD experiment, and gives a brief summary of its scientific goals.

*e-mail: Adam.Kisiel@jinr.ru

 https://doi.org/10.1051/epjconf/202225909002

2 NICA and MPD construction status

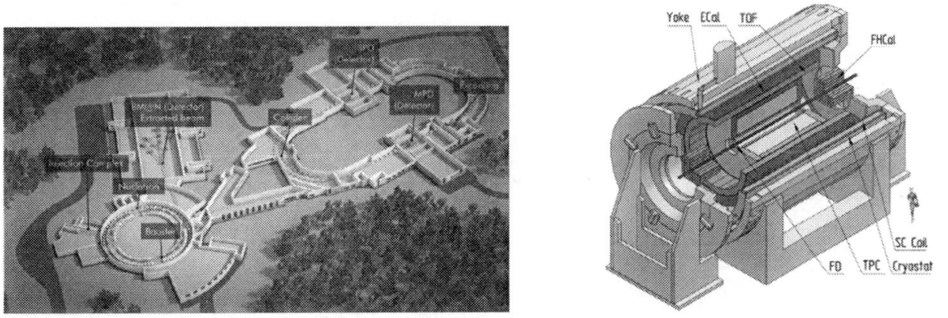

Figure 1. (Left) The schematic view of the NICA complex. (Right) The setup of the MPD apparatus in the Stage I of NICA.

The ion beams for NICA (schematically shown in Fig. 1) are provided by two injection complexes. The KRION source coupled to the HiLac linac produce light ions up to carbon. Dedicated sources with LU-20 linac produce heavier ions, including krypton, xenon and bismuth. Both sources are commissioned and operating. Nuclotron is a circular accelerator based on fast-cycling superconducting magnets, able to accelerate ions from protons up to bismuth with maximum kinetic energy of 3.8 GeV/u for Au ions. It is in operation since 1993. Booster is an auxiliary accelerator, needed for pre-acceleration of heavy ions up to 600 MeV/u. It has been put into operation in December of 2020. The commissioning of the full Booster and Nuclotron complex is planned for the end of 2021 and beginning of 2022. Once completed, they will provide beams of ions to the fixed-target area, where the Baryonic Matter at Nuclotron (BM@N) experiment is operating. Physics runs with beams of Fe/Kr/Xe are planned in 2022 for this experiment. In parallel the civil construction and production of magnets is progressing for the NICA collider. Installation of equipment is planned in stages through 2021 and 2022, the commissioning is expected in late 2022 and first beams will be provided in 2023. NICA is a racetrack-shaped collider, with 503 m of circumference. Two experimental halls are located in two straight sections. NICA is designed to provide Au+Au collisions in the center-of-mass energy $\sqrt{s_{NN}}$ range of 4 to 11 GeV. Luminosity at maximum energy is 10^{27} cm^{-2}s^{-1}. NICA complex can accelerate ions from protons and deuterons (possibility for polarized beams will be added in the Stage II of project preparation) through light (C) and intermediate (Ar, Fe, Kr, Xe, and others) to heavy (Au, Bi) ions. Therefore, a complete energy vs. system size scan is a possible physics program at NICA. In 2023 Bi+Bi collisions at $\sqrt{s_{NN}} = 9.2$ GeV are planned for initial NICA operation. Luminosity is expected to be approximately 10^{25} cm^{-2}s^{-1} with a few months of running time. 2024 will be dedicated for achieving Au+Au collisions with maximum energy of 11 GeV, with routine physics operation foreseen after reaching this goal.

The main physics experiment at NICA is the Multi-Purpose Detector (MPD), which will be operating at the collider. In 2018 an international scientific collaboration of MPD has been established. Currently it is composed of 42 institutes from 12 countries, as well as JINR as a host institution. The collaboration will be operating the MPD apparatus, which is shown schematically in Fig. 1 (right). Its designated position is the MPD Hall, which is located at the northern straight section of the NICA collider. Since 2020 the building has been available for MPD activities.

The construction of the apparatus began in 2020. The outer magnet yoke has been fully assembled with satisfactory mechanical precision. Uniform magnetic field up to 0.57 T (with the default setting of 0.5 T) will be provided by the superconducting solenoid, which has been manufactured by the ASG company in Italy. It has been delivered to JINR in November of 2020 and has been placed inside the magnet yoke in July of 2021. The completion of the assembly, commissioning and magnetic field measurements are foreseen for the reminder of the year. The magnet yoke will also serve as a mechanical fixing point for the dedicated carbon fiber support structure, which will house other detectors. It will be installed in 2022.

Located closest to the beam, the Time Projection Chamber (TPC), a 3.4 m wide, barrel shaped gaseous tracking detector will determine the charged particles' trajectories and their specific energy loss $\langle \mathrm{d}E/\mathrm{d}x \rangle$. Maximum event rate for the TPC is 7 kHz. All components of the TPC vessel are ready and the production of the read-out chambers is progressing at JINR. The Time-of-Flight (ToF) detector will surround the TPC. Its Multigap Resistive Plate Chambers (MRPC) will measure the arrival time of charged particles with time resolution of 50 ps. Together, TPC and TOF will allow for discrimination between pions and kaons up to p_{T} of 1.5 GeV/c and mesons from protons up to 2.5 GeV/c. Significant sample of MRPCs is already produced at JINR. TPC and ToF will be installed in 2022. The Electromagnetic Calorimeter (ECal) is the outermost central barrel detector. Its main goal is to measure the energy of photons and electrons produced in the collision. It is composed of shashlyk-type sampling calorimeters with plastic scintillators, light collected by WLS (Wave Length Shifting) fibers and MAPD (Micro-Pixel Avalanche Photo Diode) read-out. 38400 towers are arranged in the projective geometry in a barrel around ToF. The design energy resolution is 5% at 1 GeV/c momentum. ECal is produced jointly by Russian and Chinese collaborators. Roughly half of ECal will be installed in 2022 with the remaining modules added in the following years.

In the forward region, the Fast Forward Detector (FFD) based on fast Cherenkov modules, placed 140 cm from the interaction point around the beam pipe on either side, will provide level-0 trigger and start time information for the ToF. The Forward Hadronic Calorimeter (FHCal) will be located near the magnet end caps. On each side it will contain 44 sampling calorimeter modules, capable of delivering information on the energy deposited by spectator nucleons and fragments. It will allow for the determination of the event centrality as well as the orientation of the event plane at large rapidities. FFD and FHCal will be installed in 2022. The overall commissioning of the MPD detector is foreseen for late 2022 and will include calibration runs with cosmic muons. A dedicated MPD Cosmic Ray Detector (MCORD) will be used for triggering in this task. It may also be added to the MPD setup later, possibly installed on the outside of the magnet yoke.

3 The MPD experiment physics performance and prospects

The QCD phase diagram is shown schematically in the left panel of Fig. 2. The MPD experiment aims to characterize this diagram, in particular in the area of moderately large temperatures and large baryochemical potential. This region of the diagram has not been fully explored so far. The investigation will be carried out by performing an collision energy scan in the full NICA range of $\sqrt{s_{\mathrm{NN}}}$ from 4 to 11 GeV. The accelerator complex also allows for collision system size scan. This gives possibility for a doubly differential investigation of the onset of deconfinement. In the region of temperatures strongly exceeding the value of 160 MeV (at which the deconfinement is expected to occur), as well as baryochemical potential approaching zero the experiments at RHIC (STAR, PHENIX, BRAHMS and PHOBOS) and LHC (ALICE, CMS and ATLAS) have performed an extensive search for deconfined matter. The production of Quark Gluon Plasma was observed at these conditions, and the transition back to hadronic matter was determined to be of the cross-over type. Going towards larger baryon

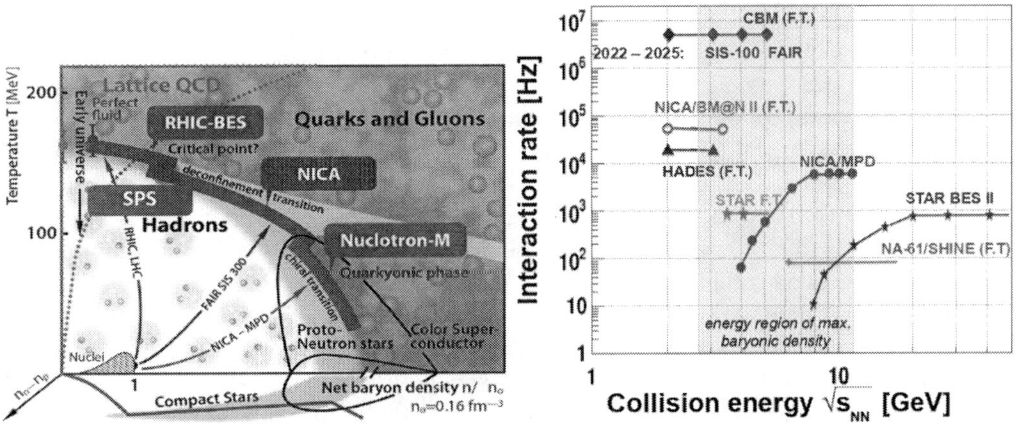

Figure 2. (left) The schematic view of the QCD Phase diagram. (right) The collision energies and expected interaction rates of currently ongoing and planned experimental programs.

densities theoretical calculations point towards the existence of first-order phase transition between QGP and hadronic matter, as well as the existence of the critical end-point in the diagram. The search for this point remains one of the major goals of the heavy-ion collision program at collision energies below 100 GeV. The experiments at SPS, the Beam Energy Scan (BES) program conducted by STAR at RHIC, as well as investigations at lower energies at GSI/FAIR and elsewhere participate in this effort. So far no conclusive experimental evidence for the existence of this critical point has been found.

MPD at NICA will provide unique coverage of collision energies and collision system sizes at large intensities with respect to other facilities operating at similar energies, as shown in the right panel of Fig. 2. Moreover, the MPD experiment is offering collider geometry in this region. This provides large acceptance, which varies only slightly with the change of the collision energy. Therefore, NICA will provide competitive experimental conditions for the exploration of the phase diagram area of interest. The $\sqrt{s_{NN}}$ range of 4-11 GeV for heavy-ion collisions allows to explore the area of maximum baryon density. The discovery potential at NICA is unique and complementary to similar scientific programs at lower energies at FAIR as well as at higher energies at SPS and RHIC BES.

The first-day physics program at MPD will focus on several key observables. The total charged track multiplicity will be measured in the TPC. In combination with the recording of the energy deposit in FHCal they enable determination of the event centrality. A Modified Wounded Nucleon, also known as MC-Glauber [1] will be employed to associate these experimentally determined quantities to the geometrical parameters, which are necessary for theory comparisons, such as the values of the impact parameter, number of participants and spectators. The careful study of nuclear fragments will also be performed. The energy density reached in the collision can be inferred from the multiplicity density per unit of pseudorapidity [2]. Its measurement will show, whether densities needed for the creation of QGP are reached in the collisions at NICA. The transverse energy density, which is related to the internal pressure in the extremely dense matter produced in collisions of heavy-ions [3] will also be measured when the particle identification will be employed.

The particle/anti-particle production ratio for identified hadrons is directly related to key aspects of the system evolution, such as the charge transport mechanism in the collision. The ratio can be measured for particle types identified in MPD, such as $p/\bar{p}$, K^+/K^-, and

π^+/π^-. Provided sufficient statistics is collected, the ratios can also be determined for heavier baryons, for example $\Lambda/\bar{\Lambda}$. After extrapolation to full 4π acceptance, total yields of various particle types, including the ones reconstructed via their decay topology will be determined. The consistent and large acceptance in η and p_{T} gives precise transverse momentum spectra for all those particle types. Through statistical model fits to this data, the freeze-out temperature and baryochemical potential of the produced system will be extracted, placing various datasets on the phase diagram of QCD, allowing the study of variation of their location with collision system size, collision energy and centrality. Secondary vertex reconstruction, three-dimensional tracking of trajectories curved in a magnetic field as well as a combination of particle identification techniques will allow for the reconstruction of particles with non-zero strangeness, including multi-strange baryons such as Ξ and Ω, as well as hypernuclei. This is instrumental in the determination of the degree of thermalization in the system, the appearance of the deconfined state and the search for signatures of the critical point.

Strongly decaying resonances, such as $\rho(770)$, $\mathrm{K}^*(892)$, $\phi(1020)$, $\Sigma(1385)$, and $\Lambda(1520)$ will be reconstructed with the invariant mass technique. Their abundances are sensitive to the existence and duration of the hadronic rescattering phase [4, 5], as their lifetime is comparable with the duration of this phase, in which they can undergo rescattering and regeneration. In thermal models they will also provide independent constraints of the fits.

The system created in heavy-ion collisions experiences extreme pressure gradients, which are converted into significant acceleration and collective behavior of matter. The efficiency of this conversion is directly sensitive to fundamental transport properties of matter, such as shear and bulk viscosity, as well as to the Equation of State of QCD matter. The detailed differential study of flow is therefore of utmost importance. The radial flow manifesting in common velocity for all particles pointing away from the central axis of the source modifies strongly the inclusive transverse momentum spectra of particles with different mass, from which its measurement can be inferred. The spectator nucleons are moving close to the created system in the initial stage after the collision, preventing the flow of matter in their direction. This produces the so-called directed flow v_1. At RHIC BES energies [6] it has intriguing non-monotonic behavior, which will be investigated in MPD. For non-central collisions in plane transverse to the beam direction the cross-section of the system has an almond shape, leading to the development of elliptic flow v_2. Its measurement, allows for determination of crucial properties of the system, and may reveal signatures of changes to the speed of sound c_s near the phase transition, resulting in the softening of the EOS.

The non-monotonic increase in the fluctuations of conserved quantities is expected in the vicinity of the critical point. Event-by-event fluctuations of total event multiplicity, mean transverse momentum and conserved charges, as well as multiplicities of identified hadrons will be measured in MPD. The advantage of the collider experiment is its large acceptance (at least $|\eta| < 1.2$ and $0.2 < p_{\mathrm{T}} < 3.0$ for identified particles), which is also constant with respect to collision energy. This presents favorable conditions for fluctuation measurements, that is low systematic uncertainties, some of which can additionally cancel out in the comparison between measurements at varying $\sqrt{s_{\mathrm{NN}}}$.

The geometrical size of the system created in heavy-ion collisions as well as important aspects of its dynamics are accessible via the femtoscopy technique. It is based on two-particle correlation measurements, routinely carried out for pairs of identical mesons and baryons. Their study as a function of collision energy, centrality and pair type is planned in MPD. In particular, it has been argued [7], that a large ratio of system sizes in the transverse plane can be viewed as a signature of the existence of the first order phase transition. Recently the correlation measurements for pairs of baryons have been used for the determination of strong interaction between them. This is especially important for baryons with non-zero strangeness, where details of such interaction have crucial influence on the properties of dense nuclear matter, found in the cores of neutron stars. Baryon production will be significantly enhanced at NICA energies, making MPD particularly well suited for such measurements.

The ECal will measure electromagnetic probes at NICA. The study of direct photons at low to intermediate p_T will give information on the initial temperature of the system [8–10]. Identification of π^0 and η mesons as a function of centrality and p_T, will also be possible. A clean sample of electrons will be obtained by combining the energy measurement from ECal with tracking and charged particle rejection from TPC and ToF. This will allow a study of light vector mesons, as well as extensive investigation of the di-lepton production at a wide range of invariant mass, as well as the total transverse momentum of the pair. NICA offers a favorable conditions for such studies, as background is expected to be significantly reduced due to suppressed charm production.

4 Summary

The NICA Accelerator Complex has been briefly introduced with a report on the status of construction and commissioning of the new elements. The ion sources, as well as Nuclotron and Booster are operational. First physics run of the collider is expected in 2023. The activities in the MPD Hall are ongoing with magnet assembly well advanced. Production of subdetector components for TPC, ToF, ECal, FHCal and FFD is ongoing. Commissioning of the MPD setup expected to start in 2022.

The physics program of MPD for initial datataking has been briefly discussed. Measurements of collision centrality and event plane, together with tracking and particle identification are foreseen. Characterization of the particle production in the soft regime is the main focus, with an important addition of electromagnetic probes. MPD will perform a search for the signatures of the deconfined matter, the existence and nature of the phase transition to ordinary hadronic matter, as well as possible existence of a critical point in the phase diagram of the QCD matter. The international collaboration of MPD is preparing for first physics runs in 2023.

Acknowledgments

Author is supported by the Polish NCN grant 2017/27/B/ST2/01947, WUT ID-UB and the grants of the Polish Plenipotentiary for JINR.

References

[1] C. Loizides, J. Nagle, P. Steinberg (2014), [SoftwareX1-2,13(2015)]
[2] K. Aamodt et al. (ALICE Collaboration), Phys. Rev. Lett. **105**, 252301 (2010)
[3] R. Sahoo, A.N. Mishra, N.K. Behera, B.K. Nandi, Adv. High Energy Phys. **2015**, 612390 (2015), [Erratum: Adv. High Energy Phys. **2017**, 4517153 (2017)]
[4] S. Acharya et al. (ALICE), Phys. Rev. **C99**, 064901 (2019)
[5] B.I. Abelev et al. (STAR), Phys. Rev. Lett. **97**, 132301 (2006)
[6] L. Adamczyk et al. (STAR), Phys. Rev. Lett. **120**, 062301 (2018)
[7] P. Batyuk, I. Karpenko, R. Lednicky, L. Malinina, K. Mikhaylov, O. Rogachevsky, D. Wielanek, Phys. Rev. C **96**, 024911 (2017)
[8] A. Adare et al. (PHENIX), Phys. Rev. Lett. **104**, 132301 (2010)
[9] A. Adare et al. (PHENIX), Phys. Rev. **C91**, 064904 (2015)
[10] J. Adam et al. (ALICE), Phys. Lett. **B754**, 235 (2016), 1509.07324

https://doi.org/10.1051/epjconf/202225909003

Prospects for the NA60+ experiment at the CERN SPS

Alessandro De Falco[1,*] for the NA60+ Collaboration.

[1]Università/INFN Cagliari, Italy

Abstract. A new heavy-ion experiment on fixed target, NA60+, has been proposed at the CERN SPS for data taking in the next years. Its main goals will be focused on precision studies of thermal dimuons, heavy quark and strangeness production in Pb-Pb collisions at center-of-mass energies ranging from 5 to 17 GeV, which will provide a unique opportunity to investigate the region of the QCD phase diagram at high baryochemical potential ($\mu_B \sim 200 - 400$ MeV). The key points of the NA60+ very broad and ambitious physics program will be described.

Heavy-ion collisions at low center-of-mass (CM) energies provide a unique tool to investigate the QCD phase diagram at large values of the baryochemical potential μ_B. An energy scan at the SPS, with CM energies ranging from 5 to 17 GeV per nucleon, will allow the study of the phase diagram in the range $\mu_B \sim 200 - 400$ MeV (fig. 1), where key information can be obtained about the presence of a critical point, the order of the phase transition at large μ_B, the chiral symmetry restoration effects and the temperature at which the onset of the deconfinement takes place.

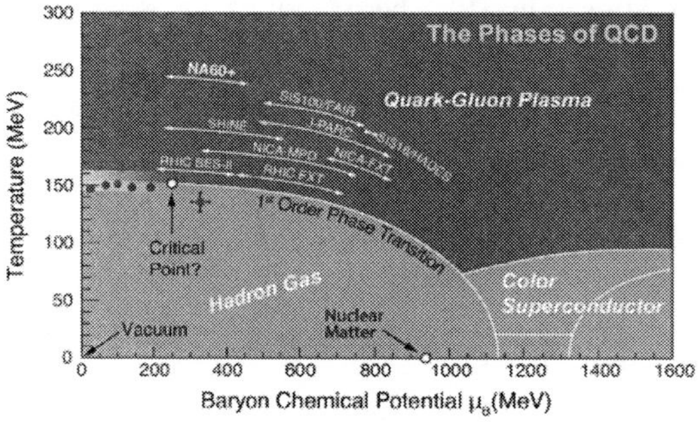

Figure 1. The QCD phase diagram (courtesy of Thomas Ullrich)

The NA60+ experiment has been proposed [1] to study hard and electromagnetic probes in this μ_B region. Such measurements would address many yet uncovered points in the understanding of the QGP phase. Hard probes may bring information on the onset of deconfinement via the measurement of J/ψ suppression as a function of the CM energy, while open charm states would allow to determine the transport properties of the Quark-Gluon Plasma (QGP) at large values of μ_B. Electromagnetic probes, and in particular dileptons, would give

*e-mail: alessandro.de.falco@ca.infn.it

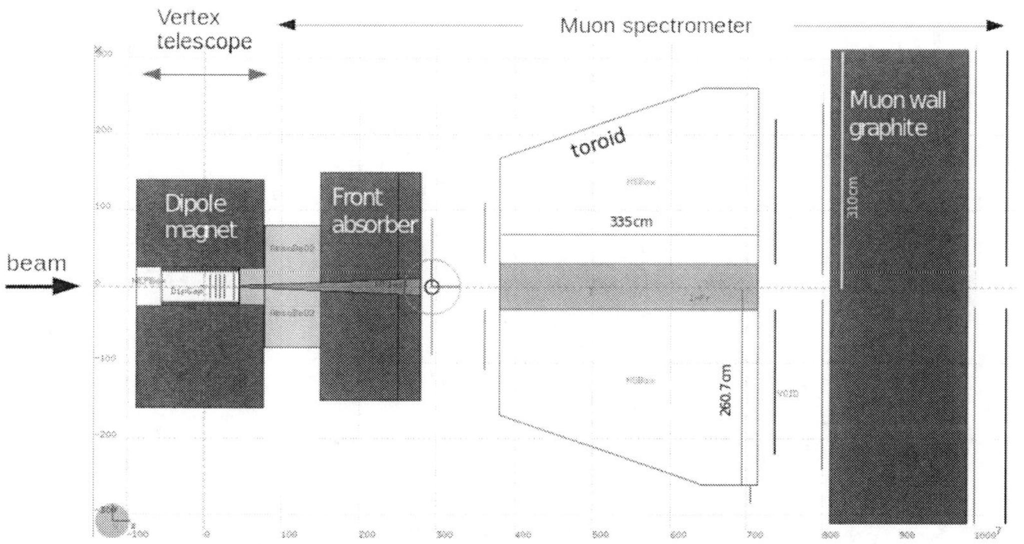

Figure 2. Layout of the proposed NA60+ apparatus in the low energy setup.

insight on the temperature of the system via the measurement of the thermal dimuon mass spectrum, on chiral symmetry restoration effects and on the order of the phase transition [2, 3].

The detector concept of NA60+ inherits the design from its predecessor NA60, and is based on a muon spectrometer coupled with a vertex telescope. Muons are detected by a magnetic spectrometer, composed of a toroidal magnet and four tracking stations, placed after a hadron absorber. A new magnet, which should provide a field $B \cdot R \sim 0.2 - 0.5$ Tm is being designed and a small-scale prototype has been constructed. The tracking stations will be equipped with about 330 GEM modules, each one measuring 50×110 cm^2, overlapping in both horizontal and vertical coordinates by 10 cm. The GEM module design profits from the experience developed for the CMS muon system and the ALICE TPC upgrades. A trigger system based on two RPC stations is placed after an additional graphite absorber. The width of the front absorber, as well as the position of the tracking stations will be changed depending on the beam energy, such that the detector keeps a good acceptance around mid-rapidity. The low-energy ($E_{\mathrm{beam}}/A = 20-40$ GeV) setup is reported in fig. 2. At high energy, $E_{beam}/A = 158$ GeV, the muon spectrometer will be moved on rails by 3.3 m and the absorber thickness will be increased to 4.6 m. The presence of the hadron absorber provides the muon identification. However, it introduces a deterioration in the muon momentum resolution, due to multiple scattering and fluctuations in the energy loss. This loss in resolution can be recovered by matching the muons reconstructed in the muon spectrometer with the charged tracks measured in the vertex telescope (VT), both in coordinates and momentum space. The VT can also be used for multiplicity measurements and for the study of hadronic decay channels of charm and strange hadrons, as a stand-alone detector. The VT consists of a dipole magnet and a set of pixel planes for tracking. The choice considered for the magnet is the CERN MEP48 dipole, that provides a field $B = 1.47$ T and an angular coverage of 21^0. The tracking system will consist of a set of (5 or 10) planes made of four large area Monolithic Active Pixel Sensors (MAPS) obtained by means of the stitching technology. The pixel dimensions are of about 15×15 μm^2. The sensor thickness is of the order of 20 μm. Moreover, the mechanical supports and cooling are moved to the border. Therefore, the material budget in

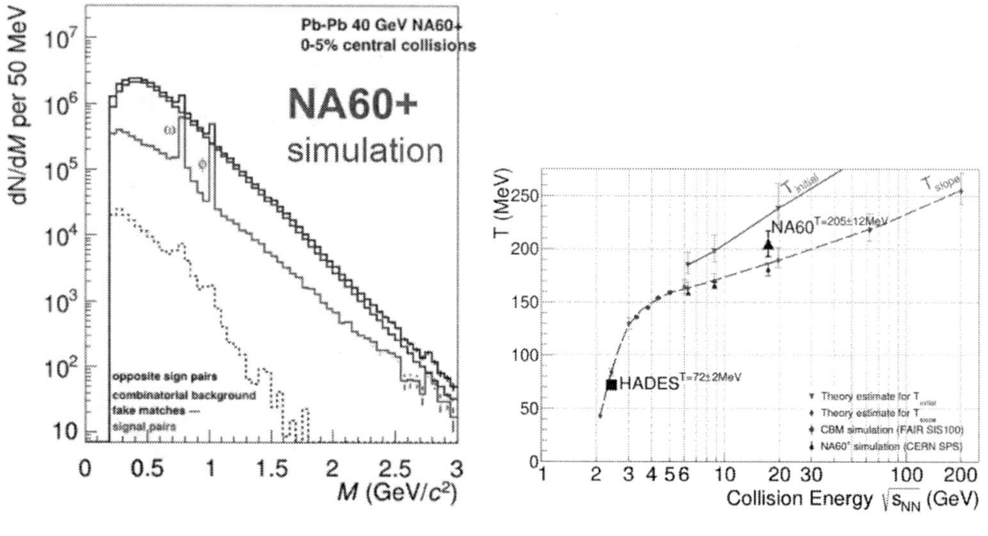

Figure 3. Left: simulated dimuon mass spectrum in central Pb-Pb collisions at $E = 40$ GeV per nucleon. Right: medium temperature evolution vs $\sqrt{s_{NN}}$ in central Pb-Pb collisions using [2]. Measurements from NA60 [4] and HADES [5].

the sensitive area is very small, lower than $0.1\% X_0$, reducing the effect of multiple scattering in the VT. The expected spatial resolution is 5μm or better, with a gain of a factor of 2 with respect to the hybrid technology.

A high intensity beam is required for the measurement of hard and electromagnetic probes. The experiment is foreseen to use the H8 beam line at the CERN SPS, which can provide an intensity of 10^7 Pb ions per 20 s spill. The interaction rate is thus expected to be one order of magnitude higher with respect to other experiments in the same μ_B range. With a typical run time of one month per energy, a very significant statistics can be reached for each of the proposed measurements.

The physics performances of NA60+ were studied performing fast simulations of the signals with semi-analytical tracking based on the Kalman filter, while the simulation of the background was performed with FLUKA. The obtained opposite sign dimuon mass spectrum is shown in the left panel of fig. 3. The mass resolution for $M = M_\omega$ is ~ 7 MeV/c^2, about one third of the NA60 resolution at the same mass. Besides the signal, composed of muon pairs, two sources of background are visible: the combinatorial background and the fake matches contribution. The former is due to uncorrelated pairs of muons mainly originated from pion and kaon decays. The latter is due to incorrect matches between muons reconstructed in the muon spectrometer and tracks reconstructed in the vertex telescope. Both contributions can be evaluated using event mixing techniques and subtracted to the opposite sign mass spectrum. The signal mass spectrum resulting after background subtraction is dominated by the hadronic cocktail for $M < 1.5$ GeV/c^2. In this region, a precision measurement of the ρ spectral function can be performed, complementing the NA60 measurement in In-In at top SPS energy with results at lower energy. In the region $1 < M < 1.5$ GeV/c^2, a dimuon enhancement due to the chiral mixing between the ρ and its chiral partner a_1 via 4π states can be observed [3]. For $M > 2$ GeV/c^2, after subtracting the Drell-Yan and open charm contributions, the mass spectrum is dominated by thermal dimuons, that can be fitted to a thermal distribution of the form $dN/dM \propto M^{-3/2} \exp(-M/T_S)$. The parameter T_S represents a space-time average of the thermal temperature over the fireball evolution and can be determined with a precision of the order of 10 MeV. Measurements of T_S vs $\sqrt{s_{NN}}$ at low energy ($<$ 10 GeV) may allow to discover a plateau in the caloric curve (fig. 3, right) that would be present in case of a phase transition of the first order [2].

Charmonium suppression was extensively studied at the SPS, where the NA50 collaboration observed a suppression of $\sim$ 30% of the J/ψ meson yield in central Pb-Pb collisions at the top SPS energy ($\sqrt{s_{NN}}$ = 17.3 GeV) that could not be ascribed to cold nuclear matter effects alone [6]. The suppression is qualitatively consistent with the melting of the less bound χ_c and $\psi(2S)$ states in a deconfined medium, which would lead to a suppression of the J/ψ produced from the decays of these particles. At present, no measurement is available at lower energy. NA60+ aims to extend the measurements down to $E_{\rm lab}/A$ = 40 GeV in order to search for the onset of the deconfinement. By correlating the measurement with the corresponding determination of the temperature obtained from thermal dimuons, one could obtain the temperature at which the χ_c and $\psi(2S)$ melt. With one month data taking at an intensity of 5×10^5 Pb ions per second, a number of J/ψ in the range $1.5 - 20 \times 10^4$, depending on the beam energy, may be reconstructed. Also cold nuclear matter effects have to be determined precisely, therefore, data taking with p-A collisions are mandatory at each CM energy. A 15-days long data taking with a beam intensity of $3 \cdot 10^8$ protons/s on 7 nuclear targets is planned to measure the J/ψ production cross section as a function of the mass number.

Open charm measurements can be performed reconstructing the decays of charmed hadrons into two or three charged hadrons, using the vertex telescope as a stand-alone detector. The huge combinatorial background can be reduced by applying geometrical selections on the displaced decay vertex topology, profiting from the fact that the $c\tau$ for these decays is about $60 - 310\mu$m, and therefore the decay vertices are displaced by a few hundred μm from the interaction point. A high resolution on the vertex reconstruction is needed. The MAPS technology can provide a signal to background ratio $\sim$ 10 times higher than the corresponding value obtainable with hybrid pixel sensors. As of today, no open charm measurement has been performed below the top SPS energy. In one month data taking, more than $3 \cdot 10^6$ D^0 can be reconstructed in central collisions at $\sqrt{s_{NN}}$ = 17.3 GeV, allowing for a precise determination of the yield and elliptic flow as a function of p_T, rapidity and centrality. At $\sqrt{s_{NN}}$ = 10.6 GeV due to the lower production cross section, the number of reconstructed D^0 is expected to be lower by about an order of magnitude. Still, the measurement will be feasible with a statistical precision at the level of the percent.

Finally, strangeness measurements in the hadronic decay channels will allow to explore the low multiplicity region for the understanding of strangeness enhancement with multiplicity. The high statistics expected to be collected will guarantee a high p_T-reach, the study in narrow centrality bins and an extention to multistrange hadrons of the elliptic flow measurements.

The broad physics program of the NA60+ experiment provides a strong motivation for its realization. The goal is to start data taking in 2027, with LHC run 4.

References

[1] T. Dahms, E. Scomparin, G. Usai (NA60+ Collaboration), Tech. rep., CERN, Geneva (2019), `https://cds.cern.ch/record/2673280`
[2] R. Rapp, H. van Hees, Phys. Lett. B **753**, 586 (2016), 1411.4612
[3] M. Dey, V.L. Eletsky, B.L. Ioffe, Phys. Lett. B **252**, 620 (1990)
[4] R. Arnaldi et al. (NA60), Eur. Phys. J. C **59**, 607 (2009), 0810.3204
[5] S. Harabasz (HADES), Nucl. Phys. A **982**, 771 (2019)
[6] B. Alessandro et al. (NA50), Eur. Phys. J. C **39**, 335 (2005), hep-ex/0412036

EPJ Web of Conferences 259, 09004 (2022) https://doi.org/10.1051/epjconf/202225909004
SQM 2021

Charm and multi-charm baryon measurements via strangeness tracking with the upgraded ALICE detector

David Dobrigkeit Chinellato for the ALICE Collaboration[1,2,*]

[1]UNICAMP
[2]CERN

Abstract.
We present a new method for detection of multiply charmed baryons via their decays into strange baryons, using 'strangeness tracking'. This method makes use of the state-of-the-art upgraded silicon detectors in ALICE during Runs 3, 4 and beyond will enable the novel possibility of tracking strange hadrons directly before they decay, leading to a very significant improvement in impact-parameter resolution. In this work, we will discuss how this new technique will be crucial to distinguish secondary strange baryons originating from charm decays from primary strange baryons. This is a particularly interesting possibility for the Ω^- baryon coming from $\Omega_c^0 \rightarrow \Omega^-\pi^+$ decays, since there is no other relevant feeddown source for Ω^-. This, in turn, means that the main Ω^- background for the Ω_c measurement will point most accurately to the primary vertex, unlike pions or protons from other charm baryon decays.
We will illustrate the achievable performance of strangeness tracking for the Run 3 configuration of ALICE with the upgraded Inner Tracking System, which is fully instrumented with silicon pixel detectors. Moreover, we will discuss the potential of this technique in a future experiment with an extensive silicon tracking detector with a first layer very close to the interaction point. Finally, we will also cover other potential major applications of strangeness tracking, including measurements of hypernuclei such as the ${}^{3}_{\Lambda}$H.

1 Introduction

A fundamental ingredient of the ALICE physics programme for the new decade is a comprehensive study of charm and multi-charm baryon production. Because charm is exclusively produced in initial hard scatterings, such measurements may provide unique insight into the QGP medium as well as hadronization from proton-proton to lead-lead collisions. However, the reconstruction of charmed hadrons poses a significant challenge. In particular, charm particles decay into strange particles, in which case topological reconstruction can serve as a powerful particle identification method. However, due to the large extrapolation distances, weak decay reconstruction is usually also associated with less than optimal pointing resolution to the primary vertex.

*e-mail: daviddc@ifi.unicamp.br

We propose to overcome this difficulty using a novel reconstruction technique called 'strangeness tracking', which is made possible by the next-generation detectors of the experiment. In this method, we use high-resolution silicon detectors very close to the primary vertex to measure the hits of a weakly decaying hadron prior to its decay and combine them with the information from the decay daughters to significantly improve pointing resolution. In this context, the upgraded ALICE inner tracking system, the ITS2 [1], is ideal for a first implementation of such a method. This is because the ITS2 has an innermost silicon pixel layer at a radius of 22 mm, as opposed to 39 mm for the previous silicon pixel detectors. As a consequence, the two multi-strange baryons Ξ^- and Ω^-, reconstructed in the channels $\Xi^- \to \pi^- + \mathrm{p}$ and $\Omega^- \to \mathrm{K}^- + \mathrm{p}$ channels and having a proper lifetime of 4.9 and 2.5 cm, respectively, are significantly more likely to leave hits in the innermost layers of the ITS2. An example of a reconstructed Ω^- decay obtained with a simulation of the full geometry of the upgraded ALICE detector and using GEANT3 propagation is shown in Fig. 1.

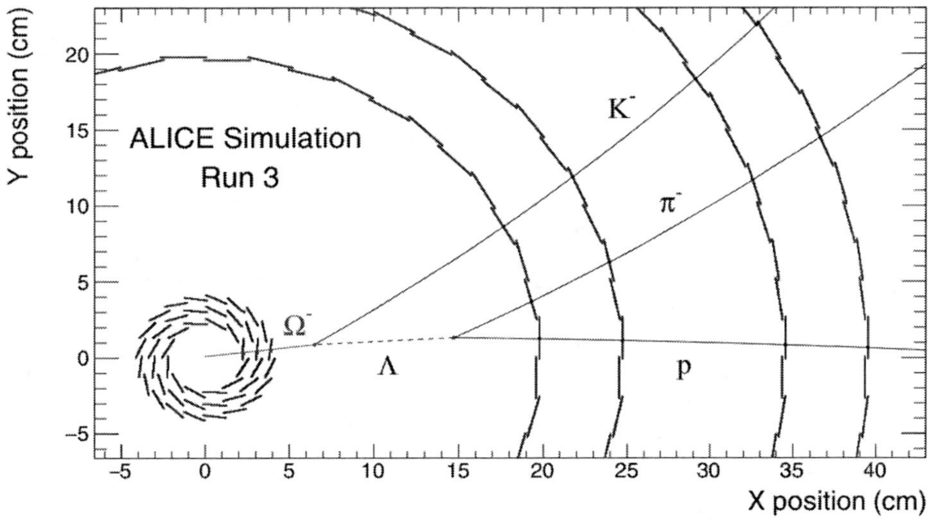

Figure 1. Event display of an $\Omega^- \to K^- + \Lambda$ decay in the upgraded ALICE detector in Run 3. The Ω^- hits in the inner layers of the ALICE ITS2 are highlighted in red and are used in the reconstruction of the weak decay when the strangeness tracking technique is employed.

2 Feasibility of strangeness tracking

Algorithmically, the task of performing strangeness tracking involves a precise enough reconstruction of the decay daughters such that the back propagation of a Ξ^- or Ω^- candidate is sufficiently constraining for the association of the correct inner layer hits to this trajectory. This establishes a crucial performance constraint: in order to ensure a low fake hit association rate, the hit density in a given layer, ρ_{hits}, should obey the relation $1/\rho_{hits} \gg \delta_{\mathrm{search}}$, where δ_{search} is the search window that is used when propagating the cascade trajectory inward to that layer. The parameter δ_{search} has been calculated in simulations and is below 0.05 mm^2 for typical Ω^- and Ξ^- inwards propagation in the innermost layer, while hit densities are not expected to exceed 50 cm^{-2} even in central Pb–Pb collisions. Because $1/\rho_{hits} = 2$ mm$^2 \gg 0.05$ mm^2, no major difficulties are expected when implementing strangeness tracking for Run 3 data.

 https://doi.org/10.1051/epjconf/202225909004

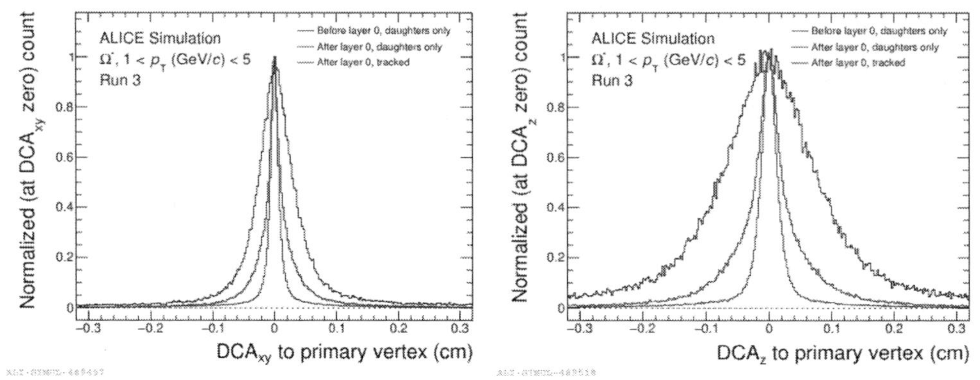

Figure 2. Distance of closest approach to the primary vertex for Ω^- baryons reconstructed in proton-proton collisions at 13 TeV with the upgraded ALICE detector in Run 3 in three different configurations, as obtained with a full detector simulation.

In this study, Monte Carlo association of the cascade hits will be employed to illustrate the expected gains of the strangeness tracking method in an idealised scenario.

3 Reconstruction improvements

The direct tracking of weakly decaying hadrons is expected to have a large impact in the pointing resolution, which can, for instance, be readily quantified by investigating the effect on the distribution of distance of closest approach (DCA) of Ω^- baryons to the primary vertex. This is shown in Fig. 2 for the DCA_{xy} and the DCA_z, which are measured in the transverse plane and in the longitudinal direction, respectively. Three distributions are shown: (a) one in which the Ω^- decayed prior to the first layer and has therefore only been reconstructed using daughter information, (b) a second curve in which the Ω^- decayed after the first layer but no direct measurement was performed, illustrating the significant worsening of the DCA resolution and (c) a third curve in which Ω^- baryons decay after the first layer and direct detection information is added to the daughter information. The resulting improvement in DCA is of a factor of approximately 4 in both transverse and longitudinal directions, showcasing the potential of the method.

When used for the detection of Ω^- from Ω_c^0 decays, strangeness tracking is crucial to isolate differences in the pointing to the primary vertex between these two sources, which will be due to the lifetime of the Ω_c^0, which has $c\tau = 80\ \mu$m/c, as illustrated in Fig. 3. While for the ITS2 the differences between primary and secondary Ω^- baryons are still not too evident, this is expected to improve significantly with further upgrades to the ALICE apparatus, which will enable a better DCA to primary vertex resolution for primary Ω^-; such upgrades will be discussed in the last section of this document.

Similar tests were also performed for the Ξ^- baryon. The DCA resolution increase is comparable to the Ω^- one and strangeness tracking also outperforms the situation in which the Ξ^- decays before the first silicon layer and therefore all daughters are tracked as best as possible. Moreover, early tests with hypertriton reconstruction also indicate a significant improvement in pointing performance that may also serve to greatly reduce the background in hypertriton analyses, as DCA selections help in removing background from secondary deuterons.

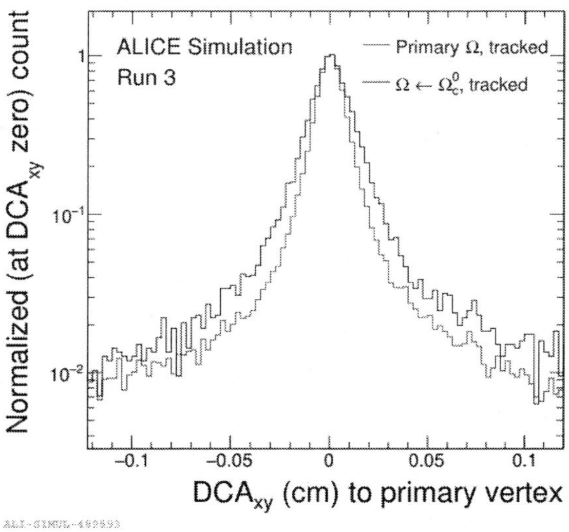

Figure 3. Distance of closest approach to the primary vertex for primary Ω^- baryons and Ω^- from $\Omega_c^0 \rightarrow \Omega^- + \pi^+$ decays as reconstructed in proton-proton collisions at 13 TeV with the upgraded ALICE detector in Run 3.

4 Outlook

As studied in this work, strangeness tracking promises to dramatically improve the pointing resolution of weakly decaying particles such as the Ξ^-, Ω^- and even hypernuclei. However, further upgrades will enable even better performance. Notably, the upgrade of the ITS scheduled for the third long shutdown of the LHC will bring the innermost layers even closer to the interaction vertex with the ITS3 [2] and reduce material budget significantly, further improving both the performance and the expected fiducial for strangeness tracking. Internal testing of strangeness tracking in that setup has already begun and shows significant gains.

Further in the future, a next-generation, all-silicon apparatus has been proposed as a replacement for the TPC-based ALICE experiment [3]. For this new setup, a design is being considered that has a retractable silicon layer that is positioned only 5 mm away from the beam axis during datataking and would serve as the ultimate strangeness tracking tool. A cornerstone of the physics programme of this setup would be the detection of the multi-charm baryons Ξ_{cc}^{++}, Ω_{cc}^{+} and Ω_{ccc}^{++}, all of which have decay channels into multi-strange baryons. Also in this case, strangeness tracking studies are ongoing and the ability to separate signal from background will be reported very soon.

References

[1] B. Abelev, J. Phys. G. **41**, 087002 (2014)
[2] L. Musa, W. Riegler, CERN-LHCC-2019-018
[3] D. Adamová, arXiv:1902.01211

Theory progress at Strange Quark Matter 2021

Björn Schenke[1,*]

[1]Physics Department, Brookhaven National Laboratory, Upton, NY 11973, USA

Abstract. I review recent theory progress reported at the 19th International Conference on Strangeness in Quark Matter (SQM), and discuss open questions to be addressed by the coming editions of SQM.

1 Introduction

The focus of the "Strange Quark Matter" (SQM) conference series is the understanding of strangeness and heavier flavor production and evolution in strongly interacting matter, as generated in heavy ion collisions or present in neutron stars. The conference also addresses related questions, such as that of the detailed structure of the phase diagram of quantum chromo dynamic (QCD) matter, including the potential existence of a critical point. Furthermore, the process of hadronization, the conversion of free quarks and gluons into color neutral hadrons, as well as the determination of QCD transport coefficients were major topics discussed at this year's conference. In this proceeding I will focus on new theory developments and important open questions that remain for future SQM conferences.

2 Equation of state and phase diagram

The QCD equation of state is well determined from lattice QCD for zero baryon chemical potential μ_B. For finite baryon chemical potential, however, the sign problem prevents direct evaluation using standard Monte Carlo methods. Different approaches to reach increasingly large μ_B are being followed, for example that of Taylor expansion around $\mu_B = 0$. This method suffers from slow convergence and bad signal to noise ratio for the higher order expansion coefficients. Recent work was presented that reorganizes the expansion scheme [1] and can determine thermodynamic observables up to $\mu_B = 3.5T$. One should note that singularities in the complex μ_B plane will make the expansions fail, and there are various indications that this happens at $\mu_B/T \gtrsim 1.5$ [2]. Understanding the nature of the singularity responsible for such a breakdown is a topic of great interest. It will need detailed quantitative studies involving careful finite-volume scaling analyses.

Besides lattice QCD, model calculations also allow to explore the equation of state at finite chemical potentials. Results from a chiral mean field model were presented [3], and it was found that results for susceptibilities were dependent on the repulsive interactions between hadrons, in particular the hard core repulsion of hyperons. Comparison to lattice data where available, also lead to the conclusion that the excluded volume for different particles matters [4], in particular that strange baryons have smaller excluded volumes than non-strange ones [3].

*e-mail: bschenke@bnl.gov

The presence of strange baryons, or hyperons, affect the equation of state, and in turn can affect neutron star properties. In fact, the "hyperon puzzle" states that when including hyperons, the mass-radius relation of neutron stars becomes incompatible with experimental observations. There are many suggestions for a way out, including modified hyperon interactions, effects of meson condensates or quark matter, or even dark matter effects or modified gravitational theories [5]. New machine learning techniques were presented as another way to constrain the neutron star equation of state [6]. From this method, the presence of a first order transition could not be excluded. Deep learning was also trained to determine the phase transition order in a linear sigma model [7], which in the future could lead to new methods to apply to experimental data.

The search for the QCD critical point relies heavily on observations of fluctuations of conserved charges, for which extrema are expected if the system passes close to the critical point. Model improvements to allow for a better comparison of fluctuation observables with experiment were presented [8]. This included hydrodynamics together with an excluded volume hadron resonance gas model, matched to lattice QCD susceptibilities. Proton cumulants in the experimental acceptance were first calculated in the grand-canonical limit, then corrections for exact baryon number conservation were applied. This method allows for example a direct comparison of net-proton cumulants, which turn out to be significantly different from net-baryon cumulants. It remains an open question whether the experimentally observed extrema in these observables as a function of beam energy indicate any criticality. Effects of non-equilibrium (in the initial state [9] and the evolution [10]) on fluctuation observables were also addressed, and can be of great significance.

New lattice QCD calculations of electric charge and strangeness cumulant ratios were presented [11]. They allow the extraction of freeze-out chemical potentials by comparison to experimental data. Freeze-out was a concern in several more presentations, all of which agreed that strange particles freeze out at a higher temperature than non-strange ones [12–14].

3 Transport coefficients and medium effects

Besides the diffusion coefficients for strange and heavy quarks, which we will focus on below, the properties of the bulk medium, characterized by the shear and bulk viscosity to entropy density (or enthalpy) ratios were also addressed. In particular the temperature and chemical potential dependence of these quantities is a matter of great current interest. A minimum of the effective $(\eta/s)(\sqrt{s})$ was found by analysis of the RHIC beam energy scan data using a scaling function [15], the shear viscosity to enthalpy ratio was computed in a hadron resonance gas model [16] as a function of T and μ_B, and effects of trajectories in the phase diagram were studied. Also, $(\eta/s)(T)$, $(\zeta/s)(T)$, and the electrical conductivity were computed within a quasiparticle model [17].

Lattice QCD provides a powerful tool to compute heavy quark diffusion coefficients. Both spatial and momentum diffusion coefficients for charm and bottom quarks were determined using both hadronic and gluonic correlators [18, 19]. Alternatively, one can use effective theories to determine diffusion coefficients for heavy flavors, such as those based on chiral and heavy-quark spin-flavor symmetries within the imaginary-time formalism [20]. Including thermal and off-shell effects, it is found within such model that the D-meson spatial diffusion coefficient matches smoothly to the latest results of lattice-QCD calculations [21].

All transport coefficients are important inputs for calculations of the bulk medium evolution and in-medium evolution of heavy-quarks. There was some focus on transport theory with results presented for comparisons of microscopic transport simulations and hydrodynamics. This study indicates that charm quarks only approach local thermal equilibrium at small transverse momentum, even though they acquire an elliptic flow that is comparable to light-quark hadrons also at larger p_T [22]. Transport simulations were also performed at fixed η/s, which can be compared to hydrodynamics, pushed far off equilibrium, and can incorporate different hadronization prescriptions, to compare e.g. microscopic implementations to Cooper-Frye [23]. Extensions to the same transport descriptions to include vorticity and electromagnetic fields were also presented [24]. Such solutions to the Boltzmann equation were also used to describe both the evolution of the medium and heavy quarks within, using fluctuating initial conditions. Diffusion coefficients for charm and bottom quarks consistent with lattice QCD results, allow to describe the experimental data for R_{AA} and v_2 for example [25].

Antikaon properties in nuclear matter were also computed using a unitarized coupled-channel model in dense and hot matter (or G-matrix approach) based on the SU(3) meson-baryon chiral Lagrangian. This can be implemented in the parton hadron string dynamics (PHSD) transport code and observables, including momentum anisotropies, calculated [26]. These calculations show, by comparison to data from the KaoS, FOPI and HADES Collaborations, that medium modifications of (anti)kaon properties are necessary to explain the data. Also effects of the strong magnetic field on heavy quark transport were investigated, yielding mainly a directional anisotropy [27].

The behavior of quarkonium in the medium was studied by solving a Lindblad equation for the evolution of the heavy-quarkonium reduced density matrix, derived using potential non-relativistic QCD and the formalism of open quantum systems [28, 29]. In combination with hydrodynamic simulations for the medium, this calculation leads to results for the suppression and elliptic flow of the $\Upsilon(1S)$, $\Upsilon(2S)$, and $\Upsilon(3S)$, as functions of centrality and transverse momentum, in very good agreement with the experimental data [29].

Medium modifications of bottomonium spectral functions, namely the mass shift and broadening, were also computed on the lattice [30]. While masses are barely modified, the widths of the $\Upsilon(1S)$, $\Upsilon(2S)$, and $\Upsilon(3S)$ increase significantly with temperature. These lattice results were also used as input in a deep learning analysis to extract the heavy quark potential in the quark gluon plasma [31], and the imaginary part turned out to be quite different from that of the hard thermal loop potential used in [29]. As for heavy quarks, also for charmonium states effects of strong electromagnetic fields, as well as vorticity fields, were considered and studied via the two-body Schroedinger equation [32]. Both mass and shape of the charmonium were significantly affected by the fields.

There has been a continued difficulty to simultaneously describe the medium suppression, quantified by R_{AA} and the elliptic anisotropy of high momentum probes, both light and heavy. At this conference, the Dynamical Radiative and Elastic Energy Loss Approach (DREENA) was used with both Bjorken expansion and a 3D hydrodynamic medium to simultaneously describe the two observables in heavy ion collisions at LHC [33, 34]. The authors give the proper description of the parton-medium interaction as the main ingredient of the framework that is responsible for its success. Somewhat in contrast, in previous studies the importance of fluctuations for the resolution of the $R_{AA} - v_2$ puzzle was highlighted [35]. It remains to be seen what the main ingredient is and what the DREENA framework will give when fluctuations are included.

4 Hadronization

One interesting experimental observation is strangeness enhancement in high multiplicity systems as measured by the ALICE Collaboration [36, 37]. Different theoretical models with very different underlying physics pictures are able to describe the experimental data. This includes PYTHIA with "rope hadronization" (at least for p+p collisions) [38], a (hydrodynamic) core+corona model [39], and thermal (statistical hadronization) model with canonical suppression included [40, 41]. It remains an open question as to which picture reflects nature most accurately, in particular for the smaller systems considered.

New results from the above mentioned statistical hadronization model, including an extension to include charm degrees of freedom were also presented [42]. It was shown that the measured multiplicities of single charm hadrons in heavy ion collisions at LHC energies can be well described with the same thermal parameters as for the light quark flavor hadrons. Production probabilities for doubly- and triply-charmed hadrons are predicted to exhibit a characteristic enhancement hierarchy. The model also allows to determine transverse momentum spectra by combining it with blast wave models [42, 43].

An interesting observation was presented that indicates a non-universality of hadronization when comparing $e^+(p) + e^-$ collisions with p+p collisions. It appears that charm quarks are "redistributed" from mesons to baryons as one goes from collisions involving electrons to p+p collisions [44]. Again, a variety of theoretical models, ranging from PYTHIA with "enhanced color reconnections", to a statistical hadronization model with additional excited baryon states, to a combined coalescence and fragmentation model, can describe this enhancement of the charmed baryon yield [45]. The latter model also does best in describing the heavier baryon states, such as Ξ_c^0 and Ξ_c^+.

After hadronization, hadrons and resonance states interact, decay, and recombine. This is often described using so called hadronic afterburners. Their effects were studied in detail in [46]. The afterburner leads to a suppression of most resonances (enhancement of some), an increase in their mean transverse momentum, and a suppression of their v_2 at small p_T. Similar results were also found by comparing EPOS with EPOS+UrQMD [47, 48].

Finally the production of light nuclei was addressed within a three-fluid dynamic model with coalescence and UrQMD [49], as well as within the Parton-Hadron-Quantum-Molecular Dynamics model [50]. Within the latter it was found that clusters are formed shortly after elastic and inelastic collisions have ceased, and behind the front of the expanding energetic hadrons, which allows them to survive.

5 Polarization and spin

The large global angular momentum (of e.g. $L = 10^7\hbar$ in 2.76 TeV collisions at LHC for impact parameter $b = 7$ fm) in heavy ion collisions generates an enormous vorticity of the order of 10^{21} s^{-1}. Spin orbit coupling leads to polarization of produced particles. For the hyperon, the decay products of its weak decay carry information on its polarization in their angular distribution. The phase space averaged polarization measured in this way at RHIC is well described by models. Calculations using AMPT were presented and able to describe the global polarization of Λ, Ξ^-, and Ω^- hyperons for 200 GeV collisions at RHIC [51]. Predictions for lower energies were also made. The ordering $P_{\Omega^-} > P_{\Xi^-} > P_\Lambda$ was found, where P_X is the polarization of the respective particle X.

It is puzzling that the angle differential polarization has the opposite sign in experiment compared to calculations. An attempt to remedy this situation by including a previously neglected shear induced polarization was presented [52, 53]. It leads to an improvement only if in addition the Λ inherits the strange quark's polarization and that spin polarization is frozen from the hadronization on. Further developments concerning the spin polarization were presented, including new proposals on how best to measure the hyperon polarization, specifically under consideration of the different rest frames involved [54]. Predictions for the collision energy dependence of Λ polarization were also presented, indicating a maximum at $\sqrt{s} \approx 7.7$ GeV [55]. Also the effect of vorticity on the spin alignment of vector mesons was addressed, in particular using a quark coalescence model [56, 57]. Effective vector meson fields and vorticity fields turn out to be important for the spin alignment in this model. There remain many open questions concerning spin in heavy ion collisions. What effect does rotation have on strongly interacting matter (equation of state, phase structure, vector mesons)? Can we develop a complete spin transport theory or spin hydrodynamics to understand all aspects of spin in heavy ion collisions? The field will be working towards answering these and other questions discussed in this short overview by the next SQM.

6 Conclusions

Besides the tremendous experimental advances [58], the theoretical progress towards understanding of strange and heavy flavor production reported at this edition of SQM has been significant. As discussed in this brief summary, many open questions remain and new ones are emerging as we learn more both from experiment and new theoretical developments.

Acknowledgments

B.P.S. is supported under DOE Contract No. DE-SC0012704.

References

[1] S. Borsányi, Z. Fodor, J.N. Guenther, R. Kara, S.D. Katz, P. Parotto, A. Pásztor, C. Ratti, K.K. Szabó, Phys. Rev. Lett. **126**, 232001 (2021), 2102.06660
[2] S. Mondal, S. Mukherjee, P. Hegde (2021), 2106.03165
[3] A. Motornenko, S. Pal, A. Bhattacharyya, J. Steinheimer, H. Stoecker, Phys. Rev. C **103**, 054908 (2021), 2009.10848
[4] J. Goswami, F. Karsch, C. Schmidt, S. Mukherjee, P. Petreczky, Acta Phys. Polon. Supp. **14**, 251 (2021), 2011.02812
[5] L. Tolos, L. Fabbietti, Prog. Part. Nucl. Phys. **112**, 103770 (2020), 2002.09223
[6] Y. Fujimoto, K. Fukushima, K. Murase, JHEP **03**, 273 (2021), 2101.08156
[7] L. Jiang, L. Wang, K. Zhou, Phys. Rev. D **103**, 116023 (2021), 2103.04090
[8] V. Vovchenko, V. Koch, C. Shen (2021), 2107.00163
[9] T. Dore, J. Noronha-Hostler, E. McLaughlin, Phys. Rev. D **102**, 074017 (2020)
[10] M. Kitazawa, G. Pihan, N. Touroux, M. Bluhm, M. Nahrgang, Nucl. Phys. A **1005**, 121797 (2021), 2002.07322
[11] D. Bollweg, J. Goswami, F. Karsch, S. Mukherjee, C. Schmidt (2020), 2010.15501
[12] P. Alba, V.M. Sarti, J. Noronha-Hostler, P. Parotto, I. Portillo-Vazquez, C. Ratti, J.M. Stafford, Phys. Rev. C **101**, 054905 (2020), 2002.12395
[13] G. Inghirami, T. Reichert, M. Bleicher (2021), 2106.04543
[14] J. Chen, J. Deng, Z. Tang, Z. Xu, L. Yi (2020), 2012.02986
[15] R. Lacey, these proceedings (2021)
[16] E. McLaughlin, J. Rose, T. Dore, P. Parotto, C. Ratti, J. Noronha-Hostler (2021), 2103.02090
[17] V. Mykhaylova, C. Sasaki, Phys. Rev. D **103**, 014007 (2021), 2007.06846
[18] H.T. Ding, O. Kaczmarek, A.L. Kruse, R. Larsen, L. Mazur, S. Mukherjee, H. Ohno, H. Sandmeyer, H.T. Shu, Nucl. Phys. A **982**, 715 (2019), 1807.06315
[19] L. Altenkort, A.M. Eller, O. Kaczmarek, L. Mazur, G.D. Moore, H.T. Shu, Phys. Rev. D **103**, 014511 (2021), 2009.13553
[20] G. Montaña, A. Ramos, L. Tolos, J.M. Torres-Rincon, Phys. Lett. B **806**, 135464 (2020)
[21] J.M. Torres-Rincon, G. Montaña, A. Ramos, L. Tolos (2021), 2106.01156

[22] C. Ding, W.Y. Ke, L.G. Pang, X.N. Wang, Chin. Phys. C **45**, 074102 (2021)
[23] G. Galesi, these proceedings (2021)
[24] S. Plumari, these proceedings (2021)
[25] M.L. Sambataro, these proceedings (2021)
[26] T. Song, L. Tolos, J. Wirth, J. Aichelin, E. Bratkovskaya, Phys. Rev. C **103**, 044901 (2021), 2012.05589
[27] M. Kurian, these proceedings (2021)
[28] S. Delorme, these proceedings (2021)
[29] N. Brambilla, M.A. Escobedo, M. Strickland, A. Vairo, P.V. Griend, J.H. Weber (2021), 2107.06222
[30] R. Larsen, S. Meinel, S. Mukherjee, P. Petreczky, Phys. Lett. B **800**, 135119 (2020)
[31] S. Shi, K. Zhou, J. Zhao, S. Mukherjee, P. Zhuang (2021), 2105.07862
[32] S. Chen, J. Zhao, P. Zhuang, Phys. Rev. C **103**, L031902 (2021), 2005.08473
[33] D. Zigic, I. Salom, J. Auvinen, M. Djordjevic, M. Djordjevic, Phys. Lett. B **791**, 236 (2019), 1805.04786
[34] D. Zigic, B. Ilic, M. Djordjevic, M. Djordjevic, Phys. Rev. C **101**, 064909 (2020)
[35] J. Noronha-Hostler, B. Betz, J. Noronha, M. Gyulassy, Phys. Rev. Lett. **116**, 252301 (2016), 1602.03788
[36] J. Adam et al. (ALICE), Nature Phys. **13**, 535 (2017), 1606.07424
[37] S. Acharya et al. (ALICE), Eur. Phys. J. C **80**, 167 (2020), 1908.01861
[38] R. Nayak, S. Pal, S. Dash, Phys. Rev. D **100**, 074023 (2019), 1812.07718
[39] Y. Kanakubo, Y. Tachibana, T. Hirano, Phys. Rev. C **101**, 024912 (2020), 1910.10556
[40] S. Acharya et al. (ALICE), Phys. Rev. C **99**, 024906 (2019), 1807.11321
[41] S. Wheaton, J. Cleymans, Comput. Phys. Commun. **180**, 84 (2009), hep-ph/0407174
[42] A. Andronic, P. Braun-Munzinger, M.K. Köhler, A. Mazeliauskas, K. Redlich, J. Stachel, V. Vislavicius, JHEP **07**, 035 (2021), 2104.12754
[43] S. Harabasz, W. Florkowski, T. Galatyuk, t. Ma Lgorzata Gumberidze, R. Ryblewski, P. Salabura, J. Stroth, Phys. Rev. C **102**, 054903 (2020), 2003.12992
[44] S. Acharya et al. (ALICE) (2021), 2105.06335
[45] S. Acharya et al. (ALICE) (2020), 2011.06078
[46] D. Oliinychenko, C. Shen (2021), 2105.07539
[47] S. Kiselev (ALICE), J. Phys. Conf. Ser. **1690**, 012100 (2020)
[48] A.G. Knospe, C. Markert, K. Werner, J. Steinheimer, M. Bleicher, Phys. Rev. C **93**, 014911 (2016), 1509.07895
[49] M. Kozhevnikova, Y.B. Ivanov, I. Karpenko, D. Blaschke, O. Rogachevsky, Phys. Rev. C **103**, 044905 (2021), 2012.11438
[50] S. Gläßel, V. Kireyeu, V. Voronyuk, J. Aichelin, C. Blume, E. Bratkovskaya, G. Coci, V. Kolesnikov, M. Winn (2021), 2106.14839
[51] H. Li, X.L. Xia, X.G. Huang, H.Z. Huang (2021), 2106.09443
[52] B. Fu, S.Y.F. Liu, L. Pang, H. Song, Y. Yin (2021), 2103.10403
[53] S.Y.F. Liu, Y. Yin, JHEP **07**, 188 (2021), 2103.09200
[54] W. Florkowski, R. Ryblewski (2021), 2102.02890
[55] Y. Guo, J. Liao, E. Wang, H. Xing, H. Zhang (2021), 2105.13481
[56] X.L. Sheng, Q. Wang, X.N. Wang, Phys. Rev. D **102**, 056013 (2020), 2007.05106
[57] X.L. Sheng, L. Oliva, Q. Wang, Phys. Rev. D **101**, 096005 (2020), 1910.13684
[58] F. Bellini, these proceedings (2021)

Off-of-equilibrium effects on Kurtosis Along Strangeness-Neutral Trajectories

Travis Dore[1], *Jamie* Karthein[2], *Debora* Mroczek[1], *Paolo* Parotto[3], *Jacquelyn* Noronha-Hostler[1], and *Claudia* Ratti[1]

[1]Department of Physics, University of Illinois at Urbana-Champaign, Urbana, IL 61801, USA
[2]Department of Physics, University of Houston, Houston, TX 77204, USA
[3]University of Wuppertal, Department of Physics, Wuppertal D-42119, Germany

Abstract. The Beam Energy Scan program at the Relativistic Heavy Ion Collider (RHIC) is searching for the QCD critical point. The main signal for the critical point is the kurtosis of the distribution of proton yields obtained on an event by event basis where one expects a peak at the critical point. However, its exact behavior is still an open question due to out-of-equilibrium effects and uncertainty in the equation of state. Here we use a simplistic hydrodynamic model that enforces strangeness-neutrality, selecting trajectories that pass close to the critical point. We vary the initial conditions to estimate the effect of out-of-equilibrium hydrodynamics on the kurtosis signal.

1 Introduction

As the search for the QCD critical point [1–5] continues, it becomes increasingly important to have a thorough understanding of its signals, including what factors influence them and how strong those influences are. One such signal is non-monotonic behavior of the kurtosis of the equilibrium net baryon distribution [6, 7], specifically, a large peak in the vicinity of the critical region signifying large fluctuations of net baryon number. Since there is no first principles way to calculate the QCD equation of state (EoS) at finite baryon densities due to the sign problem, one must rely on phenomenological methods and models.

One such implementation is done in Ref. [8] in which a critical region with the expected 3D Ising universality class is smoothly matched to the Lattice QCD EoS at zero net baryon density and uses a Taylor expansion for the non critical contribution to the pressure at finite net baryon densities. The size, shape, and strength of the critical region can be selected through the use of multiple numerical parameters, each of which produce a realization of a QCD EoS with a critical point that can be constrained against data and theoretical consistency checks (e.g. $0 \leq c_s^2 \leq 1$). The aforementioned work is an extension of a previous one [9] which now has the ability to capture 'heavy-ion' trajectories through the phase diagram by enforcing both net strangeness neutrality as well as finite electric charge, instead of $\mu_B = \mu_S = 0$. The difference in the isentropes produced (which are trajectories in the phase diagram of constant entropy to baryon number ratio, s/ρ_B) by the two EoS's can be seen in Fig. 1.

The EoS developed in that work is used here, with the parameterization being the same as that of the paper [8]. In this work, we use this EoS in conjunction with a simplified

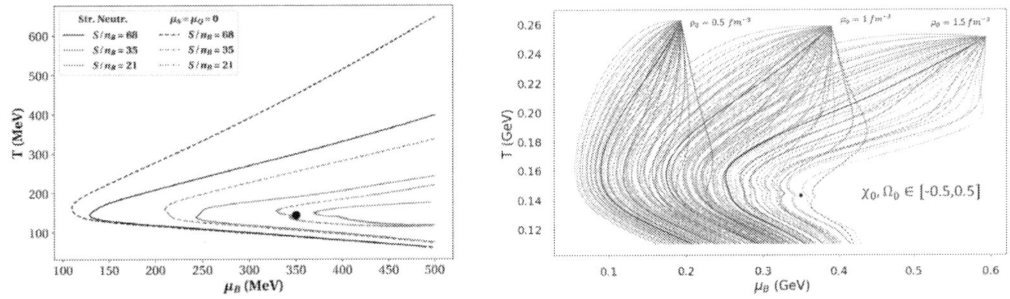

Figure 1. Left: A comparison of isentropes from an EoS realization which enforces $\mu_B = \mu_S = 0$ vs. one that enforces heavy-ion trajectories. Right: Demonstration of the ability of the system to pass through a large portion of the phase diagram depending on what its initial out-of-equilibrium initial conditions are.

hydrodynamic model to begin exploring out of equilibrium effects on the kurtosis at freeze-out. The value of the kurtosis at freeze-out is intimately related to the fluctuations in baryon number from event to even when the system freezes out [7, 10–12], thus directly tying it to an observable. The work here does not use a realistic hydrodynamic model and cannot make quantitative predictions. However, it does give a qualitative idea of the kinds of effects one might expect to encounter in realistic model to data comparison.

2 Out-of-Equilibrium Effects

It has been shown in previous works [13, 14] that out of equilibrium effects can have an impact on the (T, μ_B) trajectories explored by the system throughout its evolution. This can have a direct influence on searches for the critical point by possibly pushing the system towards or away from the critical point on an event by event basis, and even affecting the path the system takes through the critical region. In Fig. 1, we demonstrate the ability of a system to pass through a large portion of the phase diagram, using the same energy density, three different baryon densities, and many different initial values for shear-stress tensor and the bulk pressure.

It is also expected that in the vicinity of the critical region the bulk viscosity, ζ/s, should scale with the correlation length to the third power, i.e. $\zeta/s \sim (\xi/\xi_0)^3$ where ξ_0 is a scaling constant picked to scale the correlation length appropriately in the critical region [15]. In order to incorporate as many critical effects as possible, we implement a bulk viscosity that has this critical scaling, the exact form of which, can be seen in the next section.

3 Hydrodynamic Model

As a starting point for future work, we use a simplified Bjorken expanding model [16] to obtain qualitative result and a basic understanding of the sought after effects. Here we use the DNMR equations of motion detailed in [13]. These equations of motion require initial conditions for π^η_η - the rapidity eigenvalue of the shear stress tensor, Π - the bulk pressure, the energy density ϵ, and ρ - the baryon density. Second order transport coefficients are the same as in [17, 18]. We use a temperature dependent shear viscosity ($\eta T/w$ where $w = e + p$ is the enthalpy), given by an excluded volume calculation below the transition temperature

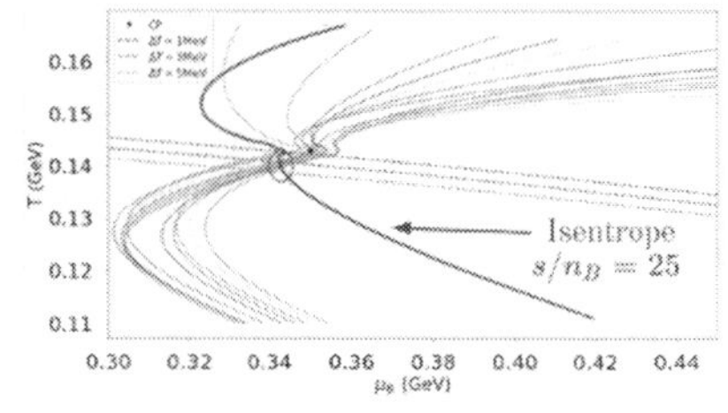

Figure 2. Trajectories of different runs that pass through the critical region

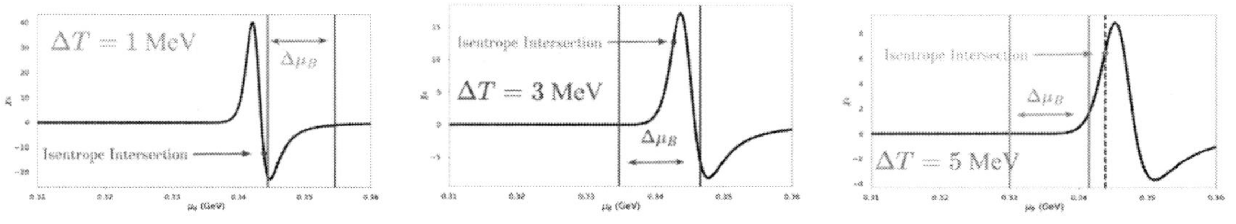

Figure 3. The fourth moment of the equilibrium net baryon distribution at fixed values of temperature. The vertical lines are bands around possible values of the kurtosis seen including out of equilibrium effects. The dots are where the isentropes intersect the parabolas of constant temperature.

[19, 20] (using the PDG16+ [21]) that is matched onto a QCD motivated parameterization in the deconfined phase [22, 23]. The form of the non-critical and critical bulk viscosities are

$$\frac{\zeta T}{w} = 36 \times \frac{1/3 - c_s^2}{8\pi} \tag{1}$$

$$\left(\frac{\zeta T}{w}\right)_{CS} = \frac{\zeta T}{w}\left[1 + \left(\frac{\xi}{\xi_0}\right)^3\right] \tag{2}$$

4 Effects On Probed Kurtosis

In Fig. 2, one can see a variety of different hydrodynamic (T, μ_B) trajectories that the system can pass through and still go through the critical region. Here, we study which value of the kurtosis is actually probed by these runs, and check this at three different values of temperature after the critical point, specifically, 1, 3, and 5 MeV afterwards. In Fig. 3 we show which different values of χ_4 are probed in our simulations. In this work, χ_4 is used as a substitute for the kurtosis since the two are closely related.

It can be seen in Fig. 3 that out of equilibrium effects lead to a spread in χ_4 probed on an event by event basis. The vertical lines for each figure represent the bounds given by out of equilibrium effects in our work, the dots show where the isentrope probes χ_4 at a given (T, μ_B). It can be seen in the bottom panel that the isentrope probes a region outside the bounds probed by out of equilibrium effects in our study.

5 Conclusion

In this work we have explored some consequences that out-of-equilibrium effects could have on the search for the critical point. In particular, we look at the fourth moment of the equilibrium net baryon distribution because of its relation to the kurtosis. We found that, for a given EoS, the value of kurtosis you probe at freeze-out is heavily dependent on the out-of-equilibrium initial conditions of the system. This can lead to different values of the kurtosis being probed on an event by event basis.

This material is based upon work supported by the the US-DOE Nuclear Science Grant No. DE-SC0019175; the National Science Foundation under grants no. PHY-1654219, PHY-1560077, and DGE – 1746047; and the DFG grant SFB/TR55.

References

[1] Y. Aoki, Z. Fodor, S.D. Katz, K.K. Szabo, Phys. Lett. B **643**, 46 (2006), hep-lat/0609068
[2] S. Borsanyi, Z. Fodor, C. Hoelbling, S.D. Katz, S. Krieg, C. Ratti, K.K. Szabo (Wuppertal-Budapest), JHEP **09**, 073 (2010), 1005.3508
[3] G. Endrodi, Z. Fodor, S.D. Katz, K.K. Szabo, JHEP **04**, 001 (2011), 1102.1356
[4] R. Bellwied, S. Borsanyi, Z. Fodor, J. Günther, S.D. Katz, C. Ratti, K.K. Szabo, Phys. Lett. B **751**, 559 (2015), 1507.07510
[5] J.N. Guenther (2020), 2010.15503
[6] D. Mroczek, A.R. Nava Acuna, J. Noronha-Hostler, P. Parotto, C. Ratti, M.A. Stephanov, Phys. Rev. C **103**, 034901 (2021), 2008.04022
[7] M.A. Stephanov, Phys. Rev. Lett. **107**, 052301 (2011), 1104.1627
[8] J.M. Karthein, D. Mroczek, A.R. Nava Acuna, J. Noronha-Hostler, P. Parotto, D.R.P. Price, C. Ratti, Eur. Phys. J. Plus **136**, 621 (2021), 2103.08146
[9] P. Parotto, M. Bluhm, D. Mroczek, M. Nahrgang, J. Noronha-Hostler, K. Rajagopal, C. Ratti, T. Schäfer, M. Stephanov, Phys. Rev. C **101**, 034901 (2020), 1805.05249
[10] M.A. Stephanov, K. Rajagopal, E.V. Shuryak, Phys. Rev. Lett. **81**, 4816 (1998), hep-ph/9806219
[11] M.A. Stephanov, K. Rajagopal, E.V. Shuryak, Phys. Rev. D **60**, 114028 (1999), hep-ph/9903292
[12] M.A. Stephanov, Phys. Rev. Lett. **102**, 032301 (2009), 0809.3450
[13] T. Dore, J. Noronha-Hostler, E. McLaughlin, Phys. Rev. D **102**, 074017 (2020), 2007.15083
[14] L. Du, X. An, U. Heinz (2021), 2107.02302
[15] A. Monnai, S. Mukherjee, Y. Yin, Phys. Rev. C **95**, 034902 (2017), 1606.00771
[16] J. Bjorken, Phys. Rev. D **27**, 140 (1983)
[17] G.S. Denicol, C. Gale, S. Jeon, A. Monnai, B. Schenke, C. Shen, Phys. Rev. C **98**, 034916 (2018), 1804.10557
[18] D. Bazow, U.W. Heinz, M. Strickland, Comput. Phys. Commun. **225**, 92 (2018), 1608.06577
[19] J. Noronha-Hostler, J. Noronha, C. Greiner, Phys. Rev. C **86**, 024913 (2012), 1206.5138
[20] E. McLaughlin, J. Rose, T. Dore, P. Parotto, C. Ratti, J. Noronha-Hostler (2021), 2103.02090
[21] P. Alba et al., Phys. Rev. D **96**, 034517 (2017), 1702.01113
[22] N. Christiansen, M. Haas, J.M. Pawlowski, N. Strodthoff, Phys. Rev. Lett. **115**, 112002 (2015), 1411.7986
[23] A. Dubla, S. Masciocchi, J.M. Pawlowski, B. Schenke, C. Shen, J. Stachel, Nucl. Phys. A **979**, 251 (2018), 1805.02985

QCD Critical Point and Net-Proton Number Fluctuations at RHIC-STAR

Yu Zhang[1,2,*]
for the STAR Collaboration

[1]Central China Normal University
[2]Lawrence Berkeley National Laboratory

Abstract. In the search of QCD phase boundary and critical point, higher-order cumulants of conserved quantities are proposed as promising observables and have been studied extensively both experimentally and theoretically. In this paper we present cumulant ratios up to 6th-order of net-proton number distributions in Au+Au collisions at $\sqrt{s_{\mathrm{NN}}}$= 7.7 - 200 GeV from STAR Beam Energy Scan program phase I and $\sqrt{s}$ = 200 GeV $p + p$ collisions. The results are compared with various models and Lattice QCD calculations.

1 Introduction

In heavy-ion collision physics it is predicted that under a very high temperature and baryon density a deconfined quark-gluon plasma (QGP) phase can be created and studying the QCD phase structure is one of the main goals. A phase diagram in terms of temperature and baryon chemical potential (μ_{B}) is usually used to explore the QCD phase structure. Regarding the phase transition between the QGP phase and hadronic phase, first principle Lattice QCD calculation [1] at μ_{B} = 0 MeV suggests a smooth crossover transition. While at large μ_{B}, various QCD-based models predict first order phase transition [2]. Thermodynamically there should be an end point of the first order phase boundary which is called QCD critical point. The possible QCD critical point and the first order phase boundary have been investigated both experimentally [3–5] and theoretically.

Higher-order cumulants of conserved quantities like net-baryon (B), net-charge (Q) and net-strangeness (S) are proposed as promising observables to search for the QCD critical point and the first order phase boundary. Higher-order cumulants are sensitive to the correlation length (ξ) [6] and are directly related to susceptibility (χ) of the system [7]. It is predicted that the fourth-order fluctuations will exhibit a non-monotonic energy dependence [8–10] when passing through the critical region. For 5th- and 6th-order cumulants recent calculations from Lattice QCD [11] and the functional renormalisation group approach (FRG) [12] show that they will be negative due to the crossover transition between QGP and hadronic phase. At high baryon density region, on the other hand, they are also sensitive to the first order phase boundary [13].

*e-mail: zhang_yu@mails.ccnu.edu.cn

2 Data and Analysis

The data of Au+Au collisions at $\sqrt{s_{NN}}$ = 7.7 - 200 GeV and $p + p$ collisions at $\sqrt{s}$ = 200 GeV are collected in RHIC Beam Energy Scan program phase I. Protons and antiprotons are identified by the Time Projection Chamber (TPC) and Time of flight (TOF) detectors at rapidity window $-0.5 < y < 0.5$ and transverse momentum window $0.4 < p_T < 2.0$ GeV/c.

The centrality is determined using charged particle multiplicity within $|\eta| < 1.0$ excluding protons and antiprotons to avoid auto-correlation effect [14]. The centrality bin width correction [14] is applied to suppress initial volume fluctuation effect. Cumulants are calculated at each multiplicity bin and then their weighted averages are taken for each centrality bin. The weight is the number of events at the corresponding multiplicity bin. Detector efficiency correction [15] in cumulant calculations are done by assuming binomial detector efficiency. Statistical uncertainties of cumulants are estimated by Bootstrap [16] and Delta methods [17].

3 Results

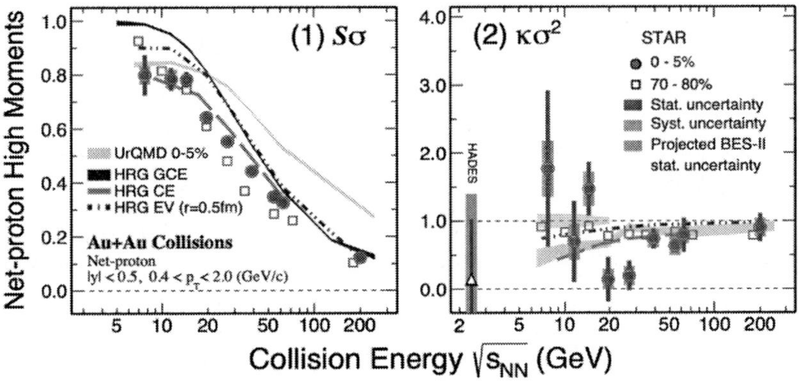

Figure 1. Energy dependence of $S\sigma$ and $\kappa\sigma^2$ of net-proton number distributions in Au+Au collisions at $\sqrt{s_{NN}}$ = 7.7 - 200 GeV. The calculations from different conditions (GCE, EV, and CE) of hadron resonance gas model (HRG) and the hadronic transport UrQMD model are shown as black, red, blue bands, and a gold band respectively.

Figure 1 shows energy dependence of $S\sigma$ and $\kappa\sigma^2$ of net-proton distributions from 0-5% and 70-80% centrality bins within $|y| < 0.5$ and $0.4 < p_T < 2.0$ GeV/c in Au+Au collisions at $\sqrt{s_{NN}}$ = 7.7 - 200 GeV [20, 21]. The $S\sigma$ (left panel) shows a decreasing trend with the increase of collision energy both in central and peripheral collisions. The decreasing trend can be qualitatively described by HRG [18] and UrQMD [19] models. The $\kappa\sigma^2$ (right panel) shows a non-monotonic energy dependence in central collisions while the results for peripheral collisions show monotonic energy dependence. The non-monotonic trend in central collisions can not be described by different conditions (GCE, EV, and CE) of HRG and UrQMD models.

Figure 2 shows energy dependence of C_5/C_1 and C_6/C_2 of net-proton distributions from 0-40% and 70-80% centrality bins within $|y| < 0.5$ and $0.4 < p_T < 2.0$ GeV/c in Au+Au collisions at $\sqrt{s_{NN}}$ = 7.7 - 200 GeV. It is suggested from Lattice QCD and FRG calculations that 5th- and 6th-order cumulants show negative signs while calculations from UrQMD and HRG models are consistent with either zero or unity. In UrQMD and HRG models, no phase transition physics is implemented. The measurements of BES-I data are shown as blue circles

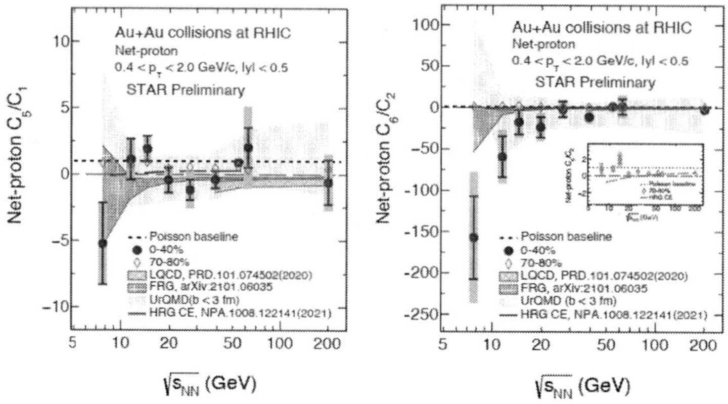

Figure 2. Energy dependence of C_5/C_1 and C_6/C_2 of net-proton number distributions in Au+Au collisions at $\sqrt{s_{NN}}$ = 7.7 - 200 GeV. The calculations from FRG model, Lattice QCD, hadronic transport UrQMD model, and Hadron Resonance Gas (HRG) model are shown as red, green, and yellow bands and red lines respectively.

for 0-40% and red diamonds for 70-80%. The cumulant ratio C_5/C_1 (left panel) deviates from zero with less than 2σ significance. In peripheral collisions C_5/C_1 shows positive sign for all energies. The ratio C_6/C_2 (right panel) for 0-40% decreases with negative sign with less than 2σ significance when decreasing energy while it shows positive sign in peripheral collisions (70-80%) for all energies.

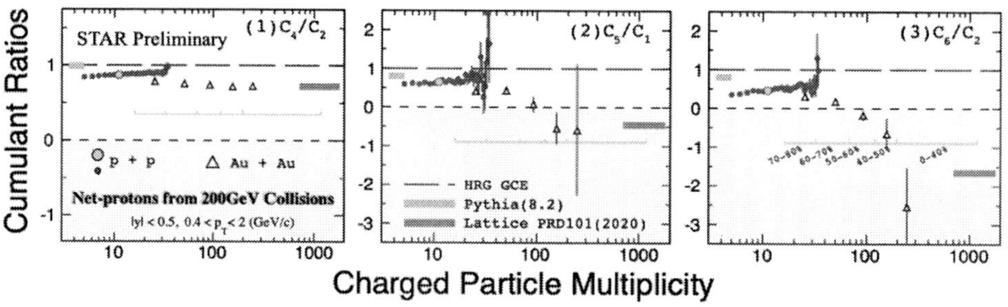

Figure 3. Multiplicity dependence of C_4/C_2, C_5/C_1, and C_6/C_2 of net-proton number distributions in $p + p$ collisions at $\sqrt{s}$ = 200 GeV. The calculations of HRG model, Pythia (version 8.2), and Lattice QCD are shown as black dashed line, yellow, and red bands respectively.

Figure 3 shows multiplicity dependence of net-proton results of C_5/C_1 and C_6/C_2 within $|y| < 0.5$ and $0.4 < p_T < 2.0$ GeV/c in $p + p$ collisions at $\sqrt{s}$ = 200 GeV. We see that the cumulant ratios (C_4/C_2, C_5/C_1, and C_6/C_2) from $p + p$ collisions fit into the multiplicity dependence of results from Au+Au collisions which are shown with triangles. The cumulant ratios C_5/C_1 and C_6/C_2 are negative for 0-40% in Au+Au collisions and positive at peripheral Au+Au collisions and $p + p$ collisions. Pythia [22] calculation using version 8.2 of $p + p$ collisions at $\sqrt{s}$ = 200 GeV is positive as shown with yellow bands. The Lattice QCD calculation [11] at $\sqrt{s_{NN}}$ = 200 GeV is negative as shown with red bands. Compared with calculations from various models, it is suggested that the negative sign for central Au+Au collisions at $\sqrt{s_{NN}}$ = 200 GeV is due to a smooth crossover transition between partonic and hadronic phases.

4 Summary

In this proceedings, we report the measurements of net-proton cumulant ratios up to 6th-order in Au+Au collisions at $\sqrt{s_{NN}}$ = 7.7 - 200 GeV and $p + p$ collisions at $\sqrt{s}$ = 200 GeV from STAR. With results from 200 GeV $p + p$ collisions and the energy dependence of C_4/C_2, C_5/C_1, and C_6/C_2 from the BES-I data sets, and the comparison with LQCD calculations, we conclude: 1) QCD matter is indeed created in the 200 GeV central (0-5%) Au+Au collisions at RHIC; 2) non-monotonic energy dependence of C_4/C_2 is observed from the most central (0-5%) Au+Au collisions [20, 21]. Future results from BES-II and STAR fixed-target experiment $\sqrt{s_{NN}}$ = 3 GeV data sets will allow to answer if QCD critical point exists in the covered energy region. In 2023 to 2025, STAR plans to collect 15 to 20 billion events of Au+Au collisions at $\sqrt{s_{NN}}$ = 200 GeV. This will allow us to perform more precise measurements of higher-order cumulants, maybe even up to 8th-order.

Acknowledgments

This work was supported by the National Key Research and Development Program of China (Grant No. 2020YFE0202002 and 2018YFE0205201), the National Natural Science Foundation of China (Grant No. 11828501, 11890711 and 11861131009) and China scholarship council (No. 201906770055).

References

[1] Y. Aoki et al., *Nature* **443**, 675(2006).
[2] M. Stephanov, K. Rajagopal, E. Shuryak *Phys. Rev. D* **60**, 114028(1999).
[3] L. Adamczyk et al. (STAR Collaboration), *Phys. Rev. Lett.* **112**, 032302(2014).
[4] L. Adamczyk et al. (STAR Collaboration), *Phys. Rev. Lett.* **113**, 092301(2014).
[5] L. Adamczyk et al. (STAR Collaboration), *Phys. Lett. B* **785**, 551(2018).
[6] M. Asakawa, S. Ejiri, and M. Kitazawa, *Phys. Rev. Lett.* **103**, 262301(2009).
[7] S. Ejiri, F. Karsch, and K. Redlich *Phys. Lett. B* **633**, 275(2006).
[8] M. A. Stephanov, *Phys. Rev. Lett.* **107**, 052301(2011).
[9] B.-J. Schaefer and M. Wagner, *Phys. Rev. D* **85**, 034027(2012).
[10] J. Chen et al., *Phys. Rev. D* **95**, 014038(2017) and references therein.
[11] A. Bazavov et al., *Phys. Rev. D* **101**, 074502(2020).
[12] W. Fu et al., arXiv:2101.06035 [hep-ph].
[13] A. Bzdak and V. Koch, *Phys. Rev. C* **100**, 051902R(2019).
[14] X. Luo et al., *J. Phys. G: Nucl. Part. Phys.* **40**, 105104(2013).
[15] X. Luo, *Phys. Rev. C* **91**, 034907(2015).
[16] Efron, B. et al (1994). https://doi.org/10.1007/978-1-4899-4541-9.
[17] X. Luo, *J. Phys. G: Nucl. Part. Phys.* **39**, 025008(2012).
[18] P. Garg et al, *Phys. Lett. B* **726**, 691(2013).
[19] M. Bleicher et al., *J. Phys. G* **25**, 1859(1999).
[20] J. Adam et al. (STAR Collaboration), *Phys. Rev. Lett.* **126**, 092301(2021).
[21] M. S. Abdallah et al. (STAR Collaboration), *Phys. Rev. C* **104**, 024902(2021).
[22] T. Sjöstrand et al. *JHEP05* (2006) **026**, *Comput. Phys. Comm.* **178** (2008) 852.

Coupled baryon, electric charge and strangeness fluctuations in heavy-ion collisions

Grégoire Pihan[1,*], *Marcus* Bluhm[1], and *Marlene* Nahrgang[1]

[1]SUBATECH UMR 6457 (IMT Atlantique, Université de Nantes, IN2P3/CNRS), 4 rue Alfred Kastler, 44307 Nantes, France

Abstract. We present for the first time quantitative results for the coupled dynamics of second order fluctuations in the three conserved charges of QCD based on stochastic diffusion equations for a Bjorken-type expanding hadronic medium. We show that the fluctuations deviate from local equilibrium expectations which can result in important phenomenological consequences for the interpretation of experimental data.

1 Introduction

Fluctuation observables in heavy-ion collisions test the constituents and the transport properties of strongly interacting matter and may signal phase transitions or the chemical freeze-out. The fluctuations in the three conserved charges of QCD, net-baryon number B, net-electric charge Q and net-strangeness S, change through diffusion [1]. Their evolution is coupled as the matter constituents typically carry more than just one charge. In this work we study the coupled fluctuation dynamics in the hadronic phase employing stochastic diffusion equations. These are expressed in terms of the Milne coordinates proper-time τ and spatial rapidity y to model a rapidly Bjorken-type expanding medium. By using realistic values for the diffusion coefficients we investigate the competition between coupled diffusion and rapid expansion and find that the fluctuation observables are necessarily driven out of equilibrium. This may have important phenomenological consequences for the interpretation of experimental data.

2 Coupled conserved charge diffusion in the hadronic phase

The dynamics of the net-charges $i = B, Q, S$ is influenced by the mixing of different diffusive currents entering the stochastic diffusion equations

$$\partial_\tau n_i = \sum_{j=B,Q,S} \partial_y \left(\frac{\kappa_{ij}}{\tau} \partial_y \left(\frac{\mu_j}{T} \right) \right) + \partial_y \xi_i \,. \tag{1}$$

Here, κ_{ij} represents the charge diffusion coefficient matrix of the hadronic medium. The fluctuation-dissipation balance requires that the noise satisfies besides the standard white-noise self-correlation also cross-correlations between the different charges introduced by the non-vanishing off-diagonal elements of κ_{ij}. These are given, for $Y = (\tau, y)$, by

$$\langle \xi_i(Y)\, \xi_j(Y') \rangle = 2\frac{\kappa_{ij}}{\tau} \delta^{(2)}(Y - Y') \,. \tag{2}$$

*e-mail: Pihan@subatech.in2p3.fr

We note that charge coupling may not only be established by means of the current mixing and the noise cross-correlations but also via the underlying equation of state for which μ_j may depend on $\mu_{i\neq j}$. In [2], the deterministic diffusion of the net-charge densities n_i was studied. It was found that the charge couplings affect the density profiles non-trivially. The current work presents the first approach to study the coupled dynamics of conserved charge fluctuations quantitatively by means of the set of stochastic diffusion equations (1) and (2).

3 Numerical results

The diffusion coefficient matrix κ_{ij} as function of temperature T and chemical potential μ_j has been calculated within kinetic theory in [3] for a hadron resonance gas composed of the 19 lightest hadron species. Consequently, we employ the same underlying model in order to determine the spatio-temporal profile of the different μ_j in Eqs. (1) from the charge density profiles assuming a homogeneous T. For the latter, the temporal evolution is modeled Hubble-like viz $T(\tau) = T(\tau_0)\tau_0/\tau$ with $\tau_0 = 1$ fm/c and $T(\tau_0) = 500$ MeV. Initially, we set $n_i(\tau, y) = 0$ fm^{-2} which implies $\mu_j = 0$ MeV for the chemical potential profiles. In Fig. 1 (left panel), we show the T-dependence of κ_{ij} at $\mu_B = 0$ MeV. In addition, the initial local variances and co-variances are also set to vanish, clearly representing a far from equilibrium situation. The imposed balance between fluctuation and dissipation allows the system to approach equilibrium, but based on the strength of the relevant κ_{ij} a rapid expansion may compete with this process. In Fig. 1 (right panel), we show the expected T-dependence of the equilibrium fluctuations of the hadronic medium.

3.1 Equilibration process

The success of the fluid dynamic modeling of heavy-ion collisions suggests that the created bulk matter is locally near equilibrium. For the given realistic values of κ_{ij} [3] it is, thus, important to understand how quickly equilibrium can be achieved starting from our far-from-equilibrium initial conditions. To this end, Eqs. (1) and (2) are solved at fixed T for a static system. In this way we can create an equilibrated initial state for a subsequent dynamic expansion as well as provide estimates for the equilibrium fluctuations realized at a given T. In Fig. 2, we show the equilibration process for the local variance of the net-baryon density, σ^2_{BB}, in the limit B not coupled to Q, S (left panel) for different T and for the fully

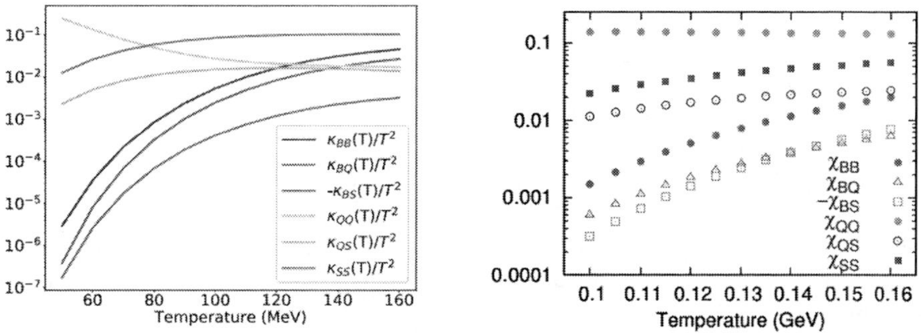

Figure 1. Left: Temperature dependence of the diffusion coefficients κ_{ij} at $\mu_B = 0$ MeV for a hadron resonance gas composed of the 19 lightest hadron species (taken from [3]). Right: Corresponding equilibrium diagonal and off-diagonal susceptibilities per unit of rapidity as function of T at $\mu_B = 0$ MeV.

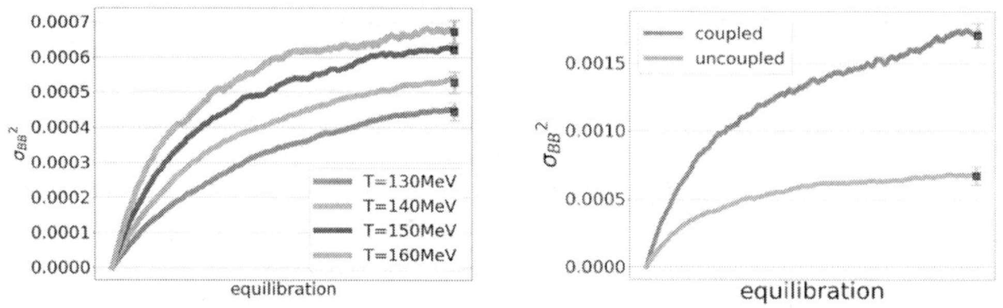

Figure 2. Time-evolution of σ_{BB}^2 during the equilibration process for B not coupled to Q, S at different T (left) and in comparison to the fully coupled system (right) at $T = 160$ MeV. The squares show our values for the equilibrium fluctuations with errors estimated through the late-time slope of the curves.

coupled system (right panel) in comparison to the uncoupled situation for $T = 160$ MeV. One observes that for the given values of κ_{ij} equilibrium is only slowly reached (the waited equilibration time is of $O(20)$ fm/c), in particular for smaller T in line with the T-dependence of κ_{BB}. Moreover, our equilibrium estimates for σ_{BB}^2 behave qualitatively similar to χ_{BB}, see Fig. 1. The coupling between the different charges leads to an even slower approach of local equilibrium, where the size of the equilibrium fluctuations is increased compared to the uncoupled situation.

3.2 Dynamic expansion and cooling

Starting from the created equilibrium initial conditions at $T = 160$ MeV, see Sec. 3.1, we can study the dynamics of the fluctuation observables as $T(\tau)$ cools down. In Fig. 3, we show the time-evolution of the local variances σ_{BB}^2 and σ_{SS}^2 for the net-baryon (left panel) and net-strangeness (right panel) density from $T = 160$ MeV to $T = 100$ MeV. In comparison to our estimates for the equilibrium fluctuations at different T (squares) one observes that for the κ_{ij} given in [3] the conserved charge fluctuations lag behind and are quickly driven out of local equilibrium in the rapidly expanding system. Deviations in σ_{BB}^2 from our equilibrium expectations are found to be stronger in the coupled system as the simultaneous nonequilibrium evolution in Q and S affects the dynamics of the fluctuations in B. In addition, we find that σ_{SS}^2 follows the equilibrium situation better, mostly because κ_{SS} is the largest diffusion coefficient in the considered temperature range, cf. Fig. 1.

4 Discussion, conclusions and outlook

We have seen that the coupling between the conserved charges in QCD impacts the diffusion dynamics of their fluctuations σ_{BB}^2 and σ_{SS}^2 in the hadronic phase significantly. Charge coupling does not only increase the size of fluctuations with possible consequences for the search of the QCD critical point, it also affects to which extent local equilibrium can be achieved and maintained. Starting from an initial condition near local equilibrium at $T = 160$ MeV we have shown, using realistic diffusion coefficients κ_{ij} [3], that it is impossible to maintain equilibrium locally in a Bjorken-type expanding medium. While diffusion still has some impact, the rapid expansion wins and the system is necessarily driven out of equilibrium in the hadronic phase. Charge coupling accelerates this nonequilibrium evolution.

The observed behavior impacts our interpretation of fluctuation measurements: We find that, as a consequence of the diffusive dynamics in the hadronic phase, final fluctuations in the

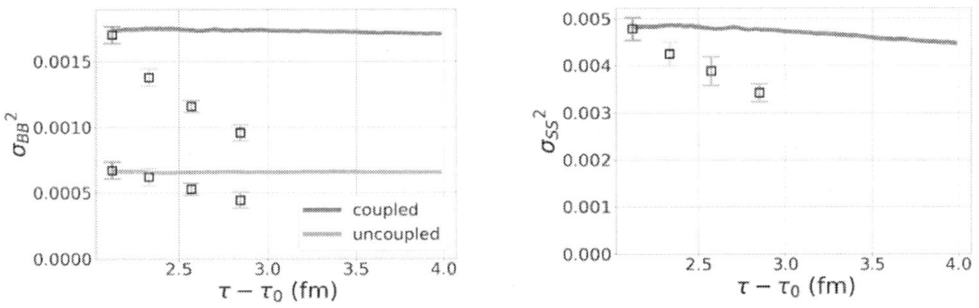

Figure 3. Nonequilibrium time-evolution of σ^2_{BB} (left) and σ^2_{SS} (right) in the hadronic phase between $T = 160$ MeV and $T = 100$ MeV for the fully coupled conserved charge dynamics in comparison with our equilibrium estimates (squares) for both fluctuation observables. For σ^2_{BB} we contrast, in addition, the corresponding results obtained in the limit B not coupled to Q, S.

net-baryon number only portray a snapshot of the initial (equilibrium) situation. Moreover, while final net-strangeness fluctuations might still reflect chemical freeze-out conditions these are necessarily over-estimated by the comparison with equilibrium models. This shows that previous determinations as in [4, 5] based on fluctuation data have to be reassessed. Our study also indicates that the critical fluctuation signals from the QCD phase transition [6] may survive the rapid expansion of the hadronic medium.

In this work, we have not discussed fluctuations in the net-electric charge, which require the largest equilibration time, nor cross-correlations between the different charges. We leave their presentation and a detailed discussion to the future [7].

Acknowledgment

We thank the authors of [3] for stimulating discussions and providing us with the data of κ_{ij}. This work received support from the program "Etoiles montantes en Pays de la Loire 2017".

References

[1] M. Asakawa, M. Kitazawa and B. Müller, Phys. Rev. C **101** no.3, 034913 (2020) [arXiv:1912.05840 [nucl-th]].

[2] J. A. Fotakis, M. Greif, C. Greiner, G. S. Denicol and H. Niemi, Phys. Rev. D **101** no.7, 076007 (2020) [arXiv:1912.09103 [hep-ph]].

[3] M. Greif, J. A. Fotakis, G. S. Denicol and C. Greiner, Phys. Rev. Lett. **120** no.24, 242301 (2018) [arXiv:1711.08680 [hep-ph]].

[4] P. Alba, W. Alberico, R. Bellwied, M. Bluhm, V. Mantovani Sarti, M. Nahrgang and C. Ratti, Phys. Lett. B **738**, 305-310 (2014) [arXiv:1403.4903 [hep-ph]].

[5] M. Bluhm and M. Nahrgang, Eur. Phys. J. C **79** no.2, 155 (2019) [arXiv:1806.04499 [nucl-th]].

[6] M. Kitazawa, G. Pihan, N. Touroux, M. Bluhm and M. Nahrgang, Nucl. Phys. A **1005**, 121797 (2021) [arXiv:2002.07322 [nucl-th]].

[7] G. Pihan, M. Bluhm and M. Nahrgang, in preparation.

Spectral reconstruction details of a gradient-flowed color-electric correlator

Luis Altenkort[1,*], *Alexander M.* Eller[2], *Olaf* Kaczmarek[1], *Lukas* Mazur[3], *Guy D.* Moore[2], and *Hai-Tao* Shu[4]

[1]Fakultät für Physik, Universität Bielefeld, D-33615 Bielefeld, Germany
[2]Institut für Kernphysik, Technische Universität Darmstadt, D-64289 Darmstadt, Germany
[3]Paderborn Center for Parallel Computing, Paderborn University, D-33098 Paderborn, Germany
[4]Institut für Theoretische Physik, Universität Regensburg, D-93040 Regensburg, Germany

Abstract. In a recently published work we provide a proof-of-concept of a novel method to extract the heavy quark momentum diffusion coefficient from color-electric correlators on the lattice using gradient flow. The transport coefficient can be found in the infrared limit of the corresponding spectral function which is reconstructed through perturbative model fits of the correlator data. In this proceedings report we want to give more detailed insights into the systematic uncertainties of this procedure and compare our results with other studies.

1 Introduction

Experimental data from heavy ion collisions provide evidence for a significant collective motion of heavy quarks (charm, bottom) with the bulk of the medium [1, 2]. From a hydrodynamical standpoint this is unexpected, as the kinetic equilibration time τ_{kin} for heavy participants of the evolution should be about a factor of M/T larger than that of the light ones [3]. In a hot medium with temperature T the behavior of heavy quarks (HQ) with mass M can be described by a nonrelativistic Langevin approach, as long as $M \gg \pi T$ [3, 4]. From this description multiple a priori unknown coefficients emerge with direct phenomenological interpretations. One of these coefficients is the HQ momentum diffusion coefficient κ, which quantifies the averaged squared momentum transfer that is exerted on a heavy quark by the hot medium per unit time. In the nonrelativistic limit it is directly related to other coefficients such as the spatial diffusion coefficient D and the drag coefficient η_D, as well as to the kinetic equilibration time τ_{kin}. In this way a determination of κ also determines τ_{kin}. However, even for high temperatures perturbative series for D or κ are ill-behaved [5] and a first principles determination from lattice QCD is necessary to provide insight into the observed behavior of heavy quarks. The coefficient κ is also a crucial input for simulation models [6].

Diffusion physics is encoded in the infrared limit of real-time spectral functions of conserved currents [7–9]. In the limit $M \gg \pi T$ one can construct a color-electric correlator whose spectral function encodes κ through [10]

$$\kappa = \lim_{\omega\to 0} 2T\frac{\rho(\omega)}{\omega}, \quad G(\tau) = \int_0^\infty \frac{\mathrm{d}\omega}{\pi}\,\frac{\cosh(\omega(\tau-\beta/2))}{\sinh(\omega\beta/2)}\,\rho(\omega). \quad \beta = 1/T. \tag{1}$$

*Speaker, e-mail: altenkort@physik.uni-bielefeld.de

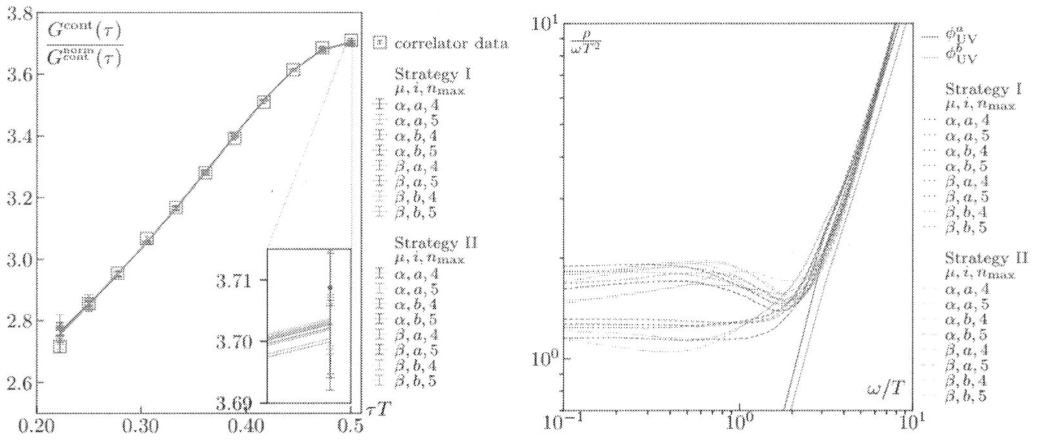

Figure 1. Left: Model correlators obtained from integration of the model spectral functions. The labels distinguish between different models (see Eq. (4)). Right: Model spectral functions from fits of the correlator data for all considered models. Statistical errors are hidden for better visibility; for $\omega = 0$ they can be found in Fig. 2, all others can be found in the data publication [11].

On a Euclidean space-time lattice the spectral function $\rho(\omega)$ is not directly accessible; instead on has to invert the above integral relation using the Euclidean correlation function $G(\tau)$. This is an ill-posed problem as it is indeterminate by definition; nevertheless it is possible to get meaningful results from this approach through various ameliorating techniques. $G(\tau)$ is a purely gluonic color-electric correlation function of Euclidean time τ that can be constructed by utilizing Heavy Quark Effective Theory [10]. It reads

$$G(\tau) = -\frac{1}{3}\sum_{i=1}^{3} \frac{\langle \mathrm{Re}\left[\mathrm{tr}\left[U(\beta,\tau)\ gE_i(\mathbf{0},\tau)\ U(\tau,0)\ gE_i(\mathbf{0},0)\right]\right]\rangle}{\langle \mathrm{Re}\left[\mathrm{tr}[\ U(\beta,0)\ ]\right]\rangle}, \tag{2}$$

where $U(b,a)$ is a temporal Wilson line connecting a and b, E_i is the color-electric field operator and g is the coupling. The main advantage of using this gluonic color-electric correlator instead of genuine hadronic correlators is that its spectral function is expected to have a smooth infrared limit which simplifies the spectral reconstruction significantly.

2 Spectral reconstruction

One way to make the inversion of Eq. (1) less indeterminate is to constrain the form of the spectral function based on additional theoretical knowledge. The idea is to build a model spectral function that obeys the correct functional form in the low- and high-frequency limits with a general interpolation in between. The true spectral function of Eq. (1) is then substituted with this general model that has a finite number of parameters. Given some set of these parameters, one can perform the integral over the model spectral function and obtain the corresponding model correlator, G^{model}, which can be compared to the original correlator data, G^{cont}, and its statistical error, δG^{cont}. In this way one can define a weighted sum of squares through

$$\chi^2 \equiv \sum_{\tau} \left[\frac{G^{\mathrm{cont}}(\tau) - G^{\mathrm{model}}(\tau)}{\delta G^{\mathrm{cont}}(\tau)}\right]^2. \tag{3}$$

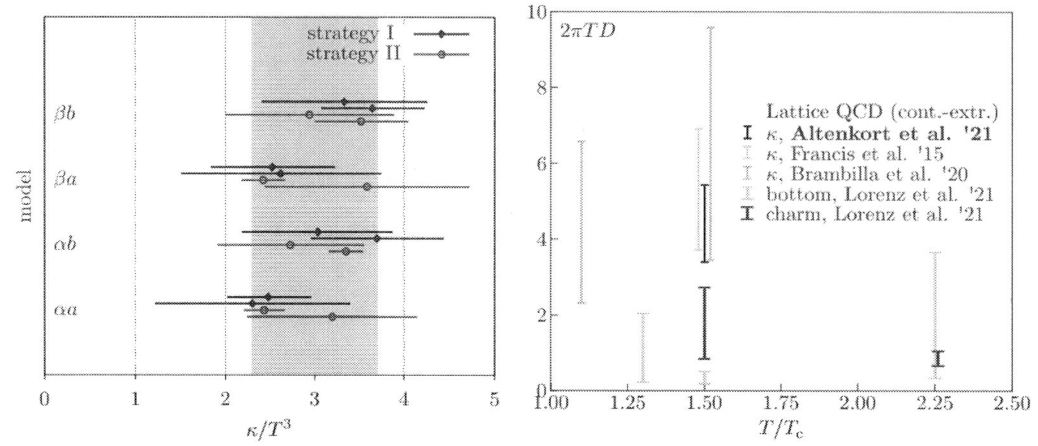

Figure 2. Left: Heavy quark momentum diffusion coefficient κ/T^3 from all considered spectral function model fits. The grey band indicates the final range we cite (see Section 3). Right: Comparison of the heavy quark spatial diffusion coefficient $2\pi TD$ from different studies [12–15]. Results from the heavy quark limit using the color-electric correlator ("κ" in legend) are converted using $D = 2T^2/\kappa$. Also shown are recent results obtained from vector meson correlators for charm and bottom quarks.

The task is then to fit the parameters of the model spectral function such that Eq. (3) is minimized. If the model for the spectral function was appropriate, then the model correlator should be in agreement with the correlator data. Note that this approach is only reasonable if the spectral function is expected to be devoid of intricate structures or peaks, which is a reasonable assumption in this case [10].

The spectral function of the color-electric correlator (Eq. (2)) can be modeled through

$$\rho^{(\mu,i)}_{\text{model}}(\omega) \equiv \left[1 + \sum_{n=1}^{n_{\max}} c_n e_n^{(\mu)}(y)\right] \sqrt{[\phi_{\text{IR}}(\omega)]^2 + [\phi^{(i)}_{\text{UV}}(\omega)]^2}, \tag{4}$$

where the term in brackets is responsible for the interpolation (see [12] for notation and details) between the infrared part ϕ_{IR} and ultraviolet part ϕ_{UV}. For the interpolation and the UV part, different functional forms are reasonable; the superscripts $\mu \in \{\alpha, \beta\}$, $i \in \{a, b\}$ and $n_{\max} \in 4, 5$ depict these (see [12] for detailed explanations). The infrared part is modeled by the simplest functional form that is consistent with Eq. (1), which reads $\phi_{\text{IR}}(\omega) \equiv \frac{\kappa\omega}{2T}$ [13]. In the end we normalize with the temperature and so κ/T^3 is simply one of the free parameters. For the fit we follow two different strategies, which are explained in detail in [12]. In summary we obtain 16 equally reasonable models that are distinguished by the parameters (μ, i), $n_{\max}$ and two fit strategies.

3 Results and comparison

All our theoretically motivated models yield model correlators that agree with the original correlator data as can be seen from the left panel of Fig. 1. The χ^2/d.o.f. of the fits lie in the range 3.4 . . . 5.4. On the right panel of Fig. 1 the resulting spectral functions are shown. Here, the estimate for the momentum diffusion coefficient κ for each spectral function can be read off at the intercept with the y-axis. All of the intercepts are shown in the left panel of Fig. 2 with a grey band that indicates the final range we cite for the HQ momentum diffusion coefficient, which is $\kappa/T^3 = 2.31 \ldots 3.70$. In the nonrelativistic limit ($M \gg \pi T$) we can use the Einstein relation $D = 2T^2/\kappa$ to convert this to $2\pi TD = 3.40 \ldots 5.44$. We can also convert to the kinetic equilibration time $\tau_{\text{kin}} = \eta_D^{-1} = (1.63 \ldots 2.61)\left(\frac{T_c}{T}\right)^2\left(\frac{M}{1.5\text{GeV}}\right)\text{fm/c}$. In comparison with light degrees of freedom of the medium (see, for example, [16]), the results suggest a rather fast thermalization, which seems to fit the experimental observations of a large collective motion of heavy quarks with the medium.

In the right panel of Fig. 2 we compare to other studies [13–15]. It agrees with both values obtained from previous studies that use the multi-level method (instead of gradient flow) and similar spectral reconstruction methods of the same continuum-extrapolated color-electric correlator obtained in the heavy quark limit. However, for all results from the heavy quark limit there is still some friction with recent results from hadronic correlators of charm and bottom quarks with physical masses [15]. Recently it has been worked out that the leading finite mass correction to the momentum diffusion coefficient is encoded in a color-magnetic correlator [17], which can be computed using a similar strategy. We expect this correction to loosen the tension between the two methods; however, there are still open questions in the renormalization of the color-magnetic correlator that need to be solved first [18].

4 Acknowledgements

The authors acknowledge support by the Deutsche Forschungsgemeinschaft (DFG, German Research Foundation) through the CRC-TR 211 'Strong-interaction matter under extreme conditions'– project number 315477589 – TRR 211. The computations in this work were performed on the GPU cluster at Bielefeld University.

References

[1] ALICE Collaboration, Phys. Rev. Lett. **120**, 102301 (2018)
[2] ALICE Collaboration, Journal of High Energy Physics **2018** (2018)
[3] G.D. Moore, D. Teaney, Physical Review C **71** (2005)
[4] A. Beraudo, A. De Pace, W. Alberico, A. Molinari, Nuclear Physics A **831**, 59 (2009)
[5] S. Caron-Huot, G.D. Moore, Physical Review Letters **100** (2008)
[6] N. Brambilla et al., JHEP **05**, 136 (2021), 2012.01240
[7] R. Kubo, M. Toda, N. Hashitsume, *Statistical Physics II*, Springer Series in Solid-State Sciences (Springer, 1978)
[8] D. Forster, *Hydrodynamic Fluctuations, Broken Symmetry, and Correlation Functions*, Advanced Book Classics (Perseus Books, 1990)
[9] H.B. Meyer, The European Physical Journal A **47**, 86 (2011)
[10] S. Caron-Huot, M. Laine, G.D. Moore, JHEP **04**, 053 (2009), 0901.1195
[11] L. Altenkort, A.M. Eller, O. Kaczmarek, L. Mazur, G.D. Moore, H.T. Shu, Bielefeld University, doi:10.4119/unibi/2949617 (2020)
[12] L. Altenkort, A.M. Eller, O. Kaczmarek, L. Mazur, G.D. Moore, H.T. Shu, Phys. Rev. D **103**, 014511 (2021)
[13] A. Francis, O. Kaczmarek et al., Phys. Rev. D **92**, 116003 (2015), 1508.04543
[14] N. Brambilla et al. (TUMQCD Collaboration), Phys. Rev. D **102**, 074503 (2020)
[15] A.L. Lorenz, Ph.D. thesis, Bielefeld University (2021)
[16] S. Schlichting, D. Teaney, Ann. Rev. Nucl. Part. Sci. **69**, 447 (2019), 1908.02113
[17] A. Bouttefeux, M. Laine, JHEP **12**, 150 (2020), 2010.07316
[18] M. Laine, JHEP **06**, 139 (2021), 2103.14270

Flavor hierarchy at the chemical freeze-out from UrQMD

Tom Reichert[1,4,*], *Gabriele* Inghirami[2], and *Marcus* Bleicher[1,2,3,4]

[1]Institut für Theoretische Physik, Goethe Universität Frankfurt, Max-von-Laue-Strasse 1, D-60438 Frankfurt am Main, Germany
[2]GSI Helmholtzzentrum für Schwerionenforschung GmbH, Planckstr. 1, 64291 Darmstadt, Germany
[3]John von Neumann-Institut für Computing, Forschungzentrum Jülich, 52425 Jülich, Germany
[4]Helmholtz Forschungsakademie Hessen für FAIR (HFHF), GSI Helmholtzzentrum für Schwerionenforschung, Campus Frankfurt, Max-von-Laue-Str. 12, 60438 Frankfurt, Germany

Abstract. We introduce a novel approach based on elastic and inelastic scattering rates to extract the hyper-surface of the chemical freeze-out from a hadronic transport simulation. We use the Ultra-relativistic Quantum Molecular Dynamics (UrQMD) model to extract the chemical freeze-out hyper-surface of pions and kaons in the energy range from $E_{\mathrm{lab}} = 1.23A$ GeV to $\sqrt{s_{\mathrm{NN}}} = 7.7$ GeV. By employing a coarse-graining procedure, we can extract the local temperature T and baryo-chemical potential μ_{B} on the chemical freeze-out surface and compare them to results from statistical model analysis. We find good agreement between the pion chemical freeze-out line extracted from the simulation and the freeze-out line from the statistical model extracted from data. In addition the simulations also hint towards the existence of a flavor hierarchy similar to the one observed in recent lattice QCD calculations.

1 Introduction

There is an ongoing debate about whether the phase transition line is connected to the chemical freeze-out line, i.e. the stage at which flavor changing reactions cease and the chemical yields of the hadrons are fixed. On the one hand there is the thermal model utilizing a relativistic hadron gas described by a volume (V), a temperature (T) and the chemical potentials (μ_i) which successfully describes final state hadron yields over a broad range energies [1]. This finding is usually interpreted as thermal/chemical equilibrium being formed via statistical hadronization from the Quark-Gluon-Plasma (QGP). On the other hand there are *ab initio* lattice calculations investigating the temperature dependence of susceptibility ratios pointing towards a separation in the chemical decoupling temperatures of light and strange quarks [2–4].

In this article, we aim to contribute to the discussion of the "flavor hierarchy" question by extracting the microscopic chemical freeze-out hyper-surface $\Sigma_\mu^{\mathrm{cfo}}(x^\nu)$ from a dynamical hadronic transport simulation without the need to fit measured particle yield. Extending our previous calculations for pions (as the dominant light hadron species) [5], this article addresses in addition the chemical freeze-out of strangeness via K mesons. As originally pointed out by Pomeranchuk, the chemical or kinetic freeze-out of a system is determined

*e-mail: treichert@itp.uni-frankfurt.de

by the competition between the scattering rate τ_{scat} and the expansion rate τ_{exp} [6]. We thus aim to trace every final state hadron back to the space-time point of its first creation. Finally, we employ a coarse-graining procedure to infer the local temperature and baryo-chemical potential.

2 Modeling chemical freeze-out in UrQMD

The present study employs the Ultra-relativistic Quantum Molecular Dynamics (UrQMD v3.4) [7, 8] transport model in cascade mode. UrQMD is a hadron cascade model that simulates the dynamical evolution of heavy ion collisions by propagating the individual hadrons, modeling their interactions via the excitation of color flux-tubes (strings) and by further elastic and inelastic scatterings. A transition to a deconfined stage is not explicitly included in the cascade mode employed here.

Our procedure is based on the fact that during the hadronic phase (pseudo-elastic) reactions like $\Delta \leftrightarrow \pi N$ or $N^* \leftrightarrow K\Lambda$ effectively do not change the π or K yield over the lifetime of the intermediating resonance. Through investigation of the full time evolution of the UrQMD output, we can find every final state hadron and trace it back to the space-time point of its first creation. This defines the chemical freeze-out hyper-surface. For the details of the procedure we refer the reader to Ref. [5], where the chemical freeze-out hyper-surface of pions has been determined with UrQMD.

To relate thermodynamic quantities to the hyper-surface Σ_μ^{cfo}, we coarse-grain the UrQMD simulations [9–11]. For the present calculation, a large ensemble of events is placed on a 4-dimensional grid of resolution $\Delta t = 0.25$ fm/c and $\Delta x, y, z = 1$ fm. In each cell, the average energy-momentum tensor $T^{\mu\nu}$ and the average net-baryon current j_B^μ is calculated. After a Lorentz-boost, their time-components represent the local energy density and the local net-baryon density in the Local Rest Frame (LRF). Finally, the energy and net-baryon density can be used to extract the local T and μ_B via interpolation of a Hadron Resonance Gas EoS [12].

3 Results

The results are obtained by analyzing central Au+Au collisions from $E_{lab} = 1.23A$ GeV to $\sqrt{s_{NN}} = 7.7$ GeV calculated with the UrQMD model. The simulations are used as input to extract the chemical freeze-out coordinates and calculate the fields ($T(t, \vec{r})$, $\mu_B(t, \vec{r})$) with the aforementioned coarse-graining procedure. Central events are selected via an impact parameter cut of $b_{max} = 3.4$ fm at all investigated energies. We focus on the chemical freeze-out of π and K which finally decouple in the volume with $|z_{kin}| \leq 5$ fm.

3.1 Freeze-out of pions versus kaons

Let us start with the distributions of the chemical freeze-out times (left), temperatures (middle) and baryo-chemical potential (right) shown in Fig. 1 for kaons (solid lines) and pions (dotted lines) in central Au+Au collisions from $E_{lab} = 1.23A$ GeV to $\sqrt{s_{NN}} = 7.7$ GeV calculated with the UrQMD model. In general, the freeze-out time, temperature and μ_B distributions of pions and of kaons are all peaked indicating that chemical decoupling indeed happens at a rather specific stage of the system evolution, hence the extraction of the chemical freeze-out hyper-surface is reasonable. At very low energies the chemical freeze-out times of the pions and kaons are both peaked but separated by 2 fm/c, however with increasing energy the shift between the distributions vanishes. This observation is expected a) due to the

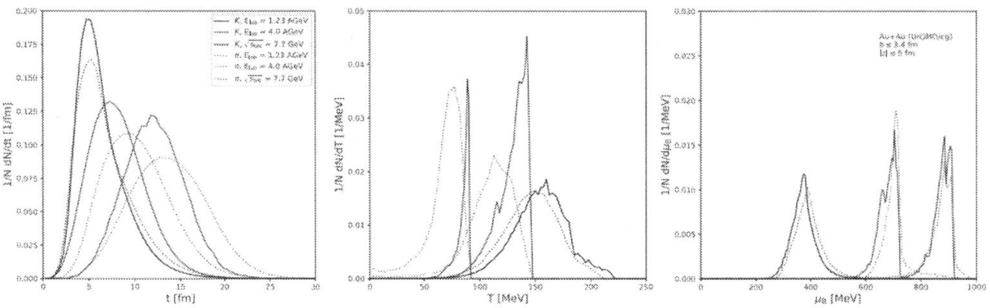

Figure 1. [Color online] Distributions of the chemical freeze-out times (left), of the chemical freeze-out temperatures (middle) and the chemical freeze-out baryo-chemical potentials (right) for kaons (solid) and pions (dotted) in central Au+Au collisions from $E_{\mathrm{lab}} = 1.23A$ GeV to $\sqrt{s_{\mathrm{NN}}} = 7.7$ GeV calculated with the UrQMD model.

fact that at low energies (low temperatures) strangeness production is canonically suppressed (also in transport simulations [13]) and furthermore b) due to the small K absorption cross section on nuclear matter. This (low energy) separation in the average decoupling time is reflected in the extracted temperatures: pions decouple during a broader time span, thus at temperatures ranging from 40 MeV to 100 MeV while kaons, to the contrary, decouple in a narrow temperature slice around 80-90 MeV. This separation again vanishes with increasing energy. The average $\langle T \rangle$ moreover approaches 150 MeV independently of the hadron species at the largest collision energy lining up with statistical model fits to measured hadron yields. The average baryo-chemical potential of both the pions and the kaons decreases with increasing energy due to the production of anti-baryons at higher energies. Here, the separation between π and K becomes small.

3.2 Thermodynamics on the hyper-surface

We are now in the position to visualize our results in the QCD phase diagram. A meaningful T, μ_{B} tuple is obtained by averaging the distributions shown in the previous section. Fig. 2 shows the T, μ_{B} combinations extracted from the chemical freeze-out of pions (black triangles up) and of kaons (red circles) and compares them to well established statistical models fits to experimental hadron yields (blue crosses) [1, 14–18]. In addition we also show results from Ref. [19] in which different ensembles of experimental hadron yields are fitted with the GCE as orange symbols and results from [20] (green symbols) in which the authors used strangeness susceptibilities to calculate flavor hierarchy at chemical freeze-out.

We observe that our results using pions are in very good agreement with the known statistical model fits to the available experimental data [14–18, 21]. Our results on the chemical freeze-out hyper-surface of kaons show generally a larger temperature speaking for an earlier freeze-out. As already discussed, in our simulation (driven by scattering rates) the separation in chemical freeze-out temperature between light and strange flavors $\Delta T \equiv T^K - T^\pi$ decreases with increasing energy from 20 MeV at HADES to 10 MeV at medium RHIC energies, different from the results in [4, 19, 20].

4 Summary

We employed a novel method to extract the microscopic chemical freeze-out hyper-surface from a dynamical hadronic transport simulation to investigate the flavor hierarchy. For this,

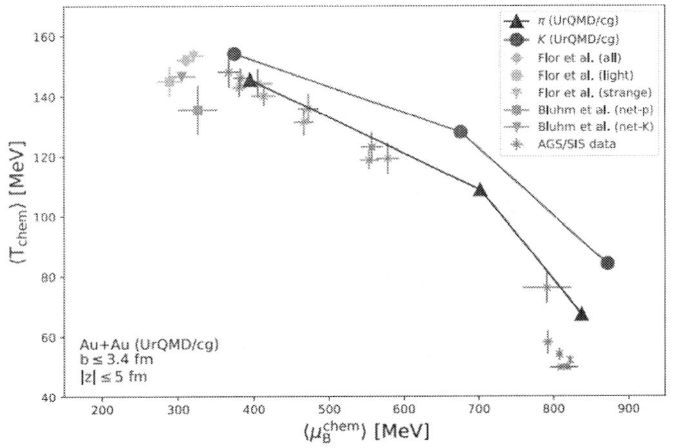

Figure 2. [Color online] The chemical freeze-out line (T, μ_B) of pions (black) and kaons (red) in central Au+Au collisions from $E_{lab} = 1.23A$ GeV to $\sqrt{s_{NN}} = 7.7$ GeV calculated with the UrQMD model. Also shown are fits to experimental hadron yields from AGS [14, 21] and SIS [15–18] as well as chemical freeze-out estimates using kaon fluctuations from [20] (green symbols) and using different hadron ensembles in thermal fits from [19] (orange symbols).

we used the Ultra-relativistic Quantum Molecular Dynamics (UrQMD) model in version 3.4 to calculate the chemical freeze-out hyper-surface of pions and of kaons from central Au+Au collisions in the energy range from $E_{lab} = 1.23A$ GeV to $\sqrt{s_{NN}} = 7.7$ GeV. The local temperature and baryo-chemical potential on this hyper-surface were extracted by coarse-graining the UrQMD output. Our calculations reveal that the statistical model data as well as flavor hierarchy in the chemical decoupling of pions and kaons can be well described by hadronic transport calculations. In contrast to previous results in the literature which investigated this phenomenon at very high energies (LHC, top RHIC), we observed that flavor separation increases with increasing energy at low collision energies (GSI, low RHIC).

The authors thank Jamie Karthein and Marcus Bluhm for fruitful discussion in the aftermath of the "Strangeness in Quark Matter 2021" conference. This work was supported in the framework of COST Action CA15213 THOR and we acknowledge support from STRONG-2020, financed by the Horizon 2020 program of the European Community. Computational resources were provided by the Center for Scientific Computing (CSC) of the Goethe University. G. Inghirami acknowledges funding by the Deutsche Forschungsgemeinschaft (DFG, German Research Foundation) – Project number 315477589 – TRR 211.

References

[1] A. Andronic, P. Braun-Munzinger, K. Redlich, J. Stachel, Nature **561**, 321 (2018), 1710.09425

[2] R. Bellwied, S. Borsanyi, Z. Fodor, S.D. Katz, C. Ratti, Phys. Rev. Lett. **111**, 202302 (2013), 1305.6297

[3] P. Alba, W. Alberico, R. Bellwied, M. Bluhm, V. Mantovani Sarti, M. Nahrgang, C. Ratti, Phys. Lett. B **738**, 305 (2014), 1403.4903

[4] J.M. Stafford, P. Alba, R. Bellwied, V. Mantovani-Sarti, J. Noronha-Hostler, P. Parotto, I. Portillo-Vazquez, C. Ratti, Springer Proc. Phys. **250**, 367 (2020), 1912.12968
[5] T. Reichert, G. Inghirami, M. Bleicher, Eur. Phys. J. A **56**, 267 (2020), 2007.06440
[6] G. Inghirami, T. Reichert, M. Bleicher (2021), 2106.04543
[7] S.A. Bass et al., Prog. Part. Nucl. Phys. **41**, 255 (1998), nucl-th/9803035
[8] M. Bleicher et al., J. Phys. G **25**, 1859 (1999), hep-ph/9909407
[9] G. Inghirami, P. Hillmann, B. Tomášik, M. Bleicher, J. Phys. G **47**, 025104 (2020), 1909.00643
[10] P. Huovinen, M. Belkacem, P.J. Ellis, J.I. Kapusta, Phys. Rev. C **66**, 014903 (2002), nucl-th/0203023
[11] S. Endres, H. van Hees, J. Weil, M. Bleicher, Phys. Rev. C **91**, 054911 (2015), 1412.1965
[12] D. Zschiesche, S. Schramm, J. Schaffner-Bielich, H. Stoecker, W. Greiner, Phys. Lett. B **547**, 7 (2002), nucl-th/0209022
[13] O. Fochler, S. Vogel, M. Bleicher, C. Greiner, P. Koch-Steinheimer, Z. Xu, Phys. Rev. C **74**, 034902 (2006), nucl-th/0601062
[14] F. Becattini, M. Gazdzicki, A. Keranen, J. Manninen, R. Stock, Phys. Rev. C **69**, 024905 (2004), hep-ph/0310049
[15] J. Cleymans, H. Oeschler, K. Redlich, Phys. Rev. C **59**, 1663 (1999), nucl-th/9809027
[16] F. Becattini, J. Cleymans, A. Keranen, E. Suhonen, K. Redlich, Phys. Rev. C **64**, 024901 (2001), hep-ph/0002267
[17] R. Averbeck, R. Holzmann, V. Metag, R.S. Simon, Phys. Rev. C **67**, 024903 (2003), nucl-ex/0012007
[18] G. Agakishiev et al. (HADES), Eur. Phys. J. A **47**, 21 (2011), 1010.1675
[19] F.A. Flor, G. Olinger, R. Bellwied, Phys. Lett. B **814**, 136098 (2021), 2009.14781
[20] M. Bluhm, M. Nahrgang, Eur. Phys. J. C **79**, 155 (2019), 1806.04499
[21] L. Bravina, A. Faessler, C. Fuchs, Z.D. Lu, E.E. Zabrodin, Phys. Rev. C **66**, 014906 (2002), nucl-th/0206014

 https://doi.org/10.1051/epjconf/202225910006

Thermal abundance of hyperons from a coupled channel model

Pok Man Lo[1,*]

[1]Institute of Theoretical Physics, University of Wroclaw, PL-50204 Wrocław, Poland

Abstract. I demonstrate how the thermal abundances of $S = -1$ strange baryons can be computed based on the density of states extracted from a coupled-channel model.

1 Introduction

The S-matrix formulation of statistical mechanics [1, 2] offers a unique theoretical framework to examine how interactions among constituents could dictate the bulk properties of the medium. The essential connection is provided by the density of states (DoS) [3, 4], which can be expressed in terms of the S-matrix by

$$\begin{aligned} B(E) &= \frac{1}{2}\mathrm{Im}\,\mathrm{Tr}\left[S^{-1}\frac{d}{dE}S - \left(\frac{d}{dE}S^{-1}\right)S \right] \\ &= \frac{d}{dE}\,\mathrm{Im}\,\ln\det S(E). \end{aligned} \tag{1}$$

The thermal partition function is given by an integral of the DoS with the appropriate Boltzmann weight. For example the thermal abundance due to the interaction can be computed as

$$\Delta n_a = \int \frac{dE}{2\pi}\, B(E)\, n_a^{(0)}(E, T), \tag{2}$$

where $n_a^{(0)}(E = m_a, T)$ denotes the particle density of an ideal gas of particles with mass m_a. Other thermal (and fluctuation) observables can be analogously computed [5, 6].

Writing the DoS in terms of the S-matrix makes possible to incorporate precise hadron properties, e.g. spectra, production mechanisms and decays, obtained from analyzing volumes of experimental data with different theoretical approaches, into the construction of thermal model. The S-matrix language also naturally describes coupled-channels effects: overlapping resonances, inelasticity, and various threshold effects as channels open up. These phenomena of hadron physics, observed in most known resonances, go beyond what can be captured with the compilation of an appropriate list of resonances, or by introducing some general repulsions among constituents. In this regard, nature is far more creative.

*e-mail: pokman.lo@uwr.edu.pl

Of course, determining the full N-body, multi-channel S-matrix of an interacting system is in general very difficult (if not impossible). This is reflected, e.g., in the poorly constrained couplings and / or the various background channels introduced to satisfy unitarity. The positive aspect is that such model would be systematically improved with data from further scattering experiments. For now we shall take these limitations as features of the approach and explore the perspectives they give when applied to studying thermal system.

2 DoS of coupled-channel hyperon system

A coupled-channel approach [7] is mandatory to describe the $S = -1$ hyperon system: the lowest channels open up in a similar energy range, and most resonances decay into multiple channel. To tackle a thermal system, prescription (1) remains valid, but the S-matrix should be formally understood as a matrix acting in the open-channel space, i.e. an $N_{\text{chan.}} \times N_{\text{chan.}}$ matrix. See table 1 for the list of channels considered.

The trace operation (Tr) in Eq. (1), originated from the thermal trace in constructing the partition function, enforces a summation over the $N_{\text{chan.}}$ channels. The effective spectral function describing a multichannel system reads [4, 6, 8]

$$B(M) = \sum_{a=1}^{N_{\text{chan.}}} B_a(M), \tag{3}$$

where

$$B_a(M) = \frac{1}{2}\,\text{Im}\left[S^{-1}\frac{d}{dM}S - \left(\frac{d}{dM}S^{-1}\right)S \right]_{aa}. \tag{4}$$

Here $[\ldots]_{aa}$ means the a–th diagonal matrix element.

The channel spectral function $B_a(M)$ describes the (energy-dependent) component of the full $B(M)$ when projected into a specific interaction channel a. It generalizes the prescription of a branching ratio ($Br(X \to a)$), and incorporates the energy dependence from resonant and non-resonant interactions [3]. The channel yield can thus be readily computed using (2) with $B \to B_a$. A display of the effective spectral functions, in the partial wave (PW) channel: S01, S11, P01, and P11 are shown in Fig. 1.

Few comments are in order:

1. These plots depict in detail how channel opens up above the threshold and the contribution to the overall DoS. For example, S01 is dominated by $\pi\Sigma$ before $\bar{K}N$ opens up. Furthermore, when a specific component dominates a resonance peak it is natural to expect a substantial branching fraction. For example, the first peak in P11 corresponds to a 2-star $\Sigma(1560)$, predicted to have strong $\pi\Lambda$ and $\pi\Sigma^*(1385)$ components. However, branching fractions thus defined are strongly energy dependent, and generally the lineshapes are not well approximated by a sum of Breit-Wigners.

2. The list of identified resonances, even with full details about their widths, is not enough to reconstruct the S-matrix. One way to understand this is to recognize the role of roots in determining the DoS [4]: they encode details of non-resonant interactions.

3. These S-matrix elements are used as input to determine the thermal hyperon yields [8] and to understand LQCD results [6]. For the former, it yields about ≈ 10–30% increase in the Λ yield compared to the standard HRG model. The relatively large uncertainty comes from the poorly constrained unitary backgrounds, which is estimated to make up around 10% of the total yield at $T = 155$ MeV (LHC conditions).

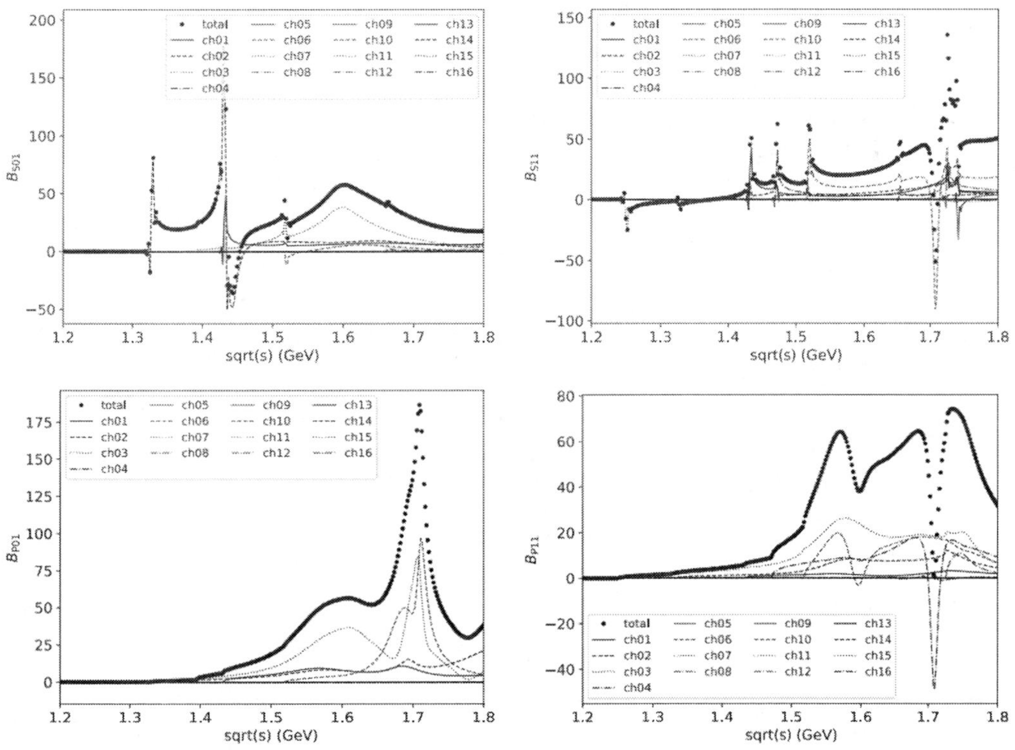

Figure 1. The effective spectral function (with its channel components) as a function of center of mass energy: for S01, S11, P01, and P11 partial waves. See Ref. [8] for details.

channel	elastic	channel	quasi-elastic	channel	unitarity
1	$\bar{K}N$	6	$\bar{K}_1^* N$	15	$\pi\pi\Lambda$
2	$\pi\Sigma$	7	$[\bar{K}_3^* N]_-$	16	$\pi\pi\Sigma$
3	$\pi\Lambda$	8	$[\bar{K}_3^* N]_+$		
4	$\eta\Lambda$	9	$[\pi\Sigma(1385)]_-$		
5	$\eta\Sigma$	10	$[\pi\Sigma(1385)]_+$		
		11	$[\bar{K}\Delta(1232)]_-$		
		12	$[\bar{K}\Delta(1232)]_+$		
		13	$[\pi\Lambda(1520)]_-$		
		14	$[\pi\Lambda(1520)]_+$		

Table 1. The list of interaction channels included in the coupled-channel PWA describing the $|S| = 1$ hyperon system by the Joint Physics Analysis Center (JPAC) Collaboration [7]. See Ref. [8] for its thermal model application.

3 Going further

The decoupling of dynamics (density of states) and thermo-statistical information is a remarkable feature of the S-matrix formulation of statistical mechanics. In principle, one can make progress in understanding the thermal medium by successively improving the S-matrix input: including relevant channels, extension to N > 2 scatterings, etc., thus working toward building an accurate virial/cluster expansion.

This is in contrast to the standard imaginary time formulation: interactions (diagrams) are intertwined with temperatures at each step of the Matsubara sum, and one is tasked to compute the in-medium modification of masses and widths. Within a low density expansion [9, 10] for some simple cases it is possible to relate the two approaches: the real part of the T-matrix, folded with a thermal distribution integral, provides a shift in the real part of the self-energy; while the imaginary part of T-matrix can be related to resonances [3, 11]. A more elaborate study of the S-matrix interpretations of the in-medium effects will be reported elsewhere.

4 acknowledgments

I thank the organizer for the kind invitation to the interesting meeting. I am also grateful to Igor Strakovsky and Christoph Hanhart for the encouraging remarks and many fruitful discussions. This work was supported by the Polish National Science Center (NCN) under the Opus grant no. 2018/31/B/ST2/01663.

References

[1] R. Dashen, S.K. Ma, H.J. Bernstein, Phys. Rev. **187**, 345 (1969)
[2] R. Venugopalan, M. Prakash, Nuclear Physics A **546**, 718 (1992)
[3] P.M. Lo, Eur. Phys. J. C **77**, 533 (2017), 1707.04490
[4] P.M. Lo, Phys. Rev. D **102**, 034038 (2020)
[5] P.M. Lo, B. Friman, K. Redlich, C. Sasaki, Phys. Lett. B **778**, 454 (2018), 1710.02711
[6] C. Fernández-Ramírez, P.M. Lo, P. Petreczky, Phys. Rev. C **98**, 044910 (2018)
[7] C. Fernandez-Ramirez, I. Danilkin, D. Manley, V. Mathieu, A. Szczepaniak, Phys. Rev. D **93**, 034029 (2016), 1510.07065
[8] J. Cleymans, P.M. Lo, K. Redlich, N. Sharma, Phys. Rev. C **103**, 014904 (2021), 2009.04844
[9] A. Schenk, Nucl. Phys. B **363**, 97 (1991)
[10] A. Schenk, Phys. Rev. D **47**, 5138 (1993)
[11] P.M. Lo, F. Giacosa, The European Physical Journal C **79** (2019)

The Role of Strange Quasiquarks in Transport Properties of the QGP

Valeriya Mykhaylova[1,*]

[1]Institute of Theoretical Physics, University of Wrocław, PL-50204 Wrocław, Poland

Abstract. We study the role of dynamical quarks in the transport properties of the deconfined hot medium utilizing the quasiparticle kinetic approach. The system is then composed of the quasiparticle excitations, whose interactions are encoded in their dynamical masses with temperature dependence specified by the effective coupling extracted from lattice QCD thermodynamics. Evaluating the temperature and flavor profiles of the shear and bulk viscosities, as well as the electrical conductivity we examine how different particle species, in particular strange quarks, modify the transport parameters of the deconfined matter.

1 Introduction

One of the major aims of the experimental and theoretical studies of the quark-gluon plasma (QGP) is to reveal its dynamical and transport properties, quantified by various transport parameters.

The shear viscosity η measures the resistance of the fluid against momentum modifications during its longitudinal motion [1], while the bulk viscosity ζ indicates the energy dissipation caused by the expansion of the QGP [2]. Their dimensionless ratios to the entropy density, i.e. η/s, ζ/s are the major input in the hydrodynamic equations [3]. The other important parameter is the electrical conductivity σ, since the strong electric and magnetic fields are expected to emerge in the non-central heavy-ion collisions. A precise determination of transport coefficients as functions of temperature and chemical potential is one of the main steps towards understanding the non-trivial evolution of strongly interacting matter.

For a comprehensive study of the bulk properties of the deconfined matter conducted within the quasiparticle model (QPM), we refer the reader to [1, 2, 4]. In this work, we summarize our numerical results on transport coefficients at finite temperature and vanishing chemical potential. We assess the influence of dynamical quasiparticles on the evolution of different coefficients, focusing particularly on the role of the strange quark flavor. In Sec. 2, we briefly discuss the main assumptions of the QPM. In Sec. 3, we present the results for transport coefficients derived in kinetic theory under the relaxation time approximation. A summary and conclusions are given in Sec. 4.

2 Quasiparticle Model

We examine transport properties of the QGP within the well-grounded quasiparticle model [1, 5]. In this aproach the QGP is described as a viscous fluid composed of quasiparticles, whose dynamical masses $m_i(T)$ arise as they propagate through the system,

*e-mail: valeriya.mykhaylova@uwr.edu.pl

$m_i^2(T) = (m_i^0)^2 + \Pi_i(T)$. The effective mass depends on the bare particle mass m_i^0 and the gauge-independent self-energy Π_i defined as [5]

$$\Pi_{l,s}(T) = 2\left(m_{l,s}^0\sqrt{\frac{G(T)^2}{6}T^2} + \frac{G(T)^2}{6}T^2\right), \qquad \Pi_g(T) = \left(3 + \frac{N_f}{2}\right)\frac{G(T)^2}{6}T^2. \qquad (1)$$

The effective running coupling $G(T)$ is deduced from the equation of state calculated in lattice gauge theory for the QGP with 2 + 1 quark flavors and the gluon plasma in pure Yang-Mills theory. This way, the $G(T)$ incorporates the non-perturbative dynamics near the QCD phase transition and reproduces the perturbative behavior in the very high-temperature regime [1, 2].

3 Transport parameters in kinetic theory

Assuming that a system slightly deviates from thermal equilibrium, we employ the expressions for the transport parameters determined from kinetic theory under the relaxation time approximation. The derivation and detailed analysis are presented in [1, 5, 6] for the shear viscosity η, in [2, 5] for the bulk viscosity ζ, and in [2, 4, 7] for the electrical conductivity σ.

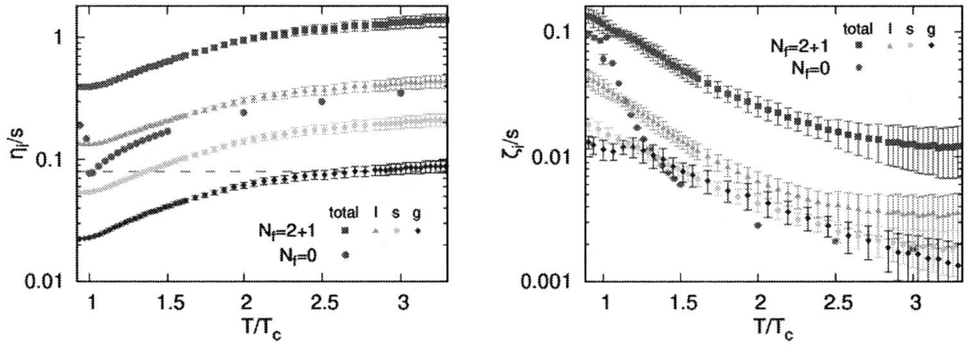

Figure 1. Shear (left) and bulk (right) viscosities to the total entropy density ratios as functions of T/T_c. The individual contributions from gluons (diamonds), light (triangles) and strange (pentagons) quarks are shown along with the total ratios for the QGP with $N_f = 2 + 1$ (squares) and $N_f = 0$ (circles). In the left panel, the Kovtun-Son-Starinets (KSS) bound of $1/4\pi$ is indicated by the dashed line.

In Fig. 1 we present the total specific shear and bulk viscosities of the hot QCD medium, along with the individual contributions of each quasiparticle sector in case of $N_f = 2 + 1$. The hierarchy in η_i/s follows the ordering in the effective masses [1], while in ζ_i/s this behavior is disturbed. Although strange quarks and gluons are described by different statistical distributions, there appears a quantitative resemblance between ζ_s/s and ζ_g/s ratios. We find that the similarity arises due to a convolution of the factors entering the expression for the bulk viscosity [2]. Overall, the light quarks bring the main impact to the total shear and bulk viscosities, while the heaviest quasiparticles are the most effective ones in equilibrating momentum degradations within the QGP.

Compared to the pure Yang-Mills theory, the presence of dynamical quarks in QCD significantly increases η/s and ζ/s ratios. Further, one clearly observes a smooth behavior of the specific viscosities in the vicinity of the QCD crossover, and a strong non-monotonicities around T_c in pure Yang-Mills theory due to the first-order phase transition [1].

Fig. 2 (left) illustrates the bulk to shear viscosity ratios ζ_i/η_i obtained by dividing each curve presented on the left side of Fig. 1 by the corresponding curve on the right. We have utilized the total ζ/η ratio in QCD and pure Yang-Mills theory to study the impact of quarks onto the deviation of the deconfined matter from conformal invariance [2]. Here, we observe an interesting overlap of the ζ_s/η_s component coming from strange quarks, with the total ζ/η ratio. Thus, the strange quark sector quantitatively resembles the bulk to shear viscosity ratio for the QGP. The contribution from the light quark sector slightly underestimates the total ζ/η ratio, while the ζ_g/η_g is much larger due to the suppressed gluonic term in the shear viscosity, see left panel of Fig. 1.

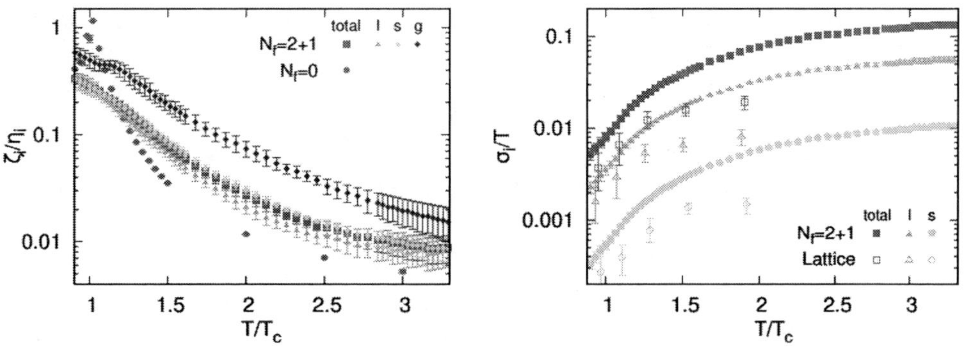

Figure 2. Left: Bulk to shear viscosity ratio for the QGP with $N_f = 2 + 1$ (squares), shown along with the individual contributions from light quarks (triangles), strange quarks (pentagons) and gluons (diamonds). Alongside, the ratio for pure gluon plasma (circles) is presented. Right: The scaled electrical coductivity σ_i/T for the QGP (full squares) along with the light (sum of up and down, full triangles) and strange (full pentagons) quark components. The corresponding lattice data is deduced from [8] for the total electrical conductivity (open squares) and for light and strange quark contributions (open triangles and open pentagons, respectively).

In the right panel of Fig. 2 we present the total scaled electrical conductivity σ/T along with the contributions from light and strange quasiparticle sectors. The light-quark contribution is larger than that of the strange quarks since it contains electric charges of up and down quarks and since the strange quark component is suppressed by the larger effective mass. The corresponding lattice data [8] near T_c is rather compatible with the QPM result, whereas the discrepancy between them emerges at $T \simeq 1.5\,T_c$ and increases gently with temperature. This can be attributed to the fact that the lattice setup includes the pion mass $M_\pi = 384(4)$ MeV, heavier than the physical one used in our model-building.

For a detailed comparison of the total σ/T ratio with the results computed in other approaches, we refer a reader to [2, 4].

4 Conclusions

We have examined the influence of different dynamical particles, especially the strange quarks, on transport coefficients of the deconfined matter. The transport parameters were computed within kinetic theory under the relaxation time approximation for a medium approached by the quasiparticle model (QPM). The dynamical masses of quasiparticles are characterized by the effective temperature-dependent running coupling deduced from the entropy density obtained by lattice simulations.

Assuming that the transport parameters are characterized by the common relaxation time for each quasiparticle species, we computed the temperature and flavor dependence of the specific shear η/s and bulk ζ/s viscosities, and of the scaled electrical conductivity σ/T.

We juxtaposed the total results for the QGP with $N_f = 2 + 1$ with the transport parameters of the gluon plasma in pure Yang-Mills theory. Above the (pseudo)critical temperature, all transport parameters of the pure gluon plasma are much smaller than that for the QGP with dynamical quarks. Thus, the presence of dynamical quarks significantly changes the transport properties of the deconfined matter.

For the QGP with quark quasiparticles, we found that the major contribution to the transport coefficients comes from the light-quark sector, while the components from strange quarks and gluons are suppressed by their larger effective masses. While for the specific shear viscosity η/s we observed a clear ordering in the contributions from different quasiparticles, for the bulk viscosity ζ/s we found that the strange quarks and gluons contribute to the total ratio in almost the same amount due to the convolution of their observables, such as effective masses and relaxation times. The bulk to shear viscosity ratio and its components reveal that in the QPM, the ζ_s/η_s ratio coming from the strange quark sector quantitatively resembles the total ζ/η ratio of the QGP with 2 + 1 quark flavor.

Finally, the individual contributions from light and strange quarks to the electrical conductivity σ/T were confronted with the total ratio and the corresponding lattice data. We found our results to be qualitatively consistent with the recent lattice QCD outsomes [8].

Acknowledgments

The author thanks Chihiro Sasaki for a scientific advisory and fruitful collaboration. This work was partly supported by the Polish National Science Center (NCN) under the Opus grant no. 2018/31/B/ST2/01663.

References

[1] V. Mykhaylova, M. Bluhm, K. Redlich and C. Sasaki, Phys. Rev. D **100**, no.3, 034002 (2019) doi:10.1103/PhysRevD.100.034002.

[2] V. Mykhaylova and C. Sasaki, Phys. Rev. D **103**, no.1, 014007 (2021) doi:10.1103/PhysRevD.103.014007.

[3] P. Huovinen, P. Kolb, U. W. Heinz, P. Ruuskanen and S. Voloshin, Phys. Lett. B **503**, 58-64 (2001) doi:10.1016/S0370-2693(01)00219-2.

[4] V. Mykhaylova, Eur. Phys. J. ST **229**, no.22-23, 3487-3496 (2020) doi:10.1140/epjst/e2020-000116-9.

[5] M. Bluhm, B. Kämpfer and K. Redlich, Nucl. Phys. A **830**, 737C-740C (2009) doi:10.1016/j.nuclphysa.2009.10.065.

[6] C. Sasaki and K. Redlich, Phys. Rev. C **79**, 055207 (2009) doi:10.1103/PhysRevC.79.055207.

[7] L. Thakur, P. K. Srivastava, G. P. Kadam, M. George and H. Mishra, Phys. Rev. D **95**, no. 9, 096009 (2017) doi:10.1103/PhysRevD.95.096009.

[8] G. Aarts, C. Allton, A. Amato, P. Giudice, S. Hands and J. I. Skullerud, JHEP **02**, 186 (2015) doi:10.1007/JHEP02(2015)186.

Resonance production in Pb+Pb collisions at 5.02 TeV

Dmytro Oliinychenko[1,*] and *Chun* Shen[2,3]

[1]Institute for Nuclear Theory, University of Washington, Seattle, WA, 98195, USA
[2]Department of Physics and Astronomy, Wayne State University, Detroit, Michigan 48201, USA
[3]RIKEN BNL Research Center, Brookhaven National Laboratory, Upton, NY 11973, USA

Abstract. The yields, mean transverse momenta, and flow of K^{*0}, ρ^0, $\Lambda(1520)$ resonances provide an evidence of a late stage hadronic rescattering in ultra-relativistic central heavy ion collisions [1]. Using hydrodynamic + hadronic afterburner simulations of Pb+Pb collisions at 5.02 TeV we achieve a reasonable description of resonance yields and spectra as a function of collision centrality. We demonstrate that the measurements of $\Lambda(1520)$'s mean transverse momentum allow to constrain the unknown branching ratios of $\Sigma^* \to \Lambda(1520)\pi$ decays. Hadronic dynamics leads to an enhanced $\Delta(1232)$ production in central collisions.

Ultra-relativistic collisions of Pb nuclei at the energy $\sqrt{s_{\rm NN}} = 5.02$ TeV per nucleon pair are conducted at the Large Hadron Collider. During a time of order 10 fm/c (= 10^{-22} s), these collisions create a highly compressed and heated quark-gluon plasma, which undergoes an explosive expansion, cools down, and turns into hadrons. The dynamics is simulated as an expansion of a fluid droplet using relativistic hydrodynamics. The later dilute-stage phase is modeled by a hadronic afterburner, in which hadrons undergo a chain of elastic and inelastic collisions and resonance formations and decays. If a resonance R is produced at the quark-gluon plasma hadronization, it does not necessarily survive through the hadronic rescattering. It can collide with other hadrons, forming a higher mass resonance, which may later decay into other products than original resonance (schematically $R + X \to R' \to A + B$). It may as well decay, and after one of the decay products scatters, the resonance R cannot be recovered experimentally. On the other hand, a new resonance of the same type can be formed in a hadronic scattering. Therefore, the measured yields and spectra of resonances can be sensitive to the hadronization and the subsequent various hadronic reactions. The sensitivity of resonance production to the hadronic rescattering stage has indeed been observed in multiple simulations within different approaches: RQMD [2], UrQMD [3–7], PHSD [8–10], and hybrid EPOS + UrQMD [11–13]. Overall, the results on resonance production from these simulations agree with experimental measurements by STAR Collaboration in pp and AuAu collisions at 200 GeV [14, 15] and ALICE Collaboration in pp, pPb, and PbPb at 2.76 and 5.02 TeV [16–19]. In this work, we attempt to gain a qualitative understanding of how the large network of reactions in the hadronic afterburner describes the experiments.

We employ an open-source 3-dimensional relativistic viscous fluid dynamic code `MUSIC v3.0` [20–22] to simulate the hydrodynamic expansion of the collision fireball. The detailed model setup for the hydrodynamic stage was discussed in Ref. [1]. Thermal hadrons are generated on a constant energy density hypersurface $\epsilon_p = 0.2\,\mathrm{GeV/fm}^3$ (corresponding in our

*e-mail: oliiny@uw.edu

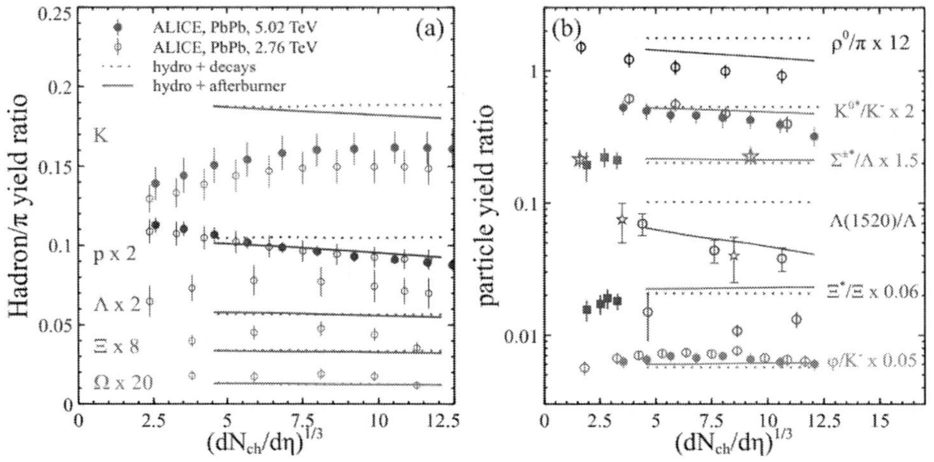

Figure 1. Midrapidity yield ratios of stable hadrons (a) and selected resonances (b) in pPb, PbPb, and AuAu collision. Ratios are shown as a function of a charged particle multiplicity per unit of pseudorapidity, $dN_{ch}/d\eta$, at $\eta = 0$. Dotted lines correspond to hydrodynamics simulation and resonance decays, full lines stand for the simulation with account of hadronic rescattering after hydrodynamical stage. The results of simulations are compared to experimental data from PbPb collisions at 5.02 TeV (full circles) [24, 25], PbPb collisions at 2.76 TeV (open circles) [16–18, 26–29], pPb collisions at 5.02 TeV (squares) [19], and AuAu collisions at 200 GeV (stars) [15].

case to temperature $T \approx 145$ MeV) and inserted into the SMASH hadronic hadronic transport simulations [23]. Switching on and off $2 \to 2$ inelastic reactions, such as $NN \to N\Delta$, $NN \to NN^*$, $NN \to N\Delta^*$ (N^* and Δ^* denote all nucleon- and delta-resonances) and strangeness exchange reactions, we found that the resonance yields are almost unchanged. Therefore, one can think of a hadronic afterburner in terms of multiple resonance excitations and decays, with at most a 10% correction for yields of baryon resonances from annihilation reactions. In SMASH, all the resonance formations and decays respect the detailed balance principle: matrix elements of forward and reverse reactions are identical. In SMASH multi-particle decays are artificially substituted by a chain of reversible $2 \to 1$ decays, for example $\omega \leftrightarrow \rho\pi$, $\rho \leftrightarrow \pi\pi$. Our simulation [1] provides a reasonable description of pion, kaon, proton, and resonance yields as a function of centrality, and a somewhat worse description of strange baryons, see Fig. 1.

Some resonances, such as ϕ, $\Sigma(1385)$, $\Xi(1530)$, are almost unaffected by the hadronic stage. For example, the $\phi(1020)$ has a long vacuum lifetime $c\tau = 46.4\pm0.14$ fm/c, and at temperatures below 150 MeV its mean free path in hadronic matter exceeds $\hbar c/\Gamma_{col} = 10$ fm [30]. In transport simulations, the mean free path is substantially larger because most of the reactions in [30] are usually not implemented. The regeneration rate $K^+K^- \to \phi(1020)$ is small because, by detailed balance principle, it is proportional to the width of ϕ. In summary, even in central Pb+Pb collisions, the fireball is transparent to ϕ, and the regeneration is negligible. This is also the case for $\Xi(1530)$, which has a large mean free path in the hadronic medium and low regeneration cross-section. Indeed, one can see in Fig. 1 (b) that the flat trend of Ξ^*/Ξ could describe data, while the model overshoots the value of Ξ^*/Ξ by almost a factor of 2. This is also the case for EPOS + UrQMD model [12]. Based on this overestimation, we conjecture that $\Xi(1530)$ interacts with pions and forms higher mass Ξ^* states, which are reactions not included in UrQMD or SMASH. Such reactions could decrease the yield of $\Xi(1530)$. Unlike for $\phi(1020)$ and $\Xi(1530)$ the fireball is not completely transparent

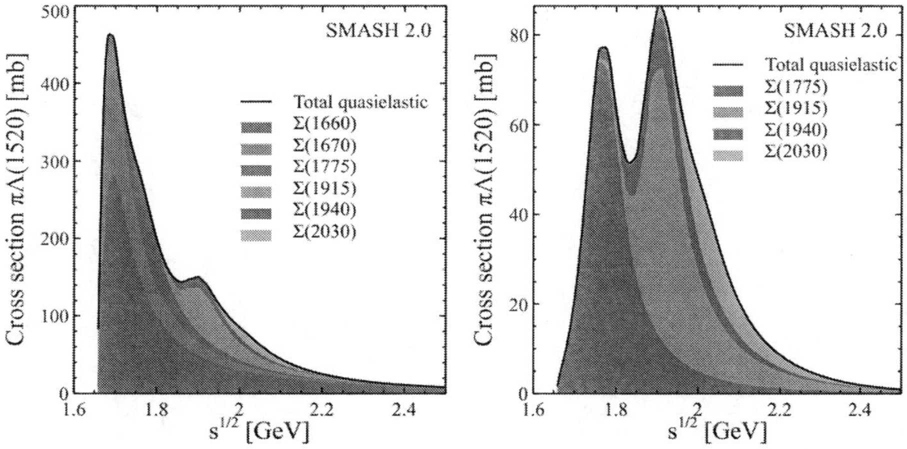

Figure 2. Cross sections of $\Lambda(1520)\pi \to \Sigma^*$ in `SMASH 2.0`. Left panel: $\Sigma(1660)$ and $\Sigma(1670)$ included; right panel: $\Sigma(1660)$ and $\Sigma(1670)$ not included.

for $\Sigma(1385)$. However, the rescattering and regeneration processes are balanced, and overall the afterburner does not affect $\Sigma(1385)$ yield.

For the resonances ρ^0, K^{*0}, and $\Lambda(1520)$ the mean free paths at $T = 145$ MeV are below 3 fm. The fireball is opaque for them, which justifies the application of a partial chemical equilibrium thermodynamic (PCE) model [31], which explains the ρ^0, K^{*0}, and $\Lambda(1520)$ yields rather well by assuming an isentropic expansion, where the yields of stable hadrons including decay contributions from resonances are fixed. However, we observe that the PCE model cannot explain the results of our simulations when a broader set of resonances is compared [1]. To better understand the microscopic origin of resonance suppression, we look in more detail at $\Lambda(1520)$, for which the suppression is the largest. We find that in SMASH $\Lambda(1520)$ has a very large interaction cross-section with pions, forming Σ^* resonances, $\Lambda(1520)\pi \to \Sigma^*$. The total $\Lambda(1520)\pi \to \Sigma^*$ cross-section is shown in Fig. 2, in SMASH it reaches a tremendous value of 460 mb. The partial cross-sections such as $\Lambda(1520)\pi \to \Sigma(1660)$ or $\Lambda(1520)\pi \to \Sigma(1775)$ are proportional to the branching ratio of the corresponding Σ^* state into $\Lambda(1520)\pi$ [23]. Many of these branching ratios are not known experimentally, in particular $\Lambda(1520)\pi \to \Sigma(1660)$ and $\Lambda(1520)\pi \to \Sigma(1670)$. After setting the latter to zero we obtain a much smaller cross-section of around 80 mb, see Fig. 2. We find that the mean transverse momentum $\langle p_T \rangle$ of $\Lambda(1520)$ is rather sensitive to this cross-section, therefore by measuring the $\langle p_T \rangle$ one can constrain the branching ratios of $\Sigma^* \to \Lambda(1520)\pi$.

Our final observation in this study is that some resonances are not suppressed in central collisions but rather enhanced. In particular, the $\Delta(1232)$ is enhanced by at least 15%. The branching ratios of Δ^* resonances into Δ are well-constrained experimentally, which reduces the possible uncertainty of this observation. We also checked that switching on and off $2 \to 2$ reactions, many of which involve $\Delta(1232)$, does not change our result – the Δ is still significantly enhanced. It would be interesting to verify the $\Delta(1232)$ enhancement experimentally in $\Delta^{++} \to p\pi^+$ or in $\Delta^+ \to pe^+e^-$ decay channels, which may however turn to be challenging.

D.O. was supported the U.S. Department of Energy (DoE) under Grant No. DE-FG02-00ER4113. C.S. was supported in part by the National Science Foundation (NSF) under grant number PHY-2012922 and by the U.S. DoE under contract number DE-SC0021969, DE-SC001346, and within the framework of the Beam Energy Scan Theory (BEST) Topical Collaboration.

References

[1] D. Oliinychenko, C. Shen (2021), 2105.07539
[2] S.C. Johnson, B.V. Jacak, A. Drees, Eur. Phys. J. C **18**, 645 (2001), nucl-th/9909075
[3] M. Bleicher, J. Aichelin, Phys. Lett. B **530**, 81 (2002), hep-ph/0201123
[4] S. Vogel, J. Aichelin, M. Bleicher, J. Phys. G **37**, 094046 (2010), 1001.3260
[5] J. Steinheimer, J. Aichelin, M. Bleicher, EPJ Web Conf. **36**, 00002 (2012)
[6] S. Vogel, M. Bleicher, *Resonance absorption and regeneration in relativistic heavy ion collisions*, in *43rd International Winter Meeting on Nuclear Physics* (2005), nucl-th/0505027
[7] T. Reichert, P. Hillmann, A. Limphirat, C. Herold, M. Bleicher, J. Phys. G **46**, 105107 (2019), 1903.12032
[8] E.L. Bratkovskaya, EPJ Web Conf. **36**, 00004 (2012)
[9] A. Ilner, D. Cabrera, C. Markert, E. Bratkovskaya, Phys. Rev. C **95**, 014903 (2017), 1609.02778
[10] A. Ilner, J. Blair, D. Cabrera, C. Markert, E. Bratkovskaya, Phys. Rev. C **99**, 024914 (2019), 1707.00060
[11] A.G. Knospe, C. Markert, K. Werner, J. Steinheimer, M. Bleicher, Phys. Rev. C **93**, 014911 (2016), 1509.07895
[12] K. Werner, A.G. Knospe, C. Markert, B. Guiot, I. Karpenko, T. Pierog, G. Sophys, M. Stefaniak, M. Bleicher, J. Steinheimer, EPJ Web Conf. **171**, 09002 (2018), 1812.06330
[13] A.G. Knospe, C. Markert, K. Werner, J. Steinheimer, M. Bleicher (2021), 2102.06797
[14] J. Adams et al. (STAR), Phys. Rev. C **71**, 064902 (2005), nucl-ex/0412019
[15] B.I. Abelev et al. (STAR), Phys. Rev. Lett. **97**, 132301 (2006), nucl-ex/0604019
[16] S. Acharya et al. (ALICE), Phys. Rev. C **99**, 064901 (2019), 1805.04365
[17] B.B. Abelev et al. (ALICE), Phys. Rev. C **91**, 024609 (2015), 1404.0495
[18] S. Acharya et al. (ALICE), Phys. Rev. C **99**, 024905 (2019), 1805.04361
[19] D. Adamova et al. (ALICE), Eur. Phys. J. C **77**, 389 (2017), 1701.07797
[20] B. Schenke, S. Jeon, C. Gale, Phys. Rev. C **82**, 014903 (2010), 1004.1408
[21] J.F. Paquet, C. Shen, G.S. Denicol, M. Luzum, B. Schenke, S. Jeon, C. Gale, Phys. Rev. C **93**, 044906 (2016), 1509.06738
[22] G.S. Denicol, C. Gale, S. Jeon, A. Monnai, B. Schenke, C. Shen, Phys. Rev. C **98**, 034916 (2018), 1804.10557
[23] J. Weil et al., Phys. Rev. C **94**, 054905 (2016), 1606.06642
[24] S. Acharya et al. (ALICE), Phys. Rev. C **101**, 044907 (2020), 1910.07678
[25] S. Acharya et al. (ALICE), Phys. Lett. B **802**, 135225 (2020), 1910.14419
[26] B. Abelev et al. (ALICE), Phys. Rev. C **88**, 044910 (2013), 1303.0737
[27] B.B. Abelev et al. (ALICE), Phys. Rev. Lett. **111**, 222301 (2013), 1307.5530
[28] B.B. Abelev et al. (ALICE), Phys. Lett. B **728**, 216 (2014), [Erratum: Phys.Lett.B 734, 409–410 (2014)], 1307.5543
[29] J. Song (ALICE), Nucl. Phys. A **967**, 920 (2017)
[30] L. Alvarez-Ruso, V. Koch, Phys. Rev. C **65**, 054901 (2002), nucl-th/0201011
[31] A. Motornenko, V. Vovchenko, C. Greiner, H. Stoecker, Phys. Rev. C **102**, 024909 (2020), 1908.11730

Production of strange particles in jets and underlying events in pp collisions at $\sqrt{s} = 13$ TeV with ALICE

Pengyao Cui[1,*] for the ALICE Collabration

[1]Central China Normal University

Abstract. The production of strange (K^0_S, Λ) and multi-strange ($\Xi^\pm$ and $\Omega^\pm$) particles in jets and the underlying event in pp collisions at $\sqrt{s} = 13$ TeV has been studied with ALICE at the LHC. The p_T-differential density of the particles produced in a jet is compared to the inclusive distribution. The corresponding $(\Lambda + \overline{\Lambda})/2K^0_S$ and $(\Xi^- + \overline{\Xi}^+)/(\Lambda + \overline{\Lambda})$ ratios in jets as a function of p_T are also reported in this work. The p_T-differential density distribution of hadrons associated with jets decreases slower than the one reported for inclusive particle production. The ratios measured in jets show a clear difference when compared to the inclusive ones at the intermediate p_T range. The results provide an opportunity to test the strange and multi-strange particle production mechanisms with hard scattering.

1 Introduction

The transverse momentum (p_T) dependence of the baryon-to-meson yield ratio in hadronic and nuclear collisions is sensitive to the collective expansion of the system, the partonic recombination into hadrons, the jet fragmentation, and hadronization. A significant enhancement of such ratio has been observed at intermediate-p_T ($2 < p_T < 6$ GeV/c) in high multiplicity pp and p–Pb collision events with respect to lower multiplicity events [1]. However, the origin of the enhancement still remains an open question. The ALICE collaboration has furthermore investigated production of the V^0 particles (K^0_S and Λ) in jets reconstructed in pp collisions at $\sqrt{s} = 7$ TeV and p–Pb collisions at $\sqrt{s_{NN}} = 5.02$ TeV [2]. In this contribution, we explore the connection between the baryon-to-meson ratio enhancement and jet production via the measurement of the p_T-differential spectrum of strange and multi-strange particles (K^0_S, Λ, and Ξ) in pp collisions at $\sqrt{s} = 13$ TeV done both inclusively and within energetic jets. The results set new constraints on the particle production mechanisms in jets and provide new insight into the understanding of the origin of flow-like correlations observed in small colliding systems.

2 Analysis Strategy

This analysis is based on 240 million pp collision events at $\sqrt{s} = 13$ TeV collected by the ALICE detector [3] at the LHC. The event selection used the minimum bias trigger which was defined by time coincidence of signals coming from the V0 arrays which cover the forward

*e-mail: pengyao.cui@cern.ch

and backward rapidity. Track reconstruction is based on spatial data points measured by the Inner Tracking System (ITS) and Time Projection Chamber (TPC) detectors. ITS is a silicon tracker providing necessary spatial resolution required for reconstruction of primary interaction vertices and secondary vertices of hadron decays. Besides tracking, TPC provides in addition also information about particle type obtained through measurements of $\mathrm{d}E/\mathrm{d}x$.

Charged particle tracks can be measured in the pseudorapidity range $|\eta| < 0.9$ and their p_{T} is constrained from below by 0.15 GeV/c. These tracks are used to reconstruct charged-particle jets, which are used to tag a hard process. The charged-particle jets are reconstructed with the anti-k_{T} algorithm [6] for $R = 0.4$ using the E-recombination scheme. Pseudorapidity of the anti-k_{T} jets is constrained to $|\eta_{\mathrm{jet}}| < 0.35$. In order to reconstruct (multi-)strange particles the following decay channels are used: $\mathrm{K^0_S} \rightarrow \pi^+ + \pi^-$, $\Lambda \rightarrow \mathrm{p} + \pi^-$, and $\Xi^- \rightarrow \Lambda + \pi^- \rightarrow \mathrm{p} + \pi^- + \pi^-$.

The strange particles were matched with a jet when the angular distance within the η–φ plane between the particles and the jet axis is smaller than the resolution parameter, $\Delta R(\text{particle, jet}) < R$. In some cases particles from the underlying event (UE) are matched to the jet. This component is subtracted from the p_{T} differential density spectra on a statistical basis. The UE contribution is estimated by perpendicular cone (PC) yields. The PC indicates the cone in the η–φ space located at the perpendicular direction with respect to the jet axis in azimuth at the same η. The density of strange particles in a jet cone (JC) or a perpendicular cone is defined as:

$$\rho^{\mathrm{str}}(p_{\mathrm{T}}) = N^{\mathrm{str}}(p_{\mathrm{T}})/A^{\mathrm{str}}, \tag{1}$$

where N^{str} is the number of strange particles and A^{str} is the acceptance in pseudorapidity and azimuthal angle. The density distributions were measured as a function of particle p_{T}. The distributions were normalized per selected event and bin width in p_{T} to obtain, p_{T} differential density $\mathrm{d}\rho/\mathrm{d}p_{\mathrm{T}}$. The density of particles coming from jet fragmentation is estimated by JE = JC − UE = JC − PC, where JE, JC, and UE denote the corresponding p_{T}-differential densities.

3 Results

Figure 1 shows the p_{T}-differential densities for $\mathrm{K^0_S}$ and $\Xi^- + \overline{\Xi}^+$. As expected the density distributions, which are corresponding to jet fragmentation, are considerably harder than the inclusive case. Let us point out that the p_{T}-differential density distribution of inclusive hadrons is lower than that of the PC selection since the latter is obtained from events which contain jets with $p^{\mathrm{ch}}_{\mathrm{T,\,jet}} > 10$ GeV/c.

The $(\Lambda + \overline{\Lambda})/2\mathrm{K^0_S}$ and $(\Xi^- + \overline{\Xi}^+)/(\Lambda + \overline{\Lambda})$ ratio of the p_{T}-differential densities corresponding to jets with $p^{\mathrm{ch}}_{\mathrm{T,\,jet}} > 10$ GeV/c and underlying events are presented in Figure 2. The data suggest that the inclusive and the UE ratios manifest an enhancement at $p_{\mathrm{T}} \sim 3 - 4$ GeV/c. The ratios corresponding to charged-particle jets are significantly lower than those for the inclusive and UE case at low and intermediate p_{T}. In addition, the ratios which are measured in jets are approximately independent of p_{T} in the region above 2 GeV/c. This suggests that the enhancement which is seen in the baryon-to-meson and baryon-to-baryon ratios at intermediate p_{T} is not driven by the jet fragmentation. In Figure 2, these ratios are compared to the PYTHIA 8 simulation with rope model [5]. It is shown that the PYTHIA can simulate the ratio $(\Lambda + \overline{\Lambda})/2\mathrm{K^0_S}$ well but fails to reproduce the $(\Xi^- + \overline{\Xi}^+)/(\Lambda + \overline{\Lambda})$ which carries information about multi-strange particles. The $(\Xi^- + \overline{\Xi}^+)/(\Lambda + \overline{\Lambda})$ ratio obtained with PYTHIA 8 soft QCD shows a strong increase, inconsistent with the data. The $(\Xi^- + \overline{\Xi}^+)/2\mathrm{K^0_S}$ and $(\Omega^- + \overline{\Omega}^+)/2\mathrm{K^0_S}$ ratios were also studied with the PYTHIA 8 rope model, see Figure 3. It can be seen that PYTHIA in the soft QCD mode predicts similar strong increase as observed above.

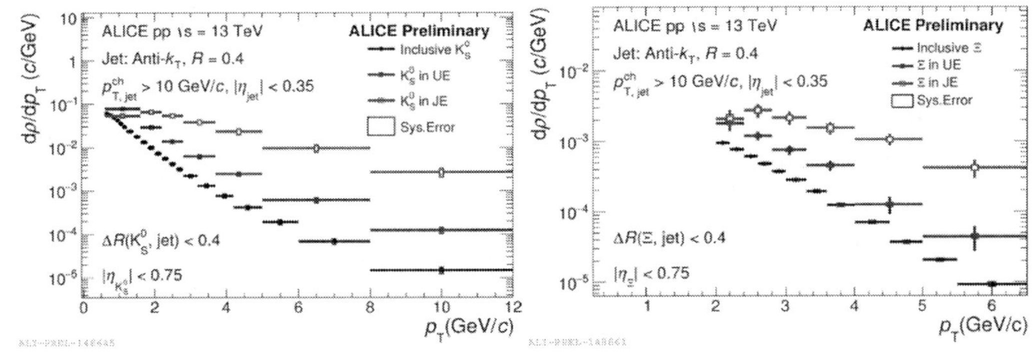

Figure 1. (color online) p_T-differential density of K^0_S (left) and $\Xi^- + \overline{\Xi}^+$ (right) in pp collisions at $\sqrt{s} = 13$ TeV. The black points represent particles from minimum bias events, the blue points represent particles within a cone perpendicular to the jet, which are associated with the underlying event, and the red points represent the particles from the jet fragmentation.

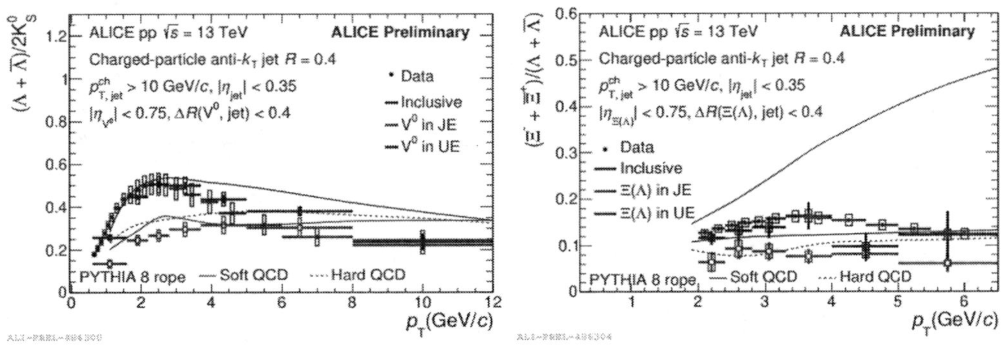

Figure 2. (color online) Ratio of p_T differential densities for $(\Lambda + \overline{\Lambda})/2K^0_S$ (left) and $(\Xi^- + \overline{\Xi}^+)/(\Lambda + \overline{\Lambda})$ (right) as a function of particle p_T in pp collisions at $\sqrt{s} = 13$ TeV. The black points correspond to the ratio evaluated for particles from minimum bias events, the blue points correspond to the ratio obtained for particles which come from a cone that is perpendicular to the jet, and the red points represent the ratio from the jet fragmentation. Charged-particle jets with $p^{ch}_{T,jet} > 10$ GeV/c were reconstructed with the anti-k_T algorithm with $R = 0.4$. The ratios are compared to the corresponding calculation by PYTHIA 8 simulations.

4 Conclusions

The the p_T dependent spectra of K^0_S, $\Lambda + \overline{\Lambda}$ and $\Xi^- + \overline{\Xi}^+$ within jets are considerably less steep than in the case of inclusive particles. In contrast to the inclusive distributions, the $(\Lambda + \overline{\Lambda})/2K^0_S$ and $(\Xi^- + \overline{\Xi}^+)/(\Lambda + \overline{\Lambda})$ ratios measured within jets do not exhibit enhancement at intermediate p_T. The enhancement of baryon to meson ratio in high multiplicity events thus might be attributed to soft processes taking place during the collisions. The $(\Xi^- + \overline{\Xi}^+)/(\Lambda + \overline{\Lambda})$ ratio measured in jets is likewise almost p_T independent. Hence, the enhanced production of multi-strange hadrons in high-multiplicity events [4] may also be explained by the soft processes. The measured ratios are compared with PYTHIA 8 rope model predictions. The

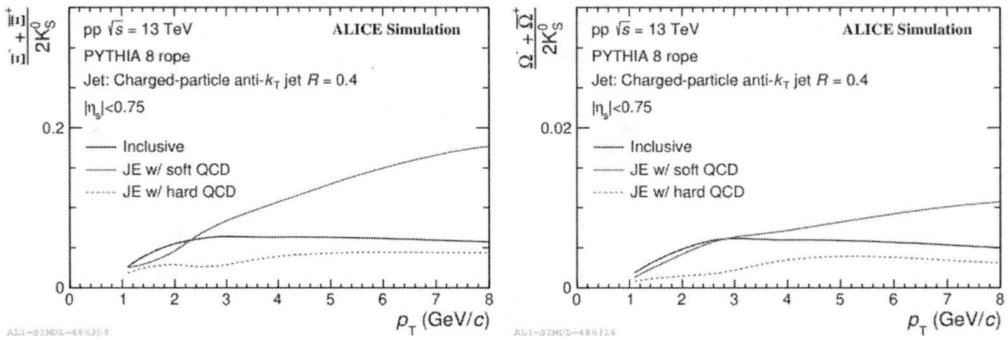

Figure 3. (color online) The $(\Xi^- + \overline{\Xi}^+)/2K^0_S$ (left) and $(\Omega^- + \overline{\Omega}^+)/2K^0_S$ (right) ratios obtained from PYTHIA 8 simulations.

PYTHIA 8 can describe the $(\Lambda + \overline{\Lambda})/2K^0_S$ but fails to reproduce the ratios that involve multi-strange particles ($\Xi^{\pm}$ and $\Omega^{\pm}$) with soft QCD.

Acknowledgment

This work is partly supported by the National Natural Science Foundation of China with Grant No. 11875143 and the National Key Research and Development Program of China under Grant No. 2016YFE0100900.

References

[1] ALICE Collaboration, B. B. Abelev et al., "Multiplicity dependence of Pion, Kaon, Proton and Lambda Production in p–Pb Collisions at $\sqrt{s_{NN}}$ =5.02 TeV" *Phys. Lett.* B728 (2014) 25–38, arXiv:1307.6796 [nucl - ex].

[2] ALICE Collaboration, S. Acharya et al., "Production of Λ and K^0_S in jets in p–Pb collisions at $\sqrt{s_{NN}}$ = 5 TeV and pp collisions at $\sqrt{s}$ = 7 TeV", arXiv: 2105.04890 [hep - ex].

[3] ALICE Collaboration, B. B. Abelev et al., "Performance of the ALICE Experiment at the CERN LHC," *Int. J. Mod. Phys.* A29 (2014) 1430044, arXiv:1402.4476 [nucl - ex].

[4] ALICE Collaboration, J. Adam *et al.*, "Enhanced production of multi-strange hadrons in high-multiplicity proton-proton collisions", *Nature Phys.* **13** (2017) 535–539, arXiv: 1606.07424 [nucl - ex].

[5] T. Sjöstrand et al., "An introduction to PYTHIA 8.2", *Comput.Phys.Commun.* **191** (2015), arXiv: 1410.3012 [hep-ph].

[6] M. Cacciari, G. P. Salam and G. Soyez, "The anti-k_t jet clustering algorithm," JHEP **04**, 063 (2008) doi:10.1088/1126-6708/2008/04/063 [arXiv:0802.1189 [hep-ph]].

EPJ Web of Conferences **259**, 10010 (2022) https://doi.org/10.1051/epjconf/202225910010
SQM 2021

Conserved charge fluctuations at vanishing net-baryon density from Lattice QCD

Jishnu Goswami[1,*], *Frithjof* Karsch[1], *Swagato* Mukherjee[2], *Peter* Petreczky[2], and *Christian* Schmidt[1]

[1]Fakultät für Physik, Universität Bielefeld, D-33615 Bielefeld, Germany
[2]Physics Department, Brookhaven National Laboratory, Upton, NY 11973, USA

Abstract. We present here continuum extrapolated results for all 2nd order cumulants using the most resent results obtained by the HotQCD collaboration in (2+1)-flavor QCD. We constrain the applicability of various HRG models by presenting a detailed comparison of our results based on different sets of hadron spectra as well as with virial expansion based model calculations. A comparison with our lattice QCD results for 2nd order cumulants with models that parametrize repulsive interactions among baryons and anti-baryons in a hadron resonance gas through a single excluded volume parameter (EVHRG) is also shown.

1 Introduction

The partition function of quantum chromodynamics (QCD) depends on the temperature (T) and the chemical potentials, μ_B, μ_Q and μ_S corresponding to three conserved charges namely baryon number (B), electric charge (Q) and strangeness (S). Fluctuations of these conserved charges can be calculated from first principles Lattice QCD formalism. On the other hand two large experimental programs at RHIC in USA and LHC in Switzerland measure the fluctuations of conserved charges either from event by event collisions or from particle yields. In those experimental programs the temperature and chemical potential at the time of freeze out is determined by using hadron resonance gas (HRG) models. Therefore it is essential to constrain the applicability of HRG models by a detailed comparison with lattice QCD, which is the fundamental theory of strong interactions [1].

In these proceedings we present all the 2nd order cumulants calculated by the HotQCD collaboration at vanishing chemical potential in (2+1) flavor QCD and show a detailed comparison with various HRG models. We show that when comparing HRG models with QCD the details of the spectrum play an important role for observables dominated by strange particles. For instance, a non interacting point like HRG model with established resonances from tables provided by the particle data group (PDG) (3 and 4 star states) [2] works poorly in the strangeness sector. Thus to improve the baryon-strangeness correlations additional resonances from PDG tables (1 and 2 star) and quark models[3, 4] are required. Baryon-electric charge correlations show some striking robust feature close to the pseudo-critical transition temperature of QCD which we will discuss in detail in the following section.

*e-mail: jishnu@physik.uni-bielefeld.de

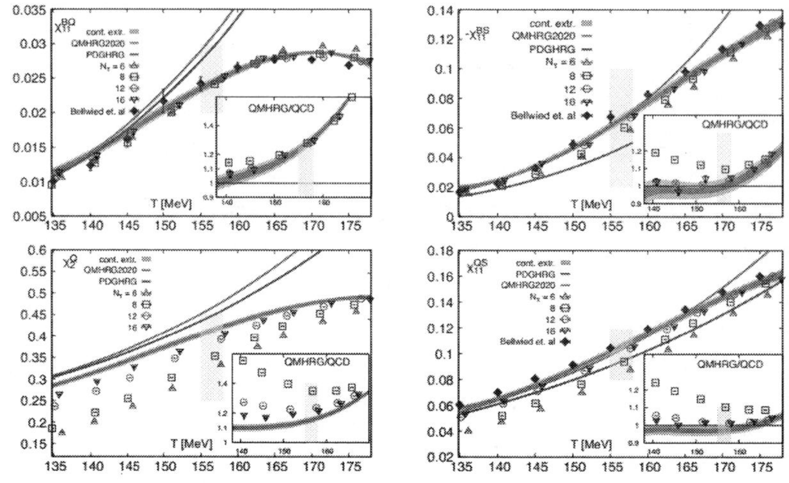

Figure 1. Four independent 2nd order cumulants versus temperature calculated on lattices with different temporal extent N_τ. Lines show results from HRG model calculations using point-like, non-interacting resonances and the hadron spectrum list QMHRG2020 [7]. Also shown are results from Bellwied et al. [6]. The insets show the ratio of continuum extrapolated lattice QCD results and HRG model calculations based on QMHRG2020. The yellow band corresponds to the crossover temperature, T_{pc}.

We also show results from model calculations that include additional interactions through excluded volume terms (EVHRG) and a virial expansion based formalism, respectively.

2 2nd order cumulants and QMHRG2020

In our calculations we have used highly improved staggered quarks (HISQ) with two light (u,d) and one strange (s) flavors. The pressure can be written as

$$P/T^4 = \frac{1}{VT^3} \ln Z(T, V, \vec{\mu}), \tag{1}$$

where $\vec{\mu} = (\mu_u, \mu_d, \mu_s)$ and Z is the QCD partition function,

$$Z = \int \mathcal{D}U \, \det[M(m_u, \mu_u)]^{1/4} \det[M(m_d, \mu_d)]^{1/4} \, \det[M(m_s, \mu_s)]^{1/4} \, e^{-S_G(U)} \tag{2}$$

The fluctuations and correlations at vanishing chemical potentials can be obtained as

$$\chi^{BQS}_{lmn} = \left. \frac{\partial P^{l+m+n}/T^4}{\partial(\mu_B/T)^l \partial(\mu_Q/T)^m \partial(\mu_S/T)^n} \right|_{\vec{\mu}=0}, \tag{3}$$

where derivatives with respect to conserved charge chemical potentials, (μ_B, μ_Q, μ_S), have been expressed in terms of the flavor chemical potentials $\vec{\mu}$ [5]. Since in (2+1)-lattice QCD calculations one uses degenerate light quark masses, $m_u = m_d$, the diagonal susceptibilities, χ_2^B and χ_2^S can be obtained using the relations $\chi_2^S = 2\chi_{11}^{QS} - \chi_{11}^{BS}$ and $\chi_2^B = 2\chi_{11}^{BQ} - \chi_{11}^{BS}$.

In Fig.(1) we present continuum extrapolations for four independent 2nd order cumulants, and a comparison with other lattice calculations is also shown [6]. In HRG model calculations with point like non-interacting resonance the non-strange baryons, strange baryons, pions and kaons provide the major contributions to χ_{11}^{BQ} , χ_{11}^{BS} , χ_2^Q and χ_{11}^{QS} respectively.

 https://doi.org/10.1051/epjconf/202225910010

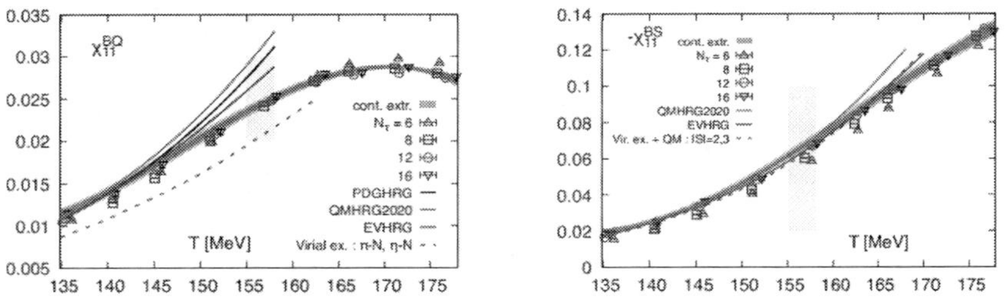

Figure 2. Continuum extrapolated results for χ_{11}^{BS} (left) and χ_{11}^{BQ} (right). Shown is a comparison with HRG model calculations based on QMHRG2020 as discussed in the text. Also shown are results obtained with excluded volume HRG models, using an excluded volume parameter $b = 1$ fm^3, and 2nd order virial expansions [13, 15], respectively.

We compare these observables with HRG model calculations based on only established resonances (PDGHRG) and it is clear that for all the 2nd order observables PDGHRG works poorly at $T > 140$ MeV. We define QMHRG2020 by including less established resonances from PDG tables (1-star and 2-star) and additional resonances from Quark Model (QM) calculations. In [7] QMHRG2020 is discussed in great detail. QMHRG2020 describes the lattice data of χ_{11}^{BS} and χ_{11}^{QS} quantitatively better than 1% in the temperature range $T < 156$ MeV indicating the abundance of strange particles in nature. However, at the same time the additional non-strange baryons included in QMHRG2020 have little effect on χ_{11}^{BQ} correlations. Although, QMHRG2020 provides better than 5% level agreement in the temperature range $T \leq 150$ MeV, it still exhibits an almost 20% deviation close to the pseudo-critical temperature, T_{pc}. Since T_{pc} [8] is close to the freeze-out temperature [9] measured by the ALICE collaboration, quantifying the agreement/deviation between HRG model calculations and QCD close to T_{pc} is the most relevant for understanding the thermal properties of the dense, strongly interacting medium created in heavy ion collision experiments.

Finally we note that electric charge fluctuations, χ_2^Q, deviate about 10% close to T_{pc} from HRG model calculations (see [7] for further discussion). Also the temperature derivatives of all cumulants calculated in HRG models are quite different from QCD at $T > 156$ MeV indicating that a hadronic description of the thermal medium does not seem to be appropriate for temperatures beyond T_{pc}.

3 Net baryon-electric charge and baryon-strangeness fluctuations at T_{pc}

Preliminary results for χ_{11}^{BQ} and χ_{11}^{BS} close to the T_{pc} have been presented by us already in [10]. In Fig.(2) (left), excluded volume HRG (EVHRG) calculations considering a hard sphere radius of $b = 1$ fm^3 [11, 12] are shown. It is evident that one needs $b > 1$ fm^3 for a proper description of χ_{11}^{BQ}. This, however is in conflict with the nice agreement obtained for χ_{11}^{BS}. Thus one needs at least two distinct b values for the non-strange baryons and strange baryons for a better agreement with lattice data. Furthermore, we also show the χ_{11}^{BQ} correlations calculated from 2nd virial coefficient [13], which falls below the Lattice data. For a better quantitative comparison, one needs information from higher order virial coefficients which is not known but has been modeled using LQCD calculations [14]. It is not shown here. The virial expansion based calculations of χ_{11}^{BS} works better but currently cannot treat multi-strange baryons which in Fig.(2) have been added as a non-interacting QMHRG contributions

[15]. Since, χ_2^Q and χ_{11}^{QS} are dominated by pions and kaons respectively, EVHRG doesn't play any significant role here.

4 Conclusions

The qualitative features of 2nd order cumulants of conserved charge fluctuations and correlations, calculated in lattice QCD, are reasonably well described by point like non-interacting HRG models up to the pseudo-critical temperature for the QCD transition. We also present calculation based on 2nd order virial expansions. We specify that for a better quantitative agreement of χ_{11}^{BQ} with Lattice QCD results higher order expansions are needed. We furthermore compared our lattice QCD results for 2nd order cumulants with model calculations that parametrize repulsive interactions among baryons and anti-baryons in a hadron resonance gas through a single excluded volume parameter (EVHRG). We point out that such an approach is not sufficient to describe all 2nd order cumulants simultaneously. At least independent excluded volume parameter for strange and non-strange baryons would be needed already for the description of 2nd order cumulants. Furthermore, all the HRG-based calculations seem to fail to describe lattice results for $T > T_{pc}$ already at the level of 2nd order cumulants.

Acknowledgements

This work was supported by the Deutsche Forschungsgemeinschaft (DFG) - project number 315477589 - TRR 211, the European Union H2020-MSCA-ITN-2018-813942 (EuroPLEx), and the U.S. Department of Energy, Office of Science, Office of Nuclear Physics through the Contract No. DE-SC0012704 and the Offices of Nuclear Physics and Advanced Scientific Computing Research within the framework of Scientific Discovery through Advance Computing award "Computing the Properties of Matter with Leadership Computing Resources".

References

[1] for recent reviews see: J. N. Guenther, Eur. Phys. J. A **57**, no.4, 136 (2021); O. Philipsen, PoS **LATTICE2019** (2019), 273 ; H. T. Ding, F. Karsch and S. Mukherjee, Int. J. Mod. Phys. E **24** (2015) no.10, 1530007

[2] P. A. Zyla *et al.* [Particle Data Group], PTEP **2020** (2020), 083C01

[3] S. Capstick and N. Isgur, Phys. Rev. D **34**, 2809 (1986) ; R. N. Faustov and V. O. Galkin, Phys. Rev. D **92**, no.5, 054005 (2015) [arXiv:1507.04530 [hep-ph]]

[4] D. Ebert, R. N. Faustov and V. O. Galkin, Phys. Rev. D **79** (2009), 114029

[5] A. Bazavov *et al.* [HotQCD], Phys. Rev. D **86**, 034509 (2012)

[6] R. Bellwied *et al.* Phys. Rev. D **101**, no.3, 034506 (2020)

[7] D. Bollweg *et al.* [arXiv:2107.10011 [hep-lat]]

[8] A. Bazavov *et al.* [HotQCD], Phys. Lett. B **795** (2019), 15

[9] J. Stachel, *et. al.*, J. Phys. Conf. Ser. **509** (2014), 012019

[10] J. Goswami *et. al.*, Acta Phys. Polon. Supp. **14**, 251 (2021)

[11] V. Vovchenko *et al.* Phys. Rev. Lett. **118** (2017) no.18, 182301;
V. Vovchenko, *et al.* Phys. Lett. B **775** (2017), 71-78

[12] J. M. Karthein, V. Koch, C. Ratti and V. Vovchenko, [arXiv:2107.00588 [nucl-th]].

[13] P. M. Lo *et. al.*, Phys. Lett. B **778** (2018), 454 [arXiv:1710.02711 [hep-ph]].

[14] A. Andronic *et. al.*, Phys. Lett. B **792**, 304-309 (2019)

[15] C. Fernández-Ramírez, P. M. Lo and P. Petreczky, Phys. Rev. C **98** (2018), 044910

Net-particle number fluctuations in a hydrodynamic description of heavy-ion collisions

Volodymyr Vovchenko[1,*], *Volker* Koch[1], and *Chun* Shen[2,3]

[1]Nuclear Science Division, Lawrence Berkeley National Laboratory, Berkeley, CA 94720, USA
[2]Department of Physics and Astronomy, Wayne State University, Detroit, Michigan 48201, USA
[3]RIKEN BNL Research Center, Brookhaven National Laboratory, Upton, NY 11973, USA

Abstract. Utilizing viscous hydrodynamic simulations of heavy-ion collisions, we study the behavior of cumulants of (net-)(anti)proton number distributions at RHIC beam energy scan energies, incorporating non-critical contributions like baryon conservation and excluded volume. The experimental data on net-proton cumulants at $\sqrt{s_{\rm NN}} > 20$ GeV are consistent with simultaneous effects of global baryon conservation and repulsive interactions in baryon sector, whereas the data at lower collision energies show possible indications for sizable attractive interactions among baryons. We discuss the behavior of factorial cumulants in addition to the ordinary cumulants, and also address the quantitative difference between proton and baryon number cumulants.

1 Introduction

As a sensitive probe of the QCD phase structure [1, 2], fluctuations of (net-)proton number are among the most valuable observables in the beam energy scan program at RHIC [3], as well as in other heavy-ion experiments [4–6]. The recent experimental data of the STAR Collaboration [7, 8] indicate a possible non-monotonic energy dependence of the net-proton kurtosis, which, if confirmed, may indicate presence of the QCD critical point [9].

A quantitative theoretical description of fluctuations is challenging. The bulk observables in heavy-ion collision are commonly well described by relativistic hydrodynamics [10–12]. Event-by-event fluctuations, however, are affected by several additional mechanisms like baryon number conservation [13, 14], the smearing of fluctuations due to momentum cuts [15], volume fluctuations [16, 17], and other effects. The two main issues mentioned above have recently been addressed at the LHC energies in Refs. [18], utilizing the approximate boost invariance of heavy-ion collisions at those energies. Here these considerations are extended to RHIC-BES energies.

2 Method

The details of our simulations setup can be found in Ref. [19]. We study 0-5% Au-Au collisions at various collision energies from the RHIC-BES program utilizing viscous hydrodynamics simulations within MUSIC [20] from Ref. [21]. The particlization of the QCD

*e-mail: vovchenko@lbl.gov

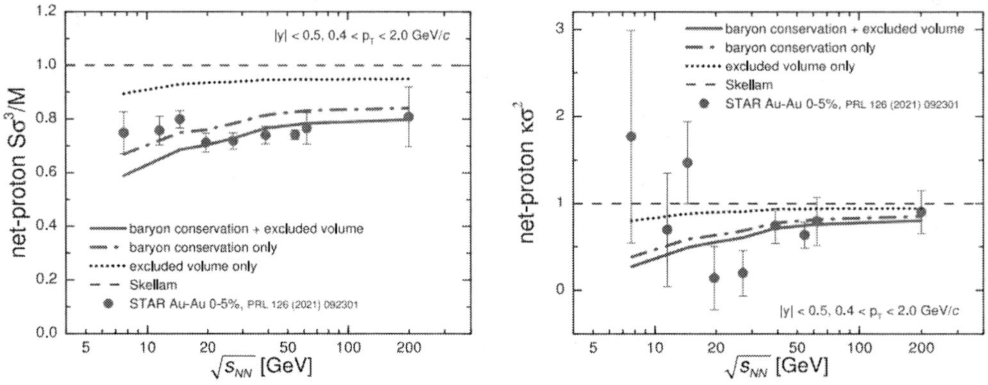

Figure 1. Collision energy dependence of the net-proton cumulant ratios $\kappa_3/\kappa_1 \equiv S\sigma^3/M$ (left) and $\kappa_4/\kappa_2 \equiv \kappa\sigma^2$ (right) in 0-5% Au-Au collisions at RHIC BES energies in a non-critical scenario compared with the STAR measurements [19, 24].

fluid takes place at a constant "switching" energy density of $\varepsilon_{\mathrm{sw}} = 0.26$ GeV/fm^3 [21]. The cumulants of proton and baryon number are calculated at particlization analytically via a generalized Cooper-Frye procedure which incorporates effects of the excluded volume and baryon conservation. This is performed in two steps:

1. The proton and baryon number cumulants within the experimental acceptance are calculated in the grand-canonical limit, i.e. neglecting the exact global conservation of baryon number.

2. The calculated cumulants are corrected for baryon conservation.

In the first step the emission of particles from each hypersurface can be calculated independently, given the grand-canonical limit of these calculations. The joint baryon-proton cumulants $\kappa_{n,m}^{B^\pm,p^\pm,\mathrm{gce}}(\Delta p_{\mathrm{acc}})$ of accepted particles correspond to the sum of contributions from each element, where each term is calculated by (i) calculating the distribution of emitted (anti)baryons using the grand-canonical susceptibilities of the excluded volume HRG model [22], then (ii) folding with the binomial distribution to model the momentum acceptance, and (iii) folding with another binomial distribution to separate protons from all baryons. The cumulants $\kappa_{n,m}^{B^\pm,p^\pm,\mathrm{gce}}(x_i; \overline{\Delta p_{\mathrm{acc}}})$ of particles outside the acceptance are calculated in the same way.

In the second step, a correction for baryon number conservation is performed. For this purpose we utilize a generalized subensemble acceptance method (SAM) of Ref. [23]. The SAM expresses the joint (net-)(anti-)proton number/net-baryon number cumulants inside the acceptance subject to global baryon conservation in terms of the grand-canonical joint cumulants both inside and outside the acceptance, i.e. it provides the mapping

$$\kappa_{n,m}^{B,p,\mathrm{ce}}(\Delta p_{\mathrm{acc}}) = \tilde{\mathcal{S}}\left[\kappa_{n,m}^{B,p,\mathrm{gce}}(\Delta p_{\mathrm{acc}}), \kappa_{n,m}^{B,p,\mathrm{gce}}(\overline{\Delta p_{\mathrm{acc}}})\right]. \qquad (1)$$

3 Results

3.1 Net proton cumulant ratios

Figure 1 shows the collision energy dependence of the net proton cumulant ratios $\kappa_3/\kappa_1 \equiv S\sigma^3/M$ and $\kappa_4/\kappa_2 \equiv \kappa\sigma^2$ in comparison with the experimental data of the STAR Collaboration [24]. Both the baryon conservation and excluded volume suppress the two ratios relative to the baseline of unity, and lead to a monotonic collision energy dependence. It is clear that baryon conservation has a larger influence at all energies compared to the excluded volume, however, both effects are necessary to obtain a quantitative description of the $S\sigma^3/M$ data at $\sqrt{s_{\rm NN}} \geq 20$ GeV. At lower energies the data indicate a smaller suppression of $S\sigma^3/M$ than predicted by the calculation. As for $\kappa\sigma^2$, the STAR data show possible indications for a non-monotonic collision energy dependence which is not observed in the baseline calculation, however, more precise data at the lowest energies are required to make a robust conclusion.

3.2 Factorial cumulants

In addition to the ordinary cumulants, we also explore the (anti)proton correlation functions (factorial cumulants) $\hat{C}_k$, which probe genuine multi-particle correlations and thus should be sensitive probes of the critical behavior [15]. We find that the high-order factorial cumulants, $\hat{C}_3/\hat{C}_1$ and $\hat{C}_4/\hat{C}_1$, indicate the presence of only mild multi-particle correlations among protons in the non-critical scenario. The second scaled factorial cumulant of protons, $\hat{C}_2/\hat{C}_1$, shown in Fig. 2, indicates sizable negative two-particle correlations. The calculation results agree with the experimental data at $\sqrt{s_{\rm NN}} \geq 20$ GeV but overestimate the strength of negative correlations at lower collision energies. This may either indicate sizable volume fluctuations at those energies, or presence of attractive interactions among protons.

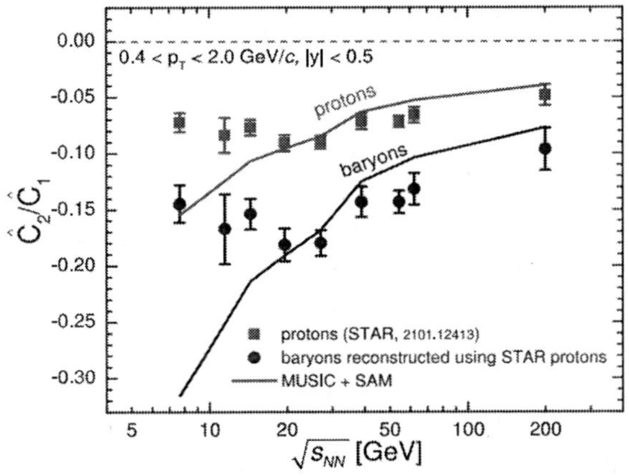

Figure 2. Collision energy dependence of the second scaled factorial cumulant $\hat{C}_2/\hat{C}_1$ of protons (red line) and baryons (black line) in 0-5% central Au-Au collisions. The red squares depict the experimental data of the STAR Collaboration for protons [8] while the black circles correspond to baryons reconstructed from the proton data using the binomial distribution.

3.3 Protons versus baryons

Our setup allows one to explore the quantitative differences between proton and baryon number fluctuations. It has been argued that baryon cumulants can be reconstructed from the measured proton ones via a binomial unfolding [25], which essentially corresponds to an additional efficiency correction. In particular, the second scaled factorial cumulant of baryons reads $\hat{C}_2^B/\hat{C}_1^B \approx 2\hat{C}_2^p/\hat{C}_1^p$, the corresponding experimental data of the STAR Collaboration based on this relation is shown in Fig. 2 by black symbols. The model calculations of two-particle baryon correlations agree well with the reconstructed STAR data for baryons at $\sqrt{s_{NN}} \geq 20$ GeV, underlying the importance of the unfolding procedure when performing comparisons to theoretical calculations of baryon fluctuations.

4 Summary

We analyzed proton number fluctuations within the hydrodynamic description of heavy-ion collisions in a non-critical scenario. The experimental data of the STAR Collaboration at $\sqrt{s_{NN}} > 20$ GeV are consistent with simultaneous effects of global baryon conservation and repulsive interactions in baryon sector while the data at lower collision energies show possible indications for sizable attractive interactions among baryons. Our calculations also reveal essential quantitative differences between proton and baryon number cumulants.

References

[1] M.A. Stephanov, K. Rajagopal, E.V. Shuryak, Phys. Rev. Lett. **81**, 4816 (1998), hep-ph/9806219
[2] M.A. Stephanov, K. Rajagopal, E.V. Shuryak, Phys. Rev. D **60**, 114028 (1999), hep-ph/9903292
[3] A. Bzdak, S. Esumi, V. Koch, J. Liao, M. Stephanov, N. Xu, Phys. Rept. **853**, 1 (2020), 1906.00936
[4] M. Gazdzicki, P. Seyboth, Acta Phys. Polon. B **47**, 1201 (2016), 1506.08141
[5] S. Acharya et al. (ALICE), Phys. Lett. B **807**, 135564 (2020), 1910.14396
[6] J. Adamczewski-Musch et al. (HADES), Phys. Rev. C **102**, 024914 (2020), 2002.08701
[7] J. Adam et al. (STAR), Phys. Rev. Lett. **126**, 092301 (2021), 2001.02852
[8] M. Abdallah et al. (STAR) (2021), 2101.12413
[9] M.A. Stephanov, Phys. Rev. Lett. **107**, 052301 (2011), 1104.1627
[10] C. Gale, S. Jeon, B. Schenke, Int. J. Mod. Phys. A **28**, 1340011 (2013), 1301.5893
[11] U. Heinz, R. Snellings, Ann. Rev. Nucl. Part. Sci. **63**, 123 (2013), 1301.2826
[12] C. Shen, L. Yan, Nucl. Sci. Tech. **31**, 122 (2020), 2010.12377
[13] A. Bzdak, V. Koch, V. Skokov, Phys. Rev. C **87**, 014901 (2013), 1203.4529
[14] V. Vovchenko, O. Savchuk, R.V. Poberezhnyuk, M.I. Gorenstein, V. Koch, Phys. Lett. B **811**, 135868 (2020), 2003.13905
[15] B. Ling, M.A. Stephanov, Phys. Rev. C **93**, 034915 (2016), 1512.09125
[16] M.I. Gorenstein, M. Gazdzicki, Phys. Rev. C **84**, 014904 (2011), 1101.4865
[17] V. Skokov, B. Friman, K. Redlich, Phys. Rev. C **88**, 034911 (2013), 1205.4756
[18] V. Vovchenko, V. Koch, Phys. Rev. C **103**, 044903 (2021), 2012.09954
[19] V. Vovchenko, V. Koch, C. Shen (2021), 2107.00163
[20] B. Schenke, S. Jeon, C. Gale, Phys. Rev. C **82**, 014903 (2010), 1004.1408
[21] C. Shen, S. Alzhrani, Phys. Rev. C **102**, 014909 (2020), 2003.05852
[22] V. Vovchenko, A. Pasztor, Z. Fodor, S.D. Katz, H. Stoecker, Phys. Lett. B **775**, 71 (2017), 1708.02852
[23] V. Vovchenko (2021), 2106.13775
[24] J. Adam et al. (STAR), Phys. Rev. Lett. **126**, 092301 (2021), 2001.02852
[25] M. Kitazawa, M. Asakawa, Phys. Rev. C **86**, 024904 (2012), [Erratum: Phys.Rev.C 86, 069902 (2012)], 1205.3292

Azimuthal anisotropy measurement of multi-strange hadrons in Au+Au collision at $\sqrt{s_{NN}}$ = 27 and 54.4 GeV at STAR

Prabhupada Dixit (for the STAR collaboration)[1,*]

[1]Indian Institute of Science Education and Research, Berhampur, 760010, India

Abstract. In these proceedings, we present the second order (v_2) and the third order (v_3) azimuthal anisotropy measurements of strange and multi-strange hadrons such as K^0_S, Λ, $\bar{\Lambda}$, ϕ, Ξ^-, $\bar{\Xi}^+$, Ω^-, $\bar{\Omega}^+$ in Au+Au collisions at $\sqrt{s_{NN}}$ = 27 and 54.4 GeV at STAR. The centrality dependence of v_2 and v_3 as well as the number of constituent quark scaling (NCQ scaling) for all the particles mentioned above are discussed. The ratio of v_2 of ϕ mesons to that of $\bar{p}$ is presented at $\sqrt{s_{NN}}$ = 54.4 GeV which gives a hint of mass ordering violation at low p_T.

1 Introduction

Relativistic heavy-ion collision is a unique tool to study matter at the extreme conditions of temperature and density. The collective expansion of the medium formed in heavy-ion collision is called flow. Depending on the initial spatial configuration of the colliding system and fluctuations, there can be the presence of different orders of anisotropy in the azimuthal distributions, known as anisotropic flow. In non-central nucleus-nucleus collisions, the overlap region of two colliding nuclei looks like the shape of an almond and in this region the pressure gradient is different along different axis of the system. The difference in pressure gradient leads to momentum anisotropy which is the dominant cause for the development of elliptic flow. But the event-by-event fluctuations of the nucleons in the colliding nuclei can give rise to other types of flow such as triangular flow, rectangular flow, etc. All these flow coefficients can be studied by using Fourier series expansion of the azimuthal distribution of particles given by

$$E\frac{d^3N}{dp^3} = \frac{1}{2\pi}\frac{d^2N}{p_T dp_T dy}\left[1 + \sum_n 2v_n \cos n(\phi - \Psi_R)\right]. \quad (1)$$

Here v_2 and v_3 represent elliptic flow coefficient and triangular flow coefficient respectively. The nth order flow coefficient is given by

$$v_n = \langle \cos n(\phi - \psi_n) \rangle. \quad (2)$$

Here ϕ is the azimuthal angle of the particles and ψ_n is the nth order event plane angle. The simultaneous measurement of v_2 and v_3 is important to constrain the shear viscosity over entropy density ratio (η/s) [1]. In this study, we have used strange and multi-strange hadrons because of their small hadronic interaction cross-section and early freeze-out [2], the measured flow coefficients remain least affected in the hadronic phase.

*e-mail: dixitprabhupada471@gmail.com

2 Analysis method

The nth order event plane is given by

$$\psi_n = \frac{1}{n}\tan^{-1}\Big(\frac{\sum_i w_i \sin(n\phi_i)}{\sum_i w_i \cos(n\phi_i)}\Big), \tag{3}$$

where ϕ_i is the azimuthal angle of the particle track and w_i is the weight factor equal to p_T $\times$ ϕ-weight. The factor ϕ-weight is the correction factor for azimuthal acceptance of the detector. Due to the finite resolution of the event plane, the v_n needs to be corrected by event plane resolution given by

$$R_n = \sqrt{\langle \cos n(\psi_A - \psi_B)\rangle}. \tag{4}$$

Here ψ_A and ψ_B are two sub-event planes in -1< η <-0.05 and 0.05 < η <1 regions respectively. In Eq. 2, the particle's ϕ and the event plane ψ_n are taken from opposite η region.

Invariant mass method [3] is used to calculate v_n. A detailed procedure to apply this method can be found in Ref. [8].

3 Results and discussion

Figure 1 shows v_2 and v_3 plotted as a function of p_T for K^0_S, Λ, ϕ, Ξ^-, Ω^- at $\sqrt{s_{NN}}$ = 54.4 GeV in 0-80% centrality. In low p_T both v_2 and v_3 show a mass ordering which is due to the radial flow of the system [4], but at p_T above 2 GeV/c a separation between baryons and mesons is observed which can be explained by the quark recombination model of hadronization [5]. A

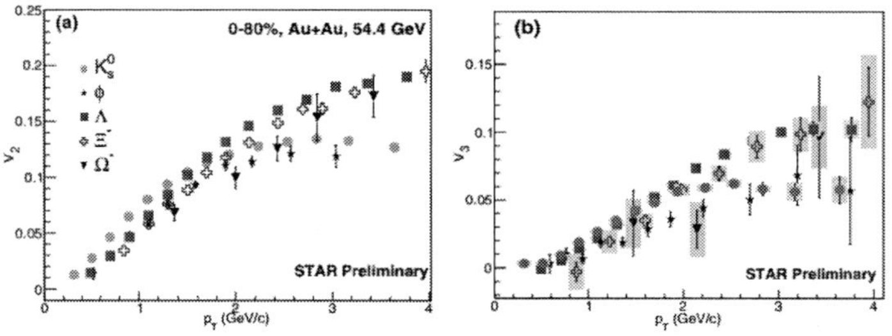

Figure 1. Panel-a shows v_2 and panel-b shows v_3 as a function of p_T for K^0_S, Λ, ϕ, Ξ^-, Ω^- in 0-80% centrality. The vertical lines represent the statistical error bars and the shaded boxes represent the systematic error bars.

centrality dependence study for both v_2 and v_3 is shown in Fig. 2 and Fig. 3 in three different centrality classes, 0-10%, 10-40% and 40-80%. The elliptic flow v_2 shows a strong centrality dependence because of initial spatial anisotropy dominance, whereas v_3 does not because it arises mostly due to event-by-event fluctuations in the initial state.

The NCQ scaled v_2 is studied as a function of NCQ scaled transverse kinetic energy for 0-80% centrality at $\sqrt{s_{NN}}$ = 27 and 54.4 GeV. The NCQ scaled v_2 for all the particles shown in Fig. 4 is found to fall on a single curve which suggests that the collectivity is developed during the QGP stage of the collision. The same result is found for v_3 also as shown in Fig. 5.

The ratio of ϕ meson v_2 to $\bar{p}$ v_2 is calculated as a function of p_T as shown in Fig. 6. Since the mass of the ϕ meson (1.019 GeV/c^2) is greater than that of $\bar{p}$ (0.938 GeV/c^2), according to mass ordering at low p_T the ratio should have been less than one. But we observe a hint of mass ordering violation at low p_T which can be explained by larger hadronic rescattering effect on $\bar{p}$ compared to ϕ meson [6]. A similar result is found in previous STAR measurement in minimum biased Au+Au collisions at $\sqrt{s_{NN}}$ = 200 GeV [7].

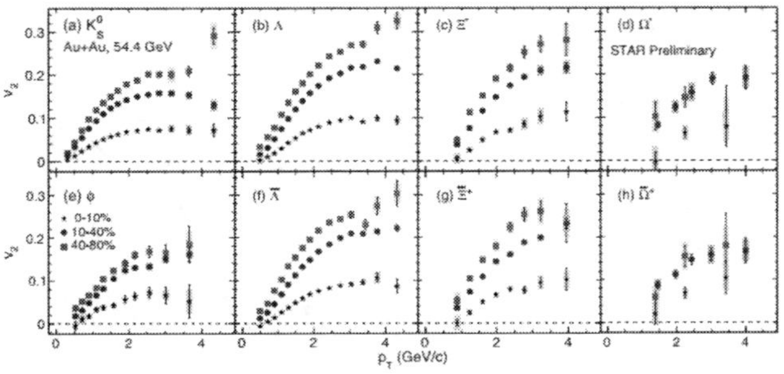

Figure 2. v_2 is plotted as a function of p_T for 0-10%, 10-40% and 40-80% centrality classes. The vertical lines represent the statistical error bars and the shaded boxes represent the systematic error bars.

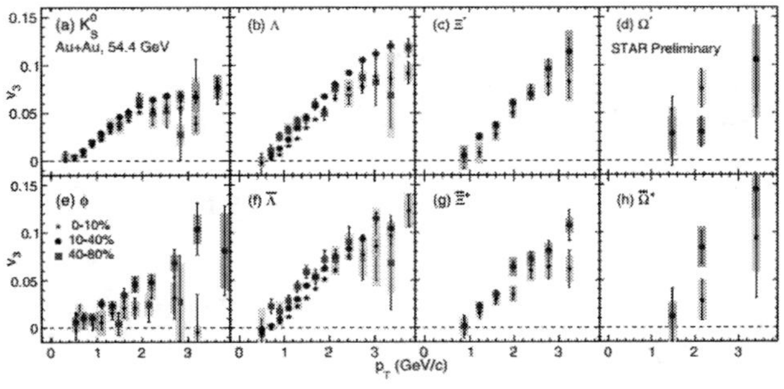

Figure 3. v_3 is plotted as a function of p_T for 0-10%, 10-40% and 40-80% centrality classes. The vertical lines represent the statistical error bars and the shaded boxes represent the systematic error bars.

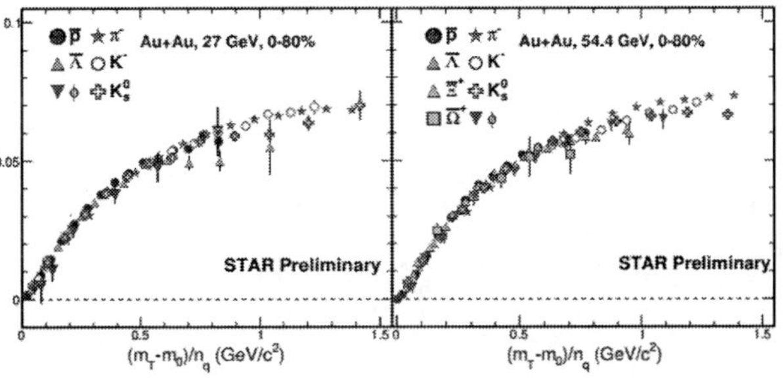

Figure 4. NCQ scaled v_2 is plotted as a function of NCQ scaled transverse kinetic energy for 0-80% centrality for $\sqrt{s_{NN}}$ = 27 (left) and 54.4 GeV (right).

4 Summary

Measurements of v_2 and v_3 as a function of p_T of strange and multi-strange hadrons, K^0_S, Λ, $\bar{\Lambda}$, ϕ, Ξ^-, $\bar{\Xi}^+$, Ω^-, $\bar{\Omega}^+$ is presented in Au+Au collisions at midrapidity for 0-80% centrality. A mass ordering is observed in the low p_T. At high p_T (> 2 GeV/c), we observe a baryon to meson separation. A strong centrality dependence is observed for v_2, but the centrality dependence is absent for v_3. The NCQ scaling is found to hold for all the particles at $\sqrt{s_{NN}}$

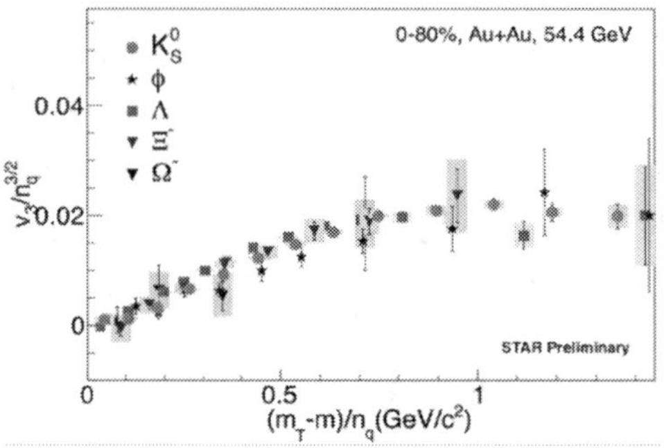

Figure 5. NCQ scaled v_3 is plotted as a function of NCQ scaled transverse kinetic energy for 0-80% centrality for $\sqrt{s_{NN}}$ = 54.4 GeV. The vertical lines represent the statistical error bars and the shaded boxes represent the systematic error bars.

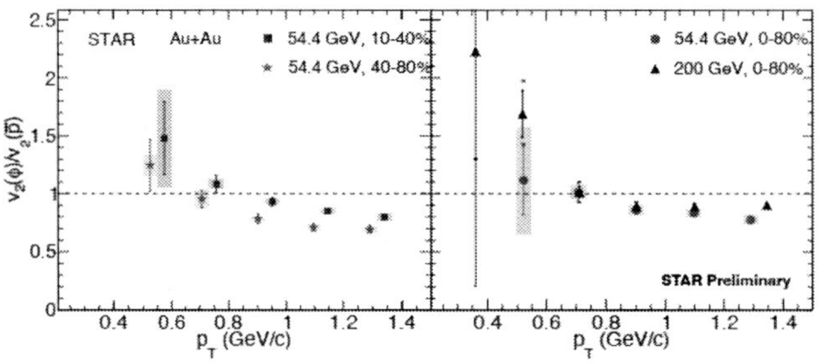

Figure 6. The left panel shows the ratio of ϕ meson v_2 to $\bar{p}$ v_2 in Au+Au collision at $\sqrt{s_{NN}}$ = 54.4 GeV in 10-40% and 40-80% centralities. The right panel shows the comparison of the ratio with 200 GeV data points for minimum bias events. The vertical lines represent the statistical error bars and the shaded boxes represent the systematic error bars.

= 27 and 54.4 GeV. The ratio of v_2 of ϕ meson to that of $\bar{p}$ gives a hint of mass ordering violation at low p_T.

References

[1] Chun Shen et al., J. Phys. G: Nucl. Part. Phys. **38**, 124045 (2011)
[2] A. Shor, Phys. Rev. Lett. **54**, 1122 (1985)
[3] N. Borghini and J.-Y. Ollitrault, Phys. Rev. C **70**, 064905 (2004)
[4] D. Teaney, J. Lauret, and E. V. Shuryak, Phys. Rev. Lett. **86**, 4783 (2001)
[5] D. Molnar and S. A. Voloshin, Phys. Rev. Lett. **91**, 092301 (2003)
[6] T. Hirano et al., Phys. Rev. C **77**, 044909 (2008)
[7] L. Adamczyk et al. (STAR), Phys. Rev. Lett. **116**, 062301 (2016)
[8] P. Dixit, arXiv:2103.04646 [hep-ex]

NCQ scaling of $f_0(980)$ elliptic flow in 200 GeV Au+Au collisions by STAR and its constituent quark content

Jie Zhao[1,⋆] (for the STAR collaboration)

[1] *Department of Physics and Astronomy, Purdue University*

Abstract. Searching for exotic state particles and studying their properties have furthered our understanding of quantum chromodynamics (QCD). The $f_0(980)$ resonance is an exotic state with relatively high production rate in relativistic heavy-ion collisions, decaying primarily into $\pi\pi$. Currently the structure and quark content of the $f_0(980)$ are unknown with several predictions from theory being a $q\overline{q}$ state, a $qq\overline{q}\overline{q}$ state, a $K\overline{K}$ molecule state, or a gluonium state. We report the first $f_0(980)$ elliptic flow (v_2) measurement from 200 GeV Au+Au collisions at STAR. The transverse momentum dependence of v_2 is examined and compared to those of other hadrons (baryons and mesons). The empirical number of constituent quark (NCQ) scaling is used to investigate the constituent quark content of $f_0(980)$, which may potentially address an important question in QCD.

1 Introduction

Searching for exotic state particles and studying their properties have furthered our understanding of quantum chromodynamics (QCD). Currently the structure and quark content of $f_0(980)$ are unknown with several predictions being a $q\overline{q}$ state, a $qq\overline{q}\overline{q}$ state, a $K\overline{K}$ molecule state, or a gluonium state [1–6]. In contrast to the vector and tensor mesons, the identification of the scalar mesons is a long-standing puzzle [7]. Previous preliminary experimental measurements [8] on the yield of $f_0(980)$ at RHIC and theoretical calculation [9] suggest that it could be a $K\overline{K}$ stat. In this analysis, the empirical number of constituent quark (NCQ) scaling [10–12] is used to investigate the constituent quark content of $f_0(980)$ [13].

2 Experiment setup and data analysis

The data reported here are from Au+Au collisions at a nucleon-nucleon center-of-mass energy of 200 GeV, collected by the STAR experiment [14] at Brookhaven National Laboratory in 2011, 2014 and 2016. A total of 2.4 billion minimum-bias (MB) events are selected for this analysis. The main subsystem used for the data analysis is the Time Projection Chamber (TPC) [15] with 2π azimuthal coverage at mid-rapidity. The TPC dE/dx is used to select $\pi^{\pm}$ candidate with $0.2 < p_T < 5.0$ GeV/c.

The $\pi^+\pi^-$ are used to reconstruct the $f_0(980)$. The combinatorial background subtraction is based on the mixed-event technique and the like-sign method [16]. The acceptance-corrected like-sign

⋆e-mail: zhao656@purdue.edu

pairs [16, 17] are used to subtract the combinatorial background after being normalized to unlike-sign pairs in the invariant mass (m_{inv}) range beyond 1.5 GeV/c^2 . Figure 1 (left) shows the background subtracted $\pi^+\pi^-$ invariant mass distribution. The resonance peaks are parametrized with the relativistic Breit-Wigner function [18, 19]. The total fit function is given by:

$$f(m_{inv}) = \left(\sum_{X=f_0,\rho^0,f_2} \frac{A_X m_{inv} m_X \Gamma(X)}{(m_{inv}^2 - m_X^2)^2 + m_X^2 \Gamma(X)^2} \right) \times PS + bg(m_{inv}) \tag{1}$$

where $\Gamma(X) = \frac{\Gamma_X m_X}{m_{inv}} \left(\frac{m_{inv}^2 - 4m_\pi^2}{m_X^2 - 4m_\pi^2} \right)^{J+1/2}$ [18, 19], $PS = \frac{m_{inv}}{\sqrt{m_{inv}^2 + p_T^2}} \exp\left(-\frac{\sqrt{m_{inv}^2 + p_T^2}}{T} \right)$ is the phase space correction taking into account the $\pi\pi$ scattering during the hadronic phase [19–22], and $bg(m_{inv})$ is a third order polynomial function to describe the residual background. m_X and Γ_X are the mass and width of the corresponding resonances. Γ_{ρ^0} is set to 160 MeV, and m_{f_2} and Γ_{f_2} are set according to the PDG values [7]. T is the kinetic freeze-out temperature, set to 120 MeV [20]. A_{f_0}, A_{ρ^0}, A_{f_2}, m_{f_0}, Γ_{f_0}, and m_{ρ^0} are free parameters.

The event-plane method [23] is used to study the elliptic flow (v_2) of $f_0(980)$. The event-plane is reconstructed by all charged particles in the TPC with pseudorapidity $|\eta| < 1$ and transverse momentum $0.2 < p_T < 5.0$ GeV/c. For each $\pi\pi$ pair, the two π candidates are removed from the event-plane reconstruction to avoid auto-correlation. The event-plane resolution is calculated by the correlation between two randomly divided sub-events from the full TPC [23]. Wide centrality bin effect is corrected by weighting the event-plane resolution with the $f_0(980)$ yield in each narrow centrality bin of 10% size [24]. Figure 1(right) shows the $f_0(980)$ yield as function of the azimuthal angle difference between the $\pi\pi$ pair (ϕ) and the event-plane direction (Ψ) in an example p_{T} bin.

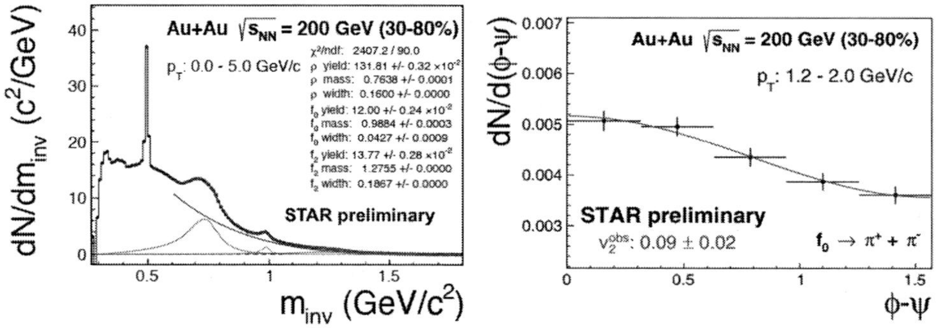

Figure 1. (Color online) (Left) The background subtracted $\pi^+\pi^-$ invariant mass distribution over the p_T range of $0 < p_T < 5.0$ GeV/c in 30–80% Au+Au collisions at $\sqrt{s_{NN}}$=200 GeV. The red line is the result of fit. The pink, green, violet lines represent the resonance peaks of the relativistic Breit-Wigner function. The solid blue line represents the residual background using a third order polynomial function. (Right) $f_0(980)$ yield as function of $\phi - \Psi$ in a given p_T bin. Errors are statistical. The red line represents a fit ($\propto (1 + 2v_2^{obs}\cos(2\phi - 2\Psi))$) to the data.

Figure 2 shows $f_0(980)$ v_2 as a function of p_{T} in 30-80% centrality Au+Au collisions at $\sqrt{s_{NN}}$ = 200 GeV. Results are compared with other identified particles: $\pi, K, p, K_s^0, \Lambda, \Xi, \Omega, \phi$ [24]. In the low p_{T} region, the $f_0(980)$ v_2 seems to follow the mass ordering. In the higher p_{T} region, the $f_0(980)$ v_2 seems closer to the baryon band.

Figure 3 shows the number of constituent quark (n_q) scaled v_2 as function of the n_q scaled p_{T} (left) and $m_T - m_0$ (right). Here the $f_0(980)$ is assumed to have either 2 quarks or 4 quarks. The data are

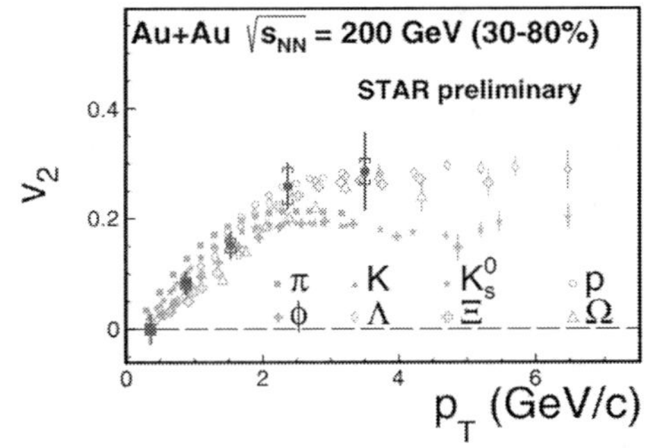

Figure 2. (Color online) $f_0(980)$ v_2 as a function of $p_{\rm T}$ in 30-80% centrality Au+Au collisions at $\sqrt{s_{NN}}$ = 200 GeV. Statistical uncertainties are shown by the vertical bars and systematic uncertainties are shown by the caps. Results of other particles are taken from Ref. [24].

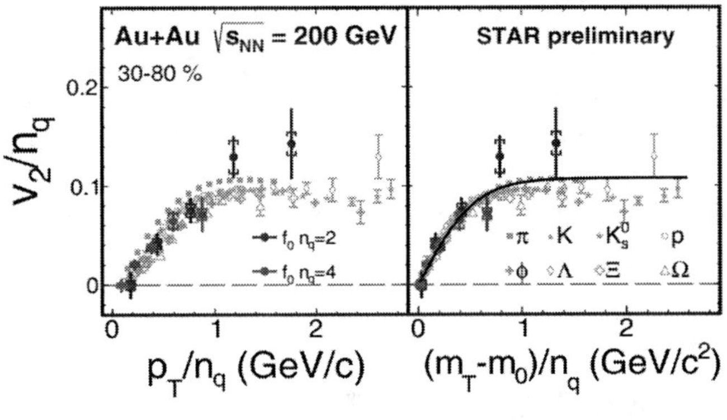

Figure 3. (Color online) $f_0(980)$ v_2 divided by n_q as a function of $p_{\rm T}/n_q$ (left) and $(m_T - m_0)/n_q$ (right) in 30-80% centrality Au+Au collisions at $\sqrt{s_{NN}}$ = 200 GeV. Results of other particles are taken from Ref. [24]. Black line in the right panel represents a fit to results of other particles using a NCQ scaling inspired function (Eq. 2).

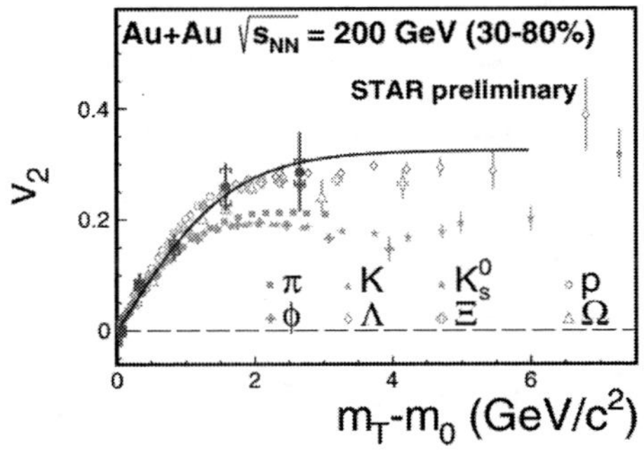

Figure 4. (Color online) $f_0(980)$ v_2 as a function of $m_T - m_0$ in 30-80% centrality Au+Au collisions at $\sqrt{s_{NN}}$ = 200 GeV. The blue curve represents the NCQ inspired fit (Eq. 2), where the only free parameter is the n_q of $f_0(980)$ and all other parameters are fixed according to the fit in the right panel of the Fig. 3.

compared to the fit of other particles [24] using a NCQ scaling inspired function [25]:

$$f_{v_2}(n_q) = \frac{a n_q}{1 + \exp(-((m_T - m_0)/n_q - b)/c)} - d n_q. \tag{2}$$

The 2-quarks (4-quarks) scaled $f_0(980)$ v_2 seems to deviate from the fit, above (below) the fit by $\sim 1\sigma$ for the last one or two points at high $(m_T - m_0)/n_q$.

Figure 4 shows $f_0(980)$ v_2 as a function of $m_T - m_0$ with a fit according to the function shown in Eq. 2. In the fit, only the n_q of $f_0(980)$ is treated as a free parameter and all other parameters are fixed according to the fit in the right panel of the Fig. 3. This NCQ scaling fit of the $f_0(980)$ v_2 yields $n_q = 3.0 \pm 0.7$ (stat) $\pm$ 0.5 (syst).

With the current uncertainty, our result is not able to determine whether $f_0(980)$ is a $q\bar{q}$, $qq\bar{q}\bar{q}$, $K\bar{K}$ molecule, gluonium state, or produced through $\pi\pi$ coalescence. It could also be given by some combined states as well. Future measurements, e.g. the $f_0(980)$ yields, could also provide different aspect to understand it.

3 Summary

Preliminary results on the $f_0(980)$ v_2 in 30-80% centrality Au+Au collisions at $\sqrt{s_{\rm NN}}$ = 200 GeV are presented. In the low $p_{\rm T}$ region ($p_{\rm T}$<2 GeV/c), the $f_0(980)$ v_2 seems to follow the mass ordering. In the higher $p_{\rm T}$ region ($p_{\rm T}$>2 GeV/c), the $f_0(980)$ v_2 seems closer to the baryon band. A NCQ scaling inspired function was used to fit the $f_0(980)$ v_2. The extracted quark content of $f_0(980)$ is n_q = 3.0 ± 0.7 (stat) ± 0.5 (syst). More data are needed to understand whether $f_0(980)$ is a $q\bar{q}$, $qq\bar{q}\bar{q}$, $K\bar{K}$ molecule, gluonium state, or produced through $\pi\pi$ coalescence. Our study indicates that heavy-ion collisions can be a useful place to examine the quark content of scalar mesons. The isobar data taken in 2018 at RHIC and the 8-fold increase in Au+Au data expected in 2023-2025 would provide more insights.

Acknowledgments This work was partly supported by the U.S. Department of Energy (Grant No. de-sc0012910).

References

[1] J.D. Weinstein, N. Isgur, Phys. Rev. D **27**, 588 (1983)
[2] J.D. Weinstein, N. Isgur, Phys. Rev. D **41**, 2236 (1990)
[3] F. Kleefeld, E. van Beveren, G. Rupp, M.D. Scadron, Phys. Rev. D **66**, 034007 (2002)
[4] V. Baru et al., Phys. Lett. B **586**, 53 (2004)
[5] M. Ablikim et al. (BESIII), Phys. Rev. D **92**, 052003 (2015)
[6] N.N. Achasov et al., Phys. Rev. D **103**, 014010 (2021)
[7] P.A. Zyla et al. (Particle Data Group), Prog. Theor. Exp. Phys. **083C01** (2020)
[8] P. Fachini, J. Phys. G **30**, S735 (2004)
[9] S. Cho et al. (ExHIC), Phys. Rev. Lett. **106**, 212001 (2011)
[10] D. Molnar, S.A. Voloshin, Phys. Rev. Lett. **91**, 092301 (2003)
[11] S.S. Adler et al. (PHENIX), Phys. Rev. Lett. **91**, 182301 (2003)
[12] J. Adams et al. (STAR), Phys. Rev. Lett. **92**, 052302 (2004)
[13] A. Gu, T. Edmonds, J. Zhao, F. Wang, Phys. Rev. C **101**, 024908 (2020)
[14] K.H. Ackermann et al. (STAR), Nucl. Instrum. Meth. **A499**, 624 (2003)
[15] M. Anderson et al., Nucl. Instrum. Meth. **A499**, 659 (2003)
[16] L. Adamczyk et al. (STAR), Phys. Rev. C **92**, 024912 (2015)
[17] L. Adamczyk et al. (STAR), Phys. Rev. Lett. **113**, 022301 (2014)
[18] C. Adler et al. (STAR), Phys. Rev. Lett. **89**, 272302 (2002)
[19] J. Adams et al. (STAR), Phys. Rev. Lett. **92**, 092301 (2004)
[20] E.V. Shuryak, G.E. Brown, Nucl. Phys. A **717**, 322 (2003)
[21] P.F. Kolb, M. Prakash, Phys. Rev. C **67**, 044902 (2003)
[22] R. Rapp, Nucl. Phys. A **725**, 254 (2003)
[23] A.M. Poskanzer, S.A. Voloshin, Phys. Rev. C **58**, 1671 (1998)
[24] L. Adamczyk et al. (STAR), Phys. Rev. Lett. **116**, 062301 (2016)
[25] X. Dong, S. Esumi, P. Sorensen, N. Xu, Z. Xu, Phys. Lett. B **597**, 328 (2004)

https://doi.org/10.1051/epjconf/202225910014

Production of light nuclei in small collision systems with ALICE

Chiara Pinto[1,2,*] on behalf of the ALICE Collaboration

[1]Dipartimento di Fisica e Astronomia "E. Majorana", Università degli Studi di Catania, Via S. Sofia 64, Catania, Italy
[2]INFN Sezione di Catania, Via S. Sofia 64, Catania, Italy

Abstract. The energy densities reached in high-energy hadronic collisions at the LHC allow significant production of light (anti)nuclei. Their production yields have been measured as a function of p_T and charged-particle multiplicity in different collision systems and at different center-of-mass energies by ALICE. One of the most interesting results obtained from such a large variety of experimental data is that the dominant production mechanism of light (anti)nuclei seems to depend solely on the event charged-particle multiplicity. Evidence for this comes from the continuous evolution of the deuteron-to-proton and ^{3}He-to-proton ratios with the event multiplicity across different collision systems and energies. The characterization of the light nuclei production mechanism is complemented by measurements of their production yields in jets and in the underlying event. In this paper, recent results on light nuclei production in small collision systems are shown and discussed in the context of the statistical hadronization and coalescence models. In addition, recent results on the deuteron production in jets and new preliminary results on its production in the underlying event measured in pp collisions at $\sqrt{s}$ = 13 TeV are discussed.

1 Physics motivation

The production mechanism of light (anti)nuclei is under intense debate in the heavy-ion physics community. Currently, the measured production yields can be described by two classes of models: the thermal-statistical models and the coalescence ones. In the Statistical Hadronization Model (SHM) [1], hadrons are produced by a thermally and chemically equilibrated source and their abundances are fixed at the chemical freeze-out. This model provides a good description of the measured hadron yields in central A–A collisions [2]. However, the mechanism of hadron production and the propagation of loosely-bound states through the hadron gas phase are not addressed by this model. On the other hand, the production of light (anti)nuclei can be modelled via the coalescence of protons and neutrons that are close by in phase space at the kinetic freeze-out and match the spin, thus forming a nucleus [3]. The key parameter of the coalescence models is the coalescence parameter B_A, which is related to the production probability of the nucleus via this process and can be calculated from the overlap of the nucleus wave function and the phase space distribution of the constituents via the Wigner formalism [4].

*e-mail: chiara.pinto@cern.ch

2 Ratio of nucleus and proton integrated yields

Light (anti)nuclei are identified using the detectors of the central barrel, which cover the pseudorapidity window $|\eta| < 0.9$. Specifically, the Time Projection Chamber (TPC) specific energy loss ($\mathrm{d}E/\mathrm{d}x$) allows the separation from other particles of deuterons in the low transverse momentum region and of nuclei with Z=2 in the full p_{T}. The use of the Time of Flight (TOF) information complements the light nuclei identification in the high p_{T} region. In order to extract the light (anti)nucleus integrated yields, p_{T} spectra are extrapolated to the unmeasured regions by means of a fit with a Lévy-Tsallis function [5].
The ratio between the measured yields of nuclei and that of protons is sensitive to the particle production mechanism. In Fig. 1 the ratio between deuteron (left panel), ^{3}H and ^{3}He (right panel) measured yields and proton yields as a function of the mean charged-particle multiplicity density ($\langle \mathrm{d}N_{\mathrm{ch}}/\mathrm{d}\eta_{\mathrm{lab}} \rangle$) measured in pp, p–Pb and Pb–Pb collisions [6–11] is compared to the expectations of the models. These nuclei-to-proton yield ratios increase smoothly with the multiplicity, reaching constant values in Pb–Pb collisions. The two ratios show a similar trend with $\langle \mathrm{d}N_{\mathrm{ch}}/\mathrm{d}\eta_{\mathrm{lab}} \rangle$, suggesting that the production mechanism depends solely on the event charged-particle multiplicity density. The observed evolution of the d/p ratio is well described by the coalescence approach across all multiplicities. For high $\langle \mathrm{d}N_{\mathrm{ch}}/\mathrm{d}\eta_{\mathrm{lab}} \rangle$, the coalescence calculations and the canonical statistical model (CSM) expectations are close and both describe the behavior observed in data. On the other hand, the ratio to protons for nuclei with $A = 3$ is described only qualitatively by both models. Therefore, in the current state it is not possible to discern between the production mechanisms of light nuclei. In order to further investigate the phenomena underlying the production of light nuclei, it is interesting to study the small system properties by means of new observables, such as the underlying event (UE) activity [12].

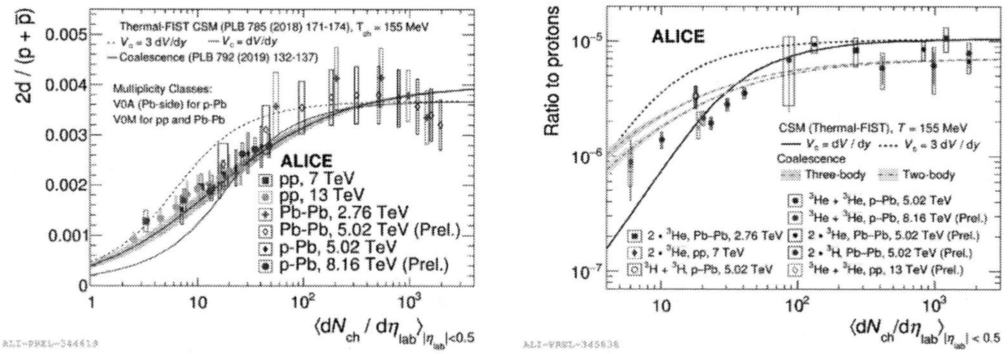

Figure 1. Deuteron (left panel), (anti)^{3}He and (anti)^{3}H (right panel) over proton integrated yield ratios as a function of the mean charged-particle multiplicity density for different collision systems and energies. Lines show the statistical uncertainties whereas boxes represent the systematic ones. The results are compared to the expectations of SHM and coalescence models described in the legends.

3 Underlying Event activity

ALICE recently released a new analysis performed on pp data collected at a center-of-mass energy $\sqrt{s} = 13$ TeV, aiming at understanding the possible role played by the underlying event activity in the light nuclei production. The transverse plane is divided into three azimuthal regions, identified by the relative position with respect to the highest-p_{T} hadron (trigger particle). The three equal-size regions, $\pi/3$ wide, are: the one around the trigger particle (towards), the one back-to-back to it (away), and the one transverse to both of them, where the UE contribution is dominant. The activity in the UE can be quantified using the self-normalised charged-particle multiplicity in the transverse region $R_\mathrm{T} = \frac{N_{ch,\mathrm{T}}}{\langle N_{ch,\mathrm{T}} \rangle}$ [12]. Only events with at least one leading charged particle with $p_\mathrm{T} \geq 5$ GeV/c are considered, since the particle density in the transverse region reaches a plateau at LHC energies. Small R_T values correspond to an event topology dominated by the hard scattering, while large R_T values correspond to events characterized by large UE activity.

4 Coalescence parameters

The coalescence parameter B_A is given by the ratio between the invariant yield of the nucleus with mass number A and that of protons, defined as:

$$B_A = \left(\frac{1}{2\pi p_\mathrm{T}^A}\frac{\mathrm{d}^2 N_A}{\mathrm{d}y\mathrm{d}p_\mathrm{T}^A}\right) \Bigg/ \left(\frac{1}{2\pi p_\mathrm{T}^\mathrm{p}}\frac{\mathrm{d}^2 N_\mathrm{p}}{\mathrm{d}y\mathrm{d}p_\mathrm{T}^\mathrm{p}}\right)^A, \tag{1}$$

where the labels A and p indicate the nucleus and the proton, respectively, and $p_\mathrm{T}^\mathrm{p} = p_\mathrm{T}^A/A$.

The coalescence parameter B_A has been measured as a function of p_T/A in different collision systems. In Fig. 2, B_2 in pp collisions at $\sqrt{s} = 13$ TeV and B_3 in p–Pb collisions at $\sqrt{s_\mathrm{NN}} = 5.02$ TeV are shown for several multiplicity classes and in the minimum bias (MB) case. B_2 in pp collision is flat with p_T/A in all multiplicity classes, in agreement with the predictions of a simple coalescence model, which takes into account only momentum space distributions and not space-time correlations [13]. The rising trend of B_A observed in the MB case can be explained as a consequence of the hardening of the proton spectra with increasing multiplicity in addition to hard scattering effects at high p_T. B_3 in p–Pb collisions is slightly increasing with p_T and it is known from previous measurements that the B_A parameters in Pb–Pb collisions also increase as a function of transverse momentum [6]. The behavior of B_3 with p_T/A in p–Pb collisions cannot be explained by simple coalescence hypotheses, as investigated in Ref. [11]. Therefore, the observed results, in addition to what is known from earlier measurements, suggest that more sophisticated coalescence models have to take into account the volume dependence in order to explain the data.

In Fig. 3, B_2 measured in pp collisions at $\sqrt{s} = 13$ TeV in the transverse and towards regions, for several R_T classes, are compared with the expectations of Pythia 8.3 simulations. The B_2 parameters as a function of R_T are flat with p_T/A in all the azimuthal regions, in agreement with the predictions of a simple coalescence picture. However, similar values of B_2 are found in the transverse and towards regions, against any naive expectations that would expect a larger B_2 in the towards region, where nucleons are closer due to the presence of a jet, with respect to what happens in the transverse region. Pythia 8.3 simulations include deuteron production via coalescence considering the reaction cross sections [14]. The p_T-dependence and the R_T ordering are well reproduced by Pythia simulations. However, Pythia fails in reproducing the magnitude of the coalescence parameter, therefore further tuning of the Monte Carlo parameters are needed to reproduce the data.

The obtained results suggest that deuterons are mostly produced in the underlying event. The B_2 values are close in all azimuthal regions indicating that the predominant contribution in the deuteron production is due to the underlying event particle production. In support of this argument, the recent results of deuteron production in jets [15] have shown that the fraction of deuterons produced in the jet is 8–15%, increasing with increasing p_T, while the majority of the deuterons are produced in the underlying event.

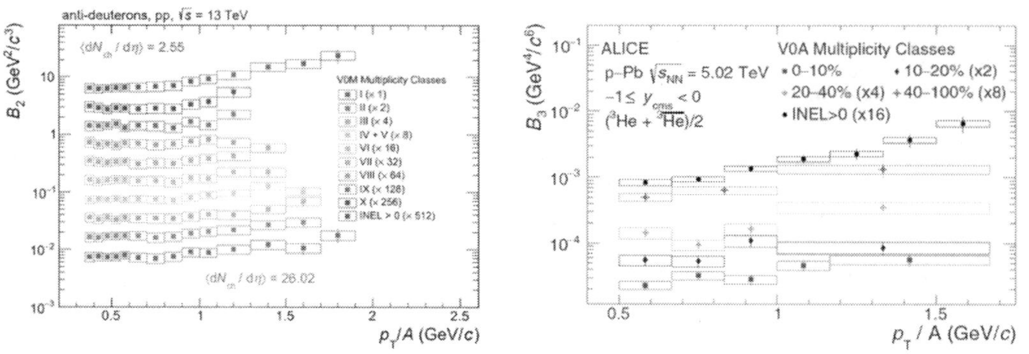

Figure 2. B_2 as a function of p_T/A in pp collisions at $\sqrt{s}$ = 13 TeV (left panel) [10] and B_3 in p–Pb collisions at $\sqrt{s_{NN}}$ = 5.02 TeV (right panel) [11].

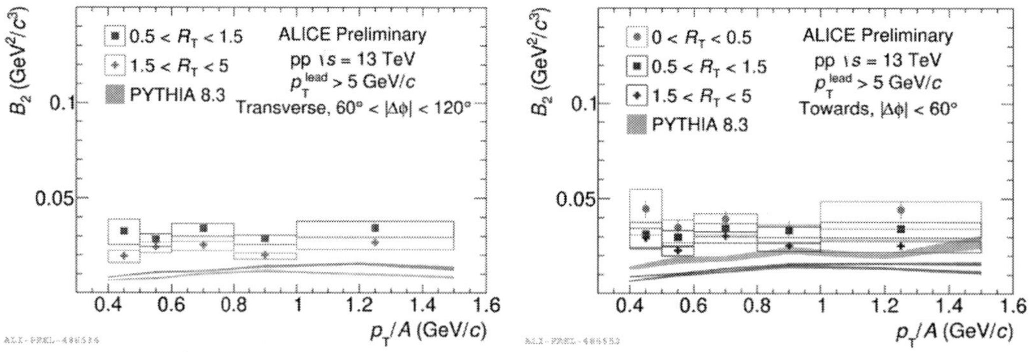

Figure 3. B_2 as a function of p_T/A in pp collisions at $\sqrt{s}$ = 13 TeV, for several R_T classes, in the transverse region on the left and in the towards region on the right.

References

[1] A.Andronic, P.Braun-Munzinger, J.Stachel, H.Stöcker, Phys.Lett.B 697 203-207(2011).
[2] A.Andronic, P.Braun-Munzinger, K.Redlich, J.Stachel, Nature 561 321–330 (2018).
[3] Butler, S. T. and Pearson, C. A., Phys. Rev. 129 836 (1963).
[4] R. Scheibl and U. W. Heinz, Phys. Rev. C59 1585-1602 (1999).
[5] C. Tsallis, J Stat Phys 52 479–487 (1988).
[6] ALICE Collaboration, Phys. Rev. C93 024917 (2016).
[7] ALICE Collaboration, Phys. Rev. C97 024615 (2018).
[8] ALICE Collaboration, Phys. Lett. B 794 50 (2019).
[9] ALICE Collaboration, Phys. Lett. B 800 135043 (2020).
[10] ALICE Collaboration, Eur. Phys. J. C 80 9, 889 (2020).
[11] ALICE Collaboration, Phys.Rev.C 101 4, 044906 (2020).
[12] ALICE Collaboration, JHEP 04 192 (2020).
[13] J. I. Kapusta, Phys. Rev. C 21 4, 1301 (1980).
[14] L.A. Dal, A.R. Raklev, Phys.Rev.D 91 12 (2015), Erratum: Phys.Rev.D 92, 069903 (2015), Erratum: Phys.Rev.D 92, 089901 (2015).
[15] ALICE Collaboration, Phys. Lett. B 819 136440 (2021).

Finite chemical potential equation of state for QCD from an alternative expansion scheme

Paolo Parotto[1,*], *Szabolcs* Borsányi[1], *Zoltan* Fodor[2,1,3,4], *Jana N.* Guenther[5], *Ruben* Kara[1], *Sandor D.* Katz[6], *Attila* Pásztor[6], *Claudia* Ratti[7], and *Kalman K.* Szabó[1,4]

[1]University of Wuppertal, Department of Physics, Wuppertal D-42119, Germany
[2]Pennsylvania State University, Department of Physics, State College, PA 16801, USA
[3]Inst. for Theoretical Physics, ELTE Eötvös Loránd University, Pázmány P. sétány 1/A, H-1117 Budapest, Hungary
[4]Jülich Supercomputing Centre, Forschungszentrum Jülich, D-52425 Jülich, Germany
[5]Aix Marseille Univ., Université de Toulon, CNRS, CPT, Marseille, France
[6]Eötvös University, Budapest 1117, Hungary
[7]Department of Physics, University of Houston, Houston, TX 77204, USA

Abstract. The Taylor expansion approach to the equation of state of QCD at finite chemical potential struggles to reach large chemical potential μ_B. This is primarily due to the intrinsic difficulty in precisely determining higher order Taylor coefficients, as well as the structure of the temperature dependence of such observables. In these proceedings, we illustrate a novel scheme [1] that allows us to extrapolate the equation of state of QCD without suffering from the poor convergence typical of the Taylor expansion approach. We continuum extrapolate the coefficients of our new expansion scheme and show the thermodynamic observables up to $\mu_B/T \leq 3.5$.

1 Introduction

The phase diagram of Quantum Chromodynamics (QCD) is a field of major interest from both a theoretical and experimental standpoint. At vanishing baryon density, it is known from lattice QCD [2] that the chiral/deconfinement transition is a smooth crossover. At finite baryon density lattice QCD faces a sign problem, due to the introduction of a real chemical potential that makes the action complex. Although new techniques that allow to directly simulate the theory at finite chemical potential exist [3, 4], they cannot be applied to large scale QCD simulations yet.

Several results have been published that exploit simulations at zero or imaginary chemical potential, and rely on an analytical continuation to extrapolate to real chemical potential. The most straightforward method is the Taylor expansion, whereby the first few μ_B-derivatives of thermodynamic quantities are calculated [5, 6]. Notable successes of analytic continuation methods include the determination of the QCD transition line, whose μ_B-dependence is now known up to the next-to-leading order [7, 8].

At large temperatures, resummed perturbation theory has provided a quantitative description of the chemical potential dependence of several observables, which agree with lattice

*Speaker: parotto@uni-wuppertal.de

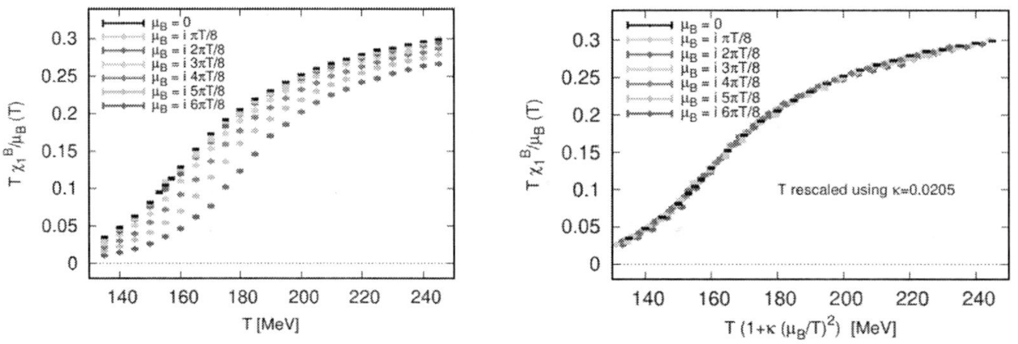

Figure 1. (Left panel): Normalized baryon density at imaginary baryon chemical potentials. The points at $\mu_B = 0$ (black) show the second baryon susceptibility $\chi_2^B(T)$. (Right panel): Same curves as in the upper panel, with a temperature rescaled in accordance to Eq. (1) with $\kappa = 0.0205$.

calculations [9, 10]. Similarly, functional methods provide an alternative approach to the study of the phase diagram [11]. However, lattice QCD remains the major *first-principle* tool of investigation, especially in the transition region of QCD.

From an experimental standpoint, heavy-ion collisions are capable of exploring the phase structure of strongly interacting matter. Systems created in such experiments are modeled through hydrodynamic simulations, for which the equation of state (EoS) of QCD is a crucial input. At vanishing chemical potential it has been known for several years [12–14]. Despite enormous computational efforts aimed at extending these results to higher μ_B by adding more terms in the Taylor series, at the moment even the sixth μ_B-derivative of the QCD pressure is only available with modest precision from lattice simulations [5, 6].

In this contribution, we describe a new [1], alternative scheme for the extrapolation of the QCD equation of state to finite density, which displays improved convergence over the Taylor series, leading to wider coverage in μ_B and to more precise results for the thermodynamic observables.

2 The alternative scheme

We start by *observing* that, at imaginary chemical potentials, the normalized baryon density $\chi_1^B/\hat{\mu}_B$ presents the same structure at all chemical potentials (including $\hat{\mu}_B = 0$, where the limit $\lim_{\mu_B\to 0}\chi_1^B/\hat{\mu}_B = \chi_2^B$), as can be seen in the left panel of Fig.1. This is reflected in the fact that a μ_B-dependent rescaling of the temperatures makes all the curves collapse onto each other, as shown in the right panel of Fig. 1. In particular, a single T-independent parameter can encode such rescaling quite well according to the following equality:

$$\frac{\chi_1^B(T,\hat{\mu}_B)}{\hat{\mu}_B} = \chi_2^B(T',0)\ , \qquad T' = (1+\kappa\hat{\mu}_B^2)\ . \tag{1}$$

Although quite suggestive, Eq.(1) cannot serve as an alternative expansion scheme as is. Thus, we include higher order terms in the expansion, and allow the expansion parameters to depend on the temperature:

$$T' = T\left(1+\kappa_2(T)\hat{\mu}_B^2+\kappa_4(T)\hat{\mu}_B^4+O(\hat{\mu}_B^6)\right)\ . \tag{2}$$

In essence, the expansion scheme we present amounts to an expansion in $\Delta T = T' - T$ *inside* the argument of χ_2^B, and corresponds to a re-organization of the different terms in the

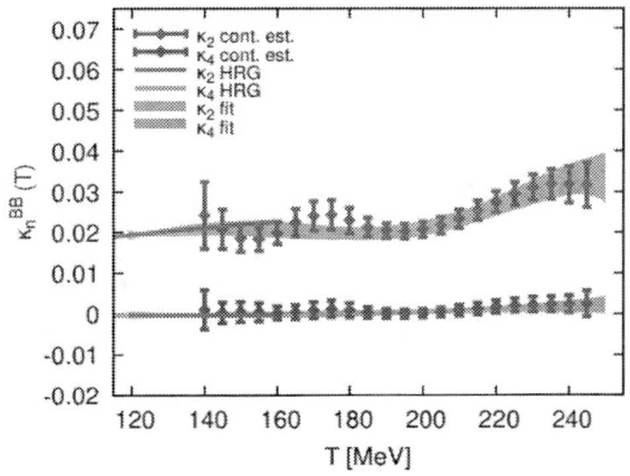

Figure 2. Continuum extrapolated result for the expansion parameters $\kappa_2(T)$ and $\kappa_4(T)$. HRG results are shown up to $T = 160\,\mathrm{MeV}$ (in green for κ_2, orange for κ_4, respectively). The bands show correlated polynomial fits.

Taylor series. One can visualize this as an expansion along lines of constant (normalized) baryon density $\chi_1^B\hat{\mu}_B$, while a Taylor expansion is carried out at constant temperature. Working out such expansions, and equating equal order terms, one finds:

$$\kappa_2(T) = \frac{1}{6T}\frac{\chi_4^B(T)}{\chi_2^{B\prime}(T)}\,, \qquad \kappa_4(T) = \frac{1}{360\,\chi_2^{B\prime}(T)^3}\left(3\,\chi_2^{B\prime}(T)^2\chi_6^B(T) - 5\,\chi_2^{B\prime\prime}(T)\chi_4^B(T)^2\right)\,. \quad (3)$$

These relations can be straightforwardly extended to higher orders, and could in principle utilized to determine the coefficients $\kappa_n(T)$. However, for $n > 2$, such coefficients would still suffer from severe cancellations between competing terms. Thus, exploiting simulations at imaginary chemical potentialts $\hat{\mu}_B = i\,n\,\pi/8$, with $n = 0, ..., 8$, we calculate the proxy quantity:

$$\Pi(T,\hat{\mu}_B^2) = \frac{T' - T}{T\hat{\mu}_B^2} = \kappa_2(T) + \kappa_4(T)\hat{\mu}_B^2 + O(\hat{\mu}_B^4)\,, \qquad (4)$$

at different temperatures, and on three different lattices with $N_\tau = 10, 12, 16$. We then perform a combined fit in $\hat{\mu}_B^2$ and $1/N_\tau^2$, from which we extract κ_2 and κ_4 at each temperature. The results for these coefficients are shown in Fig. 2, along with polynomial fits of both $\kappa_n(T)$ which take into account the full statistical and systematic correlations between different temperatures. At low temperature, we also show results from the hadron resonance gas (HRG) model, which is in good agreement with our results. Notably, we find a large separation in the values of $\kappa_2(T)$ and $\kappa_4(T)$, which hints at improved convergence of the expansion.

We reconstruct the baryon density at finite real chemical potential by employing Eq. (1), then obtain the pressure from it by simply integrating over the chemical potential. We show the baryon density and the pressure in the left and right panels of Fig. 3, respectively. We also compare with HRG results at low temperature, finding good agreement at all chemical potentials.

Crucially, the errors are well under control up to $\hat{\mu}_B = 3.5$, and no unphysical, nonmonotonic behavior is observed. For the baryon density, we compare the results with a simplified scenario in which we set $\kappa_4 \equiv 0$. We find that the inclusion of a non-zero κ_4 increases the error on the final results, but does not change its prediction. This hints at a better convergence of our series expansion, compared to the Taylor method.

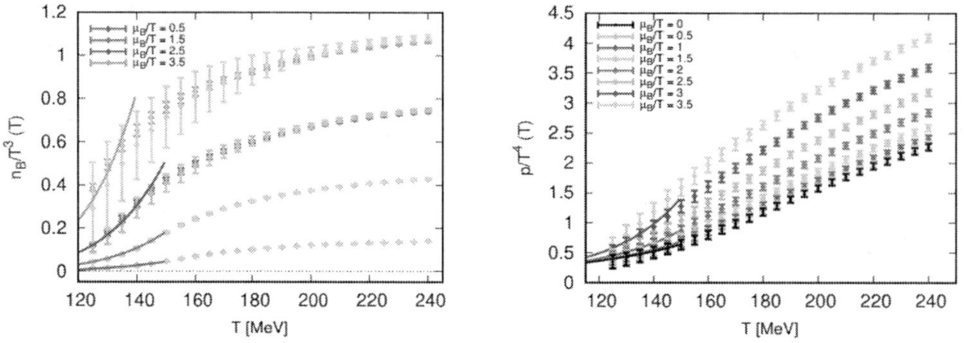

Figure 3. Baryon density (left) and pressure (right) at increasing values of $\hat{\mu}_B$. With solid lines we show the results from the HRG model. For the baryon density we also show in darker shades the results obtained by setting $\kappa_4 \equiv 0$.

Acknowledgments

This project was funded by the DFG grant SFB/TR55. The project also received support from the BMBF Grant No. 05P18PXFCA. This work was also supported by the Hungarian National Research, Development and Innovation Office, NKFIH grant KKP126769. A.P. is supported by the J. Bolyai Research Scholarship of the Hungarian Academy of Sciences and by the ÚNKP-20-5 New National Excellence Program of the Ministry for Innovation and Technology. The project leading to this publication has received funding from Excellence Initiative of Aix-Marseille University - A*MIDEX, a French "Investissements d'Avenir" programme, AMX-18-ACE-005. This material is based upon work supported by the National Science Foundation under grants no. PHY-1654219 and by the U.S. DoE, Office of Science, Office of Nuclear Physics, within the framework of the Beam Energy Scan Topical (BEST) Collaboration. This research used resources of the Oak Ridge Leadership Computing Facility, which is a DOE Office of Science User Facility supported under Contract DE-AC05-00OR22725. The authors gratefully acknowledge the Gauss Centre for Supercomputing e.V. (www.gauss-centre.eu) for funding this project by providing computing time on the GCS Supercomputer HAWK at HLRS, Stuttgart. Part of the computation was performed on the QPACE3 funded by the DFG ind hosted by JSC. C.R. also acknowledges the support from the Center of Advanced Computing and Data Systems at the University of Houston.

References

[1] S. Borsányi, Z. Fodor, J. N. Guenther, R. Kara, S. D. Katz, P. Parotto, A. Pásztor, C. Ratti and K. K. Szabó, Phys. Rev. Lett. **126**, no.23, 232001 (2021) doi:10.1103/PhysRevLett.126.232001 [arXiv:2102.06660 [hep-lat]].

[2] Y. Aoki, G. Endrodi, Z. Fodor, S. D. Katz and K. K. Szabo, Nature **443**, 675-678 (2006) doi:10.1038/nature05120 [arXiv:hep-lat/0611014 [hep-lat]].

[3] D. Sexty, Phys. Rev. D **100**, no.7, 074503 (2019) doi:10.1103/PhysRevD.100.074503 [arXiv:1907.08712 [hep-lat]].

[4] M. Giordano, K. Kapas, S. D. Katz, D. Nogradi and A. Pasztor, JHEP **05**, 088 (2020) doi:10.1007/JHEP05(2020)088 [arXiv:2004.10800 [hep-lat]].

[5] S. Borsanyi, Z. Fodor, J. N. Guenther, S. K. Katz, K. K. Szabo, A. Pasztor, I. Portillo and C. Ratti, JHEP **10**, 205 (2018) doi:10.1007/JHEP10(2018)205 [arXiv:1805.04445 [hep-lat]].

[6] A. Bazavov, D. Bollweg, H. T. Ding, P. Enns, J. Goswami, P. Hegde, O. Kaczmarek, F. Karsch, R. Larsen and S. Mukherjee, *et al.* Phys. Rev. D **101**, no.7, 074502 (2020) doi:10.1103/PhysRevD.101.074502 [arXiv:2001.08530 [hep-lat]].

[7] A. Bazavov *et al.* [HotQCD], Phys. Lett. B **795**, 15-21 (2019) doi:10.1016/j.physletb.2019.05.013 [arXiv:1812.08235 [hep-lat]].

[8] S. Borsanyi, Z. Fodor, J. N. Guenther, R. Kara, S. D. Katz, P. Parotto, A. Pasztor, C. Ratti and K. K. Szabo, Phys. Rev. Lett. **125**, no.5, 052001 (2020) doi:10.1103/PhysRevLett.125.052001 [arXiv:2002.02821 [hep-lat]].

[9] R. Bellwied, S. Borsanyi, Z. Fodor, S. D. Katz, A. Pasztor, C. Ratti and K. K. Szabo, Phys. Rev. D **92**, no.11, 114505 (2015) doi:10.1103/PhysRevD.92.114505 [arXiv:1507.04627 [hep-lat]].

[10] N. Haque and M. Strickland, Phys. Rev. C **103**, no.3, 031901 (2021) doi:10.1103/PhysRevC.103.L031901 [arXiv:2011.06938 [hep-ph]].

[11] N. Dupuis, L. Canet, A. Eichhorn, W. Metzner, J. M. Pawlowski, M. Tissier and N. Wschebor, Phys. Rept. **910**, 1-114 (2021) doi:10.1016/j.physrep.2021.01.001 [arXiv:2006.04853 [cond-mat.stat-mech]].

[12] S. Borsanyi, G. Endrodi, Z. Fodor, A. Jakovac, S. D. Katz, S. Krieg, C. Ratti and K. K. Szabo, JHEP **11**, 077 (2010) doi:10.1007/JHEP11(2010)077 [arXiv:1007.2580 [hep-lat]].

[13] S. Borsanyi, Z. Fodor, C. Hoelbling, S. D. Katz, S. Krieg and K. K. Szabo, Phys. Lett. B **730**, 99-104 (2014) doi:10.1016/j.physletb.2014.01.007 [arXiv:1309.5258 [hep-lat]].

[14] A. Bazavov *et al.* [HotQCD], Phys. Rev. D **90**, 094503 (2014) doi:10.1103/PhysRevD.90.094503 [arXiv:1407.6387 [hep-lat]].

 https://doi.org/10.1051/epjconf/202225910016

Charm and Bottom quarks dynamics in heavy-ion collisions: R_{AA}, anisotropic flows v_n and their correlations to the bulk.

Maria Lucia Sambataro[1,2,*], *Salvatore* Plumari[1,2], *Yifeng* Sun[2], *Vincenzo* Minissale[1,2], and *Vincenzo* Greco[1,2]

[1]Department of Physics and Astronomy 'E. Majorana', University of Catania, Via S. Sofia 64, 1-95123 Catania, Italy
[2]Laboratori Nazionali del Sud, INFN-LNS, Via S. Sofia 62, I-95123 Catania, Italy

Abstract. We describe the propagation of heavy quarks (HQs), charm and bottom, in the quark-gluon plasma (QGP) by means of a full Boltzmann transport approach including event-by-event fluctuations within a coalescence plus fragmentation hadronization. The non-perturbative dynamics of the interaction between HQs and plasma particles have been taken into account through a Quasi-Particle Model (QPM). We show that the resulting charm in-medium evolution is able to correctly predict simultaneously not only the experimental data for the average D-mesons $R_{AA}(p_T)$ and $v_{2,3}(p_T)$ at LHC energies but also the extension of the analysis to the event-shape engineeering tecnique that classify events according to magnitude of the second-order harmonic reduced flow vector q_2. In the same scheme we show predictions for $R_{AA}(p_T)$ of electrons from semi-leptonic B-mesons decays at top LHC energies. Our results entail a determination of D_s which is consistent with the lattice QCD calculations.

1 Introduction

Heavy quarks (HQs), namely charm and bottom, are excellent probes of the system created in a ultra-Relativistic Heavy Ion Collision (uRHIC). They are produced out-of-equilibrium in the early stage of the collisions by pQCD process and being $M_{HQ} >> T$, they are expected to thermalize slower in the Quark-Gluon Plasma (QGP) with respect to their light counterparts so that they can conserve memory of the history of the plasma evolution. One of the main observable in HQs sector that it has been also extensively used as a probe of QGP, is the nuclear suppression factor R_{AA} which is defined as the ratio between the spectra of heavy flavor hadrons measured in nucleus-nucleus collisions with the same spectra in proton-proton collisions [1, 2]. Another set of key observables are the anisotropic flows that are characterized by the magnitude of the coefficient v_n in the free expansion of the azimutal particle distribution[3, 4]. For a smooth matter distribution, all odd v_n coefficients are zero by symmetry but the event-by-event fluctuation in initial state entail the presence of non-zero odd harmonic coefficients, i.e. triangular flow v_3 that is an asimmetry triangularly shaped. Therefore, due to event by event fluctuations, collisions belonging to the same centrality class can give rise to systems with different initial eccentricity and hence different flow harmonics

*e-mail: sambataro@lns.infn.it

for the final hadron distributions. We have developed an event-by-event transport approach that incorporates initial state fluctuations to study the collective flows in Pb + Pb collisions and the correlations between initial geometry and final collective flows [5, 6]. This issue can also be addressed to the Event-Shape-Engineering (ESE) tecnique consisting in selecting events in the same centrality class with different average elliptic anisotropy of final-state particles, selected according to the magnitude of the second-order harmonic reduced flow vector $q_2 = |\vec{Q}_2|/\sqrt{M}$ where $Q_2 = \sum_{j=1}^{M} e^{ij\phi_j}$ and M is the multiplicity of charged particles [7, 8]. This tecnique is adopted to study the correlation between the flow coefficients of heavy quark mesons and soft hadrons. Within this approach, we study the $v_n(p_T)$ of D mesons making a comparisons with the available experimental data and we extend our transport approach to study the bottom dynamics showing the comparison between our results on $R_{AA}(p_T)$ of electrons from semileptonic B-mesons decay and the available ALICE experimental data.

2 Transport equation for charm and bottom quarks in the QGP

In our approach, we describe both the bulk and HQs evolution by solving relativistic Boltzmann equations developed to perform studies of the dynamics of heavy-ion collisions [9–12]:

$$\begin{aligned} \{p_k^\mu \partial_\mu + m^*(x)\partial_\mu m^*(x)\partial_p^\mu\} f_k(x, p_k) &= C[f_q, f_g](x, p_k) \\ p^\mu \partial_\mu f_Q(x, p) &= C[f_q, f_g, f_Q](x, p) \end{aligned} \tag{1}$$

where $f_k(x, p)$ is the on-shell phase space one-body distribution function of the k parton and $C[f_q, f_g, f_Q](x, p)$ is the relativistic Boltzmann-like collision integral. In order to take into account the non-perturbative effects in HQs scattering, we evaluate $C[f_q, f_g, f_Q](x, p)$ within a quasi-particle model (QPM) approach in which the interaction is encoded in the quasi-particle masses that behave like massive constituents of free gas plus a background field interaction. It has been shown that QPM can reproduce the lQCD equation of state: pressure, energy density and interaction measure $T^\mu_\mu = \epsilon - 3P$ [13, 14] . In our model the collision integral $C[f_q, f_g](x, p_k)$ is gauged to viscous hydrodynamics allowing to construct a relativistic transport approach at fixed $\eta/s \approx 0.1$. For more details see Ref.s [5, 15]. For Pb + Pb collisions at $\sqrt{s} = 5.02 TeV$, the initial conditions of plasma particles in the r-space are given by a Monte-Carlo Glauber model in order to take into account the initial event-by-event fluctuations while we distrubute light partons in p-space using a Boltzmann-Juttner distribution up to $p_T = 2\ GeV$ and we include mini-jet production distributed at larger momenta according pQCD calculation at NLO [16]. Regarding HQs distribution in p-space, we use the Fixed Order + Next-to-Leading Log (FONLL) calculations [17]. Finally, in this paper we have considered a hybrid model of coalescence plus fragmentation hadronization [18] in order to determine the final D meson and B meson spectra and therefore the final $R_{AA}(p_T)$ and $v_n(p_T)$.

3 Results

We discuss now the comparison of the results for the D-meson elliptic flow $v_2(p_T)$, also evaluated in ESE selection, and triagular flow $v_3(p_T)$ with the available ALICE experimental data. In the left panel of Fig.1, we show our results for the D meson $v_{2,3}(p_T)$ at mid-rapidity for Pb + Pb collisions at $\sqrt{s} = 5.02\ TeV$ at 0-10% centrality class. Within our approach we find a finite $v_2(p_T)$ (red solid line) and $v_3(p_T)$ (green dashed line) of D mesons that are qualitatively in agreement with the experimental data. In the right panel of Fig.1, the $v_2(p_T)$ of D mesons at midrapidity for different q_2 selections is shown as function of p_T.

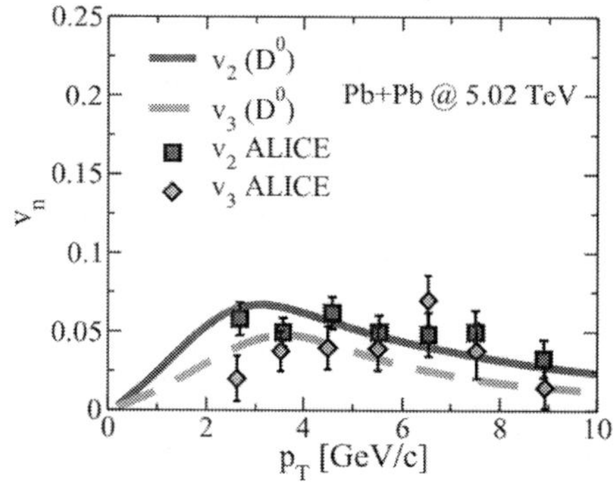

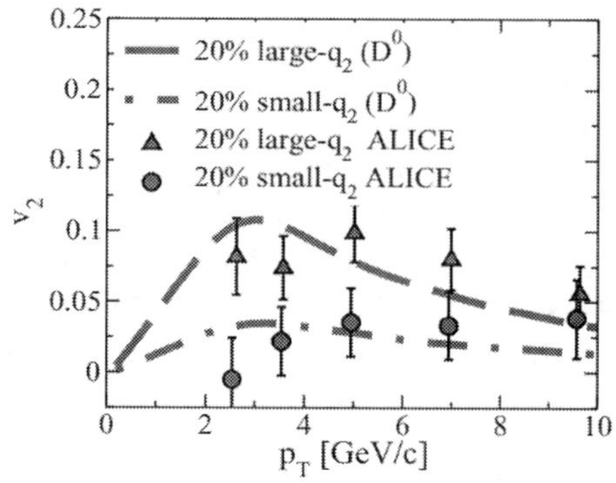

Figure 1. Left panel: D meson $v_2(p_T)$(red solid line) and $v_3(p_T)$(green dashed line) as function of p_T at $0-10\%$ centrality class for Pb + Pb $\sqrt{s}$ = 5.02 TeV. Right panel: D meson $v_2(p_T)$ at large (red dashed line) and small (red dot-dashed line) q_2 selection as function of p_T in the same centrality class. Experimental data have been taken from Ref.s [7, 8].

As shown in the right panel of Fig.1, the 20% large q_2 selection gives an elliptic flow that is larger with respect to the unbiased one (red solid line in the left panel of Fig.1). As one should expect, a large q_2 selection corresponds to a fireball with an initial larger eccentricity ϵ_2 that produces, after the fireball expansion, a corresponding larger final elliptic flow. In a similar way a small q_2 selection corresponds to smaller initial ϵ_2 and smaller final v_2. Our results show a significant difference of v_2 in the two q_2 selections of about 50% in agreement with the ALICE experimental data. Notice that the space diffusion coefficient $2\pi TD_s$ extracted with our approach is in satisfying agreement with the lQCD data within the present systematic uncertainties, for more details see Ref. [19]. In the same scheme of boltzmann transport approach with hybrid coalescence plus fragmentation approach for the hadronization process, we extend this analysis to bottom quark dynamics. Notice that the results shown in this section have been obtained with an interaction between bottom quarks and bulk that is the same to the one of charm quarks which gives a spacial diffusion coefficient in agreement with lQCD data.

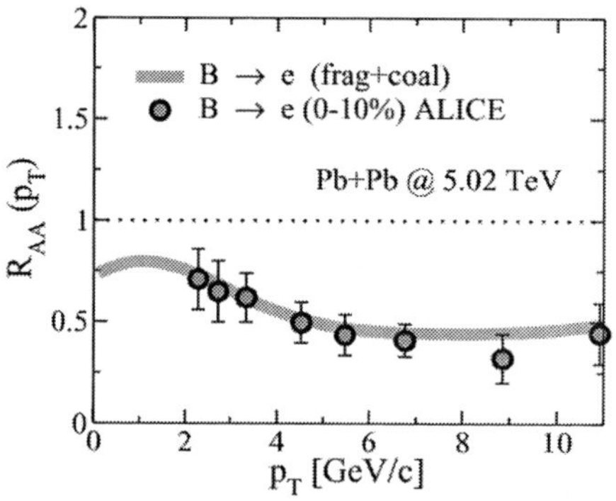

Figure 2. Nuclear modification factor $R_{AA}(p_T)$ of electrons from semi-leptonic B-mesons decays at $0-10\%$ centrality class for Pb + Pb at $\sqrt{s}$ = 5.02 TeV. Experimental data have been taken from Ref. [20].

In the Fig.2 we show the comparison between our result for $R_{AA}(p_T)$ of electrons from semi-leptonic B-mesons decays at $0-10\%$ centrality class and the available ALICE experimental data for Pb + Pb at $\sqrt{s} = 5.02\ TeV$. In order to obtain the electrons spectra we have implemented in our code the decay channel of $B \to e$ taking into account the semileptonic decay matrix weigthed by the different branching ratios of the decay. As shown in Fig. 2, our result are in good agreement with the ALICE experimetal data suggesting a strong coupling with collectively expanding fireball for bottom quark.

4 Conclusions

We have studied the charm and bottom quark propagation in QGP at LHC energies within a relativistic Boltzmann transport approach including non-perturbative effects of interaction by means of QPM approach. The hadronization process has been described by means of a hybrid model of coalescence plus fragmentation. We have studied the D meson $v_2(p_T)$ and $v_3(p_T)$ and the D meson $v_2(p_T)$ evaluated with the ESE technique. Our results show that with our approach we can simultaneously reproduce the elliptic flow of D mesons for different q_2 selections within the current experimental uncertainties. The same transport approach has been applied in order to study the bottom quark dynamics showing an $R_{AA}(p_T)$ in agreement with experimental data. The spatial diffusion coefficient D_s of charm and bottom quarks extracted from D and B mesons $R_{AA}(p_T)$ is in good agreement with the lQCD data within the still large uncertainties.

References

[1] B.I. Abelev, et al., Phys. Rev. Lett. 98, 192301 (2007)
[2] J. Adam, et al., JHEP 03, 081 (2016).
[3] A. Adare, et al., Phys. Rev. Lett. 98, 172301 (2007).
[4] B.B. Abelev, et al., Phys. Rev. C90(3), 034904 (2014).
[5] S. Plumari, G. L. Guardo, F. Scardina, V. Greco, Phys.Rev.C 92 (2015) 5, 054902.
[6] S. Plumari, G. Coci, V. Minissale, S. K. Das, Y. Sun and V. Greco, Phys.Lett.B 805 (2020) 135460.
[7] S. Acharya et al. (ALICE), Phys.Lett.B 813, 136054 (2021).
[8] S. Acharya et al. (ALICE), JHEP 02, 150 (2019).
[9] S. K. Das, F. Scardina, S. Plumari, and V. Greco, Phys. Lett. B 747, 260 (2015).
[10] M. L. Sambataro, S. Plumari, V. Greco, Eur.Phys.J.C 80 (2020) 12, 1140.
[11] M. Ruggieri et al., Phys. Rev. C 92, 064904 (2015).
[12] A. Gabbana et al., Phys.Rev.C 101 (2020) 6, 064904.
[13] S. Plumari, W. M. Alberico, V. Greco and C. Ratti, Phys. Rev. D 84, 094004 (2011).
[14] S. Borsanyi et al., JHEP 1011, 077 (2010).
[15] M. Ruggieri, F. Scardina, S. Plumari and V. Greco, Phys.Rev. C89,054914 (2014).
[16] V. Greco, C. M. Ko and P. Levai, Phys. Rev. Lett. 90, 202302 (2003).
[17] M. Cacciari, S. Frixione, N. Houdeau, M. L. Mangano, P. Nason and G. Ridolfi, JHEP 10 137 (2012).
[18] S. Plumari, V. Minissale, S.K. Das, G. Coci and V. Greco, Eur.Phys.J.C 78 (2018) 4, 348.
[19] F. Scardina, S. K. Das, V. Minissale, S. Plumari and V. Greco, Phys.Rev.C 96 (2017) 4, 044905.
[20] R. Arnaldi, 10th International Conference on Hard and Electromagnetic Probes of High-Energy Nuclear Collisions, online (2020).

Learning Langevin dynamics with QCD phase transition

Lingxiao Wang[1], *Lijia* Jiang[1,2], and *Kai* Zhou[1,*]

[1]Frankfurt Institute for Advanced Studies, Ruth Moufang Strasse 1, D-60438, Frankfurt am Main, Germany
[2]Institute of Modern Physics, Northwest University, 710069, Xi'an, China

Abstract. In this proceeding, the deep Convolutional Neural Networks(CNNs) are deployed to recognize the order of QCD phase transition and predict the dynamical parameters in Langevin processes. To overcome the intrinsic randomness existed in a stochastic process, we treat the final spectra as image-type inputs which preserve sufficient spatiotemporal correlations. As a practical example, we demonstrate this paradigm for the scalar condensation in QCD matter near the critical point, in which the order parameter of chiral phase transition can be characterized in a $1+1$-dimensional Langevin equation for σ field. The well-trained CNNs accurately classify the first-order phase transition and crossover from σ field configurations with fluctuations, in which the noise does not impair the performance of the recognition. In reconstructing the dynamics, we demonstrate it is robust to extract the damping coefficients η from the intricate field configurations.

Phase transition and critical phenomena are extensively observed in various many-body systems. The orders of phase transition or critical exponents, they can be well-extracted in equilibrium systems. However, in a non-equilibrium scenario, the situation becomes more complicated especially for a stochastic process emerges in quantum systems. The intrinsic randomness breaks the deterministic description of the dynamics, which hinders our further understanding to those exotic non-equilibrium systems, e.g., heavy-quark diffusion in the Quark-Gluon Plasma(QGP) [1, 2]. The Langevin dynamics is a general description of a non-equilibrium system as a stochastic differential equation, in which the degree of freedom typically are the collective variables changing very slowly, compared to the other microscopic variables in the system. In QCD phase transitions, one challenging of both theoretical and experimental sides is if we can identify the order of the phase transition and the critical exponent of the system from the configurations generated from such a stochastic process, which is significant for the critical-point search [3]. Deep learning with a hierarchical structure of artificial neural networks is emerging as a novel tool to uncover structure in complex data and to efficiently describe it with a minimal number of parameters [4]. Nowadays, the deep learning method is also utilized in the field of physics research [5, 6], such as recognizing phase transitions in heavy ion collisions(HICs) [7, 8] and improving the computation of Lattice QCD [9–11]. Meanwhile, the model-free prediction on state evolution has been discussed with machine learning for chaotic dynamical systems [12]. It makes recognizing phase transitions in stochastic processes feasible.

In recent study [13], we proposed a deep CNN model to detect phase transition and dynamics in stochastic processes as Fig. 1 shown. In the problem set-up, the raw observations

*e-mail: zhou@fias.uni-frankfurt.de

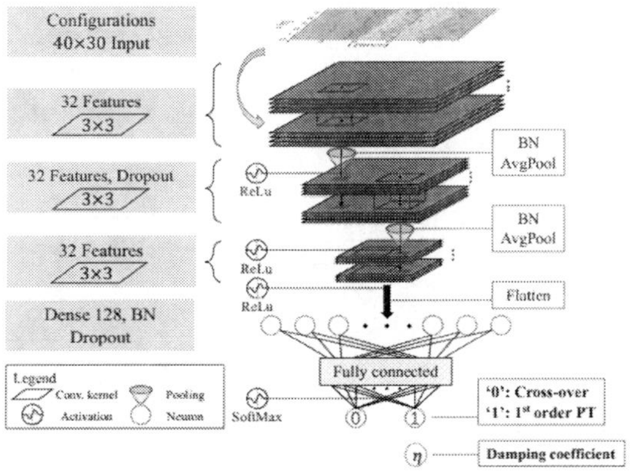

Figure 1. The architecture of the deep CNNs for recognizing the phase order and predicting the damping coefficient from the σ field configurations. Details of the network structure can be found in Ref. [13]

(configurations) for the system are nothing but the stochastic time series data. With regard to the effective inputs to the machine, a proper choice is the final-state particle spectra. Thus, we feed the event-by-event spatiotemporal scalar field configurations in final stage to the neural network to identify the dynamics, including the phase order and dynamical parameter as labels. As a practical example of Langevin dynamics with QCD phase transitions, we derive the dynamic equation from a linear sigma model(LSM). The effective potential from the LSM presents a scenario in which the crossover locates at the small chemical potential region and first-order phase transition occurs at the large region [14]. In HICs, the hot and dense fireball created an extreme environment where the QCD phase transition can happen [15, 16]. To model the corresponding phase transition processes, the Langevin equation is adopted to describe the semi-classical evolution for the long wavelength mode of the σ field (for more context and details can see [13, 17, 18]),

$$\partial^{\mu}\partial_{\mu}\sigma(t,x) + \eta\partial_{t}\sigma(t,x) + \frac{\delta V_{eff}(\sigma)}{\delta\sigma} = \xi(t,x), \quad (1)$$

where η is the damping coefficient, $\xi(t,x)$ is the noise term, and the effective potential V_{eff} decides the type of phase transition in the stochastic process (see more details in Ref. [19]). The terms η and ξ are both from the interaction between σ field and the thermal background, which follow the fluctuation-dissipation theorem. In the zero-momentum mode limit, the correlation has the form $\langle\xi(t)\xi(t')\rangle \propto \eta\coth\left(\frac{m_{\sigma}}{2T}\right)\delta(t-t')$. In our calculation, η is taken as a free parameter while the noise is set as white noise, and B is set as the amplitude of the white noise. For simplicity but without loss of generality, we assume the heat bath evolves along trajectories with constant baryon chemical potential, and the temperature drops down in a Hubble-like way, $T(t)/T_0 = (t/t_0)^{-0.45}$, where T_0 ($> T_c$) is the initial temperature, and $t_0 = 1$ fm is the initial time for the evolution. With regard to the dynamical evolution of the σ field, we set the damping coefficient η to be constant across the evolution with values ranging from 1.0 to 5.5 fm^{-1}. As for the details of the numerical set-up, we simulate the evolution of σ field in 1-dimensional space with range $L = 6.0$ fm, and the spatial grid size $dx = 0.2$ fm. The duration of the evolution is 16 fm/c at most with the temporal step size $dt = 0.1$ fm/c. With the above set-ups, the σ field was evolved according to Eq. (1) on an event-by-event basis. The configurations from later episode after phase transition with 4 fm duration of the

σ field are censored as the input data-sets. Thus, the input is in $t \in [7, 11]\,\mathrm{fm}^{-1}$, or in the last 40 time-steps from the evolution, where the ambient temperature is already much lower than T_c, ensured that the potential phase transition already happened. Therefore, the prepared input configuration contains $N = 40 \times 30 = 1200$ pixels in each event.

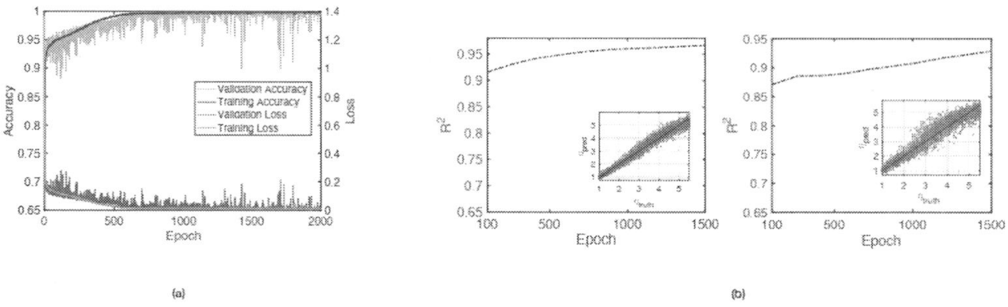

Figure 2. (a) The accuracy and loss on training and validation data sets with different noise parameters B=0.5, 1, 1.5, 2 and 2.5. (b) The training process in predicting the damping coefficient from configurations. The left panel are prediction results for the crossover scenario, the right panel are prediction results for the first-order scenario.

To train the neural networks, we prepare 10 000 events at each parameter setup with labels. First we deploy the machine on data-sets with B = 0.5 and 1, and the training history is recorded in Fig.2(a). The validation loss tends to decrease same as training loss with epoch increasing, in other words, it reaches stable without over-fitting. Moreover, the testing accuracy reaches to 99.9%, it means the performance of the well-trained machine is robust and these final-stage configurations indeed carry with phase transition information. It should be noticed that, although the configurations are totally different from the learned ones due to the intense noise, the machine can keep an accurate prediction on the phase order in each event. As for the damping coefficient, as the core dynamic parameter, it can influence not only the thermal diffusion speed of the system but also the fluctuations. Here we use events generated on both the crossover and the first-order scenarios with the damping coefficient in the range of $(1.0 - 2.5)$ and $(4.6 - 5.5)\,\mathrm{fm}^{-1}$, in each η-bin, 1000 events were collected as inputs. In the training process of Fig. 2(b), the related coefficient R^2 grows quickly from 87% to 93% for the crossover and from 91% to 97% for the first-order scenario. The evaluation on the testing data-set is shown in the insert part. It consists of orange dots which are labeled by the damping coefficients of the ground truth and predictions on testing data-set from the trained CNN. If we take a close step to see the performance of the machine. It is found that although inside the training set has no supervision in the region $(2.6 - 4.5)\,\mathrm{fm}^{-1}$, the predictions still keep a reasonable performance.

In summary, we demonstrate the framework is effective in recognizing the QCD phase transitions and extracting the Langevin dynamics from complicated configurations. The first-order or crossover-type phase transitions, encoded inside the stochastic dynamical evolution, can be detected by a well-trained machine. Although the field configurations differ totally from each other because of the noise terms and also the random initial conditions, the machine successfully learns to make an accurate prediction on the phase order for previously unseen evolution events in the testing stage. This is related to the powerful capability of the deep CNNs for extracting hidden correlations in image-type data-set, which facilitates the presented phase order identification from the field configurations.We further prepare mixed configurations with damping coefficient in the range of $(1.0 - 2.5)\text{fm}^{-1}$ and $(4.6 - 5.5)\text{fm}^{-1}$, nevertheless, the predictions to the dynamics are made for the entire range. It reveals an acceptable generalization ability when tested on configurations containing damping beyond the training set. The present method can be helpful for a broader field, like, there is a potential application in topological-dependent stochastic process [20], in which the topological charge could be extracted by the deep CNNs. A recent attempt is to detect the chiral magnetic effects in HICs with the same paradigm [21].

Acknowledgment

The authors acknowledge inspiring discussions with Horst Stoecker. The work is supported by the AI grant at FIAS of SAMSON AG, Frankfurt (L. J., L. W. and K. Z.), by the BMBF under the ErUM-Data project (K. Z.), and by the NVIDIA Corporation with the generous donation of NVIDIA GPU cards for the research (K. Z.).

References

[1] H. van Hees, M. Mannarelli, V. Greco, R. Rapp, Phys. Rev. Lett. **100**, 192301 (2008)
[2] J. Zhao, K. Zhou, S. Chen, P. Zhuang, Prog. Part. Nucl. Phys. **114**, 103801 (2020)
[3] S. Mukherjee, R. Venugopalan, Y. Yin, Phys. Rev. C **92**, 034912 (2015)
[4] Y. LeCun, Y. Bengio, G. Hinton, Nature **521**, 436 (2015)
[5] K. Albertsson, P. Altoe, D. Anderson, M. Andrews, J.P.A. Espinosa, A. Aurisano, L. Basara, A. Bevan, W. Bhimji, D. Bonacorsi et al., J. Phys.: Conf. Ser. **1085**, 022008 (2018)
[6] G. Carleo, I. Cirac, K. Cranmer, L. Daudet, M. Schuld, N. Tishby, L. Vogt-Maranto, L. Zdeborová, Rev. Mod. Phys. **91**, 045002 (2019)
[7] L.G. Pang, K. Zhou, N. Su, H. Petersen, H. Stöcker, X.N. Wang, Nature Commun. **9**, 210 (2018)
[8] Y.L. Du, K. Zhou, J. Steinheimer, L.G. Pang, A. Motornenko, H.S. Zong, X.N. Wang, H. Stöcker, Eur. Phys. J. C **80**, 516 (2020)
[9] K. Zhou, G. Endrődi, L.G. Pang, H. Stöcker, Phys. Rev. D **100**, 011501 (2019)
[10] S. Blücher, L. Kades, J.M. Pawlowski, N. Strodthoff, J.M. Urban, Phys. Rev. D **101**, 094507 (2020)
[11] G. Kanwar, M.S. Albergo, D. Boyda, K. Cranmer, D.C. Hackett, S. Racanière, D.J. Rezende, P.E. Shanahan, Phys. Rev. Lett. **125**, 121601 (2020)
[12] J. Pathak, B. Hunt, M. Girvan, Z. Lu, E. Ott, Phys. Rev. Lett. **120**, 024102 (2018)
[13] L. Jiang, L. Wang, K. Zhou, Phys. Rev. D **103**, 116023 (2021)
[14] H.S. Roh, T. Matsui, EPJ A **1**, 205 (1998)
[15] A. Bzdak, S. Esumi, V. Koch, J. Liao, M. Stephanov, N. Xu, Physics Reports (2020)
[16] N. Xu, K. Fukushima, B. Mohanty, AAPPS Bull. **31**, 1 (2021)
[17] M. Nahrgang, S. Leupold, C. Herold, M. Bleicher, Phys. Rev. C **84**, 024912 (2011)
[18] L. Jiang, S. Wu, H. Song, Nuclear Physics A **967**, 441 (2017)
[19] K. Paech, H. Stöcker, A. Dumitru, Phys. Rev. C **68**, 044907 (2003)
[20] L. Zhao, Z. Wang, X. Zhang, X. Liang, J. Xia, K. Wu, H.A. Zhou, Y. Dong, G. Yu, K.L. Wang et al., Phys. Rev. Lett. **125**, 027206 (2020)
[21] Y.S. Zhao, L. Wang, K. Zhou, X.G. Huang, ArXiv210513761 Hep-Ph Physicsnucl-Th (2021)

Early strangeness freeze-out from RHIC BES to LHC*

Jia Chen[1,2,**], *Jian* Deng[1,2], *Zebo* Tang[3], *Zhangbu* Xu[4], and *Li* Yi[1,2]

[1]Institute of Frontier and Interdisciplinary Science, Shandong University, Qingdao, Shandong, 266237, China

[2]Key Laboratory of Particle Physics and Particle Irradiation of Ministry of Education, Shandong University, Qingdao, Shandong, 266237, China

[3]State Key Laboratory of Particle Detection and Electronics, University of Science and Technology of China, Hefei, Anhui, 230026, China

[4]Brookhaven National Laboratory, Upton, New York, 11973, USA

Abstract. In this talk, we investigate the collision energy and particle species dependence of kinetic freeze-out properties in relativistic heavy ion collisions from $\sqrt{s_{\mathrm{NN}}}$ =7.7 - 200 GeV at Relativistic Heavy Ion Collider (RHIC) and 2.76 TeV at the Large Hadron Collider (LHC) with Tsallis Blast-Wave (TBW) model. Strangeness and non-strange particles show a similar radial flow, while the strange hadrons have higher temperature and smaller non-equilibrium degree. Strangeness approaches equilibrium more quickly than non-strange particles from peripheral to central collisions. The kinetic freeze-out temperature of non-strange particles in central collisions decreases from RHIC to LHC energies, while strangeness does not show this behavior. Finally we discuss the system bulk viscosity dependence on collision energy.

1 Introduction

The final transverse momentum (p_T) spectra of hadrons carry information about kinetic freeze-out properties of the hot and dense system created in relativistic heavy ion collisions. The study of energy dependence of freeze-out properties is an important tool to explore QCD phase diagram. The RHIC Beam Energy Scan (BES) program and the heavy ion program at the LHC has provided a wide energy coverage for the freeze-out properties study.

Strange hadrons serve as an excellent probe to locate the QCD phase boundary. Specifically strangeness enhancement in heavy ion collisions, compared to elementary p + p collisions, has long been suggested as one of the QGP signatures[1]. Strangeness yields lead to a deeper understanding of strangeness production mechanisms in nuclear collisions[2], and a better constrained extraction of the chemical and kinetic freeze-out parameters. The collision energy dependence of kinetic freeze-out properties for strange hadrons in heavy ion collisions is valuable for exploring QCD phase diagram.

Boltzmann-Gibbs Blast-Wave (BGBW) model [3, 4] has been traditionally used to study the produced system at kinetic freeze-out. Tsallis Blast-Wave (TBW) model was later introduced to describe the particle production for an extended p_T range in high energy collisions [5–10]. In this proceeding, we will employ blast-wave fits on particle productions from

*Presentation based on study in Phys. Rev. C **104**, 034901 (2021).
**Speaker. e-mail: jiachen@rcf.rhic.bnl.gov

RHIC BES to the LHC and focus on the strangeness and non-strangeness kinetic freeze-out comparison.

2 Analysis method

In blast-wave model, particle production is assumed to be radiated from local thermal sources. The source is boosted radially with transverse flow profile β and is boost-invariant in longitudinal direction. BGBW uses Boltzmann-Gibbs distribution for local thermal equilibrated emission source. The form of particle production in BGBW can be found in [3, 4]. For system deviated from thermal equilibrium, TBW model uses a non-extensive Tsallis statistics for emission source [9–13]. Tsallis statistics introduced a new parameter q for the degree of non-equilibrium, with values larger than 1 for deviation from equilibrium. The produced particle spectrum in TBW is

$$\frac{d^2N}{2\pi m_T dm_T dy}|_{y=0} \approx \int_{-y_b}^{+y_b} e^{\sqrt{y_b^2-y_s^2}} m_T \cosh(y_s)dy_s \times \int_0^R rdr \int_{-\pi}^{\pi} [1 + \frac{q-1}{T}E_T]^{-1/(q-1)} d\phi, \quad (1)$$

where $E_T = m_T \cosh(y_s)\cosh(\rho) - p_T \sinh(\rho)\cos(\phi)$, with y_s and y_b for source and beam rapidity.

3 Results and discussions

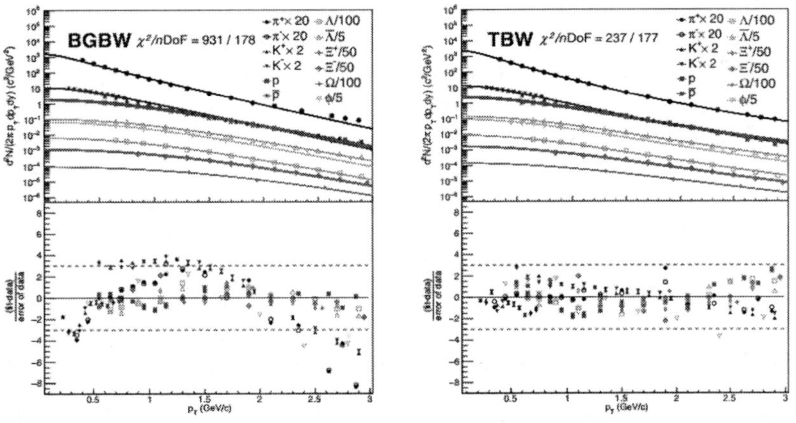

Figure 1. BGBW (left column) and TBW (right column) fits to hadron spectra (top panels) and model deviations from data divided by data uncertainties (bottom panels) in 40%-60% Au + Au collisions at $\sqrt{s_{NN}}$ =200 GeV. The solid curves represent fit functions.

The charged hadron spectra for RHIC Au + Au collision data of 7.7-200 GeV and the LHC Pb + Pb collision data of 2.76 TeV are used in this study. Spectra fit range is $p_T \leqslant$ 3 GeV/c for a comparable p_T range across all energies while focusing on the bulk properties. Figure 1 shows blast-wave fits to identified particle transverse momentum spectra in 40%-60% Au + Au collisions at $\sqrt{s_{NN}}$ =200 GeV in top panels, with corresponding deviations of those fits to experimental data divided by data uncertainties shown in the bottom panels. BGBW and TBW models are compared in the left and right column of Fig.1. TBW agrees with most data points within 3σ. With smaller χ^2/nDoF, TBW gives a better description for

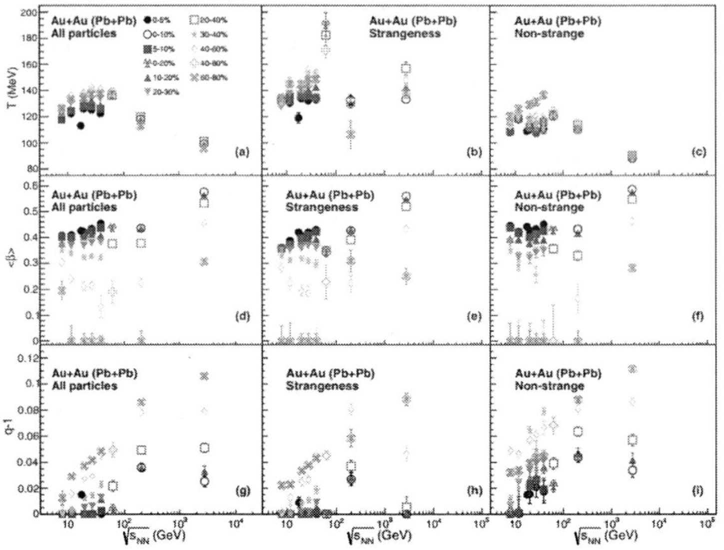

Figure 2. Collision energy dependence of the TBW extracted kinetic freeze-out parameters in different centralities for all particles (left column), strangeness (middle column), and non-strange particles (right column).

data than BGBW fit. Figure 2 shows the energy and centrality dependence of kinetic freeze-out parameters temperature T, radial flow velocity $\langle\beta\rangle$ and the degree of non-equilibrium $(q-1)$ of the TBW model. Temperature T shows an increasing trend with beam energy at RHIC BES energy, then decreases toward LHC energy. In central collisions, radial flow velocity $\langle\beta\rangle$ increases as collision energy increases. At any given collision energy, $\langle\beta\rangle$ decreases from central to peripheral collisions. The degree of non-equilibrium $(q-1)$ increases from central to peripheral and with increasing collision energy. In central collisions at RHIC BES energy, $(q-1)$ is close to 0. Smaller $\langle\beta\rangle$ with smaller $(q-1)$ at lower collision energy suggests that a weaker expansion with smaller deviation from thermal equilibrium, or energy density fluctuations at initial state either decrease or are more washed out at later stage at those lower collision energies. A higher T and smaller $\langle\beta\rangle$ for RHIC than LHC suggests the system at RHIC reaches freeze-out earlier with a lower radial flow than at LHC.

The comparison of strangeness and non-strange in the middle and right columns of Fig.2 shows that the strange hadrons have smaller non-equilibrium degree $(q-1)$ but higher temperature T, with similar radial flow $\langle\beta\rangle$ than their non-strange partners. It suggests that strangeness freeze-out earlier and approach equilibrium quickly from peripheral to central collisions than light hadrons. Non-strange hadrons exhibit a higher kinetic freeze-out temperature at RHIC than LHC. However, strange hadrons consistently show similar freeze-out temperatures for RHIC and LHC. This difference indicates that the system at RHIC spends shorter time for non-strange particles to reach kinetic freeze-out than at LHC.

Bulk viscosity measures damping for volumetric expansion in fluid. It has been suggested that the bulk viscosity increases dramatically toward the phase transition [14]. In non-equilibrium statistics framework, the dependence of temperature on the non-equilibrium degree is associated with the bulk viscosity ξ in linear or quadratic proportion [5]. We extract information about ξ from the quadratic fits of $T = T_0 + b(q-1) - d\xi(q-1)^2$ for each centrality

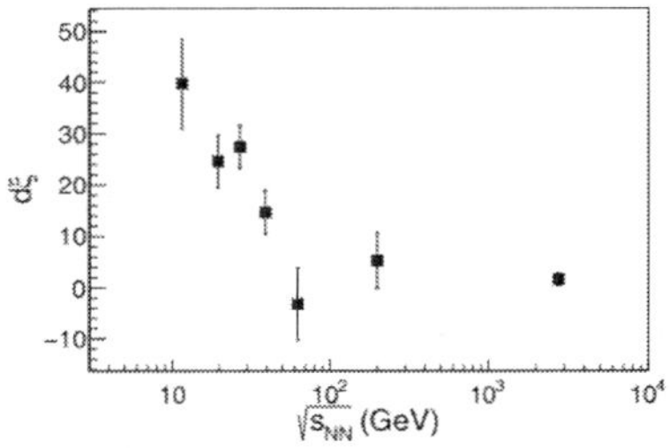

Figure 3. Collision energy dependence of bulk viscosity related parameters $d\xi$.

class at each collision energy, where $d\xi$, T_0 and b are fitting paramaters. We observed that $d\xi$, the term associated with bulk viscosity, indeed increases toward lower collision energies.

4 Conclusion

We present a study of BGBW and TBW fits on spectra of Au + Au (RHIC) collisions at $\sqrt{s_{NN}}$ =7.7 - 200 GeV, and Pb + Pb (LHC) collisions of $\sqrt{s_{NN}}$ =2.76 TeV. We find TBW describes experimental measurements better. With TBW, T shows an increasing trend with beam energy at RHIC BES, then decreases toward LHC. $\langle\beta\rangle$ and $q-1$ increase as collision energy increases. Strange particles freeze out earlier than light hadrons and approach equilibrium more quickly from peripheral to central collisions. The system at RHIC spends shorter time for non-strange particles to reach kinetic freeze-out than that at LHC. Bulk viscosity related parameter increases quickly toward lower collision energies.

This work was partly supported by the National Natural Science Foundation of China under Grants No. 11890710, No.11890713 and No.11720101001.

References

[1] J. Rafelski and B. Muller, Phys. Rev. Lett. **48** (1982), 1066 [erratum: Phys. Rev. Lett. **56** (1986), 2334]
[2] J. Adam *et al.* [STAR Collaboration], Phys. Rev. C **102**, 034909 (2020)
[3] E. Schnedermann, J. Sollfrank and U. W. Heinz, Phys. Rev. C **48**, 2462 (1993)
[4] E. Schnedermann and U. W. Heinz, Phys. Rev. C **50**, 1675 (1994)
[5] G. Wilk and Z. Wlodarczyk, Eur. Phys. J. A **40**, 299 (2009)
[6] W. M. Alberico, A. Lavagno and P. Quarati, Eur. Phys. J. C **12**, 499 (2000)
[7] T. Osada and G. Wilk, Phys. Rev. C **77**, 044903 (2008) Erratum: [Phys. Rev. C **78**, 069903 (2008)]
[8] T. S. Biro and B. Muller, Phys. Lett. B **578**, 78 (2004)
[9] K. Jiang *et al.*, Phys. Rev. C **91**, 024910 (2015)
[10] Z. Tang, Y. Xu, L. Ruan, G. van Buren, F. Wang and Z. Xu, Phys. Rev. C **79**, 051901 (2009)
[11] K. Urmossy, G. G. Barnaföldi, S. Harangozó, T. S. Biró and Z. Xu, J. Phys. Conf. Ser. **805**, no.1, 012010 (2017)
[12] M. Shao, L. Yi, Z. Tang, H. Chen, C. Li and Z. Xu, J. Phys. G **37**, 085104 (2010)
[13] Z. Tang *et al.*, Chin. Phys. Lett. **30**, 031201 (2013)
[14] F. Karsch, D. Kharzeev and K. Tuchin, Phys. Lett. B **663**, 217-221 (2008)

EPJ Web of Conferences **259**, 11002 (2022) https://doi.org/10.1051/epjconf/202225911002
SQM 2021

Elliptic flow for φ mesons measured by PHENIX

Iurii Mitrankov[1,*], *Yaroslav* Berdnikov[1], *Alexander* Berdnikov[1], and *Dmitry* Kotov[1]

[1]Peter the Great St.Petersburg Polytechnic University (SPbPU)

Abstract. The systematic study of hadronic elliptic flow in various relativistic heavy ion collisions is important for the investigation of the initial geometry influence on the quark gluon plasma characteristics. The φ meson consists of strange and antistrange quarks and has a small interaction cross section with non-strange hadrons. Therefore, φ mesons are barely affected by late hadronic stage and reflect detailed information about hot and dense matter properties. Additionally, the comparison of elliptic flow for φ mesons to those of charged hadrons will provide additional information on the flavor dependence of flow. PHENIX has measured second order azimuthal anisotropy coefficients for φ mesons in Cu+Au collisions at $\sqrt{s_{NN}}$ = 200 GeV and in U+U collisions at $\sqrt{s_{NN}}$ = 193 GeV at midrapidity ($|\eta| < 0.35$). The obtained data suggest scaling of elliptic flow for φ mesons with eccentricity of participant nucleons in Cu+Au, U+U, and Au+Au collisions. Viscous hydrodynamic model iEBE-VISHNU provides a simultaneous description of the obtained data.

1 Introduction

One of the most important goals of the current ultra-relativistic heavy ion research is the investigation of the quark gluon plasma (QGP) [1] properties. Study of second order azimuthal anisotropy has played a key role in the establishment of the QGP formation at the Relativistic Heavy Ion Collider [2]. Elliptic flow is commonly quantified by the second Fourier moment $v_2 \equiv \langle\cos 2(\phi - \Psi_{RP})\rangle$ of the azimuthal momentum distribution [3]. The detailed dependencies of v_2 on centrality, transverse momentum p_T, and particle species can lend insight on reaction dynamics and are important for defining parameters of viscous hydrodynamic, which may describe QGP behavior [4].

According to the Okubo-Zweig-Iizuka (OZI) rule φ-mesons have a relatively large mean free path, compared to the transverse size of the emitting system and those of $\pi^{\pm}$-meson [5, 6]. Also, the φ-mesons mostly decay after the QGP phase [1, 5], therefore φ kinematic properties are not affected by hadronic stage and bring detailed information of the QGP properties. Since the φ-meson mass is comparable to the (anti)protons $(p + \bar{p})/2$ mass, the comparisons of φ-meson, $\pi^{\pm}$-meson and $(p + \bar{p})/2$ provide an investigation of v_2 dependence on quark content and hadron mass. If the hadronic v_2 is proportional to the number of valence quarks n_q of the hadron, the flowing medium reflects quark degrees of freedom, otherwise hadronic stage is responsible for elliptic flow development [2].

The study of φ-meson production in heavy ion collisions, such as Cu+Au and U+U collisions, indicate additional mechanisms involved in its production in these collision systems

*e-mail: mitrankovy@gmail.com

[7, 8]. Additionally, anisotropic flow is strongly coupled to the medium density, initial geometric shape, therefore v_2 for φ-mesons was studied in asymmetric Cu+Au collisions and collisions of deformed uranium nucleus U+U.

To investigate the underlying processes behind v_2 evolution, the comparison of experimental elliptic flow v_2 for φ-mesons to theoretical predictions is needed. The iEBE-VISHNU model [9] performs event-by-event simulations for relativistic heavy-ion collisions using (2+1)D viscous hydrodynamic and hadronic cascade model. This model has proved itself valid for recent PHENIX results on elliptic and triangular flow for charged hadrons, published in the Nature Physics [10].

In this paper elliptic flow for φ-mesons in Cu+Au collisions at $\sqrt{s_{NN}}$ = 200 GeV and in U+U collisions at $\sqrt{s_{NN}}$ = 193 GeV is studied as a function of centrality and kinetic properties, i.e. transverse momentum p_T and transverse kinetic energy KE_T. The obtained results are compared to iEBE-VISHNU hydrodynamic model predictions.

2 Results and discussion

The procedure for calculating reaction plane azimuthal angle in Cu+Au collisions at $\sqrt{s_{NN}}$ = 200 GeV and in U+U collisions at $\sqrt{s_{NN}}$ = 193 GeV, φ meson v_2, and systematic uncertainties can be found elsewhere [11–13].

The left and right panels of Figure 1 compare the unscaled and scaled (respectively) φ-meson v_2 (p_T) in 0-20%, 20-40%, 40-60%, and 20-60% Cu+Au, 0-50% U+U collisions, previous 20-60% Au+Au PHENIX results, 0-30% and 30-80% Au+Au STAR results [11]. The scaling of φ-meson elliptic flow suggests that the v_2 values follow common empirical scaling with $\varepsilon_2 N_{part}^{1/3}$. The same scaling was observed in previous PHENIX results on charged hadrons v_2 in Cu+Au, Au+Au, and Cu + Cu collisions at $\sqrt{s_{NN}}$ = 200 GeV [13]. The average number of participating nucleons N_{part} and participant eccentricity of second order ε_2 were estimated using Glauber model Monte-Carlo simulation of the each collision system. Scaling with participant eccentricity of second order ε_2 represents dependence of v_2 on collision geometry. While N_{part} is proportional to the volume of the QGP, $N_{part}^{1/3}$ is proportional to the radius of the QGP.

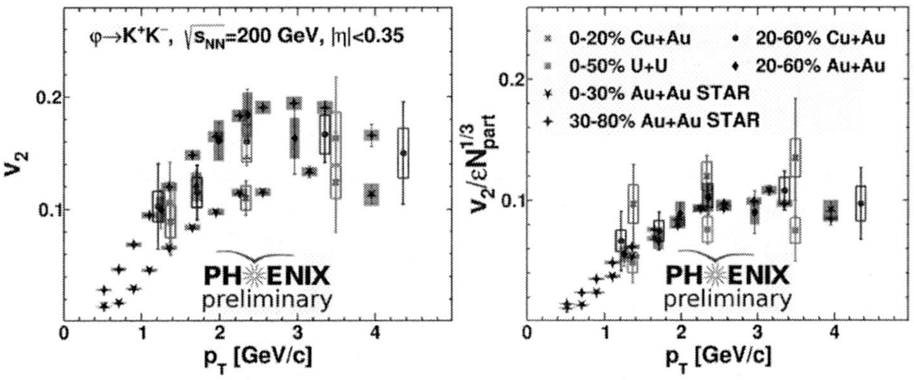

Figure 1. The elliptic flow v_2 (left) and $v_2/(\varepsilon_2 N_{part}^{1/3})$ (right) vs. p_T for φ-mesons in 0-20%, 20-40%, 40-60%, and 20-60% Cu+Au collisions, 0-50% U+U collisions, previous 20-60% Au+Au PHENIX results, 0-30% and 30-80% Au+Au STAR results. Here and below error bars and open boxes around points correspond to statistical and systematic uncertainties.

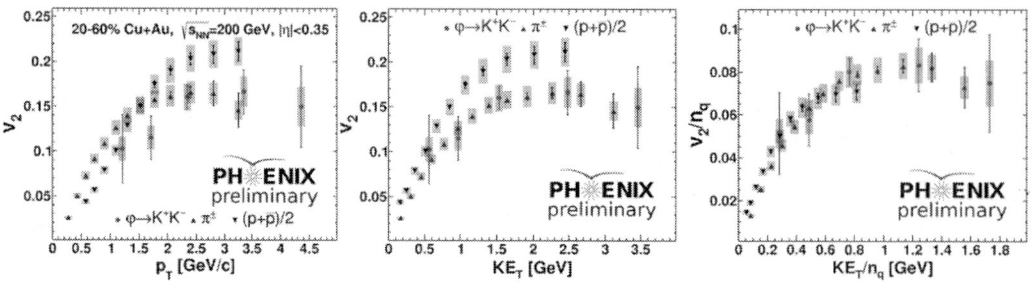

Figure 2. The comparison of elliptic flow v_2 and v_2/n_q for φ-mesons in 20-60% Cu+Au collisions to those for $\pi^{\pm}$ and $(p + \bar{p})/2$ [14] as a function of p_T (left), KE_T (middle) and KE_T/n_q (right).

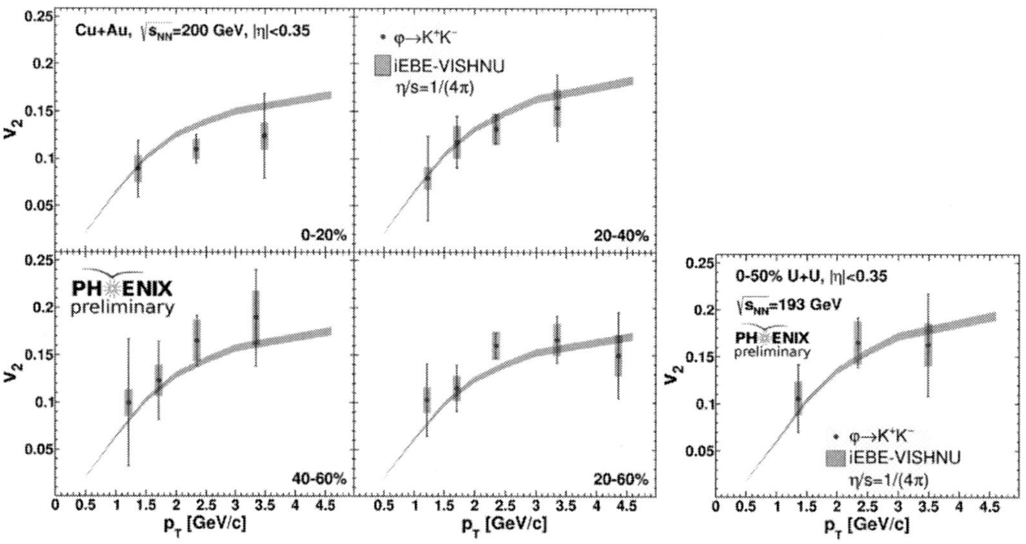

Figure 3. The comparison of measured elliptic flow v_2 (p_T) for φ-mesons in 0-20%, 20-40%, 40-60%, and 20-60% Cu+Au collisions at $\sqrt{s_{NN}}$ = 200 GeV and in 0-50% U+U collisions at $\sqrt{s_{NN}}$ = 193 GeV to iEBE-VISHNU hydrodynamic model prediction with specific viscosity $\eta/s = 1/(4\pi)$.

The example of the comparison of elliptic flow obtained for φ-mesons in 20-60% Cu+Au collisions at $\sqrt{s_{NN}}$ = 200 GeV to those for $\pi^{\pm}$ and $(p + \bar{p})/2$ [14] is shown in Figure 2. The scaling of elliptic flow for light hadrons with the number of valence quarks n_q of the hadron and transverse kinetic energy per one quark KE_T/n_q is observed (Figure 2 right panel). This result along with a smaller rescattering cross section for φ-mesons than for $\pi^{\pm}$ may indicate that elliptic flow development occurs before hadronization in the QGP phase of heavy-ion collision.

In order to better understand the physics behind v_2 development, the comparisons of measured elliptic flow for φ-mesons in 0-20%, 20-40%, 40-60%, and 20-60% Cu+Au collisions and in 0-50% U+U collisions to iEBE-VISHNU model predictions are shown in Figure 3. The iEBE-VISHNU model is based on (2+1)D viscous hydrodynamic and includes the QGP formation. The results are well described with v_2 obtained with iEBE-VISHNU calculations. A specific viscosity value of $\eta/s = 1/(4\pi)$ was used in the calculations.

3 Conclusion

Elliptic flows for φ-mesons were measured in Cu+Au collisions at $\sqrt{s_{NN}} = 200$ GeV and in U+U collisions at $\sqrt{s_{NN}} = 193$ GeV as a function of p_T, KE_T, and KE_T/n_q and compared with previous measurements in Au+Au collisions, v_2 for $\pi^{\pm}$ and $(p+\bar{p})/2$, and hydrodynamic model iEBE-VISHNU calculations.

The scaling of obtained φ-meson v_2 with second order participant eccentricity, and cube root of the number of participant nucleons is observed and confirms the previous PHENIX results. Additionally, data suggest dependence of light hadron elliptic flow on the number of valence quarks n_q of the hadron. Therefore, system size and geometry influence can be considered by normalizing factor $\varepsilon_2 N_{part}^{1/3}$, whereas hadron type dependence - by scaling with n_q. Due to all above, elliptic flow development probably occurs before hadronization in the QGP phase of heavy-ion collision.

The agreement of experimental data with iEBE-VISHNU calculations suggests that QGP behaviour can be described with (2+1)D viscous hydrodynamic with specific viscosity $\eta/s = 1/(4\pi)$.

Acknowledgments

The research is partially funded by the Ministry of Science and Higher Education of the Russian Federation under the strategic academic leadership program 'Priority 2030' (Agreement 075-15-2021-1333 dated 30.09.2021)

References

[1] K. Adcox et al. (PHENIX), Nuclear Physics A **757**, 184–283 (2005)
[2] S.S. Adler et al. (PHENIX), Phys. Rev. Lett. **94**, 232302 (2005)
[3] S. Voloshin, Y. Zhang, Z. Phys. C **70**, 665 (1996)
[4] V. Greco, C.M. Ko, P. Lévai, Phys. Rev. C **68**, 034904 (2003)
[5] A. Shor, Phys. Rev. Lett. **54**, 1122 (1985)
[6] P. Zyla et al. (Particle Data Group), PTEP **2020**, 083C01 (2020)
[7] A. Berdnikov, D. Kotov, I. Mitrankov, Journal of Physics: Conference Series **1135**, 012044 (2018)
[8] M.M. Mitrankova, Y.A. Berdnikov, A.Y. Berdnikov, D.O. Kotov, I.M. Mitrankov (PHENIX), Phys. Scripta **96**, 084010 (2021)
[9] C. Shen, Z. Qiu, H. Song, J. Bernhard, S. Bass, U. Heinz, *The iEBE-VISHNU code package for relativistic heavy-ion collisions* (2015)
[10] C. Aidala et al. (PHENIX), Nature Phys. **15**, 214 (2019)
[11] S. Afanasiev et al. (PHENIX), Phys. Rev. Lett. **99**, 052301 (2007)
[12] A. Adare et al. (PHENIX), Phys. Rev. C **88**, 064910 (2013)
[13] A. Adare et al. (PHENIX), Phys. Rev. C **94**, 054910 (2016)
[14] J. Barrette, R. Bellwied, S. Bennett et al., Phys. Rev. C **56**, 3254 (1997)

Dynamical cluster and hypernuclei production in heavy-ion collisions

Susanne Gläßel[1], *Viktar* Kireyeu[2], *Vadim* Voronyuk[2], *Jörg* Aichelin[3,4], *Christoph* Blume[1], *Elena* Bratkovskaya[5,6,7,*], *Gabriele* Coci[5], *Vadim* Kolesnikov[2], and *Michael* Winn[3]

[1]Institut für Kernphysik, Johann Wolfgang Goethe-Universität, Max-von-Laue-Str. 1, 60438 Frankfurt am Main, Germany
[2]Joint Institute for Nuclear Research, Joliot-Curie 6, 141980 Dubna, Moscow region, Russia
[3]SUBATECH, Université de Nantes, IMT Atlantique, IN2P3/CNRS, 4 rue Alfred Kastler, 44307 Nantes cedex 3, France
[4]Frankfurt Institute for Advanced Studies, Ruth Moufang Str. 1, 60438 Frankfurt, Germany
[5]GSI Helmholtzzentrum für Schwerionenforschung GmbH, Planckstr. 1, 64291 Darmstadt, Germany
[6]Institut für Theoretische Physik, Johann Wolfgang Goethe-Universität, Max-von-Laue-Str. 1, 60438 Frankfurt am Main, Germany
[7]Helmholtz Research Academy Hesse for FAIR (HFHF), GSI Helmholtz Center for Heavy Ion Physics, Campus Frankfurt, 60438 Frankfurt, Germany

Abstract. We study light cluster and hypernuclei production in heavy-ion collisions from SIS to RHIC energies based on the n-body dynamical transport approach PHQMD (Parton-Hadron-Quantum-Molecular-Dynamics). In PHQMD clusters are formed dynamically due to the interactions between baryons described on the basis of Quantum Molecular Dynamics (QMD) which allows to propagate the n-body Wigner density and n-body correlations in phase-space, which is essential for the cluster formation. The clusters are identified by the MST (Minimum Spanning Tree) or the SACA ('Simulated Annealing Cluster Algorithm') algorithm which finds the most-bound configuration of nucleons and clusters. Collisions among hadrons as well as Quark-Gluon-Plasma formation and parton dynamics in PHQMD are treated in the same way as in the PHSD (Parton-Hadron-String-Dynamics) transport approach. We study the time evolution of the cluster formation in the expanding medium and the stability of the clusters. We present a comparison of the PHQMD results for d, ^{3}He as well as for the hypernuclei with experimental data.

1 Introduction

The nature of cluster and hypernuclei production in heavy-ion collisions is one of the intriguing topics nowadays. The clusters have been observed experimentally at midrapidity from SIS up to LHC energies (cf. [1–6]). However, according to statistical models, the temperature of the surrounding matter at freeze-out is about $T \sim 100$ MeV. An understanding of how weakly-bound objects, such as clusters (with binding energies below 10 MeV), can be formed and survive in such a hot environment appears to be a puzzle, usually called "ice in the fire".

*e-mail: E.Bratkovskaya@gsi.de

The theoretical study of cluster formation - based on statistical or coalescence models - can not shed much light on the microscopic origin of such phenomena. In order to understand the dynamical processes leading to the creation of clusters at midrapidity, we studied in Ref. [7] the cluster and hypernuclei formation based on the Parton-Hadron-Quantum-Molecular-Dynamics (PHQMD) approach [8]. The PHQMD is a n-body microscopic transport approach for the description of heavy-ion collisions including cluster production. The dynamically formed clusters are identified by the SACA and MST algorithms. The PHQMD merges the collision integrals of the Parton-Hadron-String Dynamics (PHSD) approach [9, 10] with density dependent 2-body potentials between baryons, similar to the Quantum Molecular Dynamics (QMD) approach. Baryons are described by Gaussian wave functions which propagate under the influence of mutual-density-dependent 2-body forces (and not due to a mean field). As demonstrated in [8], PHQMD reproduces single-particle observables for energies $\sqrt{s_{NN}} \sim 2 \div 200$ GeV and describes the cluster formation as well.

In PHQMD clusters are formed *dynamically* due to the potential interactions between baryons. For this study we used a Skyrme density dependent static potential and a Coulomb interaction entering the quantal Hamiltonian equations of motion with 2-body potential energies. The clusters are identified by the MST [11] or the SACA [12, 13] algorithm, which finds the most-bound configuration of nucleons and clusters. In the MST algorithm only the coordinate-space information is used to define clusters: Two nucleons are considered as part of a cluster if their distance is less than $r_0 = 4$ fm in the cluster rest system obtained by a Lorentz transformation from the computational frame. Nucleons, which are more distant, do not feel anymore an attractive interaction. The MST procedure has been recently applied to study the cluster production in different transport approaches [14]. This study showed that in QMD-like approaches, where the few-body correlations are kept, more light clusters at midrapidity are produced than in mean-field or cascade models.

In spite of the advantages of accounting for the n-body interactions (on the level of 2-body potentials) compared to mean-field-based models, the semi-classical QMD models can not achieve the stability of clusters, since they cannot project the n-body density onto the quantum ground state of a cluster, which might influence the nucleon distribution inside the cluster for very late times. To overcome this problem we have fixed a cluster freeze-out time t at which the cluster production is terminated: $t = t_0 \cdot \cosh y$, where t_0 is the cluster freeze-out time at midrapidity. We call t the "physical time" because it marks identical times in the rest systems. Due to time dilatation it depends on the rapidity of the cluster rest system in the center-of-mass frame of the heavy-ion collision.

2 Cluster and hypernuclei formation in heavy-ion collisions

Here we highlight some results from our recent study [7] and we refer the readers to this reference for details and further results. We note that the MST algorithm is applied here for cluster recognition.

We start with showing the comparison between the midrapidity clusters measured by the NA49 Collaboration in heavy-ion collisions at the SPS at CERN and the PHQMD calculations. Figure 1 shows the rapidity distribution of deuterons (left)and ^{3}He (right) for central Pb+Pb collisions at the beam energy $E_{kin} = 40$ AGeV. Here we identify the clusters at $t_0 = 53$ fm/c and 67 fm/c. The shape of the rapidity distributions are well reproduced by PHQMD within the measurement uncertainties. We note also that in the whole p_T range, where experimental data are available, the transverse-momentum spectra in different rapidity bins are remarkable well reproduced by PHQMD (cf. Fig. 9 in [7])). Similar findings hold for the AGS and RHIC BES energies where the PHQMD provides a very good description of the rapidity distributions, p_T spectra and covariant coalescence function, B_A, for d, t, and ^{3}He.

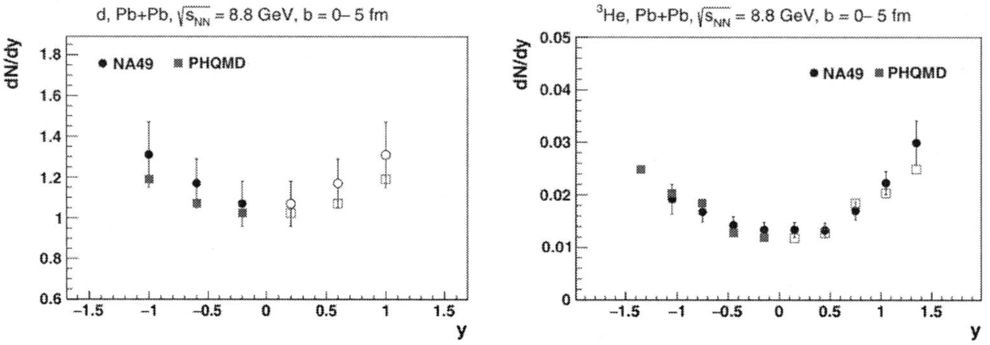

Figure 1. The rapidity distribution of deuterons (left) and ^{3}He (right) for central Pb+Pb collisions at the beam energy E_{kin} = 40 AGeV. The dots indicate the experimental data from the NA49 Collaboration [3]; the red squares show the PHQMD results taken at the physical time $t = t_0 \cosh(y)$ for t_0 = 53 fm/c for deuterons and for t_0 = 67 fm/c for ^{3}He. The figure is taken from Ref. [7].

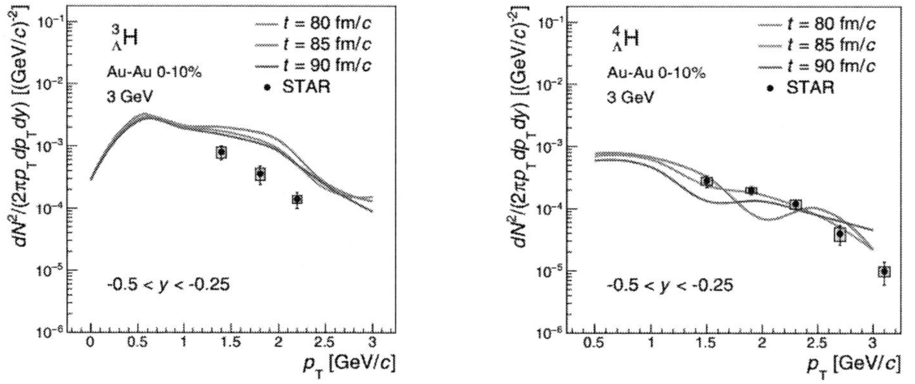

Figure 2. Transverse-momentum distribution of ${}^3_\Lambda$H (left) and ${}^4_\Lambda$H (right) for $-0.5 < y < -0.25$ in central Au+Au collisions at $\sqrt{s_{NN}}$ = 3 GeV. The filled circles indicate the preliminary experimental data from the STAR Collaboration [15]. The PHQMD results are taken at the times t = 80 fm/c (red lines), 85 fm/c (blue lines) and 90 fm/c (green lines). The figure is taken from Ref. [7].

Recently, first preliminary data of hypernuclei production at a center-of-mass energy of $\sqrt{s_{NN}}$ = 3 GeV have become public. In their fixed-target program the STAR collaboration measured for two hypernuclei, ${}^3_\Lambda$H and ${}^4_\Lambda$H, the transverse-momentum distribution $d^2N/p_T dp_T$ in different rapidity bins [15].

In Fig. 2 we show the results for ${}^3_\Lambda$H (left) and ${}^4_\Lambda$H (right) for $-0.5 < y < -0.25$. The filled circles indicate the experimental data from the STAR Collaboration [15]. The PHQMD results are taken at the times t = 80 fm/c (red lines), 85 fm/c (blue lines) and 90 fm/c (green lines), but at this low energy the time dependence of the cluster yield is weak. The calculations show that the trend of the experimental p_T spectra is well reproduced. We stress that the ΛN potential implemented in PHQMD is presently quite simple (= 2/3 V_{NN}), so we do not expect an exact quantitative agreement. We overpredict the differential yield of ${}^3_\Lambda H$, but reproduce the yield of ${}^{4}_{\Lambda}H$. This demonstrates that PHQMD in its present version is a good starting point for more sophisticated studies of hypernucleus production in this energy regime.

Since PHQMD is a microscopic transport approach and allows one to follow the history of each particle, we also studied the history of cluster production and identified when and where clusters are formed. We find a visible similarity of the cluster production at midrapidity from SIS to RHIC BES energies: the clusters are produced shortly after the collisions between the hadrons have ceased and, most important, in coordinate space they are formed behind the front of the fast expanding hadrons. This is the reason why clusters can survive. Thus, according to our PHQMD study, the "ice in the fire" puzzle can be explained by the fact that clusters come from different space regions than free nucleons. This does not exclude that they have p_T spectra compatible with those expected from a homogeneous thermal source.

Acknowledgements

The authors acknowledge support by the Deutsche Forschungsgemeinschaft (DFG, German Research Foundation): grants BR 4000/7-1 and BL 982/3-1, by the Russian Science Foundation grant 19-42-04101 and by the GSI-IN2P3 agreement under contract number 13-70. Also we thank the COST Action THOR, CA15213 and the European Union's Horizon 2020 research and innovation program under grant agreement No 824093 (STRONG-2020). The computational resources have been provided by the LOEWE-Center for Scientific Computing and the "Green Cube" at GSI, Darmstadt.

References

[1] H.H. Gutbrod, A. Sandoval, P.J. Johansen, A.M. Poskanzer, J. Gosset, W.G. Meyer, G.D. Westfall, R. Stock, Phys. Rev. Lett. **37**, 667 (1976)
[2] S. Nagamiya, M.C. Lemaire, E. Moller, S. Schnetzer, G. Shapiro, H. Steiner, I. Tanihata, Phys. Rev. C **24**, 971 (1981)
[3] T. Anticic et al. (NA49), Phys. Rev. C **94**, 044906 (2016), 1606.04234
[4] J. Adam et al. (STAR), Phys. Rev. C **99**, 064905 (2019), 1903.11778
[5] I. Arsene et al. (BRAHMS), Phys. Rev. C **83**, 044906 (2011), 1005.5427
[6] S. Acharya et al. (ALICE), Nucl. Phys. A **971**, 1 (2018), 1710.07531
[7] S. Gläßel, V. Kireyeu, V. Voronyuk, J. Aichelin, C. Blume, E. Bratkovskaya, G. Coci, V. Kolesnikov, M. Winn, Phys. Rev. C **105**, 014908 (2022), 2106.14839
[8] J. Aichelin, E. Bratkovskaya, A. Le Fèvre, V. Kireyeu, V. Kolesnikov, Y. Leifels, V. Voronyuk, G. Coci, Phys. Rev. C **101**, 044905 (2020), 1907.03860
[9] W. Cassing, E.L. Bratkovskaya, Phys. Rev. C **78**, 034919 (2008), 0808.0022
[10] W. Cassing, E.L. Bratkovskaya, Nucl. Phys. A **831**, 215 (2009), 0907.5331
[11] J. Aichelin, Phys. Rept. **202**, 233 (1991)
[12] R.K. Puri, C. Hartnack, J. Aichelin, Phys. Rev. C **54**, 28 (1996)
[13] R.K. Puri, J. Aichelin, J. Comput. Phys. **162**, 245 (2000), nucl-th/9811018
[14] V. Kireyeu, Phys. Rev. C **103**, 054905 (2021), 2103.10542
[15] Y.H. Leung (STAR) (2021), International conference on Critical Point and Onset of Deconfinement (CPOD), https://indico.cern.ch/event/985460/

System size and energy dependence of resonance production at ALICE

Vikash Sumberia[1] **(for the ALICE collaboration)***

[1]University of Jammu, Jammu and Kashmir, 180006, India

Abstract. Hadronic resonances, thanks to their relatively short lifetimes, can be used to probe the properties of the hadronic phase in ultrarelativistic heavy-ion collisions. In particular their lifetimes are exploited to investigate the interplay between particle rescattering and regeneration after hadronization. In this contribution we present recent results on $\rho(770)^0$, $K^*(892)$, $\phi(1020)$, $\Sigma(1385)^{\pm}$, $\Xi(1530)^0$ and $\Lambda(1520)$ production in pp, p–Pb, Pb–Pb and Xe–Xe collisions at LHC energies.

1 Introduction

Production properties of hadronic resonances and their comparison with those of stable hadrons are useful tools to understand the properties of the strongly-interacting matter formed in relativistic nucleus-nucleus collisions. Due to their short lifetimes (comparable to the lifetime of the hadronic phase [1]), resonances provide an insight into the dynamical evolution of the fireball. The estimated yield of the resonances may be modified even after the chemical freezeout due to (pseudo)-elastic rescattering or regeneration of their decay daughters. These processes are dominant at low p_T ($\lesssim 2$ GeV/c) [2–4]. The rescattering process supresses the observed yields with respect to what is produced at chemical freeze-out while regeneration enhances the yield of resonances [5]. Thus, the study of resonance particles can provide an estimate for the lifetime of the hadronic phase (resonance to stable hadron yield ratios), understand the partonic in-medium energy losses (R_{AA} and R_{pA}) and the mechanism of particle production. In these proceedings, we present recent results obtained by ALICE for the mesonic resonances ($\rho(770)^0$, K$(892)^{*0}$ and $\phi(1020)$) and baryonic resonances ($\Sigma(1385)^{\pm}$, $\Xi(1530)^0$ and $\Lambda(1520)$) in pp, p–Pb, Pb–Pb and Xe–Xe collisions at various centre-of-mass collision energies. Hereafter, the particles $\rho(770)^0$, K$(892)^{*0}$, $\phi(1020)$, $\Sigma(1385)^{\pm}$ and $\Xi(1530)^0$ will be represented by ρ^0, K^{*0}, ϕ, $\Sigma^{\pm}$ and Ξ^0 respectively.

2 Analysis Method

This analysis used the data recorded by the ALICE detector [6]. The main components of ALICE which are relevant to these analyses are the Inner Tracking System (ITS, for tracking and vertex finding), Time Projection Chamber (TPC, for tracking and particle identification) and V0 detector (for triggering and centrality estimation). The events selected for the analysis are required to have a reconstructed primary vertex within 10 *cm* of the nominal interaction

*e-mail: vikash.sumberia@cern.ch

point along the beam direction (at the centre of the ALICE barrel). The measurements are performed at mid-rapidity ($|y| < 0.5$ in pp, Pb–Pb and Xe–Xe collisions and $0 < y_{cm} < 0.5$ in p–Pb collisions) and are presented as a function of charge particle density (which is also measured at mid-rapidity). The invariant mass is reconstructed from the measured momenta of identified particles. The combinatorial background is estimated using the event-mixing technique. The mixed-event distribution is then normalized to the unlike-charge distribution in the mass range $1.65 < M_{pK} < 1.75$ GeV/c^2 for $\Lambda(1520)$ (for details on other analyses see these citations [7–9, 11, 12]) and then subtracted from unlike-charge distribution in each p_T bin. The residual background is removed by fitting the invariant mass distribution with the sum of a Voigtian and a polynomial function. The raw yields are calculated after background subtraction and are corrected for finite detector acceptance, tracking efficiency of daughter particles and branching ratio. Table 1 shows the lifetime values, decay modes used for reconstruction and the branching ratios of some of the measured hadronic resonances.

Table 1: Lifetimes, reconstructed decay modes and branching ratios of measured resonance particles

	$\rho(770)^0$	$K(892)^{*0}$	$\Sigma(1385)^{\pm}$	$\Lambda(1520)$	$\Xi(1530)^0$	$\phi(1020)$
Lifetime (fm/c)	1.3	4.2	5.5	12.6	21.7	46.4
Decay mode	$\pi^+ \pi^-$	$K^+ \pi^-$	$\Lambda\, \pi^{\pm}$	pK^-	$\Xi^- \pi^+$	$K^+ K^-$
B.R. (%)	100	66.6	87.0	22.5	66.7	49.2

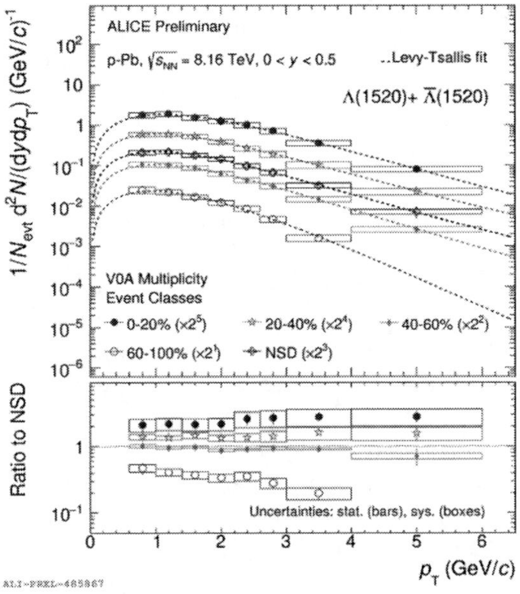

Figure 1. Corrected $\Lambda(1520)$ p_T spectrum for minimum-bias collisions (0–100%) and four multiplicity intervals (0–20%, 20–40%, 40–60% and 60–100%) in mid-rapidity p-Pb collisions at $\sqrt{s_{NN}}$ =8.16 TeV. The errors shown here are both statistical (bars) and systematic (boxes). The shaded curves represent the Levy-Tsallis fit.

3 Results and discussion

We present new results on p_T spectra for $\Lambda(1520)$ resonance in p–Pb collisions in the mid-rapidity region at $\sqrt{s_{NN}} = 8.16$ TeV in minimum-bias collisions (0-100 %) and four V0 event multiplicity classes (0–20%, 20–40%, 40–60% and 60–100%) as shown in Fig. 1. The minimum-bias spectra have been normalized to the fraction of non-single diffractive (NSD) events. We observed that p_T spectra harden with an increase in multiplicity and similar behaviour is observed for other measured resonances in pp, p–Pb, Pb–Pb and Xe–Xe collisions [7–9].

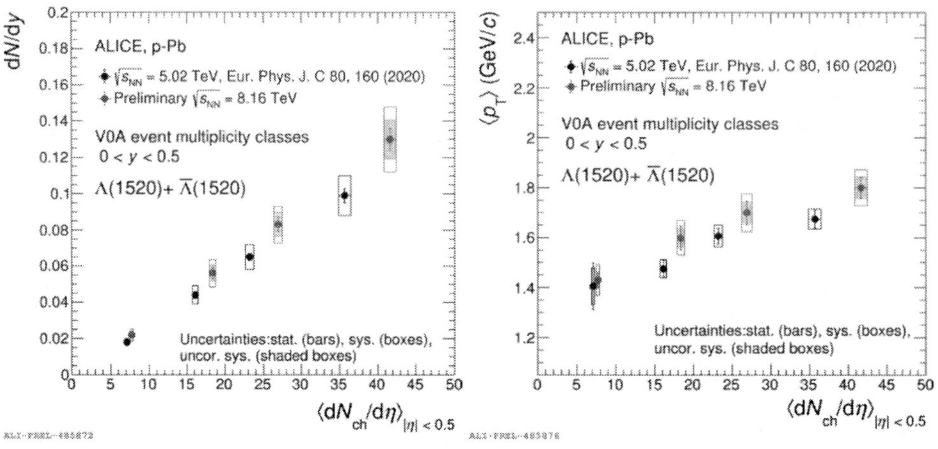

Figure 2: Integrated yield (left panel) and Average p_T (right panel) of Λ(1520) in p–Pb collisions at $\sqrt{s_{NN}}$ = 5.02 and 8.16 TeV in mid-rapidity region in various V0 event multiplicity classes. Bars show statistical errors, boxes show systematic errors and shaded boxes show uncorrelated errors.

The p_T-integrated yields of the Λ(1520) in p–Pb collisons at $\sqrt{s_{NN}}$ = 5.02 and 8.16 TeV (shown in the left panel of Fig. 2) have been measured in the mid-rapidity region in various V0 event multiplicity classes using the fitted Levy-Tsallis function to the measured p_T spectra of these resonances. Integrated yields of the resonances increase with event multiplicity and are independent of the colliding system and collision energy i.e. event multiplicity drives the resonance production. For the first time, average p_T of Λ^* in p–Pb collisions at $\sqrt{s_{NN}}$ = 8.16 TeV is measured in mid-rapidity region for various V0 event multiplicity classes (shown in the right panel of Fig. 2). The results have been compared with the previous measurements in p–Pb at $\sqrt{s_{NN}}$ = 5.02 TeV [10] and are comparable within the systematic uncertainties. The average p_T increases with event multiplicity.

The ratios of yields of resonances to those of stable hadrons is shown in the Fig. 3. We observe that the yields of ρ^0, K^{*0}, $\Sigma^{\pm}$, Λ(1520) and Ξ^0 show a suppression with an increase in the centrality of Pb–Pb collisions, while no such suppression is seen in the ϕ resonance (due to its longer lifetime compared to the lifetime of hadronic phase). Different suppression of resonances, depending on their lifetimes, provide some evidence for the extraction of a lifetime of the hadronic phase. In small system collisions (pp and p–Pb), the yield suppression with increase in event multiplicity is only observed for ρ^0 and K^{*0} resonances. This shows the dependence of the lifetime of the hadronic phase on the colliding systems i.e. the lifetime of the hadronic phase increases with the increase in the size of the colliding system.

4 Conclusions

Recent results of the hadronic resonances ($\rho(770)^0$, $K(892)^{*0}$, $\phi(1020)$, $\Sigma(1385)^{\pm}$, Λ(1520) and $\Xi(1530)^0$) in pp, p–Pb, Pb–Pb and Xe–Xe collisions obtained from the ALICE detector have been presented. The p_T spectra of these resonances get harder with the increase in multiplicity of the events. At similar event multiplicities resonance yields are independent of the colliding system or collision energy, which indicates that multiplicity drives the production

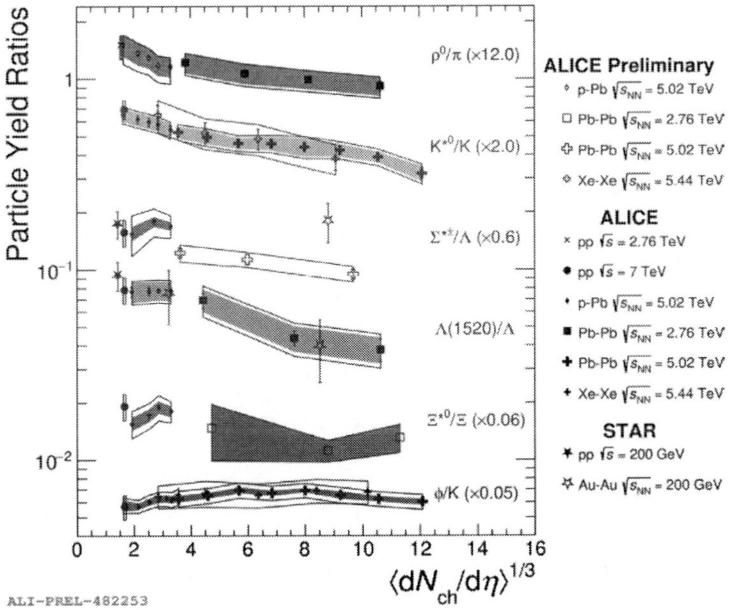

Figure 3: Ratios of integrated yields of various resonances to their respective stable particles in various systems and collision energies [13–18]. The bars and the boxes represent statistical and systematic uncertainty, respectively.

of resonance particles. The lifetime of the hadronic phase is limited (non-zero and finite) and this lifetime increases with the increase in the size of the colliding system.

References

[1] A. Aamodt *et al.* (ALICE Collaboration), Phys. Lett. B **696** , 328–37 (2011)
[2] Marcus Bleicher and Horst Stöcker, J. Phys. G **30** , S111–8 (2004)
[3] Markert C *et al.*, AIP Conf. Proc. **631**, 533–52 (2002)
[4] Badalà Angela, EPJ Conf. Proc. **96**, 01003 (2015)
[5] S. A. Bass and A. Dumitru, Phys. Rev. C **61**, 064909 (2000)
[6] Aamodt K *et al.* (ALICE Collaboration), J. Inst. **3**, S08002 i-245 (2008)
[7] S. Acharya *et al.* (ALICE Collaboration) , Phys. Lett. B **802**, 135225 (2020)
[8] S. Acharya *et al.* (ALICE Collaboration) , Phys. Lett. B **807**, 135501 (2020)
[9] D. S. D. Albuquerque (ALICE Collaboration) , Nucl. Phys. A **982**, 823–826 (2019)
[10] S. Acharya *et al.* (ALICE Collaboration) , E. Phys. J. C **80**, 160 (2020)
[11] S. Acharya et al. (ALICE Collaboration) Phys. Rev. C **99**, 064901 (2019)
[12] J. Adam et al. (ALICE Collaboration) Eur. Phys. J. C **76**, 245 (2016)
[13] S. Acharya et al. (ALICE Collaboration) Phys. Rev. C **99**, 024906 (2019)
[14] D Adamova (ALICE Collaboration) Eur. Phys. J. C **77**, 389 (2017)
[15] D Adamova (ALICE Collaboration) Eur. Phys. J. C **75**, 1 (2015)
[16] Sevil Salur (Star Collaboration) NATO Science series **166**, S665-S667 (2004)
[17] S. Acharya et al. (ALICE Collaboration) Phys. Rev. C **99**, 024905 (2019)
[18] J. Adam et al. (ALICE Collaboration) Nat. Phys. **13**, 535-539 (2017)

Multi-strange hadron production in p+p interactions at $\sqrt{s_{NN}}$=17.3 GeV

Szymon Pulawski[1,*]
for the NA61/SHINE Collaboration

[1]Institute of Physics, University of Silesia, Katowice

Abstract. The production of multi-strange hadrons in proton-proton interactions is currently studied with the NA61/SHINE experiment at the SPS. These particles are reconstructed via their weak decay typologies, exploiting the tracking and particle identification capabilities of NA61/SHINE. New measurements of rapidity and transverse momentum spectra of Ξ^- and Ξ resonances and their antiparticles are shown. The results are compared to those observed in A+A collisions as well as to model predictions.

1 NA61/SHINE facility

The data used for the analysis were recorded at the CERN SPS accelerator complex with the NA61/SHINE fixed target large acceptance hadron detector [1]. The NA61/SHINE tracking system consists of 4 large volume time projection chambers (TPCs). Two of the TPCs (VTPC1 and VTPC2) are within superconducting dipole magnets. Downstream of the magnets, two larger TPCs (MTPC-R and MTPC-L) provide acceptance at high momenta. The fifth small TPC (GAP-TPC) is placed between VTPC1 and VTPC2 directly on the beamline. The interactions were measured in the H2 beamline in the North Experimental Hall with a secondary beam of 158 GeV/*c* protons impinging on a cylindrical Liquid Hydrogen Target (LHT) of 20 cm length and 2 cm diameter. This beam was produced by 400 GeV/*c* protons hitting a Be-target. The primary protons were extracted from the SPS in a slow extraction mode with a flat top lasting about 10 seconds. Protons and other positively charged particles produced in the Be-target constitute the secondary hadron beam. Two Cherenkov counters identified the protons, a CEDAR (either CEDAR-W or CEDAR-N) and a threshold counter (THC). A selection based on signals from the Cherenkov counters identified the protons with a purity of about 99% [2]. A set of scintillation counters selects individual beam particles. Their trajectories are precisely measured by three beam position detectors (BPD-1, BPD-2, BPD-3) [1].

2 Multi-strange measurements

The presented results cover doubly strange hyperon production in inelastic p+p interactions at 158 GeV/*c* corresponding to $\sqrt{s_{NN}}$=17.3 GeV. Results are corrected for losses due to limited geometrical detector acceptance and reconstruction efficiency, as well as for secondary interactions and branching ratios to unmeasured channels.

*e-mail: szymon.pulawski@us.edu.pl

A total of 33 million minimum bias $p + p$ at 158 GeV/c events were recorded in 2009, 2010, and 2011 and analysed. Interactions in the target are selected with the trigger system and offline selection criteria.

Multi-strange hyperons are identified by their decay typologies. The $\Xi(1530)^0$ is produced in the primary interaction and decays strongly into Ξ^- and π^+. Then the Ξ^- travels for some distance, after which it decays into a Λ and a π^-. Subsequently the Λ decays into a proton and a π^-. A schematic drawing of the $\Xi(1530)^0$ decay chain is shown in Fig. 1.

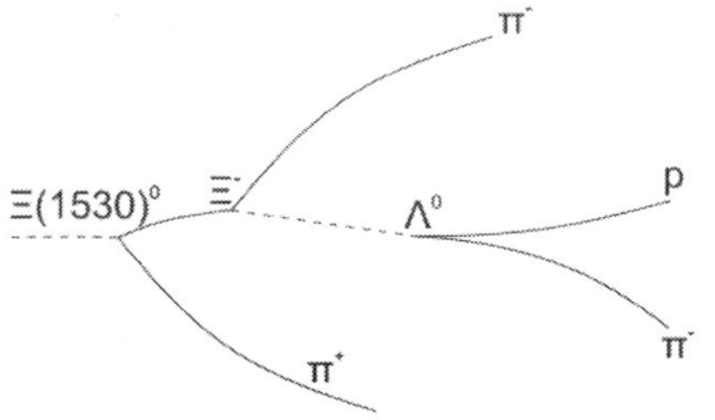

Figure 1. Schematic sketch of the $\Xi(1530)^0$ decay scheme.

2.1 Ξ production in inelastic $p + p$ collisions at 158 GeV/c

The p_T spectra in successive rapidity intervals of Ξ^- and $\overline{\Xi}^+$ measured in $p + p$ collisions at 158 GeV/c [3] are shown in Fig. 2 and scaled by appropriate factors for better visibility.

Rapidity distributions were then obtained by summing the measured transverse momentum spectra and extrapolating them into the unmeasured regions using the fitted functions. The resulting rapidity distributions are shown in Fig. 3.

The NA61/SHINE data on charged Ξ production in inelastic $p + p$ interactions are compared with predictions of the following microscopic models: Epos 1.99 [4], UrQMD 3.4 [5, 6], Ampt 1.26 [7–9], Smash 1.6 [10, 11] and Phsd [12, 13]. The model predictions are compared with the NA61/SHINE data in figs. 4. Epos 1.99 describes well the Ξ^- and $\overline{\Xi}^+$ rapidity spectra but fails on the shape of the transverse momentum distribution. The comparison of the UrQMD 3.4 calculations with the NA61/SHINE measurements reveals major discrepancies for the $\overline{\Xi}^+$ hyperons. The model output describes almost perfectly the rapidity and transverse momentum spectra of Ξ^- but strongly overestimates $\overline{\Xi}^+$ yields. Consequently also the ratio of $\overline{\Xi}^+$ to Ξ^- cannot be described by the UrQMD model, see Fig. 5(c). The Ampt, Smash and Phsd models fail in the description of both transverse momentum spectra and rapidity distributions. Ampt overestimates the Ξ^- and $\overline{\Xi}^+$ multiplicities while Smash underestimates them, both failing to describe the ratio. Phsd underestimates the Ξ^- yields and overestimates $\overline{\Xi}^+$. Obviously Phsd also fails to describe the ratio. Epos differs from the UrQMD, Ampt, Smash and Phsd models in its treatment of Pomeron-Pomeron interactions and of the valence quark remnants at the string ends.

2.2 Strangeness enhancement factors

The Ξ mean multiplicities measured by NA61/SHINE in inelastic $p + p$ interactions are used to calculate the enhancement factors of Ξs observed in centrality selected Pb+Pb, in semi-central C+C, and in Si+Si collisions as measured by NA49 [15] at the CERN SPS. The results

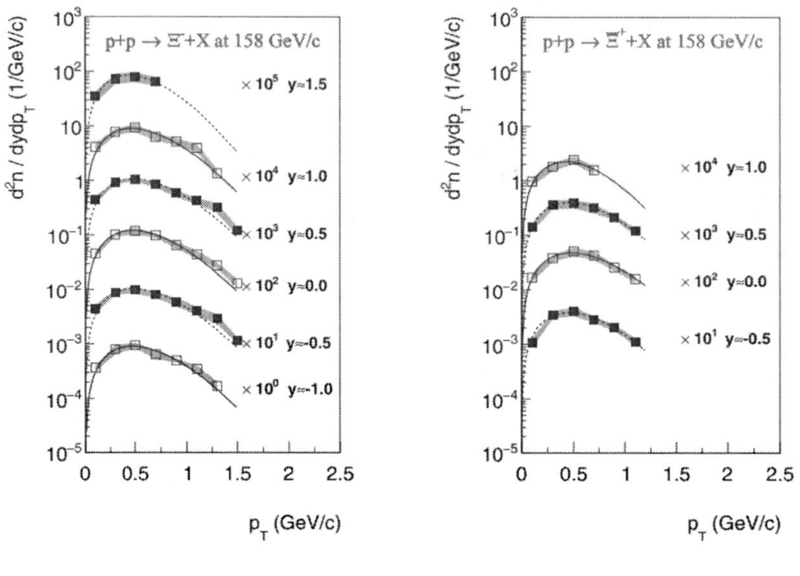

Figure 2. (Color online) Transverse momentum spectra in rapidity slices of Ξ^- (*left*) and $\overline{\Xi}^+$ (*right*) produced in inelastic $p+p$ interactions at 158 GeV/*c*. Rapidity values given in the legends correspond to the middle of the corresponding interval. Statistical uncertainties are smaller than the marker size, shaded bands show systematic uncertainties. Spectra are scaled by the given factors for better separation.

for mid-rapidity densities are shown in Fig. 6 (*left*) as a function of $\langle N_W \rangle$. The enhancement factor increases approximately linearly from 3.5 in C+C to 9 in central Pb+Pb collisions. This result is compared to data from the NA57 experiment at the SPS [16], the STAR experiment at the Relativistic Heavy Ion Collider (RHIC) [17] and the ALICE experiment at the Large Hadron Collider (LHC) [18]. The published enhancement factor reported by NA57 at the CERN SPS was computed using p+Be instead of inelastic $p + p$ interactions. Since strangeness production is already slightly enhanced in p+A collisions [19], this is not a proper reference. With the advent of the NA61/SHINE results on Ξ production in $p + p$ interactions a new baseline reference becomes available and it is used here for the recalculation of the enhancement observed in the NA57 p+Be and A+A data. The STAR Collaboration published results on multi-strange hyperon production in Au+Au collisions at $\sqrt{s_{NN}}$ from 7.7 to 39 GeV [20], however the corresponding data on $p + p$ and p+A interactions are missing. The agreement between the enhancement factors calculated using the NA49 and the NA57 A+A (p+Be) data is satisfactory. The STAR data show a slightly lower enhancement, but the enhancement observed by ALICE is significantly lower. Figure 6 (*right*) shows the rapidity densities dn/dy of $\overline{\Xi}^+$ at mid-rapidity per mean number of wounded nucleons divided by the corresponding values for inelastic $p+p$ collisions as a function of $\langle N_W \rangle$. Apart from a slightly flatter rise the overall picture remains unchanged.

2.3 $\Xi(1530)^0$ production in $p + p$ at 158 GeV/*c*

The $\Xi(1530)^0$ ($\overline{\Xi}(1530)^0$) yields are determined in 4 (4) rapidity and between 5 (3) and 6 (5) transverse momentum bins. The former is 0.5 units and the latter 0.3 GeV/*c* wide. The resulting (y, p_T) yields are presented as the function of p_T in Fig. 7. The p_T spectra in

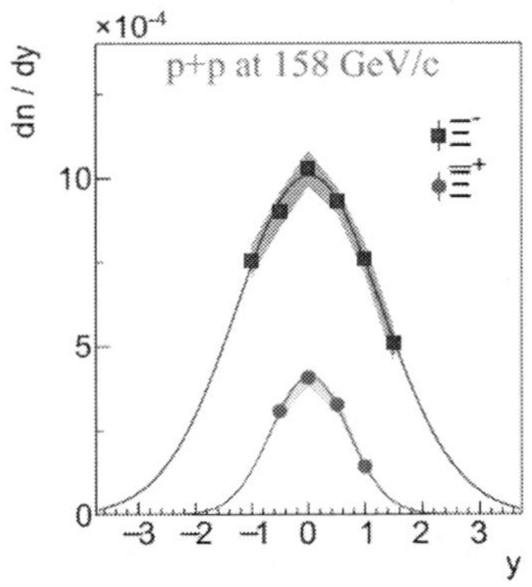

Figure 3. (Color online) Rapidity spectra of Ξ^- (blue squares) and $\overline{\Xi}^+$ (red circles) produced in inelastic $p + p$ interactions at 158 GeV/*c*. Statistical uncertainties are smaller than the marker size, shaded bands correspond to systematic uncertainties of the measurements. Curves depict Gaussian fits used to determine total mean multiplicities.

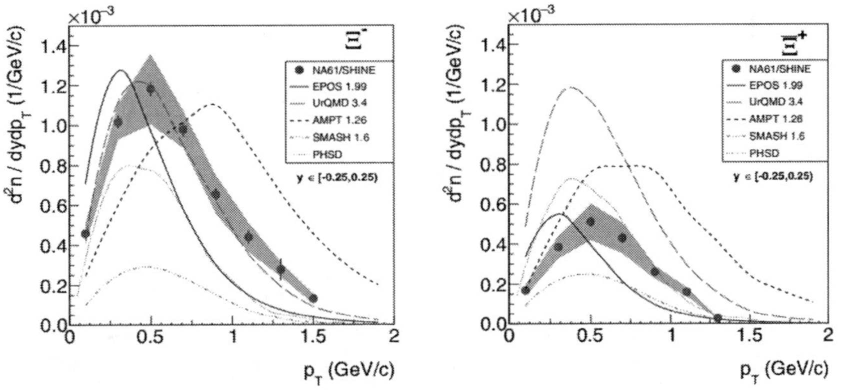

Figure 4. (Color online) Transverse momentum spectra at mid-rapidity of Ξ^- (*left*) and $\overline{\Xi}^+$ (*right*) produced in inelastic $p + p$ interactions at 158 GeV/*c*. Rapidity range is included in the legends. Shaded bands show systematic uncertainties. UrQMD 3.4 [5, 6], Epos 1.99 [4], Ampt 1.26 [7–9], Smash 1.6 [10, 11, 14] and Phsd [12, 13] predictions are shown as magenta, blue, black, gray and green lines, respectively.

successive rapidity intervals (Fig. 7) are scaled for better visibility. Statistical uncertainties are shown as error bars, and shaded bands correspond to systematic uncertainties.

The NA61/SHINE measurements of $\Xi(1530)^0$ and $\overline{\Xi}(1530)^0$ production are essential for understanding multi-strange particle production in elementary hadron interactions. The experimental results of NA61/SHINE are compared with predictions of the Epos 1.99 [4] and UrQMD 3.4 [5, 6] models in Figs. 8 and 9. Epos 1.99 describes well the $\Xi(1530)^0$ and $\overline{\Xi}(1530)^0$ transverse momentum and rapidity spectra. The comparison of the UrQMD 3.4 calculations with the NA61/SHINE measurements reveals significant discrepancies. The model strongly

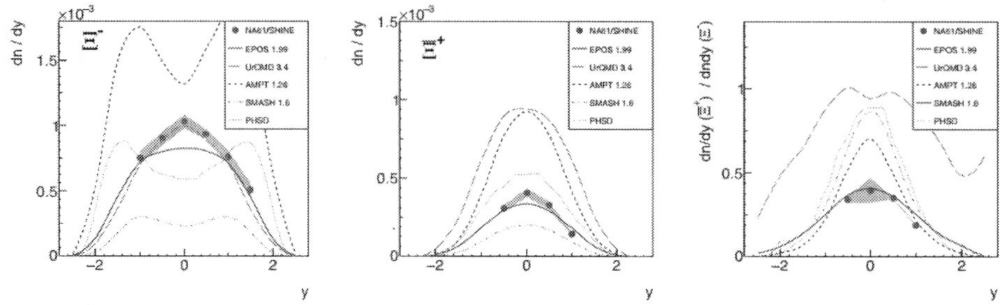

Figure 5. (Color online) Rapidity spectra of Ξ^- (*left*), $\overline{\Xi}^+$ (*middle*) and $\overline{\Xi}^+/\Xi^-$ ratio (*right*) measured in inelastic $p+p$ interactions at 158 GeV/*c*. Shaded bands show systematic uncertainties. UrQMD 3.4 [5, 6], Epos 1.99 [4], Ampt 1.26 [7–9], Smash 1.6 [10, 11, 14] and Phsd [12, 13] predictions are shown as magenta, blue, black, gray and green lines, respectively.

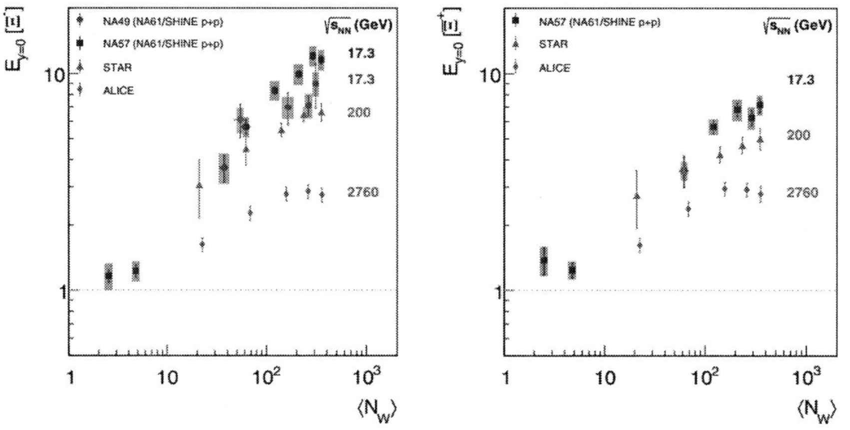

Figure 6. (Color online) The strangeness enhancement E at mid-rapidity as a function of average number of wounded nucleons $\langle N_W \rangle$ calculated as a ratio of rapidity density for Ξ^- production (*left*) and $\overline{\Xi}^+$ production (*right*) in nucleus-nucleus interactions per $\langle N_W \rangle$ divided by the corresponding value for $p+p$ interactions. Red circles – NA49 Pb+Pb at 158*A* GeV [15], blue squares - NA57 *p*+Be, *p*+Pb and Pb-Pb at the same center-of-mass energy $\sqrt{s_{NN}} = 17.3$ GeV [16], magenta triangles - STAR Au+Au at $\sqrt{s_{NN}} = 200$ GeV [17], gray diamonds - ALICE Pb+Pb at $\sqrt{s_{NN}} = 2.76$ TeV [18]. The systematic errors are represented by shaded boxes.

overestimates $\Xi(1530)^0$ and $\overline{\Xi}(1530)^0$ yields. The ratio of $\overline{\Xi}(1530)^0$ to $\Xi(1530)^0$ cannot be described by the UrQMD model but is well reproduced by Epos 1.99 (see Figs. 9 and 8).

3 HRG model in the CE formulation and $p+p$ data

The new measurements by NA61/SHINE of $\Xi(1530)^0$ and $\overline{\Xi}(1530)^0$ produced in inelastic $p+p$ interactions at 158 GeV/*c* as well as previously obtained results for π^+, π^-, K^+, K^-, p, $\overline{p}$, $K^*(892)^0$, Λ, $\phi(1020)$, Ξ^- and $\overline{\Xi}^+$ (see Refs. [2, 3, 21–25] were fitted by different variants of the Hadron Resonance Gas Model (HRG). The Canonical Ensemble with fixed $\gamma_s = 1$ and Canonical Ensemble with fitted strangeness saturation parameter γ_s configurations were

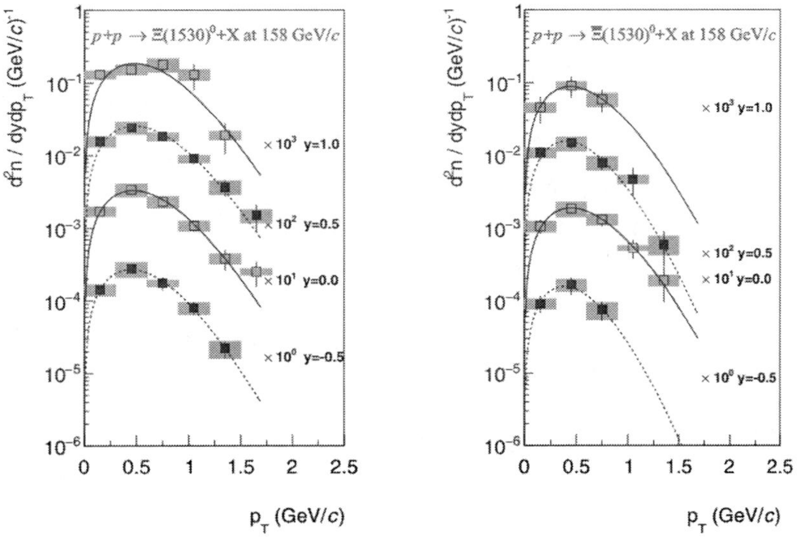

Figure 7. (Color online) Transverse momentum spectra in rapidity slices of $\Xi(1530)^0$ (*left*) and $\overline{\Xi}(1530)^0$ (*right*) produced in inelastic $p + p$ interactions at 158 GeV/*c*. Rapidity values given in the legends correspond to the middle of the corresponding interval. Statistical uncertainties are shown as vertical bars, and shaded bands show systematic uncertainties. Spectra are scaled for better visibility. Lines represent the fitted function.

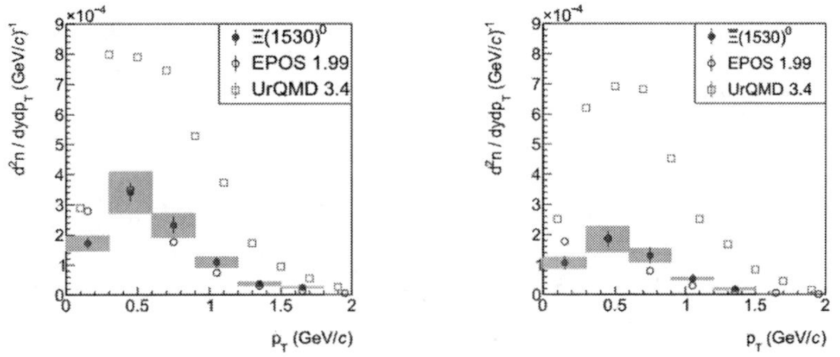

Figure 8. (Color online) Transverse momentum spectra at mid-rapidity of $\Xi(1530)^0$ (*left*) and $\overline{\Xi}(1530)^0$ (*right*) produced in inelastic $p + p$ interactions at 158 GeV/*c*. Shaded bands show systematic uncertainties. UrQMD 3.4 [5, 6] and Epos 1.99 [4] predictions are shown as magenta and blue markers, respectively.

used. Figure 10 compares the measured multiplicities of particles produced in inelastic $p + p$ interactions at 158 GeV/*c* with multiplicities of the same particles obtained from the HRG model in the CE formulation under two different model assumptions: $\gamma_s = 1$ and fitted γ_s. The software package THERMAL-FIST 1.3 [26] was used for this purpose. For the small $p + p$ system, the appropriate statistical system is the Canonical Ensemble, which has as parameters the Temperature T, the radius R of the system at chemical freezeout, and the strangeness

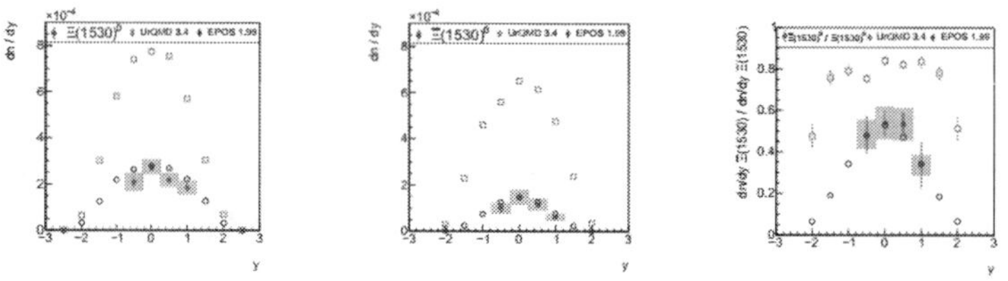

Figure 9. (Color online) Rapidity spectra of $\Xi(1530)^0$ (*left*), $\overline{\Xi}(1530)^0$ (*middle*) and $\overline{\Xi}(1530)^0/\Xi(1530)^0$ ratio (*right*) measured in inelastic $p+p$ interactions at 158 GeV/*c*. Shaded bands show systematic uncertainties. UrQMD 3.4 [5, 6] and Epos 1.99 [4] predictions are shown as magenta and blue points, respectively.

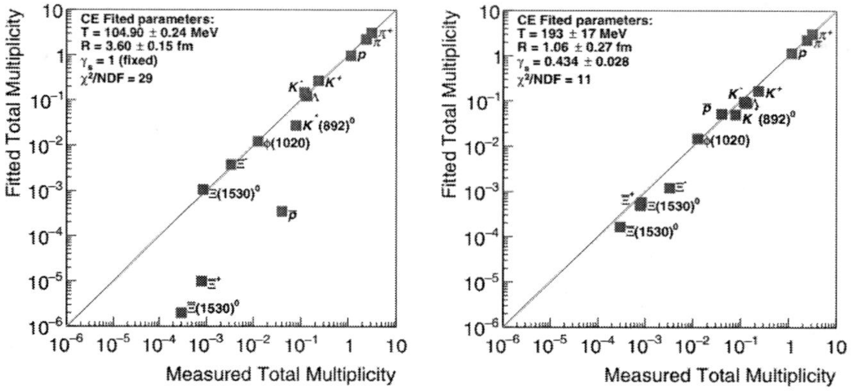

Figure 10. (Color online) Mean multiplicities of π^+, π^-, K^+, K^-, p, $\overline{p}$, $K^*(892)^0$, Λ, $\phi(1020)$, Ξ^-, $\overline{\Xi}^+$, $\Xi(1530)^0$ and $\overline{\Xi}(1530)^0$ produced in $p+p$ interactions at 158 GeV/*c* [2, 3, 21–25] measured by NA61/SHINE are compared with mean multiplicities obtained from the HRG model based on the Canonical Ensemble with fixed $\gamma_s = 1$ (*i*) and fitted γ_s (*ii*). Uncertainties of the measurement are smaller than the symbol size.

suppression factor γ_s. The results plotted in Fig. 10 show significant discrepancies for the fitted parameters T and R between the fits with $\gamma_s = 1$ and free γ_s. Moreover, the fit with fixed γ_s shows unacceptably large χ^2/NDF = 29. This demonstrates that the statistical model fails when fixing γ_s to one. The fit with free γ_s finds $\gamma_s = 0.434 \pm 0.028$ and reproduces the measurements rather well. Thus the statistical model indicates a strong suppression of strange particle production in $p+p$ collisions at CERN SPS energies.

References

[1] N. Abgrall et al. (NA61/SHINE), JINST **9**, P06005 (2014), 1401.4699
[2] N. Abgrall et al. (NA61/SHINE), Eur.Phys.J. **C74**, 2794 (2014), 1310.2417
[3] A. Aduszkiewicz et al. (NA61/SHINE), Eur. Phys. J. C **80**, 833 (2020), 2006.02062
[4] K. Werner, Nucl. Phys. Proc. Suppl. **175-176**, 81 (2008)
[5] S. Bass et al., Prog.Part.Nucl.Phys. **41**, 255 (1998), nucl-th/9803035
[6] M. Bleicher et al., J.Phys. **G25**, 1859 (1999), hep-ph/9909407
[7] Z.W. Lin, C.M. Ko, B.A. Li, B. Zhang, S. Pal, Phys. Rev. C **72**, 064901 (2005)

[8] Z.W. Lin, Phys. Rev. C **90**, 014904 (2014)
[9] B. Zhang, C.M. Ko, B.A. Li, Z. Lin, Phys. Rev. C **61**, 067901 (2000)
[10] J. Mohs, S. Ryu, H. Elfner, J. Phys. G **47**, 065101 (2020), 1909.05586
[11] J. Weil et al., Phys. Rev. C **94**, 054905 (2016), 1606.06642
[12] W. Cassing, E. Bratkovskaya, Nucl. Phys. A **831**, 215 (2009), 0907.5331
[13] W. Cassing, E. Bratkovskaya, Phys. Rev. C **78**, 034919 (2008), 0808.0022
[14] D. Oliinychenko, V. Steinberg, J. Weil, M. Kretz, H.E. (Petersen), J. Staudenmaier, S. Ryu, A. Schäfer, J. Rothermel, J. Mohs et al., *smash-transport/smash: Smash-1.6* (2019)
[15] T. Anticic et al. (NA49), Phys. Rev. C **80**, 034906 (2009), 0906.0469
[16] F. Antinori et al. (NA57), J. Phys. G **32**, 427 (2006), nucl-ex/0601021
[17] B. Abelev et al. (STAR), Phys. Rev. C **77**, 044908 (2008), 0705.2511
[18] B.B. Abelev et al. (ALICE), Phys. Lett. B **728**, 216 (2014), [Erratum: Phys.Lett.B 734, 409–410 (2014)], 1307.5543
[19] T. Susa (NA49), Nucl. Phys. A **698**, 491 (2002)
[20] J. Adam et al. (STAR) (2019), 1906.03732
[21] A. Aduszkiewicz et al. (NA61/SHINE), Eur. Phys. J. **C77**, 671 (2017), 1705.02467
[22] A. Aduszkiewicz et al. (NA61/SHINE), Phys. Rev. C **102**, 011901 (2020), 1912.10871
[23] A. Aduszkiewicz et al. (NA61/SHINE), Eur. Phys. J. C **80**, 460 (2020), 2001.05370
[24] A. Aduszkiewicz et al. (NA61/SHINE), Eur. Phys. J. C **80**, 199 (2020), 1908.04601
[25] A. Aduszkiewicz et al. (NA61/SHINE), Eur. Phys. J. **C76**, 198 (2016), 1510.03720
[26] V. Vovchenko, H. Stoecker, Comput. Phys. Commun. **244**, 295 (2019), 1901.05249

Investigating the origin of strangeness enhancement in small systems through multi-differential analyses

Francesca Ercolessi[1,*] (for the ALICE Collaboration)

[1]University and INFN, Bologna, Italy

Abstract. The main goal of the ALICE experiment is to study the physics of strongly interacting matter, including the properties of the quark-gluon plasma (QGP). The relative production of strange hadrons with respect to non-strange hadrons in heavy-ion collisions was historically considered as one of the signatures of QGP formation. However, measurements at the LHC in pp and p-Pb collisions have shown similar features to those observed in Pb-Pb, measuring an increase in the production of strange hadrons relative to pions with the charged-particle multiplicity in the event. In order to better understand the role of the event multiplicity to the observed enhancement in pp, two new complementary analyses have been performed. The first exploits the angular correlation between strange and high-p_{T} hadrons in the event to classify in-jet and out-of-jet strangeness production. The second uses the concept of the effective energy available in the event for particle production which is estimated by an anti-correlation with the energy deposited in ALICE's Zero Degree Calorimeters. The results suggest that strangeness enhancement emerges from the growth of the underlying event and is not connected to initial state properties.

1 Introduction

Strange particle production in heavy-ion collisions has proved to be an important observable to investigate the strongly interacting medium created in the collision. Strangeness enhancement was one of the first proposed signatures of quark-gluon plasma (QGP) formation [1], which is expected to lead to an increased production of strange hadrons relative to pp interactions. The enhancement was predicted to follow a hierarchy related to the strangeness content of the particle, being stronger for multi-strange hadrons. This phenomenon was observed for the first time at SPS [2], then at RHIC [3] and later at the LHC [4] at increasing collision energies. The ALICE experiment has also studied strangeness production in small collision systems, such as p-Pb and pp, by measuring the ratio of strange particle yields to pion yields as a function of the charged-particle multiplicity produced in the event. This ratio is found to increase across multiplicities, evolving smoothly within different collision systems and energies. The results can be compared to various model predictions, from a statistical hadronization description using the canonical suppression approach [5], to rope hadronization models including colour reconnection effects [7], to two-component (core-corona) models [6]. However, the mechanisms responsible for this intriguing phenomenon observed in small systems are still unclear and further studies are needed.

*e-mail: francesca.ercolessi@cern.ch

2 Results

The ALICE experiment identifies K^0_S, Λ, Ξ and Ω baryons by reconstructing their weak decay daughter tracks in the central pseudorapidity region. The main subdetectors used for particle identification (PID) and tracking include the six-layer high resolution Inner Tracking System (ITS) and the large volume Time Projection Chamber (TPC). Two forward detectors placed on both sides of the ALICE interaction point, the V0 and the Zero Degree Calorimeters (ZDC), are used to classify events in multiplicity and energy event classes, respectively. A detailed description of the ALICE apparatus and its performance can be found in [8]. The two analyses presented in these proceedings exploit multi-differential approaches to explore strangeness enhancement in pp.

The processes involved in particle production in pp collisions can be classified into hard and soft ones, depending on the momentum transfer. The relative contribution of hard and soft processes to strangeness production in pp is still not fully understood and can be studied through techniques involving two-particle correlations. In the first analysis ALICE exploits the angular correlation method to separate $\mathrm{K^0_S}$ and Ξ hadrons produced in jets (hard processes) from those produced out-of-jets (soft processes). Particles produced in the near-side jet region are characterised by a small angular separation from the leading particle of the jet, which is identified as the particle with the highest transverse momentum in the collision with $p_{\mathrm{T}} > 3$ GeV/c. The angular correlation distribution of the jet trigger particle and the associated strange hadron is studied in the $(\Delta\eta, \Delta\phi)$ plane, which is divided into a near-side-jet region, an out-of-jet region, and a full inclusive region. The near-side-jet contribution is obtained subtracting the full and out-of-jet in the $(\Delta\eta, \Delta\phi)$ corresponding region. The Ξ yields per trigger particle and per unit of $\Delta\eta\Delta\phi$ are displayed in Fig. 1 as a function of the charged-particle multiplicity produced at midrapidity.

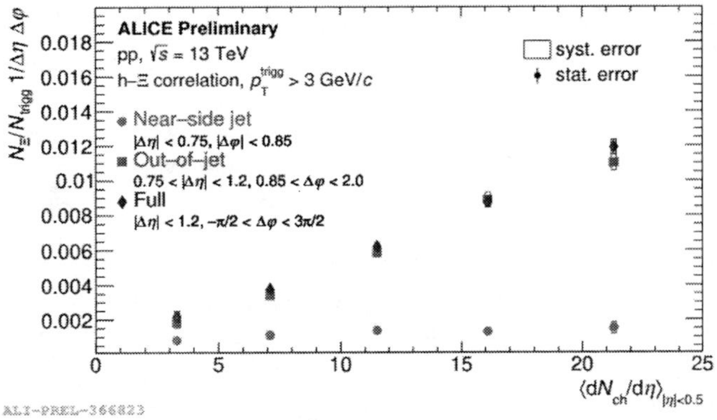

Figure 1. Ξ yields per trigger particle and per unit of $\Delta\eta\Delta\phi$ as a function of the charged-particle multiplicity produced at midrapidity.

The full and out-of-jet yields increase with multiplicity, while the near-side-jet yields show a very mild to no dependence on particle production at midrapidity. Similar results are obtained for $\mathrm{K^0_S}$ yields. These results suggest that out-of-jet (soft) processes are the dominant contribution to strange particle production in pp collisions.

The multiplicity distribution of charged hadrons is an important characteristic of the hadronic final state of any collision. But it also reflects the initial dynamics of the interaction, being strongly correlated with the energy effectively available for particle production in its initial stages (effective energy). In pp collisions the effective energy is reduced with respect to the full centre-of-mass energy due to the leading effect, which consists of a high probability to emit forward baryons with large longitudinal momenta along the direction of the incident beams. More details can be found in [9] and references therein. The effective energy can be estimated by means of ALICE's ZDC by detecting the energy deposits of the forward leading baryons in each event hemisphere: $E_{\text{effective}} = \sqrt{s} - E_{\text{leading}}$. In the second analysis the events are classified in effective energy and multiplicity percentile classes, using the information from the ZDC and V0 detectors, respectively, in order to disentangle initial and final state effects on strange hadron production. The correlation between effective energy and particle multiplicity in pp is studied using the PYTHIA 8 event generator in Fig. 2 (left), which shows that events selected using a V0 or ZDC based estimator alone are sensitive to both initial and final state. Fig. 2 (right) shows the ratio of Ξ yields to the charged-particle multiplicity in the event (self-normalized to INEL>0[1]) as a function of particle production at midrapidity selected using V0 or ZDC. Ξ hadron production is found to increase with multiplicity following a similar trend, independent of the estimator used.

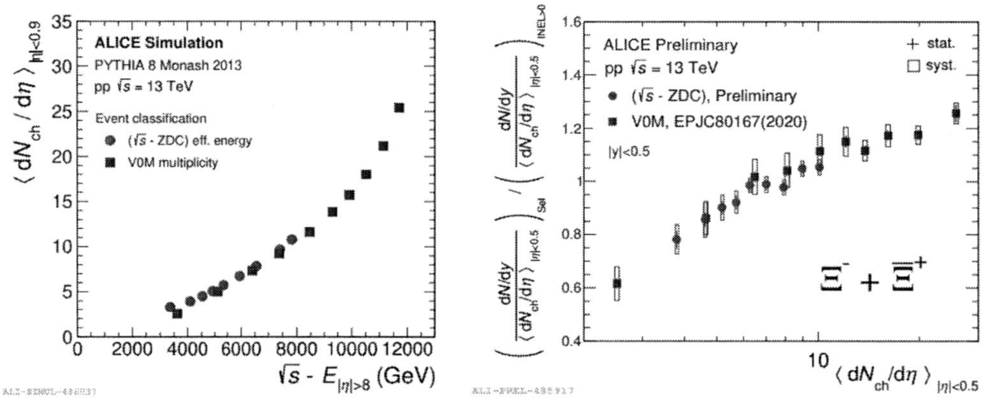

Figure 2. (left) Multiplicity and effective energy correlation in PYTHIA 8 using V0 and ZDC event classes. (right) Ratio of Ξ yields to the charged-particle multiplicity (self-normalized to INEL>0) as a function of particle production at midrapidity, using V0 and ZDC standalone selections.

In order to separate initial and final state contributions to strange particle production, V0 and ZDC combined classes are exploited. Their discriminating power in terms of effective energy and multiplicity is studied in Fig. 3 (left) via the PYTHIA 8 event generator. Events are selected using ZDC percentile selections, fixing the multiplicity through the V0 estimator in a high and low range (red points), and vice-versa using V0 percentile selections, fixing the effective energy through the ZDC estimator to a high and low range (blue points). Fig. 3 (right) shows the ratio of Ξ yields to the charged-particle multiplicity (self-normalized to INEL>0) selected using ZDC event classes, fixing the multiplicity through the V0 estimator. The points are distributed in a flat trend showing no dependence on the energy percentile estimator. The different V0 selection is responsible for the enhancement observed between the two sets of points. The results suggest that the effective energy does not play a significant role in the strangeness enhancement observed in pp collisions, which is found to be driven by the final state multiplicity.

[1] INEL>0 is a conventional event class which contains events with at least one charged particle in $|\eta| < 1$.

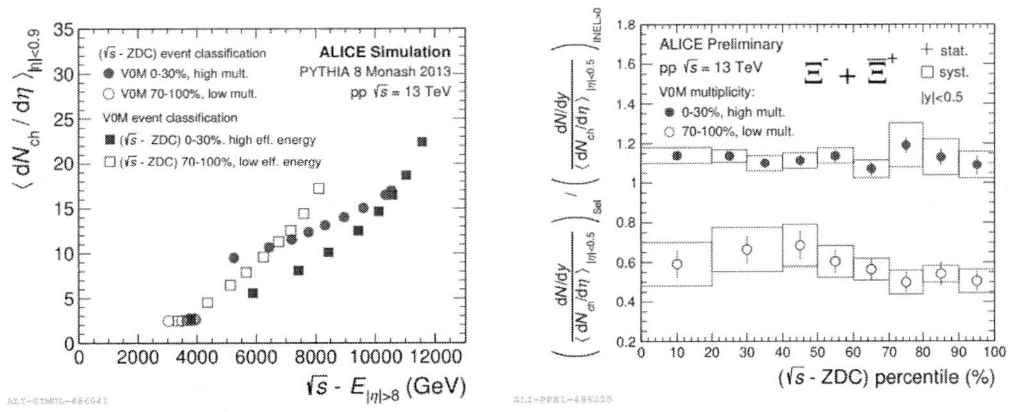

Figure 3. (left) Multiplicity and effective energy correlation in PYTHIA 8 using V0-ZDC combined classes. (right) Ratio of Ξ yields to the charged-particle multiplicity (self-normalized to INEL>0) extracted using V0-ZDC combined selections. Events are selected using ZDC event classes, fixing the multiplicity through the V0 estimator.

3 Conclusions

The ALICE Collaboration has provided a comprehensive set of results on the production of strange hadrons in pp collisions, however, the mechanisms responsible of the strangeness enhancement with multiplicity observed in small systems are still not fully understood. These preliminary results exploit a multi-differential approach to disentangle initial and final state effects on strange hadron production, suggesting a strong connection with the final particle multiplicity and no significant role of the initial effective energy. Moreover, soft out-of-jet processes are found to be the dominant contribution, with respect to in-jet hard processes, to strangeness production.

References

[1] J. Rafelski, B. Müller, Phys. Rev. Lett. **48**, 1066-1069 (1982).
[2] The WA97 Collaboration, Physics Letters B **449**, 401-406 (1999).
[3] Agakishiev, G. et al. (STAR Collaboration) Phys. Rev. Lett. **108**, 072301 (2012).
[4] Abelev, B. et al. (ALICE Collaboration) Phys. Lett. B **728**, 216 (2014).
[5] Vovchenko, V. et al., Phys. Rev. C **100**, 054906 (2019).
[6] Kanakubo, Y. et al., Phys. Rev. C **101**, 024912 (2020).
[7] Nayak, R. et al., Phys. Rev. D **100**, 074023 (2019).
[8] Aamodt, K. et al., (ALICE Collaboration) JINST **3**, S08002 (2008).
[9] Akindinov, A., et al. Eur. Phys. J. C **50**, 341–352 (2007).

Hadron production within a full transport approach with statistical hadronization mechanism at RHIC and LHC energies

Giuseppe Galesi[1,2], *Salvatore* Plumari[1,2], and *Vincenzo* Greco[1,2]

[1]Department of Physics and Astronomy, University of Catania, Via S. Sofia 64, I-95123 Catania, Italy
[2]Laboratori Nazionali del Sud, INFN-LNS, Via S. Sofia 62, I-95123 Catania, Italy

Abstract. We present for the first time results on final hadron production, with and without strangeness content, in Ultrarelativistic Heavy Ion Collisions at RHIC and LHC center of mass energies obtained combining a full 3+1D relativistic Boltzmann transport approach with a statistical hadronization mechanism. The non-perturbative interaction between quarks and gluons is described by means of a quasi-particle approach that permits to have an Equation of State close to lattice QCD. The resulting framework naturally includes both shear and bulk viscous effects. The 3+1D full transport evolution is converted to hadrons by mean of a realistic freeze-out hypersurface considering all known hadron resonances and by performing resonance decays. We present results on charged-hadron multiplicity, identified-particle spectra and identified-particle elliptic flow of π, K and p produced at RHIC and LHC energies for different centralities.

1 Introduction

Relativistic viscous hydrodynamics is able to describe very well many observables in Ultra Relativistic Heavy Ion Collisions (uRHICs) such as final hadron spectra and elliptic flow in the low p_T region [1–6]. Usually, such an approach makes use of a statistical hadronization mechanism [7–9] which converts the information contained in the energy-momentum tensor to final hadrons applying the CF (Cooper-Frye) formula [10] on a freeze-out hypersurface. This framework allows to put constraints on QGP properties: they are fixed from the beginning of a simulation and drive the dynamical evolution. Anyway, in order to transfer the viscous correction information $\delta T^{\mu\nu}$ to the final hadron particle distribution function an ansatz is needed for the deviation δf. Further, it does not allow for arbitrary far from-equilibrium initial conditions. A transport approach can self-consistently overcome both these problem because the one-particle distribution function is always known during the dynamical evolution. For these reasons we developed a transport approach at fixed η/s which permits to have direct comparison with viscous hydrodynamics and to investigate specific dissipation properties of QGP. In order to test the validity of our approach we perform statistical hadronization on a hydro-like freeze-out hypersurface at fixed temperature extracted with our code and we directly compare the resulting hadron observables with experimental data.

2 The transport equation and the statistical hadronization

In our approach, we describe the quark-gluon plasma (QGP) by solving Relativistic Transport Boltzmann (RTB) equations developed to study the dynamics of the matter created in uRHICs [11–15]:

$$\left(p^\mu \partial_\mu + \frac{1}{2}\partial_i m^2 \partial^i_{(p)}\right) f = C[f], \tag{1}$$

$C[f]$ is the collision integral for $2 \to 2$ collision processes. In our approach non-pertubative effects are encoded by using a quasi-particle model (QPM) with thermal masses determined in order to reproduce lattice QCD equation of state [16]. The collision integral is gauged to the desired η/s through the Chapmann-Enskog approximation and is computed by means of a stochastic algorithm [17, 18]. The energy-momentum tensor is evaluated during the dynamical evolution of f on a (x, y, Y_s)-grid, with Y_s the longitudinal space rapidity, according to the formula

$$T^{\mu\nu}_c = \frac{1}{N_{\text{test}} V_c} \sum_{i \in c} \frac{p^\mu_i p^\nu_i}{p^0_i}, \tag{2}$$

where N_{test} is the number of test particles per real particle, V_c the cell volume, p^μ_i the test particle 4-momentum and being the sum over all test particles in the c-cell. From that we self-consistently extract the local temperature T, the thermal mass and local flow u^μ. We then interpolate a freeze-out hypersurface Σ at fixed temperature and we give it in input to THERMINATOR 2 code [19] which performs statistical hadronization task. It is a Monte-Carlo event generator that produces final hadrons by using the following CF formula:

$$p^0_h \frac{\mathrm{d}^3 N_h}{\mathrm{d}p^3} = (2s_h + 1) \int_\Sigma \mathrm{d}\Sigma_\mu p^\mu_h f_h, \tag{3}$$

being $\frac{\mathrm{d}^3 N_h}{\mathrm{d}p^3}$ the spectrum for the h-resonance, s_h the relative spin quantum number and f_h the following equilibrium distribution:

$$f_h = \frac{1}{(2\pi)^3}\left(\exp\left(\frac{p_{h,\mu}u^\mu}{T}\right) \pm 1\right)^{-1}, \tag{4}$$

where the sign is chosen according to the statistics of the resonance. Customizations of this formula which use modified versions of the distribution function to take into account local deviation from equilibrium are of course possible but for a first test of our approach we consider Fermi-Dirac and Bose-Einstein distributions only. All known hadronic resonances from the SHARE package [20] are considered for primordial particle production on Σ and resonance decays are subsequently performed on unstable particles. In this way it is potentially possible to extract all final hadron observables to be compared with the available data from RHIC and LHC experimental collaborations in order to fix the free parameters, which are the initial central temperature T_0 for the most central collision, the freeze-out temperature T_{FO} and η/s. In particular we studied AuAu collisions at RHIC with a center of mass energy $\sqrt{s} = 200\,A\text{GeV}$ and PbPb collision at LHC with $\sqrt{s} = 2.76\,A\text{TeV}$.

3 Results

We show only results for semi-peripheral LHC collisions. We initialize the system at fixed proper time $\tau_0 = 0.25\,\text{fm/c}$. Initialization of test particles space coordinates is performed in the transverse plane by scaling particle number density or energy density according to an

OGM (Optical Glauber Model) with mixed contributions from number of wounded nucleons N_{part} and binary collisions N_{coll} and with a tilted function of rapidity. Instead, the longitudinal profile has a plateau in the mid rapidity region with gaussian tails whose parameters are set to reproduce final pseudorapidity hadron yields. In momentum space we distribute according to a 3D isotropic thermal distribution with Bjorken flow. In Figure 1 on the left we show the mid-rapidity spectra for pion, kaon and proton (filled curves) obtained in a semi-peripheral LHC collision at $b = 7.7$ fm with $T_0 = 670$ MeV, $T_{\text{FO}} = 145$ MeV and $\eta/s = 1/4\pi$, and the corresponding ALICE data [21, 22] for the centrality class 20-30 %. The pion and kaon spectra are quite well reproduced over all the p_T region shown while protons are slightly underestimated above 1.5 GeV. We attribute this fact to the missing hadronic afterburner which can have a significant effect for heavier particles, and which could be taken into account in future works by using other more sophisticated numerical code like SMASH [23]. On

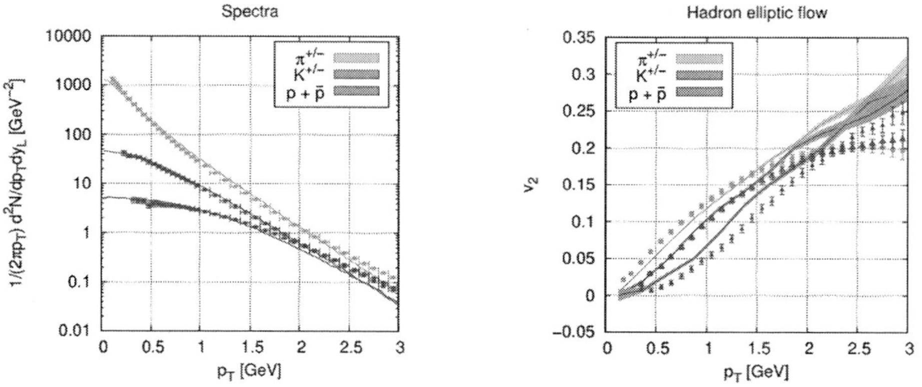

Figure 1. Left panel: Identified hadron spectra of π (orange curve), K (red line) and p+anti-p (blu curve) for PbPb collisons at $\sqrt{s} = 2.76$TeV. Circle, triangles and square point are experimental data from ALICE collaboration taken from [21]. Right panel: elliptic flow v_2 for Pb+Pb collision at $\sqrt{s} = 2.76$TeV of π, K, p for 20-30 % centrality cut and at mid-rapidity. Experimental data taken from [22]

the right of Figure 1 we show, for the same system, the differential elliptic flow caused by the initial space anisotropy. This final momentum anisotropy is due only to the shape of the freeze-out hypersurface and local flow information interpolated on it because we don't consider any non-equilibrium effect in the hadron distribution function. Nevertheless, we have a good agreement with experimental data. We have also observed a strong dependence of this quantity from the initial mixing factor of the OGM, which seems to suggest that a greater contribution should be considered from N_{coll} at LHC energies.

4 Conclusions

In conclusion, we have shown that this framework is able to reproduce quite well final hadron observables and we believe that it is still possible a further fine tuning on the free parameters to have a better agreement with experimental data [24]. This validation test opens the way to many possible applications of this code. One of the most intriguing perspective is the possibility to make a direct comparison between a microscopic hadronization model like coalescence plus fragmentation [25, 26] with a macroscopic one like the SHM presented here using an underlying common framework. Furthermore, we are planning to investigate in the next future the isotropization and hydrodynamization processes of a strongly interacting system which starts with non-equilibrium initial conditions by looking for hydrodynamic attractors of a realistic massive gas.

References

[1] P. Huovinen, P.F. Kolb, U. Heinz, P.V. Ruuskanen, S. Voloshin, Phys. Lett. B503 58-64 (2001)

[2] T. Hirano, K. Tsuda, Nucl. Phys. A715 821-824 (2002)

[3] P.F. Kolb, U. Heinz (2003)

[4] B. Schenke, S. Jeon, C. Gale, Phys. Rev. C82 014903 (2010)

[5] U. Heinz, R. Snellings, Ann. Rev. Nucl. Part. Sci. **63**, 123 (2013)

[6] M. Alqahtani, M. Nopoush, M. Strickland, Prog. Part. Nucl. Phys. **101**, 204 (2018)

[7] J. Cleymans, H. Satz, Z. Phys. C **57**, 135 (1993)

[8] F. Becattini, M. Gazdzicki, J. Sollfrank, Eur. Phys. J. C **5**, 143 (1998)

[9] W. Florkowski, W. Broniowski, M. Michalec, Acta Phys. Polon. B33: 761-769 (2001)

[10] F. Cooper, G. Frye, Phys. Rev. D **10**, 186 (1974)

[11] F. Scardina, M. Colonna, S. Plumari, V. Greco, Phys. Lett. B724 (2013) 296-300 (2012)

[12] M. Ruggieri, A. Puglisi, L. Oliva, S. Plumari, F. Scardina, V. Greco, Phys. Rev. C **92**, 064904 (2015)

[13] A. Gabbana, S. Plumari, G. Galesi, V. Greco, D. Simeoni, S. Succi, R. Tripiccione, Phys. Rev. C 101, 064904 (2020) (2019)

[14] M.L. Sambataro, S. Plumari, V. Greco (2020)

[15] Y. Sun, S. Plumari, V. Greco, Eur. Phys. J. C **80**, 16 (2020)

[16] S. Plumari, W.M. Alberico, V. Greco, C. Ratti (2011)

[17] S. Plumari, A. Puglisi, F. Scardina, V. Greco, Phys. Rev. C **86**, 054902 (2012)

[18] Z. Xu, C. Greiner, Phys. Rev. C **79**, 014904 (2009)

[19] M. Chojnacki, A. Kisiel, W. Florkowski, W. Broniowski (2011)

[20] G. Torrieri, S. Steinke, W. Broniowski, W. Florkowski, J. Letessier, J. Rafelski, Comput. Phys. Commun. **167**, 229 (2005)

[21] J.A. et al. (ALICE Collaboration), Phys. Rev. C 93 (2016)

[22] A. Collaboration, JHEP 06 (2015) 190 (2014)

[23] J. Weil et al., Phys. Rev. C **94**, 054905 (2016)

[24] G. Galesi, S. Plumari, V. Greco (In preparation)

[25] V. Greco, C.M. Ko, Acta Phys. Hung. A24 (2005) 235-240 (2004)

[26] V. Minissale, F. Scardina, V. Greco, Phys. Rev. C **92**, 054904 (2015)

Thermal model description of the particle spectra in the few-GeV energy regime

Szymon Harabasz[1,3,*], *Wojciech* Florkowski[2], *Tetyana* Galatyuk[1,3], *Małgorzata* Gumberidze[3], *Radosław* Ryblewski[4], *Piotr* Salabura[5], and *Joachim* Stroth[3,6]

[1]Institut für Kernphysik, Technische Universitä Darmstadt, 64289 Darmstadt, Germany
[2]Institute of Theoretical Physics, Jagiellonian University, 30-348 Kraków, Poland
[3]GSI Helmholtzzentrum für Schwerionenforschung GmbH, 64291 Darmstadt, Germany
[4]Institute of Nuclear Physics, Polish Academy of Sciences, 31-342 Kraków, Poland
[5]Marian Smoluchowski Institute of Physics, Jagiellonian University, 30-348 Kraków, Poland
[6]Institut für Kernphysik, Goethe-Universität, 60438 Frankfurt, Germany

Abstract. It has been demonstrated that Statistical Hadronization Model fits perfectly to particle yields at freeze-out in heavy-ion and hadron collisions at LHC, RHIC and SPS, where quark-gluon plasma is created. It is however entirely not clear if particles emitted in the few-GeV energy regime can be understood as emerging from thermalized hadronic medium. Our recent work suggests that this might be the case. By implementing appropriate fireball geometry and expansion pattern in the THERMINATOR (THERMal heavy IoN generATOR) it was possible to describe not only yields, but also the spectra of most abundant particles measured at GSI SIS18. Most of the latter are pure prediction of the model. We present details of the model and extended comparison with experimental data and discuss further developments.

1 Introduction

Models based on statistical hadronization have been very successful in describing over many orders of magnitude the multiplicities of hadrons produced in different collision processes, in particular in heavy-ion collisions (HICs), see e.g. [1] and references therein. This is possible with just a few free thermodynamic parameters (temperature, fugacities, system's volume), whose fitted values allow to relate collision energy and system size to the location in the QCD phase diagram. At the same time, the assumption of local thermodynamic equilibrium makes it possible to use concepts originating from fluid dynamics to describe the evolution of hot and dense *fireball* of QCD matter produced in HICs.

Collisions at energy of few GeV probe the phase diagram region of moderate temperature (tens of MeV) and high baryochemical potential μ_B (close to the nucleon mass). However, it is debated if at such collision energies matter does reach local thermal equilibrium. Intuitively, on one hand the number of newly produced particles is small ($\sim$ 40 pions), on the other hand the degree of baryon stopping is high and furthermore the lifetime of the fireball is relatively long. A possible way to test this is to compare the kinematic distributions of particles predicted by the thermal model accompanied with hydro-inspired description of the fireball expansion to experimental results.

*e-mail: S.Harabasz@gsi.de

2 Model and results

In this work we extend the THERMal heavy IoN generATOR (THERMINATOR) [2] to make it applicable for comparison with protons and pions spectra measured by the HADES Collaboration in 10% most central Au+Au collisions at $\sqrt{s_{\mathrm{NN}}} = 2.42$ GeV [4, 5]. In our calculation we assume that the chemical freeze-out (which we define as the moment when final abundances of particles, not just flavor content, are fixed) coincides with the kinetic freeze-out (when the momentum distributions are fixed). This assumption was successful in describing the RHIC measurements [3]. We use the Cooper-Frye formula for particle production [6]. The five parameters of the thermal distribution and the total volume we determine by fitting to the measured abundances of six particles: in addition to p and $\pi^{\pm}$ these are Λ [7] and $K^{\pm}$ [8]. We add the protons bound in light nuclear clusters to the ones measured as free, assuming implicitly that the cluster formation proceeds through the coalescence mechanism after the freeze-out. We assume the spherical symmetry of the fireball and its radial expansion following Siemens and Rasmussen [9] but with Hubble-like expansion velocity profile $\beta(r) = \tanh(Hr)$. The parameter H is determined with a fit to a p transverse mass spectrum at mid-rapidity. Shapes of all the remaining particle spectra do not enter the fit in any way and can therefore be considered as predictions of the model. Finally, we include the correct mass distribution of the most relevant resonance, $\Delta(1232)$, as a derivative of the empirical phase shift taken from ref. [10]. The details of the model and the complete set of results can be found in [11], all the parameters are also displayed in the upper row of Fig. 3.

3 Discussion

With the help of an event generator it is possible to check the influence of different assumptions on final results. In Fig. 1 we show a comparison of experimentally measured m_t spectra

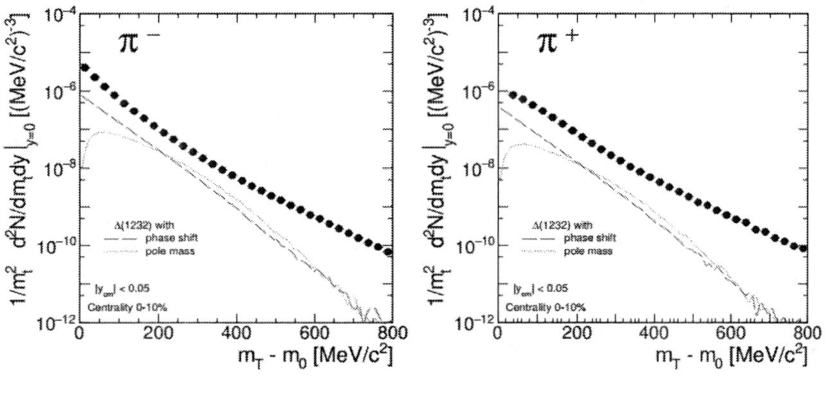

Figure 1. Transverse mass distributions of π^- (left panel) and π^+ (right panel) at mid-rapidity, measured by HADES [4] overlaid with contribution from decays of $\Delta(1232)$, calculated with mass distribution from [10] (blue solid curves) and with zero width at nominal mass (blue dashed curves).

of $\pi^{\pm}$ just to the contribution from decays of $\Delta(1232)$, either having correct mass distribution (blue dashed curves) or zero width (blue solid curves), which was the only option available in THERMINATOR by now. The difference in shape is clear. However, with relatively low temperature that we obtained as discussed above, the yield of Δ is low and the difference does not influence the final results. The situation would be different at slightly higher temperature, where Δ would not be suppressed so strongly by the thermal factor.

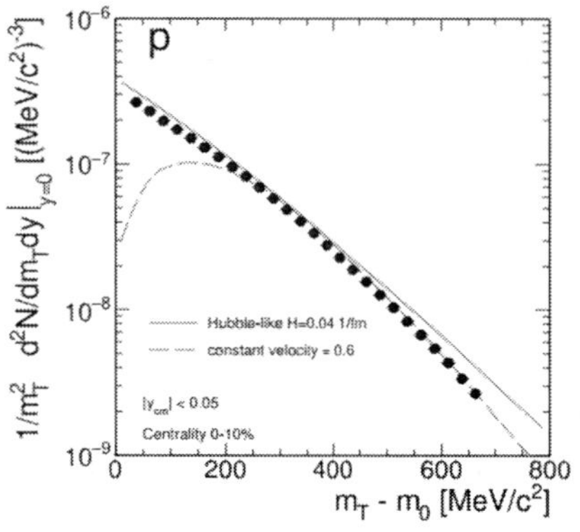

Figure 2. Mid-rapidity transverse mass distribution of protons, measured by HADES [5], compared to thermal model calculations with parameters from [11] and best fits of the fireball radial expansion velocity, assuming either Hubble-like profile $\beta(r) = \tanh(Hr)$ (blue solid curve) or constant velocity (blue dashed curve).

The role of the shape of radial expansion velocity profile is illustrated in Fig. 2. The solid line shows the best fit with Hubble-like expansion, the dashed line – with constant velocity, independent of the radial position. In the latter case the model curve tends to be more concave and has a chance to meet data points only at high m_t at the cost of missing them badly at low m_t. Such a fit gives also rise to an unexpectedly high expansion velocity of $\beta \sim 0.6$.

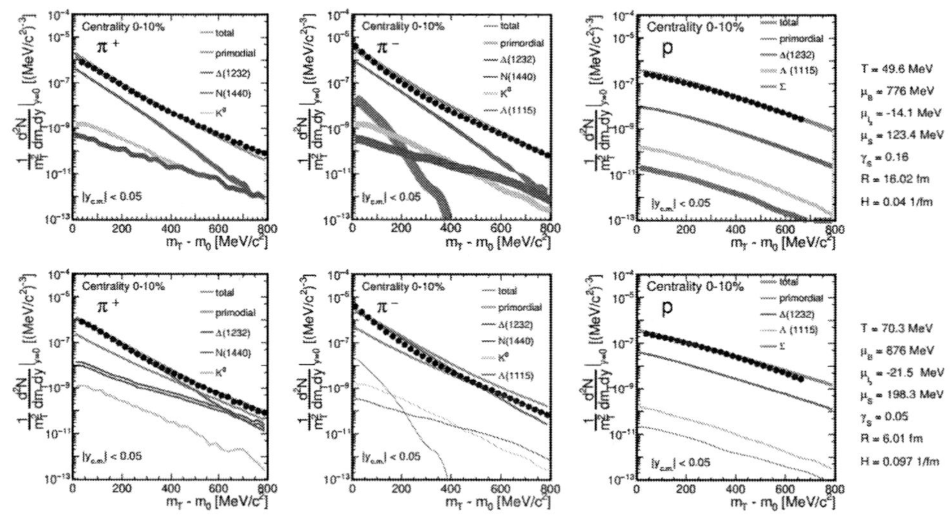

Figure 3. Transverse mass distributions of p and $\pi^{\pm}$ at mid-rapidity, measured by HADES, compared to thermal model calculations with parameters from [11] (upper row) and from [12], corresponding to the same way of treating nuclear clusters, but to the higher-temperature minimum of the thermal fit (lower row, for details see text and the reference). The respective values of the parameters are also shown.

Particle multiplicities measured by HADES have also been studied recently in ref. [12]. Several assumptions about the cluster formation have been tested there and in the setup equivalent to ours (implicit assumption of the nuclear coalescence) two degenerate minima of the thermal fit have been found. The first one results in thermal parameters consistent with the ones in [11], while the other one gives substantially higher freeze-out temperature and density. In the present work, we used THERMINATOR to generate particle spectra for the latter set of thermal parameters and the corresponding expansion coefficient H, which we extract in the same way as before. The resulting mid-rapidity m_t spectra of protons and pions are compared to the experimental data in the lower part of Fig. 3 and contrasted with the case of ref. [11] shown in the upper part of the same figure. It can be observed, that a higher temperature leads to larger relative contribution of the Δ resonance and makes the proper treatment of its spectral shape indispensable. The overall agreement of the model with experimental data is good in both cases, except that for the higher temperature it clearly overshoots π^- yield at high m_t. At the same time, the slope of pion spectra above $m_T - m_0 = 400$ MeV$/c^2$ is perhaps better reproduced with higher T. By construction, the total particle multiplicities, given by integrals of their distributions over the full phase space, agree exactly with the experiment for both sets of parameters. In all cases, particle abundances provide not sufficient information to verify the assumptions of the thermal model, especially at lower collision energies, where not many degrees of freedom are available. More comprehensive information is provided by the kinematic distributions of the particles, but in order to exclude one or another hypothesis, finer details of the model have to be considered.

4 Conclusions and outlook

Statistical hadronization together with proper treatment of the fireball expansion and resonance dynamics describes reasonably well particle spectra measured in heavy-ion collisions in the few-GeV energy regime. However there is room for further improvements. Spherical symmetry is not exact and more realistic parametrization of the fireball expansion has to be implemented. The same holds for actual inclusion of light nuclear clusters. The measured spectra of π^+ and π^- at low m_t are clearly modified by a final-state electromagnetic interaction. An event generator allows for introduction of quantum-mechanical correlations and study of various effects observed in femtoscopic measurements. A model developed in this way can be compared with recent data from STAR fixed-target program and from future facilities, like FAIR and NICA.

Acknowledgements

This work was supported in part by: the Polish National Science Center Grants No. 2016/23/B/ST2/00717, No. 2018/30/E/ST2/00432, and No. 2017/26/M/ST2/00600; TU Darmstadt, Darmstadt (Germany), Helmholtz VH-NG-823, DAAD PPP Polen 2018/57393092; Goethe-University, Frankfurt(Germany), and HIC for FAIR (LOEWE).

References

[1] A. Andronic and others, Nature **561**, 321 (2018)
[2] M. Chojnacki and others, Comput. Phys. Comm. **103**, 746-773 (2012)
[3] W. Broniowski, W. Florkowski, Phys. Rev. Lett. **87**, 272302 (2001)
[4] HADES Collaboration, Eur. Phys. J. A **56**, 250 (2020)
[5] M. Szala (HADES Collaboration), in Proceedings of the XVIII International Conference on Strangeness in Quark Matter (SQM 2019)
[6] F. Cooper, G. Frye, Phys. Rev. D **10**, 186 (1974)
[7] HADES Collaboration, Phys. Lett. B **793**, 457 (2019)
[8] HADES Collaboration, Phys. Lett. B **778**, 403 (2018)
[9] P. J. Siemens, J. O. Rasmussen, Phys. Rev. Lett. **42**, 880 (1979)
[10] P. M. Lo and others, Phys. Rev. C **96**, 015207 (2017)
[11] S. Harabasz and others, Phys. Rev. C **102**, 054903 (2020)
[12] A. Motornenko and others, arXiv:2104.06036 [hep-ph] (2021)

Hyperon Global Polarization in Nucleus-Nucleus Collisions at sub-10-GeV Beam Energy

Yu Guo[1,2], *Jinfeng* Liao[3,*], *Enke* Wang[1,2,**], *Hongxi* Xing[1,2,***], and *Hui* Zhang[1,2,****]

[1]Guangdong Provincial Key Laboratory of Nuclear Science, Institute of Quantum Matter, South China Normal University, Guangzhou 510006, China.
[2]Guangdong-Hong Kong Joint Laboratory of Quantum Matter, South China Normal University, Guangzhou 510006, China.
[3]Physics Department and Center for Exploration of Energy and Matter, Indiana University, 2401 N Milo B. Sampson Lane, Bloomington, IN 47408, USA.

Abstract. In this contribution, we report AMPT model study for the hyperon global polarization in the heavy ion collision, especially in the interesting $\hat{O}(1 \sim 10)$ GeV energy region. We find a non-monotonic trend, with the global polarization to first increase and then decrease when beam energy is lowered from 27 GeV down to 3 GeV, with a maximum polarization signal located around $\sqrt{s_{NN}} = 7.7$ GeV. In addition, local polarization patterns are also computed in the same energy regime.

1 Introduction

Where can one find the fastest spinning fluid in the world? This is a question of general interest. From a fluid dynamical point of view, the most vortical fluid droplet needs to be large enough to enable a hydrodynamic behavior while small enough to rotate fast without violating the speed of light constraint. In 2017, the STAR collaboration reported the measurement of hyperon global polarization in heavy ion collisions [1], suggesting the subatomic fireball fluid created in these collisions as the most vortical fluid with an average vorticity on the order of $10^{21\sim22}s^{-1}$. This finding was further confirmed by subsequent detailed measurements in [2–5]. In particular, the measured global spin polarization of produced hadrons can be quantitatively related to the angular momentum of the colliding system [6–10], which turns into complex fluid vorticity structures.

One notable feature of the hyperon global polarization measurement results is a strongly increasing trend from high beam energy at $\sqrt{s_{NN}} = 200$ GeV to low beam energy at $\sqrt{s_{NN}} =$ 7.7 GeV. It is tempting to ask whether such a trend would continue further into the $\hat{O}(1)$ GeV range and at which beam energy the truly most vortical fluid will be located. This beam energy region is interesting to explore also because it pertains to a likely shift from a partonic-dominant fluid to a hadronic-dominant fluid. The possible consequence of such change on the global polarization phenomenon is important to understand. In this contribution, we

*e-mail: liaoji@indiana.edu
**e-mail: wangek@scnu.edu.cn
***e-mail: hxing@m.scnu.edu.cn
****e-mail: mr.zhanghui@m.scnu.edu.cn

explore these questions by carrying out a systematic study on the beam energy dependence of hyperon global polarization phenomenon with a transport model calculation, especially in the most interesting region of $\hat{O}(1 \sim 10)$ GeV beam energy.

2 Formalism

We use the multi-phase transport (AMPT) model [18, 19] to compute the Λ hyperon spin polarization in this work. The AMPT model allows us to collect the parton space-time $x = (t, \vec{x})$ information and hadron kinematic freeze-out information from AMPT model, and calculate the vorticity by using the finite difference method after averaging a sufficient number of events. Finally we incorporate the spin polarization effect upon the hadron formation.

The rotational polarization effect on particle spin in a relativistic fluid can be determined from the thermal vorticity $\varpi_{\mu\nu}$, which is defined as [11]:

$$\varpi_{\mu\nu} = -\frac{1}{2}\left(\partial_\mu\beta_\nu - \partial_\nu\beta_\mu\right) \tag{1}$$

where $\beta_\mu = u_\mu/T$ with $T = 1/\beta$ the local temperature. A related quantity is the kinetic vorticity defined by $\Omega_{\mu\nu} = -\frac{1}{2}(\partial_\mu u_\nu - \partial_\nu u_\mu)$. Obviously $\varpi_{\mu\nu} = \beta\left\{\Omega_{\mu\nu} - \left[(\beta\partial_\mu T)u_\nu - (\beta\partial_\nu T)u_\mu\right]\right\}$. The thermal vorticity differs from the $\Omega_{\mu\nu}/T$ by terms containing gradients of temperature, $\sim (\beta\partial_\mu T) = [(\partial_\mu T)/T]$. We use the energy density ϵ to evaluate such terms via $(\partial_\mu T)/T = (\partial_\mu\epsilon)/(4\epsilon)$ [14–16]. The hyperon global polarization is then determined from the following ensemble-averaged spin 4-vector of the produced Λ from the local thermal vorticity at its formation point as [1, 11, 12, 15, 17]:

$$S^\mu = -\frac{1}{8m}\epsilon^{\mu\nu\rho\sigma}p_\nu\varpi_{\rho\sigma}, \tag{2}$$

where p^ν and m are the four-momentum and mass of the produced hyperon, respectively.

In the present work, we extend such transport model calculations to the regime of $\sqrt{s_{NN}} = (3 \sim 10)$ GeV and examining the trend of global polarization observables.

3 Results

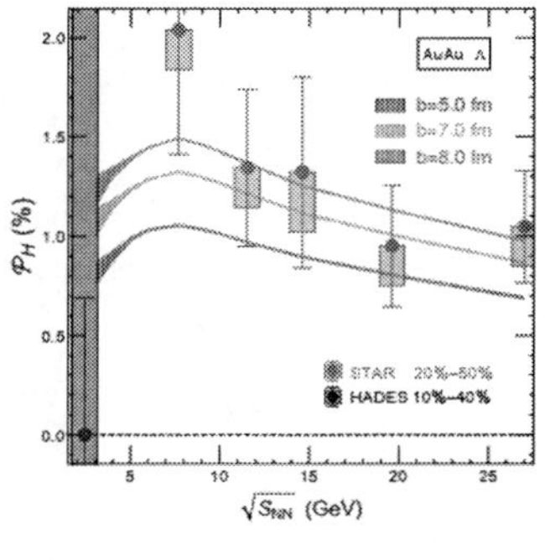

Figure 1. Λ global polarization signal as a function of collision beam energy with $|y_\Lambda| < 1.0$ and 0.4 GeV $< p_T <$ 3.0 GeV, for AuAu collisions at impact parameters b = 5 fm (red), 7 fm (green) and 8 fm (blue), with the bands representing statistical uncertainty from our simulation. Relevant measurements are also shown, with the STAR data from Ref.[1] and the HADES data from Ref. [20].

The key finding of this work is shown in Fig. 1, where the hyperon global polarization signal is computed as a function of collision beam energy, for AuAu collisions at three different impact parameters b = 5 fm (red), 7 fm (green) and 8 fm (blue). The plot includes our calculations done for $\sqrt{s_{NN}} = 3, 4, 5, 6, 7.7, 9, 10, 11.5, 14.5, 19.6, 27$ GeV. This detailed and systematic scan in beam energy has allowed us to clearly identify a non-monotonic trend, with the global polarization to first increase and then decrease when beam energy is lowered from 27 GeV down to 3 GeV. The maximum polarization signal is found to locate around $\sqrt{s_{NN}} = 7.7$ GeV. The decrease toward 3 GeV is the combined effect of the decrease in both the vorticity itself and the average energy of produced hyperons at very low energy.

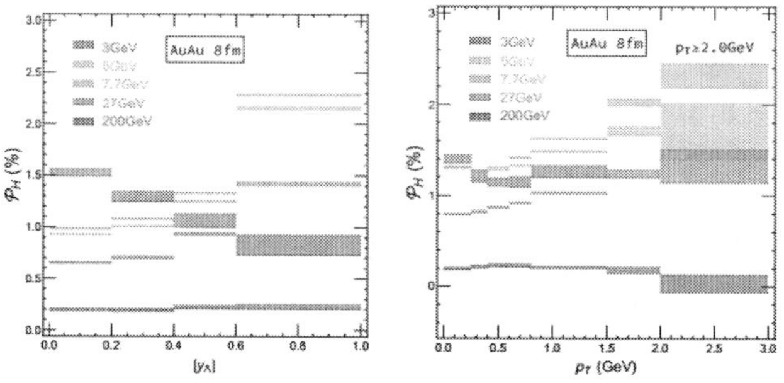

Figure 2. A global polarization signal for AuAu collisions at impact parameter $b = 8$ fm for a variety of collision beam energy from 200GeV down to 3GeV, with the bands representing statistical uncertainty from our simulation. Left: as a function of rapidity y; Right: as a function of transverse momentum p_T.

We've also computed the differential observables, with the dependence on rapidity and transverse momentum shown in the left and right panel of Fig. 2. A barely visible increase with rapidity at 200 GeV changes into a strong rising behavior at low energy down to 5 GeV, which then turns into a strong decrease with rapidity at 3 GeV. The dependence on the transverse momentum in the right panel of Fig. 2 also shows a similar pattern, in which the different beam energy dependence of vorticity and radial flow may play a role.

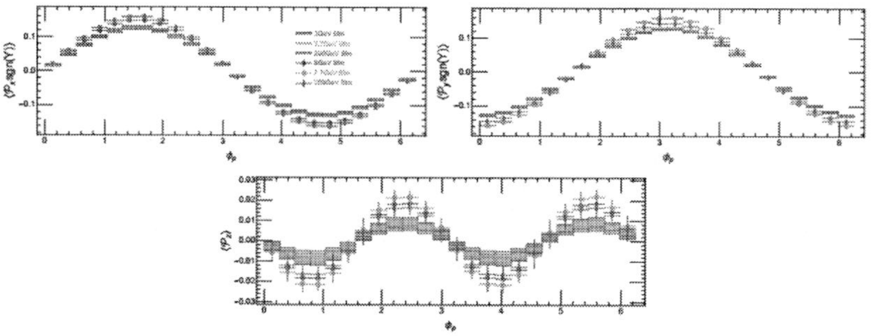

Figure 3. The dependence of hyperon local polarization components on the azimuthal angle ϕ_p for collisions with impact parameter $b = 5, 8$ fm and with beam energy $\sqrt{s_{NN}} = 3, 7.7, 200$ GeV, respectively. The bands (for $b = 5$fm) or error bars (for $b = 8$fm) represent statistical uncertainty from simulation.

In addition to global polarization, the local polarization patterns of produced hyperons provide another category of interesting observables [3, 21]. In Fig. 3, we show the dependence of all three components of the hyperon local polarization on the azimuthal angle ϕ_p with impact parameter $b = 5, 8$ fm, which are the first results of its kind for collision beam energies at 3 and 7.7 GeV. The results suggest that these azimuthal angular patterns are relatively insensitive to the collisional beam energy.

4 Summary and Discussion

In summary, we've performed a transport model (AMPT) study for the hyperon global polarization as well as local polarization patterns in the heavy ion collision. A systematic scan of the $\hat{O}(1 \sim 10)$ GeV energy region shows a non-monotonic trend with a maximum polarization signal located around $\sqrt{s_{NN}} = 7.7$ GeV. These results are of course model dependent. As the results were first reported, no experimental data between $3 \sim 7$ GeV were available yet. An agreement between this calculation and measurements could be an important confirmation of the fluid-vorticity-polarization scenario established at higher energy. On the other hand, if experimental data and the model results show discrepancy, it could be a very interesting indication of either a possible alternative mechanism of polarization generation or a possible shift of the dominant source hyperon production at those extremely low energies. In this sense, the hyperon polarization signal could offer valuable insights into such intriguing issues.

Acknowledgments

This work is supported in part by the Guangdong Major Project of Basic and Applied Basic Research (No. 2020B0301030008), by the National Natural Science Foundation of China under Grants No. 12022512, No. 12035007, No. 12047523, as well as by NSF Grant No. PHY-1913729 and the U.S. Department of Energy, Office of Science, Office of Nuclear Physics, within the framework of the Beam Energy Scan Theory (BEST) Topical Collaboration.

References

[1] STAR, L. Adamczyk *et al.*, Nature **548**, 62 (2017), arXiv:1701.06657.
[2] STAR, J. Adam *et al.*, Phys. Rev. C **98**, 014910 (2018), arXiv:1805.04400.
[3] STAR, J. Adam *et al.*, Phys. Rev. Lett. **123**, 132301 (2019), arXiv:1905.11917.
[4] STAR, J. Adam *et al.*, Phys. Rev. Lett. **126**, 162301 (2021), arXiv:2012.13601.
[5] ALICE, S. Acharya *et al.*, Phys. Rev. C **101**, 044611 (2020), arXiv:1909.01281.
[6] Z.-T. Liang and X.-N. Wang, Phys. Rev. Lett. **94**, 102301 (2005), arXiv:nucl-th/0410079, [Erratum: Phys.Rev.Lett. 96, 039901 (2006)].
[7] J.-H. Gao *et al.*, Phys. Rev. C **77**, 044902 (2008), arXiv:0710.2943.
[8] S. A. Voloshin, (2004), arXiv:nucl-th/0410089.
[9] B. Betz, M. Gyulassy, and G. Torrieri, Phys. Rev. C **76**, 044901 (2007), arXiv:0708.0035.
[10] F. Becattini, F. Piccinini, and J. Rizzo, Phys. Rev. C **77**, 024906 (2008).
[11] F. Becattini, L. Bucciantini, E. Grossi, and L. Tinti, Eur. Phys. J. C **75**, 191 (2015).
[12] F. Becattini, I. Karpenko, M. Lisa, I. Upsal, and S. Voloshin, Phys. Rev. C **95**, 054902 (2017).
[13] Y. Jiang, Z.-W. Lin, and J. Liao, Phys. Rev. C **94**, 044910 (2016), arXiv:1602.06580, [Erratum: Phys.Rev.C 95, 049904 (2017)].
[14] S. Shi, K. Li, and J. Liao, Phys. Lett. B **788**, 409 (2019), arXiv:1712.00878.
[15] H. Li, L.-G. Pang, Q. Wang, and X.-L. Xia, Phys. Rev. C **96**, 054908 (2017).
[16] X.-L. Xia, H. Li, Z.-B. Tang, and Q. Wang, Phys. Rev. C **98**, 024905 (2018).
[17] D.-X. Wei, W.-T. Deng, and X.-G. Huang, Phys. Rev. C **99**, 014905 (2019).
[18] Z.-W. Lin, C. M. Ko, B.-A. Li, B. Zhang, and S. Pal, Phys. Rev. C **72**, 064901 (2005), arXiv:nucl-th/0411110.
[19] Z.-W. Lin, Phys. Rev. C **90**, 014904 (2014), arXiv:1403.6321.
[20] HADES, F. J. Kornas, Springer Proc. Phys. **250**, 435 (2020).
[21] F. Becattini and I. Karpenko, Phys. Rev. Lett. **120**, 012302 (2018), arXiv:1707.07984.

Thermal-model-based characterization of heavy-ion-collision systems at chemical freeze-out

Jamie M. Karthein[1,2,*], *Paolo* Alba[3], *Valentina* Mantovani-Sarti[4], *Jacquelyn* Noronha-Hostler[5], *Paolo* Parotto[6], *Israel* Portillo-Vazquez[1], *Volodymyr* Vovchenko[2], *Volker* Koch[2], and *Claudia* Ratti[1]

[1]Department of Physics, University of Houston, Houston, Texas 77204, USA
[2]Nuclear Science Division, Lawrence Berkeley National Laboratory, Berkeley, CA 94720, USA
[3]Lucht Probst Assoc., GmbH, Frankfurt 60311, DE
[4]Department of Physics, Technical University Munich, Munich 80333, DE
[5]Illinois Center for Advanced Studies of the Universe, Department of Physics, University of Illinois at Urbana-Champaign, Urbana, IL 61801, USA
[6]Department of Physics, University of Wuppertal, Wuppertal 42119, DE

Abstract. We investigate the chemical freeze-out in heavy-ion collisions (HICs) and the impact of the hadronic spectrum on thermal model analyses [1, 2]. Detailed knowledge of the hadronic spectrum is still an open question, which has phenomenological consequences on the study of HICs. By varying the number of resonances included in Hadron Resonance Gas (HRG) Model calculations, we can shed light on which particles may be produced. Furthermore, we study the influence of the number of states on the so-called two flavor freeze-out scenario, in which strange and light particles can freeze-out separately. We consider results for the chemical freeze-out parameters obtained from thermal model fits and from calculating net-particle fluctuations. We will show the effect of using one global temperature to fit all particles and alternatively, allowing particles with and without strange quarks to freeze-out separately.

1 Introduction

Ultra-relativistic heavy-ion collisions provide an opportunity to study the transition from ordinary hadronic matter to the deconfined Quark-Gluon Plasma. Currently, calculations of the fundamental theory with Lattice Quantum Chromodynamics have shown there exists a crossover transition at low baryon chemical potential around T $\sim$ 156 MeV [3]. During this transition, the system traverses through chemical and kinetic freeze-out stages. The location of these freeze-out points in the phase diagram is an important characterization of heavy-ion collisions. Statistical and thermal models can be used to determine those freeze-out parameters, T_f, $\mu_{B,f}$ (and V_f), at chemical freeze-out [4–6]. Thermal fit analyses utilize the χ^2 minimization procedure for a fit to experimental data of various particle species [4, 7, 8]. Alternatively, one can determine the freeze-out parameters by utilizing net-charge fluctuations [6]. This method allows for the determination of freeze-out parameters for different particle species separately, for example for light particles and kaons [2, 6]. In these proceedings, freeze-out parameters are analyzed both via thermal fits and net-particle fluctuations.

*e-mail: jmstafford@uh.edu

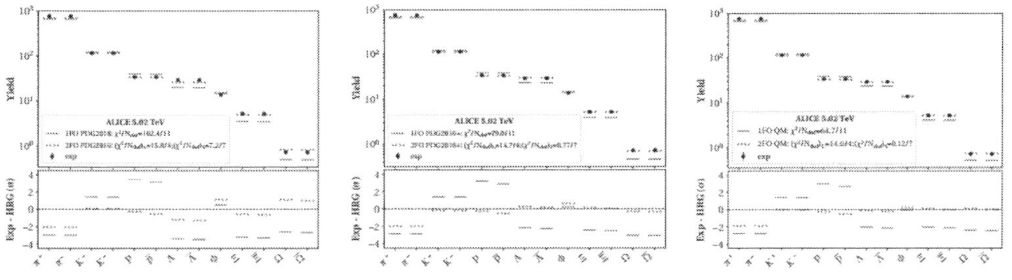

Figure 1. ALICE PbPb $\sqrt{s_{NN}} = 5.02$ GeV data for particle yields in $0-10\%$ collisions, in comparison to HRG model calculations with different resonance lists; deviations in units of experimental errors σ are shown below each panel.

2 HRG Model

This study of the chemical freeze-out stage in heavy-ion collisions is performed within the Hadron Resonance Gas (HRG) model framework. The interacting hadrons are well-described by an ideal gas of hadrons and resonances [9–11],

$$\frac{P}{T^4} = \frac{1}{VT^3} \sum_i \ln Z_i(T, V, \vec{\mu}),$$

$$\ln Z_i^{M/B} = \mp \frac{V d_i}{(2\pi)^3} \int d^3k \ln\left(1 \mp exp\left[-\left(\epsilon_i - \mu_a X_a^i\right)/T\right]\right)$$

where the index i runs over all the particles included in the HRG model, the energy for the single particle reads $\epsilon_i = \sqrt{k^2 + m_i^2}$, the conserved charges X_i are B_i, S_i, Q_i, d_i and m_i represent the degeneracy and the mass for the ith particle and V is the volume.

2.1 Thermal model fits

The HRG model has been widely employed to compare data on particle production for a wide range of energies [12–18]. Produced particle yields are calculated by adding the contribution from resonances to the primordial yields. In these proceedings, we focus on the ALICE data at $\sqrt{s_{\rm NN}} = 5.02$ TeV and $0-10\%$ centrality and perform thermal fits of the particle yields [19]. In Fig. 1 and Table 1, we show the fits for the yields for the PDG2016, PDG2016+, and Quark Model hadronic lists (described in Ref. [1]) and extract the thermal parameters (T, μ_B, V) by using the thermal fit package in FIST [5]. The FIST software allows users to choose their own particle lists, as well as data sets, in the fit.

2.2 Net-particle fluctuations

For the determination of freeze-out parameters from fluctuations, we employ the list PDG2016+ to perform the analysis of net-proton and net-charge fluctuations from Ref. [6], as well as the analysis of net-kaon fluctuations from Ref. [2] and compare to the original results. Originally, the analysis was carried out with an older PDG list, which we indicate as PDG2012. In order to determine both T and μ_B at chemical freeze-out, in general, one needs to fit two experimental quantities. However, due to the larger experimental error bars on higher order fluctuations compared to the lowest order moments, in Ref. [2] we obtained the

		T [MeV]	Volume [fm^3]	χ^2
PDG2016	Single FO	152.2 ± 1.6	5797 ± 523	102.4/12
	Light	143.2 ± 1.8	9096 ± 897	15.7/4
	Strange	169.3 ± 2.5	5797 ± 523	7.7/8
PDG2016+	Single FO	150.4 ± 1.5	6239 ± 539	79.0/12
	Light	142.5 ± 1.7	6239 ± 539	14.7/4
	Strange	164.5 ± 2.3	3087 ± 367	1.6/8
QM	Single FO	147.9 ± 1.3	6852 ± 558	64.7/12
	Light	140.9 ± 1.6	9961 ± 927	13.9/4
	Strange	158.2 ± 1.9	3949 ± 419	0.78/8

Table 1. Fit parameters for the two chemical freeze-out scenarios obtained from fits to total yields of ALICE data [19–22] for 0 − 10% centrality in PbPb collisions at $\sqrt{s_{NN}} = 5.02$ TeV, using different lists. Here, we fix the baryon chemical potential to a value 1 MeV.

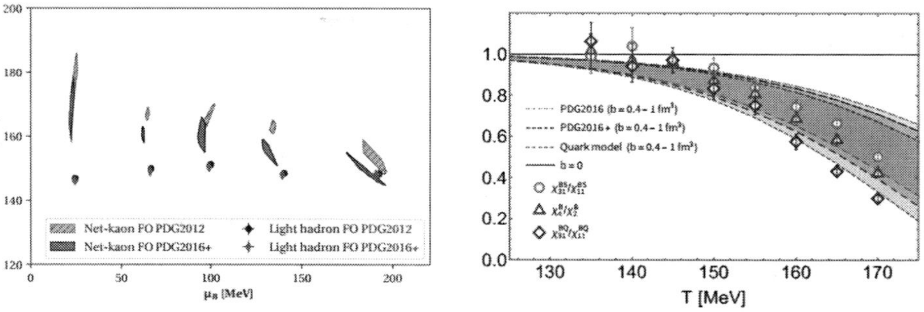

Figure 2. Left: Freeze-out parameters from light hadron and net-kaon fluctuations following the analysis from Ref. [2] with different particle lists. Right: Fourth-to-second order susceptibilities predicted to be equal in the EV-HRG model. Lattice data from Ref. [25] are shown in green with error bars, while the calculations within the EV-HRG model are shown as bands for a range of excluded volume parameter, *b*, and for different hadronic lists. The ideal HRG result corresponds to the line at unity.

freeze-out parameters as follows. Starting from the net-proton and net-charge freeze-out parameters obtained in Ref. [6], we followed the Lattice QCD isentropic trajectories determined in Ref. [23] via a Taylor-expanded equation of state. The ratio χ_1^K/χ_2^K was then calculated along these trajectories and the freeze-out points were determined by the overlap with the experimental results, which are shown in Fig. 2 in gray, while the red points correspond to the net-proton and net-charge freeze-out points from Ref. [6]. We see that this separation in temperature for the kaons and light hadrons persists for the PDG2016+ hadronic list with many more states. Furthermore, as shown in Fig. 2, we also find an indication of flavor-dependent freeze-out in the excluded volume HRG model when considering susceptibilities [24]. Around $T_c \sim 155$ MeV, we see a flavor dependence on the excluded volume parameter, *b*, namely a smaller excluded volume for strange baryons than non-strange baryons.

3 Conclusions

We have presented results on the freeze-out parameters from two different thermal analysis techniques with fits of particle yields and net-charge fluctuations. We have seen that there are two separate chemical freeze-out temperatures in heavy-ion collisions for light and strange particles. We see a better description of the experimental data when utilizing the two flavor freeze-out scenario in the thermal fits. In the case of the fluctuations analysis, we see a clear separation between the light and strange freeze-out temperatures in the phase diagram for the highest collision energies at RHIC.

References

[1] P. Alba, V.M. Sarti, J. Noronha-Hostler, P. Parotto, I. Portillo-Vazquez, C. Ratti, J.M. Stafford, Phys. Rev. C **101**, 054905 (2020), 2002.12395

[2] R. Bellwied, J. Noronha-Hostler, P. Parotto, I. Portillo Vazquez, C. Ratti, J.M. Stafford, Phys. Rev. **C99**, 034912 (2019), 1805.00088

[3] Y. Aoki, Z. Fodor, S.D. Katz, K.K. Szabo, Phys. Lett. **B643**, 46 (2006), hep-lat/0609068

[4] A. Andronic, P. Braun-Munzinger, K. Redlich, J. Stachel, Nature **561**, 321 (2018), 1710.09425

[5] V. Vovchenko, H. Stoecker, Comput. Phys. Commun. **244**, 295 (2019), 1901.05249

[6] P. Alba, W. Alberico, R. Bellwied, M. Bluhm, V. Mantovani Sarti, M. Nahrgang, C. Ratti, Phys. Lett. **B738**, 305 (2014), 1403.4903

[7] F. Bellini (ALICE), Nucl. Phys. **A982**, 427 (2019), 1808.05823

[8] L. Adamczyk et al. (STAR), Phys. Rev. **C96**, 044904 (2017), 1701.07065

[9] A. Bazavov et al. (HotQCD), Phys. Rev. D **86**, 034509 (2012), 1203.0784

[10] S. Borsanyi, Z. Fodor, C. Hoelbling, S.D. Katz, S. Krieg, C. Ratti, K.K. Szabo (Wuppertal-Budapest), JHEP **09**, 073 (2010), 1005.3508

[11] V. Vovchenko, M.I. Gorenstein, H. Stoecker, Phys. Rev. Lett. **118**, 182301 (2017), 1609.03975

[12] F. Becattini, J. Manninen, M. Gazdzicki, Phys. Rev. **C73**, 044905 (2006), hep-ph/0511092

[13] F. Becattini, M. Bleicher, T. Kollegger, T. Schuster, J. Steinheimer, R. Stock, Phys. Rev. Lett. **111**, 082302 (2013), 1212.2431

[14] J. Cleymans, H. Oeschler, K. Redlich, S. Wheaton, Phys. Rev. **C73**, 034905 (2006), hep-ph/0511094

[15] G. Torrieri, J. Rafelski, Phys. Rev. **C75**, 024902 (2007), nucl-th/0608061

[16] A. Andronic, P. Braun-Munzinger, K. Redlich, J. Stachel, J. Phys. **G38**, 124081 (2011), 1106.6321

[17] P. Braun-Munzinger, K. Redlich, J. Stachel (2003), nucl-th/0304013

[18] A. Andronic, P. Braun-Munzinger, J. Stachel, Nucl. Phys. **A772**, 167 (2006), nucl-th/0511071

[19] B.B. Abelev et al. (ALICE), Phys. Lett. **B728**, 216 (2014), [Erratum: Phys. Lett.B734,409(2014)], 1307.5543

[20] B. Abelev et al. (ALICE), Phys. Rev. **C88**, 044910 (2013), 1303.0737

[21] B.B. Abelev et al. (ALICE), Phys. Rev. Lett. **111**, 222301 (2013), 1307.5530

[22] B.B. Abelev et al. (ALICE), Phys. Rev. **C91**, 024609 (2015), 1404.0495

[23] J. Gunther, R. Bellwied, S. Borsanyi, Z. Fodor, S.D. Katz, A. Pasztor, C. Ratti, EPJ Web Conf. **137**, 07008 (2017), 1607.02493

[24] J.M. Karthein, V. Koch, C. Ratti, V. Vovchenko (2021), 2107.00588

[25] S. Borsanyi, Z. Fodor, J.N. Guenther, S.K. Katz, K.K. Szabo, A. Pasztor, I. Portillo, C. Ratti, JHEP **10**, 205 (2018), 1805.04445

Interpretation of Λ hyperon spin polarization measurements

Wojciech Florkowski[1,*] and *Radoslaw* Ryblewski[2,**]

[1]Institute of Theoretical Physics, Jagiellonian University, PL-30-348 Krakow, Poland
[2]Institute of Nuclear Physics, Polish Academy of Sciences, PL-31-342 Krakow, Poland

Abstract. A new global polarization measurement method based on the projection of the polarization vector on the properly defined system's orbital angular momentum in the Λ hyperon's rest frame is briefly described. The advantages of the new method over the standard procedure are discussed.

1 Introduction

The recent measurements of the spin polarization of particles emitted in non-central relativistic heavy-ion collisions [1–4] are quite often listed among the most interesting experimental findings in contemporary high-energy physics [5–9]. The quantum nature of these phenomena provides new insights into the properties of the hot and dense strongly-interacting matter and triggers many new theoretical developments [10–25]. Apart from theoretical challenges, the experimental procedures employed to measure the new phenomena also require better understanding [26]. In this work, we discuss the issues connected with relativistic character of the spin measurement of Λ hyperons in heavy-ion collisions and propose a new method which may improve its interpretation.

2 The weak-decay law

The magnitude P' and orientation $\hat{\boldsymbol{P}}'$ of the spin polarization vector $\boldsymbol{P}' = P'\hat{\boldsymbol{P}}'$ of the Λ hyperon is usually determined through its parity-violating weak-decay, $\Lambda \to p + \pi^-$, employing the fact that in this process the daughter proton is preferentially emitted along the direction of the Λ spin. The measurement is performed in the Λ rest frame $S'(\boldsymbol{p}_\Lambda)$ which is related to the center-of-mass (COM) frame through the (canonical) Lorentz boost [27, 28]. In the $S'(\boldsymbol{p}_\Lambda)$ frame the angular distribution of emitted protons is given by the expression

$$\frac{dN_p^{\rm pol}}{d\Omega'} = \frac{1}{4\pi}\left[1 + \boldsymbol{P}' \cdot \hat{\boldsymbol{p}}'_p\right] = \frac{1}{4\pi}\left[1 + \alpha_\Lambda P'\left(\cos(\Phi' - \phi'_p)\sin\theta'_p \sin\Theta' + \cos\theta'_p\cos\Theta'\right)\right], \quad (1)$$

where $\hat{\boldsymbol{p}}'_p = \left(\sin\theta'_p\cos\phi'_p, \sin\theta'_p\sin\phi'_p, \cos\theta'_p\right)$ and $\hat{\boldsymbol{P}}' = (\sin\Theta'\cos\Phi', \sin\Theta'\sin\Phi', \cos\Theta')$, see Fig. 1. The quantity $\alpha_\Lambda = 0.732$ in Eq. (1) is the Λ decay constant.

*e-mail: wojciech.florkowski@uj.edu.pl
**e-mail: radoslaw.ryblewski@ifj.edu.pl

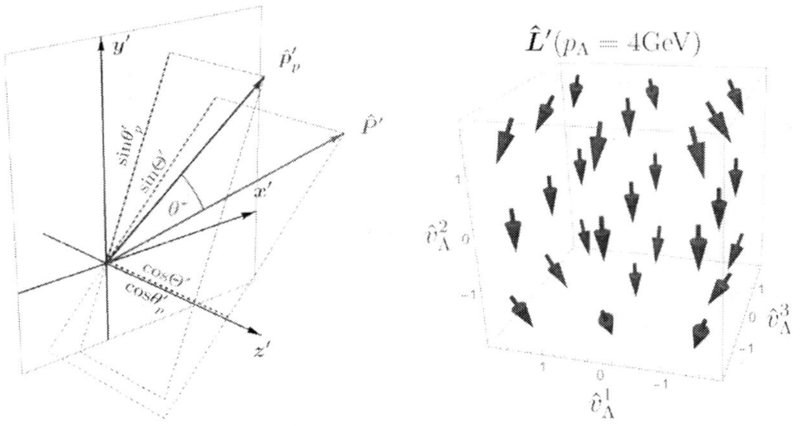

Figure 1. Left: The polarization vector $\boldsymbol{P}'$ and the proton momentum direction $\hat{\boldsymbol{p}}'_p$ as viewed in the Λ rest frame $S'(\boldsymbol{p}_\Lambda)$. The weak decay law $\Lambda \to p + \pi^-$ depends on $\cos\theta^* \equiv \cos(\Phi' - \phi'_p)\sin\theta'_p \sin\Theta' + \cos\theta'_p \cos\Theta'$, where θ^* is the angle between the two vectors, see also Eq. (1). Right: Visualization of the vectors $\boldsymbol{L}'$ for Λ with different velocity directions.

Using Eq. (1) the polarization vector can be fully determined from the averaged values of the proton's three-momentum components

$$\boldsymbol{P}' = \frac{3}{\alpha_\Lambda}\left(\langle \hat{p}'_{p,x}\rangle, \langle \hat{p}'_{p,y}\rangle, \langle \hat{p}'_{p,z}\rangle\right), \tag{2}$$

where we have used the notation $\langle \hat{p}'_{p,i}\rangle = \int \left(\frac{dN^{\rm pol}_p}{d\Omega'}\right) \hat{p}'_{p,i} \sin\theta'_p\, d\theta'_p\, d\phi'_p$.

From theoretical point of view, a particularly interesting quantity, also known as *global polarization*, is

$$P_H = \frac{8}{\pi\alpha_\Lambda}\langle \sin\phi'_p\rangle, \tag{3}$$

with $P_H = P' \sin\Theta' \sin\Phi'$ being the y-component of the polarization three-vector as seen in the Λ rest frame. This quantity has a straightforward interpretation in the non-relativistic limit (small momenta of Λ's) since, in this case, the y direction of the Λ rest frame agrees with the y direction of the COM frame, which in turn defines the direction of the global orbital angular momentum vector of the colliding system, namely

$$\hat{\boldsymbol{L}} = \frac{\boldsymbol{L}}{L} = (0, -1, 0). \tag{4}$$

Due to angular momentum conservation, the final particle-averaged angular momentum of emitted Λ's connected with spin is then expected to be directly related to the initial orbital angular momentum of the matter produced in an event. Obviously, for relativistic Λ's the interpretation of Eq. (3) is complicated by the fact that y direction seen in the Λ rest frame and the y direction in COM frame are different. The latter fact constitutes a significant problem because the measurement requires usually averaging over Λ's within a wide momentum range.

3 Correlation with global orbital angular momentum

As shown in Ref. [26] the measurement method discussed above may be considerably improved by noticing that the angular momentum vector components $L^k = -\frac{1}{2}\epsilon^{kij}L^{ij}$ are given by the spatial components of an (antisymmetric) orbital angular momentum tensor $L^{\mu\nu}$ whose Lorentz transformation properties are similar to that of the Faraday tensor [27]. In consequence, instead of the projection $\hat{\boldsymbol{L}} \cdot \boldsymbol{P}'$ one should calculate the quantity

$$\hat{\boldsymbol{L}}' \cdot \boldsymbol{P}' = \left(1 - (\boldsymbol{v}_\Lambda \cdot \hat{\boldsymbol{L}})^2\right)^{-1/2} \left(\hat{\boldsymbol{L}} \cdot \boldsymbol{P}' - \frac{\gamma_\Lambda}{\gamma_\Lambda + 1} \boldsymbol{v}_\Lambda \cdot \boldsymbol{P}' \ \boldsymbol{v}_\Lambda \cdot \hat{\boldsymbol{L}}\right), \tag{5}$$

which is the projection of the polarization vector along the direction of the total angular momentum of the system as seen in the rest frame of the Λ with three-momentum $\boldsymbol{p}_\Lambda = \boldsymbol{v}_\Lambda E_\Lambda = m_\Lambda \gamma_\Lambda \boldsymbol{v}_\Lambda$ in COM.

The main advantage of the formula (5) compared to $\hat{\boldsymbol{L}} \cdot \boldsymbol{P}'$ is that the spin vector of each Λ, irrespectively of its three-momentum $\boldsymbol{p}_\Lambda$ in COM, is projected on the same physical axis corresponding to the orbital angular momentum in COM. Hence, Eq. (5) is in our opinion the better estimate of the global polarization of the system and can be used to study the correlation between the polarization of emitted particles with the system's global angular momentum.

4 Numerical estimates

In order to quantify possible numerical differences resulting from the measurement based on Eq. (5) as compared to $\hat{\boldsymbol{L}} \cdot \boldsymbol{P}'$ one has to perform the average over the Λ's with different momenta. For that purpose we consider a simplified scenario where $\boldsymbol{P}' = P'\hat{\boldsymbol{L}}$. In this case one obtains

$$\hat{\boldsymbol{L}}' \cdot \boldsymbol{P}' = P'\left(1 - v_2^2\right)^{-1/2} \left(1 - \frac{v_2^2}{1 + \sqrt{1 - v^2}}\right) \equiv P' F_P(\boldsymbol{v}). \tag{6}$$

where (v_1, v_2, v_3) are the components of the Λ velocity in COM and $v = \sqrt{v_1^2 + v_2^2 + v_3^2}$. In the following we also assume the Fermi-Dirac form of the velocity distribution of Λ's, namely $F_T(v) = N\left[\exp\left(m_\Lambda/(T_{\rm eff}\sqrt{1-v^2})\right) + 1\right]^{-1}$, where $T_{\rm eff}$ is an effective temperature and N is an (irrelevant) normalization constant. The quantity

$$\langle \hat{\boldsymbol{L}}' \cdot \boldsymbol{P}' \rangle_{m-n} = P' \, \frac{\int_{v_{(m)}}^{v_{(n)}} dv \int d\Omega \, F_P(\boldsymbol{v}) F_T(v)}{\int_{v_{(m)}}^{v_{(n)}} dv \int d\Omega \, F_T(v)}, \tag{7}$$

where $v_{(n)} = \tanh\left[\sinh^{-1}\left(\frac{n\,\mathrm{GeV}}{m_\Lambda}\right)\right]$ gives average value of the formula (5) for Λ's with the momentum in the range (m,n) GeV. The numerical estimates using $T_{\rm eff} = 150$ MeV give: $\langle \hat{\boldsymbol{L}}' \cdot \boldsymbol{P}' \rangle_{2-3} = 0.97\,P'$, $\langle \hat{\boldsymbol{L}}' \cdot \boldsymbol{P}' \rangle_{3-4} = 0.94\,P'$, $\langle \hat{\boldsymbol{L}}' \cdot \boldsymbol{P}' \rangle_{4-5} = 0.92\,P'$, and $\langle \hat{\boldsymbol{L}}' \cdot \boldsymbol{P}' \rangle_{5-6} = 0.90\,P'$. In consequence, we expect that the effects accounted for by the newly proposed method may reach 10% for the most energetic Λ's studied at STAR.

5 Conclusions

In this work we have discussed possible interpretation issues connected with the current global spin polarization measurements in heavy-ion collisions resulting from relativistic character of the measured particles. Using the Lorentz transformation properties of the global angular momentum we have proposed a new measurement procedure which may serve as a possible solution to these problems. We have shown that the new observable may lead to quantitative differences of up to 10% of the currently measured signal.

Acknowledgements. This work was supported in part by the Polish National Science Center Grants No. 2016/23/B/ST2/00717 and No. 2018/30/E/ST2/00432.

References

[1] L. Adamczyk et al. (STAR), Nature **548**, 62 (2017), 1701.06657
[2] J. Adam et al. (STAR), Phys. Rev. C **98**, 014910 (2018), 1805.04400
[3] S. Acharya et al. (ALICE), Phys. Rev. Lett. **125**, 012301 (2020), 1910.14408
[4] M.S. Abdallah et al. (STAR) (2021), 2108.00044
[5] X.G. Huang, Nucl. Phys. A **1005**, 121752 (2021), 2002.07549
[6] F. Becattini, M.A. Lisa, Ann. Rev. Nucl. Part. Sci. **70**, 395 (2020), 2003.03640
[7] W. Florkowski, A. Kumar, R. Ryblewski, Prog. Part. Nucl. Phys. **108**, 103709 (2019), 1811.04409
[8] E. Speranza, N. Weickgenannt, Eur. Phys. J. A **57**, 155 (2021), 2007.00138
[9] S. Bhadury, J. Bhatt, A. Jaiswal, A. Kumar, Eur. Phys. J. ST **230**, 655 (2021), 2101.11964
[10] J.H. Gao, S.W. Chen, W.T. Deng, Z.T. Liang, Q. Wang, X.N. Wang, Phys. Rev. **C77**, 044902 (2008), 0710.2943
[11] F. Becattini, E. Grossi, Phys. Rev. **D92**, 045037 (2015), 1505.07760
[12] T. Hayata, Y. Hidaka, T. Noumi, M. Hongo, Phys. Rev. **D92**, 065008 (2015), 1503.04535
[13] W. Florkowski, B. Friman, A. Jaiswal, E. Speranza, Phys. Rev. **C97**, 041901 (2018), 1705.00587
[14] S. Li, M.A. Stephanov, H.U. Yee (2020), 2011.12318
[15] J. Hu, Phys. Rev. D **103**, 116015 (2021), 2101.08440
[16] J.H. Gao, Z.T. Liang, S. Pu, Q. Wang, X.N. Wang, Phys. Rev. Lett. **109**, 232301 (2012), 1203.0725
[17] R.H. Fang, L.G. Pang, Q. Wang, X.N. Wang, Phys. Rev. **C94**, 024904 (2016), 1604.04036
[18] W. Florkowski, A. Kumar, R. Ryblewski, Phys. Rev. C **98**, 044906 (2018), 1806.02616
[19] N. Weickgenannt, X.L. Sheng, E. Speranza, Q. Wang, D.H. Rischke, Phys. Rev. D **100**, 056018 (2019), 1902.06513
[20] N. Weickgenannt, E. Speranza, X.l. Sheng, Q. Wang, D.H. Rischke (2020), 2005.01506
[21] S. Bhadury, W. Florkowski, A. Jaiswal, A. Kumar, R. Ryblewski, Phys. Lett. B **814**, 136096 (2021), 2002.03937
[22] L. Tinti, W. Florkowski (2020), 2007.04029
[23] D.E. Kharzeev, D.T. Son, Phys. Rev. Lett. **106**, 062301 (2011), 1010.0038
[24] D. Montenegro, L. Tinti, G. Torrieri, Phys. Rev. **D96**, 056012 (2017), 1701.08263
[25] D. Gallegos, U. Gursoy, A. Yarom (2021), 2101.04759
[26] W. Florkowski, R. Ryblewski (2021), 2102.02890
[27] J.D. Jackson, *Classical Electrodynamics* (Wiley, 1998), ISBN 978-0-471-30932-1
[28] E. Leader, *Spin in particle physics*, Vol. 15 (2011), ISBN 978-0-511-87418-5, 978-0-521-35281-9, 978-0-521-02077-0

EPJ Web of Conferences **259**, 11012 (2022) https://doi.org/10.1051/epjconf/202225911012
SQM 2021

Studying light flavour hadrons produced in the collision of different nuclei at the LHC

Nicolò Jacazio[1,*] For the ALICE Collaboration.

[1]CERN

Abstract. The study of identified particle production as a function of event multiplicity is a key tool for understanding the similarities and differences among different colliding systems. Now for the first time, we can investigate how particle production is affected by the collision geometry in heavy-ion collisions at the LHC. In these proceedings, we report newly obtained ALICE results on charged and identified particle production in Pb–Pb and Xe–Xe collision at $\sqrt{s_{\rm NN}} = 5.02$ and $\sqrt{s_{\rm NN}} = 5.44$ TeV, respectively, as a function of transverse momentum ($p_{\rm T}$) and collision centrality. Particle spectra and ratios are compared between two different colliding systems at similar charged-particle multiplicity densities ($\langle {\rm d}N_{\rm ch}/{\rm d}\eta \rangle$), and different initial eccentricities. We find that in central collisions, spectral shapes of different particles are driven by their masses. The $p_{\rm T}$-integrated particle yield ratios follow the same trends with $\langle {\rm d}N_{\rm ch}/{\rm d}\eta \rangle$ as previously observed in other systems, further suggesting that at the LHC energies, event hadrochemistry is dominantly driven by the charged-particle multiplicity density and not the collision system, geometry or center-of-mass energy.

1 Introduction

The ultimate goal of heavy-ion physics is the study of the properties of the Quark-Gluon Plasma (QGP), a deconfined and chirally restored state of matter. The measurement of the transverse momentum ($p_{\rm T}$) spectra of identified particles provides a way into the collective properties and of particle production in the fireball created in heavy-ion collisions. The ALICE experiment [1, 2] is particularly well suited to study the production of both identified and unidentified charged particles due to its excellent tracking performance coupled with extensive particle identification (PID) capabilities over a wide range of transverse momentum. Particles are identified by combining different techniques allowing their continuous separation over a large transverse momentum interval. In 2017 ALICE recorded Xe–Xe collision at the highest energy per nucleon ever achieved in the laboratory, allowing for the first time at the LHC a quantitative comparison of different heavy-ion collision systems.

2 Results

We report results on the production of π, K, p and ϕ measured in Xe–Xe collisions at $\sqrt{s_{\rm NN}} = 5.44$ TeV as a function of centrality, as discussed in [3]. The data sample was recorded with a minimum-bias trigger. The total charge collected in the V0 detectors, a set

*e-mail: nicolo.jacazio@cern.ch

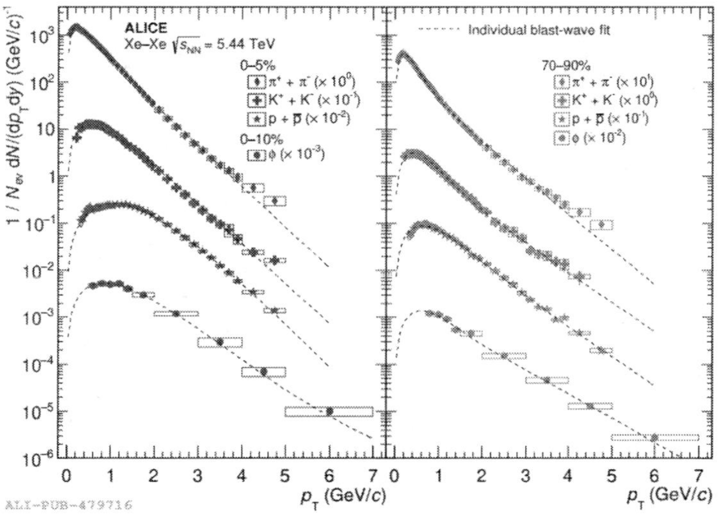

Figure 1. Transverse momentum distributions of $\pi^{\pm}$, $K^{\pm}$, $p, \bar{p}$ and ϕ as measured in central (left) and peripheral (right) Xe–Xe collisions at $\sqrt{s_{NN}}$ = 5.44 TeV.

of two scintillator detectors located in the pseudorapidity region $2.8 < \eta < 5.1$ (V0A) and $-3.7 < \eta < -1.7$ (V0C) and covering the full azimuth, was used to determine the centrality of each Xe–Xe collision defined as percentiles of the total hadronic cross section. Further details are given in [4]. Contributions from weak decays of strange particles and from particle knock-out in the material were removed with the data driven approach described in [3]. Systematic uncertainties were estimated by varying the PID technique and the selection criteria used to define the track sample. The amount of pile-up per event was reduced by selecting runs with low interaction rate and by rejecting events with more than one reconstructed vertex, resulting in a negligible effect. The evaluation of the efficiency and acceptance corrections was performed using events simulated with the HIJING [5] event generator and embedded into a detailed description of the ALICE detector through which tracks are propagated with the GEANT3 [6] transport code.

The p_T spectra of identified π, K, p and ϕ are shown for different centrality classes in Fig. 1. A visual comparison of the spectral shapes reveals that the spectra become harder with increasing centrality. This hardening is found to be mass dependent with protons being more affected than pions and experiencing a similar effect as the ϕ meson. This is consistent with the presence of a collective evolution, known also as radial flow. This effect can be better appreciated in Fig. 2 where the average transverse momentum extracted from the spectra is shown as a function of the charged particle multiplicity for each particle and compared to the results obtained in Pb–Pb collisions. The increase in average p_T is the largest for the heavier particles. In addition, it is striking to observe that in the two systems the average p_T are similar provided that they are measured at the same multiplicity and particles with similar mass i.e. p and ϕ have similar average p_T. The effect of radial flow can also be seen in the baryon/meson ratios shown in Fig. 3 as a depletion of p/π at low p_T and an enhancement at higher p_T. It is worth noting that the baryon/meson ratio exhibits an almost flat behavior only when considering particles with similar mass (i.e. p and ϕ) indicating that the hardening is driven by the mass and not by the quark content of the hadron. The p/π ratio is shown as a function of multiplicity in Fig. 4 (left) to further illustrate the effect of the radial flow, here it

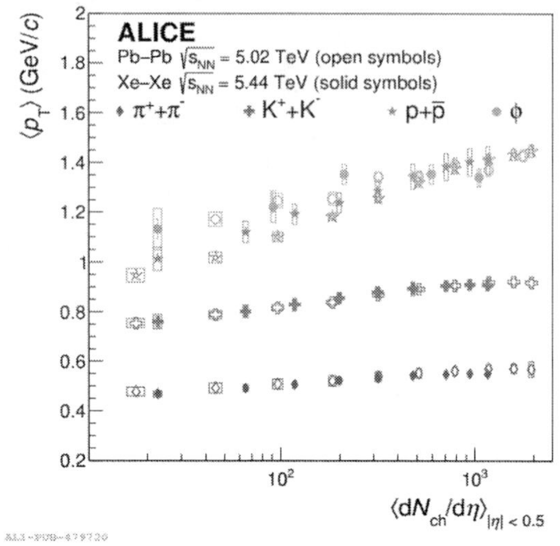

Figure 2. Average transverse momentum of $\pi^{\pm}$, $K^{\pm}$, $p, \bar{p}$ and ϕ as a function of the charged-particle multiplicity density as measured in Xe–Xe collisions at $\sqrt{s_{\mathrm{NN}}}$ = 5.44 TeV [3] and compared to Pb–Pb results at $\sqrt{s_{\mathrm{NN}}}$ = 5.02 TeV [7, 8].

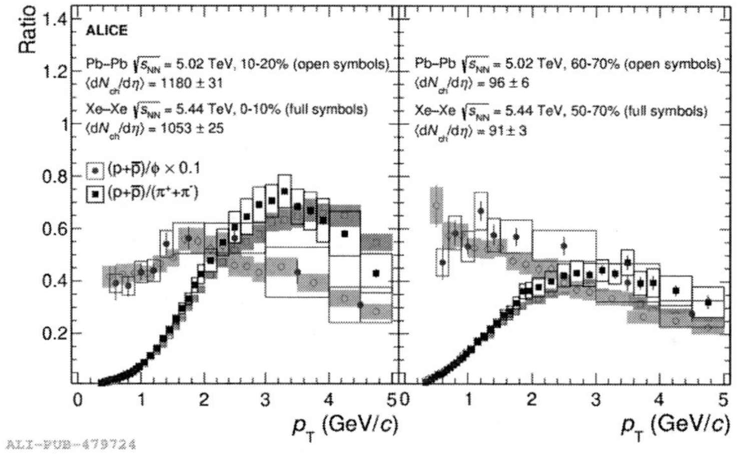

Figure 3. Left: p/ϕ and p/π as a function of the transverse momentum in 0–10% central Xe–Xe collisions at $\sqrt{s_{\mathrm{NN}}}$ = 5.44 TeV [3] compared to 10–20% central Pb–Pb collisions at $\sqrt{s_{\mathrm{NN}}}$ = 5.02 TeV [7, 8] to match the charged particle multiplicity in the two systems. The same is shown in the right panel for peripheral collisions.

can be seen that the amount of enhancement (depletion) at high p_{T} (low p_{T}) is compatible in Xe–Xe and Pb–Pb collisions when compared at similar multiplicity. Conversely, this is not true for the v_2 that exhibits a strong dependence on the centrality, indicating a dominant effect of the initial shape of the collision. Also hadrochemistry is found to be scaling well with the charged particle multiplicity as it can be seen in Fig. 4 (right). Here a general agreement of the results is found across different systems and energies. Furthermore, the results are found to be in agreement with predictions from thermal models.

3 Conclusions

In these proceedings, results on the $\pi^{\pm}$, $K^{\pm}$, $p, \bar{p}$ and ϕ production measured as a function of centrality in Xe–Xe collisions at $\sqrt{s_{\mathrm{NN}}}$ = 5.44 TeV were reported [3]. For the first time at the LHC, it was possible to compare collisions of different nuclei. A comparison of data across different collision systems indicates that the features of light hadron production scale with the event multiplicity and are independent of the colliding system, collision geometry or collision energy. In addition, the radial flow is correlated with the charged particle multiplicity, and the anisotropic flow is affected by collision geometry. It was also shown that the hadrochemistry

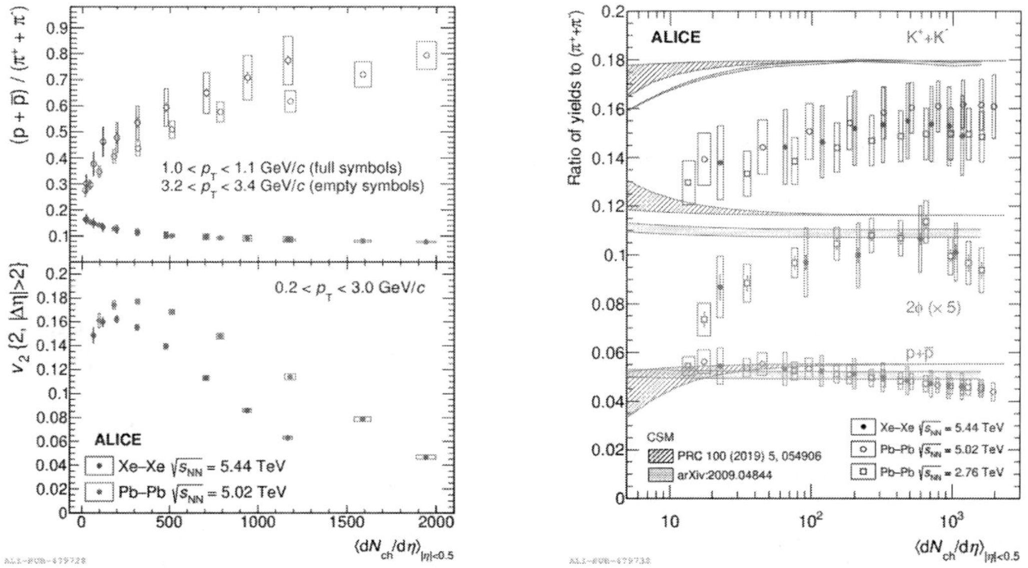

Figure 4. Left figure, top panel: p/π as a function of charged particle multiplicity density in two p_T intervals for Xe–Xe $\sqrt{s_{NN}} = 5.44$ TeV [3] compared to Pb–Pb collisions at $\sqrt{s_{NN}} = 5.02$ TeV [7, 8]. Left figure, bottom panel: the flow coefficient $v_2\{2, |\Delta\eta| > 2\}$ reported for the same collision systems [9, 10] as a function of charged particle multiplicity density. Right figure: K/π, p/π and p/π for integrated yields as a function of the charged-particle multiplicity density for Xe–Xe collisions at $\sqrt{s_{NN}} = 5.44$ TeV [3] compared to Pb–Pb collisions at $\sqrt{s_{NN}} = 2.76$ TeV [11, 12] and 5.02 TeV [7, 8]. Predictions from the canonical statistical model (CSM) are shown as bands considering different correlation volumes [13] (based on [14]) and chemical freeze-out temperatures [15].

does not depend significantly on the collision system and scales with the charged particle multiplicity.

References

[1] K. Aamodt et al. (ALICE), JINST **3**, S08002 (2008)

[2] B. Abelev et al. (ALICE), Int. J. Mod. Phys. **A29**, 1430044 (2014), 1402.4476

[3] S. Acharya et al. (ALICE), Eur. Phys. J. C **81**, 584 (2021), 2101.03100

[4] S. Acharya et al. (ALICE) (2018), ALICE-PUBLIC-2018-003

[5] X.N. Wang, M. Gyulassy, Phys. Rev. D **44**, 3501 (1991)

[6] R. Brun et al., CERN Program Library Long Write-up p. No. W5013 (1994)

[7] S. Acharya et al. (ALICE), Phys. Lett. B **802**, 135225 (2020), 1910.14419

[8] S. Acharya et al. (ALICE), Phys. Rev. C **101**, 044907 (2020)

[9] S. Acharya et al. (ALICE), JHEP **09**, 006 (2018), 1805.04390

[10] S. Acharya et al. (ALICE), Phys. Lett. **B784**, 82 (2018), 1805.01832

[11] B. Abelev et al. (ALICE), Phys. Rev. **C88**, 044910 (2013), 1303.0737

[12] B. Abelev et al. (ALICE), Phys. Rev. C **91**, 024609 (2015), 1404.0495

[13] V. Vovchenko, B. Dönigus, Private Communication (2021)

[14] V. Vovchenko, B. Dönigus, H. Stoecker, Phys. Rev. **C100**, 054906 (2019), 1906.03145

[15] J. Cleymans, P.M. Lo, K. Redlich, N. Sharma, Phys. Rev. C **103**, 014904 (2021), 2009.04844

Improved quark coalescence model for spin alignments of vector mesons

Xin-Li Sheng[1,*]

[1]Key Laboratory of Quark and Lepton Physics (MOE) and Institute of Particle Physics, Central China Normal University, Wuhan, Hubei 430079, China.

Abstract. We propose an improved quark coalescence model for spin alignment of vector mesons by spin density matrix with phase space dependence. Within this model we propose an understanding of spin alignments of vector mesons ϕ and K^{*0} in the static limit: a large positive deviation of ρ_{00} for ϕ mesons from $1/3$ may come from the electric part of the vector ϕ field, while a negative deviation of ρ_{00} for K^{*0} mesons may come from the electric part of vorticity fields. In the low-p_T region, ρ_{00} for K^{*0} mesons is proportional to p_T^2, which is qualitatively agree with experimental results.

1 Introduction

In ultra-relativistic heavy-ion collisions, a huge orbital angular momentum (OAM) can be generated in the direction perpendicular to the reaction plane. During the evolution of the fireball, the OAM is transferred to the spin polarization of quarks through the spin-orbit coupling in nonlocal scatterings [1]. The polarized quarks will then combine and form polarized hadrons and mesons in the freezeout stage [1–3]. The spin alignment of vector mesons can be measured through the angular distribution of decay daughters [4]. Here we report on our recent paper [5], where an improved quark coalescence model in phase space with spin degrees of freedom was proposed. We give a general relation between the density matrix of vector mesons and that of quarks. Using the spin polarization of quarks in vorticity and vector meson fields, predicted by the Wigner function approach [3], we derive spin alignments of vector mesons ϕ and K^{*0}. Our work propose possible explanations for the experimental results obtained by ALICE and STAR collaborations [6–8]. This work should be tested by a detailed and comprehensive simulation of vorticity tensor fields and vector meson fields in heavy ion collisions.

Notations and conventions. We suppress the time dependence of all quantities throughout this paper since we work at the formation time of hadrons. We use boldfaced symbols to represent three-vectors, e.g., $\mathbf{x}$ or $\mathbf{p}$, and use $\mathbf{x}_i$ with subscript $i = x, y, z$ for the three components. The momentum integration measure is denoted using shorthand notation $[d^3\mathbf{p}] \equiv d^3\mathbf{p}/(2\pi)^3$.

2 Spin density matrix and quark coalescence model in phase space

In Ref. [3], a quark coalescence model is constructed based on the spin density matrix in momentum space. However, we need to formulate a model that includes space-time dependence

*e-mail: xls@mail.ustc.edu.cn

because spin polarizations of quarks are inhomogeneous in heavy ion collisions. We generalize the definition of density operator by including off-diagonal elements in momentum space,

$$\rho = \sum_s \int d^3\mathbf{x} \int [d^3\mathbf{p}] w(s, \mathbf{x}, \mathbf{p}) \int [d^3\mathbf{q}] e^{-i\mathbf{x}\cdot\mathbf{q}} \left| s, \mathbf{p} + \frac{\mathbf{q}}{2} \right\rangle \left\langle s, \mathbf{p} - \frac{\mathbf{q}}{2} \right| , \tag{1}$$

where $|s, \mathbf{p}\rangle$ is the spin-momentum state of considered particles. Here $\mathbf{x}$ is the conjugate position of $\mathbf{q}$, i.e., the difference between momenta of state bases. Such a definition is consistent with the wave-packet description for a quantum particle, wherein a particle has finite momentum uncertainty and consists of plane waves with various momenta. The weight function $w(s, \mathbf{x}, \mathbf{p})$ is actually the Wigner function, which is the probability of finding a particle (described by a wave packet) with spin s, center position $\mathbf{x}$, and average momentum $\mathbf{p}$. It can be obtained by projecting the density operator onto two states with the same spin and different momenta

$$w(s, \mathbf{x}, \mathbf{p}) = \int [d^3\mathbf{q}] e^{i\mathbf{q}\cdot\mathbf{x}} \left\langle s, \mathbf{p} + \frac{\mathbf{q}}{2} \right| \rho \left| s, \mathbf{p} - \frac{\mathbf{q}}{2} \right\rangle . \tag{2}$$

For quarks with spin $1/2$, we parameterize the weight function as

$$w(q|s, \mathbf{x}, \mathbf{p}) = \frac{1}{2} f_q(\mathbf{x}, \mathbf{p}) \left[1 + s P_q(\mathbf{x}, \mathbf{p}) \right] , \tag{3}$$

where $f_q(\mathbf{x}, \mathbf{p})$ is the normalized quark distribution function, $\int d^3\mathbf{x} \int d^3[\mathbf{p}] f_q(\mathbf{x}, \mathbf{p}) = 1$, and $P_q(\mathbf{x}, \mathbf{p})$ denotes the average spin polarization at phase space point $(\mathbf{x}, \mathbf{p})$.

We note that definitions of the density operator in Eq. (1) and the weight function in Eq. (2) can be applied to any kind of particles, such as quarks/antiquarks or vector mesons. Therefore spin density matrix elements for mesons are obtained similar to Eq. (2) by putting the density matrix operator between two meson states,

$$\rho^M_{S_{z1}, S_{z2}}(\mathbf{x}, \mathbf{p}) = \int [d^3\mathbf{q}] e^{i\mathbf{q}\cdot\mathbf{x}} \left\langle M; S, S_{z1}; \mathbf{p} + \frac{\mathbf{q}}{2} \right| \rho_M \left| M; S, S_{z2}; \mathbf{p} - \frac{\mathbf{q}}{2} \right\rangle , \tag{4}$$

where M labels the flavor state of the meson, $S = 1$ for a vector meson, and $S_z = -1, 0, 1$ denote spin states in the spin quantization direction. We take the approximation that the meson's density matrix is given by the direct product of density matrices of its constitute quark/antiquark, $\rho_M = \rho_{q\bar{q}} \equiv \rho_q \otimes \rho_{\bar{q}}$. Putting $\rho_{q\bar{q}}$ into Eq. (4), we derive

$$\begin{aligned} \rho^M_{S_{z1}, S_{z2}}(\mathbf{x}, \mathbf{p}) = & \int d^3\mathbf{x}_b [d^3\mathbf{p}_b][d^3\mathbf{q}_b] e^{-i\mathbf{q}_b\cdot\mathbf{x}_b} \varphi^*_M \left(\mathbf{p}_b + \frac{\mathbf{q}_b}{2} \right) \varphi_M \left(\mathbf{p}_b - \frac{\mathbf{q}_b}{2} \right) \\ & \times \sum_{s_1, s_2} w \left(q_1 \middle| s_1, \mathbf{x} + \frac{\mathbf{x}_b}{2}, \frac{\mathbf{p}}{2} + \mathbf{p}_b \right) w \left(\bar{q}_2 \middle| s_2, \mathbf{x} - \frac{\mathbf{x}_b}{2}, \frac{\mathbf{p}}{2} - \mathbf{p}_b \right) \\ & \times \left\langle S, S_{z1} \,\middle|\, s_1, s_2 \right\rangle \left\langle s_1, s_2 \,\middle|\, S, S_{z2} \right\rangle , \end{aligned} \tag{5}$$

where we have used the fact that the spin state in a meson's wave function can be decoupled from the momentum state. The combination of spin states is controlled by the Clebsch-Gordan coefficients $\langle S, S_{z1} \,|\, s_1, s_2 \rangle$. Meanwhile, the inner product between the meson's momentum state $|M; \mathbf{p}\rangle$ and the quark-antiquark's momentum state $|\mathbf{p}_1, \mathbf{p}_2\rangle$ gives the meson wave function φ_M,

$$\left\langle \mathbf{p}_1, \mathbf{p}_2 \middle| M; \mathbf{p} \right\rangle = (2\pi)^3 \delta^{(3)} (\mathbf{p}_1 + \mathbf{p}_2 - \mathbf{p}) \varphi_M \left(\frac{\mathbf{p}_1 - \mathbf{p}_2}{2} \right) . \tag{6}$$

The simplest choice of $\varphi_M(\mathbf{k})$ is the Gaussian distribution $\varphi_M(\mathbf{k}) = N_M \exp[-\mathbf{k}^2/(2a_M^2)]$, where a_M is the average momentum width for the meson's wave function. The pre-factor N_M is determined by the normalization condition $\int [d^3\mathbf{k}] \varphi^*_M(\mathbf{k}) \varphi_M(\mathbf{k}) = 1$.

3 Spin alignments of ϕ and K^{*0}

In relativistic heavy ion collisions, there are different sources for spin polarizations of massive fermions: vorticity fields, electromagnetic fields, and mean fields of vector mesons. Here we only consider polarizations induced by vorticity and vector meson fields since contributions from electromagnetic fields are believed to be negligible [9]. For quarks and antiquarks, the polarization along the direction of global OAM, i.e., the y-direction, are given by

$$P^y_{q/\bar{q}}(\mathbf{x},\mathbf{p}) = \frac{1}{2}\omega_y \pm \frac{1}{2m_q}(\boldsymbol{\varepsilon}\times\mathbf{p})_y \pm \frac{g_V}{2m_q T}\mathbf{B}^V_y + \frac{g_V}{2m_q E_p T}(\mathbf{E}_V\times\mathbf{p})_y\,, \tag{7}$$

which are derived in the Wigner function approach [3]. We have defined the three-vector of the electric part of the thermal vorticity tensor as $\boldsymbol{\omega} = \nabla\times(\beta\mathbf{u})/2$, and the magnetic part as $\boldsymbol{\varepsilon} = -[\partial_t(\beta\mathbf{u}) + \nabla(\beta u^0)]/2$, where $\beta = 1/T$ is the inverse temperature and $u^\mu = (u^0,\mathbf{u})$ is the flow velocity. The strength tensor $F^{\mu\nu}_V$ of vector meson fields is also decomposed into the electric part $\mathbf{E}^V_i = F^{i0}_V$ and the magnetic part $\mathbf{B}^V_i = -\epsilon^{ijk}F^{jk}_V$ in analogy with normal electromagnetic fields, with $i,j,k = x,y,z$. We use g_V to represent the coupling constant between quark/antiquark and vector meson fields in the quark-meson model [10, 11]. For light quarks, i.e., u and d quarks, the major contributions are ρ and ω meson fields, while s quarks are mainly polarized by ϕ meson fields.

Substituting the average polarization (7) for s and $\bar{s}$ into Eqs. (3) and (5), we obtain the average value of the normalized spin density matrix element $\bar{\rho}_{00} \equiv \rho_{00}/(\rho_{-1,-1}+\rho_{00}+\rho_{11})$ (which is also called the spin alignment) for ϕ mesons,

$$\begin{aligned}
\left\langle\bar{\rho}^{\phi}_{00}(\mathbf{x},\mathbf{p})\right\rangle \approx& \frac{1}{3} - \frac{1}{9}\left\langle\omega_y^2\right\rangle - \frac{1}{27m_s^2}\left(\left\langle\varepsilon_x^2\right\rangle + \left\langle\varepsilon_z^2\right\rangle\right)\left\langle\mathbf{p}_b^2\right\rangle_\phi \\
&+ \frac{g_\phi^2}{9m_s^2}\left[\left\langle\left(\beta\mathbf{B}^\phi_y\right)^2\right\rangle + \left\langle\left(\beta\mathbf{E}^\phi_z\right)^2\right\rangle\left\langle\frac{\mathbf{p}^2_{b,x}}{E_{p_1}E_{p_2}}\right\rangle_\phi + \left\langle\left(\beta\mathbf{E}^\phi_x\right)^2\right\rangle\left\langle\frac{\mathbf{p}^2_{b,z}}{E_{p_1}E_{p_2}}\right\rangle_\phi\right] \\
&+ \frac{1}{36m_s^2}\left[\left\langle\varepsilon_z^2\right\rangle - g_\phi^2\left\langle\left(\beta\mathbf{E}^\phi_z\right)^2\right\rangle\left\langle\frac{1}{E_{p_1}E_{p_2}}\right\rangle_\phi\right]\mathbf{p}_x^2 \\
&+ \frac{1}{36m_s^2}\left[\left\langle\varepsilon_x^2\right\rangle - g_\phi^2\left\langle\left(\beta\mathbf{E}^\phi_x\right)^2\right\rangle\left\langle\frac{1}{E_{p_1}E_{p_2}}\right\rangle_\phi\right]\mathbf{p}_z^2\,,
\end{aligned} \tag{8}$$

where $\mathbf{p}_1 = \mathbf{p}/2 + \mathbf{p}_b$ and $\mathbf{p}_2 = \mathbf{p}/2 - \mathbf{p}_b$. Here the spin quantization direction is chosen as the y-direction. We use $\langle\cdots\rangle$ to represent the average over generating positions of ϕ mesons, while $\langle\cdots\rangle_\phi$ to represent the average on the ϕ meson's wave function. We have $\langle\cdots\rangle_\phi \equiv 8\int d^3\mathbf{x}_b d^3\mathbf{p}_b \exp\left(-\mathbf{p}_b^2/a_M^2 - a_M^2\mathbf{x}_b^2\right)(\cdots)$ if we take the Gaussian distribution approximation for the meson wave function. We find that, in Eq. (8), all contributions appear independently as positive or negative quantities. For nearly static ϕ mesons with $\mathbf{p}\simeq 0$, vorticity fields, including $\boldsymbol{\omega}$ and $\boldsymbol{\varepsilon}$, have negative contributions, while vector meson fields $\mathbf{E}_\phi$ and $\mathbf{B}_\phi$ have positive contributions. We have argued in Ref. [9] that the dominant contribution to $\bar{\rho}^{\phi}_{00}$ may possibly be from the electric part $\mathbf{E}_\phi$, which results in the positive deviation from 1/3 for ϕ mesons' spin alignment [6–8].

Let us turn to the spin alignment of vector meson K^{*0} with flavor ($d\bar{s}$). We note that d quarks will be polarized by ρ and ω meson fields, i.e., P^y_d depends on $\mathbf{E}_\rho$, $\mathbf{B}_\rho$, $\mathbf{E}_\omega$, and $\mathbf{B}_\omega$. On the other hand, $\bar{s}$ is polarized by ϕ meson fields $\mathbf{E}_\phi$ and $\mathbf{B}_\phi$. We take the approximation that different kinds of meson fields do not have large correlation in space, therefore only quadratic terms survive in the spin alignment of K^{*0},

$$\left\langle \bar{\rho}_{00}^{K^{*0}}(\mathbf{x},\mathbf{p})\right\rangle \approx \frac{1}{3}-\frac{1}{9}\left\langle \omega_y^2\right\rangle - \frac{1}{27m_s^2}\left(\left\langle \boldsymbol{\varepsilon}_x^2\right\rangle + \left\langle \boldsymbol{\varepsilon}_z^2\right\rangle\right)\left\langle \mathbf{p}_b^2\right\rangle_{K^{*0}} + \frac{1}{36m_s m_d}\left[\left\langle \boldsymbol{\varepsilon}_z^2\right\rangle \mathbf{p}_x^2 + \left\langle \boldsymbol{\varepsilon}_x^2\right\rangle \mathbf{p}_z^2\right]. \quad (9)$$

This result is obtained by substitute polarizations of d and $\bar{s}$, given by Eq. (7), into Eqs. (3) and (5). We see that contributions from vector meson fields are absent in Eq. (9). The spin alignment of K^{*0} is dominated by vorticity fields and will be smaller than $1/3$ for nearly static K^{*0}, which qualitatively agrees with experimental results obtained by ALICE and STAR [6–8]. We note that the contribution from $\boldsymbol{\varepsilon}$ for $\bar{\rho}_{00}^{K^{*0}}$ is amplified by about $2.1 \sim 2.3$ compared to that for $\bar{\rho}_{00}^{\phi}$. Therefore $\boldsymbol{\varepsilon}$ may have a sizable magnitude of contribution for $\bar{\rho}_{00}^{K^{*0}}$, but is less important for $\bar{\rho}_{00}^{\phi}$. We also notice that the spin alignment of K^{*0} is proportional to p_T^2 in the low-p_T region, which seems to be agree with the p_T dependence in Refs. [6–8].

4 Summary

We have constructed an improved quark coalescence model based on the spin density matrix in phase space with coordinate dependence. Spin alignments for vector mesons ϕ and K^{*0} are derived from spin polarizations for quarks. We propose understandings of significant positive (negative) deviations of $\bar{\rho}_{00}$ for ϕ (K^{*0}) mesons from $1/3$, which should be tested by detailed simulations of vorticity and vector meson fields in heavy ion collisions.

acknowledgments

X.-L. S. is supported in part by the National Natural Science Foundation of China (NSFC) under Grant Nos. 11935007, 11221504, 11861131009, 11890714 (a sub-grant of 11890710), 12047528, and by the Fundamental Research Funds for Central Universities in China by the UCB-CCNU Collaboration Grant.

References

[1] Z. T. Liang and X. N. Wang, Phys. Rev. Lett. **94** (2005), 102301 [erratum: Phys. Rev. Lett. **96** (2006), 039901].
[2] Z. T. Liang and X. N. Wang, Phys. Lett. B **629** (2005), 20-26.
[3] Y. G. Yang, R. H. Fang, Q. Wang and X. N. Wang, Phys. Rev. C **97** (2018) no.3, 034917.
[4] K. Schilling, P. Seyboth and G. E. Wolf, Nucl. Phys. B **15**, 397-412 (1970) [erratum: Nucl. Phys. B **18**, 332 (1970)].
[5] X. L. Sheng, Q. Wang and X. N. Wang, Phys. Rev. D **102** (2020) no.5, 056013.
[6] S. Acharya *et al.* [ALICE], Phys. Rev. Lett. **125**, no.1, 012301 (2020).
[7] S. Singha [STAR], Nucl. Phys. A **1005**, 121733 (2021).
[8] S. Kundu [ALICE], Nucl. Phys. A **1005**, 121912 (2021).
[9] X. L. Sheng, L. Oliva and Q. Wang, Phys. Rev. D **101** (2020) no.9, 096005.
[10] J. J. Sakurai, Annals Phys. **11**, 1-48 (1960).
[11] A. Zacchi, R. Stiele and J. Schaffner-Bielich, Phys. Rev. D **92**, no.4, 045022 (2015).

${}^{3}_{\Lambda}$H and ${}^{4}_{\Lambda}$H Lifetime, Yield, Directed Flow Measurements in Au+Au Collisions at $\sqrt{s_{NN}}$ = 3 GeV With the STAR Detector

Chenlu Hu[1,*] for STAR collaboration

[1]Quark Matter Research Center, Institute of Modern Physics, Chinese Academy of Sciences

Abstract. In this proceedings, the lifetime and yields of ${}^{3}_{\Lambda}$H and ${}^{4}_{\Lambda}$H in Au+Au collisions at $\sqrt{s_{NN}}$ = 3 GeV are presented. The measured yields are compared to measurements at other energies and theoretical models, and the physics implications are discussed. We also report the first observation of the ${}^{3}_{\Lambda}$H and ${}^{4}_{\Lambda}$H directed flow in 5 - 40% centrality. The directed flow of ${}^{3}_{\Lambda}$H and ${}^{4}_{\Lambda}$H are compared with those of the copiously produced particles such as p, Λ, d, t, ^{3}He, and ^{4}He. These results shed light on light hyper-nuclei production in heavy-ion collisions in the high baryon density region.

1 Introduction

As is known to all, the normal nucleus is made up of protons and neutrons. When a nucleon is replaced by a Λ hyperon ($S = -1$. Here S denotes the quantum number of strangeness), the nucleus is transformed into a hyper-nucleus which allows us to study the hyperon-nucleon (Y-N) interaction. It is well known that 2-body and 3-body Y-N interactions, especially at high baryon density, are essential for understanding the inner structure of compact stars [1-2]. Measurements of the lifetime, binding energy, decay branching ratios of hyper-nuclei can give us important information on Y-N interaction.

Anisotropic flow has been commonly used for studying the properties of matter created in high energy nuclear collisions, due to its genuine sensitivity on early stage collision dynamics [3]. The first order coefficient of the Fourier-expansion of azimuthal distribution, known as directed flow (v_1), has been analyzed for all particles ranging from the lightest pion-mesons to light nuclei in such collisions [4-5].

In this proceedings, the lifetime, yields and directed flow of ${}^{3}_{\Lambda}$H and ${}^{4}_{\Lambda}$H in Au+Au collisions at $\sqrt{s_{NN}}$ = 3 GeV will be discussed. The data was collected by the STAR experiment at RHIC with the fixed-target (FXT) setup. The gold beam of 3.85 GeV/u is collided on a thin gold target with 1% interaction probability, located at 200 cm along the beam direction from the center of the STAR Time-Projection Chamber (TPC). A total of 260M good minimum bias (MB) events were selected for this analysis.

2 Data Analysis, Results and Discussion

At the $\sqrt{s_{NN}}$ = 3 GeV collisions, the first order event plane is determined by the Event Plane Detector (EPD) [6], which is designed to measure the pattern of forward-going charged

*e-mail: huchenlu@impcas.ac.cn

particles emitted in heavy-ion collisions and covers a pseudorapidity range of $2.14 < |\eta| < 5.09$. The directed flow (v_1) discussed below is determined by the first order event plane.

2.1 Particle Reconstruction

The hyper-nuclei ${}^{3}_{\Lambda}$H and ${}^{4}_{\Lambda}$H are reconstructed with following decay channels: ${}^{3}_{\Lambda}\mathrm{H} \to {}^{3}\mathrm{He} + \pi^-$, ${}^{3}_{\Lambda}\mathrm{H} \to \mathrm{d} + \mathrm{p} + \pi^-$, ${}^{4}_{\Lambda}\mathrm{H} \to {}^{4}\mathrm{He} + \pi^-$. To assure the quality of each track, a minimum of 15 hits out of 45 hits in the TPC is required. The secondary decay topology is reconstructed by the KFParticle program which is based on a Kalman filter method [7]. In the program, the error-matrices are used to enhance the reconstruction significance. A set of cuts on topological variables are applied to the hyper-nuclei candidates to optimize the significance.

2.2 ${}^{3}_{\Lambda}$H and ${}^{4}_{\Lambda}$H Lifetime Measurements

The reconstructed ${}^{3}_{\Lambda}$H and ${}^{4}_{\Lambda}$H candidates are divided into different $L/\beta\gamma$ intervals, where L is is the decay length, β and γ are particle velocity and Lorentz factor, respectively. The raw signal counts N^{raw} for each $L/\beta\gamma$ interval are obtained from corresponding background-subtracted invariant mass spectrum using a bin counting method. The signal counts are corrected with the detector acceptance and reconstruction efficiency ($\varepsilon_{TPC} \times \varepsilon_{PID}$). The corrected hyper-nuclei counts as a function of $L/\beta\gamma$ is fitted to an exponential function ($N = N_0 e^{-L/\beta\gamma c\tau}$) to obtain the mean lifetime τ.

The lifetimes 232 ± 29(stat.) ± 37(syst.) for ${}^{3}_{\Lambda}$H (2-body decay channel) and 218 ± 8(stat.) ± 12(syst.) for ${}^{4}_{\Lambda}$H are obtained from the $\sqrt{s_{NN}} = 3$ GeV data. As shown in Fig. 1, the ${}^{4}_{\Lambda}$H measurement is the most precise measurement to date, and within uncertainties, the measured ${}^{3}_{\Lambda}$H and ${}^{4}_{\Lambda}$H lifetimes are consistent with previous measurements from ALICE [8, 9], STAR [10], HypHI [11].

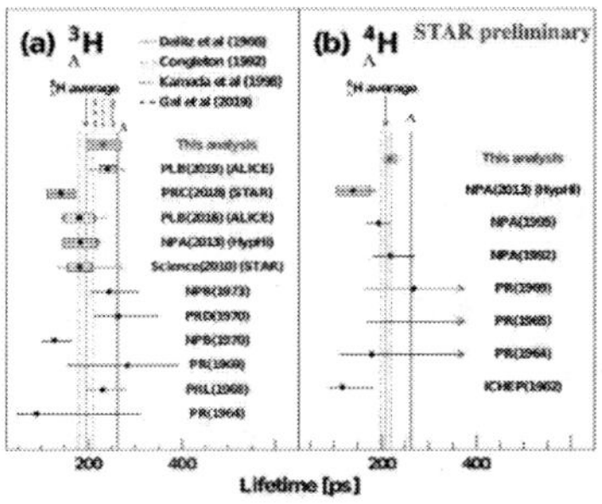

Figure 1. Measured lifetimes of ${}^{3}_{\Lambda}$H (a) and ${}^{4}_{\Lambda}$H (b) are shown comparing to previous measurements and theoretical calculations as well as the free Λ lifetime. The experimental average lifetimes and the corresponding uncertainty of ${}^{3}_{\Lambda}$H and ${}^{4}_{\Lambda}$H are also shown as orange bands.

2.3 ${}^{3}_{\Lambda}$H and ${}^{4}_{\Lambda}$H Yield Measurements

The hyper-nuclei ${}^{3}_{\Lambda}$H and ${}^{4}_{\Lambda}$H yields from their 2-body decay channels are extracted as a function of p_T and y in two centrality selections: 0–10% and 10–50%. The efficiency-corrected p_T spectra in each rapidity slice are extrapolated down to p_T=0 to obtain p_T integrated value of yields (dN/dy). Different functions (e.g blast-wave function) are used to estimate the systematic uncertainties in the unmeasured p_T regions. We have assumed branching ratios of 25% and 50% for the 2-body decay of ${}^{3}_{\Lambda}$H and ${}^{4}_{\Lambda}$H, respectively.

The ${}^{3}_{\Lambda}$H and ${}^{4}_{\Lambda}$H yields at $|y| < 0.5$ as a function of beam energy in central heavy-ion collisions are extracted and are compared to theoretical models as shown in Fig. 2. For ${}^{3}_{\Lambda}$H, the measured yield is consistent with the thermal model from GSI/Heidelberg [12]. The thermal model adopting the canonical ensemble can approximately describe the ${}^{3}_{\Lambda}$H yield both at 3 GeV and 2.76 TeV. Canonical ensemble thermal statistics is required to account for the large ϕ/K^- and ϕ/Ξ^- ratios measured at the same energy as well. We also observe that the coalescence model (DCM) [13] is consistent with the ${}^3_\Lambda$H yield while underestimating the ${}^4_\Lambda$H. On the other hand, the hybrid UrQMD overestimates both ${}^3_\Lambda$H and ${}^4_\Lambda$H yields by an order of magnitude.

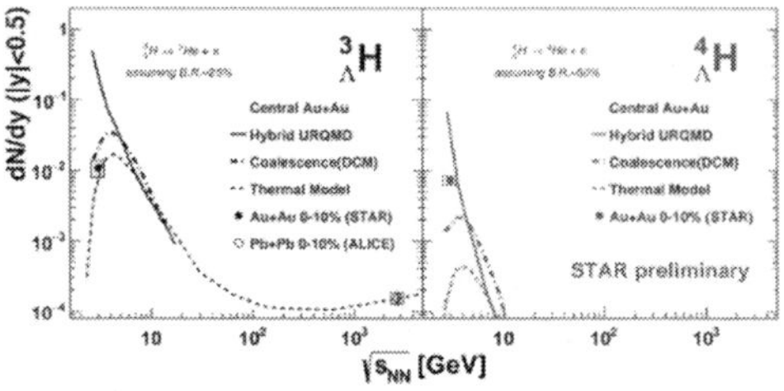

Figure 2. ${}^3_\Lambda$H (a) and ${}^4_\Lambda$H (b) yields at |y| < 0.5 as a function of beam energy in central heavy ion collisions. The symbols represent measurements while the lines represent different theoretical calculations. The data points assume a branching ratio of 25(50)% for ${}^3_\Lambda$H(${}^4_\Lambda$H) → ^{3}He(^{4}He) + π^-.

2.4 ${}^3_\Lambda$H and ${}^4_\Lambda$H Directed Flow Measurements

Directed flow of Λ hyperons, ${}^3_\Lambda$H, and ${}^4_\Lambda$H are extracted with event plane method. Figure 3 shows the v_1 for hyper-nuclei and Λ hyperons versus rapidity from the $\sqrt{s_{NN}}$ = 3 GeV Au + Au collisions. The yellow-red line is the result of linear fit to the data and is plotted in full rapidity region |y| ⩽ 0.9. For comparison, the v_1 distributions for p, d, t, ^{3}He and ^{4}He, from the events with same centrality, are shown as open symbols in the figure. Here the results of the linear fits to the light-nuclei are plotted as dashed-lines only in the positive rapidity region. As one can see, the v_1 of Λ hyperons is consistent with that of protons, and the slopes of hyper-nuclei v_1 are also similar to that of the corresponding light-nuclei with the same mass number within statistical uncertainties.

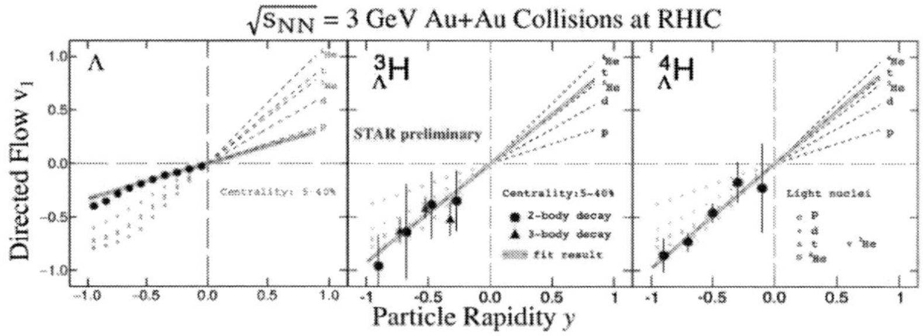

Figure 3. Hyper-nuclei v_1 as a function of rapidity from the $\sqrt{s_{NN}}$ = 3 GeV 5 – 40% mid-central Au + Au collisions at RHIC-STAR. In case of ${}^3_\Lambda$H, both 2-body (dots) and 3-body (triangles) decays are used. Results from fitting with a first-order polynomial function are shown as the yellow-red lines. The rapidity dependence of v_1 for p, d, t,^{3}He and ^{4}He are also shown as open-circles, diamonds, up-triangles, down-triangles and squares, respectively. The corresponding results of the linear fits are shown as dashed lines in the positive rapidity region.

Extracted mid-rapidity v_1 slopes, $dv_1/dy|_{y=0}$, for Λ hyperons, ${}^3_\Lambda$H, and ${}^4_\Lambda$H, are summarized in Fig. 4 as red filled-squares, as a function of particle mass. For comparison, the slopes of light-nuclei p, d, t, ^{3}He, and ^{4}He from the events with same centrality class (5-40%) in $\sqrt{s_{NN}}$ = 3 GeV Au+Au collisions are shown as open circles. The result of a linear fit to the light-nuclei is shown as the yellow-red line in the figure. Overall, hyper-nuclei v_1 slopes are consistent with that of light-nuclei which has similar mass albeit the large uncertainties in the results. The mass dependence of the v_1 slope implies that the coalescence is the dominant mechanism for hyper-nuclei production in heavy-ion collisions.

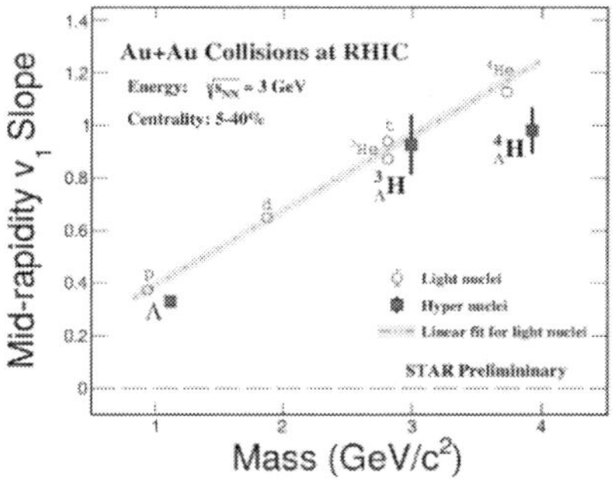

Figure 4. Mass dependence of the mid-rapidity v_1 slope $dv_1/dy|_{y=0}$ for Λ, ${}^{3}_{\Lambda}$H and ${}^{4}_{\Lambda}$H, from the $\sqrt{s_{NN}} = 3$ GeV mid-central 5-40% Au+Au collisions. Combined results of 2-body and 3-body decays are used for ${}^{3}_{\Lambda}$H while the ${}^{4}_{\Lambda}$H is only reconstructed from the 2-body decay. The slopes of light-nuclei p, d, t, ^{3}He and ^{4}He from the same collisions are shown as open circles. The yellow-red line is the result of a linear fit to the measured light nuclei v_1 slopes.

3 Summary

In summary, we reconstruct the light hyper-nuclei ${}^{3}_{\Lambda}$H and ${}^{4}_{\Lambda}$H from $\sqrt{s_{NN}} = 3$ GeV Au+Au collisions at RHIC-STAR. Lifetimes of ${}^{3}_{\Lambda}$H and ${}^{4}_{\Lambda}$H from their 2-body decay channel are measured to be 232 ± 29(stat.) ± 37(syst.) and 218 ± 8(stat.) ± 12(syst.) respectively. The ${}^{3}_{\Lambda}$H and ${}^{4}_{\Lambda}$H lifetimes are consistent with previous measurements and theoretical calculations. Meanwhile, the hyper-nuclei ${}^{3}_{\Lambda}$H and ${}^{4}_{\Lambda}$H yields at $|y| < 0.5$ as a function of beam energy in central heavy-ion collisions are reported and compared to theoretical models. We also reported the first observation of ${}^{3}_{\Lambda}$H and ${}^{4}_{\Lambda}$H directed flow v_1 from mid-central (5-40%) collisions. The rapidity dependence of their v_1 are compared with that of Λ hyperon and light nuclei p, d, t, ^{3}He and ^{4}He from the collisions with the same centrality class. It is found that, within statistical uncertainties, the mid-rapidity v_1 slope of ${}^{3}_{\Lambda}$H and ${}^{4}_{\Lambda}$H are similar to those of light nuclei with the similar mass, such as t, ^{3}He, and ^{4}He. In other words, they seem to follow the baryon mass scaling. These observations imply that coalescence of nucleons and Λ hyperons is the dominant mechanism for the light hyper-nuclei production in such collisions.

4 Acknowledgments

We thank Drs. Y. Nara and J. Steinheimer for interesting discussions and the use of simulations code of JAM and UrQMD.

References

[1] D. Gerstung, N. Kaiser, and W. Weise, Eur. Phys. J. **A56**, 175 (2020)
[2] D. Lonardoni et al., Phys. Rew. Lett. **114**, 092301 (2015)
[3] C.M. Hung and E. Shuryak, Phys. Rev. Lett. **75**, 4003 (1995)
[4] L. Adamczyk et al. (STAR Collaboration), Phys. Rev. Lett. **112**, 162301 (2014)
[5] M.S. Abdallah et al. (STAR Collaboration), Phys. Rev. **C103**, 034908 (2021)
[6] J. Adams et al. (STAR Collaboration), NIM **A968**, 163970 (2020)
[7] I. Kisel et al. (CBM Collaboration), J. Phys. Conf. Ser. **1070**, 012105 (2018)
[8] S. Acharya et al. (ALICE), Phys. Lett. B **797**, 134905 (2019), 1907.06906.
[9] J. Adam et al. (ALICE), Phys. Lett. B **754**, 360 (2016), 1506.08453.
[10] L. Adamczyk et al. (STAR), Phys. Lett. C **97**, 054909 (2018), 1710.00436.
[11] C. Rappold et al., Nucl. Phys. A **913**, 170 (2013), 1305.4871.
[12] A. Andronic, Phys. Lett. B **679**, 203 (2011), 1010.2995.
[13] J. Steinheimer, K. Gudima, A. Botvina, I. Mishustin, M. Bleicher, and H. Stocker, Phys. Lett. B **714**, 85 (2012), 1203.2547.

Measurements of p-Ξ, Λ-Λ, and Ξ-Ξ Correlation in Au+Au collisions at $\sqrt{s_{\mathrm{NN}}}$ = 200 GeV at RHIC-STAR

Moe Isshiki for the STAR Collaboration[1,*]

[1]University of Tsukuba, 1-1-1, Tenno-dai, Tsukuba, Ibaraki, 305-8571, Japan

Abstract.
The hyperon-hyperon and hyperon-nucleon interactions are important in search for exotic hadrons such dibaryon as a bound state of two baryons. Λ-Λ correlation has been measured in Au+Au collisions at $\sqrt{s_{\mathrm{NN}}}$ = 200 GeV at RHIC-STAR experiment using high statistics data recently taken, which has been found to clearly show the anti-correlation. The first measurements of p-Ξ and Ξ-Ξ correlations in Au+Au collisions have been shown. The p-Ξ correlation function shows the strong attractive interaction, which is consistent with HAL QCD calculation. Ξ-Ξ correlation has also been measured to be negative (as anti-correlation).

1 Introduction

Hyperon-hyperon (Y-Y) and hyperon-nucleon (Y-N) interactions are important for studying exotic hadrons and to understand the equation of state of neutron stars. In case of Y-Y interaction, the existence of H-dibaryon (consisting of six quarks: *uuddss*) is predicted by Jaffe in1977 [1]. Also, the Y-N interaction is interesting because N-Ξ interaction in the spin-singlet and isospin-singlet channel is most attractive according to lattice QCD calculation [2].

Study of Λ-Λ correlation has been performed in STAR [3] and ALICE experiments [4], but the interaction is still not clear due to large experimental uncertainty and complication of (partly unknown) feed-down contribution. In ALICE experiment, the attractive interaction of p-Ξ has been observed in p+p and p+Pb collisions [5, 6], which is consistent with a prediction by HAL QCD calculation. In this analysis, we presents new results of Λ-Λ correlation as well as p-Ξ correlation from high statistics data of Au+Au collisions at $\sqrt{s_{\mathrm{NN}}}$ = 200 GeV. We also present the first measurement of Ξ-Ξ correlation to study a possible bound state or weak attractive interaction of Ξ-Ξ system predicted by lattice QCD calculations [7, 8].

2 Femtoscopy analysis

A technique based on Bose-Einstein and Fermi-Dirac correlations has been used in heavy-ion collisions to probe the spatial and temporal extent of particle emitting source. Femtoscopic correlations arise due to quantum statistical and final state effects (strong interaction and Coulomb interaction) at low relative momentum of two particles. The experimental correlation function is defined as

$$C(Q_{inv}) = \frac{A(Q_{inv})}{B(Q_{inv})}, \tag{1}$$

*e-mail: misshiki@rcf.rhic.bnl.gov

 https://doi.org/10.1051/epjconf/202225911015

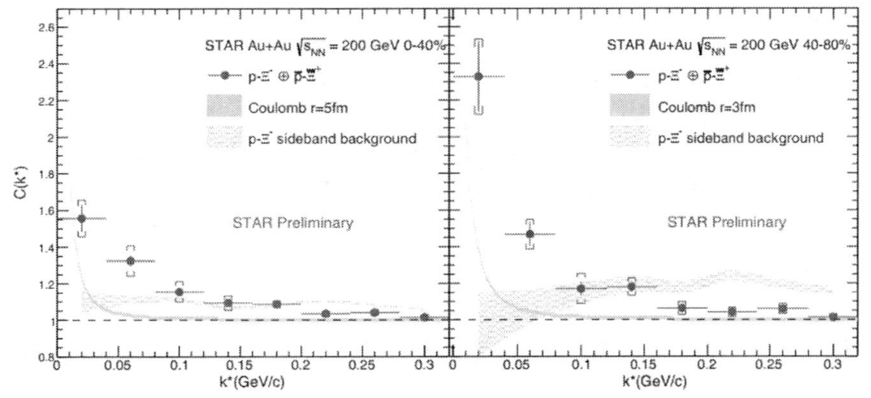

Figure 1. (Color online) p-Ξ correlation function in 0-40% and 40-80% central Au+Au collisions at $\sqrt{s_{NN}}$ = 200 GeV [10]. The orange bands show the Coulomb strength, varying the source size 5 fm (central collisions) and 3 fm (peripheral collisions). The red hatched bands show the side-band background correlation.

$$Q_{inv} = \sqrt{(p_{x1} - p_{x2})^2 + (p_{y1} - p_{y2})^2 + (p_{z1} - p_{z2})^2 - (E_1 - E_2)^2} \tag{2}$$

where $A(Q_{inv})$ is the distribution of real pairs from the same events, $B(Q_{\text{inv}})$ is the distribution of mixed pairs from two different events as background distribution, and Q_{inv} is relative momentum. Protons and daughter particles of Λ and Ξ were identified using Time Projection Chamber (TPC) and Time Of Flight (TOF) detectors, and then Λ and Ξ were identified based on decay topology and/or KFParticle package [9].

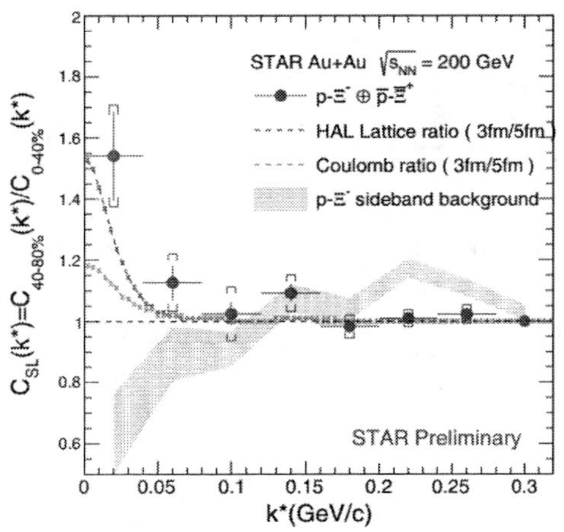

Figure 2. (Color online) System ratio of the p-Ξ correlation function, in 40-80% to that in 0-40% [10]. The green dashed line is the ratio between the source sizes 3 fm to 5 fm calculated by the HAL lattice QCD model. The yellow dashed lines shows the ratio between the source sizes 3 fm to 5 fm estimated for the Coulomb interaction. The red hatched band shows the side-band background correlation.

3 Results

3.1 Measurement of p-Ξ correlation function

Figure 1 shows the correlation function for p-Ξ pairs combined with their charge conjugate using the data taken in 2010, 2011, and 2014 [10]. The p-Ξ analysis has been done in the pair rest frame. The feed-down effect was corrected using the Theminator2 model [11], but a possible residual correlation is not considered yet. The p-Ξ correlation shows an enhancement compared to the Coulomb interaction. This is a hint for the presence of strong interaction, that can not be described by sideband background. The sideband background could come from other particle pair's correlation outside of invariant mass range. The correlation function in peripheral collisions is stronger than that in central collisions because the system size becomes smaller. Figure 2 shows the ratio of the correlation function in peripheral collision over that in central collision, which is defined as

$$C_{\mathrm{SL}}(k^*) = \frac{C(k^*)_{40-80\%}}{C(k^*)_{0-40\%}}, \tag{3}$$

where k^* is half of the relative momentum in the pair rest frame. The ratio $C_{\mathrm{SL}}(k^*)$ is more sensitive to the strong interaction because of some partial cancellation of the Coulomb interaction. Below k^*= 0.1 GeV/c, the signal has been found to be enhanced beyond the Coulomb interaction. The lattice QCD result calculated with Correlation Analysis Tool using the Schrodinger equation (CATS) model [13] shows similar behavior to the experimental data which suggests an attractive strong interaction between p and Ξ [14].

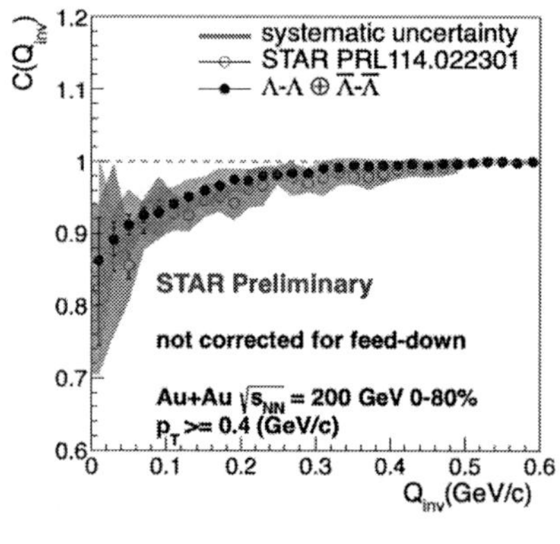

Figure 3. Correlations for Λ-Λ pairs in 0-80% central Au+Au collisions from the 2011, 2014, and 2016 data sets, requiring Λ particles to satisfy $p_T \geq$ 0.4 GeV/c, |y|<1. The filled symbols are the data of this analysis, and the open symbols are the previous results from STAR.

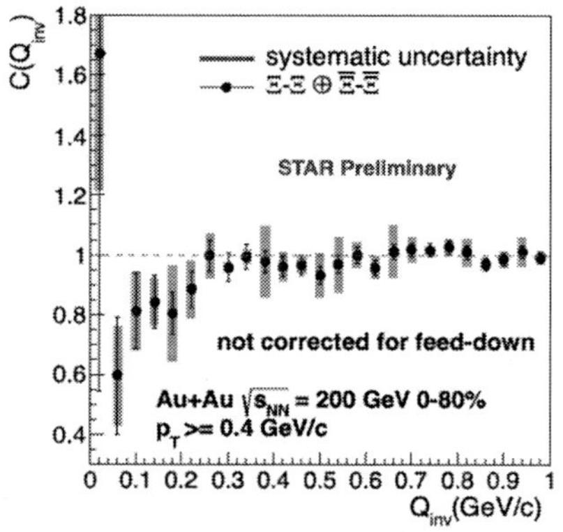

Figure 4. Correlations for Ξ-Ξ pairs in 0-80% central Au+Au collisions from the 2011, 2014 and 2016 data sets.

3.2 Measurement of Λ-Λ correlation function

Figure 3 is the correlation function for Λ-Λ ($\bar{\Lambda}$-$\bar{\Lambda}$) pairs in Au+Au collisions at $\sqrt{s_{NN}}$ = 200 GeV using the data taken in 2011, 2014 and 2016. The statistics used in this analysis is four times larger than that in the previous study [3]. This analysis result has not yet been corrected for feed-down. Feed-down means particles originating from weak decays. The feed-down particles can be misidentified as primordial particles. The new result with better statistical precision is found to be consistent with the previous result within systematic uncertainty. The correlation shows a clear anti-correlation. There is a long tail of residual correlation at high Q_{inv}, that could be coming from mini-jets and other decays. The femtoscopic fitting with Lednicky-Lyuboshitz model [15] will be applied to extract a specific scattering parameter and it would hopefully clarify the existence of a bound state of Λ-Λ system.

3.3 Measurement of Ξ-Ξ correlation function

According to NPLQCD calculation [7], the Ξ-Ξ is found to be a bound state, while HAL QCD calculation suggests that Ξ-Ξ is strongly attractive but not sufficient to form a bound state [8]. Figure 4 is the first mesurement of Ξ-Ξ ($\bar{\Xi}$-$\bar{\Xi}$) correlation in Au+Au collisions at $\sqrt{s_{\mathrm{NN}}} = 200$ GeV. The result is not corrected for feed-down. The result shows anti-correlation at $Q_{\mathrm{inv}} < 0.25$ GeV/c, which would be given by the combination of quantum statistics, strong interaction and Coulomb interaction. Feed-down and Coulomb effects need to be evaluated for further discussion. More experimental data will be taken in 2023 and 2025 at RHIC-STAR, where much better statistically significant measurement is expected.

4 summary

We have reported the first measurements of p-Ξ and Ξ-Ξ correlations in Au+Au collisions at $\sqrt{s_{\mathrm{NN}}} = 200$ GeV and also revisited Λ-Λ correlations with high statistics data. In the p-Ξ correlation analysis, strong attractive interaction is observed. The correlation function $C(k^*)$ ratio of peripheral over central collisions, $C_{\mathrm{SL}}(k^*)$ is found to be above the Coulomb interaction. This is similar to lattice QCD calculation which suggests an attractive strong interaction between p and Ξ. For the Λ-Λ correlation function, new result with high statistics data is presented and consistent with the previous result from STAR, where a clear anti-correlation is observed. The first mesurement of the Ξ-Ξ correlation seems to show the anti-correlation.

References

[1] R. L. Jaffe, Phys. Rev. Lett.38, 195 (1977), Phys. Rev. Lett. 38, 617(1977).
[2] K. Sasaki *et al.*, Nuclear Physics A 998,121737(2020).
[3] L. Adamczyk *et al.*, (STAR Collaboration), Phys. Rev. Lett. 114, 022301(2015).
[4] S. Acharya *et al.*, (ALICE Collaboration), Phys. Lett. B 797, 134822(2019)
[5] S. Acharya *et al.*, (ALICE Collaboration), Phys. Rev. Lett.123, 112002(2019).
[6] ALICE Collaboration, Nature 588, 232-238(2020)
[7] S. R. Beane *et al.*,(NPLQCD Collaboration) Phys. Rev. D 85, 054511 (2012).
[8] T. Doi *et al.*, EPJ Web Conf. 175, 05009 (2018).
[9] I. Kisel (CBM Collaboration), J. Phys. Conf. Ser.1070, 012015 (2018).
[10] K. Mi et al. STAR, talk presented at APS April Meeting 2021, L13.00007 (2021)
[11] M. Chojnacki *et al.*, arXiv:1102, 0273v1(2011).
[12] K. Morita *et al.*, Phys. Rev. C 94, 031901(2016).
[13] D.L.Miaylov *et al.*, Eur. Phys. J. C 78: 394 (2018)
[14] T. Hatsuda *et al.*, Nuclear Physics A 967, 856-859 (2017).
[15] R. Lednicky, V. L. Lyuboshitz, Sov. J. Nucl. Phys. 35, 770(1982).

Systematics in the global polarization measurements of Λ hyperons with HADES at SIS18

Frédéric Julian Kornas[1,*] for the HADES Collaboration

[1]Technische Universität Darmstadt, Schlossgartenstraße 9, 64289 Darmstadt, Germany

Abstract. The global polarization of Λ hyperons has been measured in Au+Au and Ag+Ag collisions at $\sqrt{s_{NN}}$ = 2.4 and 2.55 GeV recorded with HADES. An increase of the polarization is observed following the trend measured by the STAR Collaboration. The high statistics Ag+Ag data allowed for differential measurements of the polarization. The study of acceptance effects is very important in the fixed target setup, as the phase-space coverage is not symmetric. These studies are reported together with the evaluation procedure of the systematic uncertainties.

Introduction. In non-central heavy-ion collisions, large orbital angular momenta are created [1] which might manifest itself in a global spin polarization of the produced particles [2, 3]. Such a global polarization can be accessed using weak decays of strange hyperons, where the spin direction is transferred to the momentum direction of the decay products. The most abundantly produced hyperons in heavy-ion collisions are Λs. Measurements performed during the beam energy scan phase I by the STAR Collaboration show an increasing trend of the global polarization in Au+Au collisions down to $\sqrt{s_{NN}}$ = 7.7 GeV [4]. Further measurements by the ALICE and STAR Collaboration confirm the decreasing trend towards higher collision energies [5, 6]. The HADES experiment at SIS-18 collected high statistics data samples for the reactions Au+Au at $\sqrt{s_{NN}}$ = 2.42 and Ag+Ag at 2.55 GeV. A measurement of the global polarization of the Λ hyperons at these lower collision energies provides new insights into the evolution of the angular momentum and the collective behavior of the matter created. Theoretical predictions are not conclusive in this energy range [7–10].

Measurement technique. The measurement of the global polarization is performed using the decay channel $\Lambda \rightarrow p + \pi^-$ in which the proton is preferentially emitted in the spin direction of the Λ hyperon. Hence, the polarization P_Λ is connected to the momentum direction $\vec{p}_p^*$ of the proton in the rest frame of the Λ by $P_\Lambda = (3/\alpha_\Lambda)\langle\vec{p}_p^*\rangle$ [11]. Here $\alpha_\Lambda = 0.732 \pm 0.014$ [12] is the decay parameter quantifying the emission probability of the proton along the spin direction of the Λ hyperon and $\langle.\rangle$ denotes the average over all Λ decays. The direction of the orbital angular momentum is perpendicular to the reaction plane, spanned by the impact parameter of the colliding nuclei and the beam direction. It can be defined by a single azimuthal angle Ψ_{RP} in the laboratory frame which is estimated from a Q-vector analysis using the spectator distributions [13, 14]. The result of this procedure is the so-called event plane angle Ψ_{EP}. This can be used to express the polarization in terms of azimuthal angles by integrating over the polar angle and assuming perfect detector acceptance [11]:

$$P_\Lambda = 8/(\pi\alpha_\Lambda) \times \langle\sin(\Psi_{EP} - \phi_p^*)\rangle / R_{EP}. \quad (1)$$

*e-mail: f.kornas@gsi.de

Due to the finite resolution of the event plane reconstruction, the associated reduction of the measured correlation has to be taken into account by introducing the event plane resolution correction R_{EP} [14]. The event plane and its resolution are determined for the HADES detector [15] as described in [16]. The reconstruction of the Λ hyperons in HADES is done by identifying their its decay topology in a multi-variate analysis. The polarization signal is extracted with the invariant mass method. More details can be found in [17].

Efficiency correction. The efficiency map for the reconstruction of Λ hyperons in HADES in transverse momentum and rapidity space is shown in the left panel of Fig. 1(bottom). As only relative changes are important for the averaged polarization observable, for application the efficiencies are rescaled to the weighted mean efficiency. The Λ hyperons reconstructed in the experimental data as displayed in the left panel of Fig. 1(top) are used as weights. To minimize effects related to the finite binning, a Savitzky-Golay filter is used in addition [18].

Radial distance asymmetry (RDA) correction. A strong dependence of the extracted polarization signal has been observed on the proton radial distance R_p. The R_p is a signed value, calculated as the minimum distance of the proton track to a straight line parallel to the beam axis through the event vertex of the collision. The sign is determined by the cross product between the minimum distance vector and the track direction. The same dependence could be reproduced using Monte-Carlo simulations with polarized Λ hyperons and a multi-differential flow pattern. However, the signal distribution as a function of R_p is not symmetric around zero in the experiment in contrast to the simulations. A data driven approach has been developed to correct for this asymmetry. Therefore, the distribution is divided into a finite amount of bins which are weighted in order to restore the symmetry of the distribution. The application of the same weights to the polarization observable results in a slight increase of the extracted signal value by $\sim 8\,\%$ consistently for both, Au+Au and Ag+Ag. Variations of the correction procedure, in an extreme scenario by flattening the distribution of R_p, give consistent results within statistical uncertainties.

Acceptance effects. The phase-space coverage in fixed target experiments is usually not symmetric as displayed in the left panel of Fig. 1. To study the acceptance effects in more detail, Monte-Carlo simulations with polarized Λ hyperons are used. The flow pattern for the directed and elliptic flow has been implemented in addition according to the experimentally measured values. Different input values have been tested for the Λ polarization. For the different samples the input value could be reconstructed within statistical uncertainties as summarized in the right panel of Fig. 1. A non-zero background polarization is extracted which qualitatively matches the observations in the experimental data. It could be traced back to a mis-matching of π^- from polarized Λ hyperons with primary protons. However, the effect on the polarization extraction is negligible.

Systematic uncertainties. The systematic uncertainties are evaluated by varying single parameters in the analysis and comparing the extracted polarization signals. In order to remove statistical fluctuations the Barlow criterion [19] is used for the difference between the measured result $P_{\mathrm{M}} \pm \sigma_{\mathrm{M}}$ and the result from the varied set $P_{\mathrm{V}} \pm \sigma_{\mathrm{V}}$:

$$B_{\mathrm{crit}} = |P_{\mathrm{M}} - P_{\mathrm{V}}| \Big/ \sqrt{|\sigma_{\mathrm{M}}^2 - \sigma_{\mathrm{V}}^2|}\,. \qquad (2)$$

If $B_{\mathrm{crit}} > 1$ a variation is treated as significant beyond statistical fluctuations and added in quadrature to the overall systematic uncertainty. Otherwise it is neglected. For the particle selection, the decay topology parameters and MVA response have been varied such that the amount of reconstructed Λ hyperons differs by $\pm 15\,\%$. In general, the variations are around $2\,\%$, except for the distance of the proton track to the primary vertex where it is found to be $7.5\,\%$. Concerning the invariant mass method, the difference between determining the signal

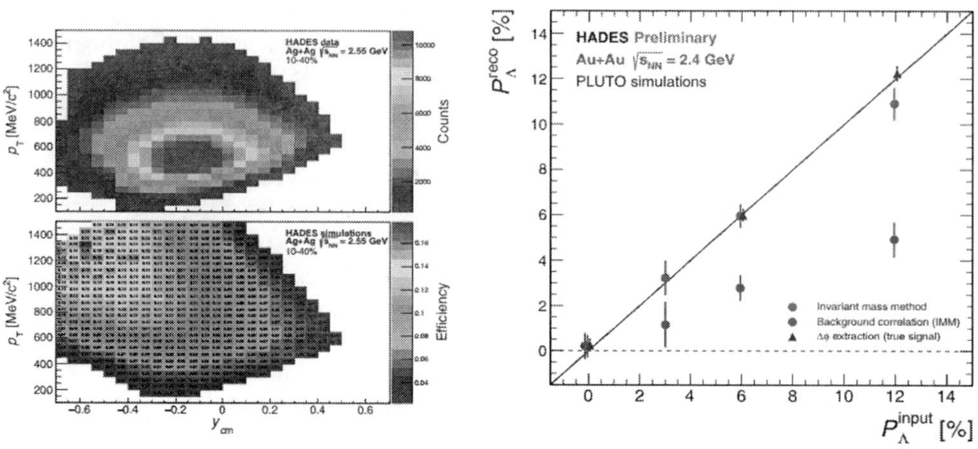

Figure 1. Left panel: Phase-space distribution (top) of the reconstructed Λ hyperons and the corresponding reconstruction efficiency (bottom) in Ag+Ag collisions at $\sqrt{s_{NN}} = 2.55\,\mathrm{GeV}$. Right panel: Extraction of the Λ polarization from Monte-Carlo simulations as a function of the input polarization using two different methods (blue and red points). The background "polarization" extracted from the invariant mass method is depicted by the green points.

and background fractions using either a direct fit to the same-event data or the mixed-event technique are found to be negligible. The same applies to variations of the invariant mass range used to fix the background shape. Different assumptions have been tested, i.e. a linear and constant background shape as a function of the invariant mass, however, the magnitude of the variation is small $\sim 1\,\%$. As a second method the $\Delta\phi$ extraction method [20] has been used, but no significant difference to the default result has been observed. The same observation has been made for the systematic variations related to the efficiency and RDA corrections. Finally, two scaling uncertainties are included in the systematic uncertainties: $2\,\%$ due to the measurement uncertainty of the decay parameter α_Λ [12] and $5\,\%$ related to the uncertainty in the determination of the event plane resolution correction R_{EP}. A summary of this evaluation procedure is shown in the left panel of Fig. 2 for both data sets. Most of these systematic uncertainties also apply to the differential study of the global polarization. However, sources that might be different for different regions of phase-space are re-evaluated for each bin separately and the overall uncertainties are plotted in the right panel of Fig. 2 for the rapidity dependence of the global Λ polarization. These are: the efficiency and RDA corrections, the $\Delta\phi$ extraction method and the constant background assumption.

In summary, the global polarization of Λ hyperons has been measured in Au+Au and Ag+Ag collisions at $\sqrt{s_{NN}} = 2.4$ and $2.55\,\mathrm{GeV}$ with the fixed-target experiment HADES. Acceptance effects have been studied using experimental data but also Monte-Carlo simulations. An effect related to the direction of the proton tracks has been studied and an experimental approach has been developed to correct for it. The appearance of a background correlation in the experimental data could be qualitatively reproduced in the simulations and is found to be insignificant for the signal extraction.

A detailed study of the systematic uncertainties has been performed using the Barlow criterion to identify relevant contributions beyond statistical fluctuations. For the Ag+Ag data, the global polarization has been extracted differentially as a function of centrality, rapidity and transverse momentum. Overall, the increasing trend towards lower collision energies observed by STAR is found to continue down to $\sqrt{s_{NN}} = 2.4\,\mathrm{GeV}$. HADES measured

Systematic on P_Λ[%]	Au+Au@1.23 *A*GeV			Ag+Ag@1.58 *A*GeV		
	Absolute	Relative	B_{crit}	Absolute	Relative	B_{crit}
D_{min}(MVA)	±0.553	±11.9%	✓	±0.042	±1.3%	×
d_v	±0.301	±6.5%	✓	±0.084	±2.6%	✓
d_1	±0.263	±5.7%	✓	±0.051	±1.6%	✓
d_2	±0.672	±14.6%	✓	±0.237	±7.5%	✓
d_3	±0.233	±5.0%	✓	±0.019	±0.6%	×
d_t	±0.341	±7.4%	✓	±0.074	±2.3%	✓
Best D_{min}(MVA)	±0.005	±0.1%	×	±0.013	±0.4%	×
Same-event mass fit	±0.000	±0.0%	×	±0.023	±0.7%	×
Fit range	±0.102	±2.2%	×	±0.032	±1.0%	×
Pol0 BG fit	±0.088	±1.9%	×	±0.035	±1.1%	✓
Efficiency	±0.611	±13.3%	✓	±0.053	±1.7%	×
$\Delta\phi$ method	±0.253	±5.5%	×	±0.053	±1.7%	×
R_{EP}	±0.138	±3.0%	✓	±0.159	±5.0%	✓
α_Λ	±0.094	±2.0%	✓	±0.063	±2.0%	✓
RDA	±0.686	±14.9%	×	±0.121	±3.8%	×
Δ_{sys}^{total}	±1.220	±26.5%	-	±0.319	±10.1%	-

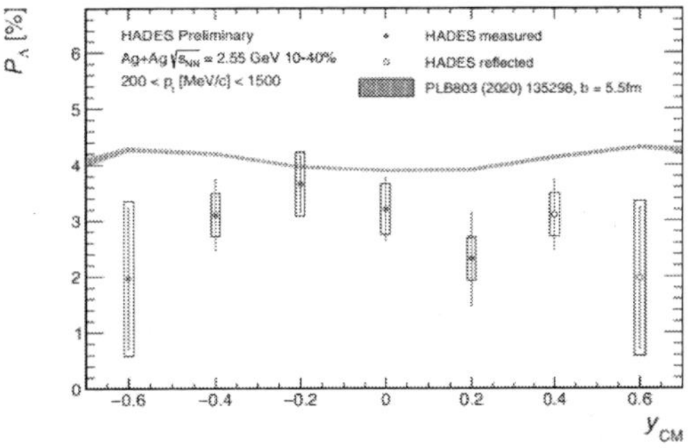

Figure 2. Left panel: Summary table of the systematic uncertainty evaluation for the integrated results in Au+Au and Ag+Ag collisions at $\sqrt{s_{NN}}$ = 2.4 and 2.55 GeV. Right panel: Global polarization measurement of Λ hyperons as a function of rapidity in the center-of-mass frame. The two data points at most negative rapidity have been reflected (open symbols) as the polarization is expected to be symmetric around midrapidity (y_{CM} = 0). The experimental data is compared to model calculations based on UrQMD [9] depicted by the blue area.

the polarization to reach a value of $P_\Lambda[\%]$ = 4.6 ± 1.0 (stat.) ± 1.2 (syst.) in Au+Au and $P_\Lambda[\%]$ = 3.2 ± 0.3 (stat.) ± 0.3 (syst.) in Ag+Ag collisions.

Acknowledgements. This work was supported by the European Union's Horizon 2020 research and innovation program under grant agreement No. 871072, the Ministry of Science and Higher Education of the Russian Federation, Project "Fundamental properties of elementary particles and cosmology" No 0723-2020-0041, and the Russian Foundation for Basic Research (RFBR) funding within the research project no. 18-02-40086.

References

[1] F. Becattini et al., Phys. Rev. C **77**, 024906 (2008).
[2] Z.-T. Liang and X.-N. Wang, Phys. Rev. Lett. **94**, 102301 (2005).
[3] S. A. Voloshin, arXiv:nucl-th/0410089 (2005).
[4] L. Adamczyk et al. (STAR), Nature **548**, 62-65 (2017).
[5] J. Adam et al. (STAR), Phys. Rev. C **98**, 014910 (2018).
[6] S. Acharya et al. (ALICE), Phys. Rev. C **101**, 044611 (2020).
[7] Y. B. Ivanov, Phys. Rev C **102**, 044904 (2020).
[8] X.-G. Deng et al., Phys. Rev. C **101**, 064908 (2020).
[9] O. Vitiuk et al., Phys. Lett. B **803**, 135298 (2020).
[10] A. Ayala et al., arXiv:2106.14379 [hep-ph] (2021).
[11] B. I. Abelev et al. (STAR), Phys. Rev. C **76**, 024915 (2007).
[12] P. A. Zyla et al. (Particle Data Group), Prog. Theor. Exp. Phys. 2020, 083C01 (2020).
[13] J.-Y. Ollitrault, Nucl. Phys. A **638**, 195-206 (1998).
[14] A. M. Poskanzer et al., Phys. Rev. C **58**, 1671-1678 (1998).
[15] G. Agakishiev et al., EPJA **41**, 243 (2017).
[16] J. Adamczewski-Musch et al. (HADES), Phys. Rev. Lett. **125**, 262301 (2020).
[17] F. J. Kornas, Springer Proc. Phys. **250**, 435-439 (2020).
[18] A. Savitzky and M.J.E. Golay, Analytical Chemistry **36**, 1627-1639 (1964).
[19] R. Barlow, *Conf. on Adv. Stat. Techniques in Particle Physics* (2002) 134-144.
[20] S. A. Voloshin et al., Landolt-Bornstein **23**, 293 (2010).

Global hyperon polarization and effects of decay feeddown

Hui Li[1,*], *Xiao-Liang* Xia[2,**], *Xu-Guang* Huang[1,2,***], and *Huan Zhong* Huang[1,3,****]

[1]Key Laboratory of Nuclear Physics and Ion-beam Application (MOE), Fudan University, Shanghai 200433, China
[2]Department of Physics and Center for Field Theory and Particle Physics, Fudan University, Shanghai 200433, China
[3]Department of Physics and Astronomy, University of California, Los Angeles, CA 90095, USA

Abstract. We study the global polarizations of Λ, Ξ^-, and Ω^- hyperons in non-central Au + Au collisions at $\sqrt{s_{NN}}$ = 7.7–200 GeV. We highlight the importance of effect of decay feeddown to the measured global polarization. With the decay contributions taken into account, the global polarization ordering $P_{\Omega^-} > P_{\Xi^-} > P_\Lambda$ can be naturally explained, which is consistent with the observation recently reported from the STAR experiment from Au+Au collisions at 200 GeV. We also extend our calculations to predict expectations from the RHIC-BES II data.

1 Introduction

The large orbital angular momentum produced by relativistic heavy-ion collisions leads to the spin polarization of final-state hadrons [1, 2]. Therefore, the spin polarization in turn provides us with a tool to detect the rotational motion of the quark-gluon plasma (QGP) created in the collision. The global polarization of Λ and $\bar{\Lambda}$ hyperons, which reflects the average vorticity strength of the QGP, has been measured by the STAR [3–5], the ALICE [6], and the HADES [7] experiments in a wide collision energy range covering $\sqrt{s_{NN}}$ = 2.4–5020 GeV. The data in the energy range of $\sqrt{s_{NN}}$ = 7.7–5020 GeV [3, 4, 6] demonstrate an increasing global Λ ($\bar{\Lambda}$) polarization with the decreasing collision energy. This trend can be reproduced by many model calculations, such as [8–12], which provides strong support to the vorticity interpretation of the global polarization.

In order to further examine the global polarization mechanism, it is important to study the spin polarization (or spin alignment) of other hadron species. Recently, the STAR experiment has measured the global polarizations of Ξ and Ω hyperons at $\sqrt{s_{NN}}$ = 200 GeV [13]. The STAR and the ALICE experiments have also measured the spin alignment of ϕ and K^{*0} mesons [14–16]. However, the observed deviation of the spin-density matrix element ρ_{00} from 1/3 does not necessarily indicate the global polarization. Local polarization distribution caused by the QGP anisotropic expansion, which does not contribute to the global polarization, can also lead to $\rho_{00} \neq 1/3$ [17]. This makes the spin alignment of vector mesons more complicated than the global polarization of hyperons.

*e-mail: lihui_fd@fudan.edu.cn
**e-mail: xiaxl@fudan.edu.cn
***e-mail: huangxuguang@fudan.edu.cn
****e-mail: huanzhonghuang@fudan.edu.cn

In this proceeding, we report our recent research [18] on the global polarizations of Λ, Ξ^-, and Ω^- hyperons within the framework based on thermal vorticity and spin transfer dynamics in decay. The study improves our understanding of the global hyperon polarization and provides predictions for future measurements at RHIC-BES energies.

2 Framework

We calculated the global polarizations of Λ, Ξ^-, and Ω^- hyperons in Au + Au collisions at RHIC energies $\sqrt{s_{\rm NN}} = 7.7$–200 GeV. The calculation can be divided into the following two procedures.

We first compute the global polarization of *primary* Λ, Ξ^-, and Ω^- hyperons. The primary hyperons are those produced by hadronization at the end of QGP fluid. Spin statistical and kinetic theory shows that their spin polarizations are determined by thermal vorticity of the fluid [19–21]:

$$P^\mu \simeq -\frac{S+1}{6m}\epsilon^{\mu\nu\rho\sigma}p_\nu\varpi_{\rho\sigma}(x), \tag{1}$$

where P^μ is the spin polarization four-vector, and S, m, p, and x are the spin, mass, momentum four-vector, and space-time coordinate four-vector of the hyperon. The thermal vorticity is defined as $\varpi_{\rho\sigma} = \frac{1}{2}(\partial_\sigma\beta_\rho - \partial_\rho\beta_\sigma)$ with $\beta_\rho = u_\rho/T$ in which u is the fluid four-velocity and T is the temperature. We extract the thermal vorticity field and the phase-space distributions of primary Λ, Ξ^-, and Ω^- hyperons from a multi-phase transport (AMPT) model [22]. With this information, the global polarization of primary hyperons can be calculated using Eq. (1). More details about the calculation can be found in [18].

Then we estimate the decay contributions to the global polarization of *inclusive* hyperons. Note that the global polarization calculated in the first procedure is of the primary hyperons. However, the experimental measurements include the hyperons produced by decay of heavier particles. Through a statistical model calculation, we find that about 80% of the final Λ's and about 40% of the final Ξ^-'s are products of decays. Following from Eq. (1), the thermal vorticity polarizes not only the primary Λ, Ξ^-, and Ω^- hyperons, but also all other hadron species created from the QGP (except spinless hadrons). When these polarized hadrons decay, their spin polarizations are transferred to secondary particles. The detailed spin-transfer rules can be found in [18, 23–25]. Taking the decay contributions into account, the global polarization of inclusive hyperons can be calculated by

$$P_{H,\rm inclusive} = \frac{N_H P_{H,\rm primary} + \sum N_{X\to H}P_{X\to H}}{N_H + \sum N_{X\to H}}, \tag{2}$$

where H is the hyperon which we want to study, X stands for particles which can decay to H, $P_{H,\rm inclusive}$ ($P_{H,\rm primary}$) is the global polarization of inclusive (primary) H hyperon, N_H is the primary yield number of H, $N_{X\to H}$ and $P_{X\to H}$ are the yield number and the global polarization of H from the decay $X \to H$, and the sum runs over all possible decays (including cascade decays) from X to H. We use the THERMUS package [26] (which is a statistical model) to calculate the primary yields of the following hyperons: Λ, $\Lambda(1405)$, $\Lambda(1520)$, $\Lambda(1600)$, $\Lambda(1670)$, $\Lambda(1690)$, Σ^0, $\Sigma(1385)$, $\Sigma(1660)$, $\Sigma(1670)$, Ξ, and $\Xi(1530)$. We then use the spin-transfer rules to simulate decay of these hyperons and obtain $N_{X\to H}$ and $P_{X\to H}$. Finally, these quantities are substituted into Eq. (2) to calculate the inclusive global polarization.

3 Results and discussion

In this section, we present and discuss our results on the global hyperon polarizations. In the current framework, we have assumed that the primary hyperons follow the same global

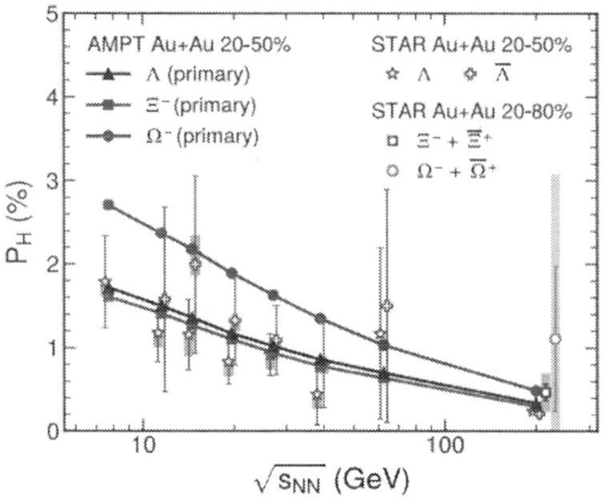

Figure 1. Global polarizations of Λ, Ξ^-, and Ω^- hyperons without (left) and with (right) the decay contributions to Λ and Ξ^- taken into account. The STAR data [3, 4, 13] of inclusive hyperon polarizations are also shown for comparison.

polarization mechanism, that is, they are polarized by the thermal vorticity through Eq. (1). Therefore, we expect that the global polarizations of the three hyperons have a similar energy dependence — they all decrease with the increase of collision energy, as demonstrated in Fig. 1.

The left panel of Fig. 1 shows that the global polarizations of primary Λ, Ξ^-, and Ω^- hyperons approximately fulfill $P_\Lambda : P_{\Xi^-} : P_{\Omega^-} \simeq 1 : 1 : 5/3$. The main reason is that Ω^- is a spin-3/2 hyperon, while Λ and Ξ^- are spin-1/2, and equation (1) shows that the global polarization is proportional to $S + 1$. In addition to the spin number, equation (1) also indicates that the global polarization depends on hyperon's four-velocity (p_ν/m) and space-time coordinate x. However, as we examined in the AMPT model, the three hyperons have very similar velocity and space-time distributions. As a result, the difference between the primary global polarizations of Λ and Ξ^- is very small. Recent hydrodynamic calculation [27] yielded similar conclusion.

The right panel of Fig. 1 shows the global polarizations of Λ, Ξ^-, and Ω^- hyperons, but with the decay contributions to Λ and Ξ^- taken into account. Compared with the primary global polarizations (Fig. 1 left), decay of heavier particles reduces the global polarization of Λ by about 10%, but increases the global polarization of Ξ^- by about 25% [18]. The Ω^- hyperon contains hardly any contribution from decays, therefore we neglect the decay contributions to the global Ω^- polarization. Finally, after we consider the decay contributions, the global polarizations of inclusive Λ and Ξ^- hyperons become separated, and we find $P_{\Omega^-} > P_{\Xi^-} > P_\Lambda$. This explains the recent experimental measurement at 200 GeV [13]. Our calculations predict that the ordering of $P_{\Omega^-} > P_{\Xi^-} > P_\Lambda$ will become more prominent from the RHIC-BES II data if there would be sufficient statistics to make the measurements.

4 Summary

In summary, we studied the global polarizations of Λ, Ξ^- and Ω^- hyperons in Au + Au collisions at RHIC energies $\sqrt{s_{NN}} = 7.7$–200 GeV, within the framework based on the thermal vorticity and spin transfer in decay. We find that the global polarizations of primary hyperons approximately fulfill $P_\Lambda : P_{\Xi^-} : P_{\Omega^-} \simeq 1 : 1 : 5/3$, and the global polarizations of inclusive hyperons are in the ordering: $P_{\Omega^-} > P_{\Xi^-} > P_{\Lambda}$. We have analyzed multiple reasons for the relations among P_{Λ}, P_{Ξ^-}, and P_{Ω^-}. We found that Ω^- has the largest global polarization, which is due to its larger spin number ($S = 3/2$). On the other hand, the decay contributions play an important role in the separation between the inclusive global polarizations of Λ and Ξ^-. Our findings can naturally explain the recent experimental data at 200 GeV [13]. We also expect more experimental measurements at RHIC-BES energies to further test our results.

Acknowledgment

This work is supported in part by NSFC under grants Nos. 11835002 and 12075061, by Shanghai Natural Science Foundation under grant No. 20ZR1404100, and by China Postdoctoral Science Foundation under grant Nos. 2018M641909 and 2019M661333.

References

[1] Z. T. Liang and X. N. Wang, Phys. Rev. Lett. **94**, 102301 (2005)
[2] S. A. Voloshin, arXiv:nucl-th/0410089 [nucl-th]
[3] L. Adamczyk *et al.* [STAR], Nature **548**, 62-65 (2017)
[4] J. Adam *et al.* [STAR], Phys. Rev. C **98**, 014910 (2018)
[5] M. S. Abdallah *et al.* [STAR], arXiv:2108.00044 [nucl-ex]
[6] S. Acharya *et al.* [ALICE], Phys. Rev. C **101**, no.4, 044611 (2020)
[7] F. Kornas [HADES], these proceedings (2021)
[8] I. Karpenko and F. Becattini, Eur. Phys. J. C **77**, no.4, 213 (2017)
[9] H. Li, L. G. Pang, Q. Wang and X. L. Xia, Phys. Rev. C **96**, no.5, 054908 (2017)
[10] Y. Xie, D. Wang and L. P. Csernai, Phys. Rev. C **95**, no.3, 031901 (2017)
[11] S. Shi, K. Li and J. Liao, Phys. Lett. B **788**, 409-413 (2019)
[12] D. X. Wei, W. T. Deng and X. G. Huang, Phys. Rev. C **99**, no.1, 014905 (2019)
[13] J. Adam *et al.* [STAR], Phys. Rev. Lett. **126**, no.16, 162301 (2021)
[14] C. Zhou [STAR], Nucl. Phys. A **982**, 559-562 (2019)
[15] S. Acharya *et al.* [ALICE], Phys. Rev. Lett. **125**, no.1, 012301 (2020)
[16] S. Singha [STAR], Nucl. Phys. A **1005**, 121733 (2021)
[17] X. L. Xia, H. Li, X. G. Huang and H. Z. Huang, Phys. Lett. B **817**, 136325 (2021)
[18] H. Li, X. L. Xia, X. G. Huang and H. Z. Huang, arXiv:2106.09443 [nucl-th]
[19] F. Becattini, V. Chandra, L. Del Zanna and E. Grossi, Annals Phys. **338**, 32-49 (2013)
[20] R. H. Fang, L. G. Pang, Q. Wang and X. N. Wang, Phys. Rev. C **94**, no.2, 024904 (2016)
[21] Y. C. Liu, K. Mameda and X. G. Huang, Chin. Phys. C **44**, no.9, 094101 (2020)
[22] Z. W. Lin, C. M. Ko, B. A. Li, B. Zhang and S. Pal, Phys. Rev. C **72**, 064901 (2005)
[23] F. Becattini, I. Karpenko, M. Lisa, I. Upsal and S. Voloshin, Phys. Rev. C **95**, no.5, 054902 (2017)
[24] F. Becattini, G. Cao and E. Speranza, Eur. Phys. J. C **79**, no.9, 741 (2019)
[25] X. L. Xia, H. Li, X. G. Huang and H. Z. Huang, Phys. Rev. C **100**, no.1, 014913 (2019)
[26] S. Wheaton and J. Cleymans, Comput. Phys. Commun. **180**, 84-106 (2009)
[27] B. Fu, K. Xu, X. G. Huang and H. Song, Phys. Rev. C **103**, no.2, 024903 (2021)

EPJ Web of Conferences **259**, 12001 (2022) https://doi.org/10.1051/epjconf/202225912001
SQM 2021

Quarkonium dynamics inside the Quark-Gluon Plasma using open quantum systems

Stéphane Delorme[1,*], *Thierry* Gousset[1], *Roland* Katz[1], and *Pol-Bernard* Gossiaux[1]

[1]SUBATECH, Université de Nantes, IMT Atlantique, IN2P3/CNRS, 4 rue Alfred Kastler, 44307 Nantes cedex 3, France

Abstract. In recent years, a significant theoretical effort has been made towards a dynamical description of quarkonia inside the Quark-Gluon Plasma (QGP), using the open quantum systems formalism. In this framework, one can get a real-time description of a quantum system (here the quarkonium) in interaction with a thermal bath (the QGP) by integrating out the bath degrees of freedom and studying the system reduced density matrix. We investigate the real-time dynamics of a correlated heavy quark-antiquark pair inside the QGP using a quantum master equation previously derived from first QCD principles in [4]. The full equation is directly resolved in 1D to lessen computing costs and is used for the first time to gain insight on the dynamics in both a static and evolving medium following a Björken-like temperature evolution.

1 Quarkonium phenomenology

At non-zero temperatures, the interaction between the heavy quark and antiquark of the pair is modified due to the presence of other color charges. As temperature increases, the density of color charges also increases, leading to stronger screening of the interaction and melting of the pairs at high temperature.[1] This screening is encoded in the real part V of a complex potential, which depends on both the distance between the quark and antiquark but also on the temperature [2]. The bound quark and antiquark system can also scatter on partons present in the medium, which may also lead to the pair dissociation. These collisional effects can be described by adding an imaginary part W to the in-medium quark-antiquark potential V, which also depends on the temperature and relative distance between the quark and antiquark. The combination of the screening of the interaction and the collisions in the medium leads to quarkonia suppression. As the energy of the collisions increases, one would expect stronger suppression of quarkonium states. However, experimental measurements at RHIC and LHC showed a more complicated pattern, where the J/Ψ suppression was much smaller than expected for some kinetics [3]. This partial disappearance of suppression is thought to be caused by recombination of initially uncorrelated charm and anticharm quarks inside the medium during the evolution and at the hadronization. This effect is also crucial for phenomenology, as there are a high number of c and $\bar{c}$ quarks in the medium. It is quite clear that a quantum treatment of the real-time dynamics of heavy quarkonia is needed to properly treat both suppression and recombination. In order to do so, one can use the open quantum system formalism. In that framework, one can treat the dynamics of a system composed by a

*e-mail: delorme@subatech.in2p3.fr

quantum subsystem (a quarkonium) in interaction with its environment (the QGP). Only the quarkonium dynamics is of interest, so the environment degrees of freedom are traced out to obtain a quantum master equation describing the real-time quarkonium dynamics.

2 Quantum master equation

The present analysis is based on the work of Blaizot & Escobedo [4]. Assuming a weak coupling between the heavy quarks and the plasma constituents and focusing on the quantum brownian motion regime (valid at high temperatures $T >> 200$ MeV), they derived coupled quantum master equations, taking into account color degrees of freedom:

$$\frac{d}{dt}\begin{pmatrix}\mathcal{D}_s\\ \mathcal{D}_o\end{pmatrix} = \mathcal{L}\begin{pmatrix}\mathcal{D}_s(\mathbf{s},\mathbf{s}',t)\\ \mathcal{D}_o(\mathbf{s},\mathbf{s}',t)\end{pmatrix}, \tag{1}$$

where $\mathcal{D}_s$ and $\mathcal{D}_o$ are respectively the singlet and octet density matrices. $\mathbf{s}$ and $\mathbf{s}'$ are respectively the distance between the quark and the antiquark and the conjugated variable. The $\mathcal{L}$ operator can be decomposed in the following way:

$$\mathcal{L} = \begin{pmatrix}\mathcal{L}_{ss} & \mathcal{L}_{so}\\ \mathcal{L}_{os} & \mathcal{L}_{oo}\end{pmatrix}, \tag{2}$$

with the various $\mathcal{L}_{ss}, \mathcal{L}_{so}, \mathcal{L}_{os}$ and $\mathcal{L}_{oo}$ operators describing transitions between color states. We can also write $\mathcal{L}$ (and the other operators) as $\mathcal{L} = \mathcal{L}_0 + \mathcal{L}_1 + \mathcal{L}_2 + \mathcal{L}_3$. $\mathcal{L}_0$ describes the free motion of the pair, $\mathcal{L}_1$ describes static screening (involving the real part of the potential V), $\mathcal{L}_2$ describes the fluctuations in the medium (involving the imaginary part of the potential W) and $\mathcal{L}_3$ describes dissipative effects in the medium (involving derivatives of W). In their approach, Blaizot & Escobedo implemented a semi-classical approximation of their equations, resulting in Langevin equations. However, their equations break positivity at the $\mathcal{L}_3$ level, which is a core property of well-defined quantum master equations. We extended their equations[1], modifying the $\mathcal{L}_3$ operator and adding a higher-order operator $\mathcal{L}_4$. With the new equations, positivity is preserved and the master equation is well-defined. The equations were then resolved in one dimension to reduce the computing cost while still being able to explore the dynamics.

3 One-dimensionnal real-time dynamics of a heavy quarkonium in the Quark-Gluon Plasma

3.1 Fixed temperature case

We consider a quark-gluon plasma with fixed temperature and take as the initial state the 1S-like eigenstate obtained from the vacuum potential V. Figure 1 shows the instantaneous projections of the singlet density matrix $\mathcal{D}_s$ on the first three eigenstates obtained from the vacuum potential V for $T = 200, 300$ and 400 MeV. The evolution starts with a transient phase where transitions from the initial 1S-like state occur, before the equilibration of the populations. At later times, all three eigenstates decay with the same rate (except for $T = 200$ MeV) but the populations do not follow a Maxwell-Boltzmann distribution at the considered temperatures. The whole evolution is characterized by the interplay between binding, diffusion and transition between states. Note that from $T = 280$ MeV, only the 1S-like state is bound with the potential used.

[1] Following a procedure described in Appendix B of [4].

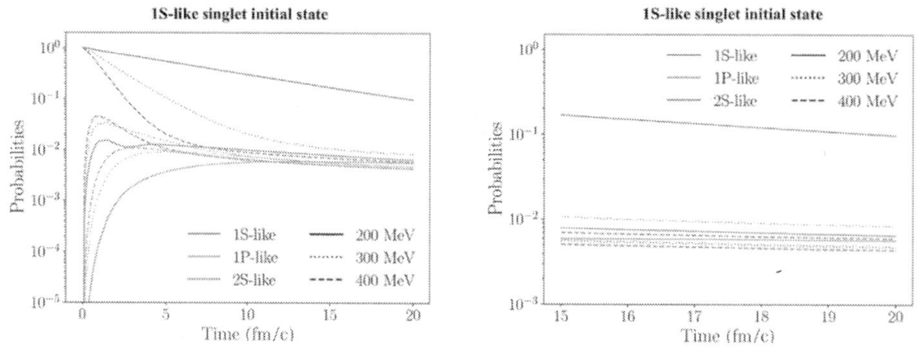

Figure 1. *Left panel:* Evolution over time of the vacuum eigenstates populations for a temperature $T = 200, 300$ and 400 MeV. *Right panel:* Zoom on the late-time evolution.

The analysis of the singlet density matrix $\mathcal{D}_s$ shows (see figure 2) the progressive delocalization of the initial state along the $s = s'$ axis. Two components are observed: 1) a remnant of the initial state is present around $s = s' = 0$ fm and 2) a semi-classical-like component on the rest of the $s = s'$ axis. The width of the distribution along the $s = -s'$ for this component is quite small, indicating that strong decoherence effects occured during the evolution of the system.

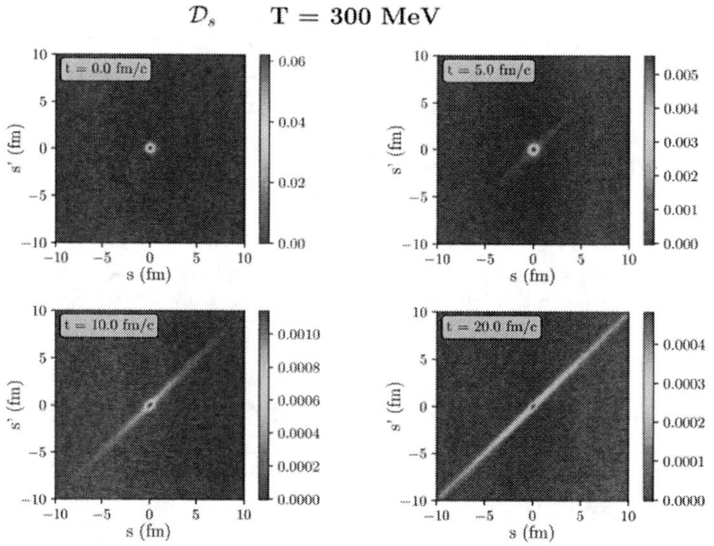

Figure 2. Evolution of the singlet density matrix $\mathcal{D}_s$ over time for a temperature $T = 300$ MeV.

3.2 Cooling medium case

To perform this study, we adopt a medium-temperature evolution following Björken's model [5], $T(t) = T_0 \left(\frac{t_0}{t_0+t}\right)^{1/3}$, where T_0 is the medium initial temperature and t_0 the initial time taken to be equal to 1 fm/c. We consider the case of a QGP medium with an initial temperature T_0 = 400 MeV and an initial 1S-like singlet state. We perform the instantaneous projection of the singlet density matrix $\mathcal{D}_s$ on the eigenstates obtained from the vacuum potential V. The results are presented in figure 3. We observe the creation at intermediate times of excited singlet bound states and the progressive re-equilibration of the populations. After around 8 fm/c, the temperature of the medium gets below T = 200 MeV, which is a limit where the model may not be valid anymore as the quantum brownian motion assumptions are not realized anymore.

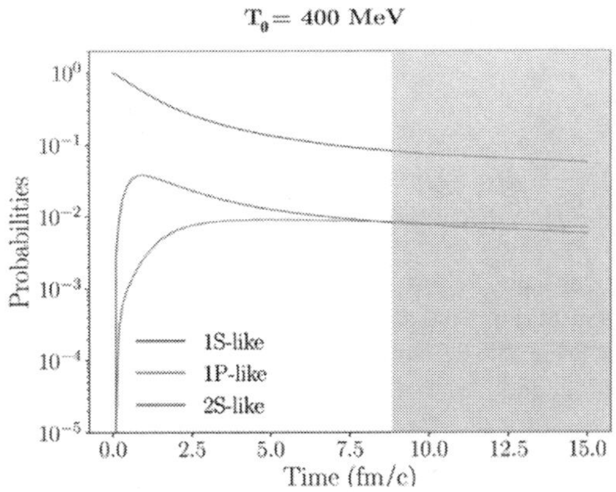

Figure 3. Evolution over time of the vacuum eigenstates populations from an initial temperature T_0 = 400 MeV. The grey band indicates the range in time where $T(t) < 200$ MeV.

4 Conclusion

We presented results on real-time quarkonium dynamics in the quark-gluon plasma described by new coupled quantum master equations. The equations involve transitions between color states as well as dissipative effects, and were directly resolved numerically for the first time in one dimension. The evolution is the result of the interplay between binding effects, diffusive effects and transitions between states. The analysis of the density matrix showed the delocalization of the initial state along a specific direction in the (s, s') space and exhibited a semi-classical-like behaviour. More details on the model and updated results will be presented in an upcoming publication.

Acknowledgements

The authors thank Jean-Paul Blaizot and Alexander Rothkopf for valuable discussions. S.D. is supported by CNRS and Région Pays de la Loire.

References

[1] T. Matsui, H. Satz, Phys. Lett. B **178**, 416 (1986)
[2] D. Lafferty, A. Rothkopf, Phys. Rev. D, **101** (2020)
[3] ALICE collaboration, Phys. Letter. B., **766**, 212-224 (2017)
[4] J.P. Blaizot, M.A. Escobedo, J. High Energ Phys. **2018**, 34 (2018)
[5] J.D. Björken, Phys. Rev. D, **27**, 140-151 (1983)

Recent J/ψ results in $p+p$ and Au+Au collisions from STAR

Kaifeng Shen[1,2,*] (for the STAR Collaboration)

[1]Department of Modern Physics, University of Science and Technology of China, Hefei 230026, China
[2]State Key Laboratory of Particle Detection and Electronics, University of Science and Technology of China, Hefei 230026, China

Abstract. In these proceedings, we present the measurements of inclusive J/ψ production at mid-rapidity region ($|y| < 1.0$) in Au+Au collisions at $\sqrt{s_{\rm NN}}$ = 54.4 GeV by the STAR experiment. The dependences of nuclear modification factor ($R_{\rm AA}$) on centrality and transverse momentum (p_T) are presented and compared to model calculations. No significant energy dependence of $R_{\rm AA}$ is found within uncertainties. We also report the distribution of $z(J/\psi) = p_T^{J/\psi}/p_T^{jet}$ and the fraction of J/ψ produced within a charged jet at mid-pseudorapidity ($|\eta| < 0.6$) with kinematic cuts of $p_T^{jet} > 10$ GeV/c and $p_T^{J/\psi} > 5$ GeV/c in $p+p$ collisions at $\sqrt{s} = 500$ GeV. Comparisons to model calculations are presented and physics implications are discussed.

1 Introduction

Heavy quarks (charm and beauty) are ideal probes to study the properties of the Quark-Gluon Plasma (QGP), because they are mainly produced via initial hard partonic scatterings and thus experience the entire evolution of the QGP created in high-energy heavy-ion collisions. Quarkonia are bound states of heavy quarks and their anti-quarks, such as charm-anticharm and bottom-antibottom, and potential holding these bound states could be screened by partons in the QGP [1]. Therefore, the cross-section of quarkonia would be reduced in heavy-ion collisions with respect to those in $p+p$ collisions scaled by the number of binary nucleon-nucleon collisions, $N_{\rm coll}$, at the same energy. Suppression of the J/ψ meson production has been observed in different collision systems at SPS, RHIC, and LHC energies [2–7]. In previous measurements by the STAR Collaboration [5], the collision energy ($\sqrt{s_{\rm NN}}$) dependence of the J/ψ suppression between 39 and 200 GeV was studied and found to be insignificant within uncertainties. In 2017, about ten times more statistics for Au+Au collisions at 54.4 GeV, compared to that used for previous STAR measurement at 62.4 GeV, was collected by the STAR experiment. This will help to understand the collision energy dependence of the J/ψ suppression with improved precision.

The production mechanism of J/ψ mesons involves both perturbative quantum chromodynamics (QCD) process - the production of charm and anti-charm pairs, and nonperturbative QCD process - the evolution of charm and anti-charm pair to a J/ψ. The latter is still not fully understood [8]. The nonrelativistic QCD (NRQCD) factorization formalism is one of the most successful approaches to describe the J/ψ production [9], which parameterizes the

*e-mail: skfwyl@mail.ustc.edu.cn

nonperturbative hadronization process with long-distance matrix elements (LDMEs). However, there exist significant differences in values of LDMEs extracted by different groups [10]. Measurements of the J/ψ production within a jet are predicated to exhibit strong distinguishing power among different models [11]. LHCb and CMS collaborations have reported their measurements of the J/ψ production within a jet [12, 13]. The observed $z(J/\psi) = p_T^{J/\psi}/p_T^{jet}$ distributions for prompt J/ψ, where $p_T^{J/\psi}$ and p_T^{jet} are the transverse momenta for J/ψ and jet, do not agree with the predictions from PYTHIA 8 [12], but can be well described by LDMEs from fits to high p_T data [14]. Measuring the J/ψ production within a jet at RHIC will provide complementary information on the J/ψ production mechanism.

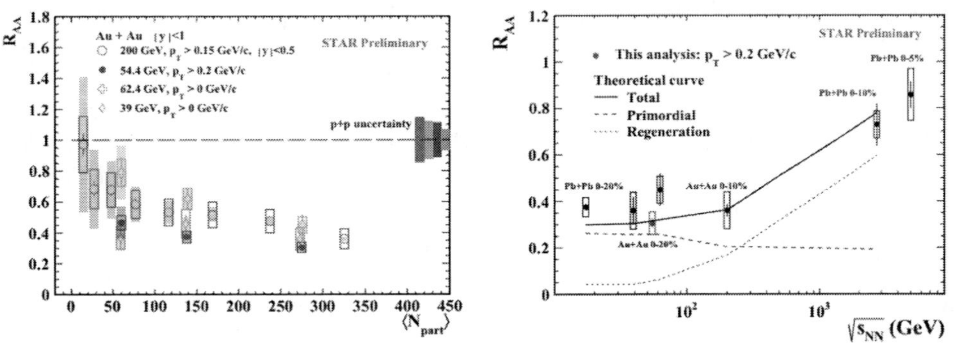

Figure 1: *Left panel*: The R_{AA} of inclusive J/ψ as a function of $\langle N_{\mathrm{part}}\rangle$ in Au+Au collisions at different collision energies at mid-rapidity [5, 6]. The error bars represent the statistical uncertainties. The boxes represent the systematic uncertainties. The shaded bands on the data points indicate the uncertainties from the nuclear overlap function $\langle T_{\mathrm{AA}}\rangle$ [15]. The bands around unity indicate the uncertainties from the $p+p$ baselines. *Right panel*: The R_{AA} of J/ψ as a function of collision energy for central collisions, in comparison with model calculations [16]. The SPS result at $\sqrt{s_{\mathrm{NN}}} = 17.2$ GeV is from [2, 3]; the STAR points at $\sqrt{s_{\mathrm{NN}}} = 39$, 62.4 and 200 GeV are from [5, 6]; the ALICE points at $\sqrt{s_{\mathrm{NN}}} = 2.76$ and 5.02 TeV are from [4, 7]. The error bars represent the statistical uncertainties and the boxes represent the systematic uncertainties, including those from $p+p$ baselines and uncertainties from $\langle T_{\mathrm{AA}}\rangle$.

2 J/ψ production in Au+Au collisions at $\sqrt{s_{\mathrm{NN}}} = 54.4$ GeV

The data sample used in this measurement is 1.3 billion minimum-bias Au+Au events at $\sqrt{s_{\mathrm{NN}}} = 54.4$ GeV collected in 2017 by the STAR experiment. A J/ψ candidate is reconstructed through its decay into an electron-positron pair, and electron candidates are reconstructed and identified using information from the Time Projection Chamber (TPC) [17], Time-of-Flight (TOF) [18], and the Barrel Electromagnetic Calorimeter (BEMC) [19].

The nuclear modification factor, R_{AA}, is defined as follows:

$$R_{\mathrm{AA}} = \frac{1}{\langle T_{\mathrm{AA}}\rangle}\frac{d^2N_{\mathrm{AA}}/dp_Tdy}{d^2\sigma_{\mathrm{pp}}^{INEL}/dp_Tdy},$$

where $d^2N_{\mathrm{AA}}/dp_Tdy$ is the J/ψ yield in A+A collisions and $d^2\sigma_{\mathrm{pp}}^{INEL}/dp_Tdy$ is the J/ψ cross section in $p+p$ collisions. The nuclear overlap function is defined as $T_{\mathrm{AA}}(\mathbf{b}) = \int T_{\mathrm{A}}(\mathbf{s})T_{\mathrm{A}}(\mathbf{s}-\mathbf{b})d^2s$, where $T_{\mathrm{A}}(\mathbf{s})$ is the transverse nucleon density and $\mathbf{b}$ is the impact parameter.

The p_T-integrated $R_{\rm AA}$ as a function of the mean number of participants $\langle N_{\rm part}\rangle$ in Au+Au collisions of different collision energies is shown in the left panel of Fig. 1. Since there are no measurements of inclusive J/ψ cross section available in p+p collisions at 39, 54.4 and 62.4 GeV, the p+p baselines are extracted from interpolations of world data [15]. The p+p baseline at 17.2 GeV is established by the data collected with lighter projectiles [20]; the p+p baseline at 200 GeV is obtained by combining STAR and PHENIX measurements [21, 22]; the p+p baseline at 2.76 TeV is obtained by interpolating the inclusive J/ψ cross-sections at mid-rapidity measured by PHENIX [23], CDF [24] and ALICE [25–27]; the p+p baseline at 5.02 TeV is measured by ALICE [28]. Suppression of the J/ψ production is observed in Au+Au collisions at 54.4 GeV with improved precision compared to previous results at 39 and 62.4 GeV. The right panel of Fig. 1 shows the $R_{\rm AA}$ as a function of collision energy for different collision systems in central collisions. There is no significant energy dependence within uncertainties up to 200 GeV. The solid line is a theoretical calculation of J/ψ $R_{\rm AA}$ from [16], in which the blue dash-dotted line represents the suppressed primordial production due to cold nuclear matter effects (CNM) and dissociation in the QGP medium, while the red dashed line denotes the regeneration contribution. The theoretical calculation is consistent with the observed energy dependence, indicating that the J/ψ production in high-energy heavy-ion collisions is affected by several effects, such as CNM, dissociation in the QGP medium, and regeneration.

Figure 2 shows the J/ψ $R_{\rm AA}$ as a function of p_T at different collision energies (left) and centralities (right). A larger suppression is observed at lower p_T and towards more central collisions.

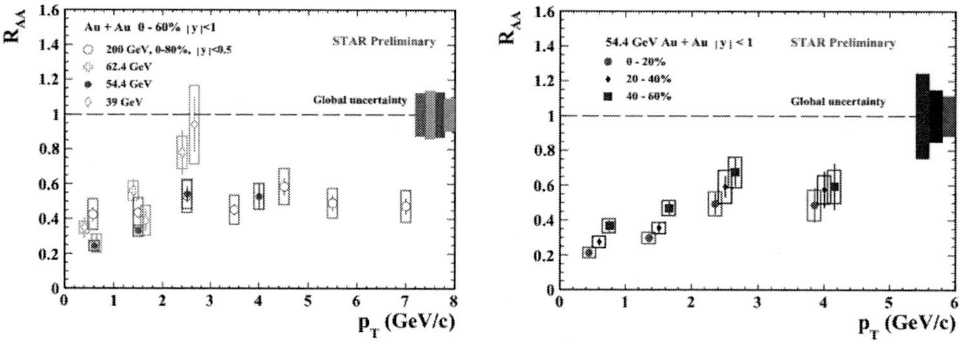

Figure 2: The inclusive J/ψ $R_{\rm AA}$ as a function of p_T in Au+Au collisions [5, 6]. The error bars represent the statistical uncertainties. The boxes represent the systematic uncertainties. The bands around unity indicate the uncertainties from the $\langle T_{\rm AA}\rangle$ and the p+p baselines.

3 J/ψ production in jets in p+p collisions at $\sqrt{s} = 500$ GeV

The data sample used in this measurement was collected from p+p collisions at $\sqrt{s} = 500$ GeV in 2011 and the integrated luminosity of this sample is 22.1 pb^{-1}. Events are triggered by the BEMC requiring an energy deposition larger than 4.3 GeV in at least one BEMC tower. The J/ψ candidates are reconstructed through their decays into electron-positron pairs, and the electron candidates are reconstructed and identified using information from the TPC and BEMC. The jet reconstruction is performed for events with a J/ψ candidate by clustering the J/ψ candidate with charged particles using the anti-k_T algorithm as implemented in the FASTJET package [30].

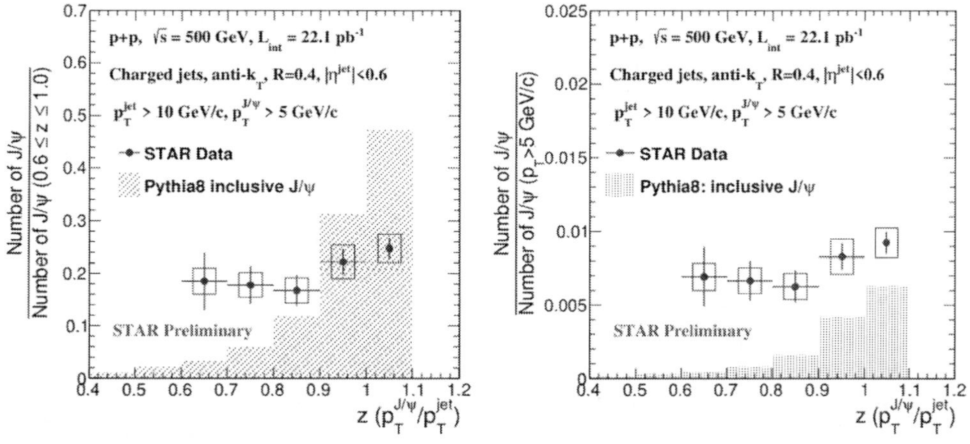

Figure 3: *Left panel*: The z distribution of inclusive J/ψ produced within a jet, normalized by the number of J/ψ with z from 0.6 to 1.0, and compared to prediction from PYTHIA 8 (gray filled histogram). The vertical lines represent statistical uncertainties and the blue boxes display systematic uncertainties. The data point for isolated J/ψ ($z = 1$) is placed at 1.05 for clarity. *Right panel*: The ratio of inclusive J/ψ produced within a jet to the total J/ψ yield [29] as a function of z, compared to PYTHIA 8 prediction.

The left panel of Fig. 3 shows the self-normalized $z(J/\psi) = p_T^{J/\psi}/p_T^{jet}$ distribution for inclusive J/ψ mesons with $p_T^{J/\psi} > 5$ GeV/c produced within a charged jet of $p_T^{jet} > 10$ GeV/c. No significant $z(J/\psi)$ dependence for $z < 1$ is observed within uncertainties. The experimental results are compared to the leading-order NRQCD-based PYTHIA 8 calculation [31], and a different trend is observed in data and PYTHIA prediction. The less isolated production of J/ψ in data than that predicted by PYTHIA 8 is similar to the LHC measurements [12, 13]. The right panel of Fig. 3 shows the ratio of the number of J/ψ within a charged jet for $p_T^{J/\psi} > 5$ GeV/c and $p_T^{jet} > 10$ GeV/c to the total number of J/ψ with $p_T^{J/\psi} > 5$ GeV/c [29]. The fraction is measured to be 3.7% ± 0.3% (stat.) ± 0.2% (sys.), significantly larger than the PYTHIA 8 predicition.

4 Summary

In this contribution, the recent results on J/ψ production in p+p and Au+Au collisions from STAR are discussed. Newly measured $R_{\rm AA}$ for J/ψ meson at $\sqrt{s_{\rm NN}} = 54.4$ GeV is shown as a function of $\langle N_{\rm part} \rangle$ and p_T with improved precision compared to previous measurements at 39 and 62.4 GeV. There is no significant energy dependence of $R_{\rm AA}$ in central collisions from 17.2 to 200 GeV within uncertainties. The first measurement of J/ψ production within a charged jet in p+p collisions at RHIC ($\sqrt{s} = 500$ GeV) is shown and compared to PYTHIA 8 predictions. The J/ψ production is less isolated in data compared to that in PYTHIA 8 and more probable to be produced in jets.

Acknowledgements

This work was funded by the National Natural Science Foundation of China under Grant Nos. 11720101001 and 11775213, Anhui Provincial Natural Science Foundation under Grant Nos. 1908085J02.

References

[1] T. Matsui, H. Satz, Phys. Lett. B **178**, 416 (1986)
[2] L. Kluberg, Eur. Phys. J. C **43**, 145 (2005)
[3] M.C. Abreu et al. (NA50), Phys. Lett. B **477**, 28 (2000)
[4] B.B. Abelev et al. (ALICE), Phys. Lett. B **734**, 314 (2014), 1311.0214
[5] L. Adamczyk et al. (STAR), Phys. Lett. B **771**, 13 (2017), 1607.07517
[6] J. Adam et al. (STAR), Phys. Lett. B **797**, 134917 (2019), 1905.13669
[7] X. Bai (ALICE), Nucl. Phys. A **1005**, 121769 (2021), 2001.11925
[8] N. Brambilla et al., Eur. Phys. J. C **71**, 1534 (2011), 1010.5827
[9] Y.Q. Ma, R. Venugopalan, Phys. Rev. Lett. **113**, 192301 (2014), 1408.4075
[10] J.P. Lansberg, Phys. Rept. **889**, 1 (2020), 1903.09185
[11] Z.B. Kang, J.W. Qiu, F. Ringer, H. Xing, H. Zhang, Phys. Rev. Lett. **119**, 032001 (2017), 1702.03287
[12] R. Aaij et al. (LHCb), Phys. Rev. Lett. **118**, 192001 (2017), 1701.05116
[13] A.M. Sirunyan et al. (CMS), Phys. Lett. B **804**, 135409 (2020), 1910.01686
[14] R. Bain, L. Dai, A. Leibovich, Y. Makris, T. Mehen, Phys. Rev. Lett. **119**, 032002 (2017), 1702.05525
[15] W. Zha, B. Huang, R. Ma, L. Ruan, Z. Tang, Z. Xu, C. Yang, Q. Yang, S. Yang, Phys. Rev. C **93**, 024919 (2016), 1506.08985
[16] X. Zhao, R. Rapp, Phys. Rev. C **82**, 064905 (2010), 1008.5328
[17] M. Anderson et al., Nucl. Instrum. Meth. A **499**, 659 (2003), nucl-ex/0301015
[18] W. Llope, Nuclear Instruments and Methods in Physics Research Section A: Accelerators, Spectrometers, Detectors and Associated Equipment **661**, S110 (2012), x. Workshop on Resistive Plate Chambers and Related Detectors (RPC 2010)
[19] M. Beddo et al. (STAR), Nucl. Instrum. Meth. A **499**, 725 (2003)
[20] M. Abreu, J. Astruc, C. Baglin, A. Baldit, M. Bedjidian, P. Bordalo, A. Bohrani, A. Bussière, P. Busson, J. Castor et al., Physics Letters B **466**, 408 (1999)
[21] J. Adam, L. Adamczyk, J. Adams, J. Adkins, G. Agakishiev, M. Aggarwal, Z. Ahammed, N. Ajitanand, I. Alekseev, D. Anderson et al., Physics Letters B **786**, 87 (2018)
[22] A. Adare, S. Afanasiev, C. Aidala, N.N. Ajitanand, Y. Akiba, H. Al-Bataineh, J. Alexander, K. Aoki, L. Aphecetche, J. Asai et al. (PHENIX Collaboration), Phys. Rev. D **82**, 012001 (2010)
[23] A. Adare, S. Afanasiev, C. Aidala, N.N. Ajitanand, Y. Akiba, H. Al-Bataineh, J. Alexander, A. Al-Jamel, K. Aoki, L. Aphecetche et al. (PHENIX Collaboration), Phys. Rev. Lett. **98**, 232301 (2007)
[24] D. Acosta, J. Adelman, T. Affolder, T. Akimoto, M.G. Albrow, D. Ambrose, S. Amerio, D. Amidei, A. Anastassov, K. Anikeev et al. (CDF Collaboration), Phys. Rev. D **71**, 032001 (2005)

[25] B. Abelev, J. Adam, D. Adamová, A. Adare, M. Aggarwal, G. Aglieri Rinella, A. Agocs, A. Agostinelli, S. Aguilar Salazar, Z. Ahammed et al., Physics Letters B **718**, 295 (2012)

[26] K. Aamodt, A. Abrahantes Quintana, D. Adamová, A. Adare, M. Aggarwal, G. Aglieri Rinella, A. Agocs, A. Agostinelli, S. Aguilar Salazar, Z. Ahammed et al., Physics Letters B **704**, 442 (2011)

[27] K. Aamodt, A. Abrahantes Quintana, D. Adamová, A. Adare, M. Aggarwal, G. Aglieri Rinella, A. Agocs, A. Agostinelli, S. Aguilar Salazar, Z. Ahammed et al., Physics Letters B **718**, 692 (2012)

[28] S. Acharya et al. (ALICE), JHEP **10**, 084 (2019), 1905.07211

[29] J. Adam et al. (STAR), Phys. Rev. D **100**, 052009 (2019), 1905.06075

[30] M. Cacciari, G.P. Salam, G. Soyez, Eur. Phys. J. C **72**, 1896 (2012), 1111.6097

[31] T. Sjostrand, S. Mrenna, P.Z. Skands, Comput. Phys. Commun. **178**, 852 (2008), 0710.3820

Charm-baryon enhancement and charm fragmentation fractions in small systems measured with ALICE

Jianhui Zhu[1,2,*] for the ALICE Collaboration

[1]Institute of Particle Physics, Central China Normal University, 152 Luoyu Road, 430079 Wuhan, China
[2]GSI Helmholtz Centre for Heavy Ion Research, Planckstraße 1, 64291 Darmstadt, Germany

Abstract. Recent measurements of charm-baryon production at midrapidity by the ALICE collaboration show baryon-to-meson yield ratios significantly higher than those measured in e^+e^- collisions, suggesting that the charm fragmentations are not universal across different collisions systems. Thus, measurements of charm-baryon production are crucial to study the charm quark hadronisation in proton–proton (pp) collisions. In proton–lead (p–Pb) collisions, the measurements of charm baryons provide important information about cold nuclear matter effects and help to understand how the possible presence of collective effects could modify the production of heavy-flavour hadrons. In this contribution, the most recent results on open charm-hadron production in pp and p–Pb collisions measured by ALICE are discussed.

1 Introduction

The production of heavy-flavour hadrons in high-energy hadronic collisions can provide important tests of the theory of quantum chromodynamics (QCD). The production cross sections of heavy-flavour hadrons can be calculated using the factorisation approach as a convolution of three factors [1]: the parton distribution functions (PDFs) of the incoming nuclei, the hard-scattering cross section at partonic level, calculated as a perturbative series in powers of the strong coupling constant α_s, and the fragmentation functions of heavy quarks into corresponding heavy-flavour hadrons, which is an inherently non-perturbative process related to, or even driven by, the confining property of QCD. The heavy-flavour baryon-to-meson ratio is an ideal observable related to the hadronization mechanism since the contributions from parton distribution function and parton-parton scattering terms cancel in the ratio. The Λ_c^+/D^0 ratio at the LHC is enhanced with respect to predictions based on e^+e^- and ep experiments, suggesting that the charm fragmentation functions are not universal among different collision systems. Several hadronization mechanisms, such as colour reconnection (CR) beyond the leading colour approximation [2], coalescence [3, 4] and feed-down from a largely augmented set of higher mass charm-baryon states beyond the current listing of the particle data group (PDG) [5], have been proposed to explain this enhancement. The newest measurements of the charm baryons Λ_c^+, $\Sigma_c^{0,++}$, $\Xi_c^{0,+}$ and Ω_c^0 performed with the ALICE experiment will be used to verify predictions from these hadronization mechanisms.

*e-mail: jianhui.zhu@cern.ch

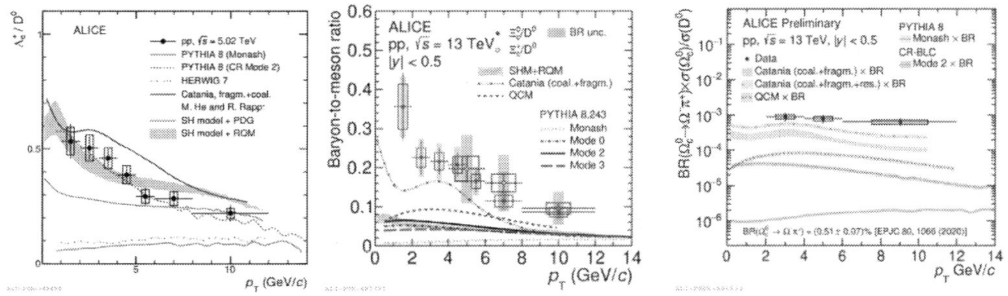

Figure 1. Left: The Λ_c^+/D^0 ratio as a function of p_T measured in pp collisions at $\sqrt{s}$ = 5.02 TeV [7]. Middle: The Ξ_c^0/D^0 and Ξ_c^+/D^0 ratios measured in pp collisions at $\sqrt{s}$ = 13 TeV [8]. Right: The $BR \times \Omega_c^0/D^0$ measured in pp collisions at $\sqrt{s}$ = 13 TeV. All the charm baryon-to-meson yield ratios are compared to theoretical calculations [2–5, 11].

2 Charm baryon-to-meson yield ratios in pp collisions

Thanks to the large data sample collected during the Run 2 period in pp collisions at $\sqrt{s}$ = 5.02 and 13 TeV, ALICE measured all the ground-state charm hadrons down to low transverse momentum (p_T), including charm mesons (D^0, D^+, D_s^+ [6]) and charm baryons (Λ_c^+ [7], $\Xi_c^{0,+}$ [8, 9], and Ω_c^0). The Λ_c^+/D^0 yield ratio is measured as a function of p_T in pp collisions at $\sqrt{s}$ = 5.02 TeV as shown in Fig. 1 (left), the Ξ_c^0/D^0 and Ξ_c^+/D^0 yield ratios are measured in pp collisions at $\sqrt{s}$ = 13 TeV as shown in Fig. 1 (middle). For Ω_c^0, the absolute decay branching ratio (BR) of $\Omega_c^0 \to \Omega^-\pi^+$ is not measured, hence the BR of $\Omega_c^0 \to \Omega^-\pi^+$ times cross section of Ω_c^0 over cross section of D^0 in pp collisions at $\sqrt{s}$ = 13 TeV is reported in Fig. 1 (right). In order to compare data with models, a theoretical calculation of $BR(\Omega_c^0 \to \Omega^-\pi^+)$ [10] is used to multiply different models. The Λ_c^+/D^0 and Ξ_c^0/D^0 ratios show a downward trend with increasing p_T. The Monte Carlo generator PYTHIA8 (Monash) [11] tuned on measurements in e^+e^- collisions largely underestimates all four charm baryon-to-meson yield ratios, providing evidence of different charm hadronisation mechanisms between e^+e^- and pp collisions. The Λ_c^+/D^0 ratio is better described by a model with colour reconnection beyond the leading colour approximation [2], a statistical hadronisation model with an augmented set of charm baryon states predicted by the relativistic quark model (RQM) [5], or a model relying on hadronisation via coalescence and fragmentation [3]. However, all the models underestimate the $\Xi_c^{0,+}/D^0$ and $BR \times \Omega_c^0/D^0$ ratios, except the Catania model [3] including charm quark hadronisation via both coalescence and fragmentation, which would indicate a partonic system similar to a quark–gluon plasma (QGP) in pp collisions.

3 Charm production and fragmentation in pp collisions

The charm fragmentation fraction $f(c \to H_c)$ shown in Fig. 2 (left) represents the probability of a charm quark hadronising into a given charm hadron. The fragmentation fraction for the Ξ_c^0 baryon is the first measurement in any collision system. An increase of about a factor of 3.3 for the fragmentation fraction for the Λ_c^+ baryon with respect to e^+e^- and ep collisions, and a corresponding decrease of about a factor of 1.2–1.4 for the D^0 meson are observed, showing that the assumption of the charm fragmentation universality (collision-system independence) is broken. Charm quarks hadronise into baryons almost 40% of the time, which is four times more often than what was measured at colliders with electron beams.

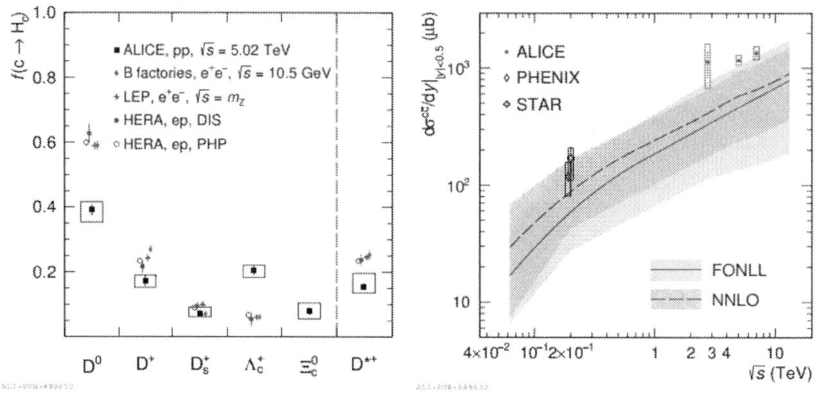

Figure 2. Left: Charm-quark fragmentation fractions into charm hadrons measured in pp collisions at $\sqrt{s}$ = 5.02 TeV in comparison with experimental measurements performed in e^+e^- and ep collisions [18]. Right: Charm production cross section at midrapidity per unit of rapidity as a function of the collision energy at the LHC [18] and RHIC [14, 15] compared to FONLL [16] and NNLO [17] calculations.

The $c\bar{c}$ production cross section per unit of rapidity at midrapidity ($d\sigma^{c\bar{c}}/dy|_{|y|<0.5}$) is calculated by summing the p_T-integrated cross sections of all measured ground-state charm hadrons (D^0, D^+, D_s^+, Λ_c^+, Ξ_c^0 and their charge conjugates). The contribution of Ξ_c^0 is multiplied by a factor of 2 in order to account for the contribution of Ξ_c^+. Since the absence of a Ω_c^0 production measurement at hadron colliders, an asymmetric systematic uncertainty is assigned assuming a contribution equal to the one of Ξ_c^0 considering the prediction of the Catania model [3]. The resulting $c\bar{c}$ production cross section per unit of rapidity at midrapidity is $d\sigma^{c\bar{c}}/dy|^{\text{pp, 5.02 TeV}}_{|y|<0.5} = 1165 \pm 44(\text{stat})^{+134}_{-101}(\text{syst})\ \mu\text{b}$. The updated fragmentation fractions obtained in pp collisions at $\sqrt{s}$ = 5.02 TeV allow the recomputation of the charm production cross sections per unit of rapidity at midrapidity in pp collisions at $\sqrt{s}$ = 2.76 and 7 TeV, which are about 40% higher than the previously published results [12, 13]. The measured $c\bar{c}$ production cross section per unit of rapidity at midrapidity together with measurements at RHIC [14, 15] are located at the upper edge of FONLL [16] and NNLO [17] predictions.

4 Charm hadronisation in p–Pb collisions

Charm hadronisation in p–Pb collisions at $\sqrt{s_{NN}}$ = 5.02 TeV is also investigated. The left panel of Fig. 3 shows the Λ_c^+/D^0 ratio as a function of p_T in pp and p–Pb collisions at 5.02 TeV. This ratio has been measured down to the p_T interval $0 < p_T < 1$ GeV/c in p–Pb collisions for the first time at the LHC. The measurements of Λ_c^+/D^0 in pp and p–Pb collisions are qualitatively consistent with each other, although a larger ratio in $3 < p_T < 8$ GeV/c and a lower ratio in $1 < p_T < 2$ GeV/c are measured in p–Pb collisions with respect to pp collisions. A clear decreasing trend with increasing p_T is obtained in both pp and p–Pb collisions for $p_T > 2$ GeV/c. The nuclear modification factor R_{pPb} of the Λ_c^+ baryon measured as a function of p_T together with R_{pPb} of non-strange D mesons are shown in the right panel of Fig. 3. There is a significant suppression for $p_T < 2$ GeV/c and enhancement for $4 < p_T < 8$ GeV/c in p–Pb collisions with respect to pp collisions, which is similar as the p_T distribution of Λ_c^+/D^0 ratio, suggesting the presence of possible radial flow effects or a further modification of the charm hadronisation mechanism in p–Pb collisions. The measurement of R_{pPb} for the Λ_c^+ baryon is compared to model calculations. The POWHEG [19] + PYTHIA6 simulations use the

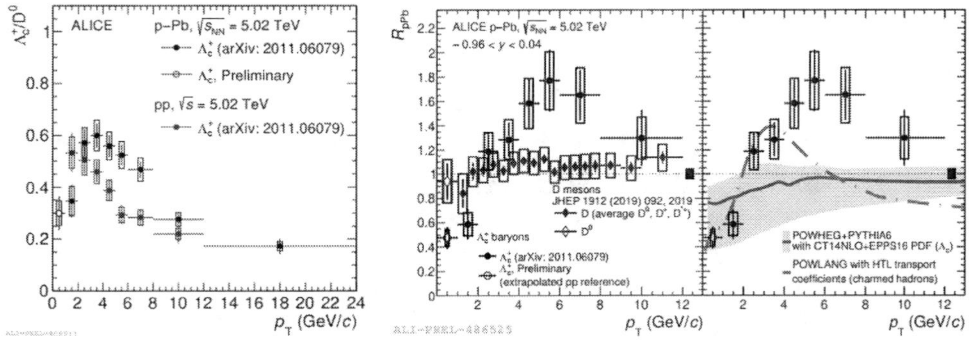

Figure 3. Left: The Λ_c^+/D^0 ratio as a function of p_T measured in pp collisions and p–Pb collisions at 5.02 TeV [7]. Right: The nuclear modification factor R_{pPb} of the Λ_c^+ baryon as a function of p_T in p–Pb collisions at $\sqrt{s_{NN}}$ = 5.02 TeV, compared to R_{pPb} of D mesons, as well as to model expectations [7].

POWHEG event generator with PYTHIA6 parton shower and EPPS16 [20] parameterisation of the nuclear modification of the PDFs. The POWLANG model [21] assumes that a hot deconfined medium is formed in p–Pb collisions. The two models capture some features of the data, but neither of them can quantitatively reproduce the data in the measured p_T interval.

5 Acknowledgements

This work was supported by the National Natural Science Foundation of China (NSFC) (No. 12105109) and the Office of China Postdoc Council (No. 20181016).

References

[1] J.C. Collins, D.E. Soper, G.F. Sterman, Nucl. Phys. B **263**, 37 (1986)
[2] J.R. Christiansen, P.Z. Skands, JHEP **08**, 003 (2015)
[3] V. Minissale, S. Plumari, V. Greco (2020), arXiv:2012.12001 [hep-ph]
[4] J. Song, H.h. Li, F.l. Shao, Eur. Phys. J. C **78**, 344 (2018)
[5] M. He, R. Rapp, Phys. Lett. B **795**, 117 (2019)
[6] S. Acharya et al. (ALICE), JHEP **05**, 220 (2021)
[7] S. Acharya et al. (ALICE) (2020), arXiv:2011.06079 [nucl-ex]
[8] S. Acharya et al. (ALICE) (2021), arXiv:2105.05187 [nucl-ex]
[9] S. Acharya et al. (ALICE) (2021), arXiv:2105.05616 [nucl-ex]
[10] Y.K. Hsiao et al., Eur. Phys. J. C **80**, 1066 (2020)
[11] P. Skands, S. Carrazza, J. Rojo, Eur. Phys. J. C **74**, 3024 (2014)
[12] B. Abelev et al. (ALICE), JHEP **07**, 191 (2012)
[13] S. Acharya et al. (ALICE), Eur. Phys. J. C **77**, 550 (2017)
[14] L. Adamczyk et al. (STAR), Phys. Rev. D **86**, 072013 (2012)
[15] A. Adare et al. (PHENIX), Phys. Rev. C **84**, 044905 (2011)
[16] M. Cacciari et al., JHEP **10**, 137 (2012)
[17] D. d'Enterria, A.M. Snigirev, Phys. Rev. Lett. **118**, 122001 (2017)
[18] S. Acharya et al. (ALICE) (2021), arXiv:2105.06335 [nucl-ex]
[19] S. Frixione, P. Nason, G. Ridolfi, JHEP **09**, 126 (2007)
[20] K.J. Eskola et al., Eur. Phys. J. C **77**, 163 (2017)
[21] A. Beraudo et al., JHEP **03**, 123 (2016)

Quarkonia as probes of the QGP and of the initial stages of the heavy-ion collision with ALICE

Ingrid McKibben Lofnes[1,*], for the ALICE Collaboration

[1]University of Bergen (Norway)

Abstract. Studies of quarkonium production in heavy-ion collisions can be used for probing QGP properties. The suppression and regeneration of bound quarkonium states is sensitive to the medium properties. Modifications of the quarkonium polarization in Pb–Pb collisions with respect to pp collisions may give further insight into the suppression and regeneration mechanisms in the QGP. Quarkonia are also sensitive to the initial stages of heavy-ion collisions, and measurements in photonuclear collisions may help constrain the nuclear gluon-distribution at low Bjorken-x. In this work, recent quarkonium measurements performed by ALICE in Pb–Pb collisions at $\sqrt{s_{\mathrm{NN}}} = 5.02$ TeV will be discussed. Preliminary measurements of the inclusive J/ψ R_{AA} measured at both forward and central rapidity will be presented. The J/ψ polarization measured for the first time in Pb–Pb collisions, as well as preliminary measurements of the coherent J/ψ photoproduction cross section, will be discussed.

The primary goal of the ALICE experiment is the study of strongly-interacting matter, known as the quark–gluon plasma (QGP), expected to be created during high-energetic heavy-ion collisions. Quarkonium production is sensitive to the medium produced in these collisions and serves as a probe of the QGP properties. Since the heavy quarks (i.e. charm and beauty) are produced in the initial hard partonic scattering, they experience the full evolution of the fireball. Bound quarkonium states immersed in a strongly interacting medium may dissociate, suppressing the quarkonium production compared to binary-scaled proton–proton (pp) collisions at the same energy. This suppression is actually a subtle interplay of several mechanisms, among them the screening of free color charges in the medium [1]. Recent theoretical developments have shown that the melting of quarkonium states is a dynamical process, where the binding of quarkonium states is weakened over time. This weakening depends on how the medium interacts with the bound states and on the time spent in the medium [2]. If the abundance of heavy quark pairs is large enough, a regeneration process may take place, where heavy quarks (re)combine either at the QGP phase boundary [3] or throughout the evolution of the fireball [4]. Recent ALICE measurements are consistent with (re)generation of charm quarks as a dominant source of the J/ψ production at low transverse momentum (p_{T}) [5, 6]. A modification of the quarkonium vector state polarization measured in Pb–Pb collisions with respect to pp collisions can shed light on quarkonium suppression and regeneration mechanisms in the QGP. In addition, quarkonia are sensitive to the initial state of the heavy-ion collisions. Looking at photoproduction of J/ψ mesons in heavy-ion collisions at the LHC, one may constrain the nuclear gluon distribution at low Bjorken-x, in the range $x \sim 10^{-5} - 10^{-2}$ [7]. The coherent photoproduction of J/ψ

*e-mail: ingrid.mckibben.lofnes@cern.ch

mesons, which takes place at very low p_T ($\sim$ 60 MeV/c), is characterized by the coherent coupling of a photon emitted by one of the nuclei while the target nucleus remains intact. Coherent photoproduction is studied in ultra-peripheral collisions, however measurements from peripheral Pb–Pb collisions show an excess yield of J/ψ at very low p_T, assumed to originate from coherent J/ψ photoproduction [8]. In these proceedings a selection of recent quarkonium measurements obtained by ALICE at both forward and central rapidity will be discussed.

The ALICE detector is capable of measuring quarkonium states down to $p_T = 0$ at both forward and central rapidity. At forward rapidity ($2.5 < y < 4$) inclusive quarkonium states are measured through the dimuon channel using the muon spectrometer. The muon spectrometer includes a front absorber for filtering muons, five tracking stations, a dipole magnet with a 3 Tm field integral, two trigger stations and an iron wall to reject punch-through hadrons and low momentum muons. At midrapidity ($|y| < 0.9$), charmonium states are reconstructed through the dielectron channel using the Inner Tracking System (ITS) composed of six cylindrical layers of silicon detectors and the Time Projection Chamber (TPC) providing track reconstruction and particle identification via the measurement of specific energy loss, dE/dx. A more detailed description of the ALICE detectors can be found in [9].

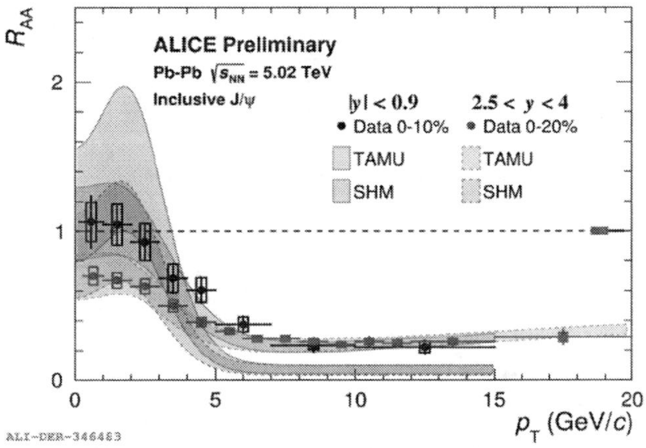

Figure 1. Preliminary measurement of the J/ψ nuclear modification factor in central Pb–Pb collisions at $\sqrt{s_{NN}} = 5.02$ TeV at forward and central rapidity in red and black. The measurements are compared to Statistical hadronisation (SHM) and transport model (TAMU) calculations [10, 11].

The medium effect on charmonium production is quantified using the nuclear modification factor (R_{AA}) calculated as the ratio between the measured yield in Pb–Pb and pp collisions at the same energy scaled by the number of binary collisions. Figure 1 shows the preliminary R_{AA} in central Pb–Pb collisions at $\sqrt{s_{NN}} = 5.02$ TeV, measured at both forward and central rapidity as a function of p_T. These measurements are obtained from a significantly larger data sample than previous measurements at the same collisions energy, allowing for a more precise and differential result [5, 6]. In the low p_T region (< 5 GeV/c) the J/ψ mesons show less suppression at midrapidity than at forward rapidity. This is consistent with the J/ψ (re)generation scenario, as a higher charm cross section is expected at midrapidity. The measurements are compared to statistical hadronisation model (SHM) [10] calculations, where the production of bound states is determined by statistical weights at

 https://doi.org/10.1051/epjconf/202225912004

the phase boundary, and transport model (TAMU) [11] calculations assuming continuous production and dissociation of charmonium states throughout the whole QGP phase. Within the current uncertainties both models describe the observed trend.

Another physical observable is the quarkonium polarization, measured through the anisotropies in the angular distribution of the decay products,

$$W(\cos\theta, \phi) \propto \frac{1}{3+\lambda_\theta} \cdot (1 + \lambda_\theta \cos^2\theta + \lambda_\phi \sin^2\theta \cos 2\phi + \lambda_{\theta\phi} \sin 2\theta \cos\phi),$$

where θ and ϕ are the polar and azimuthal angles in the adopted reference frame. The first measurement of J/ψ polarization in Pb–Pb collisions at $\sqrt{s_{NN}} = 5.02$ TeV as a function of p_T was recently reported by ALICE [12], showing a λ_θ with a maximum of 2σ deviation with respect to zero in the p_T interval from $2-4$ GeV/c. The preliminary measurement of the J/ψ polarization at forward rapidity in Pb–Pb collisions at $\sqrt{s_{NN}} = 5.02$ TeV as a function of centrality is shown in Fig. 2 for $2 < p_T < 6$ GeV/c. The polarization parameters show a flat trend in both the helicity and Collins-Soper reference frames. A non-zero λ_θ is observed in both frames, with slightly positive values in the helicity frame and slightly negative values in the Collins-Soper frame.

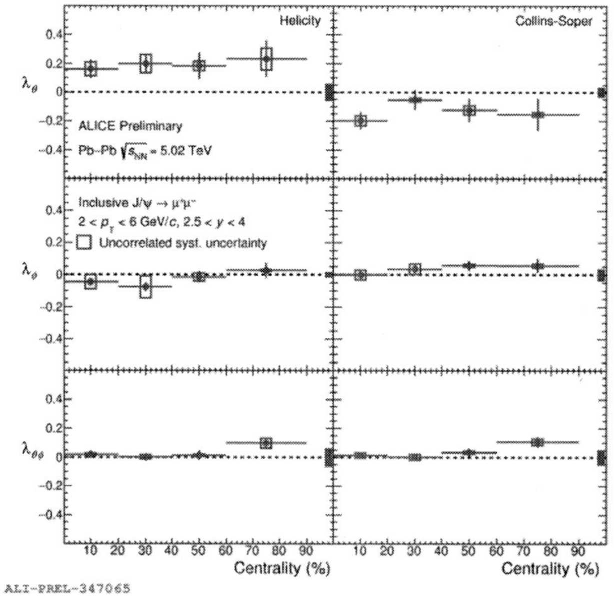

Figure 2. J/ψ polarization parameters as a function of centrality measured at forward rapidity in Pb–Pb collisions at $\sqrt{s_{NN}} = 5.02$ TeV.

The preliminary measurement of the coherent J/ψ photoproduction cross section at $\sqrt{s_{NN}} = 5.02$ TeV measured at forward rapidity is shown in Fig. 3. The measured coherent J/ψ photoproduction cross section shows an increase by a factor ~ 3 with respect to the coherently photoproduced J/ψ at $\sqrt{s_{NN}} = 2.76$ TeV [8]. The coherent photoproduction cross section is compared with models implementing a modification of the photon flux with respect to ultra-peripheral collisions (UPC). A light cone color dipole formalism was used

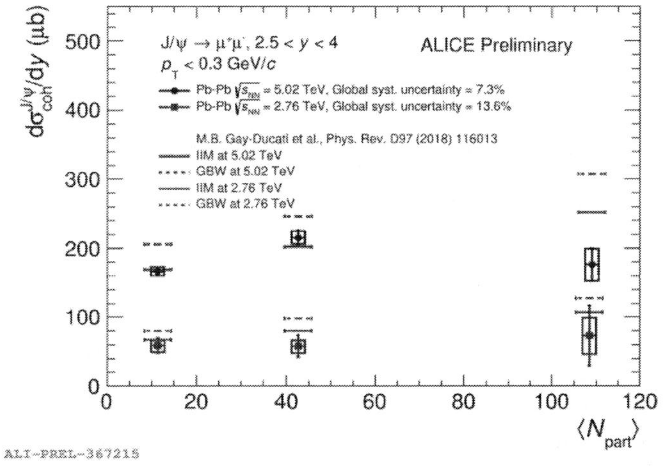

Figure 3. J/ψ coherent photoproduction cross section measured at forward rapidity in Pb–Pb collisions at $\sqrt{s_{NN}}$ = 5 and 2.76 TeV [8] in black and blue, respectively.

in the Golec-Biernat-Wusthoff (GBW) calculation, while the Iancu-Itakura-Munier (IIM) calculation is based on the Color Glass Condensate approach [13]. A qualitative agreement is observed for the most peripheral collisions, while the measurements in semicentral events deviate from the model predictions.

In summary, ALICE provides a comprehensive set of quarkonium results in heavy-ion collisions. The inclusive J/ψ R_{AA} measured at both forward and central rapidity as a function of transverse momentum is consistent with a significant contribution from (re)generation in the low p_T region. The J/ψ polarization is measured for the first time in Pb–Pb collisions, showing a flat trend for all polarization parameters as a function of centrality, with a non-zero λ_θ. Measurements of the coherent J/ψ photoproduction cross section at forward rapidity show a considerable increase in the cross section for increasing center-of-mass energy. The photoproduction cross section is well described by UPC-based models in peripheral collisions, but deviations are observed for semicentral collisions.

References

[1] T. Matsui and H. Satz, Phys. Lett. B **178** 416-422 (1986)
[2] A. Rothkopf, Phys. Rept. **858** 1-117 (2020)
[3] P. Braun-Munzinger and J. Stachel, Phys. Lett. B **490** 196-202 (2000)
[4] R. L. Thews, M. Schroedter, and J. Rafelski, Phys. Rev. C **63**, 054905 (2001)
[5] ALICE Collaboration, S. Acharya *et al.*, Phys. Lett. B **805** 135434 (2020)
[6] ALICE Collaboration, S. Acharya *et al.*, JHEP **02** 041 (2020)
[7] J. G. Contreras, Phys. Rev. C **96** 015203 (2017)
[8] ALICE Collaboration, J. Adam *et al.*, Phys. Rev. Lett. **116** 222301 (2016)
[9] ALICE Collaboration, K. Aamodt *et al.*, JINST **3** S08002 (2008)
[10] A. Andronic *et al.*, Phys. Lett. B **979** 134836 (2019)
[11] X. Du and R. Rapp, Nucl. Phys. A **943** 147-158 (2015)
[12] ALICE Collaboration, S. Acharya *et al.*, Phys. Lett. B **815** 136146 (2021)
[13] M. G. Ducati and S. Martins, Phys. Rev. D **97** 116013 (2018)

Charmonium in electromagnetic and vortical fields

Jiaxing Zhao[1,*], *Shile* Chen[1], and *Pengfei* Zhuang[1]

[1]Physics Department, Tsinghua University, Beijing 100084, China

Abstract. Due to their large mass and early production, heavy quarks (quarkonia) can be sensitive probes to investigate the fast-decaying electromagnetic and vortical fields produced in heavy-ion collisions. The non-relativistic Schrödinger-like equation for heavy quarks under strong electromagnetic fields in the rotating frame is deduced and used to construct the two-body equation for the charmonium system. The effective potential between charm and anti-charm quarks becomes anisotropic in electromagnetic and vortical fields, especially along the direction of the Lorentz force. The vorticity will affect this asymmetry and catalyze the transition from a strong interaction dominant bound state to an electromagnetic and vortical interaction controlled anisotropic bound state.

Many theoretical studies show that strong electromagnetic fields are generated in non-central relativistic heavy-ion collisions [1–5]. The maximum of the magnetic field can reach $5m_\pi^2$ in semi-central Au+Au collisions at top RHIC energy and almost $70m_\pi^2$ in semi-central Pb+Pb collisions at LHC energies [4, 5], where m_π is the pion mass. In addition, there is a finite total angular momentum $J \propto b\sqrt{s_{NN}}$ (b is the impact parameter) carried by the system of two colliding nuclei in non-central heavy-ion collisions. Although most of this total angular momentum is carried away by the spectators, there is still a sizable fraction that remains in the created quark-gluon plasma (QGP) and induces a nonzero rotational motion [6–8]. The related global polarization of Λ hyperons in relativistic heavy-ion collisions was measured by the STAR Collaboration [9]. They observed that the average vorticity of the QGP reaches $\omega \sim 10^{21}/s$ making it the most vortical fluid in nature. Both electromagnetic and vorticity fields decay fast with the expansion of the hot medium [4, 5, 8]. One needs to find a sensitive probe to characterize the electromagnetic and vortical fields. Heavy quarks and especially quarkonia are the most sensitive and effective probes due to the following reasons: The heavy quark mass is much large than the QCD cutoff, $m_c, m_b \gg \Lambda_{QCD}$, so their production can be well described by perturbative QCD; The heavy quark mass is much larger than the typical temperature of the hot medium, their mass does not change in the hot medium, and their number is conserved during the evolution; Heavy quarks are produced at a very early stage in the collision with a formation time of $\tau_c \sim 1/2m_c \approx 0.06$ fm/c for charm quark and $\tau_b \approx 0.02$ fm/c for bottom quark, hence they can feel the strongest electromagnetic and vortical fields.

So far, many interesting topics related to heavy flavor in the electromagnetic and vortical fields emerge, see review paper [10]. One example is the photoproduction of heavy flavor mesons and quarkonia in peripheral and ultra-peripheral heavy-ion collisions [11, 12] that has been observed experimentally [13]. Heavy quarks in electromagnetic and vortical fields can generate directed flow (v_1) [14, 15]. The slope of the v_1 rapidity dependence (dv_1/dy)

*e-mail: zhao-jx15@tsinghua.org.cn

of D^0 and $\bar{D}^0$ mesons is much larger than that for light hadrons [16, 17]. The charmonium states will acquire non-collective elliptic flow (v_2) in the strong magnetic field [18]. The directed flow of charmonium states is sensitive to the tilted initial energy profile that controls the magnitude of the global vorticity [19]. The static properties, e.g. mass and shape, of open/hidden heavy flavor states will be changed in electromagnetic fields [20–23]. Following these studies, we investigate charmonia in both electromagnetic and vortical fields [24].

For heavy flavor, there exists a hierarchy of scales: $m_Q \gg m_Q v \gg m_Q v^2$. Integrating out the degrees of freedom that are larger than m_Q and $m_Q v$ sequentially in the QCD Lagrangian, one obtains the non-relativistic versions, NRQCD and pNRQCD [25, 26]. A potential model is derived by neglecting the interaction between color-singlet and color-octet states in the pNRQCD framework [26]. In this case, one can employ the Schrödinger equation to study the properties of heavy quarkonia. The Schrödinger equation has been successfully applied to heavy flavor states in vacuum and at finite temperature, see the review [10]. Heavy quarks in the rotating medium in electromagnetic fields can be described by the Dirac equation in the rotating frame,

$$\left[i\gamma^\mu(\partial_\mu + iqA_\mu + \Gamma_\mu) - m\right]\Psi = 0, \tag{1}$$

where Γ_μ is the affine connection, $A_\mu(x)$ the electromagnetic field in the rotating frame, and q the charge of the quark [27]. It is straightforward to calculate the electromagnetic fields in the laboratory frame in heavy-ion collisions. One needs to transform the electromagnetic field between the rotating frame and laboratory frame. In our study, we consider a constant magnetic field along the rotating axis. Taking the symmetric gauge, the electromagnetic potential can be expressed as $A'_a = (A'_0, -By'/2, Bx'/2, 0)$ with the spacetime coordinates (t', x', y', z') in the laboratory frame. The electromagnetic potential in the rotating frame can be obtained via $A_\mu = (A_0 - (x^2 + y^2)B\omega/2, -By/2, Bx/2, 0)$ with the coordinates (t, x, y, z) in the rotating frame. In the non-relativistic limit and to first order of $1/m$, we obtain the familiar one-body Schrödinger equation,

$$\left[\frac{(\mathbf{p} - q\mathbf{A})^2}{2m} - \frac{q}{m}\mathbf{B}\cdot\mathbf{s} - qA_0 - \boldsymbol{\omega}\cdot(\mathbf{s} + \mathbf{x}\times\mathbf{p})\right]\psi = E\psi. \tag{2}$$

We can clearly see that there are no coupling terms between electromagnetic and vortical fields, which means that the rotation does not induce any electromagnetic effect. It is necessary to emphasize that the coupling term emerges if the electromagnetic fields are defined directly in the rotating frame [24].

We construct a two-body Schrödinger equation to describe a charmonium system in a rotational medium and under the impact of electromagnetic fields,

$$\left[\frac{(\mathbf{p}_a - q_a\mathbf{A}_a)^2}{2m_a} + \frac{(\mathbf{p}_b - q_b\mathbf{A}_b)^2}{2m_b} - q_a\mathbf{E}\cdot\mathbf{x}_a - q_b\mathbf{E}\cdot\mathbf{x}_b - \boldsymbol{\omega}\cdot(\mathbf{j}_a + \mathbf{j}_b) - \boldsymbol{\mu}\cdot\mathbf{B} + V\right]\psi = E\psi, \tag{3}$$

where $\mathbf{j}_i = \mathbf{s}_i + \mathbf{x}_i\times\mathbf{p}_i$ is the total angular momentum that includes spin and orbital part, $-\boldsymbol{\mu}\cdot\mathbf{B}$ is the spin magnetic moment-magnetic field coupling term, and $m_a = m_b = m$ and $q_a = -q_b = q$ are the heavy quark mass and charge. The interaction potential V between quark and anti-quark contains a spin-spin interaction term and a spin-independent part. The latter becomes the Cornell potential in vacuum. When removing the electromagnetic(vorticity) related terms, we obtain the wave equation to study quarkonium in pure vortical(electromagnetic) field. Before solving Eq. 3, we first transform the single particle coordinates and momenta to the center-of-mass and relative ones, $\mathbf{R} = (\mathbf{x}_a + \mathbf{x}_b)/2$, $\mathbf{r} = \mathbf{x}_a - \mathbf{x}_b$, $\mathbf{P} = \mathbf{p}_a + \mathbf{p}_b$, and $\mathbf{p} = (\mathbf{p}_a - \mathbf{p}_b)/2$. Different from the case with only electromagnetic fields where the total wave function can always be separated into a center-of-mass and a relative part with the help of the conserved pseudo-momentum [21, 24], it is no longer possible to find a conserved quantity to separated the two parts in a rotational field. However, considering the fact that the strength of the magnetic field in relativistic heavy-ion collisions can reach $eB \approx 70m_\pi^2$, which is much larger than the vortical field $m_Q\omega \approx m_\pi^2$ for charm quarks, we can choose the electromagnetic related parts as the main part and take the rotational related terms as perturbations. Therefore, the Hamiltonian can be expressed as,

$$\begin{aligned} H &= H_{\rm EM} + H', \\ H_{\rm EM} &\equiv \frac{\mathbf{P}_{ps}^2}{4m} + \frac{\mathbf{p}^2}{m} - \mu \cdot \mathbf{B} - q\mathbf{E}\cdot\mathbf{r} - \frac{q}{2m}(\mathbf{P}_{ps}\times\mathbf{B})\cdot\mathbf{r} + \frac{q^2}{4m}(\mathbf{B}\times\mathbf{r})^2 + V, \\ H' &\equiv -\omega\cdot(\mathbf{R}\times\mathbf{P}_{ps}) - \omega\cdot(\mathbf{l}+\mathbf{s}) + \frac{q}{2}\omega\cdot(\mathbf{R}\times(\mathbf{B}\times\mathbf{r})), \end{aligned} \tag{4}$$

where $\mathbf{P}_{ps} = \mathbf{P} + q\mathbf{B}\times\mathbf{r}/2$ is the pseudo-momentum which is the same as in pure electromagnetic fields. The contribution from the perturbation H' can be calculated through the standard method in quantum mechanics. The perturbation H' is relevant to the center-of-mass coordinate $\mathbf{R}$ hence the perturbative correction depends on the distance from the quarkonium to the rotating axis. That's the nature of particles in a vortical field and we can use the average value $\langle\mathbf{R}\rangle$ in our calculation. Details about the numerical calculation and results are shown in [24]. Here, we just show the effective potential in the two-body Schrödinger equation that can give us an intuitive understanding. Without the magnetic field, the potential (Cornell potential in

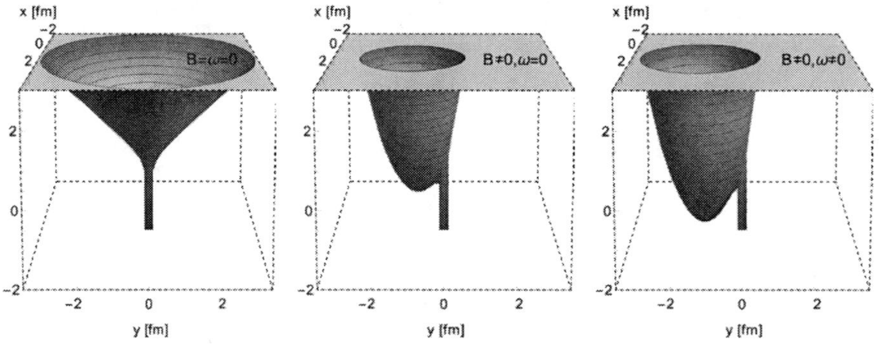

Figure 1. Spin-independent vacuum interaction potential between charm and anti-charm quark in $x-y$ plane with $z=0$ with/without magnetic and/or vortical fields. The unit of potential is GeV.

vacuum) is a symmetric function in the $x-y$ plane, but the symmetry is broken when the magnetic field is turned on. When the magnetic field is strong enough, a new potential well forms and the $c\bar{c}$ bound state may tunnel from the strong interaction well to the magnetic well, as shown in Fig. 1. This is also known as Lorentz ionization [20]. The vortical field will deepen (shallow) the magnetic well. The vorticity will catalyze the transition from a strongly to an electromagnetically and rotationally interacting bound state.

In summary, we derived the non-relativistic wave equation under strong electromagnetic fields in the rotating frame. We constructed the two-body Schrödinger equation to investigated the static properties of charmonium states in strong electromagnetic and vortical fields. Since one usually defines the electromagnetic fields in the laboratory frame in heavy-ion collisions, there are no coupling terms between the magnetic and vortical fields in the Dirac and Schrödinger equations. The effective potential between charm and anti-charm quarks becomes anisotropic in electromagnetic and vortical fields, especially along the direction of the

Lorentz force. The vorticity will increase (decrease) this asymmetry property. With increasing fields, the $c\bar{c}$ bound state which is confined by strong interaction in vacuum is gradually tunneled to an electromagnetic and vortical-field controlled anisotropic state. It is possible to realize this state in high-energy nuclear collisions. These anisotropic charmonium states will have a very different dynamic dissociation in the quark-gluon medium.

Acknowledgement: The work is supported by Guangdong Major Project of Basic and Applied Basic Research No. 2020B0301030008 and the NSFC under grant Nos. 11890712, 12075129, and 12047535.

References

[1] D. E. Kharzeev, L. D. McLerran and H. J. Warringa, Nucl. Phys. A **803**, 227-253 (2008).
[2] V. Skokov, A. Y. Illarionov and V. Toneev, Int. J. Mod. Phys. A **24**, 5925-5932 (2009).
[3] V. Voronyuk, V. D. Toneev, W. Cassing, E. L. Bratkovskaya, V. P. Konchakovski and S. A. Voloshin, Phys. Rev. C **83**, 054911 (2011).
[4] W. T. Deng and X. G. Huang, Phys. Rev. C **85**, 044907 (2012).
[5] K. Tuchin, Adv. High Energy Phys. **2013**, 490495 (2013).
[6] Z. T. Liang and X. N. Wang, Phys. Rev. Lett. **94**, 102301 (2005) [erratum: Phys. Rev. Lett. **96**, 039901 (2006)].
[7] F. Becattini, F. Piccinini and J. Rizzo, Phys. Rev. C **77**, 024906 (2008).
[8] Y. Jiang, Z. W. Lin and J. Liao, Phys. Rev. C **94**, no.4, 044910 (2016) [erratum: Phys. Rev. C **95**, no.4, 049904 (2017)].
[9] L. Adamczyk *et al.* [STAR], Nature **548**, 62-65 (2017).
[10] J. Zhao, K. Zhou, S. Chen and P. Zhuang, Prog. Part. Nucl. Phys. **114**, 103801 (2020).
[11] W. Zha, L. Ruan, Z. Tang, Z. Xu and S. Yang, Phys. Lett. B **789**, 238-242 (2019).
[12] W. Shi, W. Zha and B. Chen, Phys. Lett. B **777**, 399-405 (2018).
[13] J. Adam *et al.* [ALICE], Phys. Rev. Lett. **116**, no.22, 222301 (2016).
[14] S. Chatterjee and P. Bożek, Phys. Rev. Lett. **120**, no.19, 192301 (2018).
[15] S. K. Das, S. Plumari, S. Chatterjee, J. Alam, F. Scardina and V. Greco, Phys. Lett. B **768**, 260-264 (2017).
[16] J. Adam *et al.* [STAR], Phys. Rev. Lett. **123**, no.16, 162301 (2019).
[17] S. Acharya *et al.* [ALICE], Phys. Rev. Lett. **125**, no.2, 022301 (2020).
[18] X. Guo, S. Shi, N. Xu, Z. Xu and P. Zhuang, Phys. Lett. B **751**, 215-219 (2015).
[19] B. Chen, M. Hu, H. Zhang and J. Zhao, Phys. Lett. B **802**, 135271 (2020).
[20] K. Marasinghe and K. Tuchin, Phys. Rev. C **84**, 044908 (2011).
[21] J. Alford and M. Strickland, Phys. Rev. D **88**, 105017 (2013).
[22] S. Cho, K. Hattori, S. H. Lee, K. Morita and S. Ozaki, Phys. Rev. Lett. **113**, no.17, 172301 (2014).
[23] S. Iwasaki, M. Oka and K. Suzuki, Eur. Phys. J. A **57**, no.7, 222 (2021).
[24] S. Chen, J. Zhao and P. Zhuang, Phys. Rev. C **103**, no.3, L031902 (2021).
[25] W. E. Caswell and G. P. Lepage, Phys. Lett. B **167**, 437-442 (1986).
[26] N. Brambilla, A. Pineda, J. Soto and A. Vairo, Nucl. Phys. B **566**, 275 (2000).
[27] H. L. Chen, K. Fukushima, X. G. Huang and K. Mameda, Phys. Rev. D **93**, no.10, 104052 (2016).

Charm-baryon enhancement and charm fragmentation fractions in small systems measured with ALICE

Jianhui Zhu[1,2,*] for the ALICE Collaboration

[1]Institute of Particle Physics, Central China Normal University, 152 Luoyu Road, 430079 Wuhan, China
[2]GSI Helmholtz Centre for Heavy Ion Research, Planckstraße 1, 64291 Darmstadt, Germany

Abstract. Recent measurements of charm-baryon production at midrapidity by the ALICE collaboration show baryon-to-meson yield ratios significantly higher than those measured in e^+e^- collisions, suggesting that the charm fragmentations are not universal across different collisions systems. Thus, measurements of charm-baryon production are crucial to study the charm quark hadronisation in proton–proton (pp) collisions. In proton–lead (p–Pb) collisions, the measurements of charm baryons provide important information about cold nuclear matter effects and help to understand how the possible presence of collective effects could modify the production of heavy-flavour hadrons. In this contribution, the most recent results on open charm-hadron production in pp and p–Pb collisions measured by ALICE are discussed.

1 Introduction

The production of heavy-flavour hadrons in high-energy hadronic collisions can provide important tests of the theory of quantum chromodynamics (QCD). The production cross sections of heavy-flavour hadrons can be calculated using the factorisation approach as a convolution of three factors [1]: the parton distribution functions (PDFs) of the incoming nuclei, the hard-scattering cross section at partonic level, calculated as a perturbative series in powers of the strong coupling constant α_s, and the fragmentation functions of heavy quarks into corresponding heavy-flavour hadrons, which is an inherently non-perturbative process related to, or even driven by, the confining property of QCD. The heavy-flavour baryon-to-meson ratio is an ideal observable related to the hadronization mechanism since the contributions from parton distribution function and parton-parton scattering terms cancel in the ratio. The Λ_c^+/D^0 ratio at the LHC is enhanced with respect to predictions based on e^+e^- and ep experiments, suggesting that the charm fragmentation functions are not universal among different collision systems. Several hadronization mechanisms, such as colour reconnection (CR) beyond the leading colour approximation [2], coalescence [3, 4] and feed-down from a largely augmented set of higher mass charm-baryon states beyond the current listing of the particle data group (PDG) [5], have been proposed to explain this enhancement. The newest measurements of the charm baryons Λ_c^+, $\Sigma_c^{0,++}$, $\Xi_c^{0,+}$ and Ω_c^0 performed with the ALICE experiment will be used to verify predictions from these hadronization mechanisms.

*e-mail: jianhui.zhu@cern.ch

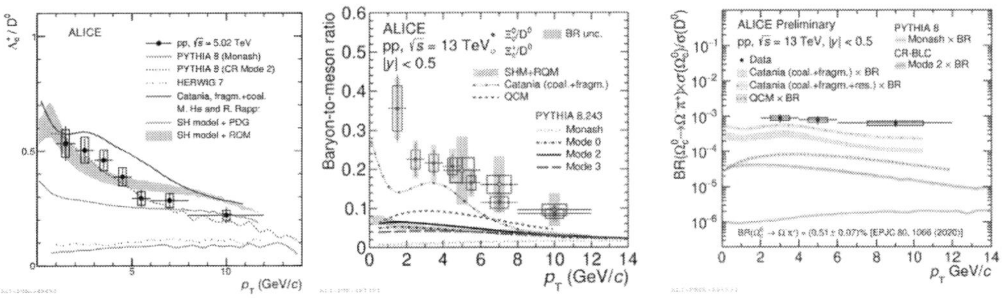

Figure 1. Left: The Λ_c^+/D^0 ratio as a function of p_T measured in pp collisions at $\sqrt{s}$ = 5.02 TeV [7]. Middle: The Ξ_c^0/D^0 and Ξ_c^+/D^0 ratios measured in pp collisions at $\sqrt{s}$ = 13 TeV [8]. Right: The $BR \times \Omega_c^0/D^0$ measured in pp collisions at $\sqrt{s}$ = 13 TeV. All the charm baryon-to-meson yield ratios are compared to theoretical calculations [2–5, 11].

2 Charm baryon-to-meson yield ratios in pp collisions

Thanks to the large data sample collected during the Run 2 period in pp collisions at $\sqrt{s}$ = 5.02 and 13 TeV, ALICE measured all the ground-state charm hadrons down to low transverse momentum (p_T), including charm mesons (D^0, D^+, D_s^+ [6]) and charm baryons (Λ_c^+ [7], $\Xi_c^{0,+}$ [8, 9], and Ω_c^0). The Λ_c^+/D^0 yield ratio is measured as a function of p_T in pp collisions at $\sqrt{s}$ = 5.02 TeV as shown in Fig. 1 (left), the Ξ_c^0/D^0 and Ξ_c^+/D^0 yield ratios are measured in pp collisions at $\sqrt{s}$ = 13 TeV as shown in Fig. 1 (middle). For Ω_c^0, the absolute decay branching ratio (BR) of $\Omega_c^0 \to \Omega^-\pi^+$ is not measured, hence the BR of $\Omega_c^0 \to \Omega^-\pi^+$ times cross section of Ω_c^0 over cross section of D^0 in pp collisions at $\sqrt{s}$ = 13 TeV is reported in Fig. 1 (right). In order to compare data with models, a theoretical calculation of $BR(\Omega_c^0 \to \Omega^-\pi^+)$ [10] is used to multiply different models. The Λ_c^+/D^0 and Ξ_c^0/D^0 ratios show a downward trend with increasing p_T. The Monte Carlo generator PYTHIA8 (Monash) [11] tuned on measurements in e^+e^- collisions largely underestimates all four charm baryon-to-meson yield ratios, providing evidence of different charm hadronisation mechanisms between e^+e^- and pp collisions. The Λ_c^+/D^0 ratio is better described by a model with colour reconnection beyond the leading colour approximation [2], a statistical hadronisation model with an augmented set of charm baryon states predicted by the relativistic quark model (RQM) [5], or a model relying on hadronisation via coalescence and fragmentation [3]. However, all the models underestimate the $\Xi_c^{0,+}/D^0$ and $BR \times \Omega_c^0/D^0$ ratios, except the Catania model [3] including charm quark hadronisation via both coalescence and fragmentation, which would indicate a partonic system similar to a quark–gluon plasma (QGP) in pp collisions.

3 Charm production and fragmentation in pp collisions

The charm fragmentation fraction $f(c \to H_c)$ shown in Fig. 2 (left) represents the probability of a charm quark hadronising into a given charm hadron. The fragmentation fraction for the Ξ_c^0 baryon is the first measurement in any collision system. An increase of about a factor of 3.3 for the fragmentation fraction for the Λ_c^+ baryon with respect to e^+e^- and ep collisions, and a corresponding decrease of about a factor of 1.2–1.4 for the D^0 meson are observed, showing that the assumption of the charm fragmentation universality (collision-system independence) is broken. Charm quarks hadronise into baryons almost 40% of the time, which is four times more often than what was measured at colliders with electron beams.

 https://doi.org/10.1051/epjconf/202225912006

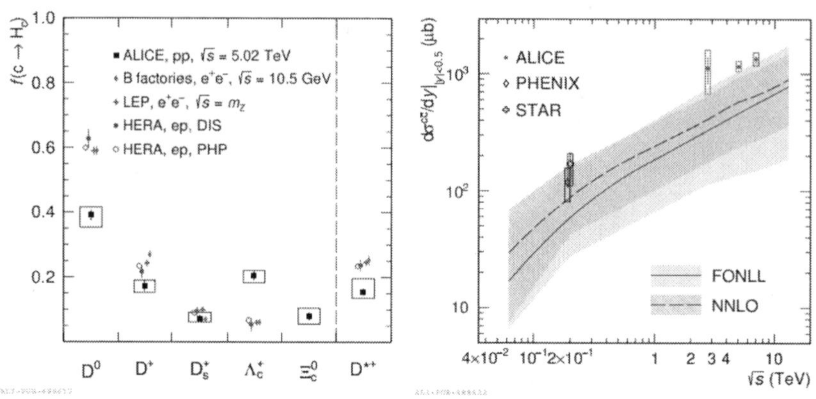

Figure 2. Left: Charm-quark fragmentation fractions into charm hadrons measured in pp collisions at $\sqrt{s}$ = 5.02 TeV in comparison with experimental measurements performed in e^+e^- and ep collisions [18]. Right: Charm production cross section at midrapidity per unit of rapidity as a function of the collision energy at the LHC [18] and RHIC [14, 15] compared to FONLL [16] and NNLO [17] calculations.

The $c\bar{c}$ production cross section per unit of rapidity at midrapidity ($d\sigma^{c\bar{c}}/dy|_{|y|<0.5}$) is calculated by summing the p_T-integrated cross sections of all measured ground-state charm hadrons (D^0, D^+, D_s^+, Λ_c^+, Ξ_c^0 and their charge conjugates). The contribution of Ξ_c^0 is multiplied by a factor of 2 in order to account for the contribution of Ξ_c^+. Since the absence of a Ω_c^0 production measurement at hadron colliders, an asymmetric systematic uncertainty is assigned assuming a contribution equal to the one of Ξ_c^0 considering the prediction of the Catania model [3]. The resulting $c\bar{c}$ production cross section per unit of rapidity at midrapidity is $d\sigma^{c\bar{c}}/dy|_{|y|<0.5}^{pp,\,5.02\,\text{TeV}} = 1165 \pm 44(\text{stat})^{+134}_{-101}(\text{syst})\ \mu b$. The updated fragmentation fractions obtained in pp collisions at $\sqrt{s}$ = 5.02 TeV allow the recomputation of the charm production cross sections per unit of rapidity at midrapidity in pp collisions at $\sqrt{s}$ = 2.76 and 7 TeV, which are about 40% higher than the previously published results [12, 13]. The measured $c\bar{c}$ production cross section per unit of rapidity at midrapidity together with measurements at RHIC [14, 15] are located at the upper edge of FONLL [16] and NNLO [17] predictions.

4 Charm hadronisation in p–Pb collisions

Charm hadronisation in p–Pb collisions at $\sqrt{s_{NN}}$ = 5.02 TeV is also investigated. The left panel of Fig. 3 shows the Λ_c^+/D^0 ratio as a function of p_T in pp and p–Pb collisions at 5.02 TeV. This ratio is measured down to $p_T = 0$ in p–Pb collisions for the first time at the LHC. The measurements of Λ_c^+/D^0 in pp and p–Pb collisions are qualitatively consistent with each other, although a larger ratio in $3 < p_T < 8$ GeV/c and a lower ratio in $1 < p_T < 2$ GeV/c are measured in p–Pb collisions with respect to pp collisions. A clear decreasing trend with increasing p_T is obtained in both pp and p–Pb collisions for $p_T > 2$ GeV/c. The nuclear modification factor R_{pPb} of the Λ_c^+ baryon measured as a function of p_T together with R_{pPb} of non-strange D mesons are shown in the right panel of Fig. 3. There is a significant suppression for $p_T < 2$ GeV/c and enhancement for $4 < p_T < 8$ GeV/c in p–Pb collisions with respect to pp collisions, which is similar as the p_T distribution of Λ_c^+/D^0 ratio, suggesting the presence of possible radial flow effects or a further modification of the charm hadronisation mechanism in p–Pb collisions. The measurement of R_{pPb} for the Λ_c^+ baryon is compared to model calculations. The POWHEG [19] + PYTHIA6 simulations use the POWHEG event generator

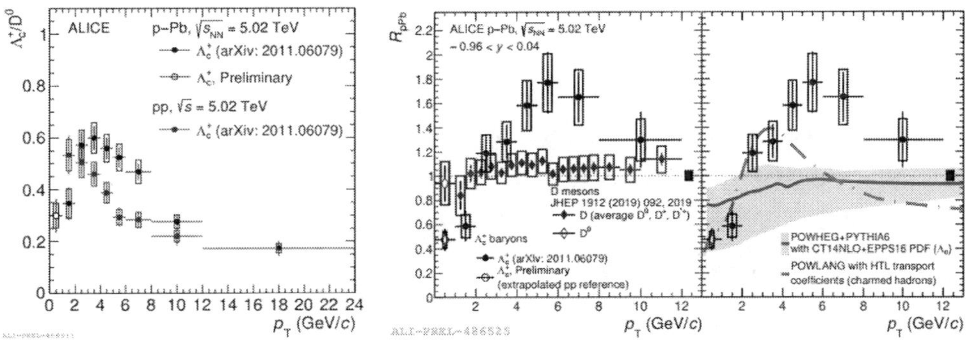

Figure 3. Left: The Λ_c^+/D^0 ratio as a function of p_T measured in pp collisions and p–Pb collisions at 5.02 TeV [7]. Right: The nuclear modification factor R_{pPb} of the Λ_c^+ baryon as a function of p_T in p–Pb collisions at $\sqrt{s_{NN}}$ = 5.02 TeV, compared to R_{pPb} of D mesons, as well as to model expectations [7].

with PYTHIA6 parton shower and EPPS16 [20] parameterisation of the nuclear modification of the PDFs. The POWLANG model [21] assumes that a hot deconfined medium is formed in p–Pb collisions. The two models capture some features of the data, but neither of them can quantitatively reproduce the data in the measured p_T interval.

5 Acknowledgments

This work is supported by the international postdoctoral exchange fellowship program of Helmholtz Association and the Office of China Postdoc Council (No. 20181016).

References

[1] J.C. Collins, D.E. Soper, G.F. Sterman, Nucl. Phys. B **263**, 37 (1986)
[2] J.R. Christiansen, P.Z. Skands, JHEP **08**, 003 (2015)
[3] V. Minissale, S. Plumari, V. Greco (2020), arXiv:2012.12001 [hep-ph]
[4] J. Song, H.h. Li, F.l. Shao, Eur. Phys. J. C **78**, 344 (2018)
[5] M. He, R. Rapp, Phys. Lett. B **795**, 117 (2019)
[6] S. Acharya et al. (ALICE), JHEP **05**, 220 (2021)
[7] S. Acharya et al. (ALICE) (2020), arXiv:2011.06079 [nucl-ex]
[8] S. Acharya et al. (ALICE) (2021), arXiv:2105.05187 [nucl-ex]
[9] S. Acharya et al. (ALICE) (2021), arXiv:2105.05616 [nucl-ex]
[10] Y.K. Hsiao et al., Eur. Phys. J. C **80**, 1066 (2020)
[11] P. Skands, S. Carrazza, J. Rojo, Eur. Phys. J. C **74**, 3024 (2014)
[12] B. Abelev et al. (ALICE), JHEP **07**, 191 (2012)
[13] S. Acharya et al. (ALICE), Eur. Phys. J. C **77**, 550 (2017)
[14] L. Adamczyk et al. (STAR), Phys. Rev. D **86**, 072013 (2012)
[15] A. Adare et al. (PHENIX), Phys. Rev. C **84**, 044905 (2011)
[16] M. Cacciari et al., JHEP **10**, 137 (2012)
[17] D. d'Enterria, A.M. Snigirev, Phys. Rev. Lett. **118**, 122001 (2017)
[18] S. Acharya et al. (ALICE) (2021), arXiv:2105.06335 [nucl-ex]
[19] S. Frixione, P. Nason, G. Ridolfi, JHEP **09**, 126 (2007)
[20] K.J. Eskola et al., Eur. Phys. J. C **77**, 163 (2017)
[21] A. Beraudo et al., JHEP **03**, 123 (2016)

Evidence for X(3872) in PbPb collisions and studies of its prompt production at $\sqrt{s_{\mathrm{NN}}} = 5.02$ TeV with CMS

Yen-Jie Lee[1,*] *for the CMS collaboration*

[1]Massachusetts Institute of Technology

Abstract. The first evidence for X(3872) production in relativistic heavy ion collisions, with a significance of 4.2 standard deviations, is reported. The X(3872) production is studied in lead-lead (PbPb) collisions at a center-of-mass energy of $\sqrt{s_{\mathrm{NN}}} = 5.02$ TeV per nucleon pair with the CMS detector. The measurement is performed in the rapidity and transverse momentum ranges $|y| < 1.6$ and $15 < p_{\mathrm{T}} < 50$ GeV/c. The prompt X(3872) to ψ(2S) yield ratio is found to be $\rho^{\mathrm{PbPb}} = 1.08 \pm 0.49(\mathrm{stat.}) \pm 0.52(\mathrm{syst.})$, to be compared with typical values of 0.1 for pp collisions. This provides new input to theoretical models of the X(3872) production mechanism, and of the nature of this exotic state.

The X(3872) is an exotic particle that was first observed in electron-position (e^+e^-) collision by Belle [1], and studied by experiments at hadron colliders [2–6] and e^+e^- [7, 8]. The quantum numbers of the X(3872) have been narrowed down by CDF [9], and later determined to be $J^{PC} = 1^{++}$ by LHCb [10]. However, the inner structure of X(3872) is still not yet fully understood. Interpretations in terms of charmonium, $\mathrm{D}^*(2010)^0\overline{\mathrm{D}}^0$ molecules [11], tetraquark states [12], and their admixture [13] have been proposed. The production and survival of the X(3872) in a quark-gluon plasma (QGP, a deconfined state of quarks and gluons [14, 15]), or in a later hadronic phase, is expected to depend upon the X(3872)'s internal structure [16–18]. Thus, the recent large data set of lead-lead (PbPb) collisions at a center-of-mass energy of $\sqrt{s_{\mathrm{NN}}} = 5.02$ TeV per nucleon pair, delivered by the Large Hadron Collider (LHC) at CERN at the end of 2018, opened new opportunities to probe the nature of this exotic state [19–22].

It is expected that in relativistic heavy ion collisions, the formation of the QGP could enhance or suppress the production of the X(3872) particle. Coalescence mechanisms could enhance the X(3872) production yield [17, 19]. These mechanisms can be modeled via the overlap of the density matrix of the constituents in an emission source with the Wigner function of the produced particle [23]. Therefore, the enhancement of the X(3872) production in the QGP would depend on the spatial configuration of the exotic state. Moreover, a longer distance between the quarks and antiquarks that constitute the state could also lead to a higher X(3872) dissociation rate, similar to that from the mechanism of quarkonium suppression in heavy ion collisions [24]. Therefore, the study of the X(3872) state in the QGP may be used as a tool to distinguish a compact tetraquark configuration with a radius ~0.3 fm from a molecular state with a radius greater than 1.5 fm [25]. Such a measurement would be complementary to the recent evidence for the radiative decay $\mathrm{X}(3872) \rightarrow \psi(2\mathrm{S})\gamma$ in proton-proton (pp) collisions reported by LHCb Collaboration [26], which does not support a pure $\mathrm{D}^*(2010)^0\overline{\mathrm{D}}$ molecular interpretation. In addition, measurements of prompt X(3872) production could

*e-mail: yenjie@mit.edu

provide an interesting test of the statistical hadronization model [27], which assumes that the produced matter is in thermodynamic equilibrium at the phase transition to hadrons [28, 29].

In this proceedings, the first evidence for X(3872) production in PbPb collisions at $\sqrt{s_{\mathrm{NN}}} = 5.02\,\mathrm{TeV}$ is reported. The PbPb sample corresponds to an integrated luminosity of $1.7\mathrm{nb}^{-1}$. The X(3872) candidates are reconstructed through the decay chain $\mathrm{X}(3872) \rightarrow \mathrm{J}/\psi(1\mathrm{S})\,\pi^+\pi^- \rightarrow \mu^+\mu^-\pi^+\pi^-$, and are measured in the $15 < p_{\mathrm{T}} < 50\,\mathrm{GeV}/c$ and $|y| < 1.6$ kinematic interval. At the LHC energies, the inclusive X(3872) yields in pp and PbPb collisions contain a significant nonprompt contribution coming from B hadron decays [6]. The nonprompt X(3872) component is related to the beauty quark energy loss and medium-modified B hadron production in heavy ion collisions, which has been studied with b-jets [30] and fully reconstructed B mesons [31]. Here, we focus on the prompt component, from charm quark fragmentation, for which the ratio ρ^i (i is pp or PbPb) between the corrected yields of X(3872) and ψ(2S) mesons (where the ψ(2S) is reconstructed with the same final-state particles in order to reduce systematic uncertainties) is presented: $\rho^i = \frac{N_i^{\mathrm{X}(3872)\rightarrow \mathrm{J}/\psi(1\mathrm{S})\pi\pi}}{N_i^{\psi(2\mathrm{S})\rightarrow \mathrm{J}/\psi(1\mathrm{S})\pi\pi}}$. The ratios in pp and PbPb collisions are connected to the nuclear modification factors $R_{\mathrm{AA}}^{\mathrm{X}(3872)}$ and $R_{\mathrm{AA}}^{\psi(2\mathrm{S})}$ (the meson yield ratio in nucleus-nucleus and pp interactions normalized by the number of inelastic nucleon-nucleon collisions) via the following relation: $\rho^{\mathrm{PbPb}} = \rho^{\mathrm{pp}} \frac{R_{\mathrm{AA}}^{\mathrm{X}(3872)}}{R_{\mathrm{AA}}^{\psi(2\mathrm{S})}}$.

The description of the CMS experiment and coordinate system can be found in Ref. [32] and the details on the analysis procedure are documented in Ref. [33]. The X(3872) and ψ(2S) signals are extracted from the $\mathrm{J}/\psi(1\mathrm{S})\,\pi^+\pi^-$ channel and $\mathrm{J}/\psi(1\mathrm{S})$ candidates are reconstructed by dimuon pairs. The raw inclusive yields of X(3872) and ψ(2S) are extracted by an extended unbinned maximum-likelihood fit. A double-Gaussian function with a common mean but independent widths is used to model each of the X(3872) and ψ(2S) peaks. For describing the combinatorial background, mostly from the random combination of a $\mathrm{J}/\psi(1\mathrm{S})$ candidate with other tracks, a 4th-order polynomial is used, which gives the best fit in terms of χ^2 per degrees of freedom. The invariant mass fits, with offline selection optimized for X(3872), are shown in the left panel of Fig. 1. The significance of the inclusive X(3872) signal against background-only hypothesis is 4.2 standard deviations, which was calculated as the square root of the logarithm of the profile likelihood ratio where the signal is zero.

The contribution from B hadron decays is subtracted from the inclusive result using the pseudo-proper decay length l_{xy}, defined as the distance in the transverse plane, L_{xy}, between the vertex formed by the 4-tracks and the primary vertex, corrected by the transverse Lorentz boost of the candidate. After nonprompt contribution subtraction [33], the ratio ρ^{PbPb} between the prompt X(3872) and ψ(2S) mesons is shown in right panel of Fig. 1, together with ρ^{pp} measured as a function of p_{T}. The pp data were measured at $\sqrt{s} = 7$ and $8\,\mathrm{TeV}$, in the $|y| < 1.2$ and $|y| < 0.75$ intervals, respectively [5, 6, 10]. The 7 TeV result was derived using the CMS Collaboration published ratio of the inclusive yields [5] and prompt fractions [5, 34]. From Fig. 1 it is clear that the prompt ρ^{pp} does not depend significantly on collision energy or rapidity. In pp collisions at $\sqrt{s} = 8\,\mathrm{TeV}$, in the kinematic range of $16 < p_{\mathrm{T}} < 22\,\mathrm{GeV}/c$, the ρ^{pp} measured by ATLAS is $0.106 \pm 0.008(\mathrm{stat.}) \pm 0.004(\mathrm{syst.})$ [6]. This is to be compared to the prompt ρ^{PbPb} measured in this proceedings, $\rho^{\mathrm{PbPb}} = 1.08 \pm 0.49(\mathrm{stat.}) \pm 0.52(\mathrm{syst.})$.

In the interval $15 < p_{\mathrm{T}} < 20\,\mathrm{GeV}/c$, the yield of the prompt ψ(2S) in PbPb collisions was reported to be significantly suppressed with respect to pp collisions, $R_{\mathrm{AA}}^{\psi(2\mathrm{S})} = 0.142 \pm 0.061(\mathrm{stat.}) \pm 0.020(\mathrm{syst.})$ [35]. This leads to an $R_{\mathrm{AA}}^{\mathrm{X}(3872)}$ central value larger than 1 (i.e., enhancement of the prompt X(3872) yield in PbPb compared to pp collisions). However, the uncertainties are such that $R_{\mathrm{AA}}^{\mathrm{X}(3872)}$ is compatible with 1 within $\sim$1 standard deviation, and with $R_{\mathrm{AA}}^{\psi(2\mathrm{S})}$ within $\sim$2 standard deviations. Thus, it is possible that in PbPb colli-

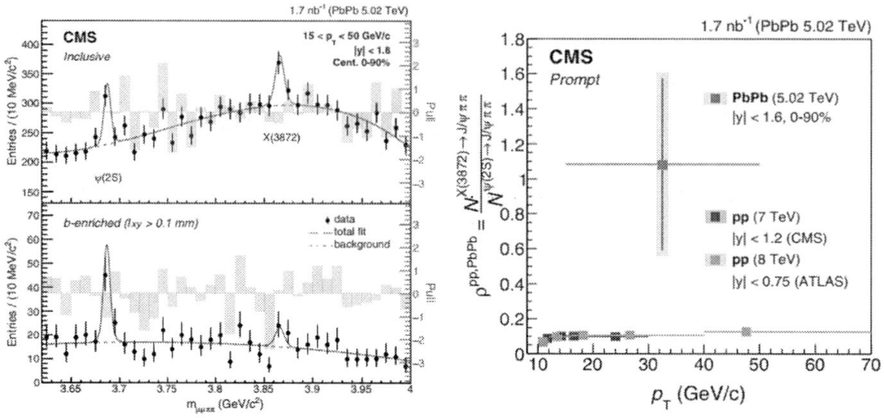

Figure 1. Left Panel: Invariant mass distribution of $m_{\mu\mu\pi\pi}$ in PbPb collisions, for the inclusive (upper) and B-enriched (bottom) samples. The vertical lines represent statistical uncertainties in the data. The pull distribution is represented by the shaded bars. Right Panel: ρ^{PbPb} in PbPb collisions at $\sqrt{s_{\mathrm{NN}}}$ = 5.02 TeV. The vertical bars (boxes) correspond to statistical (systematic) uncertainties. ρ^{pp} of prompt X(3872) over $\psi(2S)$ production in pp collisions at $\sqrt{s}$ = 8 TeV, measured by ATLAS [6], and at $\sqrt{s}$ = 7 TeV, measured by CMS [5] are also shown.

sions, the prompt X(3872) yield has either no suppression with respect to pp collisions, or as much suppression as the ψ(2S) state. The much larger data sample expected in Run 3 at the LHC will answer whether $R_{\mathrm{AA}}^{\mathrm{X(3872)}}$ is different from $R_{\mathrm{AA}}^{\psi(2\mathrm{S})}$ and significantly above unity. It may answer whether the ψ(2S) meson production (a bound state of a $q_c$and $\overline{q}_c$ quarks, with $r \sim 0.9$ fm) [36], is affected differently by the medium produced in PbPb collisions than the X(3872) state (that could be made of q_c, $\overline{q}_c$, q_u, and $\overline{q}_u$ quarks, with a radius of $r \sim 0.3$ fm or $r > 1.5$ fm), the difference in both size and quark content playing a role into their production mechanisms. It will also answer whether the X(3872) prompt state production is different in PbPb collisions compared to pp collisions. It is also of great interest to compare the PbPb results to the studies of X(3872) and ψ(2S) production in pp collisions reported by LHCb [37], where a strong suppression of X(3872) to ψ(2S) ratio is observed at high multiplicity. Naively, if the suppression gets only stronger as a function of event multiplicity, we will see even stronger suppression in PbPb collisions. The fact that an evidence of X(3872) production is reported in this analysis may indicate additional mechanism which enhances the X(3872) production in the presence of the QGP.

The question whether X(3872) is a tetraquark or a molecule cannot yet be answered, because of the statistical limitation of the data, and the disagreement among models. For example, while the AMPT transport model [19] predicts $R_{\mathrm{AA}}^{\mathrm{molecule}} >> R_{\mathrm{AA}}^{\mathrm{tetraquark}}$ with $R_{\mathrm{AA}}^{\mathrm{molecule}} > 1$, the TAMU transport model [20] predicts $R_{\mathrm{AA}}^{\mathrm{molecule}} \sim R_{\mathrm{AA}}^{\mathrm{tetraquark}}/2$ (albeit, considering only the X(3872) from regeneration processes). With the further development of the theoretical models and high precision data from various collision systems from e^+e^-, electron-ion to heavy-ion collisions, we aim to determine the internal structure of the X(3872).

In summary, the first evidence for X(3872) production in heavy ion collisions is presented using lead-lead collisions at a center-of-mass energy of $\sqrt{s_{\mathrm{NN}}}$ = 5.02 TeV per nucleon pair, recorded with the CMS detector. The significance of the inclusive X(3872) signal is 4.2 standard deviations. The ratio ρ^{PbPb} between the prompt X(3872) and ψ(2S) yields times their branching fractions into $J/\psi(1S)\pi^+\pi^-$ is found to be 1.08 ± 0.49(stat.) $\pm$ 0.52(syst.), to be compared with typical values of 0.1 for pp collisions. This result provides a unique experimental input to theoretical models of the X(3872) production mechanism, and of the nature of this exotic state.

References

[1] S.K. Choi et al. (Belle), Phys. Rev. Lett. **91**, 262001 (2003), hep-ex/0309032
[2] D. Acosta et al. (CDF), Phys. Rev. Lett. **93**, 072001 (2004), hep-ex/0312021
[3] V.M. Abazov et al. (D0), Phys. Rev. Lett. **93**, 162002 (2004), hep-ex/0405004
[4] R. Aaij et al. (LHCb), Eur. Phys. J. C **72**, 1972 (2012), 1112.5310
[5] S. Chatrchyan et al. (CMS), JHEP **04**, 154 (2013), 1302.3968
[6] M. Aaboud et al. (ATLAS), JHEP **01**, 117 (2017), 1610.09303
[7] B. Aubert et al. (BaBar), Phys. Rev. D **71**, 071103 (2005), hep-ex/0406022
[8] M. Ablikim et al. (BESIII), Phys. Rev. Lett. **112**, 092001 (2014), 1310.4101
[9] A. Abulencia et al. (CDF), Phys. Rev. Lett. **98**, 132002 (2007), hep-ex/0612053
[10] R. Aaij et al. (LHCb), Phys. Rev. Lett. **110**, 222001 (2013), 1302.6269
[11] N.A. Tornqvist, Phys. Lett. B **590**, 209 (2004), hep-ph/0402237
[12] L. Maiani, F. Piccinini, A.D. Polosa, V. Riquer, Phys. Rev. D **71**, 014028 (2005)
[13] C. Hanhart, Yu.S. Kalashnikova, A.V. Nefediev, Eur. Phys. J. A **47**, 101 (2011)
[14] J.C. Collins, M.J. Perry, Phys. Rev. Lett. **34**, 1353 (1975)
[15] E.V. Shuryak, Phys. Lett. B **78**, 150 (1978), [*Yad. Fiz.* **28** (1978) 796]
[16] S. Cho et al. (ExHIC), Phys. Rev. Lett. **106**, 212001 (2011), 1011.0852
[17] S. Cho et al. (ExHIC), Phys. Rev. C **84**, 064910 (2011), 1107.1302
[18] A. Martinez Torres, K.P. Khemchandani, F.S. Navarra, M. Nielsen, L.M. Abreu, Phys. Rev. D **90**, 114023 (2014), [Erratum: 10.1103/PhysRevD.93.059902], 1405.7583
[19] H. Zhang, J. Liao, E. Wang, Q. Wang, H. Xing, Phys. Rev. Lett. **126**, 012301 (2021)
[20] B. Wu, X. Du, M. Sibila, R. Rapp (2020), 2006.09945
[21] A. Esposito, E.G. Ferreiro, A. Pilloni, A.D. Polosa, C.A. Salgado (2020), 2006.15044
[22] E. Braaten, L.P. He, K. Ingles, J. Jiang, Phys. Rev. D **103**, L071901 (2021), 2012.13499
[23] H. Sato, K. Yazaki, Physics Letters B **98**, 153 (1981)
[24] T. Matsui, H. Satz, Phys. Lett. B **178**, 416 (1986)
[25] L.M. Abreu, K.P. Khemchandani, A. Martinez Torres, F.S. Navarra, M. Nielsen, Phys. Lett. B **761**, 303 (2016), 1604.07716
[26] R. Aaij et al. (LHCb), Nucl. Phys. B **886**, 665 (2014), 1404.0275
[27] A. Andronic, P. Braun-Munzinger, M.K. Köhler, K. Redlich, J. Stachel, Phys. Lett. B **797**, 134836 (2019), 1901.09200
[28] A. Andronic, P. Braun-Munzinger, J. Stachel, Nucl. Phys. A **772**, 167 (2006)
[29] A. Andronic, P. Braun-Munzinger, K. Redlich, J. Stachel, Nature **561**, 321 (2018)
[30] S. Chatrchyan et al. (CMS), Phys. Rev. Lett. **113**, 132301 (2014), [Erratum: Phys.Rev.Lett. 115, 029903 (2015)], 1312.4198
[31] A.M. Sirunyan et al. (CMS), Phys. Rev. Lett. **119**, 152301 (2017), 1705.04727
[32] S. Chatrchyan et al. (CMS), JINST **3**, S08004 (2008)
[33] A.M. Sirunyan et al. (CMS) (2021), 2102.13048
[34] S. Chatrchyan et al. (CMS), JHEP **02**, 011 (2012), 1111.1557
[35] A.M. Sirunyan et al. (CMS), Eur. Phys. J. C **78**, 509 (2018), 1712.08959
[36] L. Kluberg, H. Satz, *Color Deconfinement and Charmonium Production in Nuclear Collisions* (2010), 0901.3831
[37] R. Aaij et al. (LHCb), Phys. Rev. Lett. **126**, 092001 (2021), 2009.06619

Temperature dependence of the properties of open heavy-flavor mesons

Glòria Montaña[1,*], *Àngels* Ramos[1], *Laura* Tolos[2,3,4,5], and *Juan M.* Torres-Rincon[6]

[1]Departament de Física Quàntica i Astrofísica and Institut de Ciències del Cosmos (ICCUB), Facultat de Física, Universitat de Barcelona, Martí i Franquès 1, 08028 Barcelona, Spain
[2]Institute of Space Sciences (ICE, CSIC), Campus UAB, Carrer de Can Magrans, 08193, Barcelona, Spain
[3]Institut d'Estudis Espacials de Catalunya (IEEC), 08034 Barcelona, Spain
[4]Faculty of Science and Technology, University of Stavanger, 4036 Stavanger, Norway
[5]Frankfurt Institute for Advanced Studies, Ruth-Moufang-Str. 1, 60438 Frankfurt am Main, Germany
[6]Institut für Theoretische Physik, Goethe Universität Frankfurt, Max-von-Laue-Str. 1, 60438 Frankfurt, Germany

Abstract. We address the modification of open heavy-flavor mesons in a hot medium of light mesons within an effective theory approach consistent with chiral and heavy-quark spin-flavor symmetries and the use of the imaginary-time formalism to introduce the non-zero temperature effects to the theory. The unitarized scattering amplitudes, the ground-state self-energies and the corresponding spectral functions are calculated self-consistently. We use the thermal ground-state spectral functions obtained with this methodology to further calculate 1) open-charm meson Euclidean correlators, and 2) off-shell transport coefficients in the hadronic phase.

Heavy quarks, produced in large abundance in the initial stage in heavy-ion collisions at LHC and RHIC, are a unique probe of the matter formed. Open heavy-flavor mesons are created at the confinement transition and they interact with the light mesons in a hot medium at vanishing baryon density. In this paper we study open heavy-flavor mesons in hot matter, while comparing our results to lattice QCD (LQCD) data and analyzing the implications for their propagation in heavy ion collisions.

1 Effective theory at finite temperature

The scattering of open heavy-flavor ($D = \{D, D_s\}$) mesons off light mesons ($\Phi = \{\pi, K, \bar{K}, \eta\}$) in a mesonic thermal bath is described in [1, 2] with an effective Lagrangian consistent with chiral and heavy-quark spin-flavor symmetries. The s-wave scattering amplitudes are unitarized by solving the Bethe-Salpeter equation in coupled channels, $T_{ij} = V_{ij} + V_{ik}G_kT_{kj}$, where T is the unitarized amplitude, V is the interaction kernel, G is the heavy-light meson-meson ($D\Phi$) two-body propagator, and the subindices i, j, k refer to a particular $D\Phi$ channel. The thermal corrections are implemented in this model by employing the imaginary-time formalism and dressing the heavy meson in the loop with its spectral function,

*e-mail: gmontana@fqa.ub.edu

$$G_{D\Phi}(E,\vec{p};T) = \int \frac{d^3q}{(2\pi)^3} \int d\omega \int d\omega' \frac{S_D(\omega,\vec{q};T) S_\Phi(\omega',\vec{p}-\vec{q};T)}{E-\omega-\omega'+i\varepsilon} [1+f_\omega+f_{\omega'}] , \quad (1)$$

where the momentum integral is regularized with a cut-off and $f_E \equiv f(E,T)$ is the Bose-Einstein distribution function. A δ-type spectral function is considered for the light meson, as the modification of the pion with temperature is rather moderate and the effect on the heavy mesons is even smaller. The spectral function of the heavy meson is given by

$$S_D(\omega,\vec{q};T) = -\frac{1}{\pi} \text{Im}\, \mathcal{D}_D(\omega,\vec{q};T) = -\frac{1}{\pi} \text{Im} \left(\frac{1}{\omega^2 - \vec{q}^{\,2} - m_D^2 - \Pi_D(\omega,\vec{q};T)} \right) , \quad (2)$$

where the self-energy is obtained from closing the light-meson line (i.e. the pion line for a pionic bath) in the corresponding T-matrix element,

$$\Pi_D(E,\vec{p};T) = \frac{1}{\pi} \int \frac{d^3q}{(2\pi)^3} \int d\Omega \frac{E}{\omega_\Phi} \frac{f(\Omega,T) - f(\omega_\Phi,T)}{E^2 - (\omega_\Phi - \Omega)^2 + i\varepsilon} \text{Im}\, T_{D\Phi}(\Omega,\vec{p}+\vec{q};T) . \quad (3)$$

Self consistency is achieved by solving iteratively Eqs. (1) to (3).

The properties of the heavy mesons and their thermal modification are contained in their spectral functions. The temperature evolution of the masses and the widths of the D and D_s mesons in a pionic bath below the deconfining transition temperature T_c is shown in Fig. 1. The results for the open-charm vector counterparts are found in [2]. The heavy-quark spin-flavor symmetry allows us to extend the model to the open-bottom sector upon the replacement $D \to \bar{B}$. The thermal effects on the masses and widths of the $\bar{B}$ and $\bar{B}_s$ mesons are also displayed in Fig. 1, being of the same magnitude of those on the open-charm states. We observe a decrease of the masses of some tens of MeV at $T = 150$ MeV, as well as a widening of the same order.

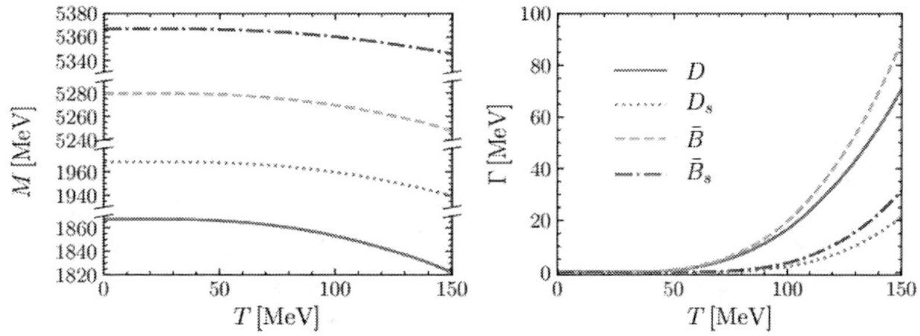

Figure 1. Thermal modification of the mass (left panel) and the width (right panel) of the open heavy-flavor ground states. (Partially taken from [2].)

The states that are dynamically generated by our model also suffer a modification at finite temperature, as can be seen by looking at the unitarized scattering amplitudes plotted in Fig. 2 for $T = 40, 100, 150$ MeV. In the sector with charm $C = 1$, spin $J = 0$ and strangeness $S = 0$ the experimental $D_0^*(2300)$ is generated with a two-pole structure at $T = 0$. These two resonances move towards lower energies and their widths increase with increasing temperatures. We note that the fitting of these structures has to be performed after background subtraction as explained in [1]. In the strange sector, where the narrow $D_{s0}^*(2317)$ is experimentally seen, the bound state found in vacuum develops a width and its mass decreases in a thermal bath. Figure 2 also shows the analogue thermal effects on the dynamically generated states in the bottom sector, where no experimental states have been reported yet.

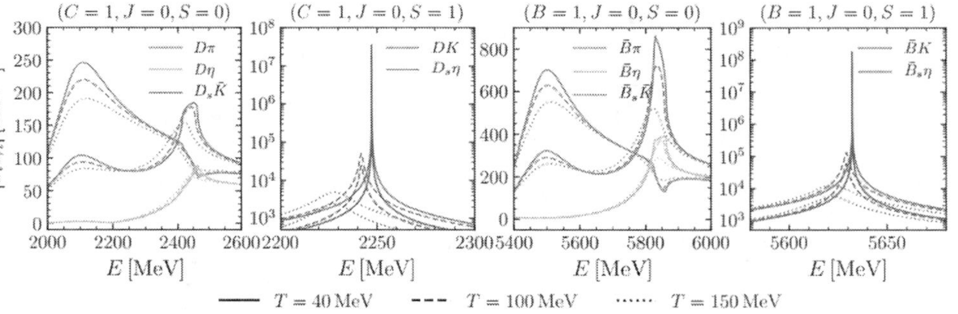

Figure 2. Diagonal elements of the T-matrix in the sectors with charm $C = 1$ (bottom $B = 1$), spin $J = 0$ and strangeness $S = 0, 1$. (Data partially taken from [1].)

2 Open-charm Euclidean correlators with EFT

Given the spectral function $\rho(\omega; T)$, the Euclidean correlator is calculated from the convolution with a temperature dependent kernel $K(\tau, \omega; T)$ as:

$$G_E(\tau; T) = \int_0^{\infty} d\omega\, K(\tau, \omega; T)\rho(\omega; T)\,, \quad K(\tau, \omega; T) = \cosh\left[\omega\left(\tau - \frac{1}{2T}\right)\right]\sinh^{-1}\left(\frac{\omega}{2T}\right). \quad (4)$$

Euclidean correlators are computed with LQCD and the inversion of Eq. (4) is required in order to extract the information about the physical states contained in the spectral function. This is an ill-posed problem and for this reason in [3] we compare the thermal EFT with LQCD results directly at the level of the Euclidean correlators. We obtain the spectral function of the ground state using Eq. (2), employing in the model the non-physical meson masses used in the LQCD calculations in [4], and we define the full spectral function

$$\rho(\omega; T) = M_D^4 S_D(\omega; T) + a\rho_{\text{cont}}(\omega; T)\,, \quad (5)$$

where the second term corresponds to the continuum of scattering states and for which we take a parametrization of the free-meson spectral function in the non-interacting limit. The parameter a weights the contribution of the continuum with respect to the ground-state.

The left panel of Fig. 3 shows the full spectral function for different values of a and temperatures below T_c, while the right panel shows the corresponding ratio of the Euclidean to the reconstructed correlators, compatible with LQCD within the errorbars for the lowest temperatures upon the inclusion of the continuum.

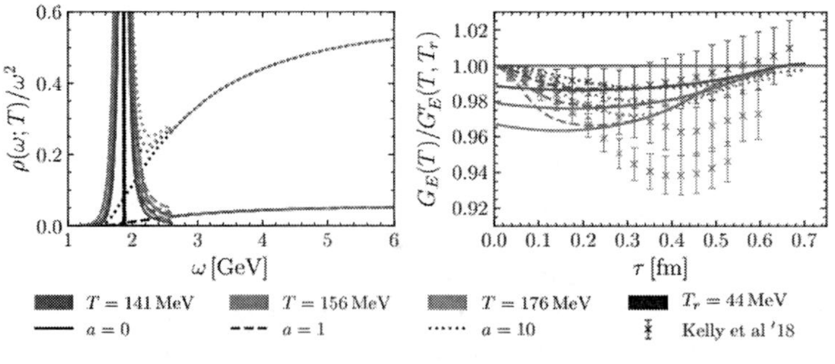

Figure 3. Full spectral functions of the D-meson (left panel), and Euclidean correlators (right panel) for various temperatures below T_c and values of the weight parameter a. (Taken from [3].)

3 Off-shell transport coefficients of D mesons

In [5] we apply the results of the EFT at non-zero temperature to the calculation of transport coefficients of an off-shell D-meson. We derive an off-shell Fokker-Planck equation for the Green's function,

$$\frac{\partial}{\partial t}G_D^<(t,k) = \frac{\partial}{\partial k^i}\left\{\hat{A}(k;T)k^i G_D^<(t,k) + \frac{\partial}{\partial k^j}\left[\hat{B}_0(k;T)\Delta^{ij} + \hat{B}_1(k;T)\frac{k^i k^j}{|\vec{k}|^2}\right]G_D^<(t,k)\right\}, \quad (6)$$

in which the transport coefficients $\hat{A}$, $\hat{B}_0$ and $\hat{B}_1$ are defined off shell, as they are calculated using spectral functions, as well as thermal scattering amplitudes. The D-meson spatial diffusion coefficient $D_s(T)$, obtained from the static limit of the $\hat{B}_0$ at the quasiparticle energy,

$$2\pi T D_s(T) = \lim_{\vec{k}\to 0}\frac{2\pi T^3}{\hat{B}_0(E_k,\vec{k};T)}, \quad (7)$$

is shown in Fig. 4 below T_c in the case of the full off-shell thermal calculation, together with the on-shell result using vacuum amplitudes. The main difference is the appearance of the so-called Landau contribution when using thermal scattering amplitudes, making it possible to smoothly match to recent results of LQCD calculations and a Bayesian analysis around T_c.

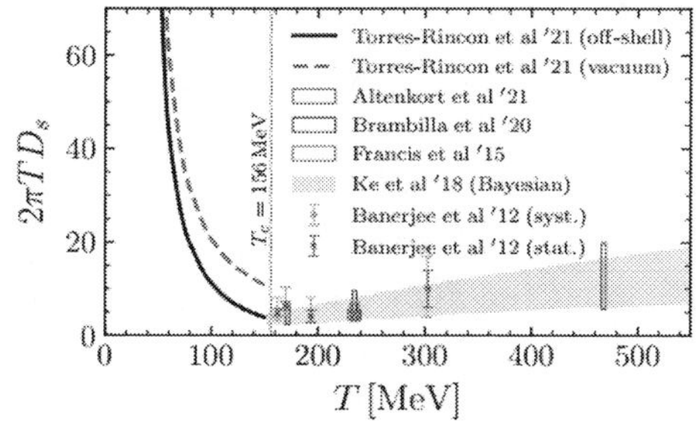

Figure 4. Spatial diffusion coefficient below and above T_c resulting from different approaches (see [5] for details). The empty boxes correspond to a range of values of $2\pi T D_s$ for the central value of T. (Taken from [5].)

Acknowledgements

G.M. acknowledges support from the FPU17/04910 Doctoral Grant from the Ministerio de Economía y Competitividad (MECD). G.M. and A.R. ackgnowledge support from the Ministerio de Economía y Competitividad (MINECO) under the project MDM-2014-0369 of ICCUB and, with additional European FEDER funds, under the contract FIS2017-87534-P. The research of L.T. has been supported by the Ministerio de Ciencia e Innovación and the European Regional Development Fund (ERDF) under contract PID2019-110165GB-I00 (Ref.10.13039/501100011033). L.T. and J.M.T.-R. acknowledge support from the Deutsche Forschungsgemeinschaft through projects no. 411563442 and no. 315477589 - TRR 211.

References

[1] G. Montaña, A. Ramos, L. Tolos, J.M. Torres-Rincon, Phys. Lett. B **806**, 135464 (2020)
[2] G. Montaña, A. Ramos, L. Tolos, J.M. Torres-Rincon, Phys. Rev. D **102**, 096020 (2020)
[3] G. Montaña, O. Kaczmarek, L. Tolos, A. Ramos, Eur. Phys. J. A **56**, 294 (2020)
[4] A. Kelly, A. Rothkopf, J.I. Skullerud, Phys. Rev. D **97**, 114509 (2018)
[5] J.M. Torres-Rincon, G. Montaña, A. Ramos, L. Tolos (2021), 2106.01156

Measurements of electrons from heavy-flavor hadron decays in 27, 54.4, and 200 GeV Au+Au collisions in STAR

Shenghui Zhang (for the STAR Collaboration)[1,2,*]

[1]University of Illinois at Chicago, Chicago, IL 60607, USA
[2]University of Science and Technology of China, Hefei 230026, China

Abstract. In these proceedings, we present the measurements of nuclear modification factors for electrons from open charm ($c \to e$) and bottom hadron decays ($b \to e$) at mid-rapidity region in Au+Au collisions at $\sqrt{s_{\rm NN}}$ = 200 GeV. The significance of the difference in nuclear modification factors between $b \to e$ and $c \to e$ is $\geq 3\sigma$. We also present the measurements of the elliptic flow (v_2) of inclusive heavy-flavor hadron decayed electrons (HFE) in Au+Au collisions at $\sqrt{s_{\rm NN}}$ = 27 and 54.4 GeV. The results show non-zero HFE v_2 in 54.4 GeV Au+Au collisions and a hint of smaller HFE v_2 in 27 GeV Au+Au collisions than those at 54.4 and 200 GeV.

1 Introduction

Heavy quarks are predominantly produced at the early stages of the relativistic heavy-ion collisions before the creation of the deconfined Quantum Chromodynamics medium, known as the Quark-Gluon Plasma (QGP). They subsequently traverse the created system throughout its evolution, and thus measurements of heavy-flavor hadron production and elliptic flow (v_2) are unique and indispensable probes of the QGP properties. Studies of the production of electrons from open charm and bottom hadron decays in Au+Au collisions serve as a valuable tool to investigate the mass hierarchy of the parton energy loss. Furthermore, measuring

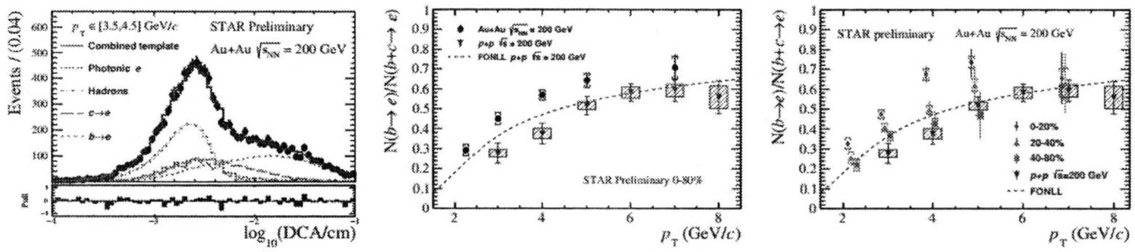

Figure 1. Left: log(DCA/cm) distribution for inclusive electrons with a template fit including $b \to e$, $c \to e$, and various background sources in 0-80% Au+Au collisions at $\sqrt{s_{\rm NN}}$ = 200 GeV. Middle: the fraction of $b \to e$ as a function of $p_{\rm T}$ in 0-80% Au+Au collisions, along with that in p+p collisions and the FONLL prediction. Right: the fraction of $b \to e$ as a function of $p_{\rm T}$ in different centralities of Au+Au collisions, along with that in p+p collisions and the FONLL prediction.

*e-mail: zhang08@mail.ustc.edu.cn

important insights into the temperature dependence of the heavy-flavor quark interactions with the QGP. In these proceedings, we present preliminary results on the production of electrons from open charm and bottom hadron decays in $\sqrt{s_{\rm NN}}$ = 200 GeV Au+Au collisions, as well as inclusive heavy-flavor electron (HFE) v_2 in $\sqrt{s_{\rm NN}}$ = 27 and 54.4 GeV Au+Au collisions.

2 Experiment and Analysis

The STAR experiment utilizes the Time Projection Chamber, the Time Of Flight detector, and the Barrel Electromagnetic Calorimeter to reconstruct charged tracks and perform particle identification. Electrons originating from charm and bottom hadron semileptonic decays are topologically separated utilizing the Heavy Flavor Tracker which provides precise identification of displaced vertices [1].

2.1 Nuclear modification factor

The log(DCA/cm) distribution, where the DCA is the 3D distance of closest approach of a candidate electron to the primary vertex, is used to separate the bottom hadron decayed electrons ($b{\to}e$) from charm hadron decayed electrons ($c{\to}e$). Such a distribution for inclusive electrons is shown in the left panel of Fig. 1, along with the template fit including $b{\to}e$, $c{\to}e$, background from photonic electrons, and the hadron contamination. The templates for $b{\to}e$ and $c{\to}e$ are obtained from a fast simulation, taking into account detector effects, coupled with the EvtGen [2] decayer, in which $D^0, D^{\pm}, D_S, \Lambda_c, B^0, B^{\pm}, B_S$, and Λ_b decays are included. The DCA distribution for $b{\to}e$ is broader than that for $c{\to}e$ because of the longer lifetime of B hadrons. The template for photonic electrons, arising from gamma conversions, π^0 and η Dalitz decays, is obtained from simulations, and constrained by the photonic electron fraction. Furthermore, hadrons misidentified as electron candidates need to be accounted for. Their template is obtained from data and the magnitude is constrained by the inclusive electron purity. The fractions of bottom-decayed electrons to the sum of bottom- and charm-decayed electrons, $f^{b\to e} = N(b{\to}e)/N(b + c{\to}e)$, as a function of transverse momentum ($p_{\rm T}$) in different centralities of 200 GeV Au+Au collisions are obtained via the template fitting and shown in the middle and right panels of Fig. 1. The fractions in Au+Au collisions are compared to the results from p+p collisions ([3] and preliminary 2012 STAR data) and the FONLL prediction [5, 6]. An enhancement of the $b{\to}e$ fraction is observed in 0-80% Au+Au collisions compared to that in p+p collisions and FONLL prediction. The enhancement is more significant in central Au+Au collisions and consistent with p+p and FONLL in peripheral collisions.

The nuclear modification factor ($R_{\rm AA}$) of bottom- and charm-decayed electrons are obtained using:

$$R_{\rm AA}^{b\to e} = \frac{f_{Au+Au}^{b\to e}}{f_{p+p}^{b\to e}} R_{\rm AA}^{\rm HFE},\ R_{\rm AA}^{c\to e} = \frac{1 - f_{Au+Au}^{b\to e}}{1 - f_{p+p}^{b\to e}} R_{\rm AA}^{\rm HFE}, \tag{1}$$

where $f^{b\to e}$ is the fraction of $b{\to}e$ in Au+Au or p+p collisions, and $R_{\rm AA}^{HFE}$ is the $R_{\rm AA}$ of the HFE defined as a ratio of HFE yield in Au+Au collisions to that in p+p collisions normalized by the average number of binary nucleon-nucleon collisions. The left panel of Fig. 2 shows $R_{\rm AA}^{b\to e}$, $R_{\rm AA}^{c\to e}$ and the ratio of $R_{\rm AA}^{b\to e}$ to $R_{\rm AA}^{c\to e}$ in 0-80% Au+Au collisions, along with a comparison to the DUKE transport model calculations that include the mass dependence of the parton energy loss [4]. From a constant fit to the ratio, the $R_{\rm AA}^{b\to e}$ is about 1.92 times larger than $R_{\rm AA}^{c\to e}$ (2.8σ significance), which is qualitatively described by the DUKE model calculation and consistent with the mass hierarchy of parton energy loss. A null hypothesis of no mass dependence for the $R_{AA}^{b\to e}$ to $R_{AA}^{c\to e}$ ratio (blue shaded curve) is constructed by using the D meson R_{AA} [7] to the $c/b\to e$ simulation, which takes into account the different decay kinematics. The p-value of the data to this curve is found to be 0.014, disfavoring the hypothesis of identical charm and bottom hadron R_{AA}.

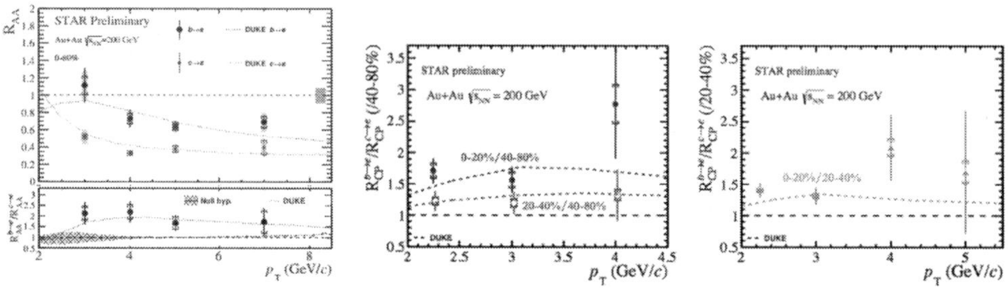

Figure 2. Left: $R_{AA}^{b\to e}$ and $R_{AA}^{c\to e}$ (top) and the $R_{AA}^{b\to e}/R_{AA}^{c\to e}$ ratio (bottom) as a function of p_T in 0-80% Au+Au collisions at $\sqrt{s_{NN}}$ = 200 GeV. The blue shaded curve represents the null hypothesis. Middle: the double ratios of $R_{CP}^{b\to e}$ to $R_{CP}^{c\to e}$ (0-20%/40-80% and 20-40%/40-80%) as a function of p_T. Right: the double ratio of $R_{CP}^{b\to e}$ to $R_{CP}^{c\to e}$ (0-20%/20-40%) as a function of p_T. Results are compared to the Duke model calculations.

The double ratio of $R_{CP}^{b\to e}$ to $R_{CP}^{c\to e}$ is obtained as follows:

$$\frac{R_{CP}^{b\to e}}{R_{CP}^{c\to e}} = \frac{f_{central}^{b\to e}}{1-f_{central}^{b\to e}}\frac{1-f_{peripheral}^{b\to e}}{f_{peripheral}^{b\to e}}. \tag{2}$$

In the middle and right panels of Fig. 2, the obtained double ratios of $R_{CP}^{b\to e}$ to $R_{CP}^{c\to e}$ are shown for 0-20%/40-80%, 20-40%/40-80% and 0-20%/20-40%, respectively. A constant fit to the double ratios is performed and the R_{CP} ratio values are found to be 1.68±0.15(stat.)±0.12(syst.) and 1.38±0.08(stat.)±0.03(syst.) for 0-20%/40-80% and 0-20%/20-40%, which are significantly different from unity at a 3.5σ and 4.4σ level, respectively.

2.2 Elliptic flow

The inclusive HFE v_2 is measured in $\sqrt{s_{NN}}$ = 27 and 54.4 GeV 0-60% Au+Au collisions in a similar way to STAR published measurements at $\sqrt{s_{NN}}$ = 200, 62.4, and 39 GeV [8]. The

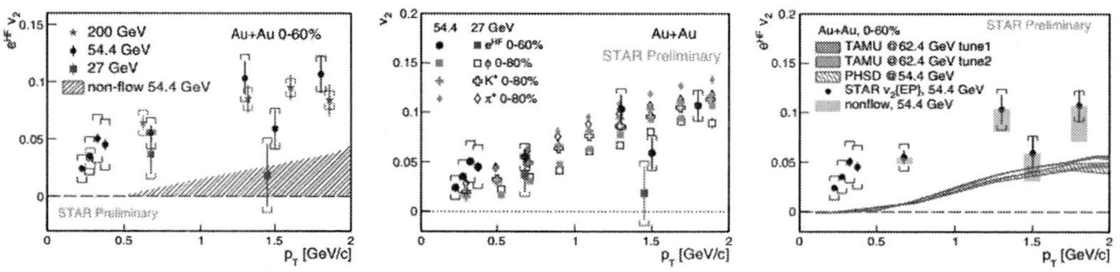

Figure 3. Left: HFE v_2 as a function of p_T in 0–60% centrality of 27 and 54.4 GeV Au+Au collisions compared with that from 200 GeV Au+Au collisions. Middle: HFE v_2 as a function of p_T compared to those of light flavor mesons in 27 and 54.4 GeV Au+Au collisions. Right: HFE v_2 as a function of p_T, compared to TAMU and PHSD calculations in 0–60% centrality of 54.4 GeV Au+Au collisions.

results are shown in left and middle panels of Fig. 3, together with a comparison to the 200 GeV Au+Au results of HFE v_2 and to those of light flavor mesons in 27 and 54.4 GeV Au+Au collisions, respectively. A non-zero v_2 is observed for HFE in 54.4 GeV Au+Au collisions, and is comparable to that at 200 GeV as well as the v_2 of light flavor mesons in 54.4 GeV Au+Au collisions, which indicates strong charm quark interactions with the medium at 54.4 GeV. There is a hint of smaller HFE v_2 in 27 GeV Au+Au collisions compared to those at 54.4 and 200 GeV and those of light flavor mesons in 27 GeV Au+Au collisions. In the right panel of Fig. 3, the HFE v_2 in 54.4 GeV Au+Au collisions is compared to the TAMU [9] and PHSD [10, 11] calculations, in which heavy quarks interact with the strongly coupled medium elastically without gluon radiation process. The TAMU and PHSD predictions are lower than the measured v_2 at low p_T and comparable to data at high p_T considering the non-flow contribution and uncertainties.

3 Summary

In these proceedings, we present the STAR measurements of R_{AA} and the double ratio of R_{CP} for open charm and bottom hadron decayed electrons in Au+Au collisions at $\sqrt{s_{NN}}$ = 200 GeV. The results indicate less suppression for electrons from bottom hadron decays compared to those from charm hadron decays with a significance larger than 3σ. They are consistent with the mass hierarchy of parton energy loss and the DUKE model calculations. We also present the measurements of HFE v_2 in Au+Au collisions at $\sqrt{s_{NN}}$ = 27 and 54.4 GeV. HFE v_2 in 54.4 GeV Au+Au collisions is non-zero and is comparable to the HFE v_2 at $\sqrt{s_{NN}}$ = 200 GeV and v_2 of light flavor mesons in 54.4 GeV Au+Au collisions, indicating that charm quarks interact strongly with the QGP medium in $\sqrt{s_{NN}}$ = 54.4 GeV Au+Au collisions. Model calculations are systematically lower than the measured HFE v_2 at low p_T and comparable to data at high p_T. There is a hint of smaller HFE v_2 in 27 GeV Au+Au collisions than those at 54.4 and 200 GeV.

Acknowledgments

This work was supported in part by Project funded by China Postdoctoral Science Foundation under Grant No. 2019M652177, National Natural Science Foundation of China with Grant No. 11890712 and 12061141008, National Key Research and Development Program of China with Grant No. 2018YFE0205200 and 2018YFE0104700, the Strategic Priority Research Program of CAS with Grant No. XDB34030000.

References

[1] G. Contin et al., Nucl. Instrum. Meth. A **907**, 60 (2018)
[2] D. J. Lange, Nucl. Instrum. Methods Phys. Res. Sect. A Accel. Spectrometers Detect. Assoc. Equip. **462**, 152–155 (2001)
[3] M. M. Aggarwal et al., Phys. Rev. Lett. **105**, 202301 (2010)
[4] S. Cao et al., Phys. Rev. C **92**, 024907 (2015)
[5] M. Cacciari et al., JHEP **1210**, 137 (2012)
[6] M. Cacciari et al., arXiv:1507.06197
[7] J. Adam et al., Phys. Rev. C **99**, 034908 (2019)
[8] L. Adamczyk et al., Phys. Rev. C **95**, 034907 (2017)
[9] M. He et al., Phys. Rev. C **91**, 024904 (2015)
[10] T. Song et al., Phys. Rev. C **92**, 014910 (2015)
[11] T. Song et al., Phys. Rev. C **96**, 014905 (2017)

Open charm and beauty measurements from small to large systems

Jonghan Park[1,*] on behalf of the ALICE Collaboration

[1]Department of Physics, Inha University

Abstract. Heavy-flavor quarks (i.e. charm and beauty) are essential probes to investigate the properties of the quark-gluon plasma (QGP) created in ultra-relativistic heavy-ion collisions. The measurement of the nuclear modification factor (R_{AA}) gives insight into the in-medium parton energy loss in heavy-ion collisions. Furthermore, the measurements of heavy-flavor elliptic flow (v_2) provide crucial information about the degree of thermalization of heavy quarks in the QGP, the path-length dependence of heavy-quark in-medium energy loss, and possible recombination effects. The higher flow harmonics, such as the triangular flow (v_3), provide further constraints on the effect of fluctuations in the initial state of the system. The measurements in proton-proton (pp) collisions allow for testing perturbative quantum chromodynamics (pQCD) calculations and are needed as a baseline for investigating the medium effects in heavy-ion collisions. In this contribution, recent results of open charm and beauty production measured with the ALICE detector are discussed.

1 Introduction

The ALICE apparatus was designed to study the quark-gluon plasma (QGP), a state of matter in which colored-partons are deconfined. Due to their large mass, heavy quarks are mainly produced in initial hard-scattering processes, which occur before the QGP formation [1]. They interact with the medium during the full evolution of the system. Therefore, they are effective probes to investigate the properties of the QGP.

While traversing the medium, heavy quarks interact with the medium constituents and lose their energy via collisional [2] and radiative processes [3]. The medium effects in heavy-ion collisions are studied measuring the nuclear modification factor (R_{AA}). The R_{AA} is defined as the ratio between the p_T-differential yield in nucleus-nucleus collisions ($d^2N_{AA}/d\eta dp_T$) and the p_T-differential production cross section in pp collisions ($d^2\sigma_{pp}/d\eta dp_T$) scaled by the average nuclear thickness function $\langle T_{AA}\rangle$, which is a quantity proportional to the average number of binary nucleon-nucleon collisions. The measurement of heavy quark production in proton-proton (pp) collisions allows for testing pQCD calculations and provides the reference for measuring the R_{AA}.

Open charm and beauty hadron production is measured with the ALICE detector via the reconstruction of hadronic and semi-leptonic decay channels at midrapidity. The D mesons are measured via the $D^0 \to \pi^+K^-$, $D^+ \to \pi^+\pi^+K^-$, $D^{*+} \to D^0\pi^+$, $D_s^+ \to \pi^+\phi \to \pi^+K^+K^-$ decay channels, and their charge conjugate. The D meson yields are extracted from an invariant

*e-mail: jonghan@cern.ch

mass analysis after applying topological and particle identification selections to enhance the signal-to-background ratio. Non-prompt D mesons are used to measure open-beauty hadron production. Exploiting machine-learning techniques based on Boosted Decision Trees, non-prompt D mesons are separated from the prompt D mesons [4]. Open heavy-flavor production is also measured by identifying the electrons from semi-electronic decays of charm and beauty hadrons. The production yield is extracted by subtracting the contamination of hadrons and of non-heavy flavor decay electrons from the inclusive sample. The production of beauty-hadron decay electrons is measured by determining statistically the fraction of beauty-decay electrons in the inclusive electron sample with an impact-parameter template fit method [5, 6].

2 Results

The production of electrons from both inclusive heavy-flavor hadron decays [7] and beauty hadron decays was measured in pp collisions at $\sqrt{s}$ = 5.02 TeV at midrapidity. The fraction of the beauty hadron decay electrons to the one from inclusive heavy-flavor hadron decays in the transverse momentum (p_{T}) region $2 < p_{\mathrm{T}} < 8$ GeV/c is shown in Fig. 1 (left). The result is described by the FONLL pQCD calculation [8–10] within the uncertainties. The measured values and the model prediction vary with p_{T}, indicating that the charm contribution is dominant at low p_{T} and the beauty contribution is increasing with p_{T}. The prompt and non-prompt D^0, D^+ and D_s^+ meson cross sections were measured in pp collisions at $\sqrt{s}$ = 5.02 TeV [4]. From the results of non-prompt D mesons, the total $\mathrm{b\bar{b}}$ production cross section was computed. Figure 1 (right) shows $\mathrm{d}\sigma_{\mathrm{b\bar{b}}}/\mathrm{d}y_{|y|=0}$ as a function of center-of-mass energy. The measurements are found to be compatible with FONLL [8–10] and NNLO [11] calculations. This result provides an important test for pQCD calculations in the beauty sector at the LHC.

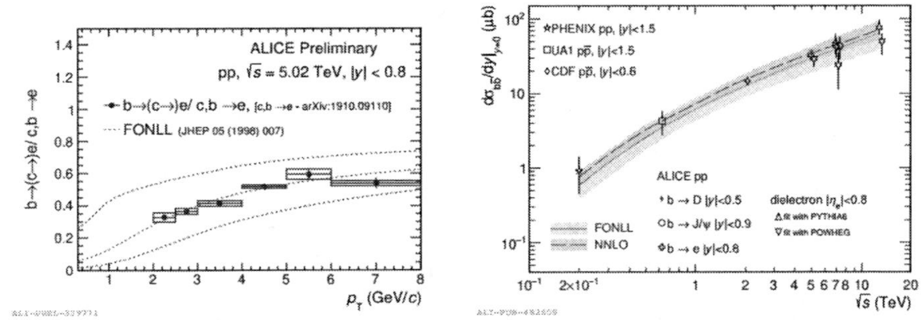

Figure 1. Left: Fraction of beauty-hadron decay electrons to heavy-flavor hadron decay electrons in pp collisions at $\sqrt{s}$=5.02 TeV. Right: Beauty production cross section per unit of rapidity at midrapidity as a function of $\sqrt{s}$ measured in pp and $\mathrm{p\bar{p}}$ collisions by different experiments [4].

The production of electrons from inclusive heavy-flavor hadron decays [7] and beauty hadron decays is also measured in Pb–Pb collisions at $\sqrt{s_{\mathrm{NN}}}$ = 5.02 TeV, as well as the production of prompt and non-prompt D^0 mesons. Figure 2 (left) shows the R_{AA} comparison between inclusive heavy-flavor hadron decays and beauty hadron decays measured in 0–10% central Pb-Pb collisions. Both R_{AA} values are smaller than unity, suggesting that the heavy-flavor production is suppressed in Pb–Pb collisions with respect to pp collisions in the intermediate p_{T} region. The R_{AA} of beauty decay electrons tends to be higher than that of inclusive heavy-flavor hadron decay electrons at low p_{T}, though they are compatible within uncertainties. At high p_T, they have very similar values. The observed trend is consistent with the expectation that the suppression of beauty hadrons is smaller than charm. Figure 2 (right) shows the R_{AA} ratio of non-prompt D^0 over prompt D^0 in the 0–10% centrality Pb–Pb collisions at $\sqrt{s_{NN}}$ = 5.02 TeV compared to several model predictions (MC@sHQ+EPOS2 [12], LGR [13, 14], TAMU [15], and CUJET3.1 [16]). The measured ratio, above unity, suggests that non-prompt D^0 mesons are less suppressed than prompt D^0 ones. This is a hint of the mass dependence of the in-medium parton energy loss. At very low p_T, the ratio decreases up to 3 GeV/c and increases from 3 GeV/c to 7 GeV/c. This behavior is mainly due to the formation of prompt D mesons via charm quark coalescence. Above 7 GeV/c, the ratio does not vary significantly with p_T within sum of statistical and systematic uncertainties.

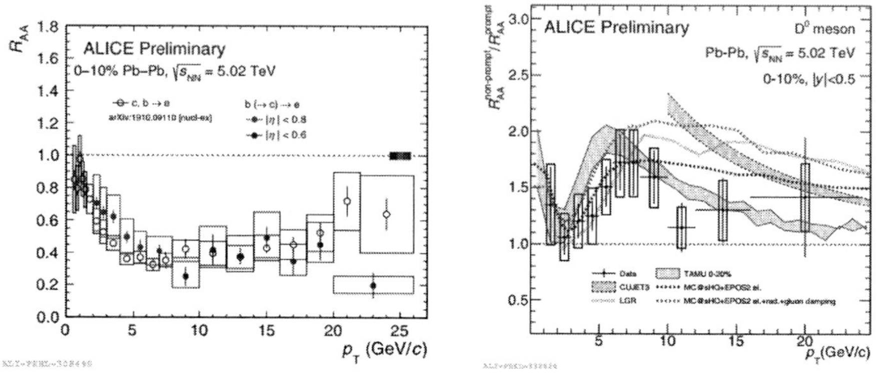

Figure 2. Left: Comparison of the R_{AA} of electrons from heavy-flavor hadron and beauty-hadron decays in Pb–Pb collisions at $\sqrt{s_{NN}}$ = 5.02 TeV. Right: Ratio of non-prompt D^0 to prompt D^0 R_{AA} in Pb–Pb collisions at $\sqrt{s_{NN}}$ = 5.02 TeV, compared to model predictions.

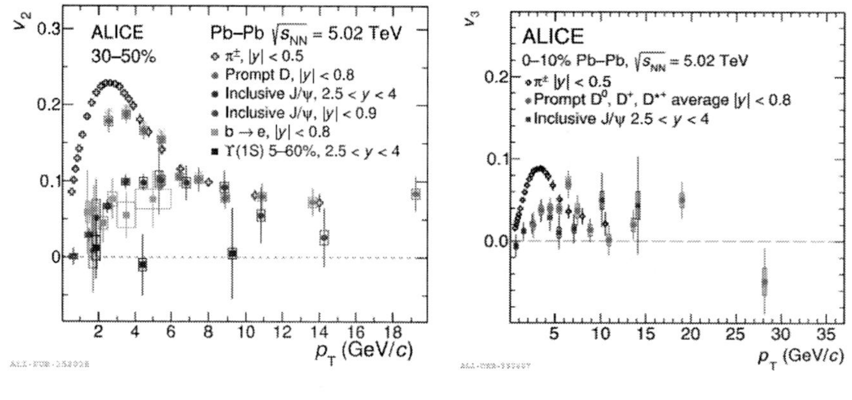

Figure 3. Anisotropic flow coefficient measurements as a function of p_T in Pb–Pb collisions at $\sqrt{s_{NN}}$ = 5.02 TeV. Left: v_2 in 30–50% central collisions. Right: v_3 in 0–10% central collisions. [17]

The azimuthal anisotropy of particles provides further insight into the interaction of heavy quarks in the QGP. It can be quantified by the Fourier coefficients of the particle azimuthal distribution defined as $v_n = \langle \cos(n(\varphi - \Psi_n)) \rangle$, where φ is particle azimuthal angle and Ψ_n is n^{th}-order asymmetry plane formed by impact parameter vector and the beam axis using V0 detector. The elliptic flow v_2, which is the dominant coefficient in semi-central collisions, is sensitive to the degree of thermalization of the heavy quarks at low p_T, while the path-length dependence of the in-medium parton energy loss becomes the main origin of a positive v_2 at high p_T. The triangular flow v_3 originates from event-by-event fluctuations in initial collision geometry. Figure 3 shows the v_2 (left) and v_3 (right) for several particle species. For v_2, the mass ordering hierarchy is seen for $p_T < 3\,\text{GeV}/c$ while the v_2 of prompt D and $\pi^\pm$ mesons are similar for $3 < p_T < 6\,\text{GeV}/c$. This supports the hypothesis that charm quarks hadronize via coalescence with light quarks from the medium. For $p_T > 6\,\text{GeV}/c$, all v_2 values converge to a similar value, except for $\Upsilon(1S)$. This suggests a similar path-length dependence of energy loss for different partons. A positive v_3 for non-strange D mesons is observed, as shown in Fig. 3 (right) indicating that the charm quarks are sensitive to the initial state event-by-event fluctuation of the created medium.

3 Summary and Outlook

In this contribution, recent results on open heavy-flavor production in pp and Pb–Pb collisions from ALICE Collaboration were presented. All the measurements are conducted utilizing Run 2 data. The results show that the pQCD calculations reasonably describe the open heavy-flavor data in pp collisions. The R_{AA} in central Pb–Pb collisions is lower than unity over a wide p_T interval, indicating that open heavy-flavor production is suppressed. The comparison between the R_{AA} of prompt and non-prompt D mesons as well as that between beauty-hadron decay electrons and the inclusive heavy-flavor hadron decay electrons suggest that the in-medium parton energy loss depends on the parton mass. The positive anisotropic flow indicates the participation of heavy quarks to the collective motion of the medium.

References

[1] Liu, Fu-Ming and Liu, Sheng-Xu, Phys. Rev. C **89**, 034906 (2014)
[2] Thoma, Markus H. and Gyulassy, Miklos, Nucl. Phys. B **351**, 491–506 (1991)
[3] Baier, R. and Dokshitzer, Yuri L. and Mueller, Alfred H. and Peigne, S. and Schiff, D., Nucl. Phys. B **484**, 265–282 (1997)
[4] S. Acharya et al. (ALICE Collaboration), JHEP **05**, 220 (2021)
[5] S. Acharya et al. (ALICE Collaboration), Phys. Rev. Lett. **126**, 162001 (2021)
[6] Barlow, Roger J. and Beeston, Christine, Comput. Phys. Commun. **77**, 219–228 (1993)
[7] S. Acharya et al. (ALICE Collaboration), Phys. Lett. B **804**, 135377 (2020)
[8] Cacciari, Matteo and Greco, Mario and Nason, Paolo, JHEP **05**, 007 (1998)
[9] Cacciari, Matteo and Frixione, Stefano and Nason, Paolo, JHEP **03**, 006 (2001)
[10] Cacciari, Matteo and Frixione, Stefano and Houdeau, Nicolas and Mangano, Michelangelo L. and Nason, Paolo and Ridolfi, Giovanni, JHEP **10**, 137 (2012)
[11] Catani, Stefano and Devoto, Simone and Grazzini, Massimiliano and Kallweit, Stefan and Mazzitelli, Javier, JHEP **03**, 029 (2021)
[12] Nahrgang, Marlene and Aichelin, Joerg and Gossiaux, Pol Bernard and Werner, Klaus, Phys. Rev. C **89**, 014905 (2014)
[13] Li, Shuang and Liao, Jinfeng, Eur. Phys. J. C **80**, 671 (2020)
[14] Li, Shuang and Xiong, Wei and Wan, Renzhuo, Eur. Phys. J. C **80**, 1113 (2020)
[15] He, Min and Fries, Rainer J. and Rapp, Ralf, Phys. Lett. B **735**, 445–450 (2014)
[16] Shi, Shuzhe and Liao, Jinfeng and Gyulassy, Miklos, Chin. Phys. C **43**, 044101 (2019)
[17] S. Acharya et al. (ALICE Collaboration), Phys. Lett. B **813**, 136054 (2021)

First measurement of the B_c^+ meson nuclear modification factor in PbPb collisions with CMS

Guillaume Falmagne[1,*], on behalf of the CMS Collaboration

[1]Laboratoire Leprince-Ringuet, CNRS, École Polytechnique, Institut Polytechnique de Paris (91120 Palaiseau, France)

Abstract. The B_c^+ meson is observed in lead-lead and proton-proton collisions at a center-of-mass energy per nucleon pair of $\sqrt{s_{NN}} = 5.02$ TeV, via the $B_c^+ \to (J/\psi \to \mu^+\mu^-)\mu^+\nu_\mu$ decay and using 2017 and 2018 data from the CMS detector. The resulting B_c^+ nuclear modification factor is measured in two bins of the trimuon transverse momentum and of the collision centrality. Less suppression is observed than for other quarkonia and most open heavy flavour mesons. This first observation of the B_c^+ meson in heavy ion collisions will shed light on the interplay of suppression and enhancement mechanisms in the production of heavy-flavor mesons in the quark-gluon plasma.

The extreme conditions of high-energy heavy-ion collisions allow quarks and gluons to move freely in a so-called quark-gluon plasma (QGP). The disappearance of quarkonia (as J/ψ or ϒ mesons) due to the screening of the heavy-quark potential at high temperature has been considered a signature of the QGP formation [1]. While the strong J/ψ suppression observed in heavy-ion collisions at lower energies [2, 3] seems consistent with screening effects, the smaller suppression reported at the Large Hadron Collider [4] points to the recombination of charm quarks in the QGP [5]. In the beauty sector, the suppression of ϒ(nS) states is sequential as expected from Debye screening [6], showing no evidence for recombination.

As the B_c^+ meson contains a charm and a beauty quark, measurements of its properties can bridge the gap between charmonia and bottomonia. The B_c^+ melting in the QGP is indeed expected to occur between the J/ψ and ϒ(1S) dissociation temperatures [7]. Heavy-quark recombination processes could also affect the B_c^+ state, with production rates typically scaling linearly with the charm multiplicity in the QGP (contrarily to a quadratic scaling for J/ψ mesons). It could lead to a large enhancement of the B_c^+ production in heavy ion collisions [7, 8] due to the rare B_c^+ production in pp collisions. This effect would mostly manifest at low transverse momentum (p_T) [7]. At large p_T, B_c^+ mesons are produced predominantly via heavy-quark fragmentation, hence probing the energy loss of a color triplet charge in the QGP (as observed for other B mesons [9, 10], J/ψ from B decays [11], and D mesons [12, 13]), whereas J/ψ mesons are mostly color octet states. Comparing the quenching of the B_c^+ meson with that of other heavy flavor mesons could therefore probe both the mass dependence of energy loss (from the dead-cone effect [14]) and its color charge dependence [15].

In this proceeding, the first observation of B_c^+ mesons in heavy ion collisions is reported. Charge-conjugate states are implicitly included in the following. We use data collected with the CMS detector in pp and PbPb collisions at $\sqrt{s_{NN}} = 5.02$ TeV, corresponding to luminosities of 302 pb^{-1} and 1.61 nb^{-1}, respectively. The B_c^+ meson is partially reconstructed from the three muons in the $B_c^+ \to (J/\psi \to \mu^+\mu^-)\mu^+\nu_\mu$ decay mode.

The results are presented in two regions of the trimuon kinematic phase space: a low-$p_T^{\mu\mu\mu}$ bin, $6 < p_T^{\mu\mu\mu} < 11$ GeV with rapidity $1.3 < |y^{\mu\mu\mu}| < 2.3$, and a high-$p_T^{\mu\mu\mu}$ bin, $11 < p_T^{\mu\mu\mu} <$

*e-mail: guillaume.falmagne@cern.ch

35 GeV with $|y^{\mu\mu\mu}| < 2.3$. The PbPb results are shown for the 0–90% centrality range, and in the 0–20 and 20–90% ranges when integrated over the two studied kinematic regions.

A detailed description of the CMS detector can be found in Ref. [16]. Events of interest are selected using a two-tiered trigger system. The first level is composed of custom hardware processors. The high-level trigger runs a fast version of the event reconstruction software. This analysis uses events selected by double-muon triggers, with the same quality criteria for PbPb collisions as in Ref. [17].

Monte Carlo (MC) simulation is used for background studies and for estimating the acceptance and efficiency of the reconstruction and selection. B_c^+ mesons are generated with BCVEGPY2.2 [18] and decayed with EVTGEN1.3. The underlying event as well as prompt and nonprompt J/ψ background samples are generated with PYTHIA8.212, tune CP5. PbPb generated events are embedded into simulated collisions generated with HYDJET1.8. All samples are passed to GEANT4 to simulate the detector response.

The $B_c^+ \to (J/\psi \to \mu^+\mu^-)\mu^+\nu_\mu$ signal decay features three muons originating from the same displaced vertex, an opposite-sign muon pair consistent with the J/ψ mass, and a trimuon invariant mass $m^{\mu\mu\mu}$ between $m_{J/\psi} + m_\mu \simeq 3.2$ GeV and $m_{B_c^+} \simeq 6.3$ GeV. Three background sources can mimic this topology. Events with a fake J/ψ, where no muon pair comes from a J/ψ decay, are estimated by interpolating the data distributions between the lower and higher dimuon mass sidebands. B decays to a true J/ψ meson and a muon (usually a misidentified hadron) are estimated with simulation, constrained by previous measurements [11]. Its normalization is left unconstrained. The rotated J/ψ sample describes combinations of a true J/ψ with a muon (usually an uncorrelated misidentified hadron) from another decay. It is estimated by rotating the data J/ψ candidates around the collision vertex before associating them with third muon candidates. In PbPb collisions, the distributions from various rotation angles are identical and averaged. In pp collisions, significant $J/\psi - \mu$ correlations are considered by allowing the rotation angles to vary in the fit.

Offline selection criteria include the event-level and single-muon identification criteria defined in Refs. [17, 19], and loose single-muon kinematic acceptance criteria. One of the opposite-sign muon pairs must have an invariant mass in the J/ψ peak region, or in the sidebands. The sidebands and peak regions are taken to be of equal width. In case two opposite-sign dimuons are in those mass regions, two corresponding trimuon candidates are kept, weighted by the probability of being a true J/ψ. Requirements are also set on five variables dealing with the quality and the displacement of the trimuon vertex, the impact of the unreconstructed neutrino on the decay topology, and the collimation of the three muons.

A Boosted Decision Tree (BDT) is then trained on the selected simulated signal and above-mentioned backgrounds, combining the discriminating power of the selection variables and of three more topological ones. For each analysis bin, three BDT intervals are set to contain fixed fractions of the expected signal. The trimuon mass templates of signal and of the three backgrounds are fitted to pp or PbPb data, simultaneously in the three BDT intervals, and in the two analysis bins ($p_T^{\mu\mu\mu}$ or centrality).

The results of the fits integrated over the two kinematic bins are shown in Figures 1 and 2. The normalizations of the fake J/ψ sample, and the PbPb rotated J/ψ sample, are data-derived. Data yields and the signal-free high-mass region help determine the other normalisations.

The fitted signal yields are corrected for the acceptance and efficiency of the reconstruction and selection, calculated with the signal MC in each analysis bin. The simulated single-muon efficiencies are corrected with a tag-and-probe method [11]. The acceptance and efficiency are evaluated iteratively by first fitting the $p_T^{\mu\mu\mu}$ spectrum measured using the original MC. This allows to correct the $p_T^{\mu\mu\mu}$ spectrum of the MC before a second run of the analysis. A last $p_T^{\mu\mu\mu}$ spectrum correction is then performed, notably improving the acceptance and efficiency estimation. The corrected yields are then divided by the (pp-equivalent) luminosities.

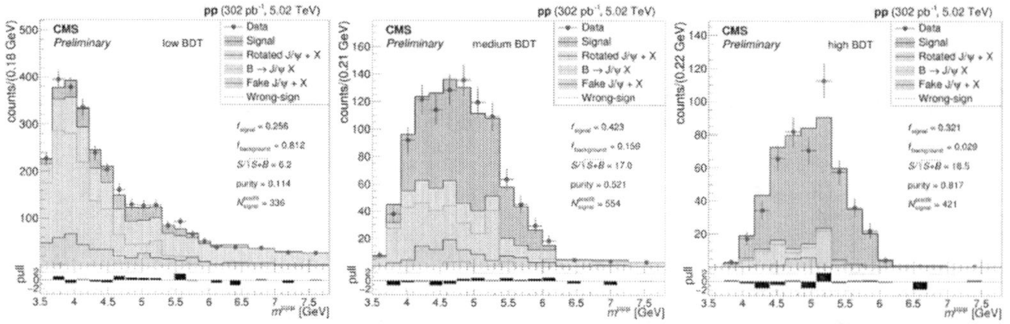

Figure 1. Template fit of the trimuon mass distributions in the three BDT bins, for the pp data sample integrated over the kinematic bins [20]. The pull between data and the fitted distributions is also shown.

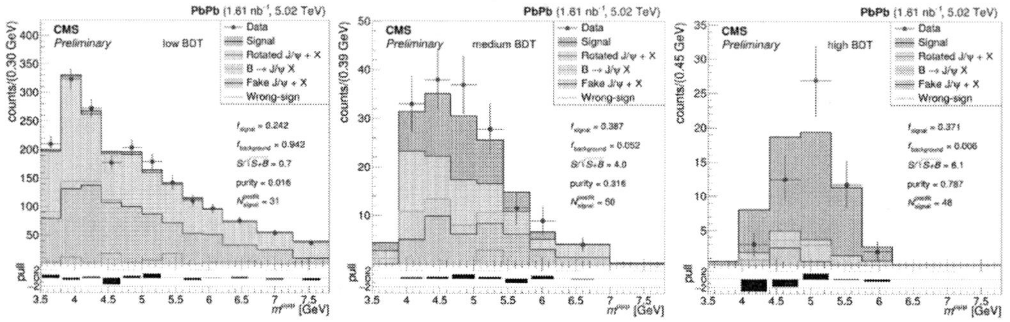

Figure 2. Template fit of the trimuon mass distributions in the three BDT bins, for the PbPb data sample integrated over the kinematic bins [20]. The pull between data and the fitted distributions is also shown.

Uncertainties arise from these sources: statistical, background shapes and normalizations, fit method, muon efficiency, B_c^+ kinematic properties, other B_c^+ decays, and overall normalization. The fit uncertainties, dominant in the PbPb $p_T^{\mu\mu\mu}$ bins, are both purely statistical and arising from background shapes and limited statistics; the latter are considered with nuisance parameters allowing variations of the background mass templates. The fit method is also varied, changing for example the $m^{\mu\mu\mu}$ or BDT bin limits or decorrelating the BDT variable from $m^{\mu\mu\mu}$. The uncertainties from the tag-and-probe, the luminosities, and the contamination from other B_c^+ decays as $B_c^+ \to J/\psi\tau^+\nu_\tau$, are sub-dominant. Since the B_c^+ kinematic spectra are not precisely known, acceptance and efficiency corrections are recalculated with toy p_T spectra obtained by fitting the measured p_T-differential yields varied within the above-mentioned uncertainties. The resulting uncertainty is the RMS of the varied acceptance and efficiency corrections or of the associated corrected yields; it dominates in the PbPb centrality bins.

The significance of the B_c^+ signal in PbPb collisions exceeds 5 standard deviations. Figure 3 shows the B_c^+ nuclear modification factor R_{PbPb}, as a function of $p_T^{\mu\mu\mu}$ (compared to open heavy flavour results from CMS) and of centrality. The latter is compared to unpublished predictions from the authors of Ref. [21] and based on similar transport model calculations on the B_c system, including correlated and uncorrelated recombination. The B_c^+ suppression is significantly lesser than for other heavy mesons [4, 6, 9, 11–13], except the B_s^0 meson [10], which may indicate that recombination is a significant B_c^+ production mechanism. The R_{PbPb} in the low-$p_T^{\mu\mu\mu}$ region stands above unity by 1 standard deviation, and above the high-$p_T^{\mu\mu\mu}$ region by 1.6 standard deviations, consistent with a softening of the p_T spectrum in the QGP. No centrality dependence is observed, contrarily to what the shown prediction suggests. The latter shows much more suppression than our result, possibly due to underestimating the recombination of B_c^+ excited states.

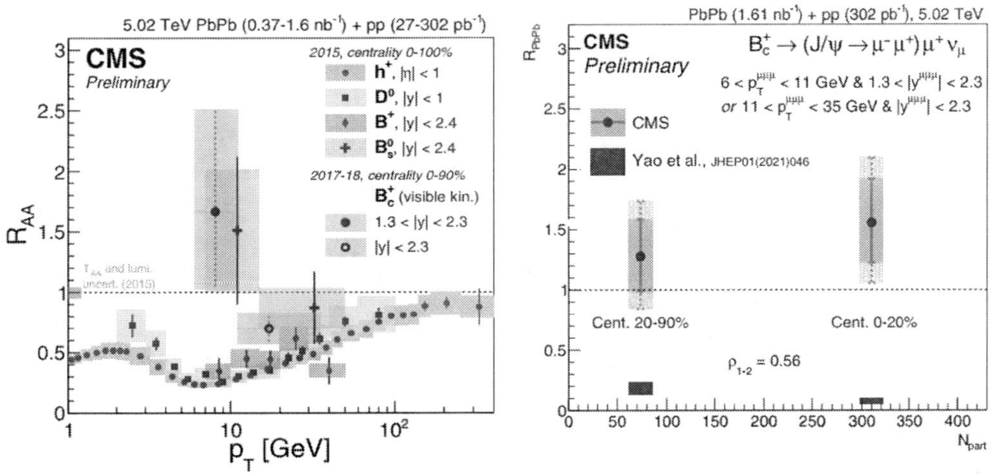

Figure 3. B_c^+ meson nuclear modification factor in $p_T^{\mu\mu\mu}$ bins corresponding to different rapidity ranges, from Ref. [20], compared to heavy-flavour measurements with CMS (*left*), and in centrality bins integrated over the studied kinematic range, compared to predictions from Yao et al. [21] (*right*). The solid and lighter rectangles, respectively, show the bin-to-bin-uncorrelated and total uncertainties.

In summary, the first measurement of the B_c^+ meson nuclear modification factor is presented, in PbPb collisions, in two bins of p_T or of centrality. This hybrid beauty-charm state can help discriminating the enhancement (possibly dominant in central events at low-p_T) and suppression (dominant at high-p_T) mechanisms in the dynamics of heavy quarks in the QGP.

References

[1] T. Matsui, H. Satz, Phys. Lett. B **178**, 416 (1986)
[2] M.C. Abreu et al. (NA50), Phys. Lett. B **477**, 28 (2000)
[3] A. Adare et al. (PHENIX), Phys. Rev. Lett. **98**, 232301 (2007), nucl-ex/0611020
[4] B.B. Abelev et al. (ALICE), Phys. Lett. B **734**, 314 (2014), 1311.0214
[5] P. Braun-Munzinger, J. Stachel, Phys. Lett. B **490**, 196 (2000), nucl-th/0007059
[6] S. Chatrchyan et al. (CMS), Phys. Rev. Lett. **109**, 222301 (2012), [Erratum: Phys.Rev.Lett. 120 (2018)199903], 1208.2826
[7] Y. Liu, C. Greiner, A. Kostyuk, Phys. Rev. C **87**, 014910 (2013), 1207.2366
[8] M. Schroedter, R.L. Thews, J. Rafelski, PRC **62**, 024905 (2000), hep-ph/0004041
[9] A.M. Sirunyan et al. (CMS), Phys. Rev. Lett. **119**, 152301 (2017), 1705.04727
[10] A.M. Sirunyan et al. (CMS), Phys. Lett. B **796**, 168 (2019), 1810.03022
[11] A.M. Sirunyan et al. (CMS), Eur. Phys. J. C **78**, 509 (2018), 1712.08959
[12] A.M. Sirunyan et al. (CMS), Phys. Lett. B **782**, 474 (2018), 1708.04962
[13] S. Acharya et al. (ALICE), JHEP **10**, 174 (2018), 1804.09083
[14] Y.L. Dokshitzer, D.E. Kharzeev, Phys. Lett. B **519**, 199 (2001), hep-ph/0106202
[15] F. Arleo, Phys. Rev. Lett. **119**, 062302 (2017), 1703.10852
[16] S. Chatrchyan et al. (CMS), JINST **3**, S08004 (2008)
[17] A.M. Sirunyan et al. (CMS), Phys. Lett. B **819**, 136385 (2021), 2006.07707
[18] C.H. Chang, et al., Comput. Phys. Commun. **174**, 241 (2006), hep-ph/0504017
[19] A. Tumasyan et al. (CMS) (2021), 2106.13235
[20] A.M. Sirunyan et al. (CMS), CMS-PAS-HIN-20-004 (2021)
[21] X. Yao, W. Ke, Y. Xu, S.A. Bass, B. Müller, JHEP **21**, 046 (2020), 2004.06746

Heavy-flavour production in small systems and evolution with multiplicity with ALICE

Shreyasi Acharya for the ALICE Collaboration[1,2,*]

[1]Variable Energy Cyclotron Centre, Kolkata, India
[2]Homi Bhaba National Institute, Mumbai, India

Abstract.
In this report, highlights on the multiplicity-dependent measurements of heavy-flavour hadrons performed by the ALICE experiment, are presented. The heavy-flavour baryon-to-meson (Λ_c^+/D^0) and strange-to-non-strange D meson (D_s^+/D^0) ratios in different charged-particle multiplicity intervals are discussed. Recent measurements of heavy-flavour self-normalised yields as a function of the self-normalised charged-particle multiplicity in pp collisions at $\sqrt{s}$ = 13 TeV are also reported. Measurement of jets containing heavy-flavour hadrons in pp and p–Pb collisions at $\sqrt{s_{NN}}$ = 5.02 TeV and the recent measurements of azimuthal correlations of heavy-flavour hadron decay electrons and charged hadrons in pp collisions at $\sqrt{s}$ = 5.02 TeV are presented.

1 Introduction

Heavy quarks are produced through hard scattering processes of partons at the early stages of an ultra-relativistic heavy-ion collision. Hence, they act as sensitive probes for exploring the properties of the hot and dense deconfined matter formed in such collisions, known as the quark-gluon plasma (QGP) [1]. Heavy-flavour measurements in pp collisions are fundamental not only as reference measurements to investigate the properties of the QGP, since the latter is not expected to be formed in small systems, but are also important since they provide precise tests for the perturbative QCD calculations. The multiplicity dependent studies in pp and p–Pb collisions are important to further understand the particle production mechanisms in such systems, and can provide an insight into the the multiple-parton-interaction and colour reconnection phenomena in the hadronisation mechanisms of heavy-flavour particles.

In ALICE [2], the open heavy-flavour hadron production is studied through their different decay channels in a wide rapidity range. At mid rapidity, they are studied either through the reconstructed hadronic decays (e.g. $D^0 \rightarrow K^-\pi^+$, $D_s^+ \rightarrow K^-K^+\pi^+$, $D^+ \rightarrow K^-\pi^+\pi^+$, $D^* \rightarrow D^0\pi^+$, $\Lambda_c^+ \rightarrow pK^-\pi^+$, $\Lambda_c^+ \rightarrow pK_s^0$) or via the measurement of electrons from the semi-leptonic decays of heavy-flavour hadron (i.e. D, B $\rightarrow$ e + X) using the particle identification (PID) information from the Time Projection Chamber (TPC) and Time-Of-Flight (TOF) detectors. The Electromagnetic Calorimeter (EMCal) is employed along with the TPC for electron identification in the transverse momentum region $p_T \gtrsim 4$ GeV/c. At forward rapidity, the open heavy-flavour hadron production is studied through the muons decayed from the heavy-flavour hadrons (i.e. D, B $\rightarrow \mu$ + X) using the Muon Spectrometer.

*e-mail: shreyasi.acharya@cern.ch

2 Heavy-flavour production as a function of charged-particle multiplicity

Figure 1 shows the ratio of the yields of D_s^+ and Λ_c^+ with respect to the D^0 yield, on the left and right panels respectively, as a function of their transverse momentum (p_T) in different charged-particle multiplicity intervals in pp collisions at $\sqrt{s} = 13$ TeV. The D_s^+/D^0 ratio does

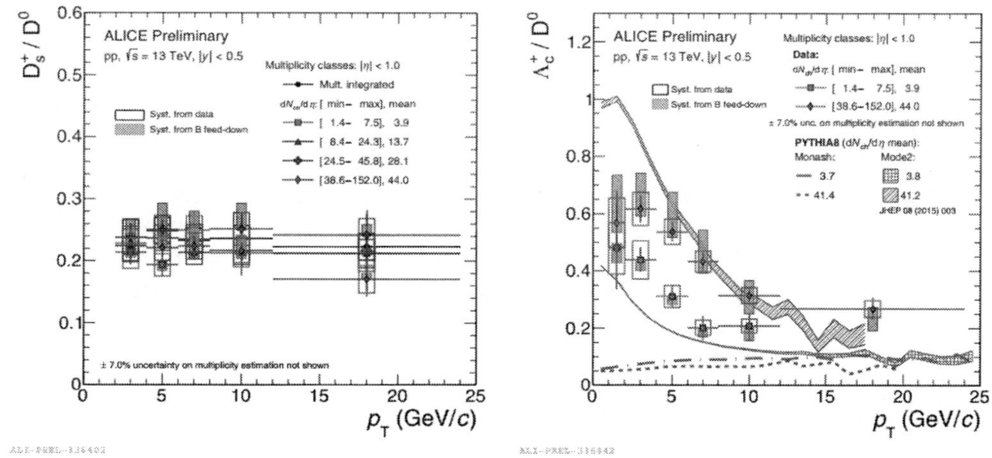

Figure 1. D_s^+/D^0 (left panel) and Λ_c^+/D^0 (right panel) ratio as a function of transverse momentum (p_T) in different charged-particle multiplicity intervals in pp collisions at $\sqrt{s} = 13$ TeV.

not show any dependence on charged-particle multiplicity within uncertainty and their abundance is compatible with the average of the p_T-integrated measurements performed in e^- e^+ collisions at LEP [3]. However, the Λ_c^+/D^0 ratio shows a significant increase from the lowest

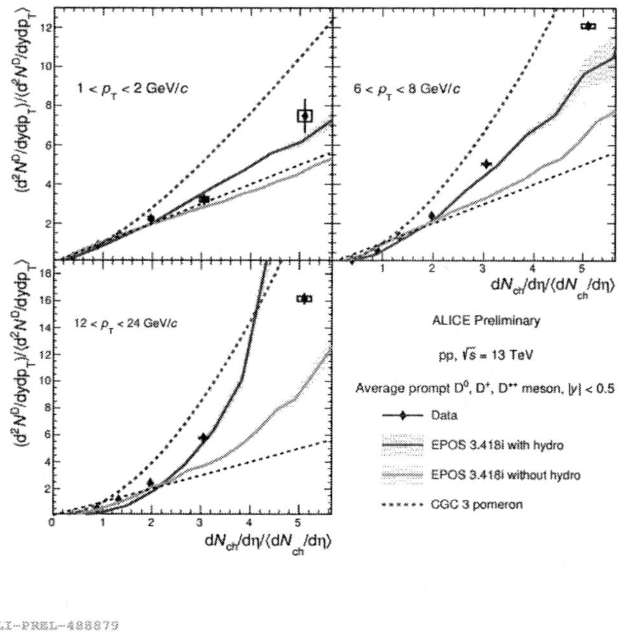

Figure 2. Self-normalised yield of D mesons as a function of the normalised charged-particle density in pp collisions at $\sqrt{s} = 13$ TeV, compared with predictions from EPOS3.418i (with and without hydrodynamics) and 3-Pomeron CGC models.

to the highest multiplicity indicating a modification of the hadronisation mechanisms from small to high multiplicity. The Λ_c^+/D^0 abundance is higher in the low-intermediate p_T region than the value measured in e^- e^+ collisions, indicating that the charm-quark fragmentation may not be universal. The Λ_c^+/D^0 measurement has also been compared with predictions from two different PYTHIA8 calculations. While predictions from default PYTHIA8 calculation (Monash tune) [4] largely underestimate the data, the predictions with colour reconnection mechanisms beyond the leading colour approximation (Mode2 tune) [5] describe the increasing trend of the baryon-over-meson ratio as a consequence of the new colour reconnection topologies included in the calculation.

Figure 2 shows the self-normalised yield of D mesons as a function of the normalised charged-particle density, in different p_T intervals, in pp collisions at $\sqrt{s} = 13$ TeV. The term 'self-normalised' implies that the yield in multiplicity intervals are normalised by the average yield calculated in minimum-bias events. A faster than linear increasing trend of the self-normalised yield is observed in all p_T intervals, with higher p_T D mesons showing a steeper increase. The D meson self-normalised yields are compared with different theoretical predictions. The yields from the EPOS3.418i model [6] without hydrodynamical calculations and from the three-pomeron color-glass condensate (CGC) model [7] underestimate and overestimate the data respectively while the predictions from EPOS3.418i model with hydrodynamical calculations tend to describe the trend in data.

3 Heavy-flavour jets, correlations and flow

The study of jets containing heavy-flavour decay hadrons provides a direct access to the heavy-quark kinematics, and allows for studying possible modifications of its fragmentation in different multiplicity environments. Figure 3 shows the p_T-differential cross sections of

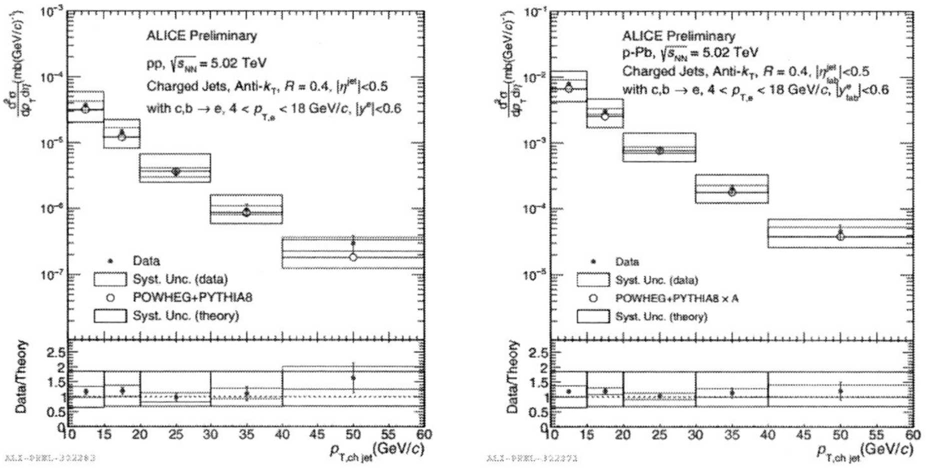

Figure 3. p_T-differential cross section of jets tagged with electrons from heavy-flavour hadron decay in pp (left panel) and p–Pb (right panel) collisions at $\sqrt{s_{NN}} = 5.02$ TeV.

jets tagged with electrons from heavy-flavour hadron decays in pp and p–Pb collisions at $\sqrt{s_{NN}} = 5.02$ TeV on the left and the right panels, respectively. The jets are identified by the presence of heavy-flavour decay electrons within $4 < p_{T,e} < 18$ GeV/c. The results have been compared with the next to leading order (NLO) perturbative QCD calculations using POWHEG + PYTHIA8 [4, 8] and are found to be in good agreement with NLO pQCD predictions.

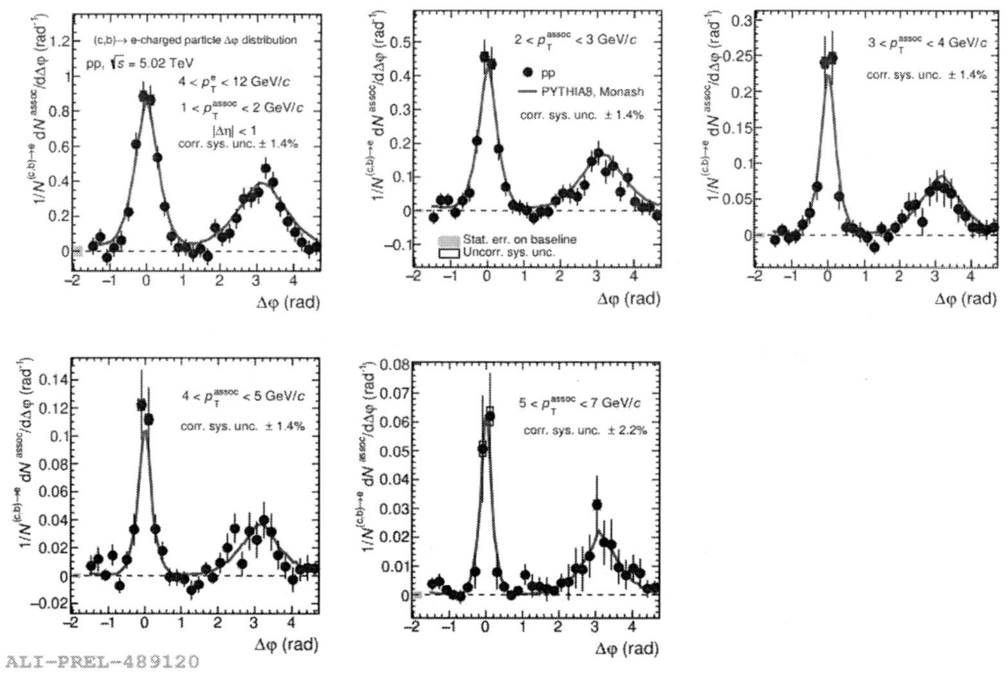

Figure 4. Azimuthal correlation between heavy-flavour hadron decay electrons and charged hadrons in pp collisions at $\sqrt{s} = 5.02$ TeV.

Figure 4 shows the azimuthal correlation between heavy-flavour hadron decay electrons and charged hadrons in pp collisions at $\sqrt{s} = 5.02$ TeV. Heavy-flavour hadron decay electrons within $4 < p_{\mathrm{T}} < 12$ GeV/c are chosen as trigger particles while charged hadrons in various intervals of p_{T} within 1 GeV/c and 7 GeV/c are chosen as associated particles as shown in the different panels of the figure. Predictions from PYTHIA8 (Monash tune) are consistent with the data. The near-side yield can be used to study modification of jet fragmentation while the away-side yield can provide information on the suppression and path length dependence of energy loss, and the correlation measurements in pp collisions will provide necessary reference for future heavy-ion measurements.

References

[1] BRAHMS Collaboration, Nucl. Phys. A 757 (2005) 1-27; PHOBOS Collaboration, Nucl. Phys. A 757 (2005) 28-101; STAR Collaboration, Nucl. Phys. A 757 (2005) 102-183; PHENIX Collaboration, Nucl. Phys. A 757 (2005) 184-283

[2] ALICE Collaboration, JINST 3 (2008) S08002

[3] L. Gladilin, Eur. Phys. J. C 75 (2015), no. 1 19, 1404.3888.

[4] T. Sjöstrand et. al., Comput. Phys. Commun. 191 (2015) 159-177

[5] J. R. Christiansen and P. Z. Skands, JHEP 08 (2015) 003

[6] K.Werner et. al., Phys. Rev. C 89, 064903 (2014)

[7] Iván Schmidt and Marat Siddikov, Phys. Rev. D 101, 094020 (2020)

[8] C. Oleari, Nucl. Phys. Proc. Suppl. 205-206 (2010) 36–41

Reactions Governing Strangeness Abundance in Primordial Universe

Cheng Tao Yang[1] and *Johann* Rafelski[1]

[1]Department of Physics, The University of Arizona, Tucson, Arizona 85721, USA

Abstract. Strangness production processes can balance natural strangeness decay in the early hadronic Universe. Comparing to the characteristic Hubble time $1/H$, the reaction rates for $\mu^{\pm}+\nu_{\mu}\to K^{\pm}$, $l^{-}+l^{+}\to\phi$, and $\pi+\pi\to K$ in sequence become slower than expansion rate at $T=33.9\,\mathrm{MeV}$, $T=25\,\mathrm{MeV}$ and $T=20\,\mathrm{MeV}$ respectively. This means that in the antibaryon annihilation epoch near to $T\simeq 40\,\mathrm{MeV}$ strangeness is in chemical equilibrium.

Overview: The present work is a first consideration of strangeness quark flavor evolution in the Universe following on the hadronization of quark-gluon plasma (QGP) near to $T_h=150\,\mathrm{MeV}$. We characterize Universe expansion dynamics during the epoch of interest $T_h\approx 150\geq T\geq 10\,\mathrm{MeV}$ and compare strangeness population influencing reactions with the characteristic Hubble time $1/H$.

We evaluate the dynamic rates employing detailed balance: the natural decay of particles concerned provides also the intrinsic strength of the inverse production reaction of interest. We outline consequences of our findings, and delimit related future research. Considering that baryons are a tiny component of all hadrons, this study is focused on the meson sector of the hadronic Universe.

We study reactions in which the particle number changes by one: An example is a particle decay into two, restored in two to one reactions in what one may call one⇔two reaction process. For these the required reaction rates are obtainable from available decay rates as described below. The two⇔two reactions often have a significantly higher reaction threshold when strangeness pair abundance is produced. Threfore these reactions are important near to the QGP hadronization temperature $T_h\simeq 150\,\mathrm{MeV}$. For discussion of the here relevant reactions, such as $KK\leftrightarrow\pi\pi$, we refer the reader to Chapter 18 in Ref.[1].

Universe expansion: Since all reaction rates we obtain are functions of T we also seek $1/H$ as a function of T. The Hubble parameter can be written in terms of energy density $\rho_{\rm tot}$ [2]; $H^2=\frac{8\pi G_N}{3}\rho_{\rm tot}(T)$, $G_{\rm N}$ is the gravitational constant. For $150\geq T\geq 10\,\mathrm{MeV}$ the Universe is radiation-dominated $\rho_{\rm tot}(T)\propto T^4$, thus we easily express $1/H$ as a function of temperature required as reference in our work.

We include in $\rho_{\rm tot}$ photons, three flavors of neutrinos, charged leptons and antileptons $e^{\pm},\mu^{\pm}$ and the lightest hadrons $\pi^0,\pi^{\pm}$. Both muons and pions are not relativistic in the considered $150\geq T\geq 10\,\mathrm{MeV}$ range. Evaluating their contribution to $\rho_{\rm tot}$ we use their actual mass and chemical equilibrium abundance: Muons are coupled through electromagnetic reactions $\mu^{+}+\mu^{-}\Leftrightarrow\gamma+\gamma$ to the photon background and retain their chemical equilibrium [3]. Because π-mesons are in thermal equilibrium [4], and are the lightest hadrons, they dominate hadronic contribution to the energy density near $T\approx 50\,\mathrm{MeV}$.

Strangeness in an expanding Universe: Regarding initial conditions: We assume that before hadronization in the QGP epoch of temperature $T>150\,\mathrm{MeV}$, the strangeness formation processes are fast enough to assure chemical equilibrium – hence the Universe undergoing a relatively slow on hadronic time scale phase transformation emerges from QGP near to chemical equilibrium abundance. In this transition the excess QGP entropy, as compared to hadron gas (HG), is absorbed in additional Universe comoving volume expansion while excess strangeness has time to reequilibrate into equilibrium HG abundance. We note that:

- The strange quark abundance in the early Universe is equal to the sum of strange baryons Λ^0, Σ^0, $\Sigma^{\pm}$ and strange mesons ϕ, K, $K^{\pm}$. We will see that in the early Universe the important sources for creation of $\phi(s\bar{s})$ and $K(q\bar{s})\overline{K}(\bar{q}s)$ are given by $l^{-}+l^{+}\to\phi$, $\rho+\pi\to\phi$, $\pi+\pi\to K,\overline{K}$, and $\mu^{\pm}+\nu\to K^{\pm}$.
- Reaction rate for $l^{-}+l^{+}\to\phi$, $\pi+\pi\to K$, and $\mu^{\pm}+\nu\to K^{\pm}$ are relatively small compared to reaction $\rho+\pi\to\phi$. However, the thermal abundance of pions and leptons π, $l^{\pm}$, and ν is relatively high compared to the more massive ρ. Therefore all these reactions compete with each other.
- The strong decay rate of $\phi\to K+K$ indicates that the particles ϕ and K are in relative chemical equilibrium.
- The strong strangeness exchange rate $K+N\to Y+\pi$ following on $\phi\to K+K$ with the small decay rates of $K\to\pi+\pi$ and $Y\to N+\pi$ suggest the possibility that strangeness can accumulate in K meson and in Y hyperons resulting in strange baryon yields above chemical equilibrium yield, a topic of future study requiring

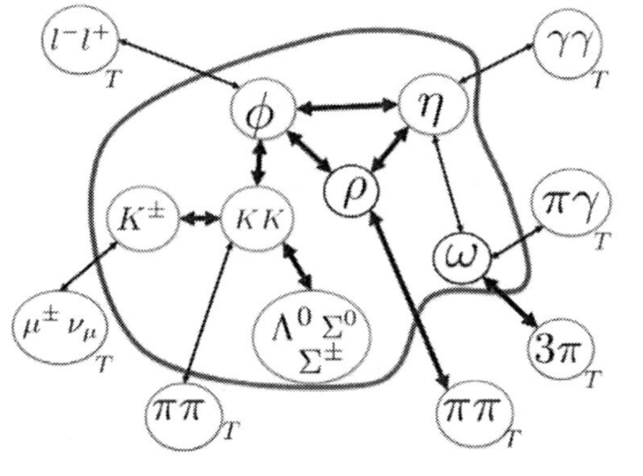

Figure 1. Principal strangeness abundance changing processes in the hadronic Universe $T < T_h = 150\,\text{MeV}$. Within the (blue) boundary we see hadronic particles expected to fall out of abundance (chemical) equilibrium. The red circles within this domain represent strangeness-carrying mesons, black non-strange mesons of importance in creation of strangeness. Particles in green circles indicating their full equilibrium (subscript T) are seen outside this (blue) boundary domain.

consideration of baryon chemical potential considered negligible in present study.

The key meson production and decay processes are illustrated in Fig. 1 including important 'hidden' $s\bar{s}$ components in η and ϕ, and multi-step reactions leading to formation of ϕ. All particles outside the (blue) boundary are in abundance (chemical) and kinetic (thermal) equilibrium in the temperature domain of interest, see Ref. [4, 5].

Strength of available reaction rates: We introduce here the typical interaction rates relevant for the discussion of strangeness in the early Universe using the particle data group data tables [6]. We have

- For ϕ meson, the partial decay rates are

$$\Gamma_{\phi\to KK} = 3.545\,\text{MeV}, \quad (1)$$
$$\Gamma_{\phi\to\rho\pi} = 0.6535\,\text{MeV}, \quad (2)$$
$$\Gamma_{\phi\to\eta\gamma} = 0.0558\,\text{MeV}, \quad (3)$$
$$\Gamma_{\phi\to ll} = 2.484\times10^{-3}\,\text{MeV}. \quad (4)$$

- For η meson, the partial decay rates are given by

$$\Gamma_{\eta\to\gamma\gamma} = 0.355\times10^{-3}\,\text{MeV}, \quad (5)$$
$$\Gamma_{\eta\to3\pi^0} = 0.516\times10^{-3}\,\text{MeV}, \quad (6)$$
$$\Gamma_{\eta\to3\pi} = 0.368\times10^{-3}\,\text{MeV}, \quad (7)$$
$$\Gamma_{\omega\to\eta\gamma} = 3.905\times10^{-3}\,\text{MeV}, \quad (8)$$
$$\Gamma_{\rho\to\eta\gamma} = 44.73\times10^{-3}\,\text{MeV}. \quad (9)$$

- For ρ meson, the decay rate is

$$\Gamma_{\rho\to\pi\pi} = 149.1\,\text{MeV}. \quad (10)$$

- For K meson, the decay rate for $K_S \to \pi+\pi$ and charge pion decay $K^\pm \to \mu^\pm\nu$ are given by

$$\Gamma_{K_S\to\pi\pi} = 7.350\times10^{-12}\,\text{MeV}, \quad (11)$$
$$\Gamma_{K^\pm\to\mu\nu} = 3.379\times10^{-14}\text{MeV}. \quad (12)$$

- For hyperons Y, the decay rate is comparable to $\Gamma_{K_S\to\pi\pi}$. We postpone discussion pending more complete study of (anti)baryon component and note here

$$\Gamma_{Y\to N\pi} \approx 6.582\times10^{-12}\,\text{MeV}. \quad (13)$$

Hadron based strangeness production: We now explore in more detail the conditions in which the nonequilibrium of strangeness arises. In order to study the strangeness nonequilibrium in the early Universe, we need to calculate the relevant dynamic reaction rates for strangeness production in detail. In the hadron sector the relevant interaction rates competing with Hubble time involving strongly interacting mesons are the reactions $\pi+\pi\leftrightarrow K$, $\pi+\pi\leftrightarrow\rho$, $\rho+\pi\leftrightarrow\phi$, and $\mu^\pm+\nu\to K^\pm$. Our results are seen in Fig. 2. We now turn to describe the methods used to obtain these results.

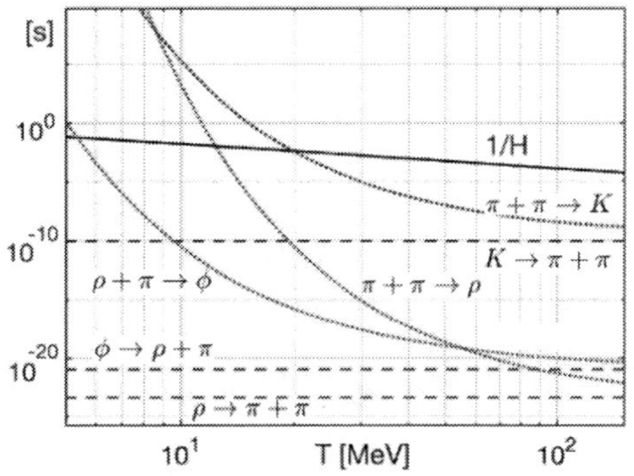

Figure 2. Hubble time $1/H$ (black line) as a function of temperature is compared to hadronic relaxation reaction times, see Eq. (20), for reactions $\pi+\pi\leftrightarrow K$(blue), $\pi+\pi\leftrightarrow\rho$ (red), $\rho+\pi\leftrightarrow\phi$(purple). The horizontal dashed lines are the corresponding natural decay lifespans.

In general, the thermal reaction rate per time and volume for reaction $1+2\to3$ is given by [1, 7]:

$$R_{12\to3} = n_1 n_2 \langle\sigma v\rangle_{12\to3} = \frac{g_1 g_2}{32\pi^4}\frac{T}{1+I_{12}}\times \quad (14)$$
$$\int_{m_3^2}^{\infty}\sigma\frac{[s-(m_1+m_2)^2][s-(m_1-m_2)^2]}{\sqrt{s}}K_1(\sqrt{s}/T)\,ds\,,$$

where the factor $1/(1+I_{12})$ is introduced to avoid double counting of indistinguishable particles; we have $I_{12}=1$ for identical particles, otherwise $I_{12}=0$. Many studies have shown that the hadronic reaction matrix element $|\mathcal{M}_{12\to3}|^2$ for hadronic reactions without a resonance phenomenon is insensitive to changes of $\sqrt{s}$ in the considered energy range [1, 7]. In this case, we assume that the transition

 https://doi.org/10.1051/epjconf/202225913001

amplitude $|\mathcal{M}_{12\to3}|^2$ is nearly constant in the energy range we consider, and the cross section σ can be written as

$$\sigma = \frac{2\pi/(g_1g_2)\sum_{\text{spin}}|\mathcal{M}_{12\to3}|^2}{2\sqrt{[s-(m_1+m_2)^2][s-(m_1-m_2)^2]}}\times \int d^4p_3\delta_0(p_3^2-m_3^2)\delta^4(p_1+p_2-p_3)\,. \quad (15)$$

In this case, the thermal reaction rate per time and volume can be written as

$$R_{12\to3} = \frac{1}{16\pi^3}\frac{T}{1+I}\sum_{spin}|\mathcal{M}_{12\to3}|^2\times \quad (16)$$

$$\frac{\sqrt{[m_3^2-(m_1+m_2)^2][m_3^2-(m_1-m_2)^2]}}{2m_3}K_1(m_3/T)\,.$$

On the other hand, the vacuum decay rate for particle 3 can be written as

$$\frac{1}{\tau_3^0} = \frac{|p|}{8\pi m_3^2}\left(\frac{1}{1+I_{12}}\right)\left(\frac{1}{g_3}\sum_{spin}|\mathcal{M}_{12\to3}|^2\right), \quad (17)$$

$$|p| = \frac{1}{2m_3}\sqrt{m_1^4+m_2^4+m_3^4-2m_1^2m_2^2-2m_1^2m_3^2-2m_2^2m_3^2}\,. \quad (18)$$

Hence, using the vacuum decay rate for the thermal reaction rate per unit time and volume can be written as

$$R_{12\to3} = \frac{g_3}{2\pi^2}\left(\frac{T^3}{\tau_3^0}\right)\left(\frac{m_3}{T}\right)^2 K_1(m_3/T)\,. \quad (19)$$

In order to compare the reaction time with Hubble time $1/H$, it is also convenient to define the relaxation time for the process $1+2\to3$ as follow

$$\tau_{12\to3} \equiv \frac{dn_1/d\Upsilon_1}{R_{12\to3}} = \frac{n_1^{\text{eq}}}{R_{12\to3}}\,, \quad (20)$$

where n_1^{eq} is the thermal equilibrium number density of particle 1, and the freeze-out temperature for the given reaction $1+2\to3$ can be estimated by considering the condition $\tau_{12\to3} = 1/H$.

Lepton and photon based strangeness production: The relevant interaction rates competing with Hubble time involving lepton background in the Universe are seen in Fig. 3. As example consider the reaction $K^\pm \leftrightarrow \mu^\pm + \nu_\mu$. For the decay modes of charged kaon $K^\pm$, we have

$$K^+ \longrightarrow \mu^+ + \nu_\mu, \quad K^- \longrightarrow \mu^- + \overline{\nu}_\mu\,, \quad (21)$$

where the lifetime of $K^\pm$ is given by $\tau_{K^\pm} = 1.238\times10^{-8}$ sec and the mass $m_{K^\pm} = 493.677$ MeV. Hence the charged kaons in the cosmic plasma can be produced by the inverse of the decay processes, and we have

$$\mu^+ + \nu_\mu \longrightarrow K^+, \quad \mu^- + \overline{\nu}_\mu \longrightarrow K^-\,. \quad (22)$$

From Eq.(19) the thermal interaction rate per unit volume for the process $\mu^\pm + \nu \leftrightarrow K^\pm$ can be written as

$$R_{\mu\nu\leftrightarrow K^\pm} = \frac{g_{K^\pm}}{2\pi^2}\left(\frac{T^3}{\tau_{K^\pm}}\right)\left(\frac{m_{K^\pm}}{T}\right)^2 K_1(m_{K^\pm}/T)\,. \quad (23)$$

Given the thermal interaction rate per volume $R_{\mu\nu\leftrightarrow K^\pm}$, we use for the process $\mu^\pm + \nu_\mu \to K^\pm$ the definition of the equilibrium relaxation time Eq.(20) to obtain

$$\tau_{\mu\nu\to K^\pm} = \frac{g_\mu}{2\pi^2 R_{\mu\nu\leftrightarrow K^\pm}}\int_{m_\mu}^{\infty} dE\frac{E\sqrt{E^2-m_\mu^2}}{\exp(E/T)+1}\,. \quad (24)$$

In Fig.(3) the dark blue line is the interaction rate for $\mu^\pm + \nu_\mu \to K^\pm$.

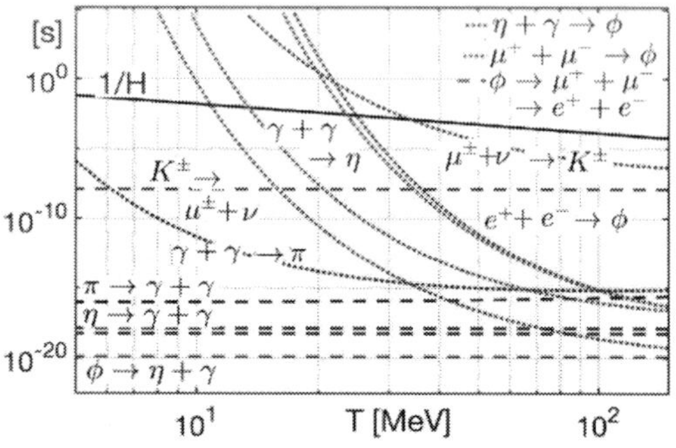

Figure 3. Hubble time $1/H$ (black line) as a function of temperature is compared to leptonic and photonic relaxation reaction times, see Eq. (20), for $\gamma+\gamma \leftrightarrow \pi$ (blue), $\gamma+\gamma \leftrightarrow \eta$ (red), $\eta+\gamma \leftrightarrow \phi$ (green), $l^+ + l^- \leftrightarrow \phi$ (brown and purple), and $\mu^\pm + \nu_\mu \longrightarrow K^\pm$(dark blue line). The horizontal dashed lines are the natural decay lifespans.

The freeze-out condition, $\tau_{\mu\nu\to K^\pm}(T_f) = 1/H(T_f)$, where the relaxation reaction time $\mu^\pm + \nu_\mu \to K^\pm$ becomes slower compared to the Universe expansion, and thus detailed balance in this reaction cannot be maintained, is near $T_f^{K^\pm} = 33.90$ MeV.

If other strangeness production reactions did not exist, strangeness would disappear as the Universe cools below $T_f^{K^\pm} = 33.90$ MeV. However, using Eq. (19) and Eq. (20) we already evaluated the relaxation time for the reactions $\pi+\pi \leftrightarrow K$, $\pi+\pi \leftrightarrow \rho$, and $\rho+\pi \leftrightarrow \phi$ as functions of temperature in the early Universe as seen in Fig. 2. The intersection of characteristic reaction times with $1/H$ occurs for $\pi+\pi\to K$ at $T = 19.85$ MeV, for $\pi+\pi\to\rho$ at $T = 12.36$ MeV, and for $\rho+\pi\to\phi$ at $T = 5.36$ MeV.

Several other leptonic and photonic reactions of relevance are shown in Fig. 3. We see now the relaxation time for $\gamma+\gamma \leftrightarrow \pi$ (blue), $\gamma+\gamma \leftrightarrow \eta$ (red), $\eta+\gamma \leftrightarrow \phi$ (green), $l^+ + l^- \leftrightarrow \phi$ (brown and purple), and $\mu^\pm + \nu_\mu \to K^\pm$(dark blue line) as a function of temperature. We see that pions remain in equilibrium since $\gamma+\gamma\to\pi$ reactions are always faster compared to $1/H$. Considering other reactions: $\gamma+\gamma\to\eta$ reaction intersects $1/H$ at $T = 13.5$ MeV, $\eta+\gamma\to\phi$ at $T = 10.85$ MeV; $e^-+e^+\to\phi$ at $T = 24.9$ MeV; $\mu^+ + \mu^- \to \phi$ at $T = 23.5$ MeV; and as noted before, $\mu^\pm + \nu_\mu \to K^\pm$ at $T = 33.9$ MeV.

Discussion: In this work we have explored reaction rates in the meson sector of strangeness. We have identified hierarchy of reactions which become slower compared to Hubble time $1/H$ involving strange particles ϕ and K. The reactions $\mu^{\pm} + \nu_{\mu} \rightarrow K^{\pm}$, $l^{-} + l^{+} \rightarrow \phi$, and $\pi + \pi \rightarrow K$ are intersecting $1/H$ at $T = 33.9\,\mathrm{MeV}$, $T = 25\,\mathrm{MeV}$, and $T = 20\,\mathrm{MeV}$, respectively.

Once these (or any other) reactions decouple from the cosmic plasma, the corresponding detailed balance is broken: Absence of the inverse decay reaction acts like a "hole" in the strangeness abundance "pot" with some strangeness decay in early Universe proceeding. However, the corresponding reverse channel of strangeness production reaction is too slow compared to the fast Universe expansion. Considering only meson content of the Universe, our results show that when the expanding Universe cools to a temperature around $T = 20\,\mathrm{MeV}$, strangeness will begin to progressively decouple but remain near-equilibrium abundance.

However, there are many different channels of strangeness production and decay so loss of one reaction channel does not end the strangeness abundance. As the Universe expands, other strangeness producing reactions become relevant; for example, $\gamma + \gamma \rightarrow \eta$ reaction intersects $1/H$ at $T = 13.5\,\mathrm{MeV}$, and $\eta + \gamma \rightarrow \phi$ at $T = 10.85\,\mathrm{MeV}$. Thus when an η is produced, we can always assume that a ϕ arises and thus considering the $\phi \leftrightarrow K+K$ reactions this generates Kaons which can decay as now several strangeness decay paths exceed in speed the corresponding channel production rate. We conclude that strangness chemical decoupling is complete once all dominant strangness formation reactions become too slow.

We have considered kinetically equilibrated distribution described by the temperature T. This certainly applies to charged particles which are embedded in the cosmic background of a dense photon-electron-positron plasma. However, many of the reactions we considered involve neutral particles; as examples consider the process $\pi^0 + \pi^0 \rightarrow K_S$, and many even more relevant reactions involving neutral strange baryons. Hadrons have relatively large strong interaction reaction cross sections. However, as the temperature decreases, their abundance is very small. Further study needs to explore the validity of kinetic equilibrium hypothesis for neutral hadrons.

This first study of strangeness production mechanisms in the early Universe, following on QGP hadronization, shows that the relevant temperature domains indicating non-trivial dynamics and freeze-out for (strange) mesons and strange (anti)baryon abundance are practically overlapping. Thus both baryon and strangeness chemical freeze-out have to be explored jointly in a dynamical model. The Universe in the range $20 \leq T \leq 60\,\mathrm{MeV}$ appears rich in physics phenomena involving strange mesons, (anti)baryons including (anti)hyperon abundances, which can enter into chemical nonequilibrium while undergoing successive freeze-out processes. This means that a quantitative study of strangeness freeze-out requires that a full account be given to strange hyperons and antihyperons. For further discussion we refer to our forthcoming work [8].

Acknowledgment: The research of CTY was supported by the US Department of Energy under Grant Contract DESC0012704 to the Brookhaven National Laboratory. We thank J. Birrell for interesting discussions.

References

[1] J. Letessier and J. Rafelski, *Hadrons and Quark-Gluon Plasma*, 397 pp, Camb. Monogr. Part. Phys. Nucl. Phys. Cosmol. **18** (2002) doi: 10.1017/CBO9780511534997, ISBN: 9780521018234 (Paperback), 9780521385367 (Hardback), 9780511037276 (Online)

[2] E. W. Kolb and M. S. Turner, *The Early Universe*, 547 pp, Front. Phys. **69**, 1 (1990), ISBN: 0201626748, 9780201626742

[3] J. Rafelski and C. T. Yang, "The muon abundance in the primordial Universe," Acta Phys. Polon. B **52** (2021), 277 [arXiv:2103.07812 [hep-ph]].

[4] I. Kuznetsova, D. Habs and J. Rafelski, "Pion and muon production in e-, e+, gamma plasma," Phys. Rev. D **78**, 014027 (2008) [arXiv:0803.1588 [hep-ph]].

[5] J. Birrell, C. T. Yang and J. Rafelski, "Relic Neutrino Freeze-out: Dependence on Natural Constants," Nucl. Phys. B **890**, 481-517 (2014) [arXiv:1406.1759 [nucl-th]].

[6] M. Tanabashi *et al.* [Particle Data Group], "Review of Particle Physics," Phys. Rev. D **98**, no. 3, 030001 (2018) doi:10.1103/PhysRevD.98.030001

[7] P. Koch, B. Muller and J. Rafelski, "Strangeness in Relativistic Heavy Ion Collisions," Phys. Rept. **142**, 167-262 (1986) doi:10.1016/0370-1573(86)90096-7

[8] C. T. Yang and J. Rafelski, "Cosmological Strangeness Abundance," [arXiv:2108.01752 [hep-ph]].

EPJ Web of Conferences **259**, 13002 (2022) https://doi.org/10.1051/epjconf/202225913002
SQM 2021

In-medium effects in strangeness production in heavy-ion collisions at (sub-) threshold energies

Taesoo Song[1,*], *Laura* Tolos[2,3,4,5], *Joana* Wirth[6,7], *Joerg* Aichelin[4,8], and *Elena* Bratkovskaya[1,9,10]

[1]GSI Helmholtzzentrum für Schwerionenforschung GmbH, Planckstr. 1, 64291 Darmstadt, Germany
[2]Institute of Space Sciences (ICE, CSIC), Campus Universitat Autonoma de Barcelona, Carrer de Can Magrans, 08193 Bellaterra, Spain
[3]Institut d'Estudis Espacials de Catalunya (IEEC), 08034 Barcelona, Spain
[4]Frankfurt Institute for Advanced Studies, Ruth-Moufang-Strasse 1, 60438 Frankfurt am Main, Germany
[5]Faculty of Science and Technology, University of Stavanger, 4036 Stavanger, Norway
[6]Excellence Cluster 'Origin and Structure of the Universe', 85748 Garching, Germany
[7]Physik Department E62, Technische Universität München, 85748 Garching, Germany
[8]SUBATECH UMR 6457 (IMT Atlantique, Université de Nantes, IN2P3/CNRS), 4 Rue Alfred Kastler, F-44307 Nantes, France
[9]Institut für Theoretische Physik, Johann Wolfgang Goethe-Universität, Max-von-Laue-Str. 1, 60438 Frankfurt am Main, Germany
[10]Helmholtz Research Academy Hesse for FAIR (HFHF), GSI Helmholtz Center for Heavy Ion Physics, Campus Frankfurt, 60438 Frankfurt, Germany

Abstract. We study the in-medium effects in strangeness production in heavy-ion collisions at (sub-)threshold energies based on the microscopic Parton-Hadron-String Dynamics (PHSD) transport approach. The in-medium modifications of the antikaon properties are described via the self-consistent coupled-channel unitarized scheme based on a SU(3) chiral Lagrangian while the in-medium modification of kaons are accounted via the kaon-nuclear potential, which is assumed to be proportional to the local baryon density. We find that the modifications of (anti)kaon properties in nuclear matter are necessary to explain the experimental data in heavy-ion collisions.

1 Introduction

In this contribution we present the highlights of our recent study [1] of the in-medium effects in the strangeness production in heavy-ion collisions at (sub-)threshold energies employing the microscopic transport approach PHSD which incorporates the in-medium description of the antikaon-nucleon interactions based on 'state of the art' many-body theory realized by the G-matrix formalism [2]. This chiral unitary approach in coupled channels incorporates the s- and p-waves of the kaon-nucleon interaction, its modification in the hot and dense medium to account for Pauli blocking effects, mean-field binding for baryons, as well as pion and kaon self-energies. For the in-medium scenario of the kaon-nucleon interaction we adopt a

*e-mail: t.song@gsi.de

repulsive potential linear in the baryon density. We then confront our theoretical results with available experimental data on (anti)kaon production at SIS energies (cf. Ref. [1] for details).

2 Chiral unitarized model for $\bar{K}N$ in hot nuclear matter

For low-energy scattering, the s-wave meson-baryon interaction from the lowest-order chiral Lagrangian which couples the octet of light pseudoscalar mesons to the octet of $1/2^+$ baryons can be written as [2]

$$V^s_{ij} \simeq -C_{ij}\frac{1}{4f^2}(k^0_i + k^0_j), \tag{1}$$

where the coefficients C_{ij}, with i, j indicating the particular meson-baryon channel, can be found explicitly in Ref. [4], f is the meson weak decay constant, and $k^0_{i(j)}$ is energy of the meson in the $i(j)$ channel in the the center-of-mass (c.m.) frame. As for p-wave meson-baryon interaction $\tilde{V}^p_{ij}$, one can find the details in Refs. [2, 3]. The transition amplitudes can be obtained by means of a unitarization coupled-channel procedure based on solving the Bethe-Salpeter equation by using the tree level contributions to the s- and p-wave meson-baryon scattering as the kernel of the equation. Within the on-shell factorization [4, 5], the Bethe-Salpeter equation is given in matrix notation as

$$T = [1 - VG]^{-1}V, \tag{2}$$

where V is the kernel (potential), and the G is a diagonal matrix accounting for the loop function of a meson-baryon propagator. Nuclear matter effects at finite temperature are introduced through the modification of the meson-baryon propagators where all mesons and baryons in the intermediate loops interact with the nucleons of the Fermi sea, and their properties are modified with respect to those in vacuum in addition to the Pauli blocking, as described in Refs. [2, 6]. Once the transition amplitudes in free space and in matter are evaluated, the corresponding cross sections can be obtained by

$$\frac{d\sigma_{ij}}{d\Omega}(\sqrt{s}) = \frac{1}{16\pi^2}\frac{M_i M_j}{s}\frac{q_j}{q_i}\left\{|T^s_{ij} + (2T^p_{ij+} + T^p_{ij-})\cos\theta|^2 + |T^p_{ij+} - T^p_{ij-}|^2\sin^2\theta\right\}, \tag{3}$$

where $M_i(M_j)$ is the incoming (outgoing) baryon mass, $q_i(q_j)$ is the modulus of the on-shell c.m. three-momentum of the incoming (outgoing) meson-baryon pair, and θ is the scattering angle in that frame. The self-energy of the antikaon in either s- or p-wave are obtained by convoluting the transition amplitudes with the nucleon Fermi distribution at a given temperature, for example,

$$\Sigma^L_{\bar{K}}(k;T) = 4\int\frac{d^3p}{(2\pi)^3}n_N(\vec{p},T)\bar{T}^L_{\bar{K}N\to\bar{K}N}(P;T), \tag{4}$$

where P is the total energy and momentum of the $\bar{K}N$ pair in the nuclear medium rest frame, k stands for the energy and momentum of the $\bar{K}$ meson also in this frame, and $\bar{T}^L$ indicates the spin and isospin averaged scattering amplitude for a given partial wave ($L = 0$ or $L = 1$). The spectral function can be then reconstructed as

$$A_{\bar{K}}(m^2) = \frac{-2\,\mathrm{Im}\Sigma_{\bar{K}}}{(m^2 - m^2_{\bar{K}} - \mathrm{Re}\Sigma_{\bar{K}})^2 + (\mathrm{Im}\Sigma_{\bar{K}})^2}, \tag{5}$$

where $m_{\bar{K}}$ is $\bar{K}$ pole mass in vacuum.

3 $K/\bar{K}$ production in heavy-ion collisions

For the simulations of heavy-ion collisions we use the Parton–Hadron–String Dynamics transport approach [7–11], which is a microscopic off-shell transport approach for the description of strongly interacting hadronic and partonic matter in and out-of equilibrium. The antikaon production cross section in the medium is evaluated by folding the corresponding vacuum production cross section with the in-medium spectral function in Eq. (5):

$$\sigma^*_{\bar{K}}(\sqrt{s}) = \int_0^{(\sqrt{s}-m_4)^2} \frac{dm^2}{2\pi} A_{\bar{K}}(m^2)\, \sigma_{\bar{K}}(\sqrt{s} - \Delta m_{\bar{K}}), \tag{6}$$

where m_4 is the invariant mass of the final particles different from antikaons [1], m is the off-shell mass of the $\bar{K}$ meson in the medium defined by the spectral function $A(m^2)$ according to Eq. (5); $\Delta m_{\bar{K}} = m - m_{\bar{K}}$ describes the deviation from the vacuum mass. The equations of motion for the produced off-shell antikaon are given by the Cassing-Juchem off-shell transport equations for testparticles [12] used also for the propagation of antikaons in the early HSD study [13]. On the other hand, the in-medium kaon mass m^*_K is expressed in terms of the repulsive potential $V_K = 25$ MeV (ρ/ρ_0) with ρ_0 being the nuclear saturation density [1]:

$$m^*_K \simeq m_K\Big(1 + E_K V_K / m^2_K\Big). \tag{7}$$

The cross section for kaon production is then shifted by the mass difference:

$$\sigma_K(\sqrt{s}) \to \sigma_K(\sqrt{s} - \Delta m_K) \tag{8}$$

where $\Delta m_K = m^*_K - m_K$. Since Δm_K is positive, the mass shift suppresses K production in heavy-ion collisions.

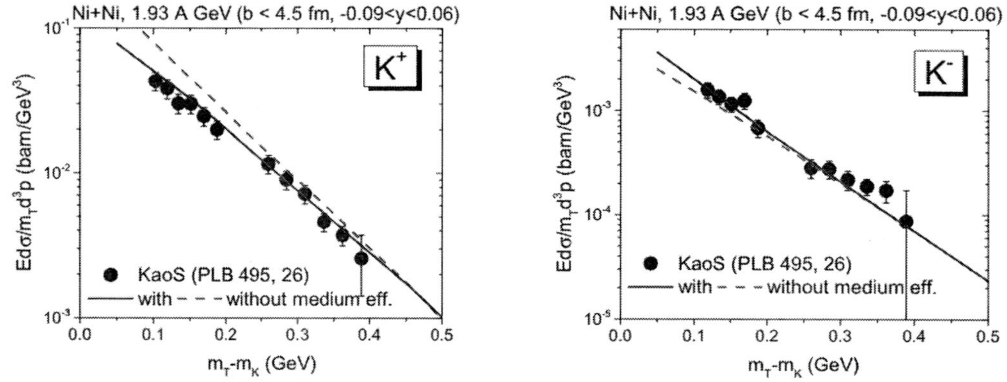

Figure 1. The PHSD results for the m_T-spectra of K^+ (left) and K^- (right) as a function of the transverse kinetic energy $m_T - m_K$ in central Ni+Ni collisions at 1.93 A GeV in comparison to the experimental data of the KaoS Collaboration [14]. The dashed lines indicate the results without medium effects for (anti)kaons, while the solid lines display the results with the medium effects. The figure is taken from Ref. [1].

Figure 1 shows the PHSD results for the m_T-spectra versus the transverse kinetic energy of K^+ and K^- at midrapidity in central Ni+Ni collisions at 1.93 A GeV in comparison to the experimental data of the KaoS Collaboration [14]. The kaon-nuclear potential suppresses the K^+ yield and pushes the K^+ spectrum to larger transverse momenta, while the antikaon potential enhances the K^- yield and pulls the p_T-spectrum toward smaller transverse momenta. The kaon potential is necessary to reproduce the experimental data on the p_T- spectra of K^+ and K^-.

4 Summary

We have investigated strangeness production in heavy ion collisions at (sub-)threshold energies within the microscopic Parton-Hadron-String Dynamics (PHSD 4.5) transport approach, extended for the incorporation of the in-medium effects for strangeness production in terms of the state-of-the-art coupled-channel G-matrix approach for the modification of antikaon properties in a dense and hot medium. We have found that the kaon-nuclear potential increases the threshold energy for the kaon production in a hot and dense medium. Consequently, the K^+ production in heavy-ion collisions is suppressed. On the other hand, the self-consistent coupled-channel unitarized scheme based on a SU(3) chiral Lagrangian leads to a broadening of the K^- spectral function without drastic changes of the pole mass due to the finite momentum. The K^- production in heavy-ion collisions is enhanced by the broad spectral width which leads to a reduction of the production threshold. Furthermore, the kaon-nuclear potential hardens the p_T-spectra of K^+ mesons while the in-medium effects in terms of self-energies within the G-matrix approach soften the K^- spectra.

Acknowledgements

The authors acknowledge support by the Deutsche Forschungsgemeinschaft (DFG, German Research Foundation): grant BR 4000/7-1, by the Russian Science Foundation grant 19-42-04101 and by the GSI-IN2P3 agreement under contract number 13-70. L.T. acknowledges support from the Deutsche Forschungsgemeinschaft (DFG, German research Foundation) under the Project Nr. 383452331 (Heisenberg Programme), Nr. 411563442 (Hot Heavy Mesons). Her research has been also supported by the Spanish Ministerio de Economía y Competitividad under contract FPA2016-81114-P and Ministerio de Ciencia e Innovación under contract PID2019-110165GB-I00. This project has, furthermore, received funding from the European Union's Horizon 2020 research and innovation program under grant agreement No 824093 (STRONG-2020). The computational resources have been provided by the LOEWE-Center for Scientific Computing and the "Green Cube" at GSI, Darmstadt.

References

[1] T. Song, L. Tolos, J. Wirth, J. Aichelin and E. Bratkovskaya, Phys. Rev. C **103**, 044901 (2021)
[2] D. Cabrera, L. Tolos, J. Aichelin and E. Bratkovskaya, Phys. Rev. C **90**, 055207 (2014)
[3] L. Tolos, A. Ramos and E. Oset, Phys. Rev. C **74**, 015203 (2006)
[4] E. Oset and A. Ramos, Nucl. Phys. A **635**, 99 (1998).
[5] J. A. Oller and U. G. Meissner, Phys. Lett. B **500**, 263 (2001)
[6] L. Tolos, D. Cabrera and A. Ramos, Phys. Rev. C **78**, 045205 (2008)
[7] W. Cassing and E. L. Bratkovskaya, Phys. Rev. C **78**, 034919 (2008)
[8] W. Cassing, Eur. Phys. J. ST **168**, 3 (2009)
[9] W. Cassing and E. L. Bratkovskaya, Nucl. Phys. A **831**, 215 (2009)
[10] E. L. Bratkovskaya, W. Cassing, V. P. Konchakovski and O. Linnyk, Nucl. Phys. A **856**, 162 (2011)
[11] O. Linnyk, E. L. Bratkovskaya and W. Cassing, Prog. Part. Nucl. Phys. **87**, 50 (2016)
[12] W. Cassing and S. Juchem, Nucl. Phys. A **672**, 417 (2000)
[13] W. Cassing, L. Tolos, E. L. Bratkovskaya, and A.Ramos, Nucl. Phys. A **727**, 59 (2003)
[14] M. Menzel *et al.* [KaoS], Phys. Lett. B **495**, 26 (2000)

Study of Charge Symmetry Breaking in A = 4 hypernuclei in $\sqrt{s_{NN}} = 3$ GeV Au+Au collisions at RHIC

Tianhao Shao[1](for the STAR Collaboration)[*]

[1]Key Laboratory of Nuclear Physics and Ion-beam Application (MOE), Institute of Modern Physics, Fudan University, Shanghai 200433, China

Abstract. In this paper, we present the measurement of the charge symmetry breaking in A = 4 hypernuclei in Au+Au collisions at $\sqrt{s_{NN}} = 3$ GeV. The signal reconstruction and binding energy measurement of ${}^4_\Lambda$H and ${}^4_\Lambda$He, including corrections and systematic uncertainty evaluation, are discussed. Combined with the energy levels of excited states, our preliminary result of Λ binding energy difference for excited states is $\Delta B_\Lambda(1^+) = -190 \pm 130$(stat.)$\pm 70$(syst.) keV which shows a negative value and its magnitude is comparable to the result of ground states $\Delta B_\Lambda(0^+) = 130 \pm 130$(stat.) $\pm$ 70(syst.) keV. These results are compared with previous measurements and theoretical model calculations.

1 Introduction

The charge symmetry of strong interactions predicts that the Λ-p and Λ-n interactions should be identical as they cannot be influenced by charge. This leads to a conclusion that the Λ binding energies of a pair mirror hypernuclei should be identical. However, in 1970's nuclear emulsion experiments measured the Λ binding energies and the binding energy difference in A = 4 mirror hypernuclei, ${}^4_\Lambda$H and ${}^4_\Lambda$He, and found a difference of $\Delta B^4_\Lambda(0^+_{g.s.}) = 350 \pm 50$ keV [1]. Such a large difference cannot be explained with the mass difference of the up and down quarks in nuclear systems. In 2015, the J-PARC E13 γ-ray spectroscopy experiment measured the transition energy from the 1^+ first excited state of ${}^4_\Lambda$He to be $1406 \pm 2 \pm 2$ keV [2]. The E13 collaboration combined the Λ binding energies of ground states from emulsion experiments of 1970s [1] with a γ-ray transition energy for ${}^4_\Lambda$H measured in 1976 [3] and their new γ-ray transition measurement for ${}^4_\Lambda$He to determine the difference in excited states to be $\Delta B^4_\Lambda(1^+_{exc}) = 30 \pm 50$ keV [2] which is much smaller than that in ground states. It was suggested that the charge symmetry breaking effect may have a large spin-dependence. In 2016, the MAMI A1 collaboration used spectrometers to provide a new measurement of the ground state Λ binding energy of ${}^4_\Lambda$H [4]. Combining their new measurement with the previous ${}^4_\Lambda$He Λ binding energy, and the measurements of the γ-ray transition energies for ${}^4_\Lambda$H [1] and ${}^4_\Lambda$He [2], they updated the estimate of the binding energy differences to be $\Delta B^4_\Lambda(0^+_{g.s.}) = 233 \pm 92$ keV and $\Delta B^4_\Lambda(1^+_{exc}) = -83 \pm 94$ keV. However many theoretical model calculations failed to reproduce the experimental results [5–9]. In 2016, the ab initio no-core shell model calculations plus a charge symmetry breaking $\Lambda - \Sigma^0$ mixing vertex of A = 4 hypernuclei got a large charge symmetry breaking in excited states and concluded

[*]e-mail: shaoth21@outlook.com

that $\Delta B_{\Lambda}^{4}(1_{exc}^{+}) \approx -\Delta B_{\Lambda}^{4}(0_{g.s.}^{+}) < 0$ [10]. Independent experiments are needed to test these calculations [11].

To study the physics of QCD matter in a high baryon density region, the STAR detector ran in fixed-target mode during the BES-II program. A stationary gold target was mounted inside the beam pipe and in two meters to the west of the center of the detector, which was in the plane of the west end-cap of the TPC detector. In collider mode, the lowest $\sqrt{s_{NN}}$ for Au+Au collisions that RHIC can effectively run is 7.7 GeV, whereas in fixed-target mode, this low energy limit can be extended down to 3 GeV. In 2018, STAR has taken about 300 million events data of Au+Au collisions at 3 GeV fixed-target mode. Model calculations predict that the production yields of hypernuclei will become larger at lower collision energies [12]. The STAR fixed target program gives us an opportunity to study the Λ binding energy of $^{4}_{\Lambda}$H and $^{4}_{\Lambda}$He in the same experiment to address the charge symmetry breaking effect.

2 Analysis details and results

In this analysis, signals of $^{4}_{\Lambda}$H and $^{4}_{\Lambda}$He are analyzed in Au+Au collisions at 3 GeV. The $^{4}_{\Lambda}$H is reconstructed via its two-body decay channel $^{4}_{\Lambda}\mathrm{H} \rightarrow {}^{4}\mathrm{He} + \pi^{-}$. The $^{4}_{\Lambda}$He is reconstructed via its three-body decay channel $^{4}_{\Lambda}\mathrm{He} \rightarrow {}^{3}\mathrm{He} + \mathrm{p} + \pi^{-}$. The decay daughters are identified mainly according to the $\langle dE/dx \rangle$ information from the Time Projection Chamber (TPC). The identification of ^{4}He and ^{3}He are done also according to the mass information from the Time Of Flight (TOF) detectors. Then the invariant mass distributions of $^{4}_{\Lambda}$H and $^{4}_{\Lambda}$He are reconstructed according to their decay topology, similar to the analysis presented in Ref. [13]. To increase the signal significance, the TMVA-BDT package [14] is used. Figure 1 shows the invariant mass distributions of $^{4}_{\Lambda}$H and $^{4}_{\Lambda}$He. The centroids and statistical uncertainties

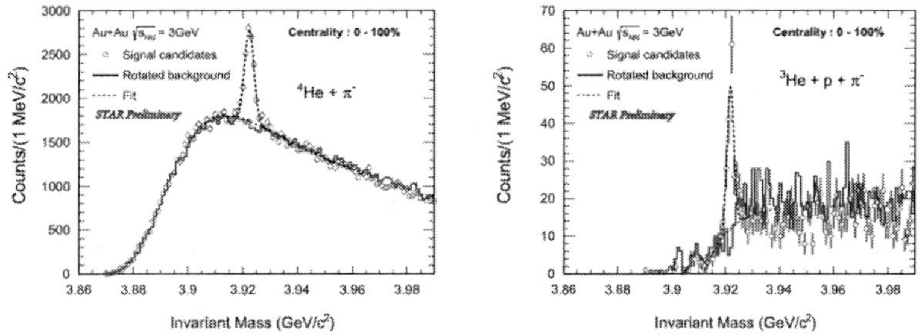

Figure 1. Invariant mass distributions for $^{4}_{\Lambda}$H and $^{4}_{\Lambda}$He reconstruction. The backgrounds are obtained by rotating ^{4}He or ^{3}He track by 180 degrees in the transverse plane. The black dashed curves represent fits with a Gaussian function plus double exponential functions.

in the ground state masses of the $^{4}_{\Lambda}$H and $^{4}_{\Lambda}$He are determined by fitting the invariant mass distributions with a Gaussian function plus double exponential functions represented by black dashed curves in Fig. 1. The mass results are

$$\mathrm{m}(^{4}_{\Lambda}\mathrm{H}) = 3922.36 \pm 0.06(\mathrm{stat.}) \pm 0.18(\mathrm{syst.})\ \mathrm{MeV/c^2}, \quad (1)$$

$$\mathrm{m}(^{4}_{\Lambda}\mathrm{He}) = 3921.70 \pm 0.12(\mathrm{stat.}) \pm 0.14(\mathrm{syst.})\ \mathrm{MeV/c^2}. \quad (2)$$

Due to the particle's energy loss in material prior the tracking region of the TPC and the precision of the magnetic field, the measured momenta of decay daughters need to be corrected. The first correction is for the particle's energy loss. This correction is done by using the STAR embedding data. Generated samples of ${}^{4}_{\Lambda}$H and ${}^{4}_{\Lambda}$He from a Monte Carlo program are inputted into a GEANT virtual STAR detector. Then the momentum loss of particles can be determined by comparing the momentum difference between MC input and detector output. The second correction is for magnetic field measurement accuracy. From previous studies of the invariant mass of known particles, it has been determined that the magnetic field of STAR detector should be scaled by 0.2%, therefore the momentum of particles are scaled with a factor 0.998 in this analysis. These two corrections have been checked in Λ invariant mass. The Λ invariant mass measured in Au+Au collisions at 3 GeV with these two corrections is consistent with the PDG mass. Four sources of systematic uncertainties have been analyzed: magnetic field accuracy, energy loss correction, BDT cut, and fit method.

The Λ binding energies of hypernuclei can be calculated using the mass of a given hypernucleus and its constituents:

$$\mathrm{B}_{\Lambda} = (\mathrm{M}_{\Lambda} + \mathrm{M}_{\mathrm{core}} - \mathrm{M}_{\mathrm{hypernucleus}})c^2, \tag{3}$$

where $\mathrm{M}_{\mathrm{core}}$ represents the mass of a triton or ^{3}He taken from CODATA [15] for ${}^{4}_{\Lambda}$H and ${}^{4}_{\Lambda}$He respectively. The Λ binding energy results are:

$$\mathrm{B}_{\Lambda}({}^{4}_{\Lambda}\mathrm{H}) = 2.24 \pm 0.06(\mathrm{stat.}) \pm 0.18(\mathrm{syst.})\ \mathrm{MeV}, \tag{4}$$

$$\mathrm{B}_{\Lambda}({}^{4}_{\Lambda}\mathrm{He}) = 2.37 \pm 0.12(\mathrm{stat.}) \pm 0.14(\mathrm{syst.})\ \mathrm{MeV}. \tag{5}$$

These results are for the ground states. The results for excited states can be obtained from the γ-ray transition energies [2, 3]. The Λ binding energy difference between ${}^{4}_{\Lambda}$H and ${}^{4}_{\Lambda}$He can be calculated:

$$\Delta\mathrm{B}_{\Lambda}(0^{+}) = 130 \pm 130(\mathrm{stat.}) \pm 70(\mathrm{syst.})\ \mathrm{keV}, \tag{6}$$

$$\Delta\mathrm{B}_{\Lambda}(1^{+}) = -190 \pm 130(\mathrm{stat.}) \pm 70(\mathrm{syst.})\ \mathrm{keV}. \tag{7}$$

In this analysis, the difference in excited states shows a negative value and its magnitude is comparable to the ground states within uncertainties. Most of theroetical calculations predict small Λ binding energy differences in both ground states and excited states. Gazda and Gal reported a large splitting in ground states and also a large value in excited states with an opposite sign and a similar magnitude, $\Delta B^{4}_{\Lambda}(1^{+}_{exc}) \approx -\Delta B^{4}_{\Lambda}(0^{+}_{g.s.}) < 0$ [10], which is slightly favored by our preliminary results.

The results in this analysis are compared to previous measurements and theoretical model calculations in Fig. 2. Due to the low statistics of ${}^{4}_{\Lambda}$He, the statistical uncertainty on the ${}^{4}_{\Lambda}$He mass drives the statistical uncertainties on the Λ binding energy differences. STAR has taken about a factor of 7 more data (about 2 billion events) at 3 GeV fixed-target Au+Au collisions in run 2021. Upgrades to the TPC and the TOF have increased the tracking and PID acceptance. The statistical uncertainties will be reduced and their expected magnitudes are shown as green shadows shown in Fig. 2.

3 Conclusions

To address the charge symmetry breaking effect in A = 4 hypernuclei, we reconstructed the invariant mass distributions of ${}^{4}_{\Lambda}$H and ${}^{4}_{\Lambda}$He in Au+Au collisions at 3 GeV taken in fixed-target mode at STAR. With the corrections for daughters' momenta the Λ binding energy difference between ${}^{4}_{\Lambda}$H and ${}^{4}_{\Lambda}$He can be determined. Using our preliminary results and the γ-ray transition energies from previous measurements, we show that the charge symmetry breaking effect in excited states has a negative value and its magnitude is comparable to that of the ground states within uncertianties. STAR has taken a factor of 7 more data at 3 GeV fixed-target in 2021. The statistical uncertainties of this analysis will be reduced in the future work.

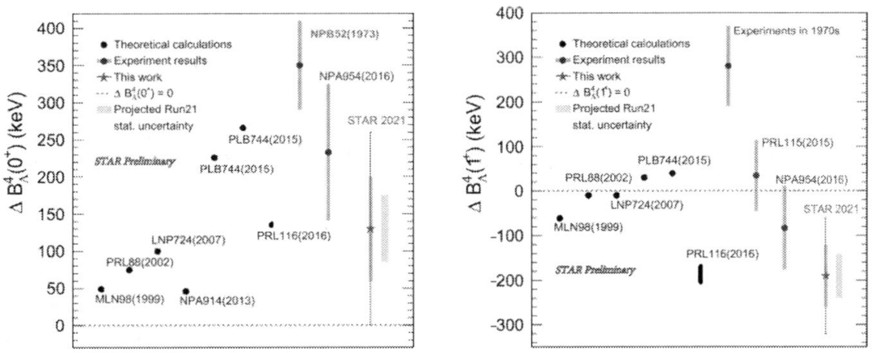

Figure 2. The Λ binding energy difference between ${}^4_\Lambda$H and ${}^4_\Lambda$He in ground states (left figure) and in excited states (right figure) compared with theoretical model calculations (black dots) [5–10] and previous measurements (blue dots) [1–4, 16]. Error bars show statistical uncertainties and shadows show the systematic uncertainties. The green shadows are projected statistical uncertainties from the STAR run 2021 3 GeV data.

4 Acknowledegments

We appreciate finance support by the National Natural Science Foundation of China under Contact No. 12025501.

References

[1] M. Juric et al., Nucl. Phys. B **52**, 1 (1973)
[2] T.O. Yamamoto et al. (J-PARC E13), Phys. Rev. Lett. **115**, 222501 (2015), 1508.00376
[3] M. Bedjidian, A. Filipkowski, J.Y. Grossiord, A. Guichard, M. Gusakow, S. Majewski, H. Piekarz, J. Piekarz, J.R. Pizzi (CERN-Lyon-Warsaw), Phys. Lett. B **62**, 467 (1976)
[4] F. Schulz et al. (A1), Nucl. Phys. **A954**, 149 (2016)
[5] A. Nogga, H. Kamada, W. Gloeckle, Phys. Rev. Lett. **88**, 172501 (2002), nucl-th/0112060
[6] J. Haidenbauer, U.G. Meissner, A. Nogga, H. Polinder, Lect. Notes Phys. **724**, 113 (2007), nucl-th/0702015
[7] A. Nogga, Nucl. Phys. A **914**, 140 (2013)
[8] S.A. Coon, H.K. Han, J. Carlson, B.F. Gibson, *Particle mixing and charge asymmetric Lambda N forces*, in *7th International Conference on Mesons and Light Nuclei 98* (1998), nucl-th/9903034
[9] A. Gal, Phys. Lett. **B744**, 352 (2015), 1503.01687
[10] D. Gazda, A. Gal, Phys. Rev. Lett. **116**, 122501 (2016), 1512.01049
[11] J. Chen, D. Keane, Y.G. Ma, A. Tang, Z. Xu, Phys. Rept. **760**, 1 (2018), 1808.09619
[12] J. Steinheimer, K. Gudima, A. Botvina, I. Mishustin, M. Bleicher, H. Stocker, Phys. Lett. B **714**, 85 (2012), 1203.2547
[13] J. Adam et al. (STAR), Nature Phys. **16**, 409 (2020), 1904.10520
[14] A. Hocker et al. (2007), physics/0703039
[15] W. Meng, A. G, K. F. G, H. W. J, N. S, X. Xing, Chinese Physics C **41**, 030003 (2017)
[16] M. Bedjidian et al. (CERN-Lyon-Warsaw), Phys. Lett. B **83**, 252 (1979)

Shear-induced spin polarization and "strange memory" in heavy-ion collisions

Shuai Y. F. Liu[1,*], *Baochi* Fu[2,3], and *Longgang* Pang[4]

[1]Quark Matter Research Center, Institute of Modern Physics, Chinese Academy of Sciences, Lanzhou 730000, China
[2]Department of Physics & State Key Laboratory of Nuclear Physics and Technology, Peking University, Beijing 100871, China
[3]Collaborative Innovation Center of Quantum Matter, Beijing 100871, China
[4]Key Laboratory of Quark & Lepton Physics (MOE) and Institute of Particle Physics, Central China Normal University, Wuhan 430079, China

Abstract. We discuss the theory of the spin polarizations induced by hydrodynamic gradients, which includes a newly discovered shear-induced polarization (SIP) term. In the phenomenological study using a hydrodynamic model, we discover that the local polarization contributed by SIP is substantial and has the "same sign" as the experimental measurements. Also, we find that the "sign" property of the local polarization observed in experiments seems to be related to "memory" effects on the polarizations of strange quarks in quark-gluon plasma.

1 Introduction

In the non-central ultra-relativistic heavy-ion collisions, a significant amount of orbital angular momentum is deposited into the quark-gluon plasma (QGP) created in these high-energy collisions. Considering the spin-orbit coupling, it has be proposed [1] that the orbital angular momentum of the rotating QGP could induce a global (spin) polarization of the hyperons. This global polarization has been observed later in the ground-breaking experiment [2] and the predictions using the thermal-vorticity formula [3–5] agree well with experimental measurements [6]. Besides the global polarization, there are proposals that the local structures of the hydrodynamic gradients [7–9] can lead to local spin polarization phenomena. However, the pattern of the local polarizations predicted [8] using thermal-vorticity formula demonstrate the "opposite sign" compared to experimental results [10, 11]. This discrepancy is sometimes called the "spin sign puzzle". On the other hand, it should be noted that the thermal-vorticity formula only has the temperature-gradient-induced and vorticity induced polarizations, while to our knowledge, the question that whether other hydrodynamic gradients, such as shear-stress tensor, could induce polarizations, has not been studied before. In the following, based on our recent works [12, 13], we will answer this question and elaborate the new formula of polarization including the newly discovered shear-induced polarization (SIP) term. Then, we will study the phenomenology of SIP, especially on local polarizations, and discuss the insights related to the "memory" effects on the strange quark polarizations in QGP phase.

*e-mail: lshphy@gmail.com

2 Theoretical formalism

The observables of the spin polarizations are closely related to the axial current in phase space $\mathcal{A}^\mu$. The expression of $\mathcal{A}^\mu$ can be obtained from a linear response theory or chiral kinetic theory, detailed in our recent works [12, 13]. Here, we will mainly discuss some extra physics and equivalent forms of $\mathcal{A}^\mu$, as well as its relations to the results in other works [14]. At first, we rewrite the final expression of $\mathcal{A}^\mu$ at momentum p^μ in Ref [13] as

$$\mathcal{A}^\mu = \frac{1}{2}n_0(1-n_0)\left\{\beta\epsilon^{\mu\nu\alpha\lambda}p_\nu\partial^\perp_\alpha u_\lambda + 2\epsilon^{\mu\nu\alpha\lambda}u_\nu p_\alpha\,\partial_\lambda\beta - 2\beta\frac{p_\perp^2}{\varepsilon_0}\epsilon^{\mu\nu\alpha\rho}u_\nu Q_\alpha^{\ \lambda}\sigma_{\rho\lambda}\right\} \tag{1}$$

where $\beta = 1/T$ is the inverse temperature and u^μ is flow velocity. The $n_0 \equiv n(\beta\varepsilon_0)$ with $n(x) = 1/(e^x+1)$ and $\varepsilon_0 = p\cdot u$. The projection of a generic vector V^μ is defined as $V^\mu_\perp = \Delta^{\mu\nu}V_\nu$ where $\Delta^{\mu\nu} = \eta^{\mu\nu} - u^\mu u^\nu$. Also, $Q^{\mu\nu} \equiv -p^\mu_\perp p^\nu_\perp/p^2_\perp + \Delta^{\mu\nu}/3$ is a quadrupole tensor and $\sigma^{\mu\nu} = \partial^{(\mu}_\perp u^{\nu)} - \Delta^{\mu\nu}\partial\cdot u/3$ is the shear strength tensor, where $\partial^{(\alpha}_\perp u^{\lambda)} \equiv (\partial^\alpha_\perp u^\lambda + \partial^\lambda_\perp u^\alpha)/2$.

The three terms in the "{...}" of Eq. (1) are vorticity-induced polarization (VoIP), T-gradient-induced polarization (TIP, spin Nernst effect) and shear-induced polarization (SIP). Since $(p^2_\perp/\varepsilon_0)\epsilon^{\mu\nu\alpha\rho}u_\nu Q_\alpha^{\ \lambda}\sigma_{\rho\lambda} \equiv \varepsilon_0^{-1}\epsilon^{\mu\nu\alpha\rho}u_\nu\, p_\rho p^\lambda\partial^\perp_{(\alpha}u_{\lambda)}$, both the TIP and SIP share a universal form $\epsilon^{\mu\nu\alpha\lambda}u_\nu p_\alpha F_\lambda$ where the "force" F_λ is proportional to either $\partial_\lambda\beta$ or $p^\alpha\partial^\perp_{(\lambda}u_{\alpha)}$ [15]. Thus, both of them can be understood as a generalized spin hall effects [16].

If we combine half of TIP with VoIP and another half of TIP with SIP, one can get another equivalent form with only two terms

$$\mathcal{A}^\mu = \frac{1}{2}n_0(1-n_0)\left\{\epsilon^{\mu\nu\alpha\lambda}p_\nu\partial_\alpha(\beta u)_\lambda - \frac{2}{\varepsilon_0}\epsilon^{\mu\nu\alpha\rho}u_\nu\, p_\rho p^\lambda\partial_{(\alpha}(\beta u)_{\lambda)}\right\}. \tag{2}$$

Note that the $\mathcal{A}^\mu$ will still stay the same if we replace all ∂_α by $\partial^\perp_\alpha$ in Eq. (2) since the superficial acceleration contributions ($(u\cdot\partial)u_\alpha$) in the first and second term in the "{...}" will cancel each other. This $\mathcal{A}^\mu$ is equivalent to the result in Ref [14] if we select the "medium rest frame" ($t^\mu = u^\mu$). The first term in the "{...}" is the thermal vorticity contribution. Analogically, $\partial_{(\mu}(\beta u)_{\nu)}$ is named as thermal-shear in Ref [14]. However, using the identity $(u\cdot\partial)u_\alpha = -\beta^{-1}\partial^\perp_\alpha\beta + O(\partial^2)$ from ideal hydrodynamic equation, we can get another equivalent form of $\mathcal{A}^\mu$ as

$$\mathcal{A}^\mu = \frac{1}{2}n_0(1-n_0)\left\{\epsilon^{\mu\nu\alpha\lambda}p_\nu\partial_\alpha(\beta u)_\lambda - \beta\frac{2}{\varepsilon_0}\epsilon^{\mu\nu\alpha\rho}u_\nu\, p_\rho p^\lambda\sigma_{\alpha\lambda}\right\}. \tag{3}$$

The $(u\cdot\partial)u_\alpha$ and the $\partial_\alpha\beta$ in the second term of Eq. (2) will cancel each other. Thus, only the shear part $\beta\sigma_{\alpha\lambda}$ out of the thermal-shear $\partial_{(\mu}(\beta u)_{\nu)}$ truly contributes to the polarization at the leading order of gradients in our framework.

With the expressions of $\mathcal{A}^\mu$, we employ the freezeout formula as that used in Ref [3]

$$P^\mu(\boldsymbol{p}) = \frac{\int d\Sigma^\alpha p_\alpha\,\mathcal{A}^\mu(x,\boldsymbol{p};m)}{2m\int d\Sigma^\alpha p_\alpha n(\beta\varepsilon_0)} \tag{4}$$

to convert the $\mathcal{A}^\mu$ into the polarization P^μ related to experimental observables.

3 Phenomenological study

We employ the 3+1-d hydrodynamics (MUSIC + AMPT initial conditions) used in Ref [17] to study the phenomenology in Au-Au collisions at $\sqrt{s_{NN}} = 200$ GeV, where the local polarizations of P_z (beam direction) and $(-P_y)$ (out-plane direction) are calculated for **two scenarios**. In the "Lambda equilibrium" scenario, we will calculate the Lambda polarization in the early hadronic phase using the mass $m = 1.116$ GeV. In the "strange memory" scenario, we will calculate the strange quark polarization P_s at the late QGP phase with an effective mass $m = 0.3$ GeV. The chemical freezeout condition is $E_f = 0.65\,\mathrm{GeV/fm^3}$ ($T_f \sim 165$ MeV) for both scenarios, where the qualitative features of the results remain the same with a different choice of E_f or T_f within the reasonable range. We will further illustrate the physics of the two scenarios later in the context of the phenomenological results.

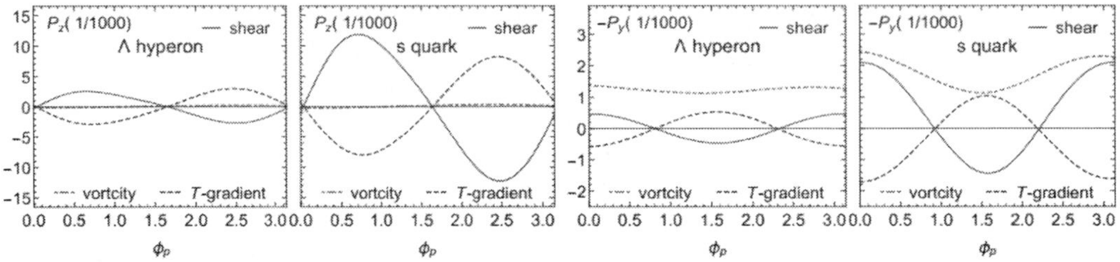

Figure 1. The P_z and $(-P_y)$ of VoIP (vorticity), TIP (T-gradient), SIP (shear) for two scenarios.

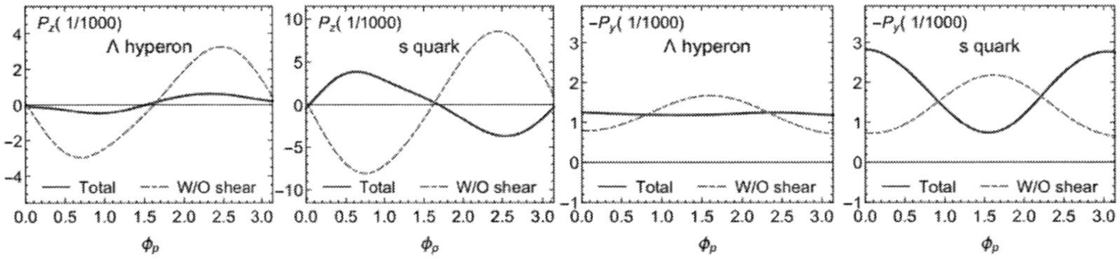

Figure 2. The P_z and $(-P_y)$ of the total (Total) and the total without SIP (W/O shear) for two scenarios. The "W/O shear"≈ "thermal-vorticity induced polarization" considering $(u \cdot \partial)u_\alpha = -\beta^{-1}\partial_\alpha^\perp\beta + O(\partial^2)$.

The Fig. 1 demonstrates how the three terms (VoIP, TIP, SIP) defined in Eq. (1) contribute to local polarizations. In both scenarios, the VoIP is the dominant contribution to the global polarizations (in y direction), but it does not strongly affect the shapes of local polarizations. The shapes of the local polarizations are mostly influenced by TIP and SIP. In all scenarios and directions, the local polarizations from TIP always has the "opposite sign" but those from SIP always has the "same sign" as compared to those observed in experiments [10, 11]. The competitions between TIP and SIP determine the sign of the total local polarizations, which are sensitive to the mass of the spin carrier because the factor $|p_\perp^2|/\varepsilon_0^2$ in SIP term (see Eq. (1)) will enhance with a smaller mass. Thus, in Fig. 2, the total local polarizations in "Lambda equilibrium" scenario have the "opposite sign", while those in "strange memory" scenario (smaller mass) have the "same sign". Note that without SIP, results from both scenarios are of the "opposite sign" and qualitatively similar. It is the SIP that makes a difference. The SIP not only seems to be essential for understanding the "spin sign puzzle" but also provides us opportunities to distinguish the physics behind these two scenarios as discussed below.

The "Lambda equilibrium" scenario assumes the Lambda polarizations are fully equilibrated in hadronic phase, representing the conventional statistical hadronization model (SHM) in the spin physics. In this scenario, we get the overall "opposite sign" (Fig. 2) and do not find any case showing the "same sign" when we repeat the calculations using many different choices of model parameters/setups in our robustness tests. This "opposite sign" behavior in conventional SHM of polarizations has also been confirmed in Ref [18]. Probably, some physics beyond the conventional SHM are important for understanding this local polarization problem. On the other hand, beside SHMs, quark-recombination models (QRMs) are also widely-used in hadronization problems, where a well-known QRM [1] derives a relation $P_\Lambda = P_s$. If we accept this QRM and neglect the evolution of polarizations in hadronic phase, P_Λ observed in experiments is just P_s at the end of the QGP phase, which is an off-equilibrium effect and can be interpreted as the Lambda's (full) "memory" on P_s. In this "strange memory" scenario illustrated above, the local polarizations demonstrate the "same sign". Definitely, the results of the two scenarios are not the final answer to the "spin sign puzzle". Instead, we would like this comparative study provides us some motivations to go beyond the conventional SHM and to develop more microscopic and dynamic models on hadronization and hadronic evolution of the spin physics in heavy-ion collisions.

4 Summary and perspective

We discover a new mechanism–shear-induced polarization (SIP) to generate spin polarization in a fluid. For the first time, the phenomenology of this SIP in heavy-ion collisions has been studied. We find that local polarization from SIP is a major contribution and always has "same sign" as those observed in the experiment. In the study of the two scenarios, we found that without SIP, none of the scenarios can have the "same sign". Including SIP effect, the local polarization still has the "opposite sign" in "Lambda equilibrium" scenario assuming equilibrated Lambda in hadronic phase. On the other hand, in "strange memory scenario" assuming that Lambda hyperon inherits and memorizes the polarization of the strange quarks, the SIP surpasses TIP and the resulting local polarizations demonstrate the "same sign".

As suggested by the study of the two scenarios, we need to develop a sophisticated hadronization model and hadronic spin transport model to obtain quantitative understandings. As discussed [13, 19, 20], the spin polarizations are potentially the probes for many interesting physics in heavy-ion collisions. Quantitative frameworks are indispensable for these tasks. To further improve our understandings, we probably also need to study the spin transports [21, 22] or spin hydrodynamics [23] in QGP, as well as the non-perturbative effects on the spin physics.

Acknowledgement: This work was supported by the Strategic Priority Research Program of Chinese Academy of Sciences Grant No. XDB34000000 as well as the NSFC under grant No. 12075007, No. 11675004, No. 11861131009 and No. 12075098.

References

[1] Z.T. Liang, X.N. Wang, Phys. Rev. Lett. **94**, 102301 (2005), [Erratum: Phys. Rev. Lett.96,039901(2006)]
[2] L. Adamczyk et al. (STAR), Nature **548**, 62 (2017)
[3] F. Becattini, V. Chandra, L. Del Zanna, E. Grossi, Annals Phys. **338**, 32 (2013)
[4] I. Karpenko, F. Becattini, Eur. Phys. J. C **77**, 213 (2017), 1610.04717
[5] H. Li, L.G. Pang, Q. Wang, X.L. Xia, Phys. Rev. C **96**, 054908 (2017), 1704.01507
[6] J. Adam et al. (STAR), Phys. Rev. C **98**, 014910 (2018), 1805.04400
[7] L.G. Pang, H. Petersen, Q. Wang, X.N. Wang, Phys. Rev. Lett. **117**, 192301 (2016)
[8] F. Becattini, I. Karpenko, Phys. Rev. Lett. **120**, 012302 (2018)
[9] S.A. Voloshin, EPJ Web Conf. **171**, 07002 (2018), 1710.08934
[10] T. Niida (STAR), Nucl. Phys. A **982**, 511 (2019)
[11] J. Adam et al. (STAR), Phys. Rev. Lett. **123**, 132301 (2019), 1905.11917
[12] S.Y.F. Liu, Y. Yin, JHEP **07**, 188 (2021), 2103.09200
[13] B. Fu, S.Y.F. Liu, L. Pang, H. Song, Y. Yin (2021), 2103.10403
[14] F. Becattini, M. Buzzegoli, A. Palermo, Phys. Lett. B **820**, 136519 (2021), 2103.10917
[15] Thanks M. Stephanov pointed out this SHE interpation of SIP to us.
[16] S.Y.F. Liu, Y. Yin (2020), 2006.12421
[17] B. Fu, K. Xu, X.G. Huang, H. Song (2020), 2011.03740
[18] F. Becattini, M. Buzzegoli, A. Palermo, G. Inghirami, I. Karpenko (2021), 2103.14621
[19] S. Ryu, V. Jupic, C. Shen (2021), 2106.08125
[20] C. Yi, S. Pu, D.L. Yang (2021), 2106.00238
[21] S.Y.F. Liu, Y. Sun, C.M. Ko, Phys. Rev. Lett. **125**, 062301 (2020), 1910.06774
[22] X.L. Sheng, N. Weickgenannt, E. Speranza, D.H. Rischke, Q. Wang, Phys. Rev. D **104**, 016029 (2021)
[23] S. Li, M.A. Stephanov, H.U. Yee (2020), 2011.12318

Event-shape studies of strangeness production in $\sqrt{s}$ = 13 TeV proton–proton collisions with ALICE

Adrian Nassirpour[1]* , on behalf of the ALICE Collaboration

[1]Lund University, Sweden

Abstract.
The ALICE Collaboration has observed that the relative fraction of strange hadrons grows strongly with multiplicity in small collision systems at LHC energies, in particular for multi-strange baryons. The origin of this effect is still under debate, and models need significant final-state interactions to accommodate the new ALICE results, requiring new tools to discriminate experimentally between the various phenomenological ideas.
In these proceedings we present a multi-differential study of identified particle spectra with respect to the event multiplicity and two different event shape observables, the unweighted transverse spherocity $S_{0}^{p_{T}=1}$, and the self-normalized, transverse charged particle density, R_{T}, which allows us to explore particle production associated with hard and soft QCD processes. ALICE has measured π, K, ϕ, and Ξ production at mid-rapidity ($|\eta| < 0.8$) as a function of $S_{0}^{p_{T}=1}$ and R_{T} in pp collisions at $\sqrt{s}$ = 13 TeV. This work reports on how these multi-differential measurements compare with predictions from PYTHIA and EPOS-LHC.

1 Introduction

Measurements in proton–proton (pp) and proton–lead (p–Pb) collisions at the Large Hadron Collider (LHC) have revealed that the relative production of strange hadrons is enhanced in high-multiplicity events, and the strength of the enhancement increases with the strange quark content. This phenomenon was historically believed to be a feature of heavy-ion collisions, interpreted as the formation of a strongly interacting medium, and its presence in pp collisions has led to new discussions regarding the physical interpretation of this signature, as several theoretical models are able to describe this enhancement by introducing various phenomenological final-state interactions.

In regard to QCD-inspired Monte-Carlo event generators, such as PYTHIA 8 [1], this observation seems to be contrary to the central assumption that pp collisions at these energies ($\sqrt{s}$ = 13 TeV) can be described as incoherent sums of parton–parton collisions. Such models require additional finale-state effects (color ropes, junctions, etc.) to describe the strangeness enhancement. In contrast, EPOS-LHC [2] is an event generator which forms a two-phase state (core-corona) consisting of a dense core of quark-gluon plasma (QGP) and a dilute corona. The strangeness enhancement in EPOS-LHC originates from a change in the relative contribution of the corona (low strangeness production) and core (high strangeness production) with multiplicity.

*e-mail: adrian.nassirpour@hep.lu.se

New, multi-differential observables are needed in order to distinguish which of these models most accurately describes hadronization in high-multiplicity small-systems collisions; specifically to understand whether the final-state multiplicity is the driving force for the enhancement of strange hadrons, or if both the multiplicity and strangeness enhancement are driven by underlying phenomena that are currently not well understood. This contribution presents results from A Large Ion Collider Experiment (ALICE) on strange particle production as functions of the transverse spherocity, $S_{\mathrm{O}}^{p_{\mathrm{T}}=1}$, and the self-normalized charged-particle density in the transverse region, R_{T}, which relate to the event topology and the underlying event activity respectively.

2 Particle production as a function of spherocity $S_{\mathrm{O}}^{p_{\mathrm{T}}=1}$

Events can be categorized according to their azimuthal event topology. The unweighted transverse spherocity, $S_{\mathrm{O}}^{p_{\mathrm{T}}=1}$, is used to quantify the azimuthal distribution of tracks, and is defined as:

$$S_{\mathrm{O}}^{p_{\mathrm{T}}=1} = \frac{\pi^2}{4} \min_{\hat{n}} \left(\frac{\Sigma_i |\hat{p_{\mathrm{T},i}}|_{p_{\mathrm{T}}=1} \times \hat{n}|}{N_{\mathrm{trks}}} \right). \tag{1}$$

$S_{\mathrm{O}}^{p_{\mathrm{T}}=1}$ is measured using all tracks in each event, N_{trks}, and yields a value between 0 and 1. $S_{\mathrm{O}}^{p_{\mathrm{T}}=1}$ is calculated using charged-particle tracks that have a transverse momentum p_{T} greater than 0.15 GeV, although in Eq. 1 the weight factor for p_{T} is set to 1 (hence "unweighted"). The $S_{\mathrm{O}}^{p_{\mathrm{T}}=1} \to 0$ limit represents azimuthal topologies with back-to-back jet structures mainly dominated by a single hard scattering, and the $S_{\mathrm{O}}^{p_{\mathrm{T}}=1} \to 1$ limit describes event shapes where particles are isotropically distributed due to multiple softer collisions. In these proceedings, the Jetty and Isotropic event samples are defined as the bottom and top 20% quantiles of the $S_{\mathrm{O}}^{p_{\mathrm{T}}=1}$ distribution, respectively. The interpretation of the results relies on the premise that the azimuthal topology reflects the main mode of particle production.

The selected events are also required to be in the top 10% high-multiplicity quantile, estimated at mid-rapidity ($|\eta| < 0.8$), as well as having at least $N_{trks} \geq 10$, in order to maintain sensitivity to the event topology. There are also $S_{\mathrm{O}}^{p_{\mathrm{T}}=1}$ studies performed where the multiplicity is measured at forward rapidity, which are discussed in further detail in Ref. [3]. π,K,p are identified via the specific energy loss $\mathrm{d}E/\mathrm{d}x$ and the time-of-flight (TOF). ϕ and Ξ yields are then extracted from invariant mass distributions of their identified decay daughters ($\phi \to \mathrm{K}^+\mathrm{K}^-$ and $\Xi^-[\Xi^+] \to \pi^-[\pi^+] + \Lambda[\bar{\Lambda}]$).

2.1 Results

The K-to-π, ϕ-to-π and Ξ-to-π ratios as a function of $S_{\mathrm{O}}^{p_{\mathrm{T}}=1}$ are all presented in Fig. 1. EPOS-LHC is able to qualitatively describe the K-to-π trend, but neither EPOS-LHC nor PYTHIA is able to describe the ϕ-to-π and Ξ-to-π ratios as a function of $S_{\mathrm{O}}^{p_{\mathrm{T}}=1}$. However, both EPOS-LHC and PYTHIA are able to qualitatively describe the double-ratios (lower panels), which suggests that the simulations capture most of the $S_{\mathrm{O}}^{p_{\mathrm{T}}=1}$ selection but overestimate/underestimate the general enhancement of strange particle production in high-multiplicity events.

Both the K/π and Ξ/π ratios highlight an overall relative enhancement of K and Ξ hadrons compared to pions in Isotropic events, and a suppression in Jetty events. This suggests that strangeness enhancement in high-multiplicity pp collisions is driven by the multiple softer collisions found in isotropic topologies, rather than hard processes. The ϕ-to-π ratio suggests that the ϕ meson production is mostly insensitive to $S_{\mathrm{O}}^{p_{\mathrm{T}}=1}$, which differentiates it from other strange particle species.

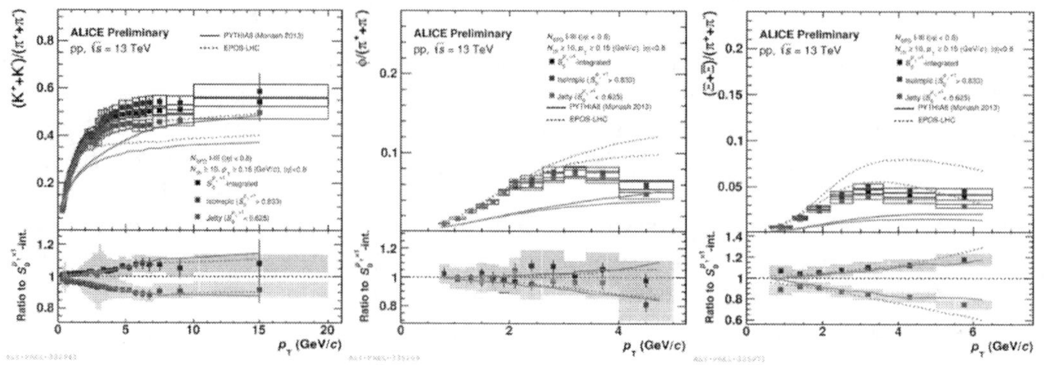

Figure 1: K-to-π (left), ϕ-to-π (middle) and Ξ-to-π (right) ratios as a function of p_{T} in bins of $S_{\mathrm{O}}^{p_{\mathrm{T}}=1}$, for Jetty (red), Isotropic (blue), and $S_{\mathrm{O}}^{p_{\mathrm{T}}=1}$-integrated (black) events. Lower panels show the ratios to $S_{\mathrm{O}}^{p_{\mathrm{T}}=1}$-integrated. The results are compared to predictions from PYTHIA8-Monash (solid lines) and EPOS-LHC (dashed lines).

3 Particle production as a function of R_{T}

The purpose of the R_{T} analysis is to estimate the strange particle production as a function of the size of the underlying event (UE), relative to a hard probe. The events are cut into different azimuthal segments relative to the azimuthal angle φ_{L} of the leading track (which has a p_{T} ≥ 5 GeV/c requirement to ensure that a hard scattering occurs).

The "Toward" region is located around the leading track, where $|\varphi - \varphi_L| \leq \frac{\pi}{3}$. The particle production in this region is dominated by jet fragmentation due to the hard scattering, in addition to the soft collisions in the UE. The "Transverse" regions are located perpendicular to the leading particle, where $\frac{\pi}{3} < |\varphi - \varphi_L| < \frac{2\pi}{3}$, and the particle production in these regions will almost be a pure sample of the UE (with possible local fluctuations due to initial-and final state radiation).

R_{T} is defined as the self-normalized charged particle density in the transverse region, $R_{\mathrm{T}} = N_{\mathrm{tracks}}^{\mathrm{Transverse}} / \left\langle N_{\mathrm{tracks}}^{\mathrm{Transverse}} \right\rangle$. With R_{T}, one is able to control the size of the UE relative to the hard scattering. Events with $R_{\mathrm{T}} < 1.0$ have a smaller-than-average UE, and vice versa for $R_{\mathrm{T}} > 1.0$. Particle momentum spectra are then measured in different R_{T} classes in order to estimate the impact of the UE on strange particle production.

3.1 Results

As seen in Fig. 2, the strange hadron yields in the Toward region show a significant dependence on R_{T}. This indicates not only that the relative strange particle production rate is proportional to R_{T}, but also that the relative strange particle production rate due to jet-fragmentation is fundamentally different from the same production rate in the UE. Furthermore, the relative yields in the high-R_{T} events in the Toward region approach the values of the (K, ϕ, Ξ)-to-π ratios found in the Transverse Region, although the R_{T} dependence in the Transverse region seems to be diminished. Notably, there is a very large modification of ϕ and Ξ hadron yields in the Toward region for events with large underlying event activity, with respect to the R_{T}-integrated spectra. This implies that at high-R_{T}, the contribution from the hard scattering is relatively small to the contribution from the UE, suggesting that the enhancement of strange hadrons at high-multiplicity pp collisions originates from the UE.

Contrasting the two models, both EPOS-LHC and PYTHIA qualitatively describe the K-to-π ratio, but neither is able to accurately predict the enhancement of ϕ and Ξ hadrons in high-R_{T} events. EPOS-LHC qualitatively reproduces the general trends going from low-

R_{T} to high-R_{T}, but seems to underestimate and overestimate the ϕ and Ξ high-R_{T} curves, respectively. On the other hand, while PYTHIA is able to describe the low-R_{T} trends in the Toward region, it is currently not capable of describing the enhancement found in the high-R_{T} events.

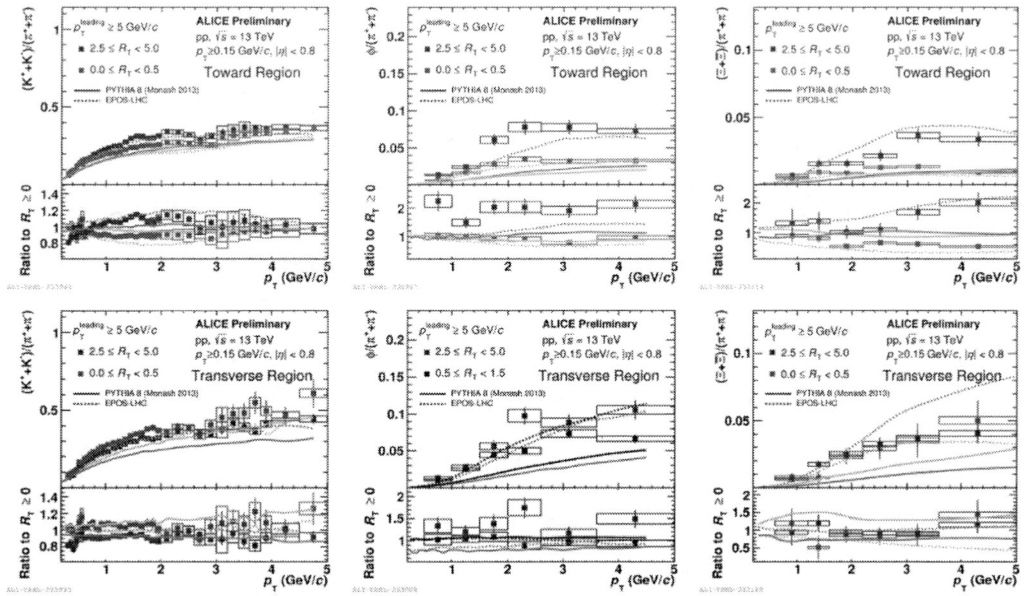

Figure 2: K-to-π (left), ϕ-to-π (middle) and Ξ-to-π (right) ratios as a function of p_{T} in bins of low-R_{T} (red, black) and high-R_{T} (blue) events. The upper/lower rows showcases the Toward and Transverse regions, respectively. The lower panels in each plot show the ratios to R_{T}-integrated. The results are compared to predictions from PYTHIA8-Monash (solid lines) and EPOS-LHC (dashed lines).

4 Conclusion

Using $S_{\mathrm{O}}^{p_{\mathrm{T}}=1}$, one is able to disentangle event topologies based on the azimuthal distribution of particles. Observations shown in Fig. 1 indicate that the enhancement of strange hadrons in high-multiplicity pp collisions is primarily driven by dynamics that arise from events with isotropic topologies. Likewise, Fig. 2 highlights how the R_{T} results support this indication. They suggest that the rate of strangeness production is proportional to the size of the UE, as the UE is produced through softer isotropic interactions. Moreover, they indicate that the rate of strangeness production in the UE outweighs the production due to jet-fragmentation. The measurements of R_{T} and $S_{\mathrm{O}}^{p_{\mathrm{T}}=1}$ suggest that strangeness enhancement in high-multiplicity pp collisions is driven by several soft collisions.

The role of the ϕ contrasting the two observables is striking to the author: the ϕ is insensitive to the $S_{\mathrm{O}}^{p_{\mathrm{T}}=1}$ measurement, unlike the K and Ξ, which have non-zero net strangeness. However, the R_{T} measurement shows a significant ϕ modification, alongside other strange particles. Narrower event criteria are being examined to gain further insight into this effect.

References

[1] T.Sjöstrand et al, *Computer Physics Communications* **191**, 159–177 (2015)
[2] T. Pierog et al, *Phys.Rev.C* **92** (2015) 3, 034906
[3] A. Nassirpour et al, *J.Phys.Conf.Ser.* **1602** (2020) 1, 012007

 https://doi.org/10.1051/epjconf/202225913006

Shear viscosity at finite baryon densities

E McLaughlin[1], J Rose[2], T Dore[3], P Parotto[4], C Ratti[5] and J Noronha-Hostler[3]

[1] Department of Physics, Columbia University, 538 W. 120th St., New York, NY 10027, USA
[2] Department of Physics, Dr. Karl-Remeis Observatory - Astronomical Institute, Bamberg D-96049, Germany
[3] Illinois Center for Advanced Studies of the Universe, Department of Physics, University of Illinois at Urbana-Champaign, Urbana, IL 61801, USA
[4] University of Wuppertal, Department of Physics, Wuppertal D-42097, Germany
[5] Department of Physics, University of Houston, Houston, TX 77204, USA

E-mail: egm2153@columbia.edu

Abstract. We use the excluded volume Hadron Resonance Gas (HRG) model with the most up-to-date hadron list to calculate $\eta T/w$ at low temperatures and at finite baryon densities ρ_B. This $\eta T/w$ is then matched to a QCD-based shear viscosity calculation of the QGP for different profiles of $\eta T/w$ across T,μ_B including cross-over and critical point transitions. When compared to ideal hydrodynamic trajectories across T,μ_B, we find that the $\eta T/w(T, \mu_B)$ profiles would require initial conditions at much larger baryon density to reach the same freeze-out point.

1. Introduction

Quark Gluon Plasma (QGP) has been observed in relativistic heavy ion collisions at both the Large Hadron Collider (LHC) and Relativistic Heavy Ion Collider (RHIC). In these relativistic heavy ion collisions at nearly zero baryon chemical potential, there is a smooth crossover phase transition from the QGP to HRG. However, experiments such as the RHIC Beam Energy Scan are looking for the existence of a critical point in the QGP to hadron gas phase transition at finite baryon densities.

From comparisons of theory to experimental data we know the shear viscosity to entropy density ratio, η/s, of the QGP at zero baryon chemical potential is known to be very small, close to the KSS bound [1]. It is not yet possible to calculate η/s directly from lattice QCD, which necessitates the use of other approaches such as the HRG model [2, 3, 4, 5], holography [1] and transport models [6]. However, in most of these models, one can only calculate $\eta T/w(T, \mu_B)$, the shear viscosity to enthalpy density ratio, in the hadronic phase (like in the cases of the HRG and transport models).

Here we present a framework [7] for connecting $\eta T/w(T, \mu_B)$ in the HRG phase to a parametrized QCD-based description of $\eta T/w(T, \mu_B)$ in the QGP phase. We construct four profiles of $\eta T/w(T, \mu_B)$ across the QCD phase diagram: two cases of a fully crossover transition and two cases with a critical point (CP), the first at $(T, \mu_B) = (143, 350)$ MeV, matching the BEST collaboration EOS [8], and the second at $(T, \mu_B) = (89, 724)$ MeV, to match the holographic prediction in [9].

2. Shear Viscosity in HRG Model

We use the HRG model with particles from the 2016 Particle Data Group (PDG16+) list [10], where we take hadrons to be point like particles and can then calculate the total pressure of the system as:

$$\frac{p(T,\mu_B,\mu_S,\mu_Q)}{T^4} = \sum_i (-1)^{B_i+1} \frac{g_i}{2\pi^2} \int_0^{\infty} p^2 \ln\left[1 + (-1)^{B_i+1} e^{(-\frac{\sqrt{p^2+m_i^2}}{T}+\tilde{\mu}_i)}\right] dp \quad (1)$$

To account for repulsive interactions in the hadron gas, we use the excluded volume approach, where we delegate a volume v to each hadron and then recalculate thermodynamic quantities taking into account this excluded volume. The shear viscosity of the HRG was then calculated at finite μ_B

$$\eta^{HRG} = \frac{5}{64\sqrt{8}} \frac{1}{r^2} \frac{1}{n} \sum_i n_i \frac{\int_0^{\infty} p^3 e^{-\sqrt{p^2+m_i^2}/T+\tilde{\mu}_i} dp}{\int_0^{\infty} p^2 e^{-\sqrt{p^2+m_i^2}/T+\tilde{\mu}_i} dp} \quad (2)$$

The $\eta T/w(T,\mu_B)$ in the HRG phase is normalized to $\eta/s = 0.08$ at $\mu_B = 0$ and splined to a parametrized $\eta/s(T)$ from [11] using a transition line across the (T,μ_B) plane [12], the final results are shown in Fig. 1. In the smooth crossover transition we see a longer flat region

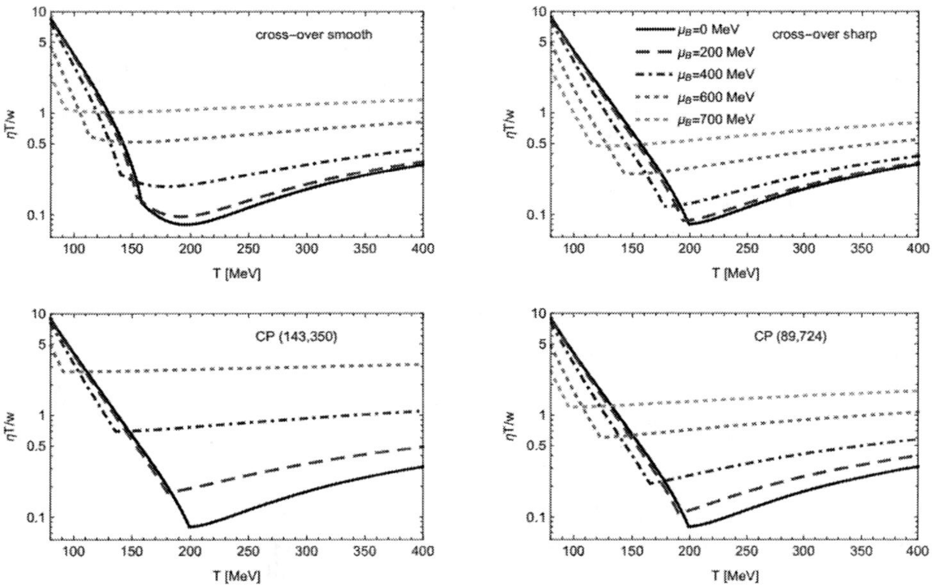

Figure 1. Temperature dependence of $\eta T/w(T,\mu_B)$ for all 4 transition profiles along slices of constant μ_B

in $\eta T/w$ that starts at around $\mu_B = 0$ and becomes more pronounced as you move to higher μ_B. The matching of the QGP to HRG occurs at a lower temperature leading to a stronger μ_B dependence in the smooth crossover transition. In comparison, in the sharp crossover the HRG to QGP matching occurs at a much higher temperature leading to $\eta T/w$ having a weaker μ_B dependence So the shape and location in the T,μ_B plane of the transition line from the deconfined to confined phase has a significant effect on how $\eta T/w$ is dependent on μ_B. For the

CP (143, 350) MeV transition there is also a strong μ_B dependence, since the critical point lies relatively close to $\mu_B = 0$, the transition line must fall steeply in temperature to pass through the critical point at (T, μ_B) = (143, 350) MeV.

3. Fluctuating Hydrodynamics

Variations in temperature and baryon chemical potential values probed by events at a specific beam energy can be caused by event-by-event fluctuations of the initial conditions, local temperature variations of the QGP, and entropy production due to the presence of viscosity. Different rapidity cuts may be used as a tool to probe these variations in T and μ_B throughout the collision space. Using the T and μ_B dependence of a single hydrodynamic event modeling a Au-Au collision at 19.6 GeV from [13], we construct the trajectory of $\eta T/w$ at central and forward space-time rapidity for three of our test cases, both crossover transitions and the (143, 350) MeV CP transition, as shown in Fig. 2 where the mean temperature and baryon chemical potential is shown in solid lines, while $\eta T/w$ from T and μ_B values ± 1 standard deviation from the mean are shown in dashed lines. In both the central and forward rapidity regions, the

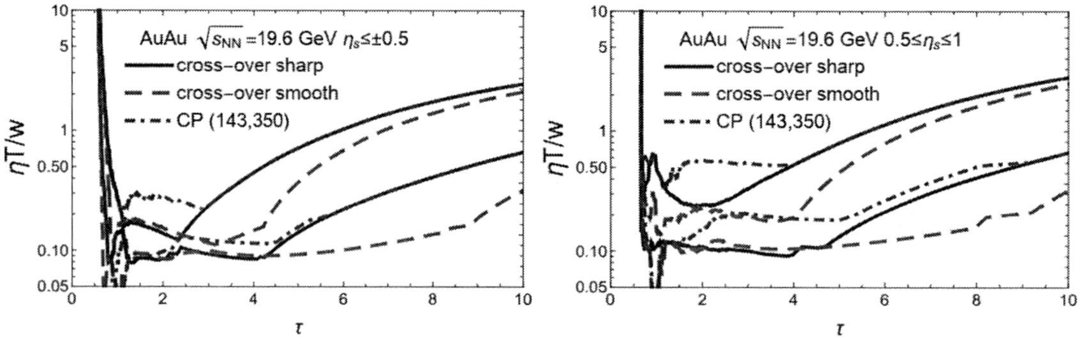

Figure 2. Approximation of $\eta T/w$ at central (left) and forward (right) rapidity probed within relativistic viscous hydrodynamics vs. time (using the (T, μ_B) from a single hydrodynamic background from Ref. [13])

different profiles have similar minimum values for $\eta T/w$. This is most likely due to the fact that the average μ_B values in these regions are still relatively small (~200-300 MeV). At forward rapidity, $\eta T/w$ at early times is slightly higher and has more variation in the ± 1 standard deviation range, so we expect that this region is not at large μ_B but farther from equilibrium.

4. Influence on $T - \mu_B$ trajectories

We present the trajectories for hydrodynamic runs following the setup in [14], beginning at T=500 MeV and various μ_B to compare the effect of using $\eta T/w(T, \mu_B)$ to ideal hydrodynamics (isentropic trajectory) or a constant value for η/s. Viscosity shifts to lower μ_B at freeze out because viscosity produces entropy as the system expands and cools. At large μ_B we see that the dependence of $\eta T/w$ on T and μ_B becomes more pronounced and the T, μ_B trajectory deviates further from both ideal hydrodynamics and viscous hydrodynamics with a constant η/s, as shown in Fig. 3. In order to achieve the same freeze out conditions as the isentropic results, viscous hydrodynamic runs would need to have initial conditions that start at a higher μ_B.

5. Conclusions

We have produced a framework for creating a functional $\eta T/w(T, \mu_B)$ that can be used an input into relativistic viscous hydrodynamic codes simulating HIC at low RHIC energies where finite

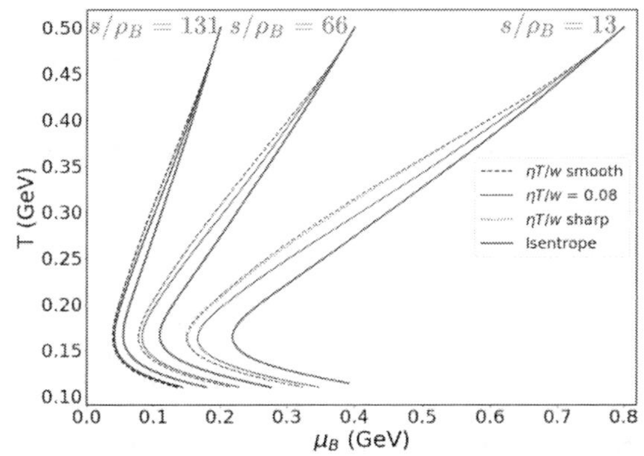

Figure 3. $\{T, \mu_B\}$ trajectories for various initial s/ρ_B using $\eta T/w = 0$, $\eta T/w = 0.08$, or $\eta T/w$ profiles.

μ_B is of interest. In our framework, the profiles of $\eta T/w$ across (T, μ_B) and minimum value of $\eta T/w$ along lines of finite μ_B are heavily determined by the transition line from QGP to HRG that one picks. The trajectories followed by a hydrodynamic system using our $\eta T/w(T, \mu_B)$ profiles push the trajectories towards a lower μ_B at freeze out when comparing to both ideal hydrodynamics ($\eta/s = 0$) and a constant $\eta/s = 0.08$. The initial conditions of these viscous runs would need to be at higher μ_B to achieve the same freeze out conditions as ideal hydrodynamics. Furthermore, to better model HIC at low energies and finite baryon denisites, future steps should be taken to compute a description of $\eta T/w$ across the full (T,μ_B, μ_S, μ_Q) space.

Acknowledgements

This material is based upon work supported by the the US-DOE Nuclear Science Grant No. DE-SC0019175; the NSF under grants no. PHY-1654219 and PHY-1560077; and the DFG grant SFB/TR55.

References

[1] Kovtun P, Son D T and Starinets A O 2005 *Phys. Rev. Lett.* **94** 111601 (*Preprint* hep-th/0405231)
[2] Noronha-Hostler J, Noronha J and Greiner C 2009 *Phys. Rev. Lett.* **103** 172302 (*Preprint* 0811.1571)
[3] Noronha-Hostler J, Noronha J and Greiner C 2012 *Phys. Rev. C* **86** 024913 (*Preprint* 1206.5138)
[4] Gorenstein M I, Hauer M and Moroz O N 2007 *Phys. Rev. C* **77** 214–220 (*Preprint* 0708.0137)
[5] Rischke D H, Gorenstein M I, Stoecker H and Greiner W 1991 *Z. Phys. C* **51** 485–490
[6] Rose J B, Torres-Rincon J M, Schäfer A, Oliinychenko D R and Petersen H 2018 *Phys. Rev. C* **97** 055204 (*Preprint* 1709.03826)
[7] McLaughlin E, Rose J, Dore T, Parotto P, Ratti C and Noronha-Hostler J 2021 (*Preprint* 2103.02090)
[8] Parotto P, Bluhm M, Mroczek D, Nahrgang M, Noronha-Hostler J, Rajagopal K, Ratti C, Schäfer T and Stephanov M 2020 *Phys. Rev. C* **101** 034901 (*Preprint* 1805.05249)
[9] Critelli R, Noronha J, Noronha-Hostler J, Portillo I, Ratti C and Rougemont R 2017 *Phys. Rev. D* **96** 096026 (*Preprint* 1706.00455)
[10] Alba P *et al.* 2017 *Phys. Rev. D* **96** 034517 (*Preprint* 1702.01113)
[11] Dubla A, Masciocchi S, Pawlowski J M, Schenke B, Shen C and Stachel J 2018 *Nucl. Phys. A* **979** 251–264 (*Preprint* 1805.02985)
[12] Borsanyi S, Fodor Z, Guenther J N, Kara R, Katz S D, Parotto P, Pasztor A, Ratti C and Szabo K K 2020 *Phys. Rev. Lett.* **125** 052001 (*Preprint* 2002.02821)
[13] Shen C and Schenke B 2019 *Nucl. Phys. A* **982** 411–414 (*Preprint* 1807.05141)
[14] Dore T, Noronha-Hostler J and McLaughlin E 2020 *Phys. Rev. D* **102** 074017 (*Preprint* 2007.15083)

Low-mass dielectron measurements with ALICE at the LHC

Raphaelle Bailhache[1,*] for the ALICE Collaboration

[1]Goethe Universitaet Frankfurt-am-Main, Germany

Abstract. Recent measurements of e^+e^- pair production in pp and p–Pb collisions at the center-of-mass energy $\sqrt{s_{NN}} = 5.02$ TeV are reported. Cold nuclear matter effects such as shadowing, as well as the possible presence of thermal radiation, are investigated in p–Pb collisions with the dielectron nuclear modification factor R_{pPb}. Furthermore, results on dielectrons at low $p_{T,ee}$ in peripheral Pb–Pb collisions at $\sqrt{s_{NN}} = 5.02$ TeV and in pp collisions at $\sqrt{s} = 13$ TeV are presented and compared to calculations.

1 Introduction

Dileptons and photons are unique tools to study the space–time evolution of the hot and dense matter created in ultra-relativistic heavy-ion collisions. They are produced during the entire history of the collision with negligible final state interaction. Therefore they carry undistorted information since the early stages of the collision. On top of the hadronic interactions, dielectrons can be produced in initial photon annihilation processes, triggered by the coherent electromagnetic (EM) fields of the incoming nuclei. The latter can be used to map the very strong, varying EM fields produced by the passing nuclei.
The data shown in this contribution were recorded with the ALICE experiment at the LHC during proton-proton (pp), proton-lead (p–Pb) and lead-lead (Pb–Pb) runs at a nucleon-nucleon center-of-mass energy $\sqrt{s_{NN}} = 5.02$ TeV and during a campaign with a reduced magnetic field of the ALICE solenoid in pp collisions at $\sqrt{s} = 13$ TeV. Electrons are tracked and identified at mid-rapidity ($|\eta_e| < 0.8$) in the central barrel with the Inner-Tracking-System (ITS), the Time Projection Chamber (TPC) and the Time-Of-Flight detector.

2 Dielectrons in pp and p–Pb collisions at $\sqrt{s_{NN}} = 5.02$ TeV

First, the measurement of the dielectron production in pp collisions at $\sqrt{s} = 5.02$ TeV is shown on the left panel of Fig.1 and constitutes an important baseline for studies in p–Pb and Pb–Pb collisions at the same $\sqrt{s_{NN}}$ [1]. The results, obtained for a transverse momentum of the electrons $p_{T,e} > 0.2$ GeV/c, are fully described by the expectations from known hadronic sources, i.e. the hadronic cocktail. Independent measurements of mesons are taken as input to compute the dielectron cross sections from light-flavour hadron and J/ψ decays, whereas the contributions from $c\bar{c} \rightarrow e^+e^-$ and $b\bar{b} \rightarrow e^+e^-$ are calculated with the POWHEG event generator and fitted to data in the dielectron invariant mass m_{ee} region $1.1 < m_{ee} < 2.7$ GeV/c^2.
The dielectron cross-section in p–Pb collisions is directly compared to the measurement in

*e-mail: raphaelle.bailhache@cern.ch

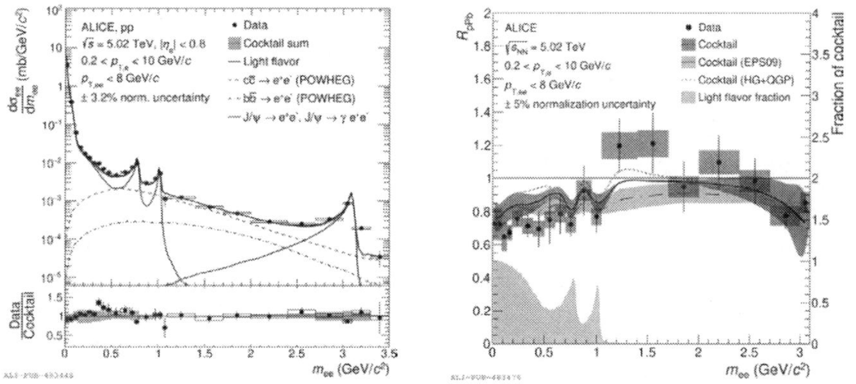

Figure 1. Left: dielectron production in pp collisions at $\sqrt{s}$ = 5.02 TeV compared to the expectations from known hadronic decays [1]. Right: dielectron nuclear modification factor in p–Pb collisions at $\sqrt{s_{\rm NN}}$ = 5.02 TeV compared to calculations [1].

pp collisions at the same $\sqrt{s_{\rm NN}}$ (= 5.02 TeV) via the dielectron nuclear modification factor $R_{\rm pPb}$ on the right panel of Fig.1 [1]. The $m_{\rm ee}$ dependence of $R_{\rm pPb}$ reflects the different scaling behaviours of the light-flavour ($m_{\rm ee}$ < 1 GeV/c^2) and heavy-flavour ($m_{\rm ee}$ > 1 GeV/c^2) productions. The results are compared to $c\bar{c}$ calculations including nuclear shadowing with the EPS09 nPDF and predictions including thermal radiation from the medium [1]. Both are disfavoured by the data in different $m_{\rm ee}$ ranges, i.e. 1.1 < $m_{\rm ee}$ < 2.7 GeV/c^2 and $m_{\rm ee}$ < 0.5 GeV/c^2, respectively. Future analyses as a function of the distance-of-closest approach (DCA) to the primary vertex of the e^+e^- pairs, as well as the charged-particle multiplicity in the event, would give the possibility to study possible prompt thermal contributions, separated from the heavy-flavour contribution.

3 Dielectrons in peripheral Pb–Pb collisions at $\sqrt{s_{\rm NN}}$ = 5.02 TeV

The preliminary results on the dielectron yield, integrated over $p_{\rm T,ee}$, in semi-peripheral (50-70%) and peripheral (70-90%) central Pb–Pb collisions at $\sqrt{s_{\rm NN}}$ = 5.02 TeV are compatible with cocktails of known hadronic sources, as can be seen on the left panel of Fig.2. The data are nevertheless also in agreement with calculations including thermal radiations from the medium [2]. For $p_{\rm T,ee}$ < 0.1 GeV/c, the vacuum hadronic cocktails clearly underestimate the measured e^+e^- yields shown on the right panel of Fig.2. The overshoot is more significant in more peripheral Pb–Pb collisions. The measured $p_{\rm T,ee}$ excess spectrum in 0.4 < $m_{\rm ee}$ < 0.7 GeV/c^2 for peripheral Pb–Pb collisions, in the left panel of Fig.3, is fairly well reproduced by predictions for the coherent photoproduction of dielectrons ($\gamma\gamma \rightarrow e^+e^-$) using the Wigner distribution formalism to take into account the impact parameter dependence of the electromagnetic fields [3]. This process leads to a $m_{\rm ee}$ continuum of very soft e^+e^- pairs in $p_{\rm T,ee}$, whose centrality dependence is much milder than that of the known hadron decays or thermal radiations in hadronic collisions. Therefore its relative contribution increases in peripheral collisions.

4 Soft dielectron production in pp collisions at $\sqrt{s}$ = 13 TeV

In the past decades, several experiments reported an anomalous soft-photon and dilepton excess in hadron-hadron collisions [4]. In pp collisions at $\sqrt{s}$ = 13 TeV, a dedicated data-

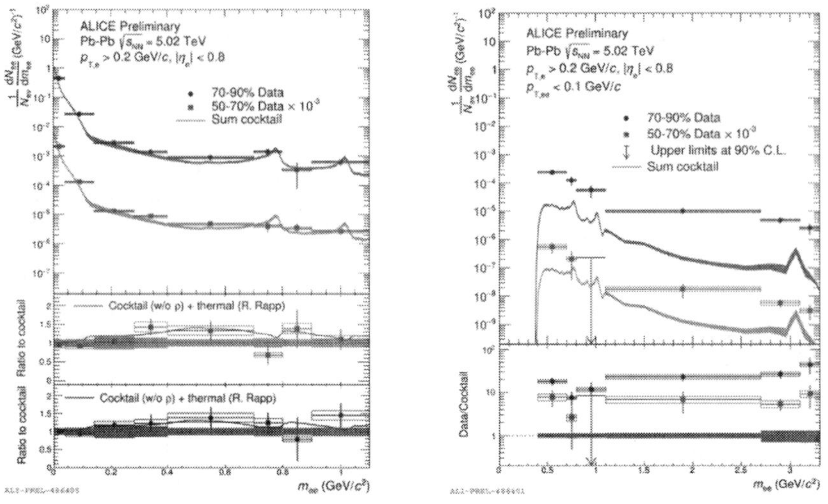

Figure 2. Dielectron yields in semi-peripheral (50-70%) and peripheral (70-90%) Pb–Pb collisions integrated over $p_{T,ee}$ (left panel) and for $p_{T,ee}$ < 0.1 GeV/c (right panel) compared to the hadronic cocktail including or not thermal radiations.

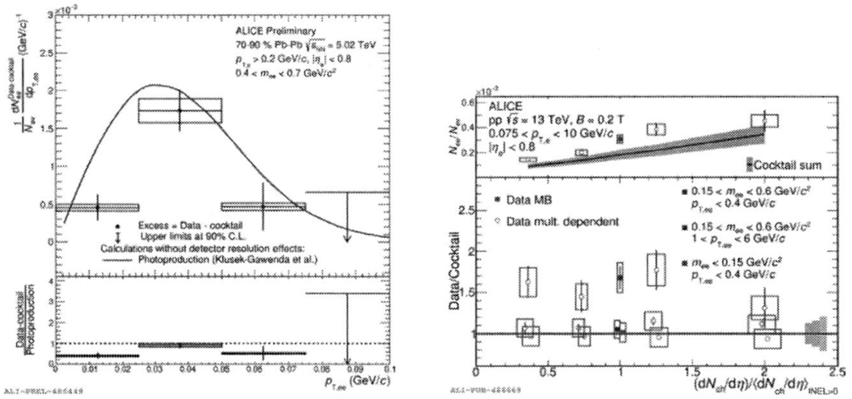

Figure 3. Left: excess dielectron yield as a function of $p_{T,ee}$ in $0.4 < m_{ee} < 0.7\,\mathrm{GeV}/c^2$ for peripheral Pb–Pb collisions compared to calculations for $\gamma\gamma \rightarrow e^+e^-$ [3]. Right upper panel: e^+e^- yield in pp collisions at $\sqrt{s}$ = 13 TeV in the excess region as a function of the event multiplicity compared with the hadronic cocktail. Right lower panel: enhancement factor data/cocktail in pp collisions at $\sqrt{s}$ = 13 TeV for three different kinematic regions [4].

taking campaign with a reduced magnetic field of the ALICE solenoid (0.2 T) allows the study of soft e^+e^- production ($p_{T,e} > 75\,\mathrm{MeV}/c$), previously inaccessible at the LHC. The measured dielectron cross section is above the expectations from the known hadronic decays by a factor 1.61 ± 0.13 (stat.) ± 0.17 (syst.) ± 0.34 (cocktail) in $0.14 < m_{ee} < 0.6\,\mathrm{GeV}/c^2$ for $p_{T,ee} < 0.4\,\mathrm{GeV}/c$ [4]. In the right upper panel of Fig.3, the multiplicity dependence of the dielectron yield is shown in orange and compared to the one of the hadronic cocktail. No clear multiplicity dependence of the data-to-cocktail ratio is observed within the uncertainties. For comparison, this ratio is also plotted for lower m_{ee} (blue) or higher (green) $p_{T,ee}$, where no excess is found. The excess of the acceptance corrected yield can be reproduced neither by

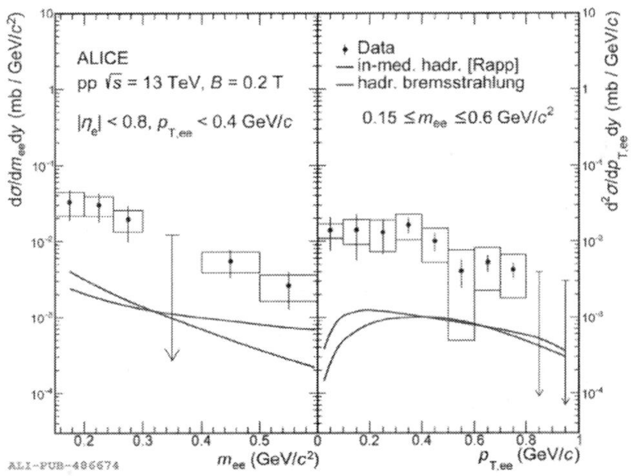

Figure 4. Dielectron excess spectra as a function of m_{ee} (left) and $p_{T,ee}$ (right) compared to calculations of bremsstrahlung from initial- and final-state hadrons, and thermal dielectron production in pp collisions at $\sqrt{s}$ = 13 TeV [4].

calculations for bremsstrahlung from initial- and final-state hadrons, nor by prediction for thermal dielectron production [4], as can be seen in Fig.4. A possible mechanism recently proposed to explain the data is a back-reaction of the QCD vacuum to non-perturbative $q\bar{q}$ pairs [5], only visible with real or virtual photons.

5 Summary and outlook

The measured dielectron cross section in pp collisions at $\sqrt{s} = 5.02$ TeV for $p_{T,e} > 0.2$ GeV/c provides a well-understood baseline for p–Pb and Pb–Pb studies at the same center-of-mass energy. In p–Pb collisions at $\sqrt{s_{NN}} = 5.02$ TeV, no significant modification of the heavy-flavour production is observed within the experimental uncertainties. Analyses as a function of the pair DCA will help to disentangle the heavy-flavour contribution from possible thermal radiations. In peripheral Pb–Pb collisions at $\sqrt{s_{NN}} = 5.02$ TeV, the measured dielectron yields clearly overshoot the hadronic cocktail for $p_{T,ee} < 0.1$ GeV/c. The excess is fairly well reproduced by calculations for coherent photoproduction of e^+e^- pairs. Analyses of high-statistics central Pb–Pb collisions are still ongoing. In pp collisions at $\sqrt{s} = 13$ TeV, the soft dielectron production ($p_{T,e} > 0.75$ GeV/c) is measured for the first time at LHC energies. An enhancement over the hadronic cocktail is found, whose multiplicity dependence is still inconclusive. Thanks to the ALICE ITS and TPC upgrades in Run 3 and 4, the DCA resolution and the readout rate will be significantly improved, leading, for example, to an increase of statistics by a factor 400 for soft-dielectron studies in pp collisions [6].

References

[1] ALICE Collaboration, Phys. Rev. C**102**, 055204 (2020)
[2] R. Rapp, Adv. High Energy Phys. **2013**, 148253 (2013)
[3] M.Klusek-Gawenda et al., Phys. Lett. B**814**, 136114 (2021)
[4] ALICE Collaboration, Phys. Rev. Lett. **127**, 042302 (2021)
[5] O. Garcia-Montero, arXiv:2104.07050
[6] ALICE Collaboration, CERN-LHCC-2020-018

Machine Learning Application for Λ Hyperon Reconstruction in CBM at FAIR

Shahid Khan[1,*], *Viktor* Klochkov[1], *Olha* Lavoryk[2], *Oleksii* Lubynets[3,4], *Ali Imdad* Khan[1], *Andrea* Dubla[3], and *Ilya* Selyuzhenkov[3,5]

[1]Eberhard Karls University of Tübingen; [2]Taras Shevchenko National University of Kyiv; [3]GSI, Darmstadt; [4]Goethe University Frankfurt; [5]NRNU MEPhI, Moscow

Abstract. The Compressed Baryonic Matter experiment at FAIR will investigate the QCD phase diagram in the region of high net-baryon densities. Enhanced production of strange baryons, such as the most abundantly produced Λ hyperons, can signal transition to a new phase of the QCD matter. In this work, the CBM performance for reconstruction of the Λ hyperon via its decay to proton and π^- is presented. Decay topology reconstruction is implemented in the Particle-Finder Simple (PFSimple) package with Machine Learning algorithms providing efficient selection of the decays and high signal to background ratio.

Introduction. Theoretical calculations predict the possibility of a first order phase transition from hadron gas to a deconfined phase of strongly interacting matter and the existence of a critical point in the region of the QCD phase diagram above 450 MeV chemical potential (μ_B) and below 180 MeV temperature [1]. The experiments at SIS-18, CERN SPS, RHIC beam energy scan and fixed target programs, and future experiments at the Facility for Antiproton and Ion Research (FAIR) and NICA facilities are dedicated to explore this region of the phase diagram. The Compressed Baryonic Matter (CBM) experiment at FAIR in Darmstadt will investigate the region at 5–8 normal baryon density ($\mu_B \approx 450 - 900$ MeV) and temperatures below 120 MeV. One of the signatures of the new phase of matter is the enhanced production of strange baryons and Λ is the most abundantly produced hyperon at FAIR SIS-100 energies. In this work, the CBM performance for reconstruction of the Λ hyperon via its decay to proton and π^- using Machine Learning algorithms to achieve an efficient selection of the decays and high signal to background ratio is presented.

CBM experiment setup. The CBM experiment consists of three main components: a tracking system inside a dipole magnet, detectors for particle identification and detectors for collision geometry determination. The tracking system consists of a Micro Vertex Detector (MVD) and a Silicon Tracking System (STS). The particle identification is provided by the Ring Imaging Cherenkov detector, Muon Chamber, Transition Radiation Detector and Time-of-Flight (TOF) wall. Geometry determination is done with the help of Projectile Spectator Detector, inner part of TOF and the tracking system. For this analysis a sample of 100*k* minimum bias Au-Au collisions at 12*A* GeV/*c* was generated using DCM-QGSM-SMM [2] (Model A) and URQMD [3] (Model B) with collision products transported through the CBM setup using GEANT4 [4].

Machine learning. In the already existing Kalman Filter Particle Finder (KFPF) package [5] for online reconstruction and selection of short-lived particles in CBM, selection

*e-mail: shahidzafarkhan@gmail.com

criteria have been manually optimized to maximize signal to background ratio for a certain collision energy and a heavy-ion event generator. The selection criteria depend on the collision energy and centrality, decay channel and detector configuration. Machine Learning (ML) algorithms can be used to adjust these criteria automatically for different data taking scenarios. ML provides an efficient non-linear and multi-dimensional selection criteria optimization. To optimize selection criteria for Λ and other particles, ML can use many variables, associated with particle candidates, provided they are not strongly correlated with the invariant mass of the particle. In this work, we study ML performance using the same variables as used by the KFPF. Multiple ML libraries were tested for bench-marking using an automated machine learning (automl) package. The XGBoost [6] model outperformed other models in terms of numerical calculation speed and efficiency. XGBoost uses decision trees and a gradient descent optimization algorithm to minimize the loss function.

Details of the Λ reconstruction in CBM can be found in [7]. Cellular automaton and Kalman Filter Particle (KFParticle) based PFSimple package [7] are used for track finding, fitting and decay kinematics reconstruction. PFSimple interfaces the mathematics of the KFParticle package and provides a convenient interface to control the reconstruction parameters. The Λ candidates constructed from proton and π^- track pairs coming from a true Λ decay are termed as signal while all other candidates are considered as background.

The list of variables used for ML includes a squared distance between the daughter tracks and primary vertex (PV) divided by its error, the closest distance between the tracks (DCA), a squared distance between daughter tracks divided by its error, and distance between PV and secondary vertex divided by its error. A set of preselection criteria are applied to the data to remove numerical artefacts. A central Au-Au collision contains mostly combinatorial background and way less signal for Λ. Since the data has imbalanced classes, the data set used to train and test the ML algorithm is over-sampled by increasing the ratio of signal (under-represented class) to background candidates. Model A is treated as simulated signal and used for efficiency calculation while Model B is treated as background (which in real data analysis is taken from experimental data).

A sample of about 1M of Λ signal from Model A in the 5σ region around the Λ peak in the invariant mass distribution and a sample of about 3M of the background from Model B on the left and right hand sides outside of the Λ peak region and below 1.3 Gev/c^2 were used for the analysis. Using background from a different model (Model B) helps to reduce over fitting and dependence of the ML result on the signal model. These samples are divided in proportion 80% to 20% into train-test samples. To tune various parameters of the algorithm so that it fits better on the training data and performs well on the test data, Bayesian optimisation [8] package is used. It divides the train data into 5 folds and searches for optimal values of parameters using parallel computing. Area Under the Curve (AUC) parameter is used to select the best model. The trained model is then applied on the test data set and it returns a probability distribution between 0 and 1 for given input variables as shown in Figure 1 (left). Candidates which are close to 0 are most probably background and vice versa for candidates close to 1. Based on this probability distribution an approximate median significance (AMS) [9] is calculated to maximise signal to background ratio.

Results and discussion. The trained and tested XGB model, along with the AMS selection criteria, is then applied on two samples for both Models A and B containing $100k$ events each. Figure 1 (right) shows Λ candidates from Model B (UrQMD) before and after XGB selection is applied. For differential analysis the kinematic phase space is divided in transverse momentum (p_T) intervals of 0.5 GeV/c wide and laboratory rapidity (y_{lab}) intervals of 0.5 step size. Efficiency of the ML selection is calculated by dividing the number of remaining true signal (true positives in the confusion matrix) in the XGB selected Λ candidates by the number of input signal for each p_T-y_{lab} interval.

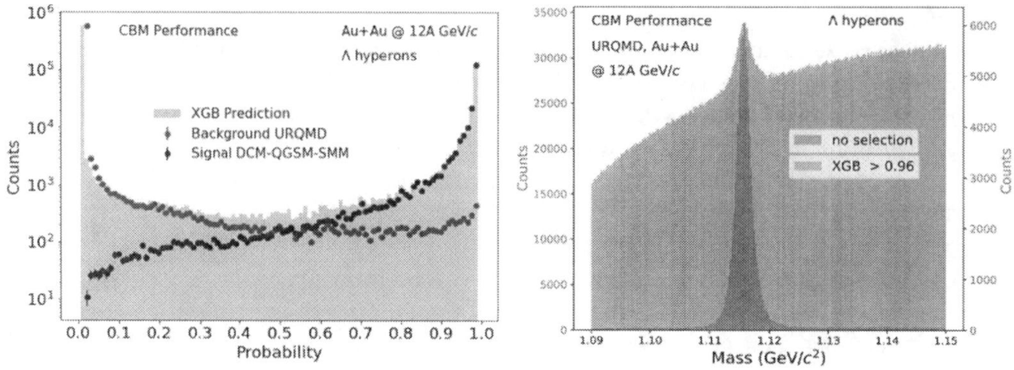

Figure 1. (left) Performance of the XGB algorithm for signal–background separation in the Λ candidates. The arrow represents the maximum AMS value at which selection was applied. (right) The invariant mass distribution of Λ candidates before (blue) and after XGB selection (red).

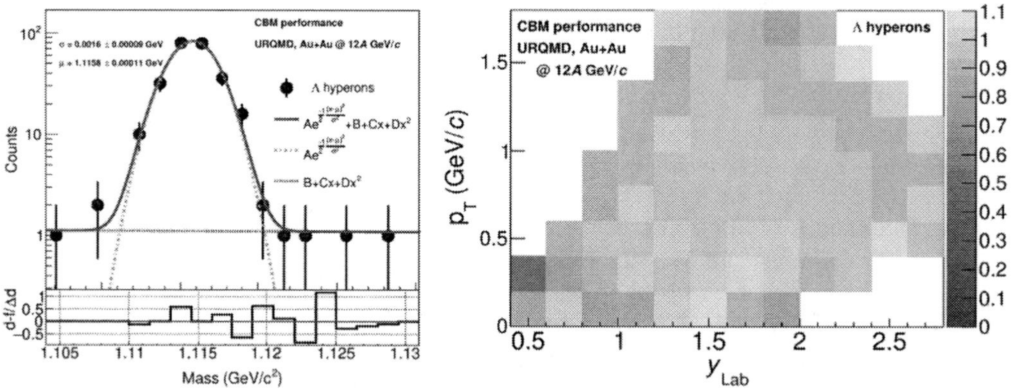

Figure 2. (left) Illustration of the procedure for Λ signal extraction after XGB selection is applied. Example is for $1.4 < p_T < 1.6$ GeV/c and $2.2 < y_{lab} < 2.4$ interval. The invariant mass yield (black circles) is fitted with the signal (Gaussian, green dotted line) and background (second order polynomial, blue line) shapes. (right) The p_T-y_{lab} distribution for the ratio of the fully corrected yield extracted after XGB selection to the simulated input.

The yield extraction is implemented as a three stage fitting procedure on Model B to each p_T-y_{lab} interval. Only the ranges which contain more than 400 candidates are considered. In stage one, the invariant mass distribution of the background, Λ candidates outside the 5σ region from the Λ peak at 1.115 GeV/c^2, is fitted with a second order polynomial (*pol*2). In the second stage, the invariant mass distribution is fitted in full range with a sum of a Gaussian ($A\exp[((x-\mu)/\sigma)^2/2]$) and *pol*2 functions. For this fit the mean (μ) and standard deviation (σ) of the Gaussian function are fixed to $\mu = 1.115$ GeV/c^2 and $\sigma = 0.0014$ GeV/c^2, while the initial values of the *pol*2 are taken from the fit at stage one. In the final stage, all parameters of the Gaussian and pol2 are released with their initial values set to the result of the fit at stage two. The number of entries under the Gaussian fit only, in the 2.5σ region around μ, are classified as Λ signal. Figure 2 (left) shows an example of the final result from the fitting procedure. The corrected Λ yield for each p_T-y_{lab} interval is extracted by dividing the signal yield from the fitting procedure with the efficiency correction factor obtained from Model A. The corrected yield divided by the simulated yield is plotted in Figure 2 (right). Figure 3 demonstrates good agreement of the corrected yield with the simulated one as a function of p_T (left) and y_{lab} (right).

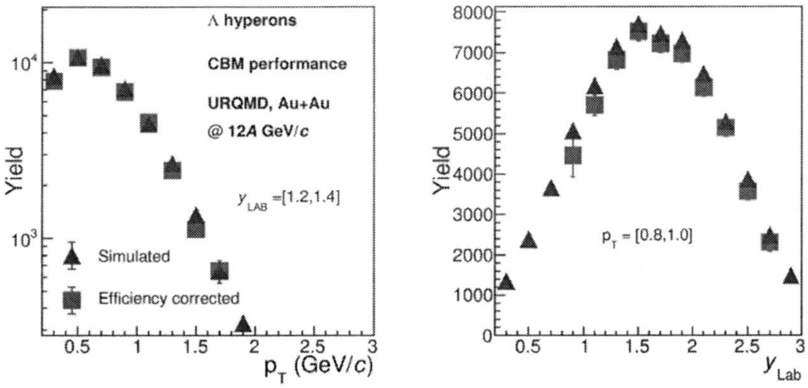

Figure 3. Yield of Λ hyperons as a function of (left) transverse momentum, p_T, for rapidity $1.2 < y_{lab} < 1.4$ and (right) rapidity for $0.8 < p_T < 1.0$.

Summary. The CBM performance for reconstruction of the Λ hyperon via its decay to proton and π^- in Au-Au collisions for the top FAIR SIS-100 beam momentum is presented. By combining the precise Kalman-Filter based reconstruction of the decay topology and machine learning algorithms a multi-differential analysis of the Λ yield in p_T and y_{lab} was performed. Reliable extraction of the Λ hyperon is demonstrated by comparison of the efficiency corrected yields after the reconstruction and ML selection with the simulated input for different heavy-ion event generators DCM-QGSM-SMM and UrQMD. In future, the method will be deployed for different collision energies and other strange and multi-strange particles and (hyper-)nuclei decays.

Acknowledgements. This work was supported by the Carlo and Karin Giersch Stiftung, HGS-HIRe Graduate School by HIC for FAIR, the Ministry of Science and Higher Education of the Russian Federation, Project "Fundamental properties of elementary particles and cosmology" No 0723-2020-0041, the Russian Foundation for Basic Research (RFBR) funding within the research project no. 18-02-40086, and the European Union's Horizon 2020 research and innovation program under grant agreement No. 871072.

References

[1] Fu et.al., Physical Review D **101**, 054032 (2020)
[2] Baznat et.al., Physics of Particles and Nuclei Letters **17**, 303 (2020)
[3] Bass et.al., Progress in Particle and Nuclear Physics **41**, 255 (1998)
[4] Allison et.al., IEEE Transactions on nuclear science **53**, 270 (2006)
[5] Zyzak, PhD thesis, 165 (2016)
[6] Chen et.al., Proceedings for "Knowledge discovery and data mining", **22**, 785 (2016)
[7] Lubynets et.al., Particles **4**, 288–295 (2021)
[8] Fernando Nogueira, https://github.com/fmfn/BayesianOptimization
[9] Adam-Bourdarios et.al., http://higgsml.lal.in2p3.fr/documentation, **9**, (2014)

EPJ Web of Conferences **259**, 13009 (2022) https://doi.org/10.1051/epjconf/202225913009
SQM 2021

Directed flow of D mesons at RHIC and LHC energy within a transport approach: non-perturbative dynamics, vorticity and electromagnetic fields

Salvatore Plumari[1,2,*], *Lucia* Oliva[3], *Yifeng* Sun[2], and *Vincenzo* Greco[1,2]

[1]Department of Physics and Astronomy 'E. Majorana', University of Catania, Via S. Sofia 64, I-95123 Catania

[2]INFN-Laboratori Nazionali del Sud, Via S. Sofia 62, I-95123 Catania, Italy

[3]Institut für Theoretische Physik, Johann Wolfgang Goethe-Universität, Max-von-Laue-Str. 1, 60438 Frankfurt am Main, Germany

Abstract. We study the propagation of charm quarks in the quark-gluon plasma (QGP) by means a relativistic Boltzmann transport (RBT) approach coupled to electromagnetic field. The interplay between these fields is responsible to generate large rapidity odd directed flow v_1 of D mesons and for a large splitting of directed flow Δv_1 between neutral D and anti-D mesons. We show that the large v_1 is generated by the longitudinal asymmetry between the bulk matter and the charm quarks and by a large non-perturbative interaction in the QGP medium.

1 Introduction

Charm quarks are the earliest charged particles appearing in the ultra relativistic Heavy Ion collisons (uRHICs). Due to their short formation time with $\tau \approx 0.1\ fm/c$ and a thermalization time which is comparable to the QGP lifetime they are good probes of both the initial stage and the subsequent evolution into a thermalized QGP. The matter produced in uRHICs is not only in the regime of extreme temperature and density but also in the regime of large vorticity due to huge initial angular momentum and intense electromagnetic (e.m.) fields that are the two possible sources for a finite directed flow v_1. Recent theoretical studies have shown that the heavy quarks can manifest a large directed flow which is one order of magnitude larger than the one of light hadrons [1, 2]. This was predicted for the average directed flow of D mesons [2] and was predicted for the first time a splitting in the directed flow between D^0 and $\bar{D}^0$ mesons induced by e.m. field [1]. In these first theoretical investigations a Langevin approach coupled to the Maxwell equations have been used. Recent experimental data from STAR collaboration [3] at top RHIC energies and from ALICE collaboration [4] at LHC energies have shown a directed flow larger than the one of light charged hadrons [5]. Recently, in Ref.s [6, 7], important features of the directed flow splitting Δv_1 induced by e.m. fields have been found suggesting that measurement of Δv_1 of D mesons and leptons form Z^0 decay and their correlations could be powerful tool to probe the initial e.m. fields in uRHICs.

*e-mail: salvatore.plumari@dfa.unict.it

2 Charm quarks transport equation in the electromagnetic field

We discuss results on the directed flow obtained within a relativistic Boltzmann transport approach developed to study the dynamics of heavy-ion collisions and recently extended to describe anisotropic flows v_n and the nuclear modification factor R_{AA} of D mesons [8–13]. The dynamical evolution of the QGP bulk (gluons and light quarks) as well as of charm quarks in the QGP bulk is described by the following equations:

$$\left[p_\mu \partial_x^\mu + q_j F_{\mu\nu}(x) p^\nu \partial_p^\mu\right] f_j(x,p) = C_{22}[f_j, f_k](x,p) \quad (1)$$

$$\left[p_\mu \partial_x^\mu + q_Q F_{\mu\nu}(x) p^\nu \partial_p^\mu\right] f_Q(x,p) = C_{22}[f_j, f_k, f_Q](x,p) \quad (2)$$

where $f_{j,k}(x,p)$ is the phase-space distribution function of quarks and gluons $j,k = g,q,\overline{q}$ while $f_Q(x,p)$ is the phase-space distribution function of charm quarks $Q = c,\overline{c}$. In the above equations the $F_{\mu\nu}$ is the electromagnetic strength tensor of the external e.m. field. On the right-hand sides C_{22} is the collision integral accounting for $2 \to 2$ scattering processes. We have employed a bulk with massive quarks and gluons given by a Quasi-Particle Model (QPM) [14] where quarks and gluons have a thermal masses $m_{g,q}(T) \propto g(T)T$ tuned to reproduce the lattice QCD thermodynamics [15]. The charm quarks interact with the bulk by means of $2 \to 2$ elastic scattering, see [12] for details of the approach used.

The e.m. field included in the transport equations are computed following Ref. [16] and the results shown have been obtained assuming a constant electrical conductivity σ_{el} of the QGP. The values of σ_{el} used are in agreement with the lattice QCD calculations [17] and correspond to typical values that can be explored in uRHICs. The total e.m. field produced by the two nuclei is obtained solving the Maxwell equations. The approach consists first to calculate the elementary e.m. fields generated by single point-like charges e located in the transverse plane and moving towards $+z$ and $-z$ respectively with speed β and then these field are convoluted with the transverse charge distribution of the nuclei ρ to get the total e.m. field. In Fig. 1 it is shown the time evolution of e.m. field for two different collision energies

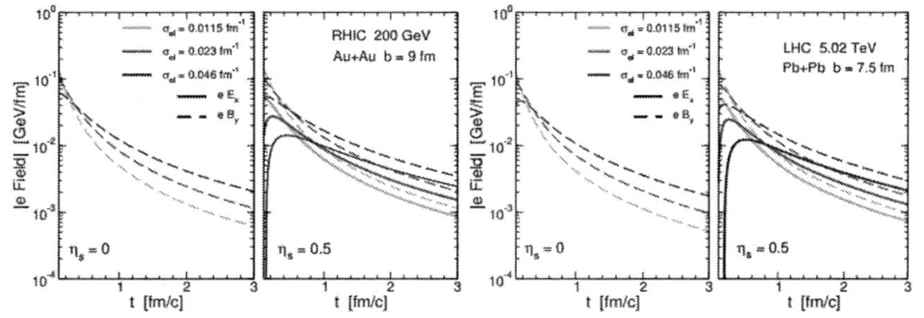

Figure 1. Time evolution of $|eE_x|$ (solid curves) and $|eB_y|$ (dashed curves) in Au+Au collisions at $\sqrt{\sigma_{NN}} = 200$ GeV with $b = 9$ fm (left panel) and Pb+Pb collisions at $\sqrt{\sigma_{NN}} = 5.02$ TeV with $b = 7.5$ fm (right panel). The different colours correspond to different values of the electric conductivity σ_{el}.

at top RHIC energies (left panel) and LHC energies (right panel). The curves are for two values of the space-time rapidity: $\eta_s = 0$ and $\eta_s = 0.5$ and for different values of the electric conductivity as computed in lattice QCD [17]. As shown, at mid-rapidity $\eta_s = 0$ the electric field vanishes due to symmetry while at forward and backward rapidity E_x become huge and comparable to B_y. The $|eB_y|$ decrease by one order of magnitude in less than 1 fm/c and the decrease is stronger for smaller electric conductivity. On the other hand, at larger η_s the time evolution of the $|eB_y|$ becomes milder.

In coordinate space the bulk it has been initialized by using the standard Glauber model assuming boost invariance along the longitudinal direction. In momentum space we have considered a mixture of Boltzmann-Jüttner distribution function up to transverse momentum $p_T^{mj} = 2GeV$ at RHIC ($p_T^{mj} = 3.5$ GeV at LHC) while mini-jet distributions as calculated by pQCD at NLO order in [18] at higher transverse momentum. The charm quark distributions are initialized in coordinate space by using N_{coll}. In momentum space we use the charm quark distribution according to the Fixed Order + Next-to-Leading Log (FONLL) calculation [19]. In a non-central collision, the system of the two incoming nuclei possesses an angular momentum J which depends on the collision energy and impact parameter. After the collision a fraction of the angular momentum is transferred to the plasma created which manifests in a nonzero vorticity of the system. In hydrodynamical simulations J has been introduced as an asymmetric initial energy density distribution in the longitudinal direction [20, 21]. We show the results obtained in [22] where similar initial conditions have been implemented in a transport approach in order to take into account the nonzero angular momentum. We show

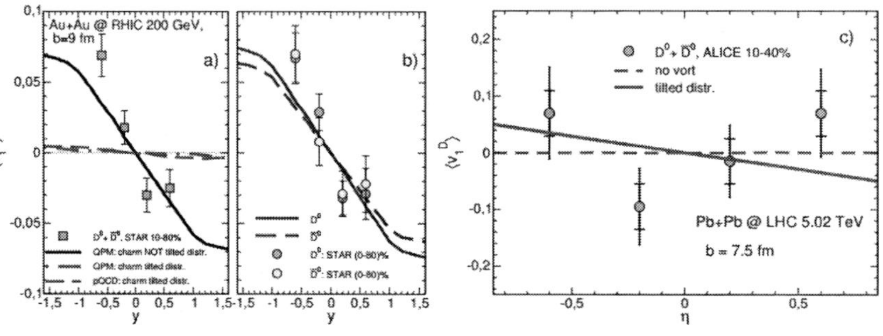

Figure 2. $v_1(y)$ of D mesons as a function of rapidity at RHIC (left) and LHC (right). In panel a) the black solid line is the case where bulk is tilted but charm distribution not while the green dot-dashed is the case where charm quark are distribuited with a tilted distribution like the bulk. The red dashed line corresponds to the case where charm quarks interact according to a pQCD interaction. In panel b) it is shown the effect of the e.m. field on the v_1 of D^0 meson (blue solid line) and $\bar{D}^0$ meson (blue dashed line). The experimental data are taken from [3]. In panel c) the solid green line is the case where the bulk has a tilted distribution while the red dashed line is the case without vorticity with a bulk forward and backward symmetric. The experimental data are taken from [4].

how the rapidity dependence of the directed flow v_1 of neutral D mesons is originated by the vorticity of the bulk matter coming from the initial tilted longitudinal distribution and the non-perturbative interaction of charm quarks. As shown in Fig. 2 a) if the initial charm quarks are initialized according to a tilted distribution, as done for the bulk, the large v_1 of the charm quarks disappears and the final v_1 is very small, see green dot-dashed lines shown in Fig. 2 a). On the other hand, the large v_1 shown by the black solid line is due to a pressure gradient of the bulk to the charm quarks toward the negative x−direction at positive η_s. The effect of this pressure gradient comes from the non-perturbative interaction of the HQ with the bulk. In fact, if one keeps the non-tilted space distribution but assume a drag and diffusion of the charm quarks according to pQCD interaction one get a very small v_1 as shown by the red dashed line in Fig. 2 a). The charm quark develops a finite v_1 only if the initial distribution of charm quark is not tilted and charm quarks interact non-perturbatively with an interaction coming from QPM. However, in addition to this dynamics there is also a motion induced by the e.m. field produced in the collision. In Fig. 2 b) it is shown the effect of the e.m. fields on the directed flow of D mesons for Au+Au collisions at top RHIC energy. The value of the electric conductivity used is $\sigma_{el} = 0.023$ fm^{-1} which is the value from lattice QCD calculation at $T = 2T_c$. As shown in Fig. 2 b) at forward rapidity positively-charged particles get a negative contribution to the v_1 and negatively-charged particles get a positive contribution to the v_1. This suggest that the overall effect is slightly dominated by the electric field. Finally, in Fig. 2 c) the results for D meson are shown in comparison with the experimental data at LHC energies. The dashed line correspond to the standard initialization while the solid line is the result with tilted initial distribution for the bulk where both simulations includes the e.m. fields. The ALICE data indicate a positive slope for the combined directed flow of D meson that is smaller than the one observed at RHIC.

3 Conclusion

The directed flow $v_1(y)$ of charm quarks it has been studied within a transport approach at both RHIC and LHC energy. We have observed that the origin of the directed flow for the D meson comes from two process: i) the rotation induced by the initial angular momentum of the spectators transferred to the bulk matter which produce a $v_1(y)$ of D^0 and $\bar{D}^0$ mesons several times larger than the one of the light mesons and ii) the interaction of HQ with the bulk matter which is non-perturbative and is responsible for the transfer of the gradient pressure of the bulk matter to the charm quarks. We have studied the splitting of $v_1(y)$ of D mesons due to the e.m. field. It has been shown that this effect is in agreement with the experimental data at RHIC energy, but due to large error bars currently it is not possible to draw any conclusions. At LHC energy theoretical models based on electromagnetic field that assume a medium at constant electric conductivity are not able to account for it.

References

[1] S.K. Das et al., Phys. Lett. B **768**, 260 (2017).
[2] S. Chatterjee, P. Bożek, Phys. Rev. Lett. **120**, 192301 (2018).
[3] J. Adam et al. (STAR), Phys. Rev. Lett. **123**, 162301 (2019).
[4] S. Acharya et al. (ALICE), Phys. Rev. Lett. **125**, 022301 (2020).
[5] L. Oliva, Eur. Phys. J. A **56**, no.10, 255 (2020).
[6] Y. Sun, S. Plumari, V. Greco, Phys. Lett. B **816**, 136271 (2021).
[7] Y. Sun, S. Plumari, V. Greco, Eur. Phys. J. C **80**, 16 (2020).
[8] M. Ruggieri, F. Scardina, S. Plumari, V. Greco, Phys. Lett. B **727**, 177 (2013).
[9] S. Plumari, G.L. Guardo, F. Scardina, V. Greco, Phys. Rev. C **92**, 054902 (2015).
[10] M. Ruggieri et al., Phys. Rev. C **92**, 064904 (2015).
[11] S.K. Das, M. Ruggieri, F. Scardina, S. Plumari, V. Greco, J. Phys. G **44**, 095102 (2017).
[12] F. Scardina et al., Phys. Rev. C **96**, 044905 (2017).
[13] S. Plumari et al., Phys. Lett. B **805**, 135460 (2020).
[14] S. Plumari, W.M. Alberico, V. Greco, C. Ratti, Phys. Rev. D **84**, 094004 (2011).
[15] S. Borsanyi et al., JHEP **11**, 077 (2010).
[16] U. Gursoy, D. Kharzeev, K. Rajagopal, Phys. Rev. C **89**, 054905 (2014).
[17] A. Amato et al., Phys. Rev. Lett. **111**, 172001 (2013).
[18] V. Greco, C.M. Ko, P. Levai, Phys. Rev. Lett. **90**, 202302 (2003).
[19] M. Cacciari et al., JHEP **10**, 137 (2012).
[20] P. Bozek, I. Wyskiel, Phys. Rev. C **81**, 054902 (2010).
[21] F. Becattini et al., Eur. Phys. J. C **75**, 406 (2015), [Erratum: Eur.Phys.J.C 78, 354 (2018)].
[22] L. Oliva, S. Plumari, V. Greco, JHEP **05**, 034 (2021).

Fixed-target program upgrade and prospects at LHCb

Edoardo Franzoso[1,2,*] for the LHCb Collaboration

[1]Università degli Studi di Ferrara, Ferrara, Italy
[2]INFN Sezione di Ferrara, Ferrara, Italy

Abstract. LHCb has the unique capability to operate in fixed-target mode to study collisions of the LHC beams on fixed targets. In Run3, the internal gas target is going to be upgraded to allow for a wider selection of target gas species and a significant increase of the rates of fixed-target collisions by up to two orders of magnitude. Along with significant data acquisition and tracking upgrades, the SMOG2 system greatly enhances the reach of LHCb's heavy ion program.

1 Fixed-target collisions in LHCb

The LHCb detector was originally conceived and designed to study heavy-flavour physics. However, today, its research program spans over a wide spectrum of high-energy physics phenomena and LHCb acts effectively as a general-purpose experiment in the forward direction with excellent tracking and particle identification performance. Since the installation of SMOG (System for Measuring overlap with gas) in 2015, LHCb offers the unique possibility to study fixed-target collisions in addition to the standard collider operations.

The SMOG system was originally designed to inject noble gases inside the primary LHC vacuum in correspondence of the LHCb vertex locator (VELO) for precise luminosity determinations through the beam-gas imaging technique [1]. The gas was injected using a Gas Feed System reaching a local pressure of 2×10^{-7} mbar. During the LHC RUN 2, the LHCb fixed-target program included runs with different targets and beam energies: between 2015 and 2018, SMOG has collected data with proton beams colliding on He, Ar and Ne; additionally, in 2018, data with Pb beams on a Ne target have been recorded.

The fixed-target configuration in LHCb opens the possibilities to perform measurements of great interest for cosmic rays and heavy-ion physics with centre-of-mass energies from 40 to 115 GeV, corresponding to beam energies ranging from 0.9 to 7 TeV. Furthermore, the forward acceptance of the LHCb spectrometer ($2 < \eta < 5$) is perfectly suited to cope with the strong boost in the lab frame of the final-state particles produced in fixed-target collisions. In the center of mass frame, particles produced at central and backward rapidities are observed, offering the possibility to study the high Bjorken-x region in the target nucleons or nuclei (large Bjorken-x can be accessed in the collider mode at much higher values of q^2, the squared transferred four-momentum.)

*e-mail: edoardo.franzoso@cern.ch

1.1 Physics results from fixed-target measurements in RUN2

During the RUN 2, LHCb performed the first fixed-target measurements with the SMOG setup. Charm production studies in samples of pAr at $\sqrt{s}$ = 86 GeV and pHe at $\sqrt{s}$ = 110 GeV have been published [2], reporting no evidence of strong intrinsic charm contribution. Moreover, the $c\bar{c}$ cross-section has been measured on pHe data very precisely in an unexplored energy region. Another interesting result is related to the secondary production of antiprotons in collisions of cosmic rays with the interstellar medium. Specifically, LHCb measured the antiproton production cross-section in the 10-100 GeV energy range [3] using a 6.5 TeV proton beam colliding on a He target, one of the main constituents of the interstellar medium. Only antiprotons originating directly from pHe collisions and decays of resonances via strong interactions were considered in the analysis. The experimental uncertainties are small compared to the spread of theoretical predictions and the results obtained are relevant for dark matter annihilation studies, and contributed to reduce the tension between the theoretical predictions and the measurements by the PAMELA and AMS-02 experiments [4].

Although originally not conceived for physics measurements, during RUN 2 SMOG has successfully demonstrated the capabilities of LHCb as a fixed-target experiment. There are, however, some drawbacks. One of those is the diffusion of the injected gas along a ± 20 m long sector of the LHC beam pipe. As a consequence, simultaneous data taking with pp collisions was not possible due to the overlapping of the luminous region and SMOG could only collect data during short dedicated runs. Additionally, the injected gas density could not be measured precisely and the only way to infer the luminosity was by exploiting the cross section of known processes, such as the single-electron scattering [3]. To overcome these constraints, an upgrade of the SMOG system is undergoing during the Long Shutdown 2 of LHC.

2 The upgrade of the fixed-target system

In general, the fixed-target program will profit from all the improvements in the detector performance expected in RUN 3. In particular, the SMOG system is undergoing a specific upgrade through the installation of a gas storage cell upstream the pp interaction point. The new system [5], called SMOG2, is shown in figure 1a.

The storage cell offers the opportunity to have a controlled environment designed specifically for a fixed-target experiment along with the possibility to inject several gas species in addition to those already available with SMOG. In particular, also H_2, D_2, N_2, O_2, Kr and Xe can in principle be injected in addition to He, Ne and Ar. SMOG2 is also coupled with a new Gas Feed System able to provide a precise density measurement of the target, hence a precise luminosity determination (not possible with the previous system). Another key point is the possibility to have a well-defined interaction region upstream of the pp interaction point (figure 1b), this feature allows easy separation of the two types of interactions related to beam-beam and beam-target collisions, respectively. A high reconstruction efficiency is expected in the fiducial volume of the cell despite the upstream position of the beam-gas primary vertices compared to the nominal collider interaction point. The possibility to run in parallel with the collider mode and to simultaneously collect beam-beam and beam-gas data for the RUN 3 can be considered since preliminary studies confirm that the performance of the measurements in the collider mode will not be affected. The last major improvement is the increase of target areal density by up to two order of magnitude, meaning a higher luminosity for fixed-target events, by assuming the same gas flow provided for SMOG.

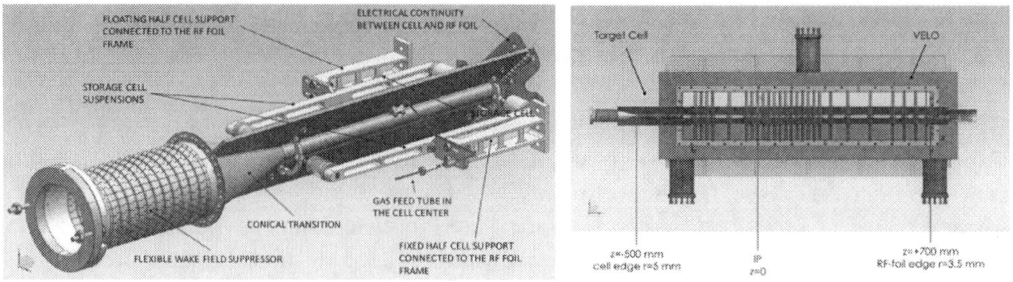

(a) (b)

Figure 1. (a) Design of the SMOG2 storage cell; (b) side view of the storage cell connected to the VELO

Table 1. Integrated luminosity and yields projection for SMOG2 in RUN3 compared with the largest SMOG sample available from RUN2

	SMOG pNe ($\sqrt{s}$ = 68 GeV)	SMOG2 pAr ($\sqrt{s}$ = 115 GeV)
Integrated luminosity	$\sim$ 100 nb^{-1}	$\sim$ 100 pb^{-1}
syst. error on J/Ψ x-sec.	6-7%	2-3%
J/Ψ yield	15k	35M
D^0	100k	350M
Λ_c yield	1k	3.5M
$\Psi(2S)$ yield	150	400k
$\Upsilon(1S)$	4	15k
Low-mass (5<$M_{\mu\mu}$<9 GeV/c^2) Drell-Yan yield	5	20k

3 Physics Prospects for RUN 3

The physics opportunities offered by the fixed-target program at LHCb are unique. The fixed-target configuration instrumented with the SMOG2 storage cell will significantly extend the studies already performed with SMOG [6]. In Table 1, a projection of the yields expected for SMOG2 is reported as compared to those measured with SMOG. Charm measurements, which are vital for studies of cold nuclear matter effects and heavy-ion physics, will benefit from the increased statistics and better luminosity determination with SMOG2. Moreover, the possibility to measure beauty production in pA collisions can be considered since a significant yield of reconstructed $\Upsilon(1S)$ is expected (Tab. 1). Also QGP-related flow observables and correlations can be studied in the full centrality range in PbA collisions.

Cosmic rays studies will highly profit from the opportunity to inject hydrogen and deuterium with the possibility to extend the study of antiproton production with the full centre of mass range available and to measure the ratios $\sigma(pHe \to \bar{p}X)/\sigma(pH \to \bar{p}X)$ and $\sigma(pD \to \bar{p}X)/\sigma(pH \to \bar{p}X)$. A deuterium target will also allow testing isospin symmetry effects. The fixed-target mode would probe the energy scale that is expected to be most relevant for the production of light anti-nuclei such as $\bar{d}$ so the opportunity to detect light anti-nuclei in LHCb is under investigation.

Looking at the nucleon structure studies, SMOG2 will be able to exploit the increased statistics to probe quark and gluon PDFs in nucleons and nuclei, especially at high-x and moderately-high q^2. Drell-Yan and weak-boson production are suitable to constraint theoretical models and reduce uncertainties on the light quark and anti-quark PDFs. Furthermore, heavy-flavour production, which is dominated by gluon-gluon interactions in high-energy hadronic collisions, will allow accessing the poorly known gluon distribution functions.

Figure 2. Storage cell installed and connected to the VELO vessel

4 Conclusions

LHCb is, as today, the only existing experiment capable to run both in fixed-target and collider mode and will continue a pioneering fixed-target program with SMOG2. The storage cell has been installed in front of the VELO as can be seen in figure 2, meanwhile, the new Gas Feed System will be ready by the end of 2021. SMOG2 will provide a wide choice of targets in combination with proton or ion beams and a significantly higher luminosity compared to SMOG, extending the LHCb fixed-target program physics reach.

References

[1] C. Barschel, Precision luminosity measurements at LHCb with beam-gas imaging. CERN-THESIS-2013-301, RWTH Aachen University (2014)

[2] LHCb collaboration, R. Aaij et al., First measurement of charm production in its fixed-target configuration at the LHC. Phys. Rev. Lett. **122**, 132002 (2019).

[3] LHCb collaboration, R. Aaij et al., Measurement of antiproton production in pHe collisions at $\sqrt{s} = 110$ GeV. Phys. Rev. Lett. **121**, 222001 (2018).

[4] Boudaud, Mathieu, et al., AMS-02 antiprotons consistency with a secondary astrophysical origin. Phys. Rev. Res. **2.2**, (2020).

[5] LHCb collaboration SMOG2, Technical Design Report (2018).

[6] PBC Report, Physics opportunities with the fixed target program of the LHCb experiment using an unpolarized gas target. LHCb-PUB-2018-015 (2018).

EPJ Web of Conferences **259**, 13011 (2022) https://doi.org/10.1051/epjconf/202225913011
SQM 2021

Estimating Microscopic Nuclear Data by Compact Star Observations

Balázs Endre Szigeti[1,2,*], *Gergely Gábor* Barnaföldi[2], *Péter* Pósfay[2], and *Antal* Jakovác[2]

[1]Institute of Physics, Eötvös University, Pázmány Péter Sétány 1/A, H-1117 Budapest, Hungary
[2]Wigner Research Centre for Physics, Konkoly-Thege Miklós út 29-33, H-1121 Budapest, Hungary.

Abstract. We studied recent observation data of pulsar masses and radii of PSR J0740+6620, PSR J0348+0432, and PSR J1614−2230 from different measurements, based on the extended version of σ-ω model. Throughout our analysis, we assumed that these pulsars are maximal-mass compact stars, thus we applied the core approximation. Based on the linear relation between the microscopic and macroscopic parameters of compact stars evaluated by our model, we estimated the average Landau mass $m_L = 752.46^{+49.1}_{-42.5}$ MeV and compressibility $K = 261.7^{+57.2}_{-28.0}$ MeV.

1 Introduction

Compact star observables are depend on the internal characteristics of the dense nuclear matter. Modification of the microscopic nuclear model and its parameters can vary the measurable macroscopic properties of a pulsar as well. In our analysis we try to approach the problem from the inverse direction and attempt to estimate the properties of the nuclear matter from compact star observables, which is made more challenging by the *masquarade problem*. It states that even large variation of the nuclear matter parameters and by exchanging of the interaction terms result in stars with very similar observable properties.

The effect of various interaction terms and the variation of the nuclear parameter values in the Lagrangian has been presented so far for the case of the extended σ-ω model [1]. Linear dependence of the mass and radius parameters on the Landau mass, m_L, compressibility, K, and nuclear asymmetry, a_{sym} was observed and tested for maximal mass (MMS) neutron star scenarios. Ordering in the strengths of these properties variation has been also obtained in general for the maximum mass star (MMS): $\Delta M_{max}(\delta m_L) \overset{10\times}{>} \Delta M_{max}(\delta K)$. Landau effective mass values we already determined by former analysis [2, 3]. We would like to investigate the compressibilty K by using recent maximal-mass pulsar observation data from PSR J0740+6620 [4–6], PSR J0348+0432 [7], and PSR J1614−2230 [8]. We estimate the mean K and m_L value including the uncertainties originating from the theory and data. The model and the analysis method was described in detail in Ref. [2, 3].

---

*e-mail: szigeti.balazs@wigner.hu

2 The Model and the Equation of State

One of the most commonly used model to describe the interior structure of the compact star is the σ-ω model, which can be naturally extended by further interactions [1, 9]. This model describes protons, electrons, and neutrons in β-equilibrium and approximates the nuclear force by introducing the σ, ω, and ρ meson with higher-order self-interaction terms for the σ meson. In our study we would like to focus on the connection between macro- and microscopical parameters, which led us to estimate nuclear parameters with high precision from pulsar data based on this model. The Lagrangian of the corresponding to the σ-ω model is,

$$\begin{aligned} \mathcal{L} &= \overline{\Psi}\left(i\partial\!\!\!/ - m_N + g_\sigma\sigma - g_\omega\omega\!\!\!/ + g_\rho\rho\!\!\!/^a\tau_a\right)\Psi + \overline{\Psi}_e\left(i\partial\!\!\!/ - m_e\right)\Psi_e - \lambda_3\sigma^3 + \lambda_4\sigma^4 \\ &+\frac{1}{2}\,\sigma\left(\partial^2 - m_\sigma^2\right)\sigma - \frac{1}{4}\,\omega_{\mu\nu}\omega^{\mu\nu} + \frac{1}{2}m_\omega^2\,\omega^\mu\omega_\mu - \frac{1}{4}\rho^a_{\mu\nu}\rho^{\mu\nu\,a} + \frac{1}{2}m_\rho^2\,\rho^a_\mu\rho^{\mu\,a}\,, \end{aligned} \tag{1}$$

where $\Psi = (\Psi_n, \Psi_p)$ is the vector of proton and neutron fields, m_N, m_σ, m_ω, m_ρ are the masses of the nucleons and σ is the scalar-, and ω, ρ are (iso-)vector mesons respectively. Moreover g_σ, g_ω, and g_ρ are the Yukawa couplings corresponding to the nucleon-meson interactions. The kinetic terms corresponding to the ω and ρ meson can be written in the following forms,

$$\omega_{\mu\nu} = \partial_\mu\omega_\nu - \partial_\nu\omega_\mu \quad \text{and} \quad \rho^a_{\mu\nu} = \partial_\mu\rho^a_\nu - \partial_\nu\rho^a_\mu + g_\rho\epsilon^{abc}\rho^b_\mu\rho^c_\nu. \tag{2}$$

We are taking the mean-field approximation at zero temperature and finite chemical potential of this model. Although in the Lagrangian we can consider the fluctuations of the ω and ρ vector meson fields, but because they are relative heavy their contribution can be neglected in the loop-integrals, only the $\omega_0 = \omega$ and $\rho_0^3 = \rho$ components gives non-zero value, due to symmetrical reasons but kinetic terms are disappeared. From this point the free energy corresponding to the model can be calculated as it is described in for example in Ref. [10]:

$$\begin{aligned} f_T &= f_F\left(m_N - g_\sigma\sigma, \mu_p - g_\omega\omega + g_\rho\rho\right) + f_F\left(m_N - g_\sigma\sigma, \mu_n - g_\omega\omega - g_\rho\rho\right) + f_F\left(m_e, \mu_e\right) \\ &+ \frac{1}{2}m_\sigma^2\sigma^2 + U_i(\sigma) - \frac{1}{2}m_\omega^2\omega^2 - \frac{1}{2}m_\rho^2\rho^2\,, \end{aligned} \tag{3}$$

where μ_p, μ_n and μ_e are the proton, neutron, and electron chemical potential respectively. The f_F term describes the free energy contribution corresponding to one fermionic degree of freedom [1]. For the fit of the free parameters of this model the nuclear saturation data were used [11], except for the Landau mass. Then after getting the optimal Landau mass value, we estimate the values for the compressibility as well. These two parameters are kept free and determined by comparing the mass radius diagrams corresponding to different values of the Landau mass to neutron star observations. The asymmetry energy plays negligible role on the mass and radius values of the maximal mass compact stars as it was pointed out in Ref. [1]. For the general relativistic description of the compact stars we assumed the usual static picture in spherically symmetric space time. We calculated the mass-radius diagram by the well-known Tolman – Oppenheimer – Volkoff equations (TOV). To focus our investigation on the effect of the nuclear matter we integrated the TOV equations, and we employed a stopping condition for the integration based on Ref. [1]: $p(r = R') = p_0$. Here, p_0 is chosen such a way that the integration stops at the core of the neutron star, so the integration does not take into account the effect of the crust and the low density EoS. To get a good approximation for p_0 we used the well known BPS nuclear equation of state which is used to describe the crust of neutron stars [12]. We used the highest pressure value for p_0 where the BPS EoS can still be considered valid.

The calculated R' in this case corresponds to the radius of the neutron star core. The mass and radius data calculated this generally can be considered a conservative approximation of the neutron star parameters, however in the case of the maximum mass star the deviations from the normal case are insignificant [1].

3 Results

Assuming maximal mass stars, we found linear dependence of both observables, the mass and radius of the maximal mass star, M_{maxM} and R_{maxM} on the m_L, where we used the saturated nuclear matter for the compressibility and asymmetry energy. The functions were fitted independently with 0.8% and 17% theoretical uncertainty, respectively, following Ref [1]:

$$M_{maxM}[\mathrm{M}_\odot] = 5.418 - 0.0043\, m_L[\mathrm{MeV}] \,, \tag{4}$$

$$R_{maxM}[\mathrm{km}] = 19.04 - 0.0104\, m_L[\mathrm{MeV}] \,. \tag{5}$$

After we fixed the Landau mass by the MMS observation we can also obtain a one-parameter linear equation between the compressibility, K and the mass and radius of the MMS compact star with $\lesssim 2\%$ and $\lesssim 14\%$ theoretical uncertainty,

$$M_{maxM}[\mathrm{M}_\odot] = 1.940 + 0.000880\, K\,[\mathrm{MeV}] \,, \tag{6}$$

$$R_{maxM}[\mathrm{km}] = 9.248 + 0.00718\, K\,[\mathrm{MeV}] \,. \tag{7}$$

One can see that the K-equations (6) and (7) have positive gradient which means that increasing compressibility makes an MMS compact star more larger and massive. Using these derived equations and pulsar observation data we obtained the optimal Landau masses, and compressibility values for each compact object in Table 1. We also determined the average of Landau masses $m_L = 752.46^{+49.1}_{-42.5}$ MeV and mean compressibility $K = 261.7^{+57.2}_{-28.0}$ MeV, which agrees well with the saturated nuclear matter data and with our recent works [2, 3]. In contrary, calculations of R_{maxM} by eqs. (5) and (7) present large uncertainties, especially in case of radius data from Ref. [6], thus core approximation might not be valid, due to the softer EoS and its relevant contribution to the crust. This result in a $\gtrsim 1$ km larger radius.

Table 1. The Landau mass, m_L and compressiblity, K values evaluated via eqs. (4)-(7) from measured pulsar masses and radii data denoted by '$*$', and assuming that these are maximal-mass neutron stars.

Ref.	Pulsar	$M_{maxM}[\mathrm{M}_\odot]$	m_L[MeV]	K[MeV]	R_{maxM}[km]
[4]	PSR J0740+6620	$2.17^{+0.11}_{-0.10}$ *	$748.39^{+63.3}_{-57.2}$	$351.8^{+115}_{-84.5}$	$11.25^{+1.06}_{-1.04}$
[7]	PSR J0348+0432	$2.01^{+0.04}_{-0.04}$ *	$785.25^{+20.0}_{-20.3}$	$206.4^{+42.7}_{-20.5}$	$10.87^{+0.82}_{-0.80}$
[8]	PSR J1614-2230	$1.97^{+0.04}_{-0.04}$ *	$794.47^{+20.1}_{-20.4}$	$170.0^{+15.5}_{-20.9}$	$10.77^{+0.82}_{-0.80}$
[6]	PSR J0740+6620	$2.07^{+0.07}_{-0.07}$ *	$778.14^{+15.3}_{-15.5}$	$278.2^{+60.9}_{-60.8}$	$11.01^{+0.46}_{-0.47}$
[6]	PSR J0740+6620	$2.64^{+1.98}_{-0.65}$	639.42^{+159}_{-125}	277.3^{+257}_{-178}	$12.39^{+1.30}_{-0.98}$ *
[5]	PSR J0740+6620	$2.08^{+0.07}_{-0.01}$ *	$769.12^{+16.9}_{-16.9}$	$285.1^{+54.8}_{-54.8}$	$11.06^{+0.41}_{-0.41}$

On the Fig. 3 we plotted the projections of the $M_{maxM}(m_L, K)$ and $R_{maxM}(m_L, K)$ curves, where the K-dependent one-parameter linear equations are determined for various m_L values. Pulsar mass and radii data from Table 1. are also plotted with color markers, where the error bars include both theoretical uncertainties from the phenomenological approach and errors from the astrophysical observation. Our results agree well with our previous results [1, 3] and also with a Bayesian analysis carried out on this extended σ-ω model [2, 3].

4 Summary

Macroscopical compact star observation parameters and microscopical superdenes nuclear matter properties has been formulated in Refs. [1–3]. This has been validated in the liu of the recent NICER [6] and NICER+XMM-Newton [5] data. We find that the concept of MMS star is valid, until the core approximation stands. Microscopic nuclear parameter values: mean Landau mass $m_L = 752.46^{+49.1}_{-42.5}$ MeV and compressibility $K = 261.7^{+57.2}_{-28.0}$ MeV were given.

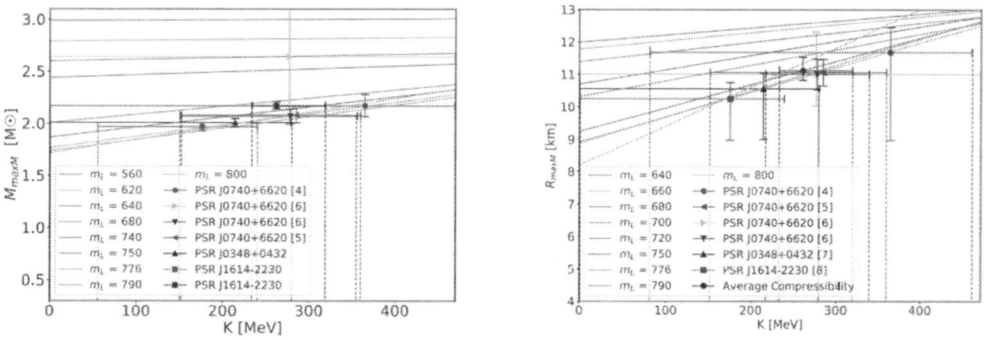

Figure 1. Left panel: mass of the maximal mass star, $M_{maxM}(m_L, K)$. Right panel: radius of the maximal mass star, $R_{maxM}(m_L, K)$. Curves at both panels are as function of K at various physical m_L values, denoted as color lines. Comparison to pulsar data of PSR J0740+6620 [4–6], PSR J0348+0432 [7], and PSR J1614−2230 [8] are also plotted, including theoretical uncertainties and data errors.

5 Acknowledgement

Authors acknowledge for NKFIH (OTKA) grants No. K123815, K135515, PHAROS (CA16214) cost action, and for the computational resources by the Wigner GPU Laboratory.

References

[1] P. Pósfay, G. G. Barnaföldi and A. Jakovác, arXiv:2004.08230 [astro-ph.HE].
[2] D. Alvarez-Castillo *et al*, Eur. Phys. J. ST **229** no.22-23, 3615-3628 (2020)
[3] G. G. Barnaföldi *et al*, Eur. Phys. J. ST **229** no.22-23, 3605-3614 (2020)
[4] H. Cromartie *et al*, Nat. Astron. **4**, 72-76 (2019)
[5] M.C. Miller *et al*, arXiv:2105.06979 [astro-ph.HE] (2021)
[6] T.E. Riley *et al*, arXiv:2105.06980 [astro-ph.HE] (2021)
[7] J. Antoniadis *et al*, Science **340**, 6131 (2013)
[8] Demorest, P.*et al*,Nature **467**, 1081-1083 (2010).
[9] P. Pósfay, G. G. Barnaföldi, and A. Jakovác, Universe **5**, 153 (2019).
[10] A. Jakovác and A. Patkós, Resummation and Renormalization in Effective Theories of Particle Physics, Lecture Notes in Physics (Springer International Publishing, 2015)
[11] J. Meng, Relativistic Density Functional for Nuclear Structure, International Review of Nuclear Physics (World Scientific Publishing Company, 2016),
[12] G. Baym, C. Pethick, and P. Sutherland, Astrophysical Journal, **170**, p.299 (1971)

https://doi.org/10.1051/epjconf/202225913012

Triple nuclear collisions – a new method to explore the matter properties under new extreme conditions

O. V. Vitiuk[1], V. M. Pugatch[1], K.A. Bugaev[2,3], P. P. Panasiuk[3], N. S. Yakovenko[1], B. E. Grinyuk[2], E. S. Zherebtsova[4,5], M. Bleicher[6], L.V. Bravina[7], A.V. Taranenko[4,5], and E.E. Zabrodin[7,8]

[1] *Institute for Nuclear Research, NAS of Ukraine, Prospekt Nauki av. 47, 03680 Kyiv, Ukraine*
[2] *Bogolyubov Institute for Theoretical Physics, NAS of Ukraine, Metrologichna str. 14-B, 03680 Kyiv, Ukraine*
[3] *Department of Physics, Taras Shevchenko National University of Kyiv, 03022 Kyiv, Ukraine*
[4] *National Research Nuclear University (MEPhI), Kashirskoe Shosse 31, 115409 Moscow, Russia*
[5] *Institute for Nuclear Research, Russian Academy of Science, 108840 Moscow, Russia*
[6] *ITP, Goethe University, Max-von-Laue-Str. 1, 60438 Frankfurt am Main, Germany*
[7] *University of Oslo, POB 1048 Blindern, N-0316 Oslo, Norway*
[8] *Skobeltsyn Institute of Nuclear Physics, Moscow State University, 119899 Moscow, Russia*

Abstract. We suggest to explore an entirely new method to experimentally and theoretically study the phase diagram of strongly interacting matter based on the triple nuclear collisions (TNC). We simulated the TNC using the UrQMD 3.4 model at the beam center-of-mass collision energies $\sqrt{s_{NN}} = 200$ GeV and $\sqrt{s_{NN}} = 2.76$ TeV. It is found that in the most central and simultaneous TNC the initial baryonic charge density is about 3 times higher than the one achieved in the usual binary nuclear collisions at the same energies. As a consequence, the production of protons and Λ-hyperons is increased by a factor of 2 and 1.5, respectively. Using the MIT Bag model equation we study the evolution of the central cell in TNC and demonstrate that for the top RHIC energy of collision the baryonic chemical potential is 2-2.5 times larger than the one achieved in the binary nuclear collision at the same time of reaction. Based on these estimates, we show that TNC offers an entirely new possibility to study the QCD phase diagram at very high baryonic charge densities.

1 Introduction

After about three decades of investigating the phase diagram of strongly interacting matter in binary nuclear (A+A) collisions it became clear that the most interesting phenomena such as the expected chiral symmetry restoration and the deconfinement phase transitions may occur at rather low center-of-mass collision energies with the thresholds $\sqrt{s_{NN}} \simeq 4-5$ GeV [1–5] for the chiral symmetry restoration phase transition and $\sqrt{s_{NN}} \simeq 9-10$ GeV [1–5] for the deconfinement one. Perhaps, the hardest lesson that the community of heavy-ion collisions learned after so many years is that in addition to the A+A collisions we need an independent and reliable source of information about the equation of state (EoS) of strongly interacting matter [6, 7]. Sharing this idea, we suggest to consider the triple nuclear collisions (TNC) [8] as an additional and independent source of information. The TNC can be done either by inserting a super-thin target into the interaction zone of two colliding

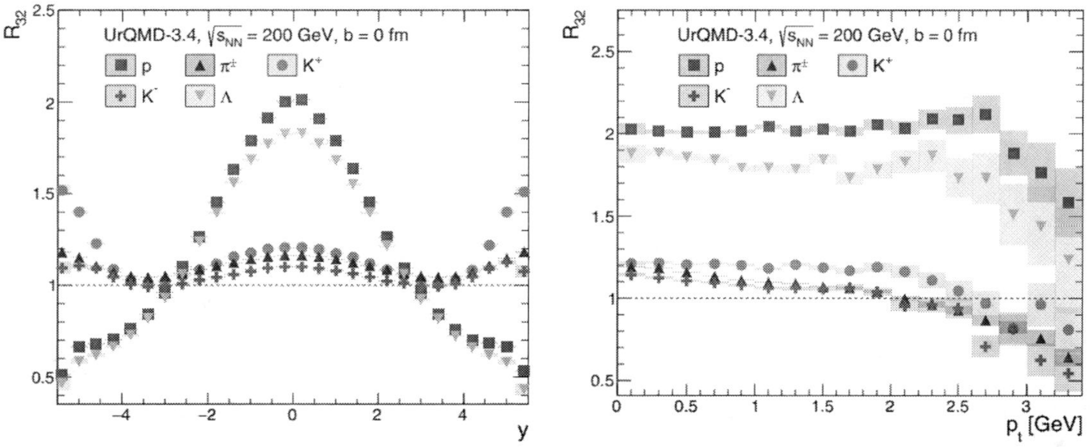

Figure 1. Left panel: The ratio of hadronic yields per rapidity unit $\frac{dN}{dy}$ expected for the most central and simultaneous Pb+Pb+Pb TNC to the one for the most central Pb+Pb collisions (i.e. 3-to-2 nuclei enhancement factor for yields) for the collision energy $\sqrt{s_{NN}}$ = 200 GeV. **Right panel:** Ratio of transversal momentum spectra of the most central and simultaneous Pb+Pb+Pb TNC to the one found for the most central Pb+Pb collisions (3-to-2 nuclei enhancement factor for p_T spectra) of hadrons obtained for the same collision energy $\sqrt{s_{NN}}$ = 200 GeV as a function of particle transverse momentum found at $|y| < 0.1$.

beams, or by making the jet target consisting of small metallic droplets of about 1-2 μm size, or by installing the third ring in a perpendicular direction to two colliding beams [8, 9]. Since the TNC is a very fresh topic, in this short work we just outline its principal advantages over the binary nuclear collisions (BNC), whereas the estimates of TNC rates and details of the experimental setup we will discuss in the separate publication [9].

2 New Extreme Conditions achieved in TNC

In order to model the TNC, we used the UrQMD 3.4 model [10, 11], which is able to rather satisfactory describe the bulk data on secondary hadron yields and their transversal momentum spectra. The primary goal of this work is to demonstrate the entirely new possibilities to study the QCD phase diagram with TNC at essentially higher baryonic and electric charge densities than the ones achieved in A+A collisions. Therefore, here we concentrate on the bulk properties of baryon production, while the question of possible phase transformations at low energies of collisions we leave for further exploration by more specialized approaches.

In order to demonstrate the strength of the effect, in Fig. 1 we show the 3-to-2 enhancement factor as a ratio of the quantity obtained by UrQMD 3.4 model for the most central Pb+Pb+Pb to the same quantity found for the most central Pb+Pb collisions at the same energy of colliding beams $\sqrt{s_{NN}}$ = 200 GeV. As one can see from the left panel of Fig. 1, the yield of protons and Λ-hyperons in the TNC is enhanced almost by a factor of 2 and, hence, this feature can be, in principle, used to detect the TNC in the event-by-event analysis. Another distinct difference between TNC and BNC can be seen in the right panel of Fig. 1. The number of pions and kaons is slightly enhanced at $p_T \leq 2$ GeV, while their number is suppressed for $p_T > 2.5$ GeV. A qualitatively similar enhancement for $p_T \leq 3.5$ GeV and suppression for $p_T > 4$ GeV we found for protons and Λ-hyperons in the TNC with the difference, that both the enhancement and the suppression of p_T spectra of these particles are

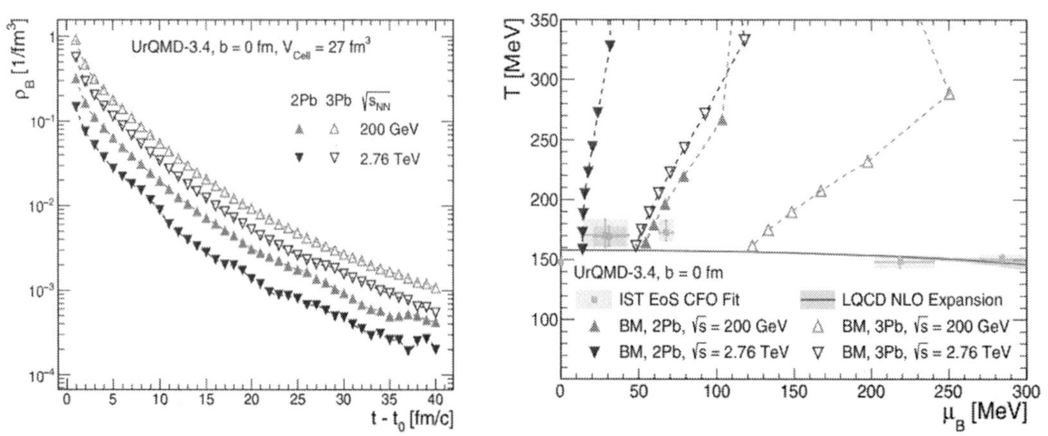

Figure 2. Left panel: The time evolution of the baryonic charge density in the central cell during the process of ordinary BNC (filled symbols) and for the TNC (empty symbols) found for $\sqrt{s_{NN}} = 200$ GeV (squares) and for $\sqrt{s_{NN}} = 2.76$ TeV (circles). The time t_0 is the moment, when the remnants of projectile nuclei have passed through the central cell. **Right panel:** The evolution of central cell parameters in the $\mu_B - T$ plane obtained with the MIT Bag Model equation of state [12]. The filled symbols correspond to the Pb+Pb collisions, while the empty ones to the Pb+Pb+Pb TNC. Collision energy $\sqrt{s_{NN}} = 200$ GeV ($\sqrt{s_{NN}} = 2.76$ TeV) points are shown by the triangles up (down). The topmost points correspond to the time $t - t_0 > 1$ fm. The curve of pseudo-critical temperature corresponds to a lattice QCD parameterization [13], while the crosses correspond to the chemical freeze-out parameters in Pb+Pb collisions found in Ref. [14].

much stronger than for pions and kaons. Therefore, the ratio of $\frac{d^2N}{dydp_T}$ for $p_T \le 2$ GeV to the same quantity, but for $p_T > 5$ GeV measured in the TNC and in the BNC will be much larger for the TNC.

This effect of the transverse momentum redistribution in TNC we observed even in the p+C+p collisions [9], when the proton-proton collisions occur inside the carbon nucleus. Because of the presence of surrounding nucleons the products of the primary p+p or p+n collisions that are moving in the transversal direction re-scatter on the nucleons of the target nucleus and lose their transverse momenta and, moreover, they produce an additional amount of secondary hadrons in comparison with the ordinary BNC. Therefore, the probability to detect the slow-moving nucleons at the midrapidity increases essentially. On the other hand, after one or two re-scatterings the products of primary collisions with the large p_T value lose the part of their momentum and increase the number of states with low p_T values. As one can see from the right panel of Fig. 1, in the TNC the effect of transverse momentum redistribution of protons and Λ-hyperons is rather strong and, hence, it can be used to identify the Pb+Pb+Pb collisions and even the p+C+p ones in the experiments.

Our simulations with UrQMD 3.4 model show us that using the TNC one can probe the new extreme conditions, namely very high baryonic and electric charge densities, even at highest RHIC energies of collisions. To demonstrate this we studied the evolution of the central cell with the size $3 \times 3 \times 3$ fm^3. The results for the most central Pb+Pb+Pb TNC for $\sqrt{s_{NN}} = 200$ GeV (RHIC) and $\sqrt{s_{NN}}$ $= 2.76$ TeV (LHC) are shown in Fig. 2. From the left panel of Fig. 2 it is easy to see that compared to the usual BNC the TNC provide an essential increase of the initial baryonic charge density: from 0.321 fm^{-3} to 0.913 fm^{-3} for $\sqrt{s_{NN}} = 200$ GeV and from 0.146 fm^{-3} to 0.567 fm^{-3} for $\sqrt{s_{NN}} = 2.76$ TeV. From these results it is clear that in the TNC one can expect a formation of about 3 times higher initial baryonic charge density for the RHIC top energy and 4 times denser ones for the LHC energy than in the BNC. The time $t_0 \simeq 10$ fm/c for the Pb+Pb+Pb TNC and $t_0 \simeq 1$ fm/c for Pb + Pb collisions correspond to the time when the remnants of target nuclei have passed through the central cell. To quantify the values of baryonic chemical potential μ_B and temperature T inside the central cell, we used the MIT Bag Model equation of state [12] for 3 colors and 3 massless quark flavors, i.e. for the pressure $p^{BM} = \frac{95}{180}\pi^2 T^4 + \frac{T^2\mu_B^2}{6} + \frac{\mu_B^4}{108\pi^2} - B_{vac}$, with a conservative value $B_{vac}^{\frac{1}{4}} = 206$ MeV [15]. Equating the baryonic charge density ρ_{cell} and the energy density ϵ_{cell} found for the central cell to the corresponding quantities of the MIT Bag Model $\rho_{BM} \equiv \frac{\partial p^{BM}}{\partial \mu_B}$ and $\epsilon_{BM} = T\frac{\partial p^{BM}}{\partial T} + \mu_B \rho_{BM} - p^{BM}$, we found the values of μ_B and T in central cell (see Fig. 2). As one can see from the right panel of Fig. 2, for the same temperature the values of μ_B are essentially larger in the TNC than in the BNC.

3 Conclusions

In this work, we briefly showed that for the same energy of colliding beams, in TNC the QCD matter can reach about 3 times higher baryonic charge density than in BNC. Apparently, this opens entirely new perspectives to study the QCD matter phase diagram under new extreme conditions and, hence, we believe that after developing the TNC concept, future experiments will be done with the TNC.

Acknowledgements

O.V.V., V.M.P. and K.A.B. acknowledge a partial support from the Program "Participation in the novel international projects in high energy and nuclear physics" launched by the Section of Nuclear Physics and Energetics of NAS of Ukraine. The work of L.V.B. and E.E.Z. was supported by the Norwegian Research Council (NFR) grant No. 255253/F50 - "CERN Heavy Ion Theory" and by the Russian Foundation for Basic Research under the grants No. 18-02-40084 and 18-02-40085. A.T received funding from the Russian Foundation for Basic Research (RFBR) under the project No. 18-02-40086. The simulations were done on the JINR supercomputer Govorun.

References

[1] K. A. Bugaev et al., Phys. Part. Nucl. Lett. **12**, 238 (2015).
[2] K. A. Bugaev et al., Eur. Phys. J. A **52** (6), 175 (2016) .
[3] W. Cassing, A. Palmese, P. Moreau, and E. L. Bratkovskaya, Phys. Rev. C **93**, 014902 (2016).
[4] A. Palmese et al., Phys. Rev. C **94**, 044912 (2016).
[5] K. A. Bugaev et al., Phys. Part. Nucl. Lett. **15**, 210 (2018).
[6] E. R. Most et al., Phys. Rev. Lett. **122** , 061101 (2019).
[7] A. Bauswein et al., Phys. Rev. Lett. **122**, 061102 (2019).
[8] K. Bugaev et al., talk at the Online «Strangeness in Quark Matter» Conference 2021, Brookhaven, May 17-22, 2021; https://indico.cern.ch/event/985652/sessions/392917/#20210521
[9] K. Bugaev et al., talk at the Offshell-2021 — The virtual HEP conference on Run4 at LHC, July 6-9, 2021; https://indico.cern.ch/event/968055/contributions/4273917/
[10] S.A. Bass et al., Prog. Part. Nucl. Phys. **41**, 255 (1998).
[11] M. Bleicher et al., J. Phys. G **25**, 1859 (1999).
[12] A. Chodos, R. L. Jaffe, K. Johnson, C. B. Thorn and V. F. Weisskopf, Phys. Rev. D **9**, 3471 (1974).
[13] S. Borsanyi *et al.*, Phys. Rev. Lett. **125**, 052001 (2020).
[14] V. V. Sagun *et al.*, Eur. Phys. J. A **54** (6), 100 (2018).
[15] C. Y. Wong, *"Introduction to high-energy heavy ion collisions,"* (Singapore, World Scientific, 1994)

CME search at STAR

Yu Hu[1,2,*]
for the STAR Collaboration

[1]Fudan University
[2]Brookhaven National Laboratory

Abstract. The hot and dense medium produced in relativistic heavy-ion collisions has been conjectured to be accompanied by an axial charge asymmetry that may lead to a separation of electric charges in the direction of the extremely strong magnetic field, also known as the Chiral Magnetic Effect (CME) [1–3]. The measurement of azimuthal correlator ($\Delta\gamma$) with respect to the spectator plane [5], gives us an opportunity to measure the possible CME fraction beyond the flow background. Preliminary results using this approach with combined Au+Au collisions at $\sqrt{s_{NN}}$ = 200 GeV and U+U at $\sqrt{s_{NN}}$ = 193 GeV show f_{CME} at (8±4±8)%. Meanwhile, the observability of CME has been conjectured to be dependent on $\sqrt{s_{NN}}$ due to changes in the lifetime of the magnetic field, the strengths of CME signal and non-CME background. At lower energies, the Event Plane Detector (EPD) installed in the year 2018 provides a unique capability for CME search. The background scenario test at Au+Au $\sqrt{s_{NN}}$ = 27 GeV by using $\Delta\gamma$ with respect to TPC and the new installed EPD shows a consistency with no-CME scenario in the current statistics. The method of the ongoing isobar blind analysis, and the latest sensitivity check with the event-by-event AVFD model on the different observables between Ru+Ru and Zr+Zr are also briefly discussed.

1 Introduction

Symmetry and it's breaking are parts of the most fundamental laws in the universe. In relativistic heavy-ion collisions, accompanied with the produced hot and dense medium, one conjecture is there could be a symmetry breaking leading to the difference in the number of right-handed and left-handed quarks [1, 2]. This imbalance leads to a separation of electric charge in the direction of the strong magnetic field (B $\sim 10^{14}$ Tesla), produced by the protons in the colliding heavy ions [3]. This phenomenon is known as the Chiral Magnetic Effect (CME). STAR (Solenoidal Tracker at RHIC) experiment is currently the only operational experiment at the Relativistic Heavy Ion Collider (RHIC), has been contributing to the CME search over many years. RHIC has collided various ion species, at many different energies over the past 20 years. In heavy-ion collisions, CME is expected to lead to charge separation across the reaction plane determined by the impact parameter and the collision direction. The reaction plane is correlated to the direction of the magnetic field [4]. To measure the charge separation quantitatively, one uses the well known observable known as the gamma correlator (γ).

*e-mail: huy17@fudan.edu.cn

To reduce the charge-independent correlation backgrounds, like from global momentum conservation, we usually use the $\Delta\gamma$, which measures the difference between the correlations from the opposite charge and the same charge. There are still many background sources that have also been identified to contribute to this observable, such as the local charge conservation and the resonance decays. We can write the $\Delta\gamma$ into 2 parts, one from the background ($\Delta\gamma_{bkg}$), another from the CME signal($\Delta\gamma_{sig}$). In principle, the $\Delta\gamma$ can be measured with respect to any plane [5, 6]: e.g. using the participate plane, which is estimated by the produced particles (Ψ_{PP}); or the spectator plane, which is estimated by the spectators neutrons (Ψ_{SP}).

2 At top RHIC energies

At the top RHIC energies, we use the Au+Au collision at 200 GeV (form year 2011, 2014, 2016) and U+U collision at 193 GeV to search for the CME. As we mentioned before, if we use the particles measured with Time Projection Chamber (TPC) to reconstruct the event plane (Ψ_{EP}) as a proxy of the Ψ_{PP}, and use the Zero Degree Calorimeters (ZDC) to reconstruct the spectator plane as a proxy of the Ψ_{SP}. The $\Delta\gamma$ with respect to Ψ_{PP}^{TPC} and Ψ_{SP}^{ZDC} contain different fractions of the CME signal and background. If we assume the the background is propositional to the elliptic flow(v_2), we will have the following relations:

$$\Delta\gamma(\Psi_{TPC}) = \Delta\gamma^{bkg}(\Psi_{TPC}) + \Delta\gamma^{sig}(\Psi_{TPC}) \quad (1)$$

$$\Delta\gamma(\Psi_{ZDC}) = \Delta\gamma^{bkg}(\Psi_{ZDC}) + \Delta\gamma^{sig}(\Psi_{ZDC}) \quad (2)$$

$$\Delta\gamma^{bkg}(\Psi_{TPC})/\Delta\gamma^{bkg}(\Psi_{ZDC}) = v_2(\Psi_{TPC})/v_2(\Psi_{ZDC}) \quad (3)$$

$$\Delta\gamma^{sig}(\Psi_{TPC})/\Delta\gamma^{sig}(\Psi_{ZDC}) = v_2(\Psi_{ZDC})/v_2(\Psi_{TPC}) \quad (4)$$

With four known variables: $\Delta\gamma(\Psi_{TPC})$, $\Delta\gamma(\Psi_{ZDC})$, $v_2(\Psi_{TPC})$, $v_2(\Psi_{TPC})$, and four unkown ones $\Delta\gamma^{bkg}(\Psi_{TPC})$, $\Delta\gamma^{bkg}(\Psi_{ZDC})$, $\Delta\gamma^{sig}(\Psi_{TPC})$, $\Delta\gamma^{sig}(\Psi_{ZDC})$, we can solve the above equations. Then we can estimate f_{CME} which represents the fraction of CME signal in the observable measured using TPC defined as:

$$f_{CME} = \Delta\gamma^{sig}(\Psi_{TPC})/\Delta\gamma(\Psi_{TPC}). \quad (5)$$

Figure 1 (left) the measurement of the f_{CME} for different run years in $20\% - 50\%$ centrality. The Au+Au and U+U combined result shows that the faction is $8 \pm 4 \pm 8\%$. The systematic uncertainties in this measurement are assessed by varying the track quality cuts and the η gap.

3 At Lower Energy

At lower energies, the Event Plane Detector (EPD) installed in the year 2018 provides a unique capability for CME search. EPD covers a pseudorapidity range of 2.1 to 5.1 on either side of the TPC. For the $\Delta\gamma$ observable, we choose particles of interest carrying charges from the TPC, and measure the event plane from the EPD. At 27 GeV, the beam rapidity ($Y_{beam} = 3.4$) falls at the center of EPD acceptance. Due to the kinematics, the inner EPD ($\eta > Y_{beam}$) detects a lot of protons, coming from spectators, stopped protons or beam fragments and they all have large directed flow which determines first-order event plane well. The outer EPD ($\eta < Y_{beam}$) detects mostly produced particles. Produced particles have large elliptic flow so we have a very well defined second order event plane. So with EPD and 27 GeV collisions we have a way to, at the same time, measure charge separation w.r.to planes of large directed flow due to spectator protons and w.r.to elliptic flow plane due to participants.

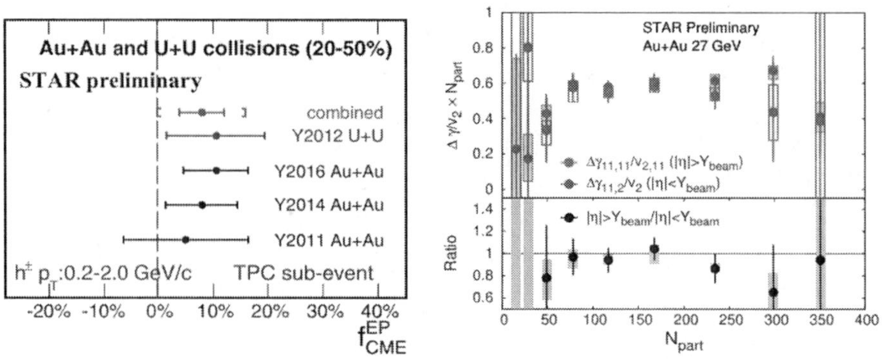

Figure 1. (Left) The fraction of the possible CME signal(f_{CME}) in $\Delta\gamma$ with respect to TPC and ZDC event planes for Au+Au collision at 200 GeV and U+U collision at 193 GeV, combined result shows $8 \pm 4 \pm 8\%$. (Right)$\Delta\gamma$ scaled by v_2 with respect to TPC-EPD-inner and TPC-EPD-outer planes to the no-CME scenario for Au+Au collision at 27 GeV.

In the background only scenario, if the background is propositional to v_2, the ratio of $\Delta\gamma/v_2$ with respect to different planes (Ψ_A, Ψ_B, Ψ_C) will be the same, in other words we expect:

$$\Delta\gamma/v_2(\Psi_A) = \Delta\gamma/v_2(\Psi_B) = \Delta\gamma/v_2(\Psi_C) = ... \tag{6}$$

Therefore, in our case we measured the $\Delta\gamma$ and v_2 with respect to the TPC-EPD-inner 1st order plane ($\eta > Y_{beam}$) and the $\Delta\gamma$ and v_2 with respect to the TPC-EPD-outer 2nd order plane ($\eta < Y_{beam}$). Figure 1(right) shows the $\Delta\gamma/v_2 \times N_{part}$ with respect to different planes. The bottom panel shows the double ratio. The results show, within the uncertainty, the ratio is consistent with unity. No significant deviation from Eq. 6 is observed.

4 Methods for blind analysis of isobar data

To better control the signal and the background in CME observables isobar program is introduced. To remove unconscious human induced biases we implement the isobar blind analysis. The $\Delta\gamma$ can be written into the signal ($\Delta\gamma^{sig}$), v_2 related background ($\Delta\gamma^{flow-bkg}$), and the non-flow driven background ($\Delta\gamma^{non-flow-bkg}$). In an ideal scenario if we can select two different systems that produce similar non-flow effects, similar flow backgrounds, then the residual difference in charge separation can be attributed to signal – isobar collisions were originally proposed with such expectations [7]. In practice with the two isobar species $^{96}_{44}$Ru+$^{96}_{44}$Ru and $^{96}_{40}$Zr+$^{96}_{40}$Zr, according to model calculations one can achieve up to 4% difference in flow background [8–11], similar non-flow determined by the multiplicity difference, and about B-field square difference of the order of $10 - 18\%$ that is relevant for a signal difference. If one can keep the systematics under control, predictions show, with two billion events collected for each species, we can achieve 5σ significance given the CME fraction in observables is at 14% level [8].

The isobar collisions are proposed to perform an experimental test of CME with the best possible control of signal and background as compared to all previous measurements. To minimize the run-to-run variation of the detector response due to the acceptance loss, efficiency changes, or the luminosity variation, we switch the collisions of two species frequently during data taking. To minimize the unconscious bias by the analysts we: 1) use the 27 GeV data which is collected in the same year to perform a closure test of our analysis, 2) mix the Ru and Zr events to freeze the analysis methods, observables, codes, and the event selection criteria, 3) perform run-by-run quanlity assurance using blind data where species information is hidden and, 4) finally perform the full analysis after unblinding the species information. The details of this analysis method can be found in Ref. [12]. One final cross check of the frozen code and response of different observables to CME signals has been recently performed using an event-by-event Anomalous-Viscous Fluid Dynamics (AVFD) simulation [13].

5 Summary

At the top RHIC energy, we measured the possible CME fraction beyond the flow background by measuring the $\Delta\gamma$ observable with respect to TPC and ZDC planes. At Au+Au 200 GeV and U+U 193 GeV collisions, the combined results indicate the CME fraction is $(8 \pm 4 \pm 8)\%$ in data for $20\% - 50\%$ centrality.

At lower energy, a test of background scenario was performed using Au+Au 27 GeV data by measuring $\Delta\gamma$ with respect to TPC and the new installed EPD systems. The results indicate no significant deviation from background scenario with the uncertainty of our measurement.

Meanwhile, for the ongoing analysis of the isobar collisions, a blind analysis is implemented to minimize unconscious bias. The sensitivity of different CME observables was checked using the event-by-event AVFD model. The results from the isobar blind analysis will soon be released by the STAR collaboration.

Acknowledgements

Y. Hu is supported by the China Scholarship Council (CSC). This work was supported in part by the National Natural Science Foundation of China under contract No. 11835002.

References

[1] Dmitri E. Kharzeev et al., Nuclear Physics A **803**, 227-253 (2008)
[2] Dmitri E. Kharzeev et al., Phys.Rev.**D61**, 1111901 (2000)
[3] Larry D. McLerran. and Vladimir V. Skokov, Nucl. Phys. A **929**,184-190 (2014)
[4] Sergei A. Voloshin, Phys. Rev. C **70**, 057901 (2004)
[5] Hao-Jie Xu et al., Chinese Phys. C **42**, 084103 (2018)
[6] Sergei A. Voloshin, Phys. Rev. C **98**, 054911 (2018)
[7] Sergei A. Voloshin, Phys. Rev. Lett. **105**, 172301 (2010)
[8] Wei-Tian Deng et al., Phys. Rev. C **94**, 041901 (2016)
[9] Björn Schenke et al., Phys. Rev. C **99**, 044908 (2019)
[10] Hao-Jie Xu et al., Phys. Rev. Lett. **121**, 022301 (2018)
[11] Hanlin Li et al., Phys. Rev. C **98**, 054907 (2018)
[12] Jaroslav Adam et al. (STAR), Nucl. Sci. Tech. **32**, 48 (2021)
[13] Shuzhe Shi et al., Annals of Physics **394**, 50-72 (2018)
[14] Subikash Choudhury et al., arXiv:2105.06044 (2021)

Low-p_T $\mu^+\mu^-$ pair production in Au+Au collisions at $\sqrt{s_{\rm NN}}$ = 200 GeV at STAR

Jian Zhou[1,2,*](for the STAR collaboration)

[1]Department of Modern Physics, University of Science and Technology of China, Hefei 230026, China
[2]State Key Laboratory of Particle Detection and Electronics, University of Science and Technology of China, Hefei 230026, China

Abstract. We report the measurements of $\mu^+\mu^-$ pair production for $p_T < 0.10$ GeV/c in 40-80% Au+Au collisions at $\sqrt{s_{\rm NN}}$ = 200 GeV at STAR. A significant enhancement with respect to the hadronic cocktail is observed. The p_T^2 and $\Delta\phi$ distributions of the excess yields are also reported and compared with model calculations. The EPA-QED calculations can describe the data very well within uncertainties.

1 Introduction

Dileptons are produced in the whole evolution of the heavy-ion collisions and escape with minimum interaction with the strongly interacting medium. Thus, dilepton measurements play an important role in the study of hot and dense nuclear matter produced in Relativistic Heavy Ion Collider (RHIC) [1, 2]. Recently, significant enhancements of e^+e^- pair at very low transverse momentum (p_T) were observed by the STAR collaboration in peripheral Au+Au collisions [3]. The excess can be explained by photon-photon interactions induced by the extremely strong electromagnetic field produced by the fast moving heavy ions. While such photon-photon interactions were traditionally studied in ultra-peripheral collisions without any nuclear overlaps [4–6], they could provide a novel probe to the Quark Gluon Plasma (QGP) created in peripheral collisions since the very-low-p_T dileptons are produced in the early stage of the collisions [7]. Recently, linearly polarized photon-photon collisions were realized that it will lead to cos(2$\Delta\phi$) and cos(4$\Delta\phi$) angular distribution which is related to vacuum birefringence [8–11]. The $\Delta\phi$ is defined as $\Delta\phi = \phi_{l^++l^-}$ - $\phi_{l^+-l^-}$, where $\phi_{l^++l^-}$ and $\phi_{l^+-l^-}$ are the azimuthal angle of the l^+l^- pair and azimuthal angle difference between l^+ and l^-, respectively. A fourth-order angular modulation has been observed from the measurements of e^+e^- pair production by STAR collaboration [12]. From model calculation in Ref. [9, 10], the cos(2$\Delta\phi$) modulation would vanish from dielectron channel, while the cos(2$\Delta\phi$) modulation will still be observed through the measurements of dimuon pair production. Measurements of $\mu^+\mu^-$ pair provide a complementary channel to investigate these phenomena and constrain the photon interaction in heavy-ion collisions.

In this proceeding, we will present invariant mass distributions of $\mu^+\mu^-$ pair production at $p_T < 0.10$ GeV/c, as well as p_T distributions. The p_T^2 and $\Delta\phi$ distributions of the excess yields will also be shown. Theoretical calculations [13–15] will be compared with data.

*e-mail: zj1994@mail.ustc.edu.cn

2 Experiment and analysis

Data are collected by the STAR in the year 2014 and about 800 million Au+Au minimum-bias (MB) good events have been used. In this analysis, the main sub-detectors used are the Time Projection Chamber (TPC) [16] and the Time of Flight (TOF) [17] detectors. The TPC is the main detector for tracking , and it can also measure the ionization energy loss (dE/dx) to provide particle identification. The TOF is used to identify particles by measuring the velocity. By combining the TPC and TOF, the muon can be identified at low momentum. The like-sign technique is used to estimate the combinatorial background, while the mixed-event technique is used to correct for the acceptance difference effect. After subtracting background from unlike-sign distribution, the raw signal can be obtained. Then the raw signal will be corrected for the detector inefficiency. Finally, a Monte-Carlo simulation is applied to evaluate the hadronic cocktail contribution [3].

3 Results and discussion

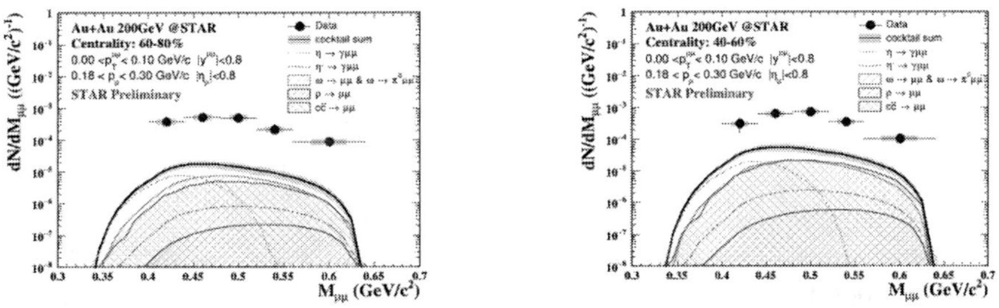

Figure 1: (color online) The $\mu^+\mu^-$ invariant mass spectra in Au+Au collisions at $\sqrt{s_{\rm NN}}$ = 200 GeV for pair p_T < 0.10 GeV/c in different centrality bins compared to the hadronic cocktail. The vertical bars on data points depict the statistical uncertainties and the systematic uncertainties are shown as gray boxes. Left: 60-80% centrality bins. Right: 40-60% centrality bins.

Figure 1 shows the $\mu^+\mu^-$ invariant mass spectra with the fiducial acceptance (0.18<p_μ<0.30 GeV/c, $|\eta_\mu|$<0.8, $|y^{\mu\mu}|$<0.8) in Au+Au collisions at $\sqrt{s_{\rm NN}}$ = 200 GeV for pair p_T < 0.10 GeV/c for 60-80% (left panel) and 40-60% (right panel) centrality bins, respectively. The invariant mass concentrates between 0.4 and 0.64 GeV/c^2 and a significant enhancement is observed with respect to the hadronic cocktail. The enhancement factor for centrality 60-80% is larger than that in semi-peripheral (40-60%) collisions. As shown in Fig. 2, the $\mu^+\mu^-$ excess mass spectra (data - cocktail) can be obtained after removing the hadronic cocktail contribution from the inclusive pair production at very low p_T (p_T < 0.10 GeV/c). Several theoretical calculations of lepton pair (l^+l^-) production from coherent photon-photon interactions in hadronic Au+Au collisions have been proposed in Ref. [15]. The excess yields are found to be consistent with EPA-QED calculations [15] within uncertainties in different centrality bins.

Figure 3 shows the p_T distributions of $\mu^+\mu^-$ pair in Au+Au collisions at $\sqrt{s_{\rm NN}}$ = 200 GeV for mass region between 0.4 and 0.64 GeV/c^2 in different centrality bins. The significant enhancement is found to concentrate below p_T $\approx$ 0.10 GeV/c. When p_T > 0.10 GeV/c, the hadronic cocktail can describe the data reasonably. The enhancement at low p_T is the evidence of photon production in hadronic heavy-ion collisions from the measurements of

Figure 2: (color online) The $\mu^+\mu^-$ excess mass spectra (data - cocktail) in Au+Au collisions at $\sqrt{s_{NN}}$ = 200 GeV for pair p_T < 0.10 GeV/c in different centrality bins compared to the model calculations.

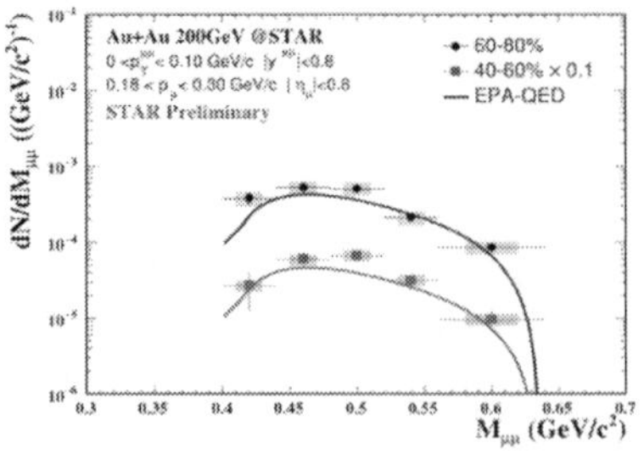

the dimuon channel. Two types of theoretical models in Ref. [15] are also compared to the results. In order to ensure the yields of data and model calculations are consistent in the same acceptance, these models are multiplied by a scale factor. The EPA-QED calculation is compatible with the data at very low p_T (p_T < 0.10 GeV/c), while the EPA calculation [15] can not describe the data very well.

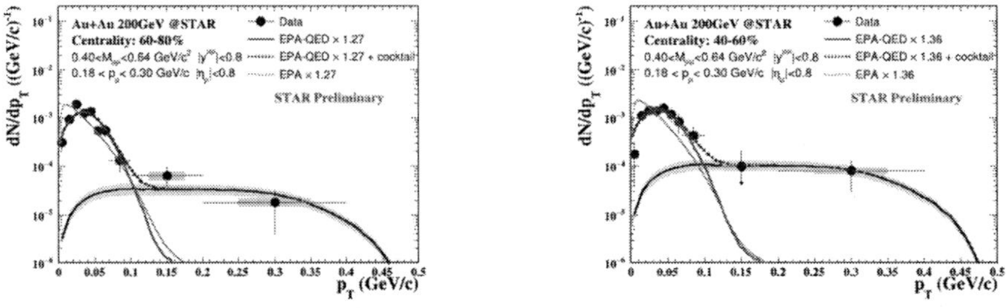

Figure 3: (color online) The p_T distributions of $\mu^+\mu^-$ pair in Au+Au collisions at $\sqrt{s_{NN}}$ = 200 GeV for mass region between 0.4 and 0.64 GeV/c^2 in different centrality bins compared to the model calculations. The gray bands depict the systematic uncertainties of the cocktails. Left: 60-80% centrality bins. Right: 40-60% centrality bins.

Figure 4 shows the p_T^2 distribution of excess yield in the mass region of 0.40-0.64 GeV/c^2 in Au+Au collisions for 60-80% centrality bin. The EPA and EPA-QED calculations are also shown in Fig. 4 as dotted and solid lines respectively. The EPA-QED curve can describe the data very well, while the EPA calculation overshoots the data at low p_T^2. The $\sqrt{< p_T^2 >}$, which characterizes the p_T broadening, is also calculated for both data and theoretical calculations. The p_T broadening is consistent with EPA-QED calculation within uncertainties.

Figure 5 shows the acceptance-corrected $\Delta\phi$ distribution of excess yield in the mass region of 0.40-0.64 GeV/c^2 for pair p_T < 0.10 GeV/c in 60-80% collisions. EPA-QED calculation is also drawn in this plot. $\Delta\phi$ distributions are fit to a function which is shown as formula in Fig. 5, where C is a constant, $A_{2\Delta\phi}$ represents the magnitude of cos($2\Delta\phi$) modulation and $A_{4\Delta\phi}$ represents the magnitude of cos($4\Delta\phi$) modulation. The cos($4\Delta\phi$) modulation is observed with 3.3σ significance which is consistent the measurements of dielectron channel in Ref. [12]. The cos($2\Delta\phi$) modulation is observed with 2.3σ significance.

4 Summary

In summary, we present the first measurements of photo-produced $\mu^+\mu^-$ pair production at very low p_T in non-central Au+Au collisions at $\sqrt{s_{NN}}$ = 200 GeV. The observed excess is

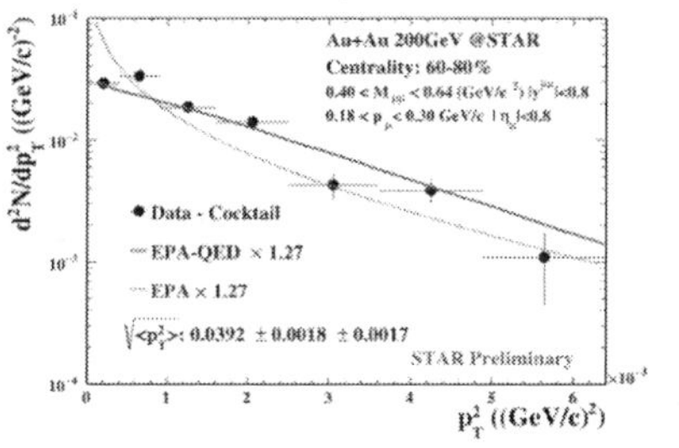

Figure 4: (color online) The p_T^2 distribution of excess yield in the mass region of 0.40-0.64 GeV/c^2 compared to the model calculations in 60-80% collisions.

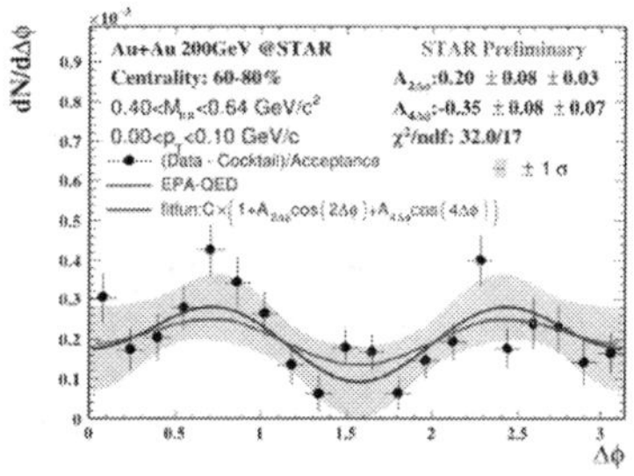

Figure 5: (color online) The $\Delta\phi$ distribution of excess yield in the mass region of 0.40-0.64 GeV/c^2 in 60-80% collisions very-low-p_T ($p_T < 0.10$ GeV/c).

found to concentrate below about $p_T \approx 0.10$ GeV/c which is consistent with EPA-QED calculations. The p_T^2 distribution in mass region of 0.4-0.64 GeV/c^2 is also presented and compared with model calculations. The broadening of p_T distribution shows agreement with EPA-QED calculation within uncertainties. The observation of a 3.3σ 4th-order azimuthal angular modulation is consistent with the measurements of dielectron channel. The 2nd-order azimuthal angular modulation with 2.3σ is indicated to exist in heavy-ion collisions.

Acknowledgments

This work was funded by the National Nature Science Foundation of China under Grant NoS. 11775213, 11720101001 and Anhui Provincial Natural Science Foundation under Grant Nos. 1908085J02.

References

[1] J. Adams *et al.* (STAR), Nucl. Phys. A **757**, 102-183 (2005)
[2] K. Adcox *et al.* (PHENIX), Nucl. Phys. A **757**, 184-283 (2005)
[3] J. Adam *et al.* (STAR), Phys. Rev. Lett. **121**,132301 (2018)
[4] B. I. Abelev *et al.* (STAR), Phys. Rev. Lett. **102**, 112301 (2009)
[5] S. Afanasiev *et al.* (PHENIX), Phys. Lett. B **679**, 321-329 (2009)
[6] B. Abelev *et al.* (ALICE), Phys. Lett. B **718**, 1273-1283 (2013)
[7] D. E. Kharzeev and H. J. Warringa, Phys. Rev. D **80**, 034028 (2009)
[8] R. P. Mignani, *et al.*, Mon. Not. Roy. Astron. Soc. **465**, no.1, 492-500 (2017)
[9] C. Li, J. Zhou and Y. J. Zhou, Phys. Lett. B **795**, 576-580 (2019)
[10] C. Li, J. Zhou and Y. J. Zhou, Phys. Rev. D **101**, no.3, 034015 (2020)
[11] W. Zha, J. D. Brandenburg, *et al.*, Phys. Rev. D **103**, no.3, 033007 (2021)
[12] J. Adam *et al.* (STAR), Phys. Rev. Lett. **127**,52302 (2021)
[13] S. R. Klein, Phys. Rev. C **97**, no.5, 054903 (2018)
[14] W. Zha, L. Ruan, Z. Tang, Z. Xu and S. Yang, Phys. Lett. B **781**, 182-186 (2018)
[15] W. Zha, J. D. Brandenburg, Z. Tang and Z. Xu, Phys. Lett. B **800**, 135089 (2020)
[16] M. Anderson, J. Berkovitz, *et al.* Nucl. Instrum. Meth. A **499**, 659-678 (2003)
[17] B. Bonner, H. Chen, *et al.*, Nucl. Instrum. Meth. A **508**, 181-184 (2003).

Studying strangeness and baryon production in small systems through Ξ–hadron correlations using the ALICE detector

Jonatan Adolfsson[1,*] (for the ALICE Collaboration)

[1]Lund University, Lund, Sweden

Abstract. These proceedings summarise recent measurements of angular correlations between the Ξ baryon and identified hadrons in pp collisions at $\sqrt{s}$ = 13 TeV using the ALICE detector. The results are compared with both string-based (PYTHIA8 with extensions) and core-corona (EPOS-LHC) models, to improve our understanding of strangeness and baryon production in small systems. The results favour baryon production through string junctions over diquark breaking, but the PYTHIA models fail at describing the relatively wide Ξ–strangeness jet peak, indicating stronger diffusion of strange quarks in data. On the other hand, EPOS-LHC is missing local conservation of quantum numbers, making it difficult to draw any conclusion about the core-corona model.

1 Introduction

Today, it is widely accepted that a quark–gluon plasma (QGP) is formed in heavy-ion collisions. What is not yet understood, however, is that several QGP key signatures have also been observed in small systems such as high-multiplicity proton–proton (pp) collisions, where the QGP is not expected to form. One such signature is the enhanced yields of multi-strange baryons, which seem to scale smoothly with system size [1].

Several phenomenological models are being developed to try to understand the origin of this. Two different approaches are being explored. In one approach, microscopic models based on colour strings are extended with new mechanisms for strange-baryon formation and other features to mimic collective behaviour. This approach is used in PYTHIA [2], where baryons in the standard configuration (Monash tune) are formed through diquark breaking. This is extended by allowing string junctions [3] and – as a further extension – rope hadronisation [4], where the latter provides strangeness enhancement. The other approach is the core-corona model, which is based on a two-phase state with a QGP-like core and a string-like corona. This approach is used in EPOS [5].

The strangeness and baryon production mechanisms predicted by these models are tested through angular correlations between the Ξ baryon (quark content dss) and other hadrons, and in particular those with opposite charge or baryon number. In the string models, strangeness is mostly formed at hadronisation, which results in short-ranged Ξ-strangeness correlations, as opposed to the core-corona model where strange quarks are largely formed in the core and can diffuse prior to hadronisation.

*e-mail: jonatan.adolfsson@hep.lu.se

2 Method

The quantity measured here is the per-trigger yield,

$$\mathbb{Y}(\Delta y, \Delta\varphi) = \frac{1}{N_{\text{trig}}} \frac{\mathrm{d}^2 N_{\text{pairs}}}{\mathrm{d}\Delta y \mathrm{d}\Delta\varphi},$$

which is the distribution of particle pairs in relative rapidity (Δy) and azimuthal angle ($\Delta\varphi$) space, divided by the number of triggers N_{trig}. To extract the part that is due to balancing charges, same-charge (or baryon number) correlations were subtracted from those of opposite charge.

This was measured for identified $\Xi - \pi$, $\Xi - \mathrm{K}$, $\Xi - \mathrm{p}$, $\Xi - \Lambda$, and $\Xi - \Xi$ pairs using approximately $8 \cdot 10^8$ minimum-bias pp events at $\sqrt{s} = 13$ TeV recorded by the ALICE detector [6]. Pions, kaons, and protons were identified via the specific energy loss in the Time Projection Chamber and the velocity in the Time-Of-Flight detectors (pseudorapidity $|\eta| < 0.8$). The Ξ and Λ baryons were reconstructed from their decay products (e.g. $\Xi^- \rightarrow \pi^- + \Lambda$, $\Lambda \rightarrow \pi^- + \mathrm{p}$), by making use of their invariant masses and various topological cuts (similar to what is done in Ref. [7]). Finally, the multiplicity dependence of the correlation function was measured by dividing the events into multiplicity classes (where the lowest percentiles correspond to the highest multiplicities) based on the response from the V0 counters in the forward regions ($-3.7 < \eta < -1.7$ and $2.8 < \eta < 5.1$).

3 Results

Figure 1 shows angular $\Xi - \pi$ correlations projected on $\Delta\varphi$ and Δy (for the near side, $|\Delta\varphi| < 3\pi/10$), for both ALICE data and models. Here, the same-sign correlations are reasonably

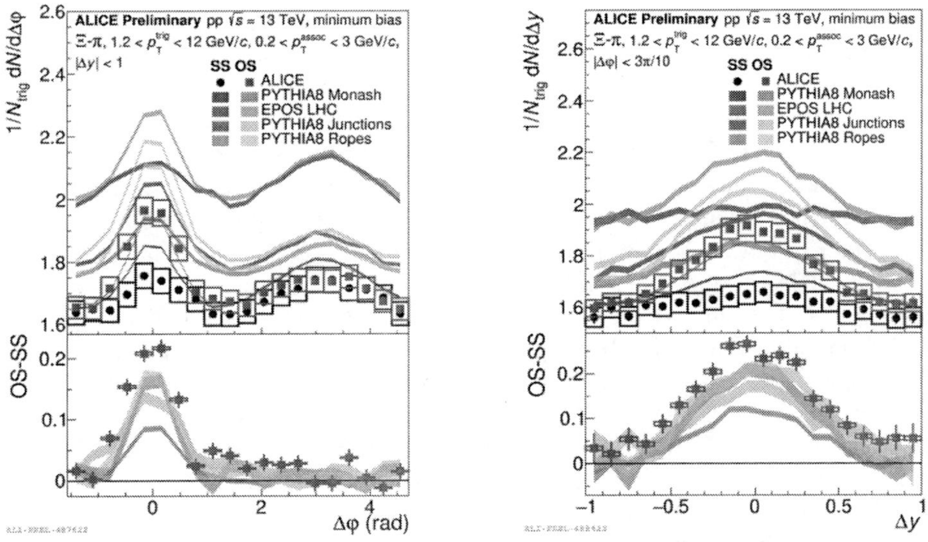

Figure 1: Per-trigger yields of $\Xi - \pi$ correlations compared with predictions from PYTHIA8 (including two extensions) and EPOS-LHC, projected onto **(left)** $\Delta\varphi$ and **(right)** Δy on the near side. The lower panels show the difference between opposite- and same-sign correlations.

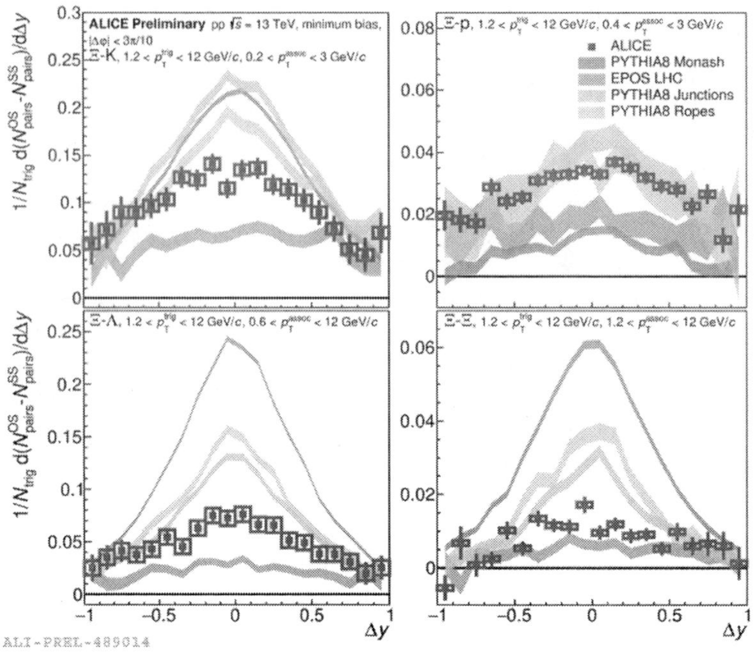

Figure 2: Differences between opposite- and same-sign (or baryon number) correlations projected onto Δy on the near side, for **(top left)** $\Xi - \mathrm{K}$ pairs, **(top right)** $\Xi - \mathrm{p}$ pairs, **(bottom left)** $\Xi - \Lambda$ pairs, and **(bottom right)** $\Xi - \Xi$ pairs.

well described by PYTHIA8 Monash, which can be attributed to good tuning of the single-particle yields, whereas there is a slight offset for the other models. Similar observations have also been made for $\Xi - \mathrm{K}$ and $\Xi - \mathrm{p}$ correlations. On the other hand, PYTHIA8 Monash underestimates the difference between opposite- and same-sign (OS-SS) correlations, which is better described by the other models or tunes. This means that these models more accurately predict the fraction of the charge from the Ξ baryons that is balanced by pions.

For the other minimum-bias results, only the near-side projections of the OS-SS difference are shown here, which are summarised in Fig. 2. Here the predictions from the different models differ significantly, with EPOS-LHC predicting a nearly flat behaviour, which is not observed in data. The reason is that local conservation of quantum numbers is not implemented in EPOS, and this is not necessarily a feature of core-corona models in general. The PYTHIA models, on the other hand, predict a stronger and narrower peak than what is observed in data, except for $\Xi - \mathrm{p}$ correlations, where no direct correlations are expected from the diquark breaking mechanism. The junction and rope extensions are better at predicting Ξ–baryon correlations, but fail at describing $\Xi - \mathrm{K}$ correlations. This favours the additional baryon production mechanism from the junction model, but more development is needed to accurately describe all observations.

Multiplicity dependent $\Xi - \pi$, $\Xi - \mathrm{K}$, and $\Xi - \Lambda$ correlations are shown in Fig. 3 (similar trends are seen also in $\Xi - \mathrm{p}$ and $\Xi - \Xi$ correlations). The shift observed in the baseline with increasing multiplicity is simply due to the increased particle yields, but a small enhancement is also observed for the OS-SS difference, along with some narrowing. A similar behaviour is predicted by PYTHIA, where this can be attributed to radial flow from colour reconnection, so this effect is likely present also in data (although possibly from a different origin). However,

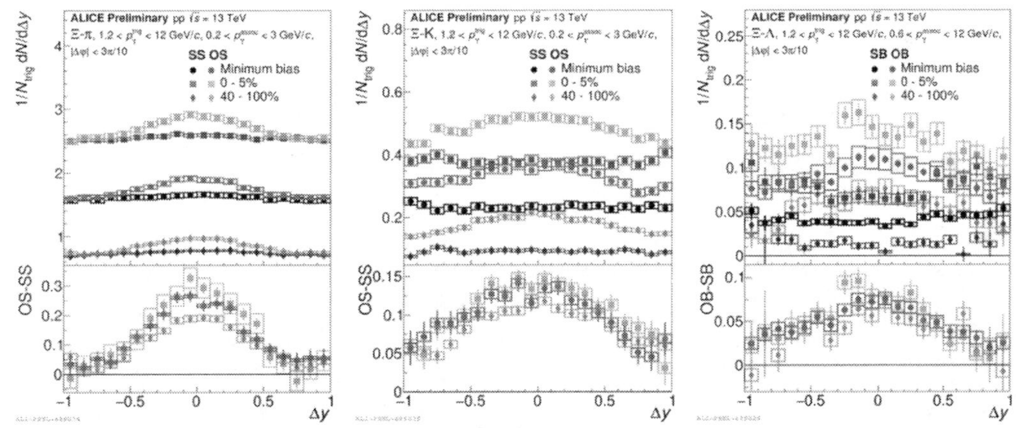

Figure 3: Per-trigger yields of **(left)** $\Xi - \pi$, **(middle)** Ξ − K, and **(right)** $\Xi - \Lambda$ correlations, measured for three different multiplicity classes and projected onto Δy on the near side. The lower panels show the differences between opposite- and same-sign correlations.

the relative difference between multiplicity classes for $\Xi - \Lambda$ correlations is weaker in data than in PYTHIA (not shown), pointing towards a competing mechanism. This could mean that the strangeness production mechanism changes with multiplicity, but more quantitative comparisons with models are needed to conclude in which way.

4 Conclusions

While the same-sign Ξ–hadron yields predicted by PYTHIA8 Monash largely describe what is observed in data, which is likely due to good tuning of the single-particle yields, the same cannot be said about the OS-SS difference. For this observable, EPOS-LHC predicts an almost flat behaviour, because local conservation of quantum numbers is not implemented in this model, and thus it is difficult to conclude anything about the accuracy of the core-corona approach. On the other hand, the PYTHIA models predict a too strong and narrow OS-SS peak for Ξ–strangeness correlations, which means that the strange quarks seem to have diffused somewhat in data prior to hadronisation. Finally, for Ξ–baryon correlations the junction extension describes the data better than the Monash tune, favouring this mechanism for producing baryons.

The multiplicity dependence seems to be dominated by radial flow, but some observations indicate a competing mechanism, possibly related to changes in the Ξ production. However, more work is required to conclude how.

References

[1] ALICE Collaboration, Nature Phys. **13**, 535-539 (2017)
[2] T. Sjöstrand *et al*, Comput. Phys. Commun. **191**, 159-177 (2015)
[3] J.R. Christiansen & P.Z. Skands, J. High Energ. Phys. **2015**, 3 (2015)
[4] C. Bierlich *et al*, J. High Energ. Phys. **2015**, 148 (2015)
[5] T. Pierog *et al*, Phys. Rev. **C92**, 034906 (2015)
[6] ALICE Collaboration, JINST **3**, S08002 (2008)
[7] ALICE Collaboration, Eur. Phys. J. **C71**, 1594 (2011)

EPJ Web of Conferences **259**, 13016 (2022) https://doi.org/10.1051/epjconf/202225913016
SQM 2021

QED Mesons, the QED Neutron, and the Dark Matter

Cheuk-Yin Wong[1]*

[1]Physics Division, Oak Ridge National Laboratory**, Oak Ridge, Tennessee 37831, USA

Abstract. Schwinger's boson solution for massless fermions in QED in 1+1D has been applied and generalized to quarks interacting in QED and QCD interactions, leading to stable and confined open-string QED and QCD boson excitations of the quark-QCD-QED system in 1+1D. Just as the open-string QCD excitations in 1+1D can be the idealization of QCD mesons with a flux tube in 3+1D, so the open-string QED excitations in 1+1D may likewise be the idealization of QED mesons with masses in the tens of MeV region, corresponding possibly to the anomalous X17 and E38 particles observed recently. A further search for bound states of quarks interacting in the QED interaction alone leads to the examination on the stability of the QED neutron, consisting of two d quarks and one u quark. Theoretically, the QED neutron has been found to be stable and estimated to have a mass of 44.5 MeV, whereas the analogous QED proton is unstable, leading to a long-lived QED neutron that may be a good candidate for the dark matter.

1 Introduction

The quark-QCD-QED vacuum is the lowest-energy state of the quark-QCD-QED system with quarks filling up the (hidden) negative-energy Dirac sea and interacting with the QCD (quantum chromodynamical) interaction and the QED (quantum electrodynamical) interaction. It is defined as the state with no valence quark above the Dirac sea and no valence antiquark as a hole below the Dirac sea. A local disturbance will generate stable collective excitations of the quark-QCD-QED system as in a fluid. There are stable and confined collective QCD excitations showing up as QCD mesons, such as π, η, and η'. The question of our central interest is whether there may also exist stable QED excitations of the quark-QCD-QED system showing up as QED mesons in the mass region of many tens of MeV, together with the related QED neutron with d-u-d quarks stabilized by the QED interaction.

Such a question may appear preposterous as one can surmise from the following debate between a Wise Guy and an Explorer. The Wise Guy objects and argues that quarks interact in QCD and QED simultaneously. When quarks and antiquarks interact, their color charges and electric charges interact simultaneously. A stable QED excitation of the quarks cannot be independent and cannot occur without the simultaneous QCD excitation. To these objections, the Explorer defends and replies that the quark currents, as well as the QCD and the QED gauge fields, are not single-element quantities. They are 3×3 color matrices with 9 matrix elements that separate naturally into $\mathbf{3} \otimes \mathbf{3}^* = \mathbf{1} \oplus \mathbf{8}$, with a color-singlet subgroup $\mathbf{1}$ and a color-octet subgroup $\mathbf{8}$ executing collective dynamics separately. In the space-time arena, there can be localized QCD excitations of the color-octet current leading to QCD mesons. There can also be independent QED excitations of the color-singlet current showing up as QED mesons at different energies. The Wise Guy objects further and says that only the non-Abelian QCD interaction can confine quarks, and the QED interaction does not confine particles such as electrons and positrons. Thereupon, the Explorer explains that light quarks can be considered approximately massless, and the application of the Schwinger's solution [1, 2] indicates that massless quarks in QED are confined in 1+1D. The confined open string in 1+1D can be considered to be the idealization of a flux tube in 3+1D. Gribov showed that massless fermions in QED in 3+1D are confined [3, 4]. Furthermore, lattice gauge calculations show that fermions in QED in 3+1D can be confined under appropriate conditions [5, 6]. Clearly, whatever the outcome of the debate may be, it is necessary to confront the predictions on the QED mesons and the QED neutron of the quark-QCD-QED system with experiments. The present investigation facilitates such a confrontation.

2 Generalization of Schwinger's solution from QED to (QED+QCD) in 1+1D

We first summarize the dynamics of the Schwinger's QED solution for massless fermions in 1+1D as reviewed in [7]. Subject to a QED gauge field disturbance A^μ with a coupling constant g_{2D} in 1+1D, the massless fermion field $\psi(x)$

*Email: wongc@ornl.gov

**This manuscript has been authored in part by UT-Battelle, LLC, under contract DE-AC05-00OR22725 with the US Department of Energy (DOE). The US government retains and the publisher, by accepting the article for publication, acknowledges that the US government retains a nonexclusive, paid-up, irrevocable, worldwide license to publish or reproduce the published form of this manuscript, or allow others to do so, for US government purposes. DOE will provide public access to these results of federally sponsored research in accordance with the DOE Public Access Plan (http://energy.gov/downloads/doe-public-access-plan), Oak Ridge, Tennessee 37831, USA satisfies the Dirac equation,

$$\gamma_\mu(p^\mu - g_{2D}A^\mu)\psi = 0. \quad (1)$$

The disturbance A^μ instructs the fermion field ψ how to move, and through the Maxwell equation,

$$\partial_\mu F^{\mu\nu} = \partial_\mu(\partial^\mu A^\nu - \partial^\nu A^\mu) = g_{2D} j^\nu = g_{2D}\bar{\psi}\gamma^\nu\psi, \quad (2)$$

the fermion field ψ in turn generates the current j^μ which instructs the gauge field A^μ how to act. A stable collective excitation of the fermion-QED system occurs, when the disturbance A^μ gives rise to the current j^μ which in turn leads to the gauge field A^μ self-consistently. By imposing the Schwinger modification factor to ensure the gauge invariance of the fermion Green's function, the fermion current $j^\mu(x)$ at the space-time point x induced by A^μ can be evaluated. After the singularities from the left and from the right cancel, the fermion current j^μ is found to relate explicitly to the perturbing QED gauge field A^μ by [1, 2]

$$j^\mu = -\frac{g_{2D}^2}{\pi}\left(A^\mu - \partial^\mu \frac{1}{\partial_\lambda \partial^\lambda}\partial_\nu A^\nu\right). \quad (3)$$

Upon substituting such a relation to the Maxwell equation (2), we get j^μ and A^μ both satisfy the Klein-Gordon equation

$$\partial_\nu\partial^\nu A^\mu + \frac{g_{2D}^2}{\pi}A^\mu = 0, \text{ and } \partial_\nu\partial^\nu j^\mu + \frac{g_{2D}^2}{\pi}j^\mu = 0, \quad (4)$$

for a boson with a mass given by $m = g_{2D}/\sqrt{\pi}$.

The Schwinger model can be generalized from QED to (QED+QCD) [8, 9]. Because of the three-color nature of quarks, the quark currents and the QED and QCD gauge fields are 3×3 color matrices which can be expanded in terms of the nine generators of the U(3) group,

$$j^\mu = \sum_{i=0}^{8} j_i^\mu t^i, \quad A^\mu = \sum_{i=0}^{8} A_i^\mu t^i, \quad t^0 = \frac{1}{\sqrt{6}}\begin{pmatrix}1 & 0 & 0\\ 0 & 1 & 0\\ 0 & 0 & 1\end{pmatrix}, \quad (5)$$

where t^0 is the generator of the U(1) color-singlet subgroup and $t^1, t^2, ..., t^8$ are the eight generators of the SU(3) color-octet subgroup. Because the current and the gauge field of each subgroup depend on each other within the subgroup, but do not depend on the current and the gauge field of the other subgroup, the two different currents and the gauge fields in their respective subgroups possess independent stable QED and QCD collective excitations at different energies. For the QCD interaction, stable QCD collective excitations can be attained in restricted variations which keep the orientation of the unit vector τ^1 in the color-octet generator space fixed while their amplitudes A_1^μ are allowed to vary. That is, with

$$j^\mu = j_0^\mu \tau^0 + j_1^\mu \tau^1, \qquad A^\mu = A_0^\mu \tau^0 + A_1^\mu \tau^1, \quad (6)$$

where $\tau^0 = t^0$, $\tau^1 = \sum_{i=1}^8 n_i t^i$, with fixed direction cosines $n_i = 2\mathrm{tr}\{\tau^1 t^i\}$ in the color-octet generator space. With the above currents and gauge fields (6) in the color-singlet and the color-octet subgroups, Schwinger's solution for QED can be trivially generalized to (QED+QCD) leading to the currents and gauge fields satisfying their corresponding Klein-Gorden equations with masses depending on their respective coupling constants [8, 9].

Consequently, there are independent collective QED and QCD excitations of the quark-QCD-QED system in 1+1D where these excitations can be described as open-string states of $q\bar{q}$ pairs. We note perceptively that in 1+1D, the open string does not have a structure, but the coupling constant g_{2D} has the dimension of a mass. In contrast in 3+1D, the flux tube has a structure with a radius R_T, but the coupling constant g_{4D} is dimensionless. The 1+1D open string can be considered an idealization of the physical meson in 3+1D with a flux tube of radius R_T, if the coupling constants g_{2D} and g_{4D} in the two different space-time dimensions are related by [8, 9]

$$g_{2D}^2 = g_{4D}^2/(\pi R_T^2) = 4\alpha/R_T^2. \quad (7)$$

Consequently, the masses of the QED and QCD mesons in 3+1D in the open-string description are approximately

$$m_{\rm QCD}^2 = 4\alpha_{\rm QCD}/\pi R_T^2, \qquad m_{\rm QED}^2 = 4\alpha_{\rm QED}/\pi R_T^2. \quad (8)$$

3 QCD and QED mesons as open strings

To get a better determination of the QCD and QED meson masses, it is necessary to take into account the flavor mixtures D_{ij} and the quark rest masses m_f. Using the method of bosonization, we obtain the semi-empirical mass formula for the neutral QCD mesons with $N_f = 3$ [9]

$$m_i^2 = \left(\sum_{f=1}^{N_f} D_{if}\right)^2 \frac{4\alpha_{\rm QCD}}{\pi R_T^2} + m_\pi^2 \sum_{f=1}^{N_f} \frac{m_f}{(m_u + m_d)/2}(D_{if})^2, \quad (9)$$

and for QED with $N_f = 2$ and isospin $(I, I_3 = 0)$,

$$m_I^2 = \left[\frac{Q_u + (-1)^I Q_d}{\sqrt{2}}\right]^2 \frac{4\alpha_{\rm QED}}{\pi R_T^2} + m_\pi^2 \frac{\alpha_{\rm QED}}{\alpha_{\rm QCD}}. \quad (10)$$

where Q_u=2/3 and Q_d=−1/3 are the charge numbers of u and d quarks respectively. With the flavor mixture and quark masses as given by the Particle Data Group [10], $\alpha_{\rm QED}$=1/137, $\alpha_{\rm QCD}$=0.68±0.08, and R_T=0.40±0.04 fm, we obtain the theoretical QCD and QED meson masses in Table 1.

Table 1. The experimental and theoretical masses of neutral, I_3=0, QCD and QED mesons, obtained with the semi-empirical mass formula (9) for QCD mesons and (10) for QED mesons.

		$[I(J^\pi)]$	Experimental mass (MeV)	Semi-empirical mass formula (MeV)
QCD meson	π^0	$[1(0^-)]$	134.9768±0.0005	134.9[‡]
	η	$[0(0^-)]$	547.862±0.017	498.4±39.8
	η'	$[0(0^-)]$	957.78±0.06	948.2±99.6
QED meson	X17	$[0(0^-)]$	16.94±0.24[#]	17.9±1.5
	E38	$[1(0^-)]$	37.38±0.71[⊕]	36.4±3.8

[‡] Calibration mass
[#] A. Krasznahorkay *et al.*, arxiv:2104.10075
[⊕] K. Abraamyan *et al.*, EPJ Web Conf 204,08004(2019)

We find from Table 1 that the open-string description of the QCD and QED mesons is a reasonable concept and the anomalous X17 [11] and E38 [12] observed recently may be QED mesons. The parent particles of the anomalous soft photons [13] may be QED mesons.

4 Schwinger's QED boson as a relativistic two-body problem in 1+1 dimensions

It is desirable to construct a phenomenological two-body model for Schwinger's QED bound state in 1+1D involving a massless fermion and an antifermion interacting in an effective two-body QED interaction Φ_{12}. The interaction Φ_{12} must be calibrated to yield Schwinger's exact solution in field theory. Neglecting spins, the relativistic two-body wave equations for the wave function Ψ for the QED interactions in 1+1 dimensions consist of two mass-shell constraints on each of the interacting particles [15, 16],

$$\mathcal{H}_1|\Psi\rangle = \left\{p_1^2 - m_1^2 - \Phi_{12}(x_{12\perp})\right\}|\Psi\rangle = 0, \quad (11a)$$

$$\mathcal{H}_2|\Psi\rangle = \left\{p_2^2 - m_2^2 - \Phi_{12}(x_{21\perp})\right\}|\Psi\rangle = 0. \quad (11b)$$

In the center-of-mass frame where the particle momentum is $p_i = (\epsilon_i, q_i)$, the above two-body wave equations become

$$\epsilon_1^2|\Psi\rangle = \left\{q_1^2 + m_1^2 + \Phi_{12}(x_{12\perp})\right\}|\Psi\rangle = 0, \quad (12a)$$

$$\epsilon_2^2|\Psi\rangle = \left\{q_2^2 + m_2^2 + \Phi_{12}(x_{12\perp})\right\}|\Psi\rangle = 0, \quad (12b)$$

and the mass of the bound states is $m = \epsilon_1 + \epsilon_2$. By requiring Φ_{12} to give the mass $m = g_{2D}/\sqrt{\pi}$ in a linearly confining potential to match Schwinger's exact solution for massless fermions in QED in 1+1D, we find [14]

$$\Phi_{12}(x_{12\perp}) = \frac{2\epsilon_1\epsilon_2}{\epsilon_1+\epsilon_2}(-Q_1Q_2)\kappa|x_1 - x_2|, \quad \kappa = \frac{g_{2D}^2}{4\pi}, \quad (13)$$

where Q_1 and Q_2 are the charge numbers and x_1 and x_2 are the spatial coordinates of the interacting particles. We shall use Φ_{12} to study the stability of the QED neutron with quarks interacting in QED.

5 The Stability of the QED Neutron

The success of the open-string description of the QCD and QED mesons leads to the search for other neutral quark systems stabilized by the QED interaction between the constituents in the color-singlet subgroup, with the color-octet QCD gauge interaction as a spectator field. Of particular interest is the QED neutron with the *d*, *u*, and *d* quarks [14]. They form a color product group of $\mathbf{3} \otimes \mathbf{3} \otimes \mathbf{3} = \mathbf{1} \oplus \mathbf{8} \oplus \mathbf{8} \oplus \mathbf{10}$, which contains a color singlet subgroup **1** where the color-singlet current and the color-singlet QED gauge field reside. In the color-singlet *d*-*u*-*d* system with three different colors, the attractive QED interaction between the *d* and *u* quarks may overwhelm the repulsive QED interaction between the two *d* quarks to stabilize the composite color-singlet QED neutron into a linear configuration. It is reasonable to treat the system in 1+1D space-time. We generalize the two-body equations of (11a) and (11b) to the three-quark system by imposing three mass-shell constraints relating the momenta, the masses, and their interactions in the form

$$\mathcal{H}_1|\Psi\rangle = \{p_1^2 - m_1^2 - [\Phi_{12}(x_{12}) + \Phi_{13}(x_{13})]\}|\Psi\rangle = 0, \quad (14a)$$
$$\mathcal{H}_2|\Psi\rangle = \{p_2^2 - m_2^2 - [\Phi_{21}(x_{21}) + \Phi_{23}(x_{23})]\}|\Psi\rangle = 0, \quad (14b)$$
$$\mathcal{H}_3|\Psi\rangle = \{p_3^2 - m_3^2 - [\Phi_{31}(x_{31}) + \Phi_{32}(x_{32})]\}|\Psi\rangle = 0, \quad (14c)$$

where Φ_{ij} is the effective two-body QED interaction (13) extracted from Schwinger's exact QED solution in 1+1D. We search for an energy minimum for the QED neutron state by using a variational wave function. It is convenient to choose a Gaussian variational wave function of the spatial dimensionless variables y_1, y_2, y_3 with standard deviations σ_1, σ_2, and σ_3 as variational parameters, for the *d*, *u*, and *d* quarks respectively,

$$\Psi(y_1, y_2, y_3) = N\exp\left\{-\frac{y_1^2}{4\sigma_1^2} - \frac{y_2^2}{4\sigma_2^2} - \frac{y_3^2}{4\sigma_3^2}\right\}, \quad (15)$$

where $y_i = \sqrt{\kappa}x_i$. The charge numbers of the quarks are $Q_1 = Q_3 = -1/3$, and $Q_2 = 2/3$. For the lowest-energy state, $\sigma_1 = \sigma_3$, and the variational parameters consist only of σ_1 and σ_2. We look for the state with the lowest composite mass $M = \epsilon_1 + \epsilon_2 + \epsilon_3$ in the variations of σ_1 and σ_2,

$$\frac{\delta^2 M(\sigma_1, \sigma_2)}{\delta\sigma_1\delta\sigma_2} = 0. \quad (16)$$

The motion of the three quarks should maintain a fixed center of mass for the composite system. The coordinates of the three quarks must satisfy the center-of-mass condition on the spatial coordinates, $\sum_{i=1}^{3}\epsilon_i y_i = 0$. The variational wave function Ψ is normalized according to

$$\int dy_1 dy_2 dy_3 |\Psi(y_1, y_2, y_3)|^2 \delta(\epsilon_1 y_1 + \epsilon_2 y_2 + \epsilon_3 y_3) = 1. \quad (17)$$

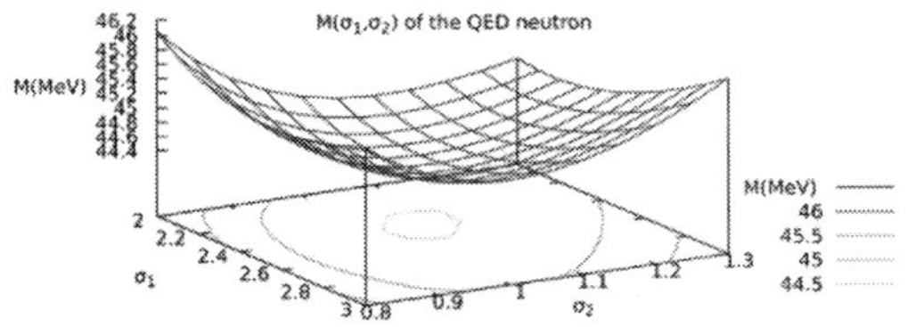

Figure 1. The mass M of the QED neutron as a function of the variation parameters σ_1, and σ_2. The QED neutron has an energy minimum at $M = 44.5$ MeV.

By the variational calculations, we find that the QED neutron is stable [14]. Its mass M has an energy minimum, M=44.5 MeV, at σ_1=2.40 and σ_2=1.05 as shown in Fig. 1, corresponding to $\sigma_1/\sqrt{\kappa}$=19.9 fm and $\sigma_2/\sqrt{\kappa}$=8.71 fm.

We study next the stability of the QED color-singlet proton with two *u* quarks and a *d* quark by carrying out variational calculations similar to those for the QED neutron. The variation over a very large range of σ_1 and σ_2 fails to find an energy minimum. Extending the range of σ will only drive the total energy of the system lower with the *u* quarks farther and farther apart without the energy turning to a minimum. An examination of the potential energies of the interaction indicates that the QED proton is unstable because of the stronger repulsion between the two u quarks in comparison with the weaker attractive interactions between the d and the u quarks. The QED proton also does not possess a continuum state with isolated quarks because the isolation of quarks is forbidden. Therefore, the QED proton does not exist as a stable bound state or a continuum state.

6 Stability of QED Neutron against Weak Decay and Dark matter

The absence of a QED proton state has an important consequence for the weak decay of the QED neutron, which could occur when a d quark decays into a u quark. Such a QED neutron weak decay would result in a possible QED proton final state, if a QED proton state could exist. Because there is no final bound or continuum QED proton state for the QED neutron to decay onto, the density of final states for the weak decay is zero. Consequently the rate for the QED neutron to decay into a QED proton is zero. The QED neutron can only decay by a baryon-number non-conserving transition which presumably has a very long life time. Therefore, the lowest energy QED neutron is a stable particle with a very long lifetime and is in fact a dark neutron. Because of its long lifetimes, the QED dark neutron may be good candidate for a part of the dark matter. Self-gravitating assemblies of QED dark neutrons may be produced by coalescence during the deconfinement-to-confinement phase transition of the quark gluon plasma in the evolution of the early Universe.

In other astrophysical frontiers, the merging of two neutron stars may lead to the production of a quark matter with deconfined quarks [17]. The coalescence of deconfined quarks during the deconfinement-to-confinement phase transition will produce QED neutrons in the post-merger environment. It has also been suggested that deconfined quark matter may be present in the core of a massive neutron star [18]. In such a neutron star, the transition region close to the core may contain QED neutron matter arising from the coalescence of deconfined quarks.

7 Conclusion and Discussions

In a system of quarks interacting in QCD and QED interactions, the current and the gauge fields reside in color-singlet QED and color-octet QCD subspaces and execute independent collective excitations. The collective excitations in the color-octet sector give rise to the QCD mesons such as π, η, and η' particles. They can be described as open-string states. In a similar manner, collective excitations of the color-singlet sector can give rise to QED mesons. The energies of the lowest QED meson states have been estimated. There are encouraging pieces of evidence for the occurrence of QED mesons as the anomalous X17 and E38 particles observed recently. On-going experiments to confirm these QED mesons are continuing.

The stability of the QED meson states leads to the study of the related QED neutron consisting of two d quarks and a u quark in QED interactions. The attractive QED interaction between the u quark and the two d quarks overwhelms the repulsion between the two d quarks to stabilize the QED neutron at an estimated mass of 44.5 MeV. The analogous QED proton has been found to be unstable, and it does not provide a bound state for the QED neutron to decay onto by way of the weak interaction. Hence the QED neutron may be stable against the weak interaction. It may have a very long lifetime and may be a good candidate for the dark matter. Because QED mesons and QED neutrons may arise from the coalescence of deconfined quarks during the deconfinement-to-confinement phrase transition in different environments such as in high-energy heavy-ion collisions, neutron-star mergers [17], and neutron star cores [18], the search of the QED bound states in various environments will be of great interest.

References

[1] J.Schwinger, *Gauge invariance and mass II*, Phys. Rev. 128, 2425 (1962).
[2] J.Schwinger, *Gauge theory of vector particles*, Trieste Lectures, 1962 (IAEA, Vienna, 1963) p. 89.
[3] V.N.Gribov, *Local confinement of charge in massless QED*, Nucl. Phys. 206, 103 (1982).
[4] V.N.Gribov, *The theory of quark confinement*,Eur.Phys. Jour.C10,91(1999),[arxiv:hep-ph/9807224]
[5] K.G.Wilson, *Confinement of quarks*, Phys. Rev. D19, 2445 (1974).
[6] G.Magnifico *et al.,Lattice QED in (3+1)dimensions at finite density with tensor networks*,[arXiv:2011.10658].
[7] C.Y.Wong,*Introduction to high-energy heavy-ion collisions*, (World Scientific,Singapore,1994).
[8] C.Y. Wong,*Anomalous soft photons in hadron production*, Phys.Rev.C81,064903(2010),[arXiv:1001.1691].
[9] C.Y.Wong, *Open string QED meson description of X17 and dark matter*,JHEP08(2020)165,[arxiv:2001.04864].
[10] PDG2019, M. Tanabashi *et al., Review of Particle Physics*, Phys. Rev. D98, 030001 (2019).
[11] A.J.Krasznahorkay *et al., Observation of anomalous internal pair creation in ^{8}Be: a possible indication of a light, neutral boson*, Phys. Rev. Lett. 116, 042501 (2016), [arXiv:1504.01527].
[12] K.Abraamyan *et al., Check of the structure in photon pairs spectra at the invariant mass of about 38* MeV/c^2, EPJ Web of Conf. 204, 08004 (2019).
[13] DELPHI Collaboration, *Direct soft photon production on the jet characteristics in hadronic Z^0 decays*, Eur. Phys. J. C67, 343 (2010), [arXiv:1004.1587].
[14] C.Y.Wong, *On the stability of the open-string QED neutron and dark matter*, [arxiv:2010.13948].
[15] H.W.Crater and P.VanAlstine,*Two body Dirac Equations*, Ann. Phys. 148, 57 (1983).
[16] H.Sazdjian,*Two-body relativistic quantum mechanics and Bethe-Salpeter equation*,J.Math.Phys.28,2618(1987).
[17] A. Bauswein *et al., Identifying a first-order phase transition in neutron-star mergers through gravitational waves*, Phys. Rev. Lett.122, 061102 (2019).
[18] E. Annala *et al., Evidence for quark-matter cores in massive neutron stars*, Nat. Phys. 16, 907 (2020)

AUTHOR INDEX

AUTHOR INDEX